D0141928

# Die Grundlehren der mathematischen Wissenschaften

in Einzeldarstellungen
mit besonderer Berücksichtigung
der Anwendungsgebiete

Band 168

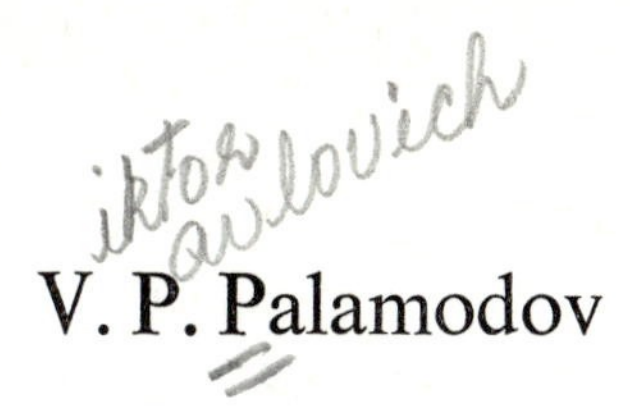

V. P. Palamodov

# Linear Differential Operators with Constant Coefficients

Translated by A. A. Brown

Springer-Verlag New York · Heidelberg · Berlin 1970

Victor P. Palamodov
Professor at the University of Moscow, U.S.S.R.

Arthur A. Brown
Acorn Park, Cambridge, Massachusetts, U.S.A.



Translation of
*Linejnye differencial'nye operatory s postojannymi koefficientami*
"Nauka", Moscow, 1967




Title No. 5151

# Preface

This book contains a systematic exposition of the facts relating to partial differential equations with constant coefficients. The study of systems of equations in general form occupies a central place. Together with the classical problems of the existence, the uniqueness, and the regularity of the solutions, we also consider the specific problems that arise in connection with overdetermined and underdetermined systems of equations: the extendability of the solutions into a wider region, the extendability of regularity, $M$-cohomology and so on. Great attention is paid to the connections and the parallels with the theory of functions of several complex variables.

The choice of material was dictated by a number of considerations. Among all the facts relating to general systems of equations, the book contains none that relate to the behavior of differential operators in spaces of slowly growing functions. Missing also are results relating to a single equation in one unknown function: the correctness of the Cauchy problem, certain theorems on $p$-convexity, and the theory of boundary values, are all set forth in other monographs (Gel'fand and Šilov [3], Hörmander [10] and Treves [4]).

The book consists of two parts. In the first, we set forth the analytic method which forms the basis for the contents of the second part, which itself is dedicated to differential equations. The first part is preceded by an introduction in which the content and methods of Part I are described. All the notes and bibliographical references are collected together in a special section.

This book was written at the suggestion of G. E. Šilov. I am grateful to him for his unfailing support. I have also benefitted greatly by my constant contacts with V. V. Grušin.

Let me make some remarks on how to use the book. For a first reading of § 1 of Chapter I, it is sufficient to limit oneself to the basic definitions. The rest of the content of this section is a detailed foundation for the argument contained in § 2. The content of § 3 of Chapter I, with the exception of the introductory section and 7°, is used only in the second part of the book, beginning with Chapter VII.

Chapter II—IV form an integrated whole. On a first reading of these, one may omit only § 4 of Chapter IV, the content of which is used, in essence, only in Chapter VII. In § 5 of Chapter IV only subsections 1°—3° need be read for an understanding of the remaining material. The general formulation of the fundamental theorem, contained in 4°—6°, is used only in § 14 of Chapter VIII.

In Chapter V, the content of §§ 1—3 represents, in essence, an exposition of more or less known facts from the theory of linear topological spaces, distributions, and Fourier transforms. In Chapter VI, 4° of § 4, 8° of § 5, and § 6 are not connected with the subsequent material, and they may be omitted on a first reading.

In the seventh chapter, §§ 11 and 12 are independent; § 13 of Chapter VIII is of auxiliary value only.

In the first part of the book, the sections are numbered independently in each chapter. In the second part, the sections are numbered sequentially. Formulae are numbered according to the following rule: A formula with the number $(a, b)$ is found in the section with the number $b$ and has the serial number $a$. The citation of a formula relating to another chapter will contain within the parenthesis the number of the chapter cited, but the citation of a formula within the same section omits the section number.

The symbol $\square$ denotes the end of a proof.

# Table of Contents

## Part II. Differential Equations with Constant Coefficients

# Introduction

The last decade has seen the completion of the foundations for what today we call the general theory of partial differential operators with constant coefficients. The general theory is distinguished from the classical in that the study of particular properties of specific operators has given way to an investigation of the structural properties of operators of general form. Perhaps the best example is provided by the study of the local properties of the solutions of homogeneous equations. Special results on the regularity of the solutions of Laplace's equation, the heat equation and certain others laid the foundation for the singling out of classes of operators having similar properties: elliptic and parabolic. These classes of operators have a common property: Every solution of the corresponding homogeneous equation is infinitely differentiable. The next step was to pose the problem of writing down all differential operators having the same property. Such operators, which we call hypoelliptic, were completely described within the class of operators with constant coefficients.

It was then observed that the regularity of solutions of hypoelliptic equations with constant coefficients is a simple consequence of a more general property which relates to all operators with constant coefficients. This general property consists in the possibility of an exponential representation and, more precisely, consists in the fact that every solution of the corresponding homogeneous equation can be written in the form of an integral with some measure over the set of exponential polynomials satisfying the same equation.

Another line, which today characterizes the general theory, had as its starting point the classical problem of the construction of fundamental solutions in the large. Within the framework of the classical theory, this problem was solved only for certain special types of operators. Then, thanks to the application of distribution theory, the following general result was obtained: For an arbitrary non-zero operator with constant coefficients, there exists in the class of distributions a fundamental solution. Moreover, it turns out that the nonhomogeneous equation corresponding to such an operator is soluble for an arbitrary right side.

The subsequent stage of development along this line was connected with the consideration of a system of equations (with constant coefficients) of general form. In this stage, the raw material for development of the general theory was provided by the corresponding portions of the theory of differential forms and functions of several complex variables. The result of their interaction was the general theorem on the solubility of an arbitrary system of equations with constant coefficients when the right-hand side satisfies the "formal condition of compatibility." This theorem on the solubility of nonhomogeneous systems, together with the already-mentioned theorem on exponential representations, provided, in its turn, a special case of the general theorem on exponential representations, which occupies an essential position in this monograph. A number of other problems in the general theory also lead to this general theorem. The second part of the monograph contains an exposition of the basic branches of the general theory of differential operators with constant coefficients, in the form of a series of consequences of the fundamental theorem on exponential representation.

The proof of the theorem on exponential representation is the main content of the first part of the monograph and also of §4 of Chapter V. In order to help the reader, we now set forth in very brief form some of the simplest cases and some of the ideas of the proof. We presuppose that the reader is familiar with the fundamentals of the theory of distributions.

## § 1. Exponential representation for an ordinary equation with one unknown function

**1°. Formulation of the problem.** An arbitrary linear differential operator with constant coefficients in $R^n$ will be written in the form

$$p(D)=\sum_{|j|\leqq m} p_j D^j, \qquad p_j \in C,$$

where

$$D^j = i^{|j|} \frac{\partial^{|j|}}{\partial \xi_1^{j_1} \dots \partial \xi_n^{j_n}}, \qquad j=(j_1, \dots, j_n), \quad |j|=j_1+\dots+j_n, \quad i=\sqrt{-1},$$

and $\xi=(\xi_1, \dots, \xi_n)$ is some fixed system of coordinates in $R^n$. Let $z=(z_1, \dots, z_n)$ be a point of the $n$-dimensional complex space $C^n$. The polynomial

$$p(z)=\sum p_j z^j, \qquad z^j = z_1^{j_1} \dots z_n^{j_n}$$

is called the characteristic polynomial of the operator $p(D)$. The algebraic variety $N \subset C^n$, formed by the roots of the polynomial $p(z)$ is also called the characteristic variety.

We choose some domain $\Omega \subset R^n$ and consider the corresponding homogeneous equation

$$p(D)\,u=0, \tag{1.1}$$

in which $u$ is assumed to be a generalized function in $\Omega$. We note that for an arbitrary point $z$, belonging to the characteristic variety $N$, the function $\exp(z, -i\xi)$ satisfies Eq. (1). In fact,

$$p(D)\exp(z, -i\xi) = p(z)\exp(z, -i\xi) = 0.$$

On the other hand, if the function $\exp(z, -i\xi)$ satisfies (1), then the point $z$ belongs to $N$. We can also find other exponential polynomials satisfying (1). First of all, we take note of Leibnitz's formula

$$p(D) f g = \sum \frac{1}{j!} D^j f p^{(j)}(D) g, \quad j! = j_1! \ldots j_n!,$$

where $p^{(j)}(z) = D_z^j p(z)$. We apply it to the functions $f = f(\xi)$ and $g = \exp(z, -i\xi)$, where $f(\xi)$ is a polynomial of order $\alpha$:

$$p(D) f(\xi) \exp(z, -i\xi) = \sum_{|j| \leqq \alpha} D^j f(\xi) p^{(j)}(z) \exp(z, -i\xi). \tag{2.1}$$

It follows from this formula that if at the point $z \in N$ all the derivatives of the polynomial $p$ up to order $\alpha$ vanish, then an arbitrary exponential polynomial of the form $f(\xi)\exp(z, -i\xi)$ satisfies (1). We note that formula (2) implies also that for the exponential polynomial $f(\xi)\exp(z, -i\xi)$ to satisfy (1), it is necessary that $z \in N$ [1].

We can now formulate in approximate terms the problem of the exponential representation of the solutions of Eq. (1): To write an arbitrary solution of this equation in the form of an integral with some measure over the set of all exponential polynomials satisfying the same equation. We now solve this problem in the most elementary case.

**2°. The case of one independent variable.** In this case Eq. (1) is an ordinary equation with constant coefficients, and the characteristic polynomial $p(z)$ has only a finite number of roots $\zeta_1, \ldots, \zeta_l$. Let $\alpha_1, \ldots, \alpha_l$ be the multiplicity of these roots; as is known, $\sum \alpha_\lambda = m$, where $m$ is the order of the polynomial $p$. From what we have said earlier, it follows that the exponential polynomial

$$\xi^j \exp(-i\zeta_\lambda \xi), \quad j = 0, \ldots, \alpha_\lambda - 1, \quad \lambda = 1, \ldots, l, \tag{3.1}$$

satisfies (1) on the whole line. We note that an arbitrary exponential polynomial satisfying (1) is a linear combination of the functions (3). Thus, the problem of the exponential representation can be formulated now in these terms: To express an arbitrary solution of (1), defined in the open set $\Omega$, in the form of a linear combination of exponential

1 In Chapter V § 7, we obtain a complete expression for the exponential polynomials satisfying (1).

polynomials of the type (3) [2]. Clearly, if this problem is to be soluble, the domain must be connected. Otherwise, we can construct a solution which is equal to distinct linear combinations of the functions (3) in the different connected components. In view of the fact that the functions (3) are linearly independent in an arbitrary open set, such a solution cannot be represented in the form of a linear combination of functions of the type (3) in the whole set $\Omega$. Therefore, we shall suppose that the region $\Omega$ is connected, that is, it is represented as an interval (finite or infinite).

**3°. Reduction of the problem.** With no essential loss of generality, we shall suppose that the interval $\Omega$ has the form $(-a, a)$ for some $a>0$. The symbol $\mathscr{D}(\Omega)$ will denote the space of all complex-valued infinitely differentiable functions on the line, of which the carriers belong to $\Omega$. By $\mathscr{D}_b$ we shall denote for every $b$, $0<b<a$, the subspace of $\mathscr{D}$ formed by all those functions with carriers belonging to the segment $[-b, b]$. The topology in the space $\mathscr{D}_b$ will be defined by the countable collection of norms

$$\|\phi\|^q = \sum_{|j| \leqq q} \sup_{\xi} |D^j \phi(\xi)|.$$

It is clear that $\mathscr{D}(\Omega) = \bigcup_{b<a} \mathscr{D}_b$. The conjugate space $\mathscr{D}'(\Omega)$, that is, the space of distributions over $\Omega$, is, by definition, the collection of all linear functionals on $\mathscr{D}(\Omega)$, having in every subspace $\mathscr{D}_b$ a restriction continuous with respect to the topology of that subspace. A functional $u \in \mathscr{D}'(\Omega)$ satisfies Eq. (1), if

$$\big(u, p^*(D)\,\phi\big) = 0 \tag{4.1}$$

for an arbitrary function $\phi \in \mathscr{D}(\Omega)$. Here $p^*(D)$ is the operator conjugate to $p(D)$; it is equal to the operator $\bar{p}(D)$, where $\bar{p}(z)$ is a polynomial obtained from the polynomial $p(z)$ by replacing each of the coefficients $p_j$ by its complex conjugate.

It is well known that the Fourier transform carries the functions $\phi \in \mathscr{D}_b$ into entire functions, for which all the norms

$$\|\psi\|_q^b = \sup_z (|z|+1)^q \exp(-b\,|\mathrm{Im}\, z|)\, |\psi(z)|, \qquad q = 0, 1, 2, \ldots \tag{5.1}$$

are finite. The space of all entire functions for which all the norms (5) are finite will be denoted by $Z^b$. We introduce in this space a topology defined by these norms. A well-known theorem states that the Fourier

2 It is well known that the classical Euler theorem on the structure of the solutions of an ordinary equation with constant coefficients provides just such an expression for all classical solutions. On the other hand, every distribution which is a solution of such an equation represents a classical solution and, consequently, the result that we have just reached, is well known. But the purpose of our argument is to provide a very simple example which will illustrate the methods that we shall apply in a much more general situation.

transform sets up a topological isomorphism of the spaces $\mathscr{D}_b$ and $Z^b$; in other words, the inverse of the Fourier transform is continuous from $Z^b$ to $\mathscr{D}_b$.

The Fourier transform of the generalized function $u \in \mathscr{D}'(\Omega)$ is a linear functional on the space $Z(\Omega) = \bigcup_{b<a} Z^b$, defined by the formula

$$(\tilde{u}, \psi) = (u, \tilde{\psi}), \qquad \psi \in Z(\Omega),$$

where $\tilde{\psi}$ is the inverse Fourier transform of the function $\psi$. From what we have said above, it follows that the functional $\tilde{u}$ is continuous on every subspace $Z^b$. From the relation $\widetilde{\bar{p}(z)\,\psi(z)} = \bar{p}(D)\,\tilde{\psi}$ it follows that Eq. (4) can be rewritten as:

$$(\tilde{u}, \bar{p}(z)\,\psi) = 0, \qquad \forall \psi \in Z(\Omega).$$

Thus, in order that the distribution $u$ in $\Omega$ should satisfy Eq. (1), it is necessary and sufficient that its Fourier transform should vanish on functions of the form $\bar{p}(z)\,\psi$, where $\psi \in (\Omega)$. The subspace consisting of such functions will be denoted by $\bar{p}\,Z(\Omega)$. We note that the polynomial $\bar{p}$ is connected with $p$ by the relationship $\bar{p}(\bar{z}) = \overline{p(z)}$, from which it follows that for every $\lambda$ the point $\bar{\zeta}_\lambda$ is a root of the polynomial $\bar{p}$ of multiplicity $\alpha_\lambda$.

We now find the Fourier transform of the exponential polynomial (3). We have

$$\begin{aligned}(\widetilde{\xi^j \exp(-i\zeta_\lambda \xi)}, \psi) \\ = (\xi^j \exp(-i\zeta_\lambda \xi), \tilde{\psi}) = \int \xi^j \exp(-i\zeta_\lambda \xi)\,\tilde{\psi}(\xi)\,d\xi = c\,D_z^j\,\psi|_{\bar{\zeta}_\lambda},\end{aligned}$$

where $c$ is a constant different from zero. Thus, the Fourier transform of the exponential polynomial $\xi^j \exp(-i\zeta_\lambda \xi)$ is (up to a constant) the $j$-th derivative of the delta-function at the point $\bar{\zeta}_\lambda$. So, our task is reduced to the following: To represent every functional on $Z(\Omega)$ that vanishes on the subspace $\bar{p}\,Z(\Omega)$ in the form of a linear combination of delta-functions at the points $\bar{\zeta}_\lambda$ and their derivatives up to certain orders.

**4°. The construction of the isomorphism.** The problem we have just formulated is essentially like a problem well known in the theory of distributions, namely, to write down the structures of distributions that are concentrated at a point. This analogy suggests the road to the solution. The fundamental step consists in proving that for an arbitrary collection of the values of the derivatives

$$D^j\,\psi(\bar{\zeta}_\lambda), \qquad j = 0, \ldots, \alpha_\lambda - 1, \quad \lambda = 1, \ldots, l,$$

we can always restore the function $\psi \in Z(\Omega)$ and this restoration is unique, up to a function belonging to the subspace $\bar{p}\,Z(\Omega)$.

For convenience, we formalize this proposition; we consider an operator $d$, which operates on functions from $Z(\Omega)$ according to the formula

$$d: \psi \to \{D^j \psi(\bar{\zeta}_\lambda),\ j=0, \dots, \alpha_\lambda - 1,\ \lambda = 1, \dots, l\} \in C^m.$$

This operator, obviously, vanishes on the subspace $\bar{p}\,Z(\Omega)$ and, therefore, can be continued to a mapping of the corresponding factor-space

$$d: Z(\Omega)/\bar{p}\,Z(\Omega) \to C^m. \tag{6.1}$$

The proposition that we wish to prove now reads as follows: The mapping $d$ is an isomorphism of linear spaces. In order to prove that it is an isomorphism, we first prove that $d$ is a monomorphism, and then that it is an epimorphism.

The assertion that $d$ is a monomorphism can be expanded as follows: Every function $\psi \in Z(\Omega)$, satisfying the conditions

$$D^j \psi(\bar{\zeta}_\lambda) = 0, \qquad j = 0, \dots, \alpha_\lambda - 1, \quad \lambda = 1, \dots, l, \tag{7.1}$$

belongs to the subspace $\bar{p}\,Z(\Omega)$. We consider the fraction $\chi = \psi/\bar{p}$. It is, clearly, analytic away from the points $\bar{\zeta}_\lambda$, $\lambda = 1, \dots, l$. On the other hand, it is analytic at any $\bar{\zeta}_\lambda$, since by (7) the function $\psi$ vanishes at $\bar{\zeta}_\lambda$ with a multiplicity not less than that of the polynomial $\bar{p}$. Thus, $\chi$ is an entire function. Since the modulus of the polynomial $\bar{p}$ does not decrease at infinity, the finiteness of the norm $\|\psi\|_q^b$ implies the finiteness of the norm $\|\chi\|_q^b$. It follows that $\chi \in Z(\Omega)$, q.e.d.

We now prove that the mapping $d$ is an epimorphism. An arbitrary vector $a \in C^m$ will be written for convenience in the form

$$a = \{a_\lambda^j,\ j = 0, \dots, \alpha_\lambda - 1,\ \lambda = 1, \dots, l\}.$$

Our problem consists in constructing a function $\psi \in Z(\Omega)$, satisfying the $m$ conditions

$$D^j \psi(\bar{\zeta}_\lambda) = a_\lambda^j, \qquad \forall j, \lambda.$$

It is easy to see that we can always find such a function $\psi$, in view of the fact that the space $Z(\Omega)$ is sufficiently ample; in particular, this space contains functions that are not identically zero, and with every such function it contains all of its derivatives and translations. Thus, the isomorphism (6) is established.

Making use of this isomorphism, we easily obtain an exponential representation. As we noted earlier, for every solution of (1) the functional $\tilde{u}$ vanishes on $\bar{p}\,Z(\Omega)$ and, therefore, can be looked at as a functional on the factor-space in (6). But because of the isomorphism (6), there corresponds to it some functional $v$ on $C^m$. The functional $v$ can be identified with a vector $\{v_\lambda^j\}$, which acts in $C^m$ according to the relation $(v, a) = \sum v_\lambda^j a_\lambda^j$. Hence

$$(\tilde{u}, \psi) = (v, d\psi) = \sum_{j,\lambda} v_\lambda^j D^j \psi(\bar{\zeta}_\lambda).$$

So we have represented $\tilde{u}$ as a linear combination of arbitrary delta-functions which are Fourier transforms of the exponential polynomial (3). Returning now to the inverse Fourier transform, we arrive at the representation of the functional $u$ that we are looking for:

$$u(\xi) = \sum_{j,\lambda} v_\lambda^j \xi^j \exp(-i\zeta_\lambda \xi). \tag{8.1}$$

## § 2. Exponential representation of the solutions of partial differential equations

**1°. General remarks.** We shall assume for simplicity that the set $\Omega$, in which we are considering Eq. (1.1), is a sphere of radius $a$ with its center at the origin of coordinates. By analogy with the case $n=1$ we construct the spaces $\mathcal{D}_b$, $Z^b$ and $Z(\Omega)$; $\mathcal{D}_b$ is the space of infinitely differentiable functions in $R^n$, with supports belonging to the closed sphere $\{\xi: |\xi| \leqq b\}$, and the space $Z^b$, which is the Fourier transform of the space $\mathcal{D}_b$, consists of all those entire functions in $C^n$ that satisfy the inequalities

$$|\psi(z)| \leqq c_q(|z|+1)^{-q} \exp(b\,|\mathrm{Im}\, z|), \qquad q = 0, 1, 2, \ldots. \tag{1.2}$$

We have $Z(\Omega) = \bigcup_{b<a} Z^b$, and by $\bar{p}\,Z(\Omega)$ we denote the subspace in $Z(\Omega)$ which is formed of functions of the type $\bar{p}\psi$, where $\psi \in Z(\Omega)$. Just as in the one-dimensional case, the problem of the exponential representation reduces to the description of the factor-space

$$Z(\Omega)/\bar{p}\,Z(\Omega). \tag{2.2}$$

In order to obtain its description, we must find suitable local conditions that will select the subspace $\bar{p}\,Z(\Omega) \subset Z(\Omega)$. Let $\bar{N}$ be the set conjugate to $N$, that is, the set of roots of $\bar{p}$. The simplest condition of the type we are looking for would be the condition

$$\psi|_{\bar{N}} = 0. \tag{3.2}$$

As we have seen, this is not sufficient for the relationship $\psi \in \bar{p}\,Z(\Omega)$ in the case $n=1$, if the polynomial has multiple roots. In the general case, the condition requiring simplicity of the roots must be altered as follows: we factor the polynomial $\bar{p}$ into the product of irreducible polynomials $\bar{p} = q_1, \ldots, q_l$ and we assume that all the polynomials $q_1, \ldots, q_l$ are distinct. We shall show that in this case $\psi \in \bar{p}\,Z(\Omega)$, provided (3) is satisfied.

**2°. Proof of the sufficiency of (3).** In $C^n$ we choose a system of coordinates $z=(z_1, \ldots, z_n)$ such that the polynomial $\bar{p}$ is normalized with respect to the variable $z_1$, that is, so that it may be written in the form

$$\bar{p}(z)=p_m z_1^m+p_{m-1}(z') z_1^{m-1}+\cdots+p_0(z'), \qquad z'=(z_2, \ldots, z_n),$$

where $m$ is its order and $p_m$ is a constant different from zero. An elementary argument will show that such a system of coordinates can always be found. Without loss of generality, we shall suppose that $p_m=1$. We may write

$$\bar{p}(z)=\prod_1^m [z_1-\zeta_j(z')],$$

where $\zeta_j(z')$ is some function of $z'$. The function

$$\mathscr{D}(z')=\prod_{j \neq i} [\zeta_i(z')-\zeta_j(z')],$$

considered as a polynomial in $z'$, is called the discriminant of $\bar{p}$ with respect to the variable $z_1$. From the conditions imposed earlier on the polynomial, it follows that the polynomial $\mathscr{D}$ is not identically zero. The set $M \subset C^{n-1}$, consisting of its roots, is called the discriminant set. In the neighborhood of an arbitrary point $z'$, not belonging to the discriminant set, all the functions $\zeta_j(z')$ are holomorphic. Accordingly, the set $\bar{N}$ is an $(n-1)$-dimensional analytic manifold in the neighborhood of an arbitrary one of its points $(z_1, z')$, where $z' \bar{\in} M$.

Suppose $\psi$ is an arbitrary function belonging to $Z(\Omega)$ and satisfying (3). We shall fix an arbitrary point $z' \in C^{n-1} \setminus M$. Since $\mathscr{D}(z') \neq 0$, all the numbers $\zeta_j(z'), j=1, \ldots, l$, are distinct. Therefore $\bar{p}(z_1, z')$ as a polynomial in $z_1$ has distinct roots. By virtue of (3) the function $\psi(z_1, z')$, considered as a function of $z_1$ alone, vanishes at these roots. Therefore the fraction $\chi=\psi/\bar{p}$ is an entire function in $z_1$. We choose some sphere $S \subset C^{n-1}$, not intersecting the set $M$. From what we have said earlier, it follows that all the functions $\zeta_j(z')$ depend analytically on $z'$ on this sphere. From this it is not difficult to conclude that the fraction $\chi=\psi/\bar{p}$ also depends analytically on $z'$ on the sphere $S$.

Thus, we have established the fact that the function $\mathscr{D}$ is analytic in all the variables $z$ over the whole space $C^n$, with the exception, perhaps, of the points of a set $\bar{N}$, which projects into $M$ (that is, which has the form $(z_1, z')$, where $z' \in M$). The set of such points has codimension 2 in $C^n$ and, therefore, because of a well-known theorem of Hartogs, is a set of removable singularities, whence it follows that the function $\chi$ is, in fact, an entire function. It is easy to verify that it satisfies the inequalities (1) and, therefore, belongs to the space $Z(\Omega)$. With this the proof is complete.

**3°. Description of the factor-space (2).** Proceeding in the same way as in the case $n=1$, we must define an operator $d$, which relates the function $\psi \in Z(\Omega)$ to its restriction on the set $\bar{N}$, and then write down the image of this operator in terms of functions defined only on the set $\bar{N}$. For this we need the concept of a function analytic on a set. As applied to the set $\bar{N}$, it reads as follows: a complex valued function $f(z)$, defined on $\bar{N}$, is said to be analytic on $\bar{N}$, if for an arbitrary point $\zeta \in \bar{N}$ there exists a function $F(z)$, holomorphic in the neighborhood of $\zeta$, which has values that on $\bar{N}$ coincide with $f(z)$ in some neighborhood of $\zeta$. It is clear that every function of the form $d\psi$, $\psi \in Z(\Omega)$, is analytic on $\bar{N}$, but not every function analytic on $\bar{N}$ has such a form, since functions belonging to $Z(\Omega)$ have bounded growth at infinity. We shall denote by $Z_{\bar{N}}(\Omega)$ the space of functions analytic on $\bar{N}$ and satisfying all of the inequalities

$$|f(z)| \leqq c_q(|z|+1)^{-q} \exp(b\,|\mathrm{Im}\, z|), \qquad q=0, 1, 2, \ldots, z \in \bar{N}, \tag{4.2}$$

for some $b<a$.

Thus, the operator $d$ acts from $Z(\Omega)$ to $Z_{\bar{N}}(\Omega)$ and vanishes on $\bar{p}\,Z(\Omega)$. Therefore we have defined a mapping

$$d\colon\ Z(\Omega)/\bar{p}\,Z(\Omega) \to Z_{\bar{N}}(\Omega). \tag{5.2}$$

We shall show that it is an isomorphism. This isomorphism provides the required description of the factor-space (2).

**4°. Proof of the isomorphism (5).** The argument of subsection 2 shows that $d$ is a monomorphism; we need only prove that it is an epimorphism. In other words, we must solve the following problem: for every function $f$ analytic on $\bar{N}$, and satisfying the inequality (4), to construct in $C^n$ an entire function $\psi$, which is a continuation of $f$, and which satisfies the analogous inequalities (1).

We now show how to solve this problem. The first step consists in constructing a "local extension". By a local extension, we mean a set of functions $\psi_\zeta(z)$, $\zeta \in C^n$, having the following properties:

a) for arbitrary $\zeta$ the function $\psi_\zeta$ is defined and holomorphic in the $\varepsilon$-neighborhood $U_\zeta$ of the point $\zeta$ ($\varepsilon$ is some number not depending on $\zeta$),

b) the functions $\psi_\zeta$ satisfy in the ensemble the inequalities

$$|\psi_\zeta(z)| \leqq c_q(|z|+1)^{-q} \exp(b\,|\mathrm{Im}\, z|), \qquad q=0, 1, 2, \ldots;\ z \in U_\zeta,\ \zeta \in C^n,$$

for some $b<a$ and

c) $$\psi_\zeta(z)|_N = f(z)$$

for $z \in \bar{N} \cap U_\zeta$.

We now outline the path leading to the construction of such a local extension. If the point $\zeta$ does not belong to $\bar{N}$, we choose the largest sphere $V_\zeta$, not intersecting $\bar{N}$, and we set $\psi_\zeta \equiv 0$ on this sphere. If $\zeta \in \bar{N}$, then we construct a set $V_\zeta$, which is equal to the product of the spheres $v_1$ and $v'$ that have the largest possible radii, not exceeding unity, and that have the following property: for all $z' \in v'$ the number of the $\zeta_1(z'), \ldots, \zeta_m(z')$, belonging to the circle $v_1$, is constant. For $\psi_\zeta$ we choose the interpolation polynomial with respect to $z_1$, which for every $z' \in v'$ takes on the same value as the function $f$ at the points $\zeta_j(z')$ belonging to $v_1$. It is not difficult to show that in the covering of $C^n$, formed by the sets $V_\zeta$, we can inscribe a covering formed by the spheres $U_\zeta = \{z : |z - \zeta| < \varepsilon\}$ for sufficiently small $\varepsilon > 0$. Restricting the function $\psi_\zeta$ to the corresponding spheres $U_\zeta$, we obtain a set which satisfies the condition a). The fulfillment of c) follows immediately from the construction of the functions $\psi_\zeta$. In order to prove that condition b) is satisfied, it is sufficient to use a well-known method of estimation of the interpolation polynomials $\psi_\zeta$ by their values at the nodes, which are values of the function $f$. The local extension is constructed.

The functions $\psi_\zeta$ which have been constructed in this first step do not solve the problem of extension, since they do not necessarily coincide on the intersections of the corresponding spheres $U_\zeta$. Our aim is now to fit these functions together, that is, to find corrections $\rho_\zeta$ to these functions, that do not violate the conditions a), b) and c), such that the "corrected" functions $\psi'_\zeta = \psi_\zeta - \rho_\zeta$ agree with one another on the pairwise intersections of the spheres $U_\zeta$.

We now look on the set of functions $\psi_\zeta$ as a null-order cochain on the covering $\{U_\zeta\}$. We consider a coboundary of this cochain, that is, for arbitrary $\zeta, \theta \in C^n$ the difference

$$\psi_{\zeta\theta} = \psi_\zeta - \psi_\theta, \tag{6.2}$$

which is defined and analytic on the intersection $U_\zeta \cap U_\theta$. It follows from c) that the function $\psi_{\zeta\theta}$ vanishes on the set $U_\zeta \cap U_\theta \cap \bar{N}$. If we modify the arguments of 2°, it is not difficult to prove that this function is divisible by $\bar{p}$, that is,

$$\psi_{\zeta\theta} = \bar{p}\,\chi_{\zeta\theta}, \tag{7.2}$$

where the function $\chi_{\zeta\theta}$ is analytic in $U_\zeta \cap U_\theta$. Estimating the quotient $\chi_{\zeta\theta}$ by the dividend $\psi_{\zeta\theta}$ and taking account of the inequality b), we arrive at the analogous inequality

$$|\chi_{\zeta\theta}(z)| \leqq c_q(|z|+1)^{-q} \exp(b\,|\mathrm{Im}\, z|), \qquad q = 0, 1, 2, \ldots; \; z \in U'_\zeta \cap U'_\theta, \tag{8.2}$$

where $\{U'_\zeta\}$ is some finer covering.

We note that the functions $\chi_{\zeta\theta}$ form a cocycle on the covering $\{U'_\zeta\}$; that is, for arbitrary $\zeta$, $\theta$ and $\eta$ we have the identity

$$\chi_{\zeta\theta}(z) + \chi_{\theta\eta}(z) + \chi_{\eta\zeta}(z) \equiv 0, \qquad z \in U'_\zeta \cap U'_\theta \cap U'_\eta. \tag{9.2}$$

In order to verify this identity, we multiply both sides by $\bar{p}(z)$. The left side vanishes because of relations (6) and (7). There remains only to note that the identity $\bar{p}\,h \equiv 0$ implies of an analytic function $h$ that $h \equiv 0$. With this, the equality (9) is proved.

The following step consists of showing that every analytic cocycle $\{\chi_{\zeta\theta}\}$, satisfying (8), is a coboundary of some analytic cochain $\{\chi_\zeta\}$ (defined, perhaps, on some finer covering), which satisfies analogous inequalities. The proof of this assertion is complicated and, therefore, we postpone it to Chapter III. We shall show that the functions

$$\psi'_\zeta = \psi_\zeta - \bar{p}\,\chi_\zeta$$

are the desired "corrections" to the local extensions. In fact, for arbitrary points $\zeta$ and $\theta$ in the intersection of the corresponding spheres, we have the equality

$$\psi'_\zeta - \psi'_\theta = \psi_\zeta - \psi_\theta - \bar{p}(\chi_\zeta - \chi_\theta) = \psi_{\zeta\theta} - \bar{p}\,\chi_{\zeta\theta} = 0.$$

The functions $\psi'_\zeta$ satisfy condition b), since the functions $\psi_\zeta$ and $\chi_\zeta$ satisfy the same condition. Therefore, the functions $\psi'_\zeta$ are restrictions of some entire function $\psi$, belonging to the space $Z(\Omega)$; the values of $\psi$ coincide on $\bar{N}$ with $f$. The extension problem is now solved and, therefore, the isomorphism (5) is established.

**5°. Exponential representation.** We now start with the isomorphism (5) and obtain an exponential representation of the solutions of Eq. (1.1). Let $u \in \mathscr{D}'(\Omega)$ be such a solution. Reasoning as we did in the case $n=1$, we represent its Fourier transform in the form

$$(\tilde{u}, \psi) = (v, d\psi), \qquad \psi \in Z(\Omega),$$

where $v$ is some functional on $Z_N(\Omega)$. By refining the arguments used in 3° and 4°, we can prove that this functional is continuous in the following sense: for arbitrary $b < a$ we can find an integer $q \geqq 0$ such that the functional $v$ is bounded in the norm

$$f \to \|f\|_q^b = \sup_N (|z|+1)^q \exp(-b\,|\mathrm{Im}\, z|)\,|f(z)|$$

on the subspace $Z_N^b \subset Z_N(\Omega)$, consisting of functions for which this norm is finite.

Let us now consider the space $C_N$, consisting of complex-valued functions continuous on $\bar{N}$ and bounded in the norm

$$\|F\| = \sup_N |F(z)|.$$

It is clear that the mapping

$$f(z) \to F(z) = f(z)(|z|+1)^q \exp(-b\,|\mathrm{Im}\, z|)$$

defines an isometric imbedding of the space $Z_{\bar{N}}^{b}$, endowed with the norm $\|\cdot\|_q^b$, in the space $C_{\bar{N}}$, endowed with the norm $\|\cdot\|$. By the Hahn-Banach theorem, the functional $v$ admits an extension $\check{v}$ to the whole space $C_{\bar{N}}$. By Riesz's theorem the functional $\check{v}$, being bounded in the norm $\|\cdot\|$, can be written in the form of an integral

$$(\check{v}, F) = \int_{\bar{N}} F \mu$$

with some complex additive measure $\mu$, the absolute value of which has a finite integral $\int |\mu|$. From this there follows finally

$$(\tilde{u}, \psi) = (v, d\psi) = \int_{\bar{N}} \psi(z)\, \mu', \tag{10.2}$$

where

$$\mu' = (|z|+1)^q \exp(-b\,|\mathrm{Im}\, z|)\, \mu. \tag{11.2}$$

Eq. (10) provides us with the required representation of the solution $u$. In order to put it into "exponential" form, we substitute in (10) $\psi(z) = \exp(z, -i\xi)$. We obtain the equation

$$u(\xi) = (\tilde{u}, \exp(z, -i\xi)) = \int \exp(z, -i\xi)\, \mu',$$

which is to be understood symbolically and, namely, as applied to the functions in $\mathscr{D}(\Omega)$

$$(u, \phi) = \int_{\bar{N}} (\exp(z, -i\xi), \phi)\, \mu'. \tag{12.2}$$

From formula (11) and from the finiteness of the integral $\int |\mu|$ it is easy to see that the righthand side of (12) converges absolutely for an arbitrary function $\phi \in \mathscr{D}_b$, and converges uniformly on every set that is bounded in $\mathscr{D}_b$.

The formula (10) is analogous to the representation (8.1) for the case when all $\alpha_\lambda = 1$. We note two essential distinctions between the formulae (8.1) and (10). The first is that the coefficients $v_\lambda^j$ in (8.1) are uniquely defined, whereas the measure $\mu'$ in (12) can be defined with a substantial degree of arbitrariness. To provide an example, we construct a measure $\mu_0$ with a compact carrier, belonging to $\bar{N}$, which upon substitution in the right side of (12) defines a null-functional. Let $\zeta \in \bar{N}$ be some point in whose neighborhood $\bar{N}$ is an $(n-1)$-dimensional analytic manifold. Within this manifold we choose some holomorphic system of coordinates in the neighborhood of $\zeta$ and we construct a sufficiently small polycylinder in this system of coordinates. Let $\gamma$ be its skeleton. The Cauchy integral of an arbitrary function analytic in $\bar{N}$, taken over $\gamma$, is equal to zero. Such an integral can be written in the form of an integral over $\bar{N}$ with respect to some measure $\mu_0$, the support of which coincides with $\gamma$. Hence,

$$\int (\exp(z, -i\xi), \phi)\, \mu_0 = \left(\int \exp(z, -i\xi)\, \mu_0, \phi\right) = 0,$$

so the measure $\mu_0$ is the one we are looking for.

Another essential distinction between (8.1) and (12) is that formula (8.1) contains all the exponential polynomials satisfying the homogeneous equation, whereas in formula (12) there are only relatively few of them. In fact, in formula (12) there appear only exponents; on the other hand for an arbitrary point $\zeta \in N$ we can construct an exponential polynomial of the form $h(\xi)\exp(\zeta, -i\xi)$ satisfying (1.1), in which the polynomial $h(\xi)$ has arbitrarily high order. For simplicity, we shall suppose that $\zeta$ is a regular point of $\bar{N}$. Let $\tau$ be a vector tangent to $\bar{N}$ at $\zeta$. We have

$$(\tau, \operatorname{grad})\, p(\zeta)=0,$$

whence

$$p(D)(\tau, \xi)\exp(\zeta, -i\xi)=0.$$

**6°. Generalization to an arbitrary convex open set.** Up to now we have assumed that the open set $\Omega$ was an open sphere. In fact, the arguments can be carried through for a substantially more general case: namely, the case when $\Omega$ is an arbitrary convex open set. The methodological reason for choosing convex open sets only is that only then can the entire functions that represent the Fourier transforms of functions in $\mathscr{D}(\Omega)$ be adequately described in terms of inequalities of the form

$$|\psi(z)| \leqq C M_\alpha(z),$$

where $\{M_\alpha(z)\}$ is some system of majorants. To show this, we choose some increasing sequence of convex compacts $K_\alpha \subset \Omega$, $\alpha=1, 2, \ldots$ such that their union is equal to $\Omega$. An entire function $\psi$ is a Fourier transform of some function $\phi \in \mathscr{D}(\Omega)$ when and only when there exists an $\alpha$ such that $\psi$ satisfies the inequalities

$$|\psi(z)| \leqq c_q(|z|+1)^{-q} \exp(\gamma_\alpha(\operatorname{Im} z)), \qquad q=0, 1, 2, \ldots,$$

where

$$\gamma_\alpha(y)=\sup_{\xi \in K_\alpha}(-y, \xi).$$

Denoting by $Z(\Omega)$ the space of such entire functions, we can repeat all the arguments of Sections 2°–5°, if everywhere we replace the majorant $\exp(b|\operatorname{Im} z|)$ by $\exp(\gamma_\alpha(\operatorname{Im} z))$.

The requirement that the region $\Omega$ be convex is not however due to methodological limitations alone; it is an essential feature of the problem. In fact, the representation (10) is, generally speaking, impossible in a non-convex open set $\Omega$. To convince ourselves of this, we have only to consider the trivial case $p \equiv 0$. In this case Eq. (1.1) is satisfied by an arbitrary distribution in $\Omega$, and in formula (12) $\bar{N}=C^n$. Therefore, the representation (10) means that every distribution in $\Omega$ is the inverse Fourier transform of some measure $\mu$ in $C^n$, such that the integral $\int|\tilde{\phi}\,\mu|$

converges for an arbitrary $\phi \in \mathscr{D}(\Omega)$. On the other hand, it is well-known that such a representation of a distribution is possible only when the set $\Omega$ is convex.

## § 3. The exponential representation of solutions of arbitrary systems

We consider a general system of differential equations with constant coefficients, in one unknown.

$$p_1(D)\,u = \cdots = p_t(D)\,u = 0, \qquad u \in \mathscr{D}'(\Omega). \tag{1.3}$$

Here $p_1(D), \ldots, p_t(D)$ are an arbitrary finite set of operators with constant coefficients, and $\Omega$ is a convex open set. By definition, the distribution $u \in \mathscr{D}'(\Omega)$ is a solution of this system if and only if it vanishes for all functions of the form

$$\bar{p}_1(D)\,\phi_1 + \cdots + \bar{p}_t(D)\,\phi_t, \qquad \phi_1, \ldots, \phi_t \in \mathscr{D}(\Omega).$$

The Fourier transform of $u$ is a functional on $Z(\Omega)$, characterized by the property that it vanishes on the subspace $(\bar{p}_1, \ldots, \bar{p}_t)\,Z(\Omega)$, consisting of functions of the form

$$\bar{p}_1(z)\,\psi_1 + \cdots + \bar{p}_t(z)\,\psi_t, \qquad \psi_1, \ldots, \psi_t \in Z(\Omega).$$

Therefore to obtain an exponential representation of the solutions of the system (1), we must study the factor-space

$$Z(\Omega)/(p_1, \ldots, p_t)\,Z(\Omega). \tag{2.3}$$

The problem discussed in 2° § 2 is a special case of the one that we have here, namely the case in which $t=1$, and the polynomial $p_1$ has no multiple divisors. For the solution of the general problem, we shall consider the characteristic variety $N$ of the system (1), that is, the set of all roots of the polynomials $p_1, \ldots, p_t$. The conjugate space $\bar{N}$ is the set of all roots of the conjugate polynomials $\bar{p}_1, \ldots, \bar{p}_t$; we shall now show how to construct the local conditions on this set that characterize the subspace $(\bar{p}_1, \ldots, \bar{p}_t)\,Z(\Omega) \subset Z(\Omega)$. This construction is as follows: there exist (a) a finite covering $\{N_\lambda, \lambda = 1, \ldots, l\}$ of the manifold $\bar{N}$, consisting of algebraic varieties, and (b) for every $\lambda$ a finite set of differential operators $d_\lambda^j(z, D)$ in $C^n$ with polynomial coefficients, such that the set of conditions

$$d_\lambda^j(z, D)\,\psi(z)|_{N_\lambda} = 0, \qquad \forall j,\ \lambda,\ \psi \in Z(\Omega),$$

is necessary and sufficient for the function $\psi$ to belong to the subspace $(\bar{p}_1, \ldots, \bar{p}_t)\,Z(\Omega)$.

In the particular case considered in the preceding section, the covering $\{N_\lambda\}$ reduces to the set $\bar{N}$ itself, and the set of operators $d_\lambda^j$ reduces to the single operator of restriction. We now carry out the construction of the operators $d_\lambda^j$ in the case when $t=1$ but there are no conditions on the polynomial $\bar{p}_1$. Let $q_1^{\alpha_1} \dots q_l^{\alpha_l}$ be a representation of it as a product of powers of irreducible polynomials, and for every $\lambda$, let $N_\lambda$ be the set of roots of $q_\lambda$. Then, we can easily verify that the operators

$$d_\lambda^j(z, D) = \frac{\partial^j}{\partial z_1^j}, \qquad j=0, \dots, \alpha_\lambda - 1, \; \lambda = 1, \dots, l,$$

are the ones we are looking for. (The polynomial $\bar{p}_1$ is supposed normalized with respect to $z_1$.) In the general case when $t>1$, the operators $d_\lambda^j$ are constructed in a somewhat more complicated way; a case typifying the general situation is considered in Chapter IV, § 4.

Later we shall introduce an operator $d$, which maps a function $\psi \in Z(\Omega)$ into the set of functions $d_\lambda^j(z, D)\, \psi(z)|_{N_\lambda}$. In order to describe the image of this operator, we generalize the notion of a function analytic on $\bar{N}$. We define a holomorphic $p$-function[3] as an arbitrary set $\{f_\lambda^j\}$, of functions $f_\lambda^j$ defined on $N_\lambda$, such that, for every point $\zeta \in \bar{N}$, there exists a function $F$ holomorphic in the neighborhood of $\zeta$ such that

$$f_\lambda^j(z) = d_\lambda^j(z, D)\, F(z)|_{N_\lambda}, \qquad \forall j, \lambda.$$

The set of all holomorphic $p$-functions, having components that satisfy the inequalities

$$|f_\lambda^j(z)| \leq c_q (|z|+1)^{-q} \exp(\gamma_\alpha(\operatorname{Im} z)), \qquad q=0, 1, 2, \dots; \; \exists \alpha,$$

is a linear space, which we denote by $Z_p(\Omega)$. The operator $d$ acts from the space $Z(\Omega)$ to $Z_p(\Omega)$ and vanishes on the subspace $(\bar{p}_1, \dots, \bar{p}_t)\, Z(\Omega)$. By the fundamental theorem of the first part of this book, this operator, in fact, sets up an isomorphism of the factor-space (2) with the space $Z_p(\Omega)$.

From this result, by analogy with § 2, we can derive an exponential representation for the solutions of the system (1)

$$u(\xi) = \sum_{j, \lambda} \int_{N_\lambda} d_\lambda^j(z, -i\xi) \exp(z, -i\xi)\, \mu_\lambda^j. \tag{3.3}$$

Here, for an arbitrary compact $K_\alpha$, the measures $\mu_\lambda^j$ can be so chosen that

$$\sum_{j, \lambda} \int_{N_\lambda} \int (|z|+1)^{-q} \exp(\gamma_\alpha(\operatorname{Im} z))\, |\mu_\lambda^j| < \infty$$

for some $q$.

Finally, we shall obtain complete generality if we consider the rectangular system of equations

$$p(D)\, u = 0 \tag{4.3}$$

3 The symbol $p$ denotes the vector $(p_1, \dots, p_t)$.

in several unknown distributions $u = (u_1, \ldots, u_s)$, where $p(D)$ is a matrix, whose elements are differential operators with constant coefficients. We shall obtain an exponential representation for the solutions of such a system, in the form of expressions similar to (3), in which the $d_\lambda^j$ will be vector differential operators. The proof of this general formula under very general conditions follows the scheme of proof of the representation in § 2. However, the technical apparatus needed for the proof is in this case considerably more complicated.

In the first chapter, we shall introduce and study families of topological modules, which will provide the language we need. In the same chapter, we collect together certain important homology theorems. In the second chapter, we prove a general theorem, which in the general case replaces the construction of the interpolation polynomials used in 4° § 2. In the third chapter, we prove that the cohomologies of analytic cochains are trivial, when the cochains are subjected to the condition of bounded growth at infinity; from this, in particular, there follows a result used without proof in 4° § 2. In the fourth chapter, we shall prove the fundamental theorem that we have just mentioned, for the case when the matrix $p$ is of general form. In Part II, we derive from the fundamental theorem an exponential representation of the solutions of the general system (4) and many other consequences that concern the differential operator $p(D)$.

PART ONE

# Analytic Methods

Chapter I

## Homological Tools

The basic content of this chapter consists of the homology theorems of § 2, in which we are concerned with families of topological modules. We shall make use of families of modules and of the results of § 2 many times in the remainder of the book. All the definitions relating to families are collected in § 1, which serves essentially as a basis for the arguments used in § 2.

In § 3, we set down a number of elementary facts concerning homology algebra. The contents of this section, with the exception of the introductory portion, subsection 2°, and subsection 7°, will be used only in the fifth chapter, beginning with § 7.

### § 1. Families of topological modules

**1°. Topological modules and mappings.** We fix our attention on some commutative ring $A$, which contains the field $C$ of complex numbers as a subring. We recall that a module over the ring $A$ or, for short, an $A$-module is a commutative group $X$ (which we shall write in additive form) in which every element $a$ of the ring $A$ defines an additive operator $x \to a\,x$, and the sum and product of elements of $A$ correspond to the sum and the composition of these operators. We shall suppose that the module is unitary, that is, that the identity operator corresponds to the unit of the ring $A$. Since $A$ contains the field $C$, every $A$-module is at the same time a linear space over $C$. The module $X$ will be said to be topological, if it is a topological linear space and if for an arbitrary element $a \in A$, the corresponding operator in $X$ is continuous.

Every $A$-module can be thought of as a topological module if we endow it with the discrete topology, that is, the topology in which every element is its own neighborhood.

Let $X$ be a topological $A$-module. Every submodule $Y$ is a topological module in the topology induced by $X$. We consider the factor-module $X/Y$, and the canonical mapping $X \to X/Y$, which carries every element of $X$ into the coset to which it belongs. We use this mapping to introduce a topology into $X/Y$, defining the open sets in the latter as the images of open sets in $X$. From now on, when we speak of submodules and factor-modules, we shall assume that their topologies have been introduced in the way just described.

Let $\phi: X \to Y$ be a mapping of $A$-modules. Several other modules are associated with the mapping: these are $\operatorname{Ker}\phi = \operatorname{kernel}\phi$, $\operatorname{Im}\phi = \operatorname{image}\phi$, $\operatorname{Coker}\phi = Y/\operatorname{Im}\phi = \operatorname{cokernel}\phi$, and $\operatorname{Coim}\phi = X/\operatorname{Ker}\phi = \operatorname{co-image}\phi$. We shall suppose that the modules $X$ and $Y$ are topological. Then, by the argument just used, all four of the associated modules are also topological. We shall say that the mapping $\phi$ is continuous if it is continuous as a mapping of the topological linear space $X$ into $Y$. In particular, the canonical mapping $X \to X/Y$ is always continuous. From now on, all the modules that we encounter will be supposed topological, and all mappings continuous (unless the contrary is explicitly stated). Therefore, the words "topological" and "continuous" will be omitted.

Let $\phi: X \to Y$ be a mapping of modules, and let $X', X'' \subset X'$ and $Y', Y'' \subset Y'$ be submodules of $X$ and $Y$ such that $\phi(X') \subset Y'$, and $\phi(X'') \subset Y''$. Restricting $\phi$ to $X'$, we obtain a mapping $\hat{\phi}: X' \to Y'$. Since the mapping $\hat{\phi}$ carries the submodule $X''$ into $Y''$, we can construct the mapping

$$\check{\phi}: X'/X'' \to Y'/Y''$$

of the corresponding cosets. The mapping $\check{\phi}$, is obviously a continuous mapping of topological modules. We shall call it the mapping associated with $\phi$.

The mapping $\phi: X \to Y$ will be called an isomorphism and will be written $X \overset{\phi}{\cong} Y$, if there exists an inverse mapping $\phi^{-1}: Y \to X$, that is, a mapping such that the compositions $\phi\phi^{-1}$ and $\phi^{-1}\phi$ represent identity mappings. We shall say that $\phi$ is a homomorphism if the associated mapping $\check{\phi}: \operatorname{Coim}\phi \to \operatorname{Im}\phi$ is an isomorphism. Clearly, every isomorphism is also a homomorphism. If the modules $X$ and $Y$ are given the discrete topology, then every mapping $\phi: X \to Y$ is a homomorphism.

A sequence of mappings

$$X \xrightarrow{\phi} Y \xrightarrow{\psi} Z$$

is said to be semi-exact (at the term $Y$), if $\operatorname{Im}\phi \subset \operatorname{Ker}\psi$. This sequence is said to be algebraically exact if $\operatorname{Im}\phi = \operatorname{Ker}\psi$. Finally, we shall say that it is exact, if it is algebraically exact, and the mappings $\phi$ and $\psi$ are homomorphisms. In particular, the whole sequence is algebraically exact, if and only if it is exact when $X$, $Y$ and $Z$ have the discrete topology.

### 2°. Many-valued mappings of modules

**Definition 1.** A many-valued mapping $\phi: X \to Y$ of topological $A$-modules is a map from $X$ into the set of subsets of $Y$, satisfying the following two conditions.

*I. Linearity.* For arbitrary $x_1, x_2 \in X$ and $a_1, a_2 \in A$, we have the inclusion relation

$$a_1 \phi(x_1) + a_2 \phi(x_2) \subset \phi(a_1 x_1 + a_2 x_2). \tag{1.1}$$

*II. Continuity.* For an arbitrary neighborhood of zero $V \subset Y$, we can find a neighborhood of zero $U \subset X$, such that for every $x \in U$ the intersection $\phi(x) \cap V$ is not empty.

It follows from Condition I that for arbitrary $a_1, a_2 \in A$

$$a_1 \phi(0) + a_2 \phi(0) \subset \phi(0).$$

So the set $\phi(0)$ is a submodule of $Y$. For arbitrary $x \in X$, the same condition implies the inclusions

$$\phi(0) + \phi(x) \subset \phi(x), \qquad \phi(x) - \phi(x) \subset \phi(0),$$

which show that the set $\phi(x)$ is a translation of the submodule $\phi(0)$, that is, $\phi(x) = \phi(0) + y$ for every element $y \in \phi(x)$. In particular, if $0 \in \phi(x)$, we have $\phi(x) = \phi(0)$.

The kernel of the mapping $\phi$ is the subset $\operatorname{Ker} \phi$ of the module $X$, consisting of all elements $x$ for which $\phi(x) = \phi(0)$. This subset is a submodule, since for arbitrary $x_1, x_2$ belonging to the subset, and $a_1, a_2 \in A$, the sets $a_1 \phi(x_1)$ and $a_2 \phi(x_2)$ contain the zero of $X$. The inclusion relation (1) implies that the set $\phi(a_1 x_1 + a_2 x_2)$ also contains the zero element and, accordingly, coincides with $\phi(0)$.

The image of the mapping $\phi$ is the set $\operatorname{Im} \phi = \phi(X)$, the union of the sets $\phi(x)$ for all $x \in X$. It is a submodule of $Y$, since for arbitrary $y_1 \in \phi(x_1)$ and $y_2 \in \phi(x_2)$, relation (1) implies the inclusion

$$a_1 y_1 + a_2 y_2 \in \phi(a_1 x_1 + a_2 x_2).$$

We shall consider also the following modules: the cokernel $\operatorname{Coker} \phi = Y/\operatorname{Im} \phi$ and the coimage $\operatorname{Coim} \phi = X/\operatorname{Ker} \phi$.

A typical example of a many-valued mapping is given by the mapping $X/Y \to X$, which carries every coset into the set of elements that belong to it. We shall call this a canonical mapping. Every ordinary, i.e., single-valued, mapping is also many-valued. From now on, unless the contrary is explicitly stated, we shall assume that all mappings are many-valued.

Let $\phi: X \to Y$ and $\psi: Y \to Z$ be two module-mappings. We define their composition. For every $x \in X$, we denote by $\psi\phi(x)$ the union $\psi(\phi(x))$ of the sets $\psi(y)$ for $y \in \phi(x)$. Let us prove that the mapping $\psi\phi$ is linear and continuous. Using the linearity of the mappings $\phi$ and $\psi$ we obtain

$$a_1 \psi(\phi(x_1)) + a_2 \psi(\phi(x_2)) \subset \psi(a_1 \phi(x_1) + a_2 \phi(x_2)) \subset \psi(\phi(a_1 x_1 + a_2 x_2))$$

for arbitrary $x_1, x_2 \in X$ and $a_1, a_2 \in A$. This verifies the linearity.

Let $W$ be a neighborhood of zero in $Z$. Since $\phi$ and $\psi$ are continuous, we can find neighborhoods of zero $V \subset Y$ and $U \subset X$ such that for an arbitrary $x \in U$ the intersection $\phi(x) \cap V$ is not empty, and for arbitrary $y \in V$ the intersection $\psi(y) \cap W$ is not empty. It follows that for arbitrary $x \in U$, the intersection $\psi(\phi(x)) \cap W$ is not empty. We have thus established that the composition $\psi\phi$ is a mapping.

### 3°. Families of topological modules

**Definition 2.** A family of topological $A$-modules will mean an arbitrary system $X = \{X_\alpha, i_\alpha^{\alpha'}\}$, consisting of the functions $X_\alpha$, defined on the set of all integers with values which are topological $A$-modules, and of the set of single-valued continuous mappings $i_\alpha^{\alpha'}: X_\alpha \to X_{\alpha'}$, defined for arbitrary pairs of integers $\alpha$ and $\alpha'$, and such that $\alpha \leqq \alpha'$, and satisfying the following conditions:

a) for arbitrary $\alpha$ the mapping $i_\alpha^\alpha$ is the identity,

and

b) for arbitrary $\alpha \leqq \alpha' \leqq \alpha''$, we have $i_{\alpha'}^{\alpha''} i_\alpha^{\alpha'} = i_\alpha^{\alpha''}$.

In some applications, we shall deal with families $\{X_\alpha, i_\alpha^{\alpha'}\}$, in which the modules $X_\alpha$ and the mappings $i_\alpha^{\alpha'}$ are defined only for $0 < \alpha \leqq \alpha'$. Such incomplete families can be completed in a trivial way by setting $X_\alpha = 0$ and $i_\alpha^{\alpha'} = 0$ for $\alpha \leqq 0$.

**Definition 3.** Let $X = \{X_\alpha, i_\alpha^{\alpha'}\}$ and $Y = \{Y_\alpha, j_\alpha^{\alpha'}\}$ be two families. A family-to-family mapping $\phi: X \to Y$ is defined as an ensemble of continuous mappings

$$\phi_\alpha: X_\alpha \to Y_{\beta(\alpha)},$$

which are defined for all integer $\alpha$, where $\beta(\alpha)$ is a nondecreasing function of $\alpha$ tending to $\pm\infty$ together with $\alpha$, and for arbitrary $\alpha < \alpha'$ the diagram

$$\begin{array}{ccc} X_{\alpha'} & \xrightarrow{\phi_{\alpha'}} & Y_{\beta(\alpha')} \\ {\scriptstyle i_\alpha^{\alpha'}}\uparrow & & \uparrow{\scriptstyle j_{\beta(\alpha)}^{\beta(\alpha')}} \\ X_\alpha & \xrightarrow{\phi_\alpha} & Y_{\beta(\alpha)} \end{array} \tag{2.1}$$

is commutative. The mappings $\phi_\alpha$ are called components of the mapping $\phi$, and the function $\beta(\alpha)$ is called the order of the mapping.

A mapping $I: X \to Y$ is said to be an identity mapping if its components are the mappings $i_\alpha^{\beta(\alpha)}$. Thus, the identity mappings can be put into one-to-one correspondence with their orders, that is, with the non-decreasing functions $\beta(\alpha)$ which tend to $\pm\infty$ together with $\alpha$. If $\beta(\alpha) \equiv \alpha$, the corresponding identity mapping is said to be a unit mapping.

Let $X = \{X_\alpha, i_\alpha^{\alpha'}\}$ be a family. The family $Y = \{Y_\alpha, j_\alpha^{\alpha'}\}$ is said to be a subfamily of the family $X$, if for every $\alpha$, $Y_\alpha$ is a submodule of $X_\alpha$ (its topology being induced by the topology of $X_\alpha$), and if the mappings $j_\alpha^{\alpha'}$ are restrictions of the mappings $i_\alpha^{\alpha'}$. Let $Y$ be a subfamily of $X$. We shall consider the sequence of factor-modules $X_\alpha/Y_\alpha$. Since for arbitrary $\alpha' > \alpha$, $i_\alpha^{\alpha'}(Y_\alpha) = j_\alpha^{\alpha'}(Y_\alpha) \subset Y_{\alpha'}$, we can define the mapping $\check{i}_\alpha^{\alpha'}: X_\alpha/Y_\alpha \to X_{\alpha'}/Y_{\alpha'}$, associated with the mapping $i_\alpha^{\alpha'}$. It is easy to see that the modules $X_\alpha/Y_\alpha$ and the mappings $\check{i}_\alpha^{\alpha'}$ form a family. This family we shall call the factor-family and denote it by $X/Y$. The canonical mappings $X_\alpha \to X_\alpha/Y_\alpha$ and $X_\alpha/Y_\alpha \to X_\alpha$ define mappings of the families $X \to X/Y$ and $X/Y \to Y$, which we shall also call canonical mappings.

Let $X = \{X_\alpha, i_\alpha^{\alpha'}\}$ and $Y = \{Y_\alpha, j_\alpha^{\alpha'}\}$ be arbitrary families and let $\phi = \{\phi_\alpha: X_\alpha \to Y_{\beta(\alpha)}\}$ be a mapping of them. Since the diagram (2) is commutative, the mapping $i_\alpha^{\alpha'}$ carries the submodule $\operatorname{Ker}\phi_\alpha$ into the submodule $\operatorname{Ker}\phi_{\alpha'}$. Therefore the submodules $\operatorname{Ker}\phi_\alpha$, with the topology induced by the modules $X_\alpha$, form a subfamily of the family $X$. We denote this subfamily by $\operatorname{Ker}\phi$.

Now we construct the image of $\phi$. For every integer $\beta$ we consider in the module $Y_\beta$ the submodule $j_{\beta(\alpha)}^{\beta}\phi_\alpha(X_\alpha)$, where $\alpha$ is the largest number such that $\beta \geqq \beta(\alpha)$. If we assign to these submodules the topology induced by the topology of the modules $Y_\beta$, we obtain a subfamily of the family $Y$, which we denote by $\operatorname{Im}\phi$ or $\phi(X)$. We shall consider also the factor-families

$$\operatorname{Coim}\phi = X/\operatorname{Ker}\phi, \qquad \operatorname{Coker}\phi = Y/\operatorname{Im}\phi.$$

Let

$$X \xrightarrow{\phi} Y \xrightarrow{\psi} Z,$$

$$\phi = \{\phi_\alpha: X_\alpha \to Y_{\beta(\alpha)}\}, \qquad \psi = \{\psi_\alpha: Y_\alpha \to Z_{\gamma(\alpha)}\}$$

be a sequence of families and mappings. The composition of the mappings will be the mapping with components $\psi_{\beta(\alpha)}\phi_\alpha: X_\alpha \to Z_{\gamma(\beta(\alpha))}$. Let us establish the validity of this definition. The function $\gamma(\beta(\alpha))$ is a non-decreasing function of $\alpha$ and tends to $\pm\infty$ together with $\alpha$, since the functions $\beta(\alpha)$ and $\gamma(\alpha)$ possess these properties by definition. The diagram

$$\begin{array}{ccccc}
X_{\alpha'} & \xrightarrow{\phi_{\alpha'}} & Y_{\beta(\alpha')} & \xrightarrow{\psi_{\beta(\alpha')}} & Z_{\gamma(\beta(\alpha'))} \\
\uparrow & & \uparrow & & \uparrow \\
X_\alpha & \xrightarrow{\phi_\alpha} & Y_{\beta(\alpha)} & \xrightarrow{\psi_{\beta(\alpha)}} & Z_{\gamma(\beta(\alpha))}
\end{array} \qquad \alpha' > \alpha$$

is commutative, since both the component squares are commutative. Accordingly, the mappings $\gamma_{\beta(\alpha)}\phi_\alpha$ do indeed form a family mapping. We note that the composition of identity mappings is again an identity mapping.

**Proposition 1**

I. *For every two identity mappings $I_1$, $I_2$ of some family $X$, we can find two other mappings $I_3$ and $I_4$ of the same family, such that $I_3 I_1 = I_4 I_2$.*

II. *Let $\phi: X \to Y$ be a family mapping. For an arbitrary identity mapping $J$ of the family $Y$, we can find two identity mappings $I: X \to X$ and $J': Y \to Y$ such that $\phi I = J' J \phi$.*

III. *Let $\phi: X \to Y$ be a mapping and $I: X \to X$ be an identity mapping. Then we can find an identity mapping $J$ of the family $Y$ such that $\phi I = J\phi$.*

*Proof.* Let us establish the first assertion. Let $\beta_i(\alpha)$ be the order of the mapping $I_i$, $i=1,2$. We write $\beta(\alpha)=\max_i \beta_i(\alpha)$. Now we extend both of the nondecreasing functions $\gamma_i$: $\beta_i(\alpha) \to \beta(\alpha)$ to all integer values of the argument while preserving the property of monotone growth. We take the functions $\gamma_i(\alpha)$ as the orders of the mappings $I_{i+2}$, $i=1,2$.

Let us prove Part II. Let $\beta(\alpha)$ and $\gamma(\alpha)$ be the orders of the mappings $\phi$ and $J$. We choose the function $\gamma(\beta(\alpha))$, which is the order of $J\phi$, to be the order of the mapping $I$, and the function $\beta(\alpha)$ as the order of $J'$.

Let us prove Part III. Let $\delta(\alpha)$ be the order of the mapping $I$. We extend the function $\beta(\alpha) \to \beta(\delta(\alpha))$ to all integer values of the argument, preserving the property of monotone growth, and we take it as the order of the mapping $J$. $\square$

### 4°. Equivalent families and mappings

**Definition 3a.** Let $X'$ and $X''$ be subfamilies of some family $X$. We shall say that the subfamily $X'$ majorizes the subfamily $X''$, and we write $X' \succ X''$, if there exists an identity mapping $I$ of the family $X$, which carries the subfamily $X''$ into $X'$.

Let $X'$, $X''$, $X'''$ be three subfamilies of $X$, satisfying the relations $X' \succ X'' \succ X'''$, that is, let there exist two identity mappings $I'$ and $I''$, which respectively carry $X''$ into $X'$ and $X'''$ into $X''$. Then the composition $I' I''$ carries $X'''$ into $X'$ and therefore $X' \succ X'''$. Thus, majorization is an ordering relation.

Let $X'$ and $X'' \prec X'$ be subfamilies of $X$, and let $I$ be an identity mapping, which carries $X''$ into $X'$. We consider the factor-family $X'/I(X'')$; an arbitrary factor-family of this form will be called a family associated with $X$.

**Definition 3b.** Let $X_1 = X_1'/X_1''$ and $X_2 = X_2'/X_2''$ be two families associated with $X$. We shall say that the first of these majorizes the second, and we write $X_1 \succ X_2$, if we have simultaneously $X_1' \succ X_2'$ and $X_1'' \succ X_2''$.

Majorization, as applied to families associated with $X$, is an ordering relation also, since the majorization of subfamilies is an ordering relation. If we make use of Proposition I, Definition 3b can be rewritten as follows: $X_1 \succ X_2$ if and only if there exists an identity mapping $I: X \to X$, such that $I(X_2') \subset X_1'$ and $I(X_2'') \subset X_1''$.

We shall say that the families $X_1$ and $X_2$ are equivalent, and we write $X_1 \sim X_2$, if we have simultaneously $X_1 \succ X_2$ and $X_1 \prec X_2$. We note that if $X' \succ X''$, the factor-family $X'/I(X'')$ is independent (up to an equivalent) of the choice of the identity mapping $I$ which carries $X''$ into $X'$. For this factor-family we shall employ an abbreviated notation $X'/X''$, which does not mean, however, that $X''$ is a sub-family of $X'$.

**Definition 4.** We shall say that two mappings $\phi, \phi': X \to Y$ are equivalent, and we write $\phi \sim \phi'$, if there exist identity mappings $J, J': Y \to Y$, such that $J\phi = J'\phi'$.

Let us validate this definition. Suppose that $\phi \sim \phi'$ and $\phi' \sim \phi''$, that is,

$$J\phi = J'\phi', \qquad J^*\phi' = J^{**}\phi''.$$

According to Proposition 1 there exist identity mappings $J_1$ and $J_2$ such that $J_1 J' = J_2 J^*$. Hence, $J_1 J\phi = J_2 J^{**}\phi''$, that is, $\phi \sim \phi''$.

**Proposition 2.** *The equivalence relation is preserved under composition.*

*Proof.* Let $\phi, \phi': X \to Y$ and $\psi, \psi': Y \to Z$ be pairs of equivalent mappings. Then by definition $J\phi = J'\phi'$ and $K\psi = K'\psi'$ where $J$, $J'$, $K$, and $K'$ are identity mappings. By Proposition 1, we can find identity mappings $K_1$, $K_1'$, such that

$$K_1 K\psi = K\psi J, \qquad K_1' K'\psi' = K'\psi' J'.$$

Then

$$K_1 K\psi\phi = K\psi J\phi = K'\psi' J'\phi' = K_1 K'\psi'\phi',$$

i.e., $\psi\phi \sim \psi'\phi'$. □

In particular, for arbitrary mappings $\phi: X \to Y$ and identity mappings $I: X \to X$, the mappings $\phi I$ and $\phi$ are equivalent.

Now suppose there is given a diagram of mappings of families, that is, some set of families $X^i$, $i \in \mathscr{I}$, and the set of mappings $\phi_i^j: X^i \to X^j$, defined for some pairs $(i,j) \in \mathscr{I} \times \mathscr{I}$. We shall say that the diagram is commutative if for any pair $(i,j) \in \mathscr{I} \times \mathscr{I}$, all the compositions of the form

$$\phi_{i*}^j \ldots \phi_{i'}^{i''} \phi_i^{i'}: X^i \to X^j$$

are equivalent. Proposition 2 implies that the diagrams remain commutative when all the mappings are replaced by equivalent mappings.

**Definition 5.** Let $X'$ and $X'' \prec X'$ be families associated with some family $X$; let $Y'$ and $Y'' \prec Y'$ be families associated with some family $Y$.

And further, let

$$\phi'\colon X' \to Y', \qquad \phi''\colon X'' \to Y''$$

be mappings of these families. Then, we shall say that the mapping $\phi'$ majorizes the mapping $\phi''$, and we write $\phi' \succ \phi''$, if there exist identity mappings $I$ and $J$ of the families $X$ and $Y$, such that the mappings $\check{I}$ and $\check{J}$ associated with them define the commutative diagram:

$$\begin{array}{ccc} X' & \xrightarrow{\phi'} & Y' \\ {\scriptstyle \check{I}}\uparrow & & \uparrow{\scriptstyle \check{J}} \\ X'' & \xrightarrow{\phi''} & Y'' \end{array} \tag{3.1}$$

that is, $\phi' \check{I} \sim \check{J} \phi''$.

It is obvious that the majorization of mappings is an ordering relation. If $X' \sim X''$, $Y' \sim Y''$, $\phi' \succ \phi''$, and $\phi'' \succ \phi'$, then we shall say that the mappings $\phi'$ and $\phi''$ are equivalent and we write $\phi' \sim \phi''$. If $X' = X''$ and $Y' = Y''$, the equivalence relationship just defined coincides with the equivalence relation set up in Definition 4. In fact let $\phi'$ and $\phi''$ be mappings that are equivalent in the sense of Definition 5. Then there exist identity mappings $J_1, J_2\colon Y' \to Y'$, such that $J_1 \phi' \check{I} = J_2 \check{J} \phi''$. By Proposition 1 we can find an identity mapping $J^*$ such that $J^* J_1 \phi' = J_1 \phi' \check{I}$. Thus, $J^* J_1 \phi' = J_2 \check{J} \phi''$, that is, $\phi'$ and $\phi''$ are mappings equivalent in the sense of Definition 4.

Let $X'$, $X''$ be families associated with $X$, and let $Y'$, $Y''$ be families associated with $Y$. We shall suppose that $X'' \prec X'$ and $Y'' \succ Y'$, that is, that there exist mappings $\check{I}\colon X'' \to X'$ and $\check{J}\colon Y' \to Y''$, associated with identity mappings. In this case, for any mapping $\phi\colon X' \to Y'$, the compositions $\phi \check{I}\colon X'' \to Y'$ and $\check{J} \phi\colon X' \to Y''$ are defined. It is clear that these compositions (up to an equivalent) do not depend on $I$ and $J$. From now on we shall make no distinction between the mappings $\phi \check{I}$ and $\check{J} \phi$ and the mapping $\phi$ itself.

Now suppose $X$ and $Y$ are arbitrary families and $X'/X''$ and $Y'/Y''$ are families associated with them. Suppose further that we are given the mapping $\phi\colon X \to Y$, satisfying the conditions $\phi(X') \prec Y'$ and $\phi(X'') \prec Y''$. Replacing $\phi$ by an equivalent mapping (if we have to) we have the inclusion relations $\phi(X') \subset Y'$ and $\phi(X'') \subset Y''$. Then we consider the mapping

$$\check{\phi}\colon X'/X'' \to Y'/Y'',$$

which has the components $\check{\phi}_\alpha\colon X'_\alpha/X''_\alpha \to Y'_\alpha/Y''_\alpha$ associated with the components of the mapping $\phi$. We shall call $\check{\phi}$ the mapping associated to $\phi$. It is easy to see that the construction we have just gone through defines $\check{\phi}$ up to an equivalence.

Let $\phi', \phi''\colon X \to Y$ be two family-mappings having the same orders. By the sum of two such mappings, we shall mean the mapping $\phi' + \phi''$:

$X \to Y$ having as components the sums of the corresponding components of $\phi'$ and $\phi''$. If $\phi'$ and $\phi''$ have different orders, their sum is the mapping $J'\phi' + J''\phi''$, where the identity mappings $J'$ and $J''$ of $Y$ are so chosen that the mappings $J'\phi'$ and $J''\phi''$ have the same orders. It is clear that the mapping $\phi' + \phi''$ (up to an equivalent) does not depend on the choice of $J'$ and $J''$. It is clear that the addition of family mappings and the multiplication of a mapping by an element of the ring $A$ does not destroy the majorization relation introduced in Definition 5.

Let $X'$ and $Y'$ be associated to $X$ and $Y$, and let $\phi$: $X' \to Y'$ be a mapping of these families. Then the families $\operatorname{Ker}\phi$ and $\operatorname{Coim}\phi$ are associated with $X'$ and, accordingly, with $X$ also; the families $\operatorname{Im}\phi$ and $\operatorname{Coker}\phi$ are associated with $Y'$ and, therefore, also with $Y$.

**Proposition 3.** *Let the mappings $\phi'$: $X' \to Y'$ and $\phi''$: $X'' \to Y''$ satisfy the relations $X' \succ X''$, $Y' \succ Y''$, $\phi' \succ \phi''$. Then we have the inequalities*

$$\begin{aligned} &\operatorname{Im}\phi' \succ \operatorname{Im}\phi'', \qquad \operatorname{Coker}\phi' \succ \operatorname{Coker}\phi'', \\ &\operatorname{Ker}\phi' \succ \operatorname{Ker}\phi'', \qquad \operatorname{Coim}\phi' \succ \operatorname{Coim}\phi''. \end{aligned} \tag{4.1}$$

*Proof.* Since the diagram (3) is commutative, there exist identity mappings $J_1$ and $J_2$ of the family $Y'$ such that $J_1\phi'\check{I} = J_2\check{J}\phi''$. Since the mapping $J_2\check{J}$ carries $\operatorname{Im}\phi''$ into $\operatorname{Im} J_2\check{J}\phi''$, the family $\operatorname{Im} J_2\check{J}\phi''$ majorizes $\operatorname{Im}\phi''$. On the other hand,

$$\operatorname{Im} J_2\check{J}\phi'' = \operatorname{Im} J_1\phi'\check{I} \subset \operatorname{Im}\phi'\check{I} \subset \operatorname{Im}\phi'.$$

Accordingly, the first of the relations (4) is proved. The second follows from the first, since $Y' \succ Y''$.

Let us now prove the third relation. Proposition 1 implies that there exist identity mappings $I_0$, $J_3$ of the families $X''$ and $Y'$ such that

$$\phi'\check{I} I_0 = J_3 J_1\phi'\check{I} = J_3 J_2\check{J}\phi''.$$

Since the mapping $\check{I} I_0$ carries $\operatorname{Ker}\phi'\check{I} I_0$ into $\operatorname{Ker}\phi'$, we have

$$\operatorname{Ker}\phi'' \subset \operatorname{Ker} J_3 J_2\check{J}\phi'' = \operatorname{Ker}\phi'\check{I} I_0 \prec \operatorname{Ker}\phi'.$$

This proves the third relation of (4). The fourth relation follows from the third. □

In particular, if the families $X'$, $X''$, $Y'$, $Y''$ and the mappings $\phi'$, $\phi''$ are equivalent, then the families (4) are pairwise equivalent.

**Proposition 4.** *Let*

$$X' \xrightarrow{\phi'} Y' \xrightarrow{\psi'} Z', \qquad X'' \xrightarrow{\phi''} Y'' \xrightarrow{\psi''} Z''$$

*be sequences of families and mappings such that*

$$X' \succ X'', \quad Y' \succ Y'', \quad Z' \succ Z'', \quad \phi' \succ \phi'', \quad \psi' \succ \psi''.$$

*Then* $\psi'\phi' \succ \psi''\phi''$. *In other words, our composition preserves the order relation of mappings.*

*Proof.* It is easy to see that the commutativity of the diagram (3) is independent of the choice of the mappings $I$ and $J$. Hence, for arbitrary mappings $\check{I}$: $X'' \to X'$, $\check{J}$: $Y'' \to Y'$ and $\check{K}$: $Z'' \to Z'$, associated with identity mappings in the diagram

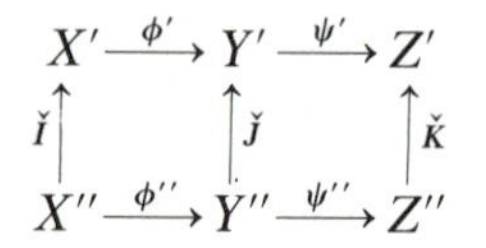

both squares are commutative. Accordingly, by Proposition 2 the entire diagram is commutative. Therefore $\psi'\phi' \succ \psi''\phi''$. ▯

**Definition 6.** The mapping $\phi$: $X \to Y$ is said to be a monomorphism if $\operatorname{Ker}\phi \sim 0$, and an epimorphism if $\operatorname{Coker}\phi \sim 0$.

By Proposition 3, this definition is invariant when the mapping $\phi$ is replaced by an equivalent one.

**Proposition 5.** *If in the sequence*

$$W \xrightarrow{\chi} X \xrightarrow{\phi} Y \xrightarrow{\psi} Z$$

*the mapping* $\phi$ *is a monomorphism, then* $\operatorname{Ker}\phi\chi \sim \operatorname{Ker}\chi$. *If* $\phi$ *is an epimorphism, then* $\operatorname{Im}\psi\phi \sim \operatorname{Im}\psi$.

*Proof.* Let $\phi$ be a monomorphism. Then, since $\operatorname{Ker}\phi \sim 0$, we can find an identity mapping $I$ of the family $X$ such that $I(\operatorname{Ker}\phi) = 0$. Now, let $w \in \operatorname{Ker}\phi\chi$, that is, $\phi\chi(w) = \phi\chi(0)$. This equation implies that we can find an element $x \in \chi(w)$ such that $\phi(x) = \phi(0)$, that is, $x \in \operatorname{Ker}\phi$. This means that $I(x) = 0$, hence, the set $I\chi(w)$ contains the zero element, and therefore coincides with $I\chi(0)$. Thus, $\operatorname{Ker}\phi\chi \subset \operatorname{Ker} I\chi$. Proposition 3 implies that $\operatorname{Ker} I\chi \sim \operatorname{Ker}\chi$, whence $\operatorname{Ker}\phi\chi \prec \operatorname{Ker}\chi$. The converse relation is obvious. The first assertion is proved and the proof of the second is carried out in a similar fashion. ▯

### 5°. Exact sequences of mappings

**Definition 7.** We shall say that the family mapping $\phi$: $X \to Y$ establishes an isomorphism, if there exists a mapping $\phi^{-1}$: $Y \to X$ such that

$$\phi\phi^{-1} = J, \quad \phi^{-1}\phi = I,$$

where $I$ and $J$ are identity mappings of the families $X$ and $Y$. We write: $X \overset{\phi}{\cong} Y$.

The equation $\phi\,\phi^{-1}=J$, implies, in particular, that the mapping $\phi$ is equivalent to a single-valued mapping. We shall refer to $\phi^{-1}$ as the inverse of $\phi$.

We shall say that the mapping $\phi$ is a homomorphism, if the associated mapping

$$\hat{\phi}\colon \operatorname{Coim}\phi \to \operatorname{Im}\phi$$

is an isomorphism. Thus, every homomorphism is equivalent to a single-valued mapping.

If the mapping $\phi$ is an isomorphism (homomorphism), then every mapping equivalent to it is also an isomorphism (homomorphism).

**Proposition 6.** *The mapping $\phi\colon X \to Y$ is a homomorphism if and only if there exists a mapping $\phi'\colon \operatorname{Im}\phi \to X/\operatorname{Ker}\phi$ such that the composition $\phi\,\phi'$ is equal to an identity mapping of the family* $\operatorname{Im}\phi$.

*Proof. Necessity.* It follows from the definition that there exists a mapping $\phi^{-1}\colon \operatorname{Im}\phi \to X/\operatorname{Ker}\phi$ such that the composition $\check{\phi}\,\phi^{-1}$ is an identity mapping. Therefore the composition of $\phi^{-1}$ and the canonical mapping $X/\operatorname{Ker}\phi \to X$ forms the desired mapping $\phi'$.

*Sufficiency.* We denote by $\phi^{-1}$ the composition of $\phi'$ and the canonical mapping $X \to X/\operatorname{Ker}\phi$. Since $\check{\phi}\,\phi^{-1}=\phi\,\phi'$, the composition $\check{\phi}\,\phi^{-1}$ is an identity mapping. Thus $\check{\phi}\,\phi^{-1}\phi \sim \check{\phi}$, that is, $\check{\phi}(\phi^{-1}\check{\phi}-j) \sim 0$ where $j$ is a unit mapping of the family $\operatorname{Coim}\phi$. Since $\check{\phi}$ is a monomorphism, Proposition 5 implies that

$$\operatorname{Ker}(\phi^{-1}\check{\phi}-j) \sim \operatorname{Ker}\check{\phi}(\phi^{-1}\check{\phi}-j) \sim \operatorname{Coim}\phi,$$

whence $\phi^{-1}\check{\phi}-j \sim 0$. □

**Definition 8.** Let

$$X \xrightarrow{\ \phi\ } Y \xrightarrow{\ \psi\ } Z \tag{5.1}$$

be a sequence of families and mappings which are equivalent to single-valued mappings. We shall say that this sequence is semi-exact or it is a complex if $\operatorname{Im}\phi \prec \operatorname{Ker}\psi$, that is, $\psi\,\phi \sim 0$. We shall say that the sequence is algebraically exact if $\operatorname{Im}\phi \sim \operatorname{Ker}\psi$. The sequence (5) will be called exact if it is algebraically exact and if the mappings $\phi$ and $\psi$ are homomorphisms.

**Proposition 7.** *Let*

$$X' \xrightarrow{\ \phi'\ } Y' \xrightarrow{\ \psi'\ } Z', \qquad X'' \xrightarrow{\ \phi''\ } Y'' \xrightarrow{\ \psi''\ } Z''$$

*be two sequences of equivalent families and mappings. If one of these sequences is semi-exact, algebraically exact, or exact, then the other sequence has the same property.*

*Proof.* By Proposition 3, $\operatorname{Im}\phi' \sim \operatorname{Im}\phi''$ and $\operatorname{Ker}\psi' \sim \operatorname{Ker}\psi''$. We need only add that if two mappings are homomorphic, then all mappings equivalent to them are also homomorphic. □

**6°. Supplementary notation.** If $X'$ and $X''$ are two subfamilies of a family $X$, we shall denote by $X'+X''$ the family formed of sums of modules of the families $X'$ and $X''$ respectively; we shall denote by $X' \cap X''$ the family formed of the intersections of such modules. It is easy to see that when $X'$ and $X''$ are replaced by equivalent families, the sums and intersections are replaced by equivalent families also.

Let $X$ be a module, and let $k>0$ be an integer. By $[X]^k$ or $X^k$ we shall denote the direct sum of $K$ summands which are each equal to $X$. For the sake of generality, we shall write $[X]^0=0$. If $\phi\colon X \to Y$ is a module mapping, then $[\phi]^k$ will denote the mapping

$$[X]^k \ni (x_1, \dots, x_k) \to (\phi(x_1), \dots, \phi(x_k)) \in [Y]^k.$$

If $X=\{X_\alpha, i_\alpha^{\alpha'}\}$ is a family of modules, then $[X]^k$ will denote the family $\{[X_\alpha]^k, [i_\alpha^{\alpha'}]^k\}$. Let

$$X \xrightarrow{\phi} Y \xrightarrow{\psi} Z \tag{6.1}$$

be a sequence of families and mappings, and let

$$[X]^k \xrightarrow{[\phi]^k} [Y]^k \xrightarrow{[\psi]^k} [Z]^k \tag{7.1}$$

be a sequence of powers of these families and mappings. Then, if $\phi$ is a homomorphism, it is clear that $[\phi]^k$ is also a homomorphism. If the sequence (6) is exact, then the sequence (7) is also exact, and so on. In order to avoid notational clumsiness, we shall often write $\phi$ and $\psi$ in sequences of the type (7) in place of $[\phi]^k$ and $[\psi]^k$.

The total effect of Sections $3°-6°$ is that we may neglect distinctions between equivalent families and between equivalent mappings.

We note that any topological module $X$ can be considered as a family if we set $X_\alpha \equiv X$ and we choose for the mappings $i_\alpha^{\alpha'}$ the identity mapping of $X$. If $X$ and $Y$ are two modules and $\phi\colon X \to Y$ is a mapping between them, then $\phi$ can be considered as a mapping of the corresponding family. In this sense, the definitions given in $3°-6°$ as applied to families and mappings of this type coincide with the corresponding definitions given in 1°.

## § 2. The fundamental homology theorem

**1°. Theorem 1.** *We consider three types of commutative diagrams (as shown immediately below) formed of families $X_i^j$ of topological modules and single-valued mappings of them.*

I. *Let us suppose that all columns and all rows of distinguished parts of these diagrams are exact. If one of the mappings $f$ or $g$ is a homomorphism, the other is a homomorphism also.*

II. *Let us note the extreme rows and columns marked with the symbols $f$, $f'$, $g$, and $g'$, and suppose, in addition to the condition stated in the first part of the theorem, that all rows and columns of our diagrams, with the exceptions of those already noted, are algebraically exact. Then the extreme rows and columns are semi-exact and we have the natural isomorphisms*

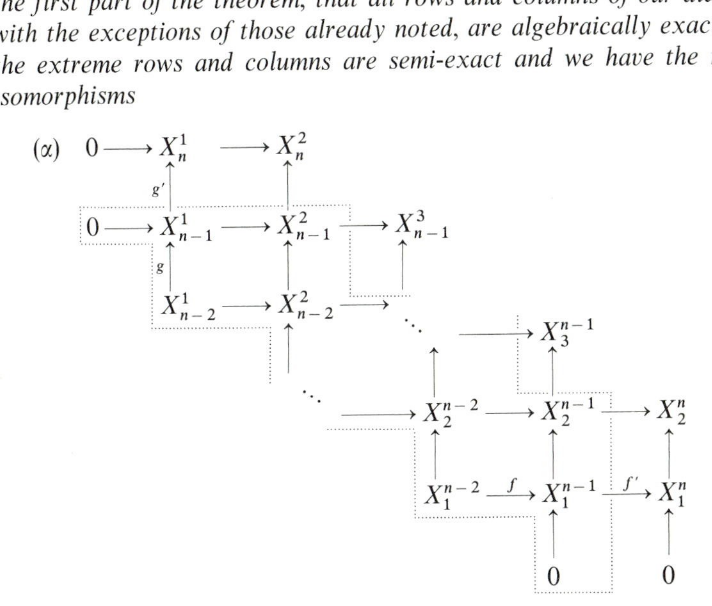


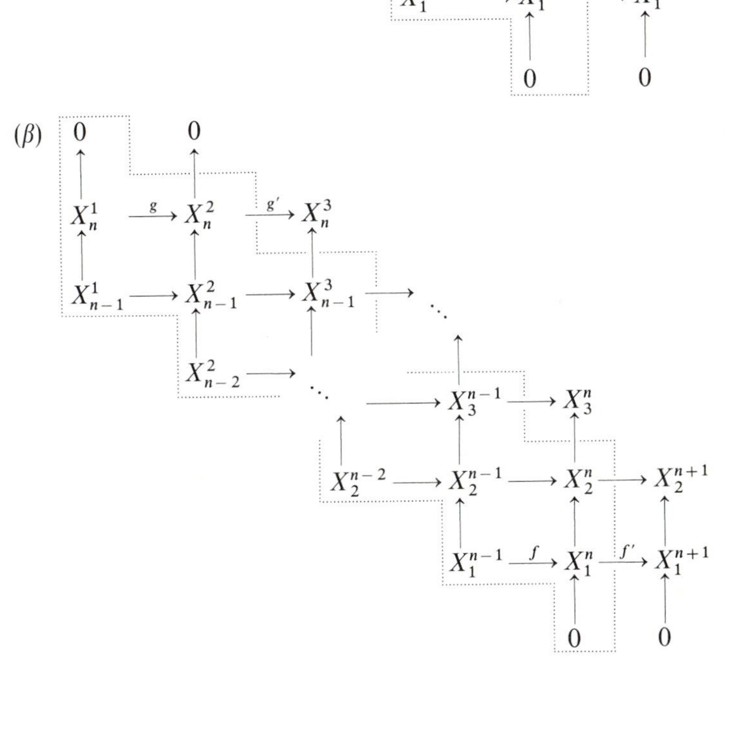

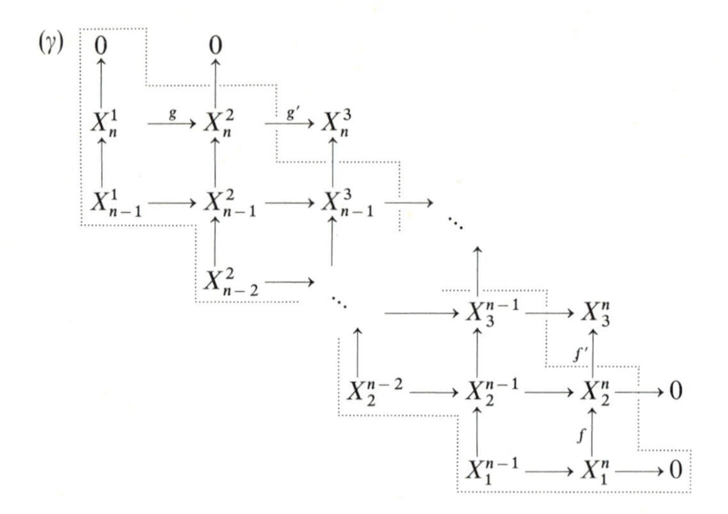

$$\operatorname{Ker} f'/\operatorname{Im} f \cong \operatorname{Ker} g'/\operatorname{Im} g. \tag{1.2}$$

We postpone the proof for

**2°. Three lemmas**

**Lemma 1.** *Let us consider the commutative diagram*

$$\begin{array}{ccccc} Z_1 & \xrightarrow{\psi_1} & Z_2 & \xrightarrow{\psi_2} & Z_3 \\ & & {\scriptstyle\chi_2}\uparrow & & \uparrow{\scriptstyle\chi_3} \\ & & Y_2 & \xrightarrow{\phi} & Y_3 \\ & & & & \uparrow \\ & & & & 0 \end{array} \tag{2.2}$$

*of mappings of families and suppose that the right-hand column is exact, that the first row is algebraically exact, and the mappings $\phi$ and $\psi_1$ are homomorphisms. Then,*

A) *The mapping $\psi_1'$: $Z_1 \to Z_2/\operatorname{Im} \chi_2$, which is an extension of the mapping $\psi_1$* (that is, it is a composition of $\psi_1$ with the canonical mapping $Z_2 \to Z_2/\operatorname{Im} \chi_2$), *is a homomorphism;*

B) *If $Z_1 \sim 0$, then the mapping $\chi_2$ is also a homomorphism.*

*Proof.* Let

$$\kappa: \quad \operatorname{Im} \psi_1 + \operatorname{Im} \chi_2 \to \operatorname{Im} \psi_1 + \operatorname{Im} \chi_2/\operatorname{Im} \chi_2 = \operatorname{Im} \psi_1',$$

$$\kappa^{-1}: \operatorname{Im} \psi_1' \to \operatorname{Im} \psi_1 + \operatorname{Im} \chi_2$$

be canonical mappings. Since the first column of (2) is semi-exact, it follows that the restriction of the mapping $\psi_2$ on $\operatorname{Im}\psi_1+\operatorname{Im}\chi_2$, in composition with some identity mapping, acts to $\operatorname{Im}\psi_2\chi_2$. Because the diagram (2) is commutative, $\operatorname{Im}\psi_2\chi_2\sim\operatorname{Im}\chi_3\phi$, whence $\psi_2(\operatorname{Im}\psi_1+\operatorname{Im}\chi_2)\prec\operatorname{Im}\chi_3\phi$; therefore, the mapping $\psi_2$ (in composition with some identity mapping) carries $\operatorname{Im}\psi_1+\operatorname{Im}\chi_2$ into $\operatorname{Im}\chi_3\phi$.

Because the right-hand column is exact, there exists a single-valued mapping $\chi_3^{-1}\colon\operatorname{Im}\chi_3\to Y_3$, which is inverse to $\chi_3$. We consider its restriction on the subfamily $\operatorname{Im}\chi_3\phi$. Since $\chi_3^{-1}\chi_3\phi\sim\phi$, the restriction carries $\operatorname{Im}\chi_3\phi$ into the subfamily $\operatorname{Im}\phi$ of the family $Y$. Since $\phi$ is a homomorphism, the inverse mapping $\phi^{-1}\colon\operatorname{Im}\phi\to Y_2$ is defined. We consider the composition $\omega=\chi_2\phi^{-1}\chi_3^{-1}\psi_2$, which acts from $\operatorname{Im}\psi_1+\operatorname{Im}\chi_2$ to $\operatorname{Im}\chi_2$. Since $\psi_2\chi_2\sim\chi_3\phi$, we have

$$\psi_2\,\omega\sim\chi_3\,\phi\,\phi^{-1}\chi_3^{-1}\psi_2\sim\psi_2.$$

So, for the mapping $\delta=(i-\omega)\,\kappa^{-1}$, which acts from $\operatorname{Im}\psi_1'$ to $Z_2$, we have $\psi_2\delta\sim 0$, that is, $\operatorname{Im}\delta\prec\operatorname{Ker}\psi_2$. Since the first row of (2) is algebraically exact, $\operatorname{Ker}\psi_2\prec\operatorname{Im}\psi_1$, and since $\psi_1$ is a homomorphism, there exists a mapping $\psi_1^{-1}\colon\operatorname{Im}\psi_1\to Z_1$, which is inverse to $\psi_1$. We now consider the composition $\psi_1^{-1}\delta\colon\operatorname{Im}\psi_1'\to Z_1$. It is clear that

$$\psi_1'\,\psi_1^{-1}\,\delta\sim\kappa\,\delta=j-\kappa\,\omega\,\kappa^{-1}$$

where $j$ is a unit mapping of the family $\operatorname{Im}\psi_1'$. Since the mapping $\omega$ acts into $\operatorname{Im}\chi_2$, the composition $\kappa\omega$ is a null mapping. Therefore, $\psi_1'\psi_1^{-1}\delta\sim j$, from which (by Proposition 6 of § 1) it follows that $\psi_1'$ is a homomorphism. This proves assertion A).

We now prove assertion B). Suppose that $Z_1\sim 0$. Since the right-hand column of (2) is exact, the mapping $\chi_3$ is an isomorphism from $\operatorname{Im}\phi$ onto $\operatorname{Im}\chi_3\phi\sim\operatorname{Im}\psi_2\chi_2$. Therefore, we may consider the composition

$$\chi\colon\ \operatorname{Im}\chi_2\xrightarrow{\psi_2}\operatorname{Im}\psi_2\chi_2\xrightarrow{\chi_3^{-1}}\operatorname{Im}\phi\xrightarrow{\phi^{-1}}Y_2.$$

Furthermore, we have

$$\psi_2\chi_2\chi=\psi_2\chi_2\phi^{-1}\chi_3^{-1}\psi_2\sim\chi_3\,\phi\,\phi^{-1}\chi_3^{-1}\psi_2\sim\psi_2,$$

that is, $\psi_2(\chi_2\chi-i)\sim 0$, where $i$ is a unit mapping of $Z_2$. Since the first row is exact, $\operatorname{Ker}\psi_2\sim 0$, and therefore, $\chi_2\chi\sim i$. It follows that $\chi_2$ is a homomorphism. ▯

**Lemma 2.** *Let us consider the commutative diagram*

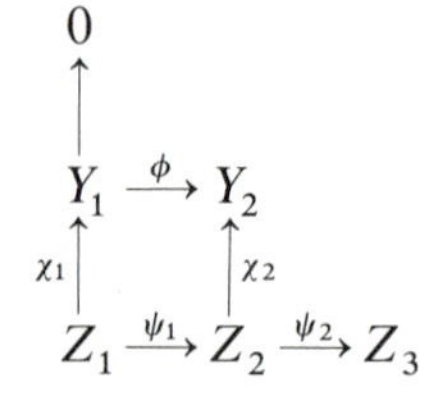


*of mappings of families and suppose that the lefthand column is exact, that the lowest row is algebraically exact, and that $\phi$ and $\psi_2$ are homomorphisms. Then*

A) *The mapping* $\psi_2'$: $\operatorname{Ker}\chi_2 \to Z_3$, *a restriction of* $\psi_2$, *is a homomorphism;*

B) *If* $Z_3 \sim 0$, *then* $\chi_2$ *is also a homomorphism.*

*Proof.* It follows from the conditions that there exists a mapping $\psi_2^{-1}$: $\operatorname{Im}\psi_2 \to Z_2/\operatorname{Ker}\psi_2$, inverse to $\psi_2$. This mapping carries the subfamily $\operatorname{Im}\psi_2'$ into the image of the canonical mapping $\operatorname{Ker}\chi_2 \to Z_2/\operatorname{Ker}\psi_2$, that is, into the family $(\operatorname{Ker}\chi_2 + \operatorname{Ker}\psi_2)/\operatorname{Ker}\psi_2$. Let $\lambda$: $Z_2/\operatorname{Ker}\psi_2 \to Z_2$ be a canonical mapping. The composition $\lambda\psi_2^{-1}$ carries $\operatorname{Im}\psi_2'$ into the sum $\operatorname{Ker}\chi_2 + \operatorname{Ker}\psi_2 \sim \operatorname{Ker}\chi_2 + \operatorname{Im}\psi_1$.

Let us now consider the sequence of mappings

$$\operatorname{Ker}\chi_2 + \operatorname{Im}\psi_1 \xrightarrow{\chi_2} \operatorname{Im}\chi_2\psi_1 \sim \operatorname{Im}\phi\chi_1 \sim \operatorname{Im}\phi \xrightarrow{\phi^{-1}} Y_1 \xrightarrow{\chi_1^{-1}} Z_1 \xrightarrow{\psi_1} \operatorname{Ker}\psi_2.$$

The composition of these mappings will be denoted by $\rho$. Since $\chi_2\phi_1 \sim \phi\chi_1$, we have $\chi_2\rho \sim \phi\chi_1\chi_1^{-1}\phi^{-1}\chi_2 \sim \chi_2$. Thus $\chi_2(i-\rho)\sim 0$, and, therefore, the mapping $(i-\rho)\lambda\psi_2^{-1}$ (up to an equivalence) acts from $\operatorname{Im}\psi_2'$ to $\operatorname{Ker}\chi_2$. Moreover

$$\psi_2'(i-\rho)\lambda\psi_2^{-1} = \psi_2\lambda\psi_2^{-1} \sim k,$$

where $k$ is a unit mapping of $Z_3$. It follows that $\psi_2'$ is a homomorphism, and with this, assertion A) is proved.

We prove assertion B). Let $Z_3 \sim 0$. Then, since the lefthand column and the lower row are algebraically exact, we have $\operatorname{Im}\chi_2 \sim \operatorname{Im}\phi$. We consider the composition of the mappings

$$\chi_2': \operatorname{Im}\chi_2 \xrightarrow{\phi^{-1}} Y_1 \xrightarrow{\chi_1^{-1}} Z_1 \xrightarrow{\psi_1} Z_2.$$

The mapping $\chi_2'$ satisfies the relations

$$\chi_2\chi_2' = \chi_2\psi_1\chi_1^{-1}\phi^{-1} \sim \phi\chi_1\chi_1^{-1}\phi^{-1} \sim l,$$

where $l$ is a unit mapping of $Y_2$; therefore, $\chi_2$ is a homomorphism. □

**Lemma 3.** *Consider the commutative diagram*

$$\begin{array}{ccc} 0 & & \\ \uparrow & & \\ Z_1 & \xrightarrow{\psi} & Z_2 \\ {\scriptstyle\chi_1}\uparrow & & \uparrow{\scriptstyle\chi_2} \\ Y_1 & \xrightarrow{\phi} & Y_2 \\ & & \uparrow \\ & & 0 \end{array} \tag{3.2}$$

*of mappings of families.*

a) *Let $\chi_1$ be an epimorphism, and let the second column be exact. Then if $\phi$ is a homomorphism, so also is $\psi$.*

b) *Let $\chi_2$ be a monomorphism and let the first column be exact. Then if $\psi$ is a homomorphism, so also is $\phi$.*

*Proof.* We establish the assertion a). Since $\chi_1$ is an epimorphism, it follows that $\operatorname{Im}\psi \sim \operatorname{Im}\psi\chi_1 \sim \operatorname{Im}\chi_2\phi$. Therefore, the mapping $\chi_2$ (up to an equivalence) establishes an isomorphism between $\operatorname{Im}\psi$ and $\operatorname{Im}\phi$. The desired mapping $\psi^{-1}$ is defined as the composition of the mappings

$$\operatorname{Im}\psi \xrightarrow{\chi_2^{-1}} \operatorname{Im}\phi \xrightarrow{\phi^{-1}} Y_1 \xrightarrow{\chi_1} Z_1.$$

We have

$$\psi\psi^{-1} = \psi\chi_1\phi^{-1}\chi_2^{-1} \sim \chi_2\phi\phi^{-1}\chi_2^{-1} \sim i,$$

where $i$ is a unit mapping of $\operatorname{Im}\psi$.

We now prove assertion b). Since $\chi_1$ is an epimorphism and the diagram (3) is commutative, we have the equivalence $\operatorname{Im}\chi_2\phi \sim \operatorname{Im}\psi\chi_1 \sim \operatorname{Im}\psi$. The desired mapping $\phi^{-1}$ is the composition of the mappings

$$\operatorname{Im}\phi \xrightarrow{\chi_2} \operatorname{Im}\psi \xrightarrow{\psi^{-1}} Z_1 \xrightarrow{\chi_1^{-1}} Y_1.$$

Since $\chi_2$ is a monomorphism, the relations

$$\chi_2\phi\phi^{-1} = \chi_2\phi\chi_1^{-1}\psi^{-1}\chi_2 \sim \psi\chi_1\chi_1^{-1}\psi^{-1}\chi_2 \sim \chi_2$$

imply that $\phi\phi^{-1} \sim j$, where $j$ is a unit mapping of $Y_2$. This proves assertion b). □

**3°. Proof of the first assertion of the theorem.** We consider the diagram ($\alpha$). Let us suppose that the mapping $f$ is a homomorphism. We then show that the mapping $g$ is also a homomorphism. In the case $n=2$, the proof follows from Assertion B) of Lemma 1. Supposing that $n>2$, we apply Lemma 1 to the portion of the diagram ($\alpha$) formed by the two lowest rows and the columns with the numbers $n-3$, $n-2$, and $n-1$. This shows that the mapping

$$X_2^{n-3} \to X_2^{n-2}/I_2^{n-2}, \qquad I_2^{n-2} = \operatorname{Im}(X_1^{n-2} \to X_2^{n-2}), \tag{4.2}$$

which is an extension of the mapping $X_2^{n-3} \to X_2^{n-2}$, is a homomorphism. We note that the family $I_2^{n-2}$ belongs to the kernel of the mapping $X_2^{n-2} \to X_3^{n-2}$. Thus, we may form the associated mapping

$$X_2^{n-2}/I_2^{n-2} \to X_3^{n-2}.$$

We now consider the diagram:

$$\begin{array}{ccccc} X_3^{n-4} & \longrightarrow & X_3^{n-3} & \longrightarrow & X_3^{n-2} \\ & & \uparrow & & \uparrow \\ & & X_2^{n-3} & \longrightarrow & X_2^{n-2}/I_2^{n-2} \\ & & & & \uparrow \\ & & & & 0 \end{array}$$

and we again apply Lemma 1. We note that all of the mappings involved are homomorphisms, and that the rows and columns are exact. Therefore the mapping

$$X_3^{n-4} \to X_3^{n-3}/I_3^{n-3},$$

which is defined in the same way as (4), is also a homomorphism. A repeated application of this argument brings us to the diagram:

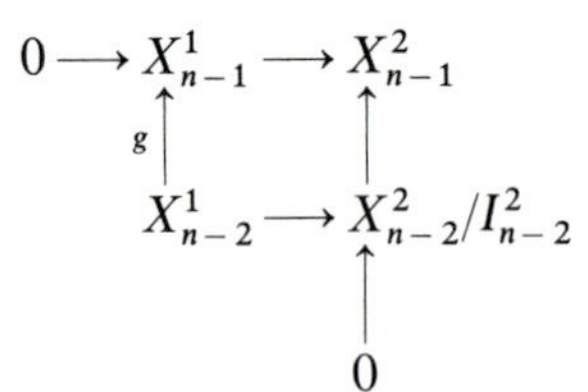

in which the rows and columns are exact, and all the mappings, with the possible exclusion of $g$, are homomorphisms. If we apply assertion B) of Lemma 1, we see that $g$ is also a homomorphism. This proves the first assertion of the theorem for the diagram ($\alpha$).

We now take up diagram ($\beta$). Suppose it is known that $f$ is a homomorphism. A repeated application of Lemma 1 leads us, as in the preceding case, to the diagram

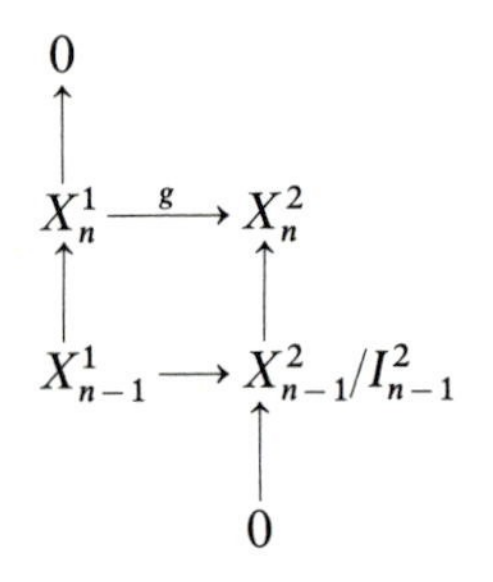

in which both columns are exact, and all the mappings, with the possible exception of $g$, are homomorphisms. If we apply the assertion a) of Lemma 3, we see that $g$ is also a homomorphism.

Let us now suppose that $g$ is a homomorphism. Applying Lemma 2 to the portion of the diagram ($\beta$) formed by the first two rows and the first three columns, we conclude that the mapping

$$K_{n-1}^2 \to X_{n-1}^3, \qquad K_{n-1}^2 = \operatorname{Ker}(X_{n-1}^2 \to X_n^2), \tag{5.2}$$

which is a restriction of the mapping $X_{n-1}^2 \to X_{n-1}^3$, is a homomorphism. We note that the image of the mapping $X_{n-2}^2 \to X_{n-1}^2$ belongs to $K_{n-1}^2$. We may therefore consider the diagram:

$$\begin{array}{ccccc}
0 & & & & \\
\uparrow & & & & \\
K_{n-1}^2 & \longrightarrow & X_{n-1}^3 & & \\
\uparrow & & \uparrow & & \\
X_{n-2}^2 & \longrightarrow & X_{n-2}^3 & \longrightarrow & X_{n-2}^4
\end{array}$$

All the mappings in this diagram are homomorphisms, and the rows and column are exact. Applying Lemma 2, we find that the mapping

$$K_{n-2}^3 \to X_{n-2}^4,$$

which is an analog of (5), is also a homomorphism. By repeated applications of this argument, we are led to the diagram:

$$\begin{array}{ccc}
0 & & \\
\uparrow & & \\
K_2^{n-1} & \longrightarrow & X_2^n \\
\uparrow & & \uparrow \\
X_1^{n-1} & \xrightarrow{f} & X_1^n \\
 & & \uparrow \\
 & & 0
\end{array}$$

in which both columns are exact, and all the mappings, with the exception, perhaps, of $f$, are homomorphisms. But assertion b) of Lemma 3 implies that the mapping $f$ is also a homomorphism. We have thus proved the first assertion of the theorem for diagram ($\beta$).

Let us now prove this assertion for diagram ($\gamma$). In order to obtain the implication "$g$ is a homomorphism implies $f$ is a homomorphism," it is sufficient to make a repeated application of assertion A) of Lemma 2 (as we did before), and then to apply assertion B) of the same lemma. The converse implication reduces to what we have proved by the symmetry of diagram ($\gamma$) with respect to its bisectors. Thus, the first assertion of the theorem is completely proved.

**4°. Proof of the second assertion of the theorem.** We shall first prove a lemma.

**Lemma 4.** *Suppose that in the diagram*

$$\begin{array}{ccc} & Z \xrightarrow{\psi} Z' \\ & \uparrow \chi \\ Y' \xrightarrow{\phi} Y & \end{array}$$

*$\chi$ is a homomorphism and $\psi\chi\phi\sim 0$. Then there exists an isomorphism*

$$[\mathrm{Im}\,\chi\cap\mathrm{Ker}\,\psi]/\mathrm{Im}\,\chi\phi\cong\mathrm{Ker}\,\psi\chi/[\mathrm{Ker}\,\chi+\mathrm{Im}\,\phi],$$

*associated with $\chi$.*

*Proof.* Since $\chi$ is a homomorphism, there exists an inverse mapping $\chi^{-1}$: $\mathrm{Im}\,\chi\to Y/\mathrm{Ker}\,\chi$. The restriction of this mapping on $\mathrm{Im}\,\chi\cap\mathrm{Ker}\,\psi$ carries the subfamily into the family $\mathrm{Ker}\,\psi\chi/\mathrm{Ker}\,\chi$. The mapping $\chi^{-1}$ establishes an isomorphism between the subfamily $\mathrm{Im}\,\chi\phi\subset\mathrm{Im}\,\chi\cap\mathrm{Ker}\,\psi$ and the image of the mapping $\mathrm{Im}\,\phi\to Y/\mathrm{Ker}\,\chi$. This image is clearly equal to the family $[\mathrm{Ker}\,\chi+\mathrm{Im}\,\phi]/\mathrm{Ker}\,\chi$. Accordingly, the mapping associated with $\chi$ establishes an isomorphism

$$\begin{aligned}[\mathrm{Im}\,\chi\cap\mathrm{Ker}\,\psi]/\mathrm{Im}\,\chi\phi &\cong(\mathrm{Ker}\,\psi\chi/\mathrm{Ker}\,\chi)/([\mathrm{Ker}\,\chi+\mathrm{Im}\,\phi]/\mathrm{Ker}\,\chi)\\ &\cong\mathrm{Ker}\,\psi\chi/[\mathrm{Ker}\,\chi+\mathrm{Im}\,\phi].\quad\square\end{aligned}$$

Let us now turn to the diagrams ($\alpha$), ($\beta$), ($\gamma$) and prove that the extreme rows and columns are semi-exact. We show, for example that $f'f\sim 0$ in the diagram ($\alpha$). The composition of the mappings $f'f$ and the monomorphism $X_1^n\to X_2^n$ is equal to the composition of the mappings

$$X_1^{n-2}\to X_2^{n-2}\to X_2^{n-1}\to X_2^n;$$

the composition of the two latter mappings is zero. Hence $f'f\sim 0$.

We now establish the isomorphism (1) for the diagram ($\alpha$). For every pair $(i,j)$, satisfying the relationship $i+j=n-1$ or $n$, there is defined a mapping $X_i^j\to X_{i+1}^{j+1}$, as the composition of the mappings

$$X_i^j\to X_i^{j+1}\to X_{i+1}^{j+1}\quad\text{or}\quad X_i^j\to X_{i+1}^j\to X_{i+1}^{j+1}. \tag{6.2}$$

These compositions are equivalent since the diagram is commutative. For any pair $(i,j)$ with $i+j=n$ or $n+1$, we consider the family $H_i^j$, associated with $X_i^j$, which is defined for $i+j=n$ by the formula

$$H_i^j=\mathrm{Ker}(X_i^j\to X_{i+1}^{j+1})/[\mathrm{Im}(X_i^{j-1}\to X_i^j)+\mathrm{Im}(X_{i-1}^j\to X_i^j)],$$

and for $i+j=n+1$ by the formula

$$H_i^j=[\mathrm{Ker}(X_i^j\to X_i^{j+1})\cap \mathrm{Ker}(X_i^j\to X_{i+1}^j)]/\mathrm{Im}(X_{i-1}^{j-1}\to X_i^j).$$

We now apply Lemma 4 to the fragment:

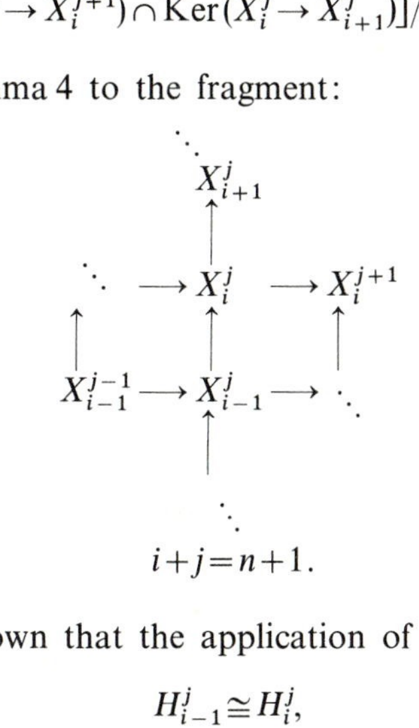

$$i+j=n+1.$$

It can be easily shown that the application of this lemma yields the isomorphism

$$H_{i-1}^j\cong H_i^j, \tag{7.2}$$

associated with the mapping $X_{i-1}^j\to X_i^j$. In the same way, we can establish the isomorphism

$$H_i^{j-1}\cong H_i^j, \tag{8.2}$$

associated with the mapping $X_i^{j-1}\to X_i^j$. Combining the isomorphisms (7) and (8), we are led to the isomorphism $H_{n-1}^1\cong H_1^{n-1}$. We note that

$$H_{n-1}^1\sim \mathrm{Ker}\, g'/\mathrm{Im}\, g, \qquad H_1^{n-1}\sim \mathrm{Ker}\, f'/\mathrm{Im}\, f,$$

and obtain (1). The construction of the isomorphism (1) for the diagrams ($\beta$) and ($\gamma$) is carried out in the same way. □

**Corollary.** *Suppose that in the diagrams* ($\alpha$), ($\beta$), *and* ($\gamma$) *all rows and columns are exact, with the exception of one extreme row or column. Then this row (or column) is also exact.*

**5°. Refinement.** I. For the sake of simplicity, we shall suppose that the diagrams ($\alpha$), ($\beta$), and ($\gamma$) are strongly commutative, that is, for arbitrary $i$ and $j$, both of the through mappings (6) coincide (they are not only equivalent as has been assumed up to now). We shall also suppose that all the rows and columns are strictly semi-exact, that is, the composition of two mappings in any rows or columns is equal to zero. We note that in the applications of Theorem 1, these conditions will be fulfilled in almost all cases.

The hypotheses of the second part of the theorem amount to the statement that there exist (generally speaking, many-valued) mappings

$$\begin{aligned} &\operatorname{Ker}(X_i^{j+1} \to X_i^{j+2}) \to X_i^j, \\ &\operatorname{Ker}(X_{i+1}^j \to X_{i+2}^j) \to X_i^j, \end{aligned} \tag{9.2}$$

which are inverse to the mappings

$$X_i^j \to X_i^{j+1}, \qquad X_i^j \to X_{i+1}^j. \tag{10.2}$$

On the other hand, the assertion of the second part of the theorem consists in the statement that there exist mappings

$$\operatorname{Ker} f'/\operatorname{Im} f \underset{J}{\overset{I}{\rightleftarrows}} \operatorname{Ker} g'/\operatorname{Im} g,$$

which set up an isomorphism. If we inspect the proof of the theorem, we can see that the mappings $I$ and $J$ can be effectively constructed, and are polynomials in the mappings (9) and (10). Therefore, the orders (in the sense of 3° § 1) of the mappings $I$ and $J$ are well-defined compositions of the orders of the mappings (9) and (10) and, therefore, depend only on the orders of these mappings, and not on the mappings (9) and (10) themselves.

II. The following result shows how far we can weaken the hypotheses of Theorem 1, and still have it continue to act in one direction (in this case, from above downward).

**Theorem 1′.** *Suppose that in the commutative diagram* (α) *all rows and columns are semi-exact and that the sequences*

$$X_i^j \to X_i^{j+1} \to X_i^{j+2}, \quad i+j=n \quad i=2, \dots, n \ (X_n^0=0),$$

*and*

$$X_{i-1}^j \to X_i^j \to X_{i+1}^j, \quad i+j=n \quad i=1, \dots, n-1 \quad (X_0^n=0),$$

*are algebraically exact. Then the bottom row of the diagram is also algebraically exact.*

The proof of this theorem is left to the reader.

### 6°. Mappings of diagrams of types (α), (β), (γ)

**Theorem 2.** *Suppose given two diagrams of type* (α) *having the same size*

$$\{X_i^j; f, f', g, g'\}, \qquad \{\mathscr{X}_i^j; \mathscr{f}, \mathscr{f}', \mathscr{g}, \mathscr{g}'\}, \tag{11.2}$$

*and satisfying the condition of Theorem* 1. *We suppose further that for an arbitrary pair* $(i, j)$ *with* $i+j=n-1$, $n$, $n+1$, $n+2$, *there are mappings* $\phi_i^j$: $X_i^j \to \mathscr{X}_i^j$, *such that the mappings* $\phi_i^j$ *commute in the ensemble with the mappings of the diagram* (11). *Then the associated mappings* $\Phi_{n-1}^1$ *and*

$\check{\phi}_1^{n-1}$ *form a commutative diagram:*

$$\begin{array}{ccc} \operatorname{Ker} \mathscr{f}'/\operatorname{Im} \mathscr{f} & \overset{\mathscr{I}}{\cong} & \operatorname{Ker} \mathscr{g}'/\operatorname{Im} \mathscr{g} \\ {\scriptstyle \check{\phi}_1^{n-1}}\uparrow & & \uparrow{\scriptstyle \check{\phi}_{n-1}^1} \\ \operatorname{Ker} f'/\operatorname{Im} f & \overset{I}{\cong} & \operatorname{Ker} g'/\operatorname{Im} g \end{array} \tag{12.2}$$

*in which $I$ and $\mathscr{I}$ are isomorphisms of type* (1). *A similar assertion holds for diagrams of type* ($\beta$) *and* ($\gamma$).

*Proof.* We shall consider only diagrams of type ($\alpha$). It follows from our hypotheses that mappings associated with $\phi_i^j$ carry kernel and image of the mapping $X_i^j \to X_i^{j+1}$ respectively into kernel and image of the mapping $\mathscr{X}_i^j \to \mathscr{X}_i^{j+1}$ and so on. Therefore, for an arbitrary pair $(i, j)$ with $i+j=n$, $n+1$, we may define the mapping $H_i^j \to \mathscr{H}_i^j$, associated with $\phi_i^j$, where $\mathscr{H}_i^j$ is a family analogous to $H_i^j$, constructed for the diagram $\{X_i^j\}$. Then, we have the commutative diagram:

$$\begin{array}{ccccc} \mathscr{H}_{i-1}^j & \cong & \mathscr{H}_i^j & \cong & \mathscr{H}_i^{j-1} \\ \uparrow & & \uparrow & & \uparrow \\ H_{i-1}^j & \cong & H_i^j & \cong & H_i^{j-1} \end{array} \qquad i+j=n+1,$$

in which the rows contain isomorphisms of type (7) and (8). Combining these diagrams, we arrive at the commutative diagram

$$\begin{array}{ccc} \mathscr{H}_1^{n-1} & \cong & \mathscr{H}_{n-1}^1 \\ \uparrow & & \uparrow \\ H_1^{n-1} & \cong & H_{n-1}^1 \end{array}$$

which coincides with (12). ▯

## § 3. Operations on modules

In this section, we introduce the functors $\otimes$ and Hom, and their derivatives[1] and we establish certain properties of these functors, limiting ourselves to the case in which the ring is a Noether ring and the first argument is a finite module. The facts which we set forth here are in general well known; they can be found for example in the text of Zarisky and Samuels [1] and Cartan and Eilenberg [1]. For the convenience of the reader, however, we supply the proofs of some of them.

---

1 We assume that the reader has at least a superficial acquaintance with the concept of functor; see, for example, Godement [1] § 1, Chapter I.

The symbol $A$ will again be used for an arbitrary commutative ring over the field $C$. All the $A$-modules that we shall encounter will be assumed to be topological, and all $A$-mappings are assumed to be continuous and single-valued.

Let $\Phi$ be an $A$-module and let $k$ be an integer. The module $\Phi^k$ will be interpreted as a module formed of columns of length $k$, the components of which belong to $\Phi$. The module $A^k$ has a canonical basis, formed of columns $e_i$, $i=1, \ldots, k$, of unit matrices of order $k$.

Let $p$: $A^s \to A^t$ be an $A$-mapping. Such a mapping is characterized by its values $p(e_i)$ on the basis elements of the modules $A^s$. We denote by $p$ a matrix of size $t \times s$, formed of the columns $p(e_i)$, $i=1, \ldots, s$. It is clear that the mapping $p$ is effectively a multiplication of the columns of $A^s$ on the left by the matrix $p$. Conversely, to every matrix $p$ of size $t \times s$ (that is, with $t$ rows and $s$ columns) with elements from $A$, there corresponds an $A$-mapping, $p$: $A^s \to A^t$, which is obtained by multiplication on the left by this matrix. We shall refer to matrices of this type as $A$-matrices.

To every $A$-matrix $p$ of size $t \times s$ and to every module $\Phi$ there corresponds a mapping $p$: $\Phi^s \to \Phi^t$, which consists in the multiplication of the columns $\Phi^s$ on the left by the matrix. The image of this mapping will be denoted by $p\Phi^s$ and the kernel by $\Phi_p$. An $A$-module $M$ will be said to be finite if it has a finite basis. For every integer $k$, for example, the module $A^k$ is finite, since it has a finite basis $\{e_i\}$. This module is, moreover, free, since its canonical basis $\{e_i\}$ is formed of elements linearly independent over $A$. All finite modules will be endowed with the discrete topology.

The free resolution of a finite module $M$ is an arbitrary exact sequence of the form

$$\cdots \to A^{s_2} \xrightarrow{p_1} A^{s_1} \xrightarrow{p_0} A^{s_0} \xrightarrow{a} M \to 0, \tag{1.3}$$

where $p_i$, $i=0, 1, 2, \ldots$ are $A$-matrices. If the ring $A$ is a Noether ring, then every finite $A$-module has at least one free resolution. From now on, we shall assume always that the ring $A$ is a Noether ring. This implies that an arbitrary submodule of a finite $A$-module is also finite. Clearly, a factor-module of a finite module is also finite.

**1°. The tensor product of modules and the functor Tor.** Let $M$ be a finite $A$-module and let (1) be a free resolution of it. Let $\Phi$ be an $A$-module. In the sequence (1), we replace the powers of the ring $A$ by the same powers of the module $\Phi$ and we consider the sequence

$$\cdots \to \Phi^{s_2} \xrightarrow{p_1} \Phi^{s_1} \xrightarrow{p_0} \Phi^{s_0} \to 0.$$

This sequence, generally speaking, is only semi-exact, since the exactness of (1) implies $p_i\, p_{i+1}=0$ for all $i$. We consider the homologies of this sequence, which are denoted by

$$M \otimes \Phi = \operatorname{Tor}_0(M, \Phi) = \Phi^{s_0}/p_0 \Phi^{s_1}$$
$$\operatorname{Tor}_i(M, \Phi) = \Phi_{p_{i-1}}/p_i \Phi^{s_{i+1}} \qquad i=1,2,\ldots. \tag{2.3}$$

In order to validate this notation, it is necessary to show that the factor-modules do, in fact, depend only on the modules $M$ and $\Phi$ and do not depend on the choice of the free resolution (1).

**Proposition 1**[2]. *Let*

$$\cdots \longrightarrow A^{\sigma_2} \xrightarrow{\pi_1} A^{\sigma_1} \xrightarrow{\pi_0} A^{\sigma_0} \xrightarrow{\alpha} M' \longrightarrow 0 \tag{3.3}$$

*be a free resolution of the finite A-module $M'$, and let $f$: $M \to M'$ be an A-mapping. There exist A-matrices $f_i$, $i=0,1,2,\ldots$, which make the diagram:*

$$\begin{array}{ccccccccc} \cdots \longrightarrow & A^{\sigma_2} & \xrightarrow{\pi_1} & A^{\sigma_1} & \xrightarrow{\pi_0} & A^{\sigma_0} & \xrightarrow{\alpha} & M' & \longrightarrow 0 \\ & {\scriptstyle f_2}\uparrow & & {\scriptstyle f_1}\uparrow & & {\scriptstyle f_0}\uparrow & & {\scriptstyle f}\uparrow & \\ \cdots \longrightarrow & A^{s_2} & \xrightarrow{p_1} & A^{s_1} & \xrightarrow{p_0} & A^{s_0} & \xrightarrow{\alpha} & M & \longrightarrow 0 \end{array} \tag{4.3}$$

*commutative.*

*If $g_i$, $i=0,1,2,\ldots$, are other A-matrices which make this diagram commutative, then these two diagrams are homotopic, that is, there exist A-matrices $h_i$: $A^{s_i} \to A^{\sigma_{i+1}}$, $i=0,1,2,\ldots$, such that*

$$f_i - g_i = \pi_i h_i + h_{i-1} p_{i-1}, \qquad i=0,1,2,\ldots\ (h_{-1}=0). \tag{5.3}$$

Since the diagram (4) is commutative, the mapping $f_i$: $\Phi^{s_i} \to \Phi^{\sigma_i}$ carries the submodule $\Phi_{p_{i-1}}$ into $\Phi_{\pi_{i-1}}$, and the submodule $p_i \Phi^{s_{i+1}}$ into $\pi_i \Phi^{\sigma_{i+1}}$. We can define the associated mapping

$$\check{f}_i\colon \Phi_{p_{i-1}}/p_i \Phi^{s_{i+1}} \to \Phi_{\pi_{i-1}}/\pi_i \Phi^{\sigma_{i+1}}, \qquad i=0,1,2,\ldots. \tag{6.3}$$

**Proposition 2.**

I. *The mappings* (6) *depend only on the module $\Phi$ and the mapping $f$.*

II. *Assume that the mapping $f$ is an isomorphism. Then, the mappings* (6) *are also isomorphisms, and their inverses are mappings associated with some A-matrices.*

III. *We again assume that $f$ is an isomorphism. Then, if some mapping $\pi_i$: $\Phi^{\sigma_{i+1}} \to \Phi^{\sigma_i}$ is a homomorphism, the mapping $p_i$: $\Phi^{s_{i+1}} \to \Phi^{s_i}$ is also a homomorphism.*

*Proof.* If $g_i$ is another matrix which renders the diagram (4) commutative, then by Proposition 1, the relation (5) is valid. But it is then clear that the mappings of the form (6) associated with the matrices $f_i$

2 The proof of this proposition can be found in Cartan and Eilenberg [1], Chapter V, and in Godement [1], Chapter I, § 5.

and $g_i$ coincide, since the restriction of the mapping $h_{i-1}p_{i-1}$ on $\Phi_{p_{i-1}}$ is null, and the component $\pi_i h_i$ carries the module $\Phi_{p_{i-1}}$ into $\pi_i \Phi^{\sigma_{i+1}}$. This proves the first assertion.

We assume that $f$ is an isomorphism. Then Proposition 1 allows us to construct $A$-matrices $\phi_i$ such that the following diagram is commutative:

$$\begin{array}{ccccccccc}
\cdots \longrightarrow & A^{s_2} & \xrightarrow{p_1} & A^{s_1} & \xrightarrow{p_0} & A^{s_0} & \longrightarrow & M & \longrightarrow 0 \\
 & \uparrow{\scriptstyle \phi_2} & & \uparrow{\scriptstyle \phi_1} & & \uparrow{\scriptstyle \phi_0} & & \uparrow{\scriptstyle f^{-1}} & \\
\cdots \longrightarrow & A^{\sigma_2} & \xrightarrow{\pi_1} & A^{\sigma_1} & \xrightarrow{\pi_0} & A^{\sigma_0} & \longrightarrow & M' & \longrightarrow 0
\end{array} \tag{7.3}$$

and, therefore, the mappings

$$\check{\phi}_i\colon \Phi_{\pi_{i-1}}/\pi_i \Phi^{\sigma_{i+1}} \to \Phi_{p_{i-1}}/p_i \Phi^{s_{i+1}},$$

associated with the matrices $\phi_i$ are defined. We shall show that these mappings are the inverses of the mappings (6). Combining diagrams (4) and (7), we obtain the commutative diagram:

$$\begin{array}{ccccccccc}
\cdots \longrightarrow & A^{s_0} & \xrightarrow{p_1} & A^{s_1} & \xrightarrow{p_0} & A^{s_0} & \longrightarrow & M & \longrightarrow 0 \\
 & \uparrow{\scriptstyle \phi_2 f_2} & & \uparrow{\scriptstyle \phi_1 f_1} & & \uparrow{\scriptstyle \phi_0 f_0} & & \| & \\
\cdots \longrightarrow & A^{s_2} & \xrightarrow{p_1} & A^{s_1} & \xrightarrow{p_0} & A^{s_0} & \longrightarrow & M & \longrightarrow 0
\end{array}$$

On the other hand, we can construct an analogous commutative diagram, in which instead of the mappings $\phi_i f_i$, we find the identity mappings $e\colon A^{s_i} \to A^{s_i}$. By Proposition 1, both of these diagrams are homotopic, that is,

$$\phi_i f_i - e^{s_i} = p_i h_i + h_{i-1} p_{i-1}, \qquad i = 0, 1, 2, \ldots, \tag{8.3}$$

with some $A$-matrices $h_i\colon A^{s_i} \to A^{s_{i+1}}$ ($h_{-1} = 0$). It now follows that the mapping

$$\check{\phi}_i \check{f}_i\colon \Phi_{p_{i-1}}/p_i \Phi^{s_{i+1}} \to \Phi_{p_{i-1}}/p_i \Phi^{s_{i+1}},$$

associated with the matrix $\phi_i f_i$ coincides with the mapping associated with the matrix $e$; i.e., it is an identity mapping. From the symmetry of the mappings $\phi_i$ and $f_i$, it follows that the composition $\check{f}_i \check{\phi}_i$ is also an identity mapping. This proves the second assertion.

We now take up the third. Suppose that for some $i \geqq 0$ the mapping $\pi_i\colon \Phi^{\sigma_{i+1}} \to \Phi^{\sigma_i}$ is a homomorphism, that is, that there is defined a many-valued mapping

$$\rho\colon \pi_i \Phi^{\sigma_{i+1}} \to \Phi^{\sigma_{i+1}},$$

inverse to $\pi_i$. We consider the mapping

$$r\colon p_i \Phi^{s_{i+1}} \xrightarrow{f_i} \pi_i \Phi^{\sigma_{i+1}} \xrightarrow{\rho} \Phi^{\sigma_{i+1}} \xrightarrow{\phi_{i+1}} \Phi^{s_{i+1}}.$$

Since the diagram (7) is commutative,

$$p_i(r-h_i)=p_i\,\phi_{i+1}\,\rho\,f_i-p_i\,h_i=\phi_i\,\pi_i\,\rho\,f_i-p_i\,h_i=\phi_i\,f_i-p_i\,h_i.$$

According to Eq. (8), the right-hand side is equal to $e+h_{i-1}\,p_{i-1}$. Since the restriction of the mapping $h_{i-1}\,p_{i-1}$ on $p_i\,\Phi^{s_{i+1}}$ is null, the mapping $p_i(r-h_i)$, on $p_i\,\Phi^{s_{i+1}}$, coincides with $e$; i.e. it is an identity mapping. Therefore the mapping $r-h_i$ is inverse to $p_i$. Accordingly, $p_i$ is a homomorphism. □

We shall assume that $M'=M$, and $f$ is an identity mapping. Then according to Proposition 2, there exist $A$-matrices, $f_i$, $i=0, 1, 2, \ldots$, such that the associated mappings (6) are isomorphisms. Then the factor-modules (2) do not depend on the choice of the free resolution (1), and this validates our notation.

Let us determine the modules (2) in the case $M=A^k$ for some integer $k$. For the free resolution of the module $A^k$, we may choose the sequence $0\to A^k\to A^k\to 0$. From the definition, we are led quickly to the isomorphisms

$$A^k\otimes\Phi\cong\Phi^k,\qquad \operatorname{Tor}_i(A^k,\Phi)=0,\qquad i\geqq 1,$$

for an arbitrary $A$-module $\Phi$. In particular, for arbitrary integers $k$ and $l$,

$$A^k\otimes A^l\cong A^{kl}.$$

If $f\colon M\to M'$ is a mapping of finite $A$-modules, then by Proposition 2 we can define the mapping $\operatorname{Tor}_i(M,\Phi)\to\operatorname{Tor}_i(M',\Phi)$, depending only on $f$ and $\Phi$. Let $\phi\colon\Phi\to\Phi'$ be a mapping of $A$-modules. It is clear that it carries $\Phi_{p_{i-1}}$ into $\Phi'_{p_{i-1}}$ and $p_i\,\Phi^{s_{i+1}}$ into $p_i\,\Phi'^{\,s_{i+1}}$. Therefore, there is an associated mapping

$$\Phi_{p_{i-1}}/p_i\,\Phi^{s_{i+1}}\to\Phi'_{p_{i-1}}/p_i\,\Phi'^{\,s_{i+1}},$$

and a corresponding mapping $\operatorname{Tor}_i(M,\Phi)\to\operatorname{Tor}_i(M,\Phi')$, both independent of the choice of the resolution (1). If we have simultaneously the mappings $f$ and $\phi$, we may consider the composition of the mappings $M\otimes\Phi\to M'\otimes\Phi\to M'\otimes\Phi'$. For this composition, one employs the special notation: $f\otimes\phi$.

In particular, if we have two mappings $p\colon A^s\to A^t$ and $q\colon A^\sigma\to A^\tau$, then their tensor product $p\otimes q\colon A^{s\sigma}\to A^{t\tau}$ corresponds to the $A$-matrix $p\otimes q$, formed as the Kronecker product of the matrices $p$ and $q$.

**Proposition 3.** *Let $M$ and $L$ be arbitrary finite $A$-modules, let* (1) *be a free resolution of the module $M$, and let*

$$\cdots\longrightarrow A^\sigma\xrightarrow{\;q\;}A^\tau\xrightarrow{\;b\;}L\longrightarrow 0$$

*be a free resolution of the module L. Then we have the isomorphisms*

$$\begin{aligned} M\otimes L &\cong A^{t\tau}/[p_0 A^{s\tau}+q A^{t\sigma}], \qquad t=s_0,\ s=s_1, \\ \operatorname{Tor}_i(M,L) &\cong [p_{i-1}A^{s_i\tau}\cap q A^{s_{i-1}\sigma}]/(p_{i-1}\otimes q)\, A^{s_i\sigma}, \qquad i\geqq 1, \end{aligned} \tag{9.3}$$

*where for brevity we have written* $p_i = p_i\otimes e^{\tau}$, $q = e^{t}\otimes q$.

*Proof.* We fix an arbitrary $i\geqq 1$ and we consider the following diagram of type $(\gamma)$:

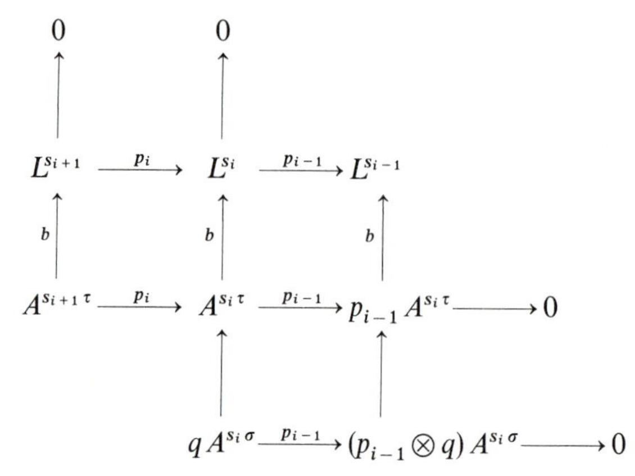

It is obviously commutative and the first and second columns and the second and third rows are exact. By Theorem 1 § 2, we conclude that the module

$$L_{p_{i-1}}/p_i L^{s_i+1} \equiv \operatorname{Tor}_i(M,L)$$

is isomorphic to the kernel of the mapping $p_{i-1}A^{s_i\tau}\xrightarrow{b} L^{s_i-1}$ factored with respect to the submodule $(p_{i-1}\otimes q)\,A^{s_i\sigma}$. This kernel is the intersection of the submodules $p_{i-1}A^{s_i\tau}\subset A^{s_{i-1}\tau}$ with the kernel of the mapping $b\colon A^{s_{i-1}\tau}\to L^{s_{i-1}}$. The latter is clearly equal to $qA^{s_i\sigma}$. Thus, the isomorphisms (9) with $i=1,2,\ldots$ have been validated.

Finally, let us consider another commutative diagram of type $(\gamma)$:

$$\begin{array}{ccccc}
0 & & 0 & & \\
\uparrow & & \uparrow & & \\
L^{s} & \xrightarrow{p} & L^{t} & \longrightarrow & 0 \\
{\scriptstyle b}\uparrow & & {\scriptstyle b}\uparrow & & \uparrow \\
A^{s\tau} & \xrightarrow{p} & A^{t\tau} & \longrightarrow & A^{t\tau}/pA^{s\tau} \longrightarrow 0 \\
& & \uparrow & & \uparrow \\
& & qA^{t\sigma} & \longrightarrow & [qA^{t\sigma}+pA^{s\tau}]/pA^{s\tau} \longrightarrow 0
\end{array}$$

It is easy to see that the second and third rows and the first and second columns are exact. Therefore, we may apply Theorem 1 § 2, whence follows the first of the isomorphisms (9). □

Since the relationship (9) is symmetric in $M$ and $L$ for the module $M \otimes L$, we have the natural isomorphism $M \otimes L \cong L \otimes M$; in other words, the tensor product is commutative. The modules $\operatorname{Tor}_i(M, L)$ have the same property for all $i \geqq 1$. The proof of this fact is left to the reader (it is suggested that he make use of the diagram (21.9) of Chapter VII).

For arbitrary finite modules $M$, $M'$ and the module $\Phi$ there is defined a natural isomorphism

$$\operatorname{Tor}_i(M \oplus M', \Phi) \cong \operatorname{Tor}_i(M, \Phi) \oplus \operatorname{Tor}_i(M', \Phi).$$

This property of the module $\operatorname{Tor}_i$ is called distributivity.

**2°. Flat modules**

**Definition 1.** An $A$-module $\Phi$ is said to be flat, if for an arbitrary exact sequence of finite $A$-modules

$$M' \xrightarrow{f} M \xrightarrow{g} M'' \tag{10.3}$$

the sequence

$$M' \otimes \Phi \xrightarrow{f \otimes I} M \otimes \Phi \xrightarrow{g \otimes I} M'' \otimes \Phi \tag{11.3}$$

(where $I$ is an identity mapping of $\Phi$) is exact.

If $\Phi$ is a flat module, then, as is easily seen, $\operatorname{Tor}_i(M, \Phi) = 0$ for all $i \geqq 1$ and for an arbitrary finite module $M$. It follows from Proposition 5 that this property is sufficient for the flatness of the module.

The simplest example of a flat module is the module $A^k$, where $k$ is an arbitrary integer. In this case, the sequence (11) takes the form:

$$[M']^k \xrightarrow{[f]^k} [M]^k \xrightarrow{[g]^k} [M'']^k.$$

Therefore, the flatness of (10) immediately implies the flatness of (11). Other examples will be met later.

We note some properties of flat modules. Let $\Phi$ be a flat module and let $p\colon A^s \to A^t$ be an $A$-matrix. Then, if the sequence $A^s \xrightarrow{p} pA^s \to 0$ is flat, the sequence

$$A^s \otimes \Phi \xrightarrow{p \otimes I} pA^s \otimes \Phi \to 0.$$

is algebraically exact. Because of the isomorphism $A^s \otimes \Phi \cong \Phi^s$ and the monomorphism $pA^s \otimes \Phi \to A^t \otimes \Phi \cong \Phi^t$, the mapping $p \otimes I$ is isomorphic to $p$. Taking account of the fact that this mapping is an epimorphism, we arrive at the isomorphism $pA^s \otimes \Phi \cong p\Phi^s$.

Let $N$ be a submodule of the finite module $M$. The identity imbedding $N \to M$ is a monomorphism, and therefore, for an arbitrary flat module

$\Phi$ the mapping $N \otimes \Phi \to M \otimes \Phi$ is also a monomorphism. Therefore the modules $N \otimes \Phi$ can be looked on as a submodule of the module $M \otimes \Phi$. We note that we have the isomorphism

$$M \otimes \Phi / N \otimes \Phi \cong M/N \otimes \Phi.$$

To prove this, it is sufficient to consider the exact sequence

$$0 \to N \to M \to M/N \to 0$$

and, subjecting it to tensor multiplication by the module $\Phi$, to make use of the fact that it is flat.

**Proposition 4.** *Let $M$ be an arbitrary finite module, let $K$ and $L$ be submodules of it, and let $\Phi$ be a flat module. From what we have said above, it follows that the module $(K \cap L) \otimes \Phi$ can be considered as a submodule of the modules $K \otimes \Phi$ and $L \otimes \Phi$. We have also the equation*

$$(K \cap L) \otimes \Phi = (K \otimes \Phi) \cap (L \otimes \Phi).$$

*Proof.* We consider the sequence

$$0 \to K \cap L \xrightarrow{\sigma} K \oplus L \xrightarrow{\rho} K + L \to 0,$$

where $\sigma$ is a mapping, carrying the element $x$ into the couple $(x, x)$, and the mapping $\rho$ carries the couple $(x, y)$ into the difference $x - y$. It is clear that this sequence is exact. Therefore, if we take the tensor product with the module $\Phi$, we arrive again at the exact sequence

$$0 \to (K \cap L) \otimes \Phi \xrightarrow{\sigma \otimes I} (K \oplus L) \otimes \Phi \xrightarrow{\rho \otimes I} (K + L) \otimes \Phi \to 0. \quad (12.3)$$

Because the tensor product is distributive, we have the isomorphism $(K \oplus L) \otimes \Phi \cong (K \otimes \Phi) \oplus (L \otimes \Phi)$. We substitute this isomorphism in (12). It is clear that the mapping $\sigma \otimes I$ goes over into the mapping $\sigma$, and the mapping $\rho \otimes I$ goes into the mapping $\rho$, that is, we obtain the following exact sequence:

$$0 \to (K \cap L) \otimes \Phi \xrightarrow{\sigma} (K \otimes \Phi) \oplus (L \otimes \Phi) \xrightarrow{\rho} (K + L) \otimes \Phi \to 0.$$

Since it is exact,

$$(K \cap L) \otimes \Phi = \operatorname{Im} \sigma = \operatorname{Ker} \rho = (K \otimes \Phi) \cap (L \otimes \Phi). \quad \square$$

### 3°. Criteria for flatness of modules

**Proposition 5.** *In order that the module $\Phi$ be flat it is necessary and sufficient that, for an arbitrary exact sequence of the form $A^s \xrightarrow{p} A^t \xrightarrow{r} A^v$, the sequence*

$$\Phi^s \xrightarrow{p} \Phi^t \xrightarrow{r} \Phi^v$$

*be exact.*

*Proof.* The necessity is obvious. We shall prove the sufficiency. To begin with, we establish a formally simpler proposition: the exactness of a sequence of the form

$$0 \to M' \xrightarrow{f} M \xrightarrow{g} M'' \to 0 \tag{13.3}$$

implies the exactness of the sequence

$$0 \to M' \otimes \Phi \xrightarrow{f \otimes I} M \otimes \Phi \xrightarrow{g \otimes I} M'' \otimes \Phi \to 0. \tag{14.3}$$

**Lemma**[3]. *The sequence of* (13) *can be included in a commutative diagram of the form:*

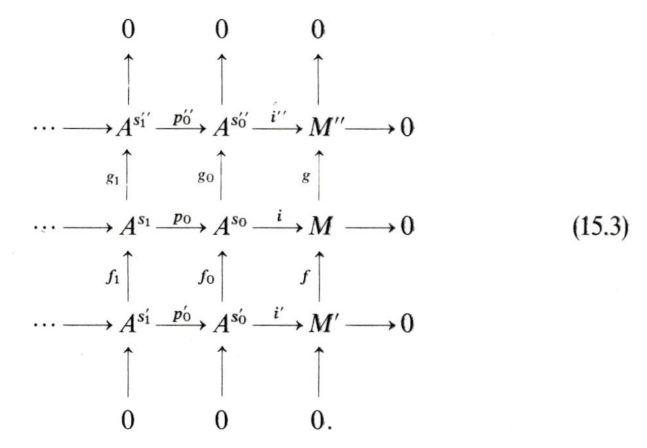

(15.3)

Subjecting diagram (15) to tensor multiplication by $\Phi$, we obtain the commutative diagram:

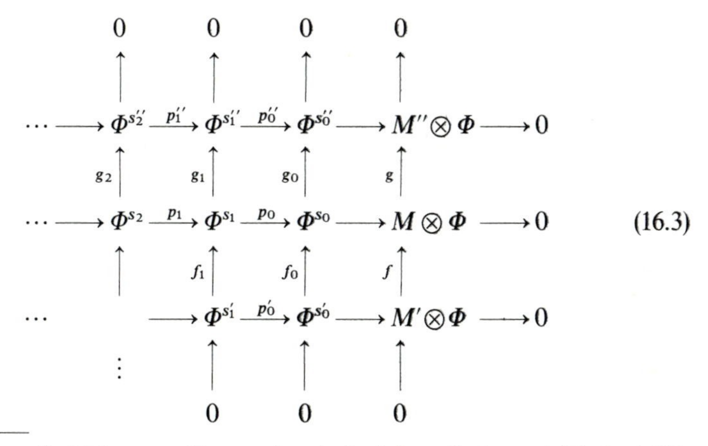

(16.3)

3 The proof of this proposition can be obtained from Cartan and Eilenberg [1], Chapter V § 2, by replacing all projective modules by free modules.

It follows from the hypothesis that all the rows, and all the columns with the exception of the right-hand column, are exact. Applying Theorem 1 of § 2, we find that the right-hand column is also exact; i.e. the exactness of (14) is now proved.

We now establish the exactness of (11) under the assumption that (10) is exact. The fact that (10) is exact implies that the sequences

$$0 \to \operatorname{Coim} f \xrightarrow{\check{f}} M \xrightarrow{\check{g}} \operatorname{Im} g \to 0,$$

$$0 \to \operatorname{Ker} f \to M' \to \operatorname{Coim} f \to 0 \to \operatorname{Im} g \to M'' \to \operatorname{Coker} g \to 0.$$

are exact. But then the proposition that we have just proved implies that the following sequences are also exact:

$$0 \to \operatorname{Coim} f \otimes \Phi \to M \otimes \Phi \to \operatorname{Im} g \otimes \Phi \to 0,$$

$$0 \to \operatorname{Ker} f \otimes \Phi \to M' \otimes \Phi \to \operatorname{Coim} f \otimes \Phi \to 0,$$

$$0 \to \operatorname{Im} g \otimes \Phi \to M'' \otimes \Phi \to \operatorname{Coker} g \otimes \Phi \to 0,$$

whence it is not difficult to derive the exactness of (11). ▯

**4°. The functors Hom and Ext.** Let $\Phi$ and $M$ be $A$-modules. We consider the set of all $A$-mappings from $M$ to $\Phi$. Such mappings can be combined by addition and multiplication with the elements of the ring $A$ in the customary way. It is clear that the set of all such mappings with these operations is an $A$-module. This module is designated by $\operatorname{Hom}(M, \Phi)$. If we are given a mapping $f\colon M \to M'$, then there is defined the mapping $\operatorname{Hom}(M', \Phi) \to \operatorname{Hom}(M, \Phi)$, which relates to every element $F \in \operatorname{Hom}(M', \Phi)$ the mapping $M \ni x \to F(f(x)) \in \Phi$. This mapping will be denoted by $\operatorname{Hom}(f, \Phi)$ or by $f^*$.

Let $k$ be an integer. We consider the module $\Phi^k$. To each of its elements $(\phi_1, \ldots, \phi_k)$ we associate the mapping

$$A^k \ni (a_1, \ldots, a_k) \to \sum a_i \phi_i \in \Phi.$$

This correspondence defines a mapping $h\colon \Phi^k \to \operatorname{Hom}(A^k, \Phi)$.

**Proposition 6.** *The mapping $h$ is an isomorphism. For an arbitrary $A$-matrix $p$, the following diagram is commutative:*

$$\begin{array}{ccc} \operatorname{Hom}(A^s, \Phi) & \xleftarrow{\operatorname{Hom}(p, \Phi)} & \operatorname{Hom}(A^t, \Phi) \\ \updownarrow h & & \updownarrow h \\ \Phi^s & \xleftarrow{p'} & \Phi^t \end{array} \tag{17.3}$$

*where $p'$ is a transposition matrix.*

*Proof.* First we note that the mapping $h$ is a one-to-one mapping. In fact, if $h(\phi_1, \ldots, \phi_k)=0$, then all the $\phi_i=0$, since the module $A^k$ consists of columns of the unit matrix.

We construct the mapping $h^{-1}$. Every mapping $F: A^k \to \Phi$ is uniquely defined by the values $F(e_\sigma)$, $\sigma=1, \ldots, k$, on the basis elements; it can be written in the form $F(a)=(F, a)$, where $F$ is a vector formed by the elements $F(e_\sigma)$. Considering this vector as an element of the module $\Phi^k$, we obtain the mapping $h^{-1}$: $\mathrm{Hom}(A^k, \Phi) \to \Phi^k$. The composition $h\,h^{-1}$ is clearly an identity mapping. Therefore, $h$ is an isomorphism.

Let us go over to diagram (17). By definition, the mapping $\mathrm{Hom}(p, \Phi)$ carries the element $F \in \mathrm{Hom}(A^t, \Phi)$ into a mapping which acts according to the formula $F'$: $a \to F(p\,a)$. Making use of the isomorphism $h$, we represent this mapping in the form

$$\big(h^{-1}(F'), a\big)=F'(a)=F(p\,a)=\big(h^{-1}(F), p\,a\big)=\big(p'\,h^{-1}(F), a\big).$$

It follows that $h^{-1}(F')=p'\,h^{-1}(F)$. □

The functor Hom can be defined in another way – by analogy with the functor of a tensor product. In the free resolution (1) of the module $M$, we replace the modules $A^k$ by the modules $\Phi^k$, and the matrices $p_k$ by the transposed matrices $p_k'$:

$$0 \to \Phi^{s_0} \xrightarrow{p_0'} \Phi^{s_1} \xrightarrow{p_1'} \Phi^{s_2} \to \cdots.$$

This sequence is semi-exact, since the exactness of (1) implies $p_i'\,p_{i-1}'=0$. The homologies of this sequence will be denoted as follows:

$$\mathrm{Ext}^0(M, \Phi)=\Phi_{p_0'}, \quad \mathrm{Ext}^i(M, \Phi)=\Phi_{p_i'}/p_{i-1}'\,\Phi^{s_{i-1}} \qquad i=1, 2, \ldots. \tag{18.3}$$

We shall show that there exists an algebraic isomorphism

$$\mathrm{Ext}^0(M, \Phi) \cong \mathrm{Hom}(M, \Phi). \tag{19.3}$$

To every mapping $F: M \to \Phi$, we set up the mapping $F'$: $A^{s_0} \to \Phi$, which is the composition of $F$ and of the mapping $a$: $A^{s_0} \to M$ from (1). The composition of the mapping $F'$ and the mapping $p_0$: $A^{s_1} \to A^{s_0}$ is clearly null, that is $\mathrm{Hom}(p_0, \Phi)(F')=0$. If we apply to $F'$ the mapping $h^{-1}$, we obtain an element $f \in \Phi^{s_0}$, and, according to Proposition 6, $p_0'\,f=0$. Conversely, to every element $f \in \Phi_{p_0'}$, the isomorphism $h$ sets up a correspondence with the mapping $F'$: $A^{s_0} \to \Phi$, which reduces to zero on the submodule $p_0\,A^{s_1}$. Such a mapping can be looked on as acting from the factor-module $A^{s_0}/p_0\,A^{s_1} \cong M$ to $\Phi$. The isomorphism (19) is then established.

Let us validate Definition (18). Let $f$: $M \to M'$ be a mapping of finite $A$-modules, and let (1) and (3) be free resolutions of these modules.

The commutativity of diagram (4) implies that the transposed matrix $f_i'$ carries $\Phi_{\pi_i'}$ into $\Phi_{p_i'}$, and carries $\pi_{i-1}'\Phi^{\sigma_{i-1}}$ into $p_{i-1}'\Phi^{s_{i-1}}$; therefore, there is defined an associated mapping

$$\tilde{f}_i': \Phi_{\pi_i'}/\pi_{i-1}'\Phi^{\sigma_{i-1}} \to \Phi_{p_i'}/p_{i-1}'\Phi^{s_{i-1}}. \tag{20.3}$$

In the same way as we proved Proposition 2, we can prove the following

**Proposition 7**

I. *The mappings* (20) *depend only on the mapping $f$ and the module $\Phi$.*

II. *We assume that the mapping $f$ is an isomorphism. Then the mappings* (20) *are also isomorphisms, and their inverses are mappings associated with certain $A$-matrices.*

III. *Let $f$ be an isomorphism. Then if for some $i$ the mapping* $\mathrm{Hom}(\pi_i, \Phi)$ *is a homomorphism, the mapping* $\mathrm{Hom}(p_i, \Phi)$ *is also a homomorphism.*

It follows from these propositions that the modules (18), up to an $A$-isomorphism, depend only on $M$ and $\Phi$, and if $f: M \to M'$ is a mapping of finite modules, then the mappings

$$f_i^*: \mathrm{Ext}^i(M', \Phi) \to \mathrm{Ext}^i(M, \Phi) \qquad i = 0, 1, 2, \ldots,$$

are uniquely defined, and $f_0^* = f^*$.

Let $\phi: \Phi \to \Phi'$ be a mapping of modules. It carries the submodule $\Phi_{p_i'}$ into $\Phi'_{p_i'}$, and carries the submodule $p_{i-1}'\Phi^{s_{i-1}}$ into $p_{i-1}'\Phi'^{s_{i-1}}$.

But this defines the associated mapping $\phi_i^*: \mathrm{Ext}^i(M, \Phi) \to \mathrm{Ext}^i(M, \Phi')$, which is easily seen to depend only on $\phi$ and $M$. The mappings $\phi_i^*$ and $f_i^*$ will be called derived (with respect to $\phi$ and $f$).

**5°. Exact sequences connecting derived mappings.** We have in mind the well-known exact sequence for the functor Ext. Let $\Phi$ be a module and

$$0 \to M' \xrightarrow{f} M \xrightarrow{g} M'' \to 0$$

an exact sequence of finite modules. Then there are defined the so-called connection mappings $\delta_i$, $i \geqq 0$, which make the sequence

$$0 \to \mathrm{Hom}(M'', \Phi) \xrightarrow{g^*} \mathrm{Hom}(M, \Phi) \xrightarrow{f^*} \mathrm{Hom}(M', \Phi) \xrightarrow{\delta_0} \mathrm{Ext}^1(M'', \Phi) \to \cdots$$
$$\to \mathrm{Ext}^i(M'', \Phi) \xrightarrow{g_i^*} \mathrm{Ext}^i(M, \Phi) \xrightarrow{f_i^*} \mathrm{Ext}^i(M', \Phi) \xrightarrow{\delta_i} \mathrm{Ext}^{i+1}(M'', \Phi) \to \cdots$$

algebraically exact.

Now suppose that $M$ is a finite module, and that

$$0 \to \Phi' \xrightarrow{\phi} \Phi \xrightarrow{\psi} \Phi'' \to 0 \tag{21.3}$$

is an algebraically exact sequence of modules. Then, there are defined mappings $\delta^i$, also called connection mappings, such that the sequence

$$0 \to \mathrm{Hom}(M, \Phi') \xrightarrow{\phi_0^*} \mathrm{Hom}(M, \Phi) \xrightarrow{\psi_0^*} \mathrm{Hom}(M, \Phi'') \xrightarrow{\delta^0} \mathrm{Ext}^1(M, \Phi') \to \cdots$$
$$\to \mathrm{Ext}^i(M, \Phi') \xrightarrow{\phi_i^*} \mathrm{Ext}^i(M, \Phi) \xrightarrow{\psi_i^*} \mathrm{Ext}^i(M, \Phi'') \xrightarrow{\delta^i} \mathrm{Ext}^{i+1}(M, \Phi') \to \cdots$$

is algebraically exact. We remark that we do not assert that the mappings $\delta^i$ are continuous mappings of modules. Their continuity will be demonstrated immediately under constraints that are additionally placed on (21).

**Proposition 8.** *If the sequence* (21) *is exact, the mappings* $\delta^i$ *are continuous.*

*Proof.* Let (1) be a free resolution of the module $M$. We consider the commutative diagram:

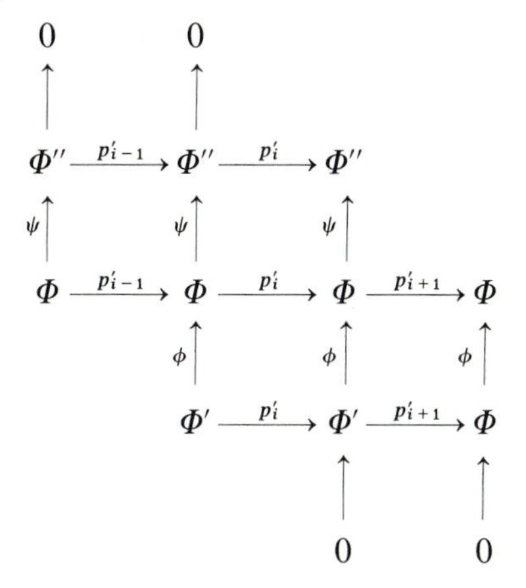

where for simplicity we denote the free module of type $\Phi^k$ by $\Phi$ alone. With the aid of this diagram, we construct the sequence of mappings

$$\begin{aligned} \Phi''_{p_i'}/p_{i-1}'\Phi'' \xrightarrow{\psi^{-1}} \Phi_{(p_i'\psi)}/[p_{i-1}'\Phi + \phi\Phi'] \xrightarrow{p_i'} \\ \xrightarrow{p_i'} [\Phi_{p_{i+1}'} \cap \phi\Phi']/\phi\, p_i'\Phi' \xrightarrow{\phi^{-1}} \Phi'_{p_{i+1}'}/p_i'\Phi'. \end{aligned} \tag{22.3}$$

The mappings $\psi^{-1}$ and $\phi^{-1}$ in this sequence are obtained by the application of Lemma 4 §2 to the triplets $(p_{i-1}', \psi, p_i')$ and $(p_i', \phi, p_{i+1}')$ and are therefore isomorphisms. Since all the mappings (22) are continuous, their composition is continuous and this is easily seen to coincide with $\delta^i$. □

If the finite module $M$ is a direct sum of the finite modules $M'$ and $M''$, we can immediately infer from the definitions the topological isomorphisms

$$\mathrm{Ext}^i(M, \Phi) \cong \mathrm{Ext}^i(M', \Phi) \oplus \mathrm{Ext}^i(M'', \Phi), \qquad i = 0, 1, 2, \ldots.$$

Similarly, if $\Phi \cong \Phi' \oplus \Phi''$, then for arbitrary $i \geqq 0$

$$\operatorname{Ext}^i(M, \Phi) \cong \operatorname{Ext}^i(M, \Phi') \oplus \operatorname{Ext}^i(M, \Phi'').$$

**6°. Injective modules**

**Definition 2.** An $A$-module $\Phi$ is said to be injective if for an arbitrary exact sequence of finite $A$-modules $M' \xrightarrow{f} M \xrightarrow{g} M''$ the sequence

$$\operatorname{Hom}(M'', \Phi) \xrightarrow{g^*} \operatorname{Hom}(M, \Phi) \xrightarrow{f^*} \operatorname{Hom}(M', \Phi).$$

is exact.

The following proposition is proved in the same way as Proposition 5.

**Proposition 9.** *The $A$-module $\Phi$ is injective if, and only if, for an arbitrary exact sequence of the form $A^s \xrightarrow{p} A^t \xrightarrow{r} A^v$ the sequence*

$$\Phi^v \xrightarrow{r'} \Phi^t \xrightarrow{p'} \Phi^s.$$

*is exact.*

**7°. The cohomology dimension of a module.** The cohomology dimension of a finite $A$-module $M$ is the smallest number $\delta = \delta(M)$ such that there exists a free resolution of this module, consisting of zeros, beginning with the $\delta+1$-st, that is, a free resolution of the form

$$0 \to A^{s_\delta} \xrightarrow{p_{\delta-1}} A^{s_{\delta-1}} \to \cdots \to A^{s_1} \xrightarrow{p_0} A^{s_0} \xrightarrow{a} M \to 0.$$

In particular, the module $M$ is free if, and only if, $\delta(M) = 0$; the equation $\delta(M) = -1$ means that $M = 0$.

If $A$ is the ring of all polynomials in $n$ variables with complex coefficients, then by Hilbert's theorem[4], an arbitrary finite $A$-module will have cohomology dimension not exceeding $n$. It follows that for an arbitrary finite $A$-module $M$ and $A$-module $\Phi$, we have $\operatorname{Ext}^i(M, \Phi) = 0$ for $i > n$.

Chapter II

# Division with Remainder in the Space of Power Series

The famous Weierstrass "preparation theorem" is a theorem on division with a remainder of formal (convergent) power series. If we are given a fixed power series $p$, every series $\phi$ can be written in the form $\phi = p\psi + \chi$, where the quotient $\psi$ and the remainder $\chi$ are power series,

4 See O. Zarisky, P. Samuel [1], Volume II.

and the remainder $\chi$ belongs to a distinguished subspace (formed of series containing only bounded powers of the distinguished variable). This latter condition defines the decomposition uniquely. In this chapter, we will establish a similar theorem on the division of vector power series by a fixed matrix $p$, formed of power series: every vector series $\phi$ can be uniquely written in the form $\phi = p\psi + \chi$, where $\psi$ and $\chi$ are also vector power series, and the series $\chi$ belongs to the distinguished subspace. We shall suppose that the matrix $p$ is formed of polynomials or analytic functions. This is not essential for the formal part of the results, but is necessary for the bounds of the operators $\phi \to \psi$ and $\phi \to \chi$. A detailed study of these operators, which we carry out here, will be extremely important in Chapter IV.

## § 1. The space of power series

**1°. Terminology and notation.** Let $s$ be an integer. The elements of the space $C^s$ will often be interpreted as columns of height $s$ with complex elements. For an arbitrary $\zeta \in C^s$, we write

$$|\zeta| = \max_\sigma |\zeta_\sigma|, \qquad \zeta = (\zeta_l, \ldots, \zeta_s).$$

Correspondingly, the linear operator $a$: $C^s \to C^t$ will be identified with the matrix $a = \{a_{\tau\sigma}\}$ of size $t \times s$ (that is, with $t$ rows and $s$ columns) which acts on the column vector $\zeta$ by multiplication on the left. We write

$$|a| = \max_{1 \leqq \tau \leqq t} \sum_{\sigma - 1}^{s} |a_{\tau\sigma}|.$$

We have, obviously, $|a\zeta| \leqq |a| \cdot |\zeta|$ for arbitrary $\zeta \in C^s$.

**Proposition 1.** *Let $a$ be a non-singular matrix of size $s \times s$. Then*

$$|a^{-1}| \leqq \frac{s\,|a|^{s-1}}{|\det a|}. \tag{1.1}$$

*Proof.* Let $A_{\tau\sigma}$ be the cofactor of the element $a_{\tau\sigma}$. It is clear, that

$$|A_{\tau\sigma}| \leqq \prod_{\tau' \neq \tau} \sum_{\sigma'} |a_{\tau'\sigma'}| \leqq |a|^{s-1}.$$

Therefore the absolute value of each element of the matrix $a^{-1}$ does not exceed $|a|^{s-1} |\det a|^{-1}$. From this follows the inequality (1). □

Sometimes it will be convenient to use the following definitions. Let $M$ and $N$ be arbitrary finite sets; a matrix of size $M \times N$ is defined as an arbitrary function $A = \{a_{\mu\nu}\}$, defined for $\mu \in M$, $\nu \in N$, with complex values in the field of complex numbers. The set of numbers $a_{\mu\nu}$ for fixed $\mu$ will be called a row of this matrix, with the index $\mu$, and the set of numbers $a_{\mu\nu}$, $\mu \in M$ will be called a column with index $\nu$. By enumerating the elements of the sets $M$ and $N$, we will convert the matrix $A$ into the customary rectangular matrix. From this as a point of departure, we proceed to define the concepts of minor, rank, and so on, for arbitrary matrices of dimension $M \times N$.

Let $A = \{a_{\mu\nu}\}$ be a matrix of dimension $M \times N$; if $M' \subset M$ and $N' \subset N$ are subsets, the set of numbers $\{a_{\mu\nu}, \mu \in M', \nu \in N'\}$ will be called a submatrix of the matrix $A$. If $A$ is a matrix of size $M \times N$, and $B$ is a matrix of size $N \times L$, we can define the product $AB$, which is a matrix of size $M \times L$.

Let us now choose an arbitrary integer $m$. By $Z^m$ we denote the subsets of the Euclidian space $R^m$, consisting of points with integer coordinates, and by $Z_+^m$, the subset of $Z^m$, consisting of points with non-negative coordinates. The sets $Z^m$ and $Z_+^m$ will be called $Z$-modules. The vectors $e_k$, $k = 1, \ldots, m$, forming the rows of the unit matrix of order $m$, will serve as a basis for the $Z$-module $Z^m$. In $Z^m$ we now consider the linear functional $i \to |i|$, which maps each point on the sum of its coordinates. The intersection of $Z_+^m$ and the set in which $|i| = k \geqq 0$, will be denoted by $\Sigma_k$.

We define an ordering relation in the lattice $Z^m$ as follows: We write $i \geqq j$, if no coordinate of the point $j$ exceeds the corresponding coordinate of the point $i$, that is, if $i - j \in Z_+^m$. We shall write $i > j$, if $i \geqq j$, and $i \neq j$.

Let $s$ be an arbitrary integer. We shall denote by $s\,Z^m$ the set consisting of all pairs $(\sigma, i)$ where $\sigma$ is an integer with values in the range from 1 to $s$, and $i \in Z^m$. More generally, if $\mathscr{I}$ is an arbitrary subset of $Z^m$, we shall mean by $s\mathscr{I}$ the subset of $s\,Z^m$ formed of pairs $(\sigma, i)$, where $\sigma = 1, \ldots, s$, and $i \in \mathscr{I}$. The set $1\mathscr{I}$ is identical with $\mathscr{I}$. Given any subset $\mathscr{I} \subset s\,Z^m$ we denote by $|\mathscr{I}|$ the number of its elements. If $\mathscr{I}$ is a subset of $s\,Z^m$, and $\mathscr{J}$ is a subset in $Z^m$, we denote by $\mathscr{I} + \mathscr{J}$ the subset of $s\,Z^m$, consisting of the pairs $(\sigma, i+j)$, where $(\sigma, i) \in \mathscr{I}$, and $j \in \mathscr{J}$.

**2°. The space of formal power series.** We now fix the Euclidian space $C^m$; the points of this space will be denoted by $\eta = (\eta_1, \ldots, \eta_m)$. By $\mathscr{S} = \mathscr{S}[\eta]$ we denote the linear space consisting of the formal power series in the variable $\eta$, with coefficients in $C$, that is, the space of formal sums of the type

$$\phi = \phi[\eta] = \sum_{i \in Z_+^m} \phi_i \eta^i, \qquad i = (i_1, \ldots, i_m), \; \eta^i = \eta_1^{i_1} \ldots \eta_m^{i_m}. \tag{2.1}$$

For an arbitrary integer $s$ we denote by $\mathscr{S}^s$ the direct sum of $s$ replicas of the space $\mathscr{S}$, in accordance with the notation of § 1 of Chapter I. We shall interpret the space $\mathscr{S}^s$ in two different ways: first, as the space of columns of type $s$, made up of elements of $\mathscr{S}$; and, on the other hand, as the space of formal power series in $\eta$, with coefficients from $C^s$. In the latter case, we shall write the elements of the space $\mathscr{S}^s$ in the form of the sum (2), to remind us that $\phi_i \in C^s$ for all $i$, or in the form

$$\phi = \sum_{s Z_+^m} \phi_{\sigma, i} \eta^{\sigma, i}, \qquad \phi_{\sigma, i} \in C,$$

where $\eta^{\sigma, i}$ is the product of $\eta^i$ and the column $e_\sigma$ with the index $\sigma$ of the unit matrix of order $s$.

For every pair $(\sigma, i) \in s Z_+^m$ we consider the linear functional $\delta_{\sigma, i}$ over $\mathscr{S}^s$, defined by the formula

$$\delta_{\sigma, i}: \phi = \sum \phi_{\tau, j} \eta^{\tau, j} \to \phi_{\sigma, i}.$$

We set $\delta_{\sigma, i} = 0$ for $i \bar{\in} Z_+^m$. If $i \in Z_+^m$, then $\delta_i$ will denote the functional over $\mathscr{S}^s$ with values in $C^s$, which maps the series (2) into its coefficient $\phi_i$. Thus, the functional $\delta_i$ may be considered as a column of height $s$, consisting of the functionals $\delta_{\sigma, i}$, $\sigma = 1, \ldots, s$.

More generally, let $\mathscr{I}$ be an arbitrary finite subset of $s Z_+^m$. By $\delta_{\mathscr{I}}$ we denote a functional on $\mathscr{S}^s$, defined by the formula

$$\delta_{\mathscr{I}}: \phi \to [\phi_{\sigma, i}, \ (\sigma, i) \in \mathscr{I}\} \in C^{|\mathscr{I}|}.$$

In other words, $\delta_{\mathscr{I}} \phi$ is a column vector, formed from the coefficients of the series $\phi$ with indices belonging to the set $\mathscr{I}$. We shall write simply $\delta_k$ for the functional $\delta_{s \Sigma_k}$.

Let $s$ and $t$ be arbitrary integers. By a functional over $\mathscr{S}^s$ with values in $C^t$, we shall mean an arbitrary finite sum of the form

$$f = f(\delta) = \sum f^{\sigma, i} \delta_{\sigma, i}, \ f^{\sigma, i} \in C^t, \tag{3.1}$$

which is defined by the formula

$$f: \phi \to \sum f^{\sigma, i} \phi_{\sigma, i}.$$

In the matrix notation Eq. (3) appears as follows: $f = \sum f^i \delta_i$, where $f^i: C^s \to C^t$ are matrices consisting of columns $f^{\sigma, i}$. The maximum value of $|i|$ for $f^i \neq 0$ will be called the order of the functional $f$, and will be denoted by $\deg f$.

The differential operator, corresponding to the functional $f$, is the expression

$$f(D) = \sum f^{\sigma, i} \frac{1}{i!} D^{\sigma, i},$$

where $D^{\sigma, i}$ is the product of the operator $D^i = \dfrac{\partial^{|i|}}{\partial \eta_1^{i_1} \ldots \partial \eta_m^{i_m}}$ and the column $e_\sigma$ with the index $\sigma$ of the unit matrix of order $s$. The following formula connects the functional $f(\delta)$ and the operator $f(D)$:

$$f(\delta)\,\phi = f(D)\,\phi|_{\eta=0}.$$

**3°. Operators in the space of formal series.** We again let $s$ and $t$ be arbitrary integers. By an operator acting from $\mathscr{S}^s$ to $\mathscr{S}^t$, we shall mean an arbitrary expression of the form

$$A = \sum_{t\,Z_+^m} \eta^{\tau, i} A_{\tau, i}(\delta), \tag{4.1}$$

where all the $A_{\tau, i}$ are functionals on $\mathscr{S}^s$ with values in $C$, as defined by the formula

$$A\colon \phi \to \sum \eta^{\sigma, i} A_{\sigma, i}(\delta)\,\phi.$$

For every $i$, the column $A_i = (A_{1,i}, \ldots, A_{s,i})$ is a functional on $\mathscr{S}^s$ with values in $C^t$. Therefore, formula (4) can be rewritten as: $A = \sum \eta^i A_i(\delta)$. But this formula can again be rewritten, if we express the functional $A_i$ in expanded form

$$A = \sum \eta^i A_i^j\,\delta_j = \sum \eta^{\tau, i} A_{\tau, i}^{\sigma, j}\,\delta_{\sigma, j},$$

where $A_i^j\colon C^s \to C^t$ are matrices, consisting of the quantities $A_{\tau, i}^{\sigma, j}$. The quantities $A_i$, $A_i^j$, and $A_{\tau, i}^{\sigma, j}$ will be called the coefficients of the operator $A$. They may be determined from the operator itself as follows:

$$A_i = \delta_i A, \qquad A_i^j = \delta_i A \eta^j, \qquad A_{\tau, i}^{\sigma, j} = \delta_{\tau, i} A \eta^{\sigma, j}.$$

In particular, the identity operator in $\mathscr{S}^s$ can be written in the form:

$$E = \sum \eta^i \delta_i = \sum \eta^{\sigma, i} \delta_{\sigma, i}.$$

We shall mean by $[\cdot]_k$ an operator in $\mathscr{S}^s$ which acts according to the formula

$$[\phi]_k = \sum_{|i| \leqq k} \eta^i \delta_i \phi.$$

The kernel of this operator will be denoted by $m_{k+1}^s$. Thus, $m_{k+1}^s$ is the space of power series in $\mathscr{S}^s$, not containing terms of order $\leqq k$.

**4°. The space of convergent power series.** For every $r > 0$, we shall consider the subspace $\mathcal{O}_r \subset \mathscr{S}$, consisting of the set of all power series that converge for $|\eta| \leqq r$. We introduce in this subspace the norm

$$\|\phi\|_r = \sup_{|\eta| \leqq r} |\sum \phi_i \eta^i|,$$

where on the right-hand side the expression $\sum \phi_i \eta^i$ is to be understood not as a formal power series but as the sum of the series. $\mathcal{O}$ will denote the union of all the spaces $\mathcal{O}_r$.

**Proposition 2.** *For arbitrary* $\phi \in \mathcal{O}_\varepsilon$, $\varepsilon > 0$, *we have the inequality*

$$|\phi_i| \leqq \left(\frac{\sqrt{m}}{\varepsilon}\right)^{|i|} \|\phi\|_\varepsilon, \quad i \in Z_+^m. \tag{5.1}$$

*Proof.* We denote by $\phi(\eta)$ the sum of the series $\sum \phi_i \eta^i$. This function is bounded on the sphere $|\eta| \leqq \varepsilon$ and is analytic within the sphere. We inscribe in the sphere the polycylinder $\left\{\eta\colon |\eta_i| \leqq \frac{\varepsilon}{\sqrt{m}}, i = 1, \ldots, m\right\}$ and we apply Cauchy's theorem, taking as the contour of integration the skeleton $\Gamma$ of this polycylinder. Then a chain of self-evident relationships

$$|\phi_i| = \left|\frac{1}{i!} D^i \phi \right|_{\eta=0}\Bigg| = \left|(2\pi)^{-n} \int_\Gamma \eta^{-i-e} \phi(\eta)\, d\eta\right| \leqq \left(\frac{\sqrt{m}}{\varepsilon}\right)^{|i|} \|\phi\|_\varepsilon, \quad e = \sum_1^s e_k$$

leads us to our goal. ▯

**5°. The spaces $\mathcal{S}$ and $\mathcal{O}$ as $\mathcal{P}$-modules.** We fix an arbitrary integer $n \geqq m$ and consider the ring $\mathcal{P} = C[z]$ of plynomials with complex coefficients in the space $C^n = C_z^n$. The space $C^m$ introduced earlier will now be looked on as a coordinate subspace in $C^n$, or more exactly, the point $\eta \in C^m$ will be identified with the point $z = (\eta_1, \ldots, \eta_m, 0, \ldots, 0)$.

Corresponding to every polynomial $f \in \mathcal{P}$ and point $z \in C^n$, we determine an operator in $\mathcal{S}$

$$f(z)\colon \phi[\eta] \to f(z+\eta)\phi[\eta], \tag{6.1}$$

whose action consists in a multiplication by the power series

$$f(z+\eta) = \sum \eta^i \frac{D^i}{i!} f(z),$$

which is the Taylor expansion of the polynomial $f$ at the point $z$ in terms of the variable $\eta$. Similarly, corresponding to every $\mathcal{P}$-matrix $p\colon \mathcal{P}^s \to \mathcal{P}^t$ (that is, a matrix of size $t \times s$ with elements from $\mathcal{P}$) and to every point $z \in C^n$, we determine an operator

$$p(z)\colon \mathcal{S}^s \ni \phi[\eta] \to p(z+\eta)\phi[\eta] \in \mathcal{S}^t,$$

which acts by multiplication by the Taylor expansion of the matrix $p$.

Let us fix the point $z$. By setting up the correspondence between the polynomial $f \in P$ and the operators (6), we have converted the space $\mathcal{S}$ into a unitary $\mathcal{P}$-module. The space $\mathcal{S}$, with this structure of a $\mathcal{P}$-module, will sometimes be denoted by $\mathcal{S}_z$. The submodule of this module consisting of elements of the subspace $\mathcal{O}$ will be denoted by $\mathcal{O}_z$. When $n = m$ the submodule will often be identified with the space of functions holomorphic in the neighborhood of $z$, where the polynomials of the ring $\mathcal{P}$ act in the usual way.

In what follows, if we do not explicitly state the contrary, the spaces $\mathscr{S}$ and $\mathscr{O}$ will be given the structures of the modules $\mathscr{S}_0$ and $\mathscr{O}_0$. We note a few well-known properties of these modules[1].

**Proposition 3.** *When $n=m$, the $\mathscr{P}$-modules $\mathscr{S}$ and $\mathscr{C}$ are flat.*

**Proposition 4.** *Let $n=m$, and let $p: \mathscr{P}^s \to \mathscr{P}^t$ be an arbitrary $\mathscr{P}$-matrix, and let $z$ be an arbitrary point in $C^n$. Then if the column $F \in \mathscr{P}^t$ belongs to $p\mathscr{S}_z^s$, it belongs to $p\mathscr{R}_z^s$, where $\mathscr{R}_z$ is the ring of all rational functions in $C^n$ which are regular at the point $z$.*

*If $F \in p\mathscr{S}_z^s$ for arbitrary $z \in C^n$, then $F \in p\mathscr{P}^s$.*

## § 2. The base sequence of matrices

From now to the end of the chapter, we shall fix an arbitrary $\mathscr{P}$-matrix $p: \mathscr{P}^s \to \mathscr{P}^t$, that is, a matrix of size $t \times s$, formed of polynomials in $C^n$, with complex coefficients.

### 1°. Construction of the base sequence

**Theorem.** *At every point $z \in C^n$ for arbitrary integers $k \geqq 0$, there exists a matrix $P_k(z, \eta)$ of size $t \times s_k$ for some $s_k \geqq 0$, consisting of polynomials in $\eta$, and having the following property.*

I. $P_k(z)\,\mathscr{S}^{s_k} = m_k^t \cap p(z)\,\mathscr{S}^s$.

II. *Let $\Delta_k(z) = \delta_k P_k(z, \eta)$[2] be a matrix of dimension $t\Sigma_k \times s_k$, formed from the columns $\delta_k P_{k,\sigma}$, where $P_{k,\sigma}$, $\sigma = 1, \ldots, s_k$ are the columns of the matrix $P_k$; and let $\rho_k(z)$ be the rank of the matrix $\Delta_k(z)$. On an arbitrary set of the form*

$$N_{k-1}^r = \{z \in C^n \colon \rho_0(z) = r_0, \ldots, \rho_{k-1}(z) = r_{k-1}\},$$

*where*

$$r = (r_0, \ldots, r_{k-1}) \in Z_+^k,$$

*the number $s_k$ is constant, and the matrix $P_k$ can be written in the form*

$$P_k(z, \eta) = p(z+\eta)\, Q_k(z, \eta), \tag{1.2}$$

*where $Q_k$ is a matrix of size $s \times s_k$, made up of polynomials in $z$ and $\eta$.*

The sequence of matrices $\Delta_k(z)$, $k = 0, 1, 2, \ldots$, will be called the base sequence. From the theorem just formulated, it is easy to deduce the following property of the base sequence. For every $k \geqq 0$ the series $\phi \in m_k^t$ belongs to the subspace $p(z)\,\mathscr{S}^s$ modulo $m_{k+1}^t$ if and only if the column $\delta_k \phi$ is a linear combination of the columns of the matrix $\Delta_k(z)$. In fact, the inclusion relation

$$\phi \in m_k^t \cap p(z)\,\mathscr{S}^s, \qquad \operatorname{mod} m_{k+1}^t$$

---

1 See, for example, O. Zarisky, P. Samuel [1], Volume II.

2 Here and in what follows the elements of the matrix $P_k(z, \eta)$ and similar matrices will be looked on as power series in $\eta$; the variables $z$ appear as parameters.

is by virtue of the first assertion of the theorem equivalent to the inclusion relation

$$\phi \in P_k(z)\,\mathscr{S}^{s_k}, \qquad \operatorname{mod} m_{k+1}^t.$$

Since the matrix $P_k$ contains no term of order less than $k$, the latter inclusion is equivalent to the statement that $\phi = P_k(z)\,\psi$, $\operatorname{mod} m_{k+1}^t$ where $\psi \in C^{s_k}$ is some vector. Operating on both sides of this equation with the functional $\delta_k$, we obtain the equivalent relationship

$$\delta_k\,\phi = \Delta_k(z)\,\psi,$$

which is what we were looking for.

*Proof of the Theorem.* We construct the matrices $P_k$ by induction on $k$. We set $P_0(z,\eta) = p(z+\eta)$; it is clear that this matrix satisfies the conditions of the theorem with $k=0$. We assume that for some $k \geqq 0$ we have constructed the matrices $P_0, \dots, P_k$, satisfying the conditions I and II. We now construct the matrix $P_{k+1}$.

We fix an arbitrary vector $r' = (r_0, \dots, r_{k-1}, \rho) \in Z_+^{k+1}$ and we postulate that the point $z$ belongs to the $N_k^{r'}$. By the hypothesis of the induction, the elements of the matrix $P_k$ on this set are polynomials in $z$ and $\eta$. On the set $t\,\Sigma_k$ we choose in an arbitrary way $\rho$ distinct elements $\alpha_1, \dots, \alpha_\rho$. We also choose arbitrarily $\rho+1$ distinct integers $\sigma_1, \dots, \sigma_\rho$, $\sigma$, each of which does not exceed $s_k$. Let $M$ be the minor of the matrix $\Delta_k$, consisting of the elements $\delta_{\alpha_i} P_{k,\sigma_j}$, $i,j = 1, \dots, \rho$. We consider the system of linear equations

$$\delta_\alpha \sum_1^\rho P_{k,\sigma_j}\,\lambda_j = (\det M)^2\,\delta_\alpha P_{k,\sigma}, \qquad \alpha = \alpha_1, \dots, \alpha_\rho, \tag{2.2}$$

in the unknowns $\lambda = (\lambda_1, \dots, \lambda_\rho)$, having $M$ for its matrix. The solution of this system will be found by the following rule not depending on the non-singularity of the matrix $M$. We consider a matrix $\check{M}$, formed of the cofactors of the elements of the matrix $M$; let $\check{M}'$ be the transpose of this matrix. We obtain the desired vector $\lambda$ by applying the matrix $(\det M)\,\check{M}'$ to the columns made up of the quantities $\delta_\alpha P_{k,\sigma}$, $\alpha = \alpha_1, \dots, \alpha_\rho$. It is clear that the elements of the vector $\lambda$ are polynomials in $z$ and they vanish when $\det M = 0$.

We remark that the quantities $\lambda_j$ which have just been determined satisfy the Eqs. (2) for all $\alpha \in t\,\Sigma_k$. In fact, if $\det M = 0$, then both sides of Eq. (2) vanish. If, however, $\det M \neq 0$, then the columns $\delta_k P_{k,\sigma_j}$, $j=1, \dots, \rho$, of the matrix $\Delta_k$ are linearly independent and therefore form a basis for all columns of this matrix, since it follows from $z \in N_k^{r'}$ that the rank of the matrix $\Delta_k$ is equal to $\rho$. Hence the column $\delta_k P_{k,\sigma}$ is a linear combination of the columns of this basis, with coefficients $\lambda_1, \dots, \lambda_\rho$, which is what we were to prove. We consider the column:

$$P^* = (\det M)^2 P_{k,\sigma} - \sum P_{k,\sigma_j}\,\lambda_j. \tag{3.2}$$

It is clearly composed of polynomials in $z$ and $\eta$, and since Eq. (2) is satisfied for arbitrary $\alpha \in t\Sigma_k$, it follows that the column $P^*$ contains no member of order less than $k+1$.

We now force $\alpha_1, \ldots, \alpha_\rho$ to run over all possible combinations of $\rho$ different elements of the set $t\Sigma_k$ in a definite order not depending on the point $z \in N_k^{r'}$. In exactly the same way, we require $\sigma_1, \ldots, \sigma_\rho$, $\sigma$ to run over all possible combinations of $\rho+1$ integers not exceeding $s_k$. In this way we obtain an ordered sequence of columns of the form (3). We add to this sequence all possible columns of the form $\eta^j P_{k,\sigma}(z, \eta)$, where $\sigma = 1, \ldots, s_k$, and $j \in \Sigma_1$; these columns are also to be ordered in a fashion independent of the point $z \in N_k^{r'}$. We denote the sequence of columns so obtained by:

$$P_{k+1,1}, \ldots, P_{k+1,s_{k+1}}.$$

For every $\sigma \leqq s_{k+1}$ $P_{k+1,\sigma}$ is the column of height $t$, formed of polynomials in $z$ and $\eta$. The matrix of dimension $t \times s_{k+1}$, formed of these columns in the established order, will be denoted by $P_{k+1}$. We shall show that this matrix satisfies the conditions of the theorem.

It is immediately obvious from the construction that on the set $N_k^{r'}$

$$P_{k+1,\sigma} = \sum_\tau P_{k,\tau} q_{k,\tau}^\sigma, \qquad \sigma = 1, \ldots, s_{k+1},$$

where $q_{k,\tau}^\sigma$ are polynomials in $z$ and $\eta$. In matrix notation

$$P_{k+1} = P_k q_k, \tag{4.2}$$

where $q_k = \{q_{k,\tau}^\sigma\}$. Therefore, from (1)

$$P_{k+1} = p Q_{k+1},$$

where $Q_{k+1} = Q_k q_k$ is a matrix formed of polynomials in $z$ and $\eta$. We have thus shown that the matrix $P_{k+1}$ satisfies condition II.

**2°. Completion of the proof of the theorem.** It remains to show that the matrix $P_{k+1}$ satisfies the first condition of the theorem. We first establish the equation

$$P_{k+1} \mathscr{S}^{s_{k+1}} = m_{k+1}^t \cap P_k \mathscr{S}^{s_k}. \tag{5.2}$$

We have from (4)

$$P_{k+1} \mathscr{S}^{s_{k+1}} = P_k q_k \mathscr{S}^{s_{k+1}} \subset P_k \mathscr{S}^{s_k}. \tag{6.2}$$

Since the columns of the matrix $P_{k+1}$ contain no term of order less than $k+1$, we have $P_{k+1} \mathscr{S}^{s_{k+1}} \subset m_{k+1}^t$, which together with (6) implies the inclusion relation $P_{k+1} \mathscr{S}^{s_{k+1}} \subset m_{k+1}^t \cap P_k \mathscr{S}^{s_k}$.

We now prove the converse inclusion. Let $\phi$ be an arbitrary element of the righthand side of (5). This means that it does not contain terms of order less than $k+1$, and can be written in the form

$$\phi=\sum_{\sigma} P_{k,\sigma} \sum_{i} \psi_{\sigma,i} \eta^i$$

for some $\psi_{\sigma,i} \in C$. The righthand side is now rewritten in the form

$$\sum_{\sigma} P_{k,\sigma} \psi_{\sigma,0} + \sum_{\sigma,|i|>0} \psi_{\sigma,i} P_{k,\sigma} \eta^i. \tag{7.2}$$

We remark that every term in the second sum belongs to the space $P_{k+1} \mathscr{S}^{s_{k+1}}$, since the columns of the matrix $P_{k+1}$ contain all the columns of the form $P_{k,\sigma} \eta^j$, where $|j|=1$.

We now transform the first sum in (7). We choose a non-singular minor $M$ of order $\rho=\rho_k=\operatorname{rank} \Delta_k$ in the matrix $\Delta_k$. Let $\sigma_1, \ldots, \sigma_\rho$ be the indices of the columns containing this minor. An arbitrary column $P_{k,\sigma}$ of the matrix $P_k$ can by virtue of (3) be expressed in terms of $P_{k,\sigma_j}$ according to the equation

$$P_{k,\sigma}=(\det M)^{-2}\left(P^* + \sum P_{k,\sigma_j} \lambda_j\right),$$

where $P^*$ is one of the columns of the matrix $P_{k+1}$. We substitute this expression in the first sum of (7), so that we reduce this sum to the form $\sum P_{k,\sigma_j} \psi_j$ for some $\psi_j \in C$, up to a linear combination of columns of the matrix $P_{k+1}$. Taking account of the fact that the sum (7) is equal to $\phi$, we obtain

$$\phi=\sum P_{k,\sigma_j} \psi_j, \quad \operatorname{mod} P_{k+1} \mathscr{S}^{s_{k+1}}. \tag{8.2}$$

Applying the functional $\delta_k$ to both sides of this equation, we conclude that

$$\sum_j \delta_k P_{k,\sigma_j} \psi_j = 0,$$

since $\phi \in m_{k+1}^t$. Since the columns $\delta_k P_{k,\sigma_j}$, $j=1, \ldots, \rho$, of the matrix $\Delta_k$ are linearly independent (in view of the choice of the minor $M$), we conclude that all $\psi_j=0$. Therefore, the righthand side of (8) vanishes, that is, $\phi \in P_{k+1} \mathscr{S}^{s_{k+1}}$, which is what we were to prove. Thus, Eq. (5) is validated.

Combining this equation with assertion I of the theorem for the matrix $P_k$, we obtain

$$P_{k+1} \mathscr{S}^{s_{k+1}} = m_{k+1}^t \cap (m_k^t \cap p \mathscr{S}^s) = m_{k+1}^t \cap p \mathscr{S}^s.$$

Therefore, the matrix $P_{k+1}$ satisfies condition I of the theorem. □

**3°. Supplementary construction.** We introduce into $Z^m$ a lexicographical ordering relationship: we write $i \in j$, if for some integer $k \leqq m$

$$i_1=j_1, \ldots, i_{k-1}=j_{k-1}, \qquad i_k > j_k,$$

where $i_\alpha$ and $j_\alpha$ are the coordinates of the points $i$ and $j$. We write $i \succeq j$, if $i \succ j$ or $i = j$. We remark that the relation $i \succ j$ is preserved under transfer in the lattice $Z^m$.

In accordance with the definitions of § 1, $t Z^m$ is the set of pairs $(\tau, i)$ where $i \in Z^m$, and $\tau$ is an integer not exceeding $t$. In this set, we also introduce an ordering relation, writing

$$(\tau, i) \succ (\sigma, j), \qquad \text{if } i \succ j \text{ or } i = j \text{ and } \tau < \sigma.$$

We also write $(\tau, i) \succeq (\sigma, j)$, if $(\tau, i) \succ (\sigma, j)$ or $(\tau, i) = (\sigma, j)$. In the case $(\tau, i) \succeq (\sigma, j)$ $((\tau, i) \succ (\sigma, j))$ we shall say that the point $(\tau, i)$ is (strongly) senior to the point $(\sigma, j)$. Let $A = (\alpha_1, \ldots, \alpha_r)$ be an arbitrary finite subset of $t Z^m$; we shall say that it is ordered by seniority if $\alpha_1 \succ \cdots \succ \alpha_r$. Let $A = (\alpha_1, \ldots, \alpha_r)$ and $B = (\beta_1, \ldots, \beta_r)$ be two subsets of $t Z^m$ with the same number of elements, and let them be ordered by seniority. We shall say that the set $A$ is strongly senior to $B$ if for some $k \leqq r$, we have the relations

$$\alpha_1 = \beta_1, \ldots, \alpha_{k-1} = \beta_{k-1}, \qquad \alpha_k \succ \beta_k.$$

We shall say that $A$ is senior to $B$, if $A$ is strongly senior to $B$ or if $A = B$.

Let us now fix an arbitrary point $z \in C^n$ and consider the base series $\Delta_k$, $k = 0, 1, 2, \ldots$, at this point (see 1°). For every $k$ the rows

$$\delta_{\tau, i} P_k = (\delta_{\tau, i} P_{k,1}, \ldots, \delta_{\tau, i} P_{k, s_k}), \qquad (\tau, i) \in t\,\Sigma_k,$$

form the matrix $\Delta_k$.

We fix an arbitrary $k$ and we consider all possible sets $\mathscr{J} \subset t\,\Sigma_k$, having the property that the corresponding rows $\delta_{\tau,i} P_k$, $(\tau, i) \in \mathscr{J}$, of the matrix $\Delta_k$ form a basis in the linear space spanned over all rows $\Delta_k$. The number of elements in each such set $\mathscr{J}$ is equal to the rank $\rho_k$ of $\Delta_k$. Ordering each of the sets $\mathscr{J}$ by seniority, we indicate the most senior by $\mathscr{I}_k$. This set is characterized by the fact that $|\mathscr{I}_k| = \rho_k$, and an arbitrary row $\delta_{\tau, i} P_k$ of $\Delta_k$ is a linear combination of rows of the same matrix with indices $\alpha \in \mathscr{I}_k$, more senior than $(\tau, i)$. The set

$$\mathscr{I}(z) = \mathscr{I} = \bigcup_k \mathscr{I}_k$$

will be called a basis at the point $z$.

We fix arbitrary integers $k \geqq K \geqq 0$. We consider the linear operator $L$: $Z^m \to Z^m$, which carries the point $i = (i_1, \ldots, i_m)$ into the point $L(i)$ with the coordinates $L_\alpha(i)$, $\alpha = 1, \ldots, m$, defined by the formulae

$$\begin{aligned} L_\alpha(i) = K^{\alpha-1}[i_1 - (K+1)\,|i|] + K^{\alpha-2}[i_2 - (K+1)\,|i|] + \cdots \\ \cdots + [i_\alpha - (K+1)\,|i|], \qquad \alpha = 1, \ldots, m-1; \\ L_m(i) = -|i|. \end{aligned}$$

It is easy to see that $L(i)=0$ implies $i=0$. Therefore, this operator generates an isomorphism of $R^m$.

The proposition which we shall now formulate defines certain properties of this operator. We introduce two pieces of notation: for every point $i=(i_1, \ldots, i_m) \in Z^m$ we denote by $|i|_-$ the sum of all the negative quantities occurring in the sequence $i_1, \ldots, i_{m-1}$, and by $|i|_+$ the sum of all the remaining coordinates of the point $i$; thus, we have $|i|=|i|_+ + |i|_-$.

**Proposition**

I. *If* $i \leqq 0$, *then* $L(i) \geqq 0$ *(i.e.,* $L_\alpha(i) \geqq 0$ *for all* $\alpha$*).*

II. *The relations* $|i| \leqq 0$, $i \succ 0$, *and* $|i|_- \geqq -K$ *imply that* $L(i)>0$.

III. *The inequalities* $|i|<0$, $|i|_+ \leqq K$ *imply that* $L(i) \geqq -|i|\, e_m$.

*Proof.* Let $i_\alpha \leqq 0$ for all $\alpha$. Then $|i| \leqq i_\alpha$, $\alpha=1, \ldots, m$, whence $i_\alpha - (K+1)\,|i| \geqq 0$ for all $\alpha$. Therefore $L_\alpha(i) \geqq 0$ for $\alpha \leqq m-1$. Moreover, it is clear that $L_m(i) \geqq 0$, and with this the first assertion is proved.

We prove the second assertion. We assume, to begin with, that $|i|=0$. The relations $i \succ 0$ and $|i|_- \geqq -K$ imply that for some $k<m$

$$i_1 = \cdots = i_{k-1} = 0, \qquad i_k > 0,$$

and the sum of the negative numbers in the sequence $i_{k+1}, \ldots, i_{m-1}$ is not less than $-K$. It follows immediately that $L_1(i) = \cdots = L_{k-1}(i) = 0$, and $L_k(i)>0$. In the case $k<\alpha<m$, we have

$$L_\alpha(i) = K^{\alpha-k}\, i_k + K^{\alpha-k-1}\, i_{k+1} + \cdots + i_\alpha.$$

Since $i_k>0$, the first term on the right side is not less than $K^{\alpha-k}$. The sum of all the negative terms on the righthand side is not less than the product of $K^{\alpha-k-1}$ by the sum of all the negative numbers in the sequence $i_{k+1}, \ldots, i_\alpha$. But this last is not less than $-K$, and therefore the whole of the negative terms in the formula for $L_\alpha(i)$ yield a sum not less than $-K^{\alpha-k}$. It follows that $L_\alpha(i) \geqq 0$. Moreover, $L_m(i)=0$, since $|i|=0$. With this the inequality $L(i)>0$ is proved.

In the case $|i|<0$, we set $i' = i - |i|\, e_m$. Clearly, we have $|i'|=0$, $i' \succ 0$ and $|i'|_- = |i|_- \geqq -K$. Therefore, from what we have already shown, $L(i')>0$. On the other hand, $L(i) = L(i') + L(|i|\, e_m) \geqq L(i')$, since by virtue of the first assertion of the proposition $L(|i|\, e_m) \geqq 0$. It follows lastly that $L(i)>0$, which is what we were to prove.

We now take up the third assertion. The inequality $|i|<0$, implies that $-K\,|i| \geqq K$, whence $|i|_+ + |i|_- - (K+1)\,|i| \geqq K$ and, therefore,

$$|i|_- - (K+1)\,|i| \geqq K - |i|_+ \geqq 0.$$

Thus for arbitrary $\alpha=1, \ldots, m-1$ the quantities $i_\alpha - (K+1)\,|i|$ are non-negative. Therefore, $L_1(i) \geqq 0, \ \ldots, \ L_{m-1}(i) \geqq 0$. To complete the proof, there remains only to note that $L_m(i) = -|i|$. □

### 4°. The fundamental lemma

**Lemma.** *For an arbitrary point* $z \in C^n$, *and arbitrary integer* $k \geqq K \geqq 0$, *the following propositions hold.*

I. $\mathscr{I}_K + \Sigma_\kappa \subset \mathscr{I}_k$, $\kappa = k - K$.

II. *For every* $j \in \Sigma_\kappa$, *we set* $\Delta_K^j = \delta_k \eta^j P_K$ *and we consider the system of equations*

$$\sum_{j \in \Sigma_\kappa} \Delta_K^j x_j = y, \tag{9.2}$$

*where* $y = \{y_i \in \mathscr{L}^t, i \in \Sigma_k\}$ *is a column of size* $t\,\Sigma_k$, *the elements of which belong to some linear space* $\mathscr{L}$. *Let* $M$ *be a non-singular minor of order* $\rho = \rho_K(z)$ *of the matrix* $\hat{\Delta}_K = \delta_{\mathscr{I}_K} P_K(z)$. *Then the quantities* $x_j \in \mathscr{L}^{sK}$ *can be so chosen that in the system (9) all the equations with indices belonging to the set* $\mathscr{I}_K + \Sigma_\kappa$, *are satisfied, and*

$$x_j = \sum R_j^i y_i, \tag{10.2}$$

*where* $R_j^i$ *is a matrix with elements that are polynomials in the elements of the matrices* $\Delta_K$ *and* $M^{-1}$; *the inequality* $R_j^i \neq 0$ *holds only under the conditions* $L(i) \geqq L(j + K e_m)$ *and*

$$|R_j^i| \leqq b^{|L(i) - L(j + K e_m)| + 1}, \tag{11.2}$$

*where*

$$b = |\Sigma_K|\, t \left(\rho \frac{(|\Delta_K| + 1)^\rho}{|\det M|}\right)^{t+1}. \tag{12.2}$$

*Proof.* First of all, we remark that among the columns of the matrix $P_k$ are contained all the columns of the form $\eta^j P_{K,\sigma}$, where $j \in \Sigma_\kappa$. We fix $j \in \Sigma_\kappa$ and we note, that

$$\delta_{\tau,i} \eta^j P_{K,\sigma} = \begin{cases} \delta_{\tau,i-j} P_{K,\sigma}, & \text{if } i \geqq j, \\ 0 & \text{otherwise}. \end{cases} \tag{13.2}$$

If the point $(\tau, i)$ runs over the set $t\,\Sigma_k$ in order of decreasing seniority, the point $(\tau, i-j)$ for $i \geqq j$ runs over the whole set $t\,\Sigma_K$, also in order of decreasing seniority. It follows that every column $\delta_k \eta^j P_{K,\sigma}$ of the matrix $\Delta_K^j$ contains as a submatrix the column $\delta_K P_{K,\sigma}$ of the matrix $\Delta_K$, and that in the remaining places, that is, the places with indices $(\tau, i)$, where $i \ngeqq j$, it has zeroes.

We shall prove the first assertion of the lemma. The points of the set $\mathscr{I}_K$, in order of decreasing seniority, will be denoted by $\alpha_\sigma = (\tau_\sigma, i_\sigma)$, $\sigma = 1, \ldots, \rho$. Let us suppose that the inclusion relation $I$ is false. Then the set $(\mathscr{I}_K + \Sigma_\kappa) \setminus \mathscr{I}_k$ is not empty; let $\alpha$ be a point belonging to this set. By definition, this point has the form $\alpha_\sigma + j$ for some $\sigma$ and $j \in \Sigma_k$ and does not belong to the set $\mathcal{I}_k$. By a property of the set $\mathcal{I}_k$ the corresponding row $\delta_\alpha P_k$ of the matrix $\Delta_k$ is a linear combination of the rows $\delta_\beta P_k$, with indices belonging to $\mathcal{I}_k$ and strongly senior to $\alpha$. Let $\lambda_\beta$ be the coefficients of this linear combination. Replacing each of the rows $\delta_\beta P_k$ by its corresponding portion $\delta_\beta \eta^j P_K$, we obtain the equation

$$\delta_\alpha \eta^j P_K = \sum_{\beta \succ \alpha} \lambda_\beta \delta_\beta \eta^j P_K.$$

By relation (13) this equation can be rewritten as:

$$\delta_{\alpha_\sigma} P_K = \sum_{\beta - j \succ \alpha_\sigma} \lambda_\beta \delta_{\beta - j} P_K.$$

But this shows that the row $\delta_{\alpha_\sigma} P_K$ of the matrix $\Delta_K$ with the index $\alpha_\sigma \in \mathcal{I}_K$ is a linear combination of rows of the same matrix with strongly senior indices. This is impossible because of the property of the set $\mathcal{I}_K$. We have obtained a contradiction which proves the inclusion relation $I$.

We now take up the proof of the second assertion. Let $M$ be a non-singular minor of the matrix $\hat{\Delta}_K$ of order $\rho$. For simplicity we shall assume that it lies in the first $\rho$ columns of the matrix. We consider the matrix $\mathcal{M}$ of size $s_K \times \rho$, of which the first $\rho$ rows consist of the matrix $M^{-1}$, and the remainder are equal to zero. We write

$$D = \Delta_K \mathcal{M}; \quad D^j = \Delta_K^j \mathcal{M}, \quad j \in \Sigma_K.$$

Let us determine the structure of these matrices. $D$ is a matrix of size $t\Sigma_K \times \rho$, and its minor, formed of rows with indices belonging to $\mathcal{I}_K$, is the unit matrix. For every $\alpha \in t\Sigma_K$ we denote by $d_\alpha$ the row with index $\alpha$. For every $\alpha$ the row $d_\alpha$ is a linear combination of the rows $d_{\alpha_\sigma}$ with indices $\alpha_\sigma \in \mathcal{I}_K$ which are senior to $\alpha$, since the matrix $\Delta_K$ has a similar property. Let $\alpha_1, \ldots, \alpha_\sigma$ be elements of the set $\mathcal{I}_K$, all senior to $\alpha$. Then only the first $\sigma$ elements of the row $d_\alpha$ can be different from zero, since $d_{\alpha_1}, \ldots, d_{\alpha_\sigma}$ are the first $\sigma$ rows of the unit matrix. Further, let $d^\sigma$, $\sigma = 1, \ldots, \rho$ be the column of the matrix $D$ with index $\sigma$. From what we have said already, it follows that all the elements of the column $d^\sigma$ with an index strongly senior to $\alpha_\sigma$ are equal to zero, and the element with index $\alpha_\sigma$ is equal to unity.

The element of the matrix $D(D^j)$, located in the column with index $\sigma$ and the row with index $\alpha$, will be denoted by $d_\alpha^\sigma(d_\alpha^{j,\sigma})$. Relation (13) implies that

$$d_{\tau,i}^{j,\sigma} = \begin{cases} d_{\tau,i-j}^\sigma, & \text{if } i \geqq j, \\ 0 & \text{otherwise.} \end{cases} \tag{14.2}$$

Thus the connection between the matrices $D$ and $D^j$ is the same as that between the matrices $\Delta_K$ and $\Delta_K^j$, and in fact, the minor of the matrix $D^j$, made up of rows with indices lying in $t\Sigma_K+j$ is equal to $D$; the remaining rows of the matrix $D^j$ are equal to zero.

Let us consider the system of equations

$$\sum_{j\in\Sigma_\kappa} D^j x'_j = y, \tag{15.2}$$

where $x'_j \in \mathscr{L}^\rho$, the solution of which yields also a solution of the system (9). In expanded form, this system reads as follows:

$$\sum d^{j,\sigma} x_{j,\sigma} = y, \quad \text{where} \quad (x_{j,1}, \ldots, x_{j,\rho}) = x'_j, \tag{16.2}$$

and $d^{j,\sigma}$ are columns of the matrix $D^j$. We now show that there exist elements $x_{j,\sigma} \in \mathscr{L}$, satisfying these equations with indices belonging to the set $\mathscr{I}_K+\Sigma_\kappa$, and that these elements can be written in the form

$$x_{j,\sigma} = \sum r^i_{j,\sigma} y_i, \tag{17.2}$$

where $r^i_{j,\sigma} \in C^t$ are rows, made up of polynomials in the elements of the matrices $\Delta_K$ and $M^{-1}$, and are different from zero only for $L(i) \geqq L(i_\sigma+j)$ and satisfy the inequalities

$$|r^i_{j,\sigma}| \leqq b_0^{|L(i)-L(i_\sigma+j)|+1}, \qquad b_0 = |\Sigma_K|\, t\, |D|^t. \tag{18.2}$$

The $x_{j,\sigma}$ will be determined by induction on the value of the sum $i_\sigma+j$. We assume that for some $i\in\Sigma_k$ we have determined all the quantities $x_{j,\sigma}$ with $i_\sigma+j \succ i$, in the forms (17)–(18) above, in such a way that all the equations in the system (16) having indices $(\tau, i') \in \mathscr{I}_K+\Sigma_\kappa$ such that $i' \succ i$ are satisfied. We remark that the values of the $x_{j,\sigma}$ not yet determined cannot disturb the fulfillment of these equations. In fact, it follows from the formula (14) and the properties of the $d^\sigma_{\tau,i}$ that the most senior, non-zero element of the column $d^{j,\sigma}$ has the index $(\tau_\sigma, i_\sigma+j)$. Therefore, if $i_\sigma+j \preceq i$, all the elements of the column $d^{j,\sigma}$ with indices $(\tau, i')$, where $i' \succ i$, are equal to zero.

We now need to find quantities $x_{j,\sigma}$ with $i_\sigma+j=i$ to satisfy all the equations of the system (16) with indices $(\tau, i) \in \mathscr{I}_K+\Sigma_\kappa$ (where $i$ is the point fixed earlier). But this means that the equation

$$\sum_{i_\sigma+j=i} d^{j,\sigma}_{\tau,i} x_{j,\sigma} = \hat{y}_{\tau,i}, \quad \text{where} \quad \hat{y}_{\tau,i} = y_{\tau,i} - \sum_{i_{\sigma'}+j' \succ i} d^{j',\sigma'}_{\tau,i} x_{j',\sigma'} \tag{19.2}$$

has to be satisfied for all $(\tau, i) \in \mathscr{I}_K+\Sigma_\kappa$. Taking account of the relation (14) we rewrite the system in the form:

$$\sum_{i_\sigma+j=i} d^\sigma_{\tau,i_\sigma} x_{j,\sigma} = \hat{y}_{\tau,i}. \tag{20.2}$$

By definition of the set $\mathscr{I}_K+\Sigma_\kappa$, there exists for each of its points $(\tau, i)$ at least one pair $(\sigma, j)$, for which $1 \leqq \sigma \leqq \rho$, and $j \in \Sigma_\kappa$ is such that $i = i_\sigma + j$, and $\tau=\tau_\sigma$. If to the given point $(\tau, i)$ there correspond several such pairs $(\sigma, j)$, we select exactly one such pair (for example, by the condition that the point $(\sigma, j)\in\rho\Sigma_\kappa$ be the most senior) and for the remaining pairs $(\sigma, j)$, we set $x_{j,\sigma}=0$. Thus, the number of undetermined quantities $x_{j,\sigma}$ in the system (20) is equal to the number of equations in the system.

Rearranging the terms on the left side of (20) in order of increasing $\tau_\sigma$, we transform the system into upper triangular form. In fact, for arbitrary $\tau<\tau_\sigma$ the element $d^\sigma_{\tau, i_\sigma}$ is equal to zero, since $(\tau, i_\sigma)\succ(\tau_\sigma, i_\sigma)=\alpha_\sigma$. On the diagonal of this triangular matrix we find the quantities $d^\sigma_{\tau_\sigma, i_\sigma}=d^\sigma_{\alpha_\sigma}$, which are equal to unity. Thus, the determinant of the matrix of the system (20) has value unity, and accordingly, the norm of the inverse matrix, because of the inequality (1.1), does not exceed $t\,|D|^{t-1}$. Therefore, the system (20) has a solution, which can be written in the form

$$x_{j,\sigma}=\rho^i_{j,\sigma}\,\hat{y}_i, \qquad \hat{y}_i=(\hat{y}_{1,i}, \ldots, \hat{y}_{t,i}), \tag{21.2}$$

where $\rho^i_{j,\sigma}$ are rows made up of polynomials in the elements of the matrix $D$, satisfying the inequalities

$$|\rho^i_{j,\sigma}|\leqq t\,|D|^{i-1}. \tag{22.2}$$

We shall verify that the $x_{j,\sigma}$ that we found has the form (17)–(18). For this, we substitute in (21) the expression (19) for the $\hat{y}_i$ in vector notation

$$\begin{aligned} x_{j,\sigma}&=\rho^i_{j,\sigma}\,y_i-\rho^i_{j,\sigma}\sum_{i_{\sigma'}+j'\succ i} d^{j',\sigma'}_i\,x_{j',\sigma'} \\ &=\rho^i_{j,\sigma}\,y_i-\rho^i_{j,\sigma}\sum_{i_{\sigma'}+j'\succ i} d^{\sigma'}_{i-j'}\sum_{i'} r^{i'}_{j',\sigma'}\,y_{i'}. \end{aligned} \tag{23.2}$$

We shall show that in the second term of the righthand side, the summation extends only over the points $i'$ satisfying the inequality $L(i')>L(i)$. In fact, the $d^{\sigma'}_{i-j'}$ are different from zero only if $i-j'\in\Sigma_K$. We conclude that $|i_{\sigma'}+j'-i|_-\geqq|j'-i|_-\geqq -K$. Since $i_{\sigma'}+j'\succ i$ and $|i_{\sigma'}+j'-i|=0$, then Proposition I implies that $L(i_{\sigma'}+j')>L(i)$. On the other hand, it follows from the induction hypothesis that $r^{i'}_{j',\sigma'}\neq 0$ only when $L(i')\geqq L(i_{\sigma'}+j')$. This inequality together with the preceding implies that $L(i')>L(i)$, whence, in particular, it follows that $i'\neq i$. Therefore, we can convert Eq. (23) to the form (17) by writing

$$r^i_{j,\sigma}=\rho^i_{j,\sigma}, \qquad r^{i'}_{j,\sigma}=-\rho^i_{j,\sigma}\sum_{i_{\sigma'}+j'\succ i} d^{\sigma'}_{i-j'}\,r^{i'}_{j',\sigma'}, \tag{24.2}$$

where $i=i_\sigma+j$, and $L(i')>L(i)$. We remark that the elements of these matrices are polynomials in the elements of the matrices $\Delta_K$ and $M^{-1}$, since the matrices $\rho^i_{j,\sigma}$ have this property, and by the induction hypothesis, so also do the matrices $r^{i'}_{j',\sigma'}$ with $i_{\sigma'}+j'\succ i$.

We estimate the norms of the matrices $r^{i'}_{j,\sigma}$. When $i'=i$, the inequality (18) follows from (22), since $t\,|D|^{t-1}\leqq b_0$, and $L(i)=L(i_\sigma+j)$. For $i'\neq i$ we make use of the second of the formulae (24). The number of terms on the right side does not exceed the number of points $j'$, for which $d^{\sigma'}_{i-j'}\neq 0$, and, therefore, does not exceed the number of points $j'$, for which $i-j'\in\Sigma_K$. But this, clearly, is not greater than $|\Sigma_K|$. Making use of this fact, and the inequalities (22) and (18), we obtain

$$|r^{i'}_{j,\sigma}|\leqq|\Sigma_K|\cdot t\cdot|D|^{t-1}\cdot|D|\cdot b_0^{|L(i')-L(i_{\sigma'}+j')|+1}\leqq b_0^{|L(i')-L(i_\sigma+j)|+1},$$

since

$$L(i_{\sigma'}+j')>L(i)=L(i_\sigma+j).$$

Thus, the formulae (17)–(18) have been proved for all pairs $(\sigma,j)$. We remark that for arbitrary $\sigma$, $i_0\succeq K\,e_m$, $|i_\sigma|=K$, and, accordingly, by virtue of Proposition I, $L(i_\sigma)\geqq L(K\,e_m)$, whence $L(i_\sigma+j)\geqq L(j+K\,e_m)$. Therefore the inequality (17) can be rewritten in vector form

$$x'_j=\sum r^i_j\,y_i, \tag{25.2}$$

where $r^i_j$ are matrices different from zero only when $L(i)\geqq L(j+K\,e_m)$ and satisfying the inequalities

$$|r^i_j|\leqq b_0^{|L(i)-L(j+K\,e_m)|+1}. \tag{26.2}$$

Since the vectors $x'_j$ satisfy the system (15), by writing $x_j=\mathcal{M}x'_j$, $j\in\Sigma_\kappa$, we obtain a solution of the system (9). It follows from (25) that the vectors $x_j$ can be written in the form (10) with $R^i_j=\mathcal{M}r^i_j$. From what we have said above, it follows that the elements of the matrices $R^i_j$ are polynomials in the elements of the matrices $\Delta_K$ and $M^{-1}$, and that the matrix $R^i_j$ is different from zero only when $L(i)\geqq L(j+K\,e_m)$. In view of the inequality (1.1) the norm of the matrix $\mathcal{M}$ does not exceed $\rho\,|\Delta_K|^{\rho-1}(\det M)^{-1}$. Hence, taking account of the fact that in the inequality (26) the exponent is always positive, we obtain the inequality (11) for the matrix $R^i_j$. □

## § 3. Stabilization of the base sequence

### 1°. Monotone sets

**Definition 1.** A subset $\mathcal{I}\subset Z^m_+$ will be said to be monotone if with every one of its points $i$ it contains all points $j\geqq i$, or, in other words, if it coincides with the union $\bigcup\{i+Z^m_+,\ i\in\mathcal{I}\}$. The set $\mathcal{I}=(\mathcal{I}^1,\dots,\mathcal{I}^t)\subset t\,Z^m_+$ will be said to be monotone if each of its components $\mathcal{I}^1,\dots,\mathcal{I}^t\subset Z^m_+$ is a monotone subset of $Z^m_+$.

Given a set $I\subset t\,Z^m_+$ and an integer $k\geqq 0$, we denote by $\mathcal{I}_k$ the intersection $\mathcal{I}\cap t\,\Sigma_k$. We can formulate the criterion of monotonicity for sets as follows: in order that the subset $\mathcal{I} \subset tZ_+^m$ be monotone, it is necessary that for arbitrary $k \geqq 0$ and arbitrary $l \geqq 0$ (and sufficient, that for all $k \geqq 0$ and $l=1$) we have the inclusion relation $\mathcal{I}_k + \Sigma_l \subset \mathcal{I}_{k+l}$. Thus, in accordance with the Fundamental Lemma of § 2, every basis set is monotone.

**Definition 2.** The source of the monotone set $\mathcal{I} \subset tZ_+^m$ is defined as the minimum subset $s(\mathcal{I})$, having the property that

$$\mathcal{I} = s(\mathcal{I}) + Z_+^m.$$

In other words, the source is the set of all points $(\tau, i) \in \mathcal{I}$, having the property that there exist no points $(\tau, j) \in \mathcal{I}$ with $j < i$. It is clear that the components of the set $s(\mathcal{I})$ are the sets $s(\mathcal{I}^1), \ldots, s(\mathcal{I}^t)$, where $\mathcal{I}^1, \ldots, \mathcal{I}^t$ are the components of the monotone set $\mathcal{I}$.

**Proposition 1.** *The source of an arbitrary monotone set $\mathcal{I} \subset tZ_+^m$ is a finite set.*

*Proof.* In view of the preceding remark, it is sufficient to limit ourselves to the case $t=1$. For $m=1$ the assertion is obvious. We shall suppose it true for an arbitrary monotone set in the lattice $Z_+^{m-1}$, and we shall show that it is true for the given set $\mathcal{I} \subset Z_+^m$.

If the set $\mathcal{I}$ is empty, the assertion that $s(\mathcal{I})$ is finite is trivially true. Let us suppose that $\mathcal{I}$ is not empty, and we denote by $i^0 = (i_1^0, \ldots, i_m^0)$ a point in $\mathcal{I}$. Let $k$ be an arbitrary integer, lying between 1 and $m$, and let $\alpha$ be an integer, satisfying the inequality $0 \leqq \alpha < i_k^0$. The intersections of the lattice $Z_+^m$, of the set $\mathcal{I}$, and of its source $s(\mathcal{I})$ respectively with the subspace $i_k = \alpha$ are the lattice $Z_+^{m-1}$, a monotone subset $\mathcal{I}'$ of this lattice, and a subset of the source $s(\mathcal{I}')$. By the induction hypothesis, the source $s(\mathcal{I}')$ is finite for arbitrary $\alpha$ and $k$. Varying $\alpha$ and $k$ within the established limits, we see that the portion of the set $s(\mathcal{I})$, lying inside the set $Z_+^m + i^0 \subset \mathcal{I}$, is finite. Of the whole set $Z_+^m + i^0$, there can belong to the source $s(\mathcal{I})$ at most the one point $i^0$. From this we conclude that the set $s(\mathcal{I})$ is finite. □

The quantity

$$K = K(\mathcal{I}) = \max \{|i|, (\tau, i) \in s(\mathcal{I})\}$$

will be called the stability constant of the monotone set $\mathcal{I} \subset tZ_+^m$. This constant has the following property. For any other monotone set $\mathcal{I}' \subset tZ_+^m$ the inclusion relation $\mathcal{I}'_k \supset \mathcal{I}_k$ for all $k \leqq K(\mathcal{I})$ implies $\mathcal{I}' \supset \mathcal{I}$. In fact, these inclusions imply that the set $\mathcal{I}'$ contains the source $s(\mathcal{I})$ and therefore, being monotone, contains the whole of $\mathcal{I}$.

**2°. Decomposition of the space $C^n$ into sets in which the rank of the base sequence is constant.** We now resolve the space $C^n$ into the sum of a finite number of sets, in each of which all the matrices $\Delta_k(z)$ depend polynomially on $z$ and have constant rank.

**Definition 3.** An algebraic stratification of the space $C^n$ is defined as an arbitrary sequence $\mathcal{N}=\{N_\nu\}$ of algebraic varieties $N_\nu\subset C^n$ such, that

$$C^n=N_0\supset N_1\supset\cdots\supset N_\nu\supset N_{\nu+1}\supset\cdots\supset N_\omega=\varnothing, \tag{1.3}$$

and for all $\nu$, $N_\nu\neq N_{\nu+1}$.

It is known that every strongly decreasing sequence of algebraic varieties is finite and, in particular, that every algebraic stratification is finite. Corresponding to the algebraic stratification $\mathcal{N}$ we set up the function $\theta(z,\mathcal{N})$, defined in $C^n$, having the following form:

$$\theta(z,\mathcal{N})=\frac{\rho(z,N_{\nu+1})}{|z|^2+1},\qquad z\in N_\nu\setminus N_{\nu+1},\ \nu=0,1,2,\ldots,$$

where for convenience we set $\rho(z,\varnothing)=1$. By a power function over the stratification $\mathcal{N}$, we shall mean an arbitrary function of the form $C\,\theta^q(z,\mathcal{N})$, where $C>0$ and $q$ is a constant.

In order not to break the thread of the argument, we postpone the study of the properties of algebraic stratification to the end of this section and take up the fundamental theorem at this point.

**Theorem.** *There exists an algebraic stratification $\mathcal{N}=\{N_\nu\}$ of the space $C^n$, having the following properties: For arbitrary $\nu\geqq 0$, on the set $N_\nu\setminus N_{\nu+1}$*

I. *the basis set $\mathcal{I}=\mathcal{I}(z)$ and therefore, the ranks of all the matrices $\Delta_k(z)$ are constant (and equal to $|\mathcal{I}_k(z)|$);*

II. *for arbitrary $k\leqq K$, where $K=K(\mathcal{I})$, we can single out in the matrix $\hat{\Delta}_k(z)=\delta_{\mathcal{I}_k}P_k(z)$ a non-singular minor $M_k(z)$ of order $|\mathcal{I}_k|$, lying in the columns with constant indices such that*

$$|\det M_k(z)|\geqq c\,\theta^q(z,\mathcal{N}) \tag{2.3}$$

*for some $c>0$ and $q$.*

*Proof.* The varieties $N_\nu$ will be constructed by induction. We set $N_0=C^n$ and we postulate that for some $\mu\geqq 0$ we have already constructed the variety $N_0,\ldots,N_\mu$, satisfying the conditions of the theorem, with $\nu=0,\ldots,\mu-1$. If the variety $N_\mu$ is empty, the theorem is proved. Supposing that $N_\mu$ is not empty, we construct the variety $N_{\mu+1}$ so as to satisfy the conditions of the theorem with $\nu=\mu$.

We begin by constructing a sequence of sets $\mathcal{I}_k\subset t\,\Sigma_k$, $k=0,1,2,\ldots$, and a sequence of algebraic varieties $L_k$, such that

$$L_0\subset L_1\subset\cdots\subset L_k\subset\cdots\subsetneqq N_\mu$$

using the following inductive construction. We assume that for some $k \geqq 0$ the sets $\mathcal{J}_0, \ldots, \mathcal{J}_{k-1}$ and the manifolds $L_0, \ldots, L_{k-1}$ have already been constructed, and for convenience we set $\mathcal{J}_{-1} = \varnothing$ and $L_{-1} = \varnothing$. Then

1) we determine $\max |\mathcal{I}_k(z)|$ for the point $z \in N_\mu \setminus L_{k-1}$, and among the sets $\mathcal{I}_k(z)$ for which this maximum is obtained, we choose the most senior, which we denote by $\mathcal{J}_k$;

2) the union of $L_{k-1}$ and of the set of points $z \in N_\mu \setminus L_{k-1}$, in which $\mathcal{I}_k(z) \neq \mathcal{J}_k$, will be denoted by $L_k$.

It is clear that $L_{k-1} \subset L_k \subsetneq N_\mu$. It remains to show that $L_k$ is an algebraic variety. By $\hat{\Delta}_k(z)$ we shall denote the submatrix of $\Delta_k(z)$ consisting of the rows with indices belonging to $\mathcal{J}_k$. We shall show that the equation $\mathcal{I}_k(z) = \mathcal{J}_k$ for points $z \in N_\mu \setminus L_{k-1}$ is equivalent to the equation

$$\operatorname{rank} \hat{\Delta}_k(z) = |\mathcal{J}_k|. \tag{3.3}$$

In fact, if $\mathcal{I}_k(z) = \mathcal{J}_k$ the rows of the matrix $\hat{\Delta}_k(z)$ (the number of these is equal to $|\mathcal{J}_k|$) are linearly independent and therefore we have Eq. (3). Conversely, if Eq. (3) is satisfied, then by 1) the rank of the matrix $\Delta_k(z)$ is equal to $|\mathcal{J}_k|$. Therefore, by its construction, the set $\mathcal{I}_k(z)$ [3] is senior to $\mathcal{J}_k$. On the other hand, by construction, the set $\mathcal{J}_k$ is the most senior of the sets $\mathcal{I}_k(z)$, for which $|\mathcal{I}_k(z)| = |\mathcal{J}_k|$, and $z \in N_\mu \setminus L_{k-1}$. It follows that $\mathcal{I}_k(z) = \mathcal{J}_k$.

The condition contrary to (3) can be written in the form

$$\det M^\alpha(z) = 0, \quad \forall_\alpha, \tag{4.3}$$

where the $M^\alpha(z)$ are all the minors of the matrix $\hat{\Delta}_k(z)$ of order $|\mathcal{J}_k|$. By 1) and 2) the sets $\mathcal{I}_0(z), \ldots, \mathcal{I}_{k-1}(z)$ are constant on $N_\mu \setminus L_{k-1}$, and therefore, the ranks of the matrices $\Delta_0(z), \ldots, \Delta_{k-1}(z)$ are constant on $N_\mu \setminus L_{k-1}$. The theorem of § 2 implies that the matrix $\Delta_k(z)$ has a constant size on this set and its elements are polynomials in $z$. Therefore, all the $\det M^\alpha(z)$ are also polynomials on the set $N_\mu \setminus L_{k-1}$. Accordingly, the system of Eq. (4) determines an algebraic subvariety $L' \subset N_\mu \setminus L_{k-1}$. It follows from 2) that $L_k$ is equal to $L' \cup L_{k-1}$ and, accordingly, is an algebraic variety, which is what we were to show. This completes the construction of the sets $\mathcal{J}_k$ and $L_k$.

At an arbitrary point $z \in N_\mu \setminus L_k$, we have $\mathcal{I}_{k-1}(z) = \mathcal{J}_{k-1}$ and $\mathcal{I}_k(z) = \mathcal{J}_k$. From this, because of the monotonicity of the set $\mathcal{I}(z)$, it follows that $\mathcal{J}_{k-1} + \Sigma_1 \subset \mathcal{J}_k$, which implies the monotonicity of the set $\mathcal{J} = \bigcup_k \mathcal{J}_k \subset t Z_+^m$. Let $K = K(\mathcal{J})$ be the stability constant of $\mathcal{J}$. We shall show that for arbitrary $k > K$ the equation $L_k = L_K$ holds. Let us suppose the contrary:

3 See 3°, § 2.

let $k$ be the smallest of the numbers for which this equation is not satisfied. In this case $L_k$ is strongly senior to $L_K$; we choose an arbitrary point $z \in L_k \setminus L_K$. It follows from 2) that $\mathscr{I}_i(z) = \mathscr{I}_i$ for all $i \leqq K$. Then by the properties of the constant $K$, it follows that $\mathscr{I}(z) \supset \mathscr{I}$. On the other hand, it follows from 1) that $|\mathscr{I}_k(z)| \leqq |\mathscr{I}_k|$, whence, employing the preceding inclusion, it follows that $\mathscr{I}_k(z) = \mathscr{I}_k$ that is, $z \bar{\in} L_k$, which contradicts the choice of the point $z$. The contradiction that we have obtained shows that $L_k = L_K$ for $k \geqq K$. Since $L_k \subset L_K$ for $k < K$, then $\mathscr{I}(z) \equiv \mathscr{I}$ on $N_\mu \setminus L_K$.

It follows in particular that on the set $N_\mu \setminus L_K$, the ranks of all the matrices $\Delta_k(z)$ are constant, and, therefore, the sizes of these matrices are constant and their elements are polynomials. For every $k \leqq K$ we select in the submatrix $\hat{\Delta}_k(z)$ a minor $M_K(z)$ which is not identically singular on the set $N_\mu \setminus L_K$ of order $|\mathscr{I}_k(z)|$, lying in the columns having a constant index on the whole set $N_\mu \setminus L_K$. By $L^0$ we denote the algebraic subvariety in $N_\mu$, defined by the polynomial equation $\prod_0^k \det M_k(z) = 0$. Setting $N_{\mu+1} = L_K \cup L^0$, we obtain the variety we are seeking. To complete the proof of the theorem there remains to establish the inequality (2). Since every polynomial of $\det M_k(z)$ fails to vanish on $N_\mu \setminus N_{\mu+1}$, this inequality follows from Corollary 2, which we obtain below. □

In what follows we shall use the symbols $\mathscr{I}$ and $K$ for the functions $\mathscr{I}(z)$ and $K(\mathscr{I}(z))$. By the theorem we have just proved, these functions are constant on all the strata $N_\nu \setminus N_{\nu+1}$.

**Corollary 1.** *For every point $z \in C^n$ and every number $k \geqq K$ the columns of the matrix $\Delta_k(z)$ are linear combinations of the columns having the form*

$$\delta_k \{\eta^j P_{K,\sigma}(z, \eta)\}, \quad j \in \Sigma_{k-K}, \; \sigma = 1, \ldots, s_K. \tag{5.3}$$

*Proof.* By the properties of the stability constant $K$, for arbitrary $k \geqq K$, we have $\mathscr{I}_k = \mathscr{I}_K + \Sigma_{k-K}$, and, accordingly, the rank of the matrix $\Delta_k(z)$ is equal to $|\mathscr{I}_K + \Sigma_{k-K}|$. On the other hand, it follows from the Fundamental Lemma of § 2 that the minor of $\Delta_k(z)$ formed from the rows with indices belonging to $\mathscr{I}_K + \Sigma_{k-K}$ and columns of the form (5), has a rank equal to the number of rows, that is, its rank is equal to $|\mathscr{I}_K + \Sigma_{k-K}|$. Hence follows the assertion. □

For this corollary those columns of the matrix $\Delta_k(z)$ $k \geqq K$ which are different from the columns listed in (5), play no role and consequently can be excluded from consideration. In what follows, we shall suppose that the matrices $\Delta_k(z)$ for $k \geqq K$ consist only of columns of the type (5). Analytically, this agreement takes the form:

$$P_k(z, \eta) = P_K(z, \eta) \, H_{k,K}(\eta), \tag{6.3}$$

where $H_{k,K}(\eta)$ is a matrix of size $s_K \times s_K |\Sigma_{k-K}|$, which is the Kronecker product of the unit matrix of order $s_K$ and the rows $\{\eta^j, j\in\Sigma_{k-K}\}$. Accordingly, to conserve Eq. (1.2) for all $k$, we write

$$Q_k(z,\eta)=Q_K(z,\eta)\,H_{k,K}(\eta) \tag{7.3}$$

for $k \geqq K$.

**3°. Properties of an algebraic stratification.** We now establish certain properties of algebraic stratifications which will be used in the proof of the theorem of this section, and some properties which will be useful in Chapter IV.

**Proposition 2**

I. *Let $f(x)$ be an arbitrary polynomial, not vanishing identically, defined in the Euclidian space $R^n$, and let $N \subset R^n$ be the set of its real roots. Then we have the inequality*

$$\frac{1}{c}(|x|^2+1)^q\,\rho(x,N) \geqq |f(x)| \geqq c\left(\frac{\rho(x,N)}{|x|^2+1}\right)^q \tag{8.3}$$

*for some positive $c$ and $q$.*

II. *Let $N$ and $M$ be algebraic varieties in $R^n$. Then*

$$\rho(x,N) \geqq c\left(\frac{\rho(x,N\cap M)}{|x|^2+1}\right)^q, \qquad x\in M, \tag{9.3}$$

*for some positive $c$ and $q$.*

*Proof.* We establish the lefthand inequality in (8). Let $x$ be an arbitrary point in $R^n$; we choose the point $y\in N$ such that $\rho(x,N)=\rho(x,y)$. Making use of Lagrange's theorem, we establish a bound for the quantity $|f(x)|=|f(x)-f(y)|$ by the product of $\rho(x,y)$ and the upper bound of $|\operatorname{grad} f|$, taken along a segment joining the points $x$ and $y$. Since $|\operatorname{grad} f(x)| \leqq C(|x|^2+1)^q$ for some $C$ and $q$, this bound leads us to the leftmost inequality in (8). The righthand inequality is well-known[4].

We pass to the second assertion. Let $\{f_\alpha(x)\}$ be a finite set of polynomials such that the set of their common real roots coincides with $N$, and let $\{g_\beta(x)\}$ be a similar collection of polynomials for the manifold $M$. It is clear that the set of roots of the polynomial $\sum|f_\alpha|^2$ is $N$, and the set of roots of the polynomial $\sum|f_\alpha|^2+\sum|g_\beta|^2$ is $N\cap M$. On the set $M$ we have the equation

$$\sum|f_\alpha|^2=\sum|f_\alpha|^2+\sum|g_\beta|^2.$$

4 See, for example, Gorin [1].

By the first assertion, the lefthand side does not exceed $C(|x|^2+1)\,\rho(x,N)$, and the righthand side is not less than

$$c\left(\frac{\rho(x,N\cap M)}{|x|^2+1}\right)^q$$

for some $C$, $c$ and $q$. The inequality (9) follows. □

**Corollary 2.** *Let $\mathcal{N}=\{N_\nu\}$ be an algebraic stratification of $C^n$, and $f(z)$ be a polynomial not vanishing on some stratum $N_\nu \setminus N_{\nu+1}$. Then*

$$|f(z)| \geqq c\,\theta^q(z,\mathcal{N}), \qquad z\in N_\nu\setminus N_{\nu+1}, \tag{10.3}$$

*for some positive $c$ and $q$.*

*Proof.* We shall consider the complex space $C^n$ as a real space $R^{2n}$. Let $N$ be the set of roots of the polynomial $f$. Applying the first assertion of Proposition 2 to $N$, and then the second assertion to $N_\nu$ and $N$, we obtain the inequalities

$$|f(z)| \geqq c\left(\frac{\rho(z,N)}{|z|^2+1}\right)^q \geqq c'\left(\frac{\rho(z,N_\nu\cap N)}{|z|^2+1}\right)^{q'}, \qquad z\in N_\nu.$$

Since by hypothesis $N_\nu \cap N \subset N_{\nu+1}$, the righthand side is not less than $c'\,\theta^{q'}(z,N)$, which implies (10). □

**Proposition 3.** *For any two algebraic stratifications $\mathcal{M}$ and $\mathcal{L}$, we can find an algebraic stratification $\mathcal{N}$ such, that*

$$c\,\theta^q(z,\mathcal{N}) \leqq \min\{\theta(z,\mathcal{M}),\theta(z,\mathcal{L})\}, \qquad z\in C^n, \tag{11.3}$$

*for some positive $c$ and $q$.*

*Proof.* Let $\mathcal{M}=\{M_\mu\}$, and $\mathcal{L}=\{L_\lambda\}$. We pose

$$N_\nu = \bigcup_{\mu+\lambda=\nu} (M_\mu \cap L_\lambda), \qquad \nu=0,1,2,\ldots.$$

It is clear that the algebraic varieties $N_\nu$, after subtracting any repetitions, form a stratification of $C^n$. We shall establish the inequality (11). The sets

$$\begin{aligned}(M_\mu\setminus M_{\mu+1})\cap(L_\lambda\setminus L_{\lambda+1})\\ =(M_\mu\cap L_\lambda)\setminus[(M_\mu\cap L_{\lambda+1})\cup(M_{\mu+1}\cap L_\lambda)]\end{aligned} \tag{12.3}$$

form a finite covering of $C^n$. Therefore, inequality (11) will be established, if it is established for each of these sets.

We choose $\mu$ and $\lambda$ arbitrarily. We have $M_\mu \cap L_\lambda \subset N_\nu$, where $\nu=\mu+\lambda$, and

$$(M_\mu\cap L_\lambda)\cap N_{\nu+1} = (M_\mu\cap L_{\lambda+1})\cup(M_{\mu+1}\cap L_\lambda).$$

It follows that the set (12) belongs to $N_\nu \setminus N_{\nu+1}$, and

$$\rho(z, N_{\nu+1}) \leqq \min\{\rho(z, M_\mu \cap L_{\lambda+1}), \rho(z, M_{\mu+1} \cap L_\lambda)\}, \quad z \in M_\mu \cap L_\lambda. \tag{13.3}$$

Considering $C^n$ as the real space $R^{2n}$, we apply the second assertion of Proposition 2 to the varieties $N = L_{\lambda+1}$ and $M = M_\mu \cap L_\lambda$. We find that on the set $M_\mu \cap L_\lambda$ we have the inequality

$$c\left(\frac{\rho(z, M_\mu \cap L_{\lambda+1})}{|z|^2+1}\right)^q \leqq \rho(z, L_{\lambda+1}).$$

In a similar fashion, we establish the inequality

$$c\left(\frac{\rho(z, M_{\mu+1} \cap L_\lambda)}{|z|^2+1}\right)^q \leqq \rho(z, M_{\mu+1}).$$

Combining these inequalities with (13), we arrive at (11). □

The stratification $\mathcal{N}$ constructed in this proposition will be called the product of the stratifications $\mathcal{M}$ and $\mathcal{L}$.

## § 4. $p$-decompositions

**1°. Formulation of the fundamental theorem.** Let us recall the notation we now need. By $n$ we have denoted an arbitrary integer; $\mathcal{P}$ is the ring of polynomials in $C^n$ with complex coefficients, and $p$ is an arbitrary matrix of size $t \times s$, formed from elements of the ring $\mathcal{P}$. Further, $m$ is an arbitrary integer not exceeding $n$, $\mathcal{S}$ is the space of formal power series in $m$ variables $\eta = (\eta_1, \ldots, \eta_m)$ with complex coefficients. For an arbitrary point $z \in C^n$ the symbol $p(z)$ denotes a linear operator, acting from $\mathcal{S}^s$ to $\mathcal{S}^t$; the action of this operator amounts to multiplication by the matrix $p(z+\eta)$, where $z+\eta = (z_1+\eta_1, \ldots, z_m+\eta_m, z_{m+1}, \ldots, z_n)$.

By $\mathcal{I} = \mathcal{I}(z)$, $K = K(\mathcal{I}(z))$ and $\mathcal{N} = \{N_\nu\}$ we denote respectively the basis set, its stability constant, and an algebraic stratification of $C^n$, constructed in § 3 for the matrix $p$. The symbol $L$ denotes a linear operator in $R^m$, constructed in § 2 with the constant $K$ equal to $K(\mathcal{I}(z))$. By $\mathcal{S}^t_{\mathcal{I}}$ we denote the subspace of $\mathcal{S}^t$ consisting of series whose coefficients vanish when their indices belong to the set $\mathcal{I}$.

**Theorem 1.** *At every point $z \in C^n$ the identity operator $E$ in $\mathcal{S}^t$ admits the decomposition*

$$E = \mathcal{D}(z) + p(z)\mathcal{G}(z), \tag{1.4}$$

*which has the following properties:*

1) *$\mathcal{D}(z)$ is a linear operator from $\mathcal{S}^t$ to $\mathcal{S}^t_{\mathcal{I}}$, vanishing on the subspace $p(z)\mathcal{S}^s$.*

2) *$\mathscr{G}(z)$ is a linear operator from $\mathscr{S}^t$ to $\mathscr{S}^s$, vanishing on the subspace $\mathscr{S}^t_{\mathscr{I}}$.*

3) *Let $\mathscr{D}(z)=\sum \eta^i \mathscr{D}_i$, and let $\mathscr{G}(z)=\sum \eta^i \mathscr{G}_i$. Then*

$$\begin{aligned} &\mathscr{D}_i \eta^\lambda \neq 0, \quad \textit{only if} \quad L(\lambda) \geqq L(i)+L(K e_m - K e_1), \\ &\mathscr{G}_i \eta^\lambda \neq 0, \quad \textit{only if} \quad L(\lambda) \geqq L(i)+L(K e_m). \end{aligned} \tag{2.4}$$

4) *For arbitrary $i$ and $\lambda$ the quantities (2) are rational functions of $z$ that do not reach infinity on any of the sets $N_\nu \setminus N_{\nu+1}$, $\nu=0,1,2,\ldots$, and*

$$\begin{aligned} &|\mathscr{D}_i \eta^\lambda| \leqq \mathscr{B}^{|L(\lambda)-L(i)-L(K e_m - K e_1)|+1}, \\ &|\mathscr{G}_i \eta^\lambda| \leqq \mathscr{B}^{|L(\lambda)-L(i)-L(K e_m)|+1}, \end{aligned} \tag{3.4}$$

*where the constant $\mathscr{B}$ satisfies the inequality*

$$\mathscr{B} \leqq C\,\theta^q(z, \mathscr{N}) \tag{4.4}$$

*for some $C$ and $q$.*

The decomposition (1) will be called a $p$-decomposition, and the operators $\mathscr{D}$ and $\mathscr{G}$ will be called $p$-operators or operators corresponding to the matrix $p$. In the future, unless we state the contrary explicitly, we shall use this decomposition in the particular case $m=n$.

**2°. Construction of a $p$-decomposition.** We fix an arbitrary point $z \in C^n$. We obtain the decomposition (1) by a sequence of transformations on the homogeneous component of the lefthand side of (1). At the minus first step our decomposition has the following form:

$$E=\sum \eta^{\tau,i} \delta_{\tau,i}=\sum \eta^{\tau,i} \mathscr{D}^{-1}_{\tau,i}+p(z+\eta)\sum \eta^i \mathscr{G}^{-1}_i, \tag{5.4}$$

where

$$\mathscr{D}^{-1}_{\tau,i}=\delta_{\tau,i}, \qquad \mathscr{G}^{-1}_i=0. \tag{6.4}$$

We now assume that for some integer $k \geqq 0$, we have carried out $k-1$ steps and have obtained the decomposition

$$\sum \eta^{\tau,i} \delta_{\tau,i}=\sum \eta^{\tau,i} \mathscr{D}^{k-1}_{\tau,i}+p(z+\eta)\sum \eta^i \mathscr{G}^{k-1}_i, \tag{7.4}$$

satisfying the following conditions:

$a_{k-1}$) The functional $\mathscr{D}^{k-1}_{\tau,i}\colon \mathscr{S}^t \to C$ is equal to zero, if $(\tau,i) \in \mathscr{I}$ and $|i| \leqq k-1$,

$b_{k-1}$) the order of the functional $\mathscr{D}^{k-1}_{\tau,i}$ does not exceed $|i|$,

$c_{k-1}$) every functional $\mathscr{G}^{k-1}_i\colon \mathscr{S}^t \to C^s$ contains the functionals $\delta_{\tau,j}$ only when $(\tau,j) \in \mathscr{I}$.

$d_{k-1}$) the order of the functional $\mathscr{G}^{k-1}_i$ does not exceed $k-1$. For the decomposition (5) this condition is clearly fulfilled for $k-1=-1$.

We now make the $k$-th step. Let $\Delta_k=\Delta_k(z)$ be a matrix belonging to the base sequence; let $\delta_k \mathscr{D}^{k-1}$ be a column of size $t\,\Sigma_k \times 1$, formed from the functionals $\mathscr{D}^{k-1}_{\tau,i}$ with $|i|=k$. We consider the system of linear equations

$$\Delta_k F = \delta_k \mathscr{D}^{k-1}, \tag{8.4}$$

where $F$ is a column formed of the unknown functionals $F_\sigma$: $\mathscr{S}^t \to C$, $\sigma = 1, \ldots, s_k$. The column $F$ will be defined in such a way that all the equations with indices $(\tau, i) \in \mathscr{I}_k$ in the system (8) will be satisfied. This can be done, since by the definition of the basis set $\mathscr{I}$ the rows of the matrix $\Delta_k$ with indices belonging to $\mathscr{I}_k$ are linearly independent. Therefore there exists a vector $F$ satisfying all of the equations of the system (8), whose components are linear combinations of the functionals $\mathscr{D}^{k-1}_{\tau,i}$ with $(\tau, i) \in \mathscr{I}_k$. We note that this vector is not uniquely defined, since the columns of $\Delta_k$ can be linearly dependent. A little later we shall eliminate this ambiguity by sharpening the choice of the vector $F$.

We set

$$\sum \eta^{\tau,i} \mathscr{D}^k_{\tau,i} = \sum \eta^{\tau,i} \mathscr{D}^{k-1}_{\tau,i} - P_k F, \qquad \mathscr{D}^k_{\tau,i} = \mathscr{D}^{k-1}_{\tau,i} - \delta_{\tau,i} P_k F. \tag{9.4}$$

Since no element of the matrix $P_k$ contains a term of order less than $k$ (with respect to $\eta$), we conclude from Eq. (9) that

$$\mathscr{D}^k_{\tau,i} = \mathscr{D}^{k-1}_{\tau,i} \qquad \text{for arbitrary } i \text{ with } |i| < k. \tag{10.4}$$

Therefore, it follows from $a_{k-1}$) that $\mathscr{D}^k_{\tau,i} = 0$ for all $(\tau, i) \in \mathscr{I}$ with $|i| < k$. We remark that the equation of the system (8) that has the index $(\tau, i)$ has the form $\delta_{\tau,i} P_k F = \mathscr{D}^{k-1}_{\tau,i}$. By construction, the vector $F$ satisfies all equations of this type with $(\tau, i) \in \mathscr{I}_k$. It therefore follows from (9) that $\mathscr{D}^k_{\tau,i} = 0$ also, if $(\tau, i) \in \mathscr{I}_k$. We have thus established the fact that the functionals $\mathscr{D}^k_{\tau,i}$ satisfy the condition $a_k$).

It follows from $b_{k-1}$) that $\deg \delta_k \mathscr{D}^{k-1} \leqq k$, whence $\deg F \leqq k$. Therefore every coefficient of the power series $P_k F$ is a functional of order not less than $k$. But since the elements of the matrix $P_k$ contain no term of order less than $k$, the coefficient of $\eta^{\tau,i}$ in this power series is a functional of order not greater than $|i|$. By $b_{k-1}$) the series $\sum \eta^{\tau,i} \mathscr{D}^{k-1}_{\tau,i}$ has the same property. It therefore follows from (9) that the same is true also for the series $\sum \eta^{\tau,i} \mathscr{D}^k_{\tau,i}$, that is, condition $b_k$) is satisfied.

The functionals $\mathscr{G}^k_i$, corresponding to the $k$-th step, are defined as follows:

$$\sum \eta^i \mathscr{G}^k_i = \sum \eta^i \mathscr{G}^{k-1}_i + Q_k F, \qquad \mathscr{G}^k_i = \mathscr{G}^{k-1}_i + \delta_i Q_k F, \tag{11.4}$$

where $Q_k$ is the matrix of the theorem in § 2. We shall verify that the conditions $c_k$) and $d_k$) are satisfied. On both sides of Eq. (7) we equate coefficients of $\eta^{\tau,i}$, where $(\tau, i)$ is an arbitrary point of the set $\mathscr{I}_k$. On the left we obtain the functional $\delta_{\tau,i}$, and on the right the sum of the functional $\mathscr{D}^{k-1}_{\tau,i}$ and a linear combination of the functionals $\mathscr{G}^{k-1}_{i'}$. In view of $c_{k-1}$), this linear combination of the functionals $\mathscr{G}_{i'}^{k-1}$ contains $\delta_{\sigma,j}$ only if $(\sigma,j)\in\mathscr{I}$. Therefore, the functional $\mathscr{D}_{\tau,i}^{k-1}$ also has this property. By construction, the functionals $F_\sigma$ are linear combinations of the functionals $\mathscr{D}_{\tau,i}^{k-1}$ with $(\tau,i)\in\mathscr{I}_k$ and therefore, they also contain $\delta_{\sigma,j}$ only if $(\sigma,j)\in\mathscr{I}$. Therefore, by Eq. (11), it follows from $c_{k-1}$) that we have $c_k$). In a similar way, $d_{k-1}$) implies $d_k$), since $\deg F\leqq k$.

Combining Eqs. (9), (11), (7), and (1.2), we obtain

$$\Sigma\,\eta^{\tau,i}\delta_{\tau,i}=\sum\eta^{\tau,i}\mathscr{D}_{\tau,i}^{k}+p(z+\eta)\sum\eta^{i}\mathscr{G}_{i}^{k}. \tag{12.4}$$

The induction is now completed and, therefore, for arbitrary $k\geqq 0$ we have constructed the decomposition (12), which satisfies the conditions $a_k$), $b_k$), $c_k$), and $d_k$).

We remark that as $k\to\infty$, the functionals $\mathscr{D}_{\tau,i}^{k}$ and $\mathscr{G}_i^k$ are stabilized. In fact, Eq. (10) implies that $\mathscr{D}_{\tau,i}^{k}=\text{const}$ as soon as $k\geqq|i|$. We set $\mathscr{D}_{\tau,i}=\mathscr{D}_{\tau,i}^{|i|}$ for all $(\tau,i)$. It further follows from (7.3) that for $k\geqq K$ the matrix $Q_k$ contains no term of order less than $k-K$ (with respect to $\eta$). The equations (11) therefore imply $\mathscr{G}_i^k=\mathscr{G}_i^{k-1}$, if $k-K>|i|$, and, accordingly $\mathscr{G}_i^k=\text{const}$ for $k-K\geqq|i|$. Setting $\mathscr{G}_i=\mathscr{G}_i^{|i|+k}$, we consider the operators

$$\mathscr{D}(z)=\sum\eta^{\tau,i}\mathscr{D}_{\tau,i},\qquad \mathscr{G}(z)=\sum\eta^{i}\mathscr{G}_i.$$

Letting $k$ tend to infinity in (12), we conclude that these operators satisfy (1).

**3°. Proof of the first and second properties.** It follows from $a_k$) that $\mathscr{D}_{\tau,i}=0$ for all $(\tau,i)\in\mathscr{I}$, and, accordingly, the image of the operator $\mathscr{D}$ belongs to $\mathscr{S}_{\mathscr{I}}^t$. Further, let $\phi$ be an arbitrary element of the space $p(z)\mathscr{S}^s$. We must show that

$$\mathscr{D}_{\tau,i}\phi=0 \tag{13.4}$$

for all $(\tau,i)$. These equations will be established by induction. We suppose that for some $k\geqq 0$, the Eqs. (13) are fulfilled for all $(\tau,i)$ with $|i|<k$ (if $k=0$, this assumption is trivially satisfied), and we show that the equations hold also for $|i|=k$.

Since $\mathscr{D}_{\tau,i}=\mathscr{D}_{\tau,i}^{k-1}$ for $|i|<k$, we have by hypothesis $\mathscr{D}_{\tau,i}^{k-1}\phi=0$ for $|i|<k$. Acting on the series $\phi$ with the decomposition (7), we obtain

$$\breve{\phi}=\phi-p(z+\eta)\sum\eta^{i}\mathscr{G}_i^{k-1}\phi=\sum\eta^{\tau,i}\mathscr{D}_{\tau,i}^{k-1}\phi.$$

Thus, the series $\breve{\phi}$ belongs to $p(z)\mathscr{S}^s$ and it does not contain terms of order less than $k$. Hence, by the theorem of § 2 it follows that the column

$$\delta_k\breve{\phi}=\delta_k\mathscr{D}^{k-1}\phi$$

is a linear combination of the columns of the matrix $\Delta_k$. Therefore the column

$$\delta_k(\mathscr{D}^{k-1}\phi - P_k F \phi)$$

is also a linear combination of columns of the matrix $\Delta_k = \delta_k P_k$. By definition of the operator $\mathscr{D}^k$, this column is equal to $\mathscr{D}^k \phi$, and, accordingly, its elements with indices belonging to $\mathscr{I}_k$ are equal to zero. Since the rows of the matrix $\Delta_k$ with indices belonging to the set $\mathscr{I}_k$ form a basis for all its rows, it follows that every column $\mathscr{D}^k \phi$ vanishes, that is, $\mathscr{D}^k_{\tau,i}\phi = 0$ for all $(\tau, i)$ for which $|i| = k$. Since $\mathscr{D}_{\tau,i} = \mathscr{D}^k_{\tau,i}$ for $|i| = k$, it further follows that we have Eq. (13) with $|i| = k$. But this establishes property 1. Property 2 follows from $c_k$).

**4°. Proof of the third property**

**Inductive Proposition.** *For arbitrary* $i, \lambda \in Z^m_+$, *with* $|i| \geqq k-1$, *we have* $\mathscr{D}^{k-1}_i \eta^\lambda \neq \delta_i \eta^\lambda$, *only if* $L(\lambda) \geqq L(i) - l^i_{k-1}$, *where*

$$l^i_{k-1} = \begin{cases} L(K e_1 - K e_m), & |i| = k-1, \\ (k-1-|i|)\, e_m, & |i| > k-1. \end{cases} \tag{$3_{k-1}$}$$

*Proof.* For $k-1 = -1$ the assertion follows from the fact that $\mathscr{D}^{-1}_i = \delta_i$. We shall demonstrate $(3_k)$, assuming that $(3_{k-1})$ is true. We note that

$$\mathscr{D}^k_i \eta^\lambda = \mathscr{D}^{k-1}_i \eta^\lambda \qquad \text{if } |\lambda| > k. \tag{14.4}$$

This relationship follows from (9), since $\deg F \leqq k$. By virtue of (14) for $|\lambda| > k$, the assertion $(3_k)$ follows immediately from $(3_{k-1})$, since $l^i_{k-1} \leqq l^i_k$.

We shall now suppose that $|\lambda| \leqq k$. We first suppose that $k < K$. We show that the inequality $L(\lambda) \geqq L(i) - l^i_k$ holds. In fact, if $|i| = k$, then

$$\lambda - i + K e_1 - K e_m \succ 0, \quad |\lambda - i + K e_1 - K e_m| \leqq 0,$$
$$|\lambda - i + K e_1 - K e_m|_- \geqq |-i|_- > -K.$$

Hence, in view of the proposition of § 2, there follows $L(\lambda - i + K e_1 - K e_m) \geqq 0$, and this, because of the linearity of the operator $L$ yields $L(\lambda) \geqq L(i) - L(K e_1 - K e_m) = L(i) - l^i_k$. If, however, $|i| > k$, we have

$$|\lambda - i| \leqq k - |i| < 0, \quad |\lambda - i|_+ \leqq |\lambda| \leqq K,$$

which again, by virtue of the proposition of § 2, leads to the inequality $L(\lambda - i) \geqq (|i| - k)\, e_m = -l^i_k$, whence $L(\lambda) \geqq L(i) - l^i_k$.

Now suppose $k \geqq K$. We make the construction of the vector $F$ more precise. To this end, we rewrite the system (8) in the form

$$\sum_j \Delta^j_K F_j = \delta_k \mathscr{D}^{k-1}, \tag{15.4}$$

where $\Delta_K^j, j \in \Sigma_{k-K}$ is the minor of the matrix $\Delta_K$ formed from the columns $\delta_k \eta^j P_{K,\sigma}(z,\eta)$, $\sigma = 1, \ldots, s_K$, and $F_j$ is a column formed from the unknown functionals $F_{j,\sigma}$, $\sigma = 1, \ldots, s_K$ which coincide up to numerical order with the functionals $F_\sigma$ in (8). By the Fundamental Lemma of § 2, we can find columns $F_j$ satisfying all the equations of this system with indices belonging to $\mathscr{I}_k$ which admit the representation

$$F_j = \sum R_j^{i'} \mathscr{D}_{i'}^{k-1}, \tag{16.4}$$

where the $R_j^{i'}$ are the matrices constructed in the Lemma, non-zero only for $L(i') \geqq L(j + K e_m)$.

Eq. (9) can now be written as:

$$\mathscr{D}_i^k - \mathscr{D}_i^{k-1} = -\delta_i \sum_j \eta^j P_k F_j = -\sum_j \delta_{i-j} P_K \sum_{i'} R_j^{i'} \mathscr{D}_{i'}^{k-1}.$$

If we apply both sides to the element $\eta^\lambda$, we obtain

$$\mathscr{D}_i^k \eta^\lambda - \mathscr{D}_i^{k-1} \eta^\lambda = -\sum_j \delta_{i-j} P_K \sum_{i'} R_j^{i'} \mathscr{D}_{i'}^{k-1} \eta^\lambda. \tag{17.4}$$

We shall show that the values of the indices $i, i', j, \lambda$, for which the terms on the righthand side are different from zero, are connected by the inequalities

$$L(\lambda) \geqq L(i') \geqq L(j + K e_m) \geqq L(i) - l_k^i. \tag{18.4}$$

Since on the righthand side of (17) $|i'| = k > k-1$, $(3_{k-1})$ implies that $\mathscr{D}_{i'}^{k-1} \eta^\lambda \neq 0$ only if $L(\lambda) \geqq L(i') + e_{m'}$ or if $i' = \lambda$. In both cases, the first inequality of (18) is fulfilled. The second follows from the fact that $R_j^{i'} \neq 0$ only if $L(i') \geqq L(j + K e_m)$. We now establish the third inequality. We note that $\delta_{i-j} P_K \neq 0$ only for $i \geqq j$. We assume that $|i| = k$. Then the equation $|j| = k - K$ implies, that

$$j - i + K e_1 \succeq 0, \quad |j - i + K e_1| = 0,$$
$$|j - i + K e_1|_- \geqq |j - i|_- = -K.$$

Hence, by the proposition of § 2, $L(j - i + K e_1) \geqq 0$, that is, $L(j + K e_m) \geqq L(i - K e_1 + K e_m)$, which leads to the third inequality of (18). Now suppose $|i| > k$. Then $|j + K e_m - i| = k - |i| < 0$, and $i \geqq j$ implies that $|j + K e_m - i|_+ \leqq |K e_m|_+ = K$. Therefore, by the same proposition $L(j + K e_m - i) \geqq (|i| - k)\, e_m = -l_k^i$, which again implies the third inequality of (18).

Comparing the left and right sides of (18), we conclude that the righthand side of (17) fails to vanish only when $L(\lambda) \geqq L(i) - l_k^i$. Accordingly, only in this case do we have $\mathscr{D}_i^k \eta^\lambda \neq \mathscr{D}_i^{k-1} \eta^\lambda$. This fact, together with $(3_{k-1})$, implies $(3_k)$, since $l_k^i \geqq l_{k-1}^i$. □

Since $\mathscr{D}_i = \mathscr{D}_i^{k-1}$ for $|i| = k-1$, $(3_{k-1})$ implies the first assertion of the theorem for the operator $\mathscr{D}$.

We shall prove this assertion for the operator $\mathscr{G}$. To this end, it is sufficient to show that for arbitrary $k$

$$\mathscr{G}_i^k \eta^\lambda \neq 0 \quad \text{only if} \quad L(\lambda) \geqq L(i + K e_m). \tag{$3'_k$}$$

Suppose $k < K$. From $\mathrm{d}_k$) it follows that $\mathscr{G}_i^k \eta^\lambda$ differs from zero only for $|\lambda| \leqq k$. We shall show that for arbitrary $\lambda$ satisfying this inequality, and for arbitrary $i$, the inequality

$$L(\lambda) \geqq L(i + K e_m). \tag{19.4}$$

holds.

Every coordinate of the point $L(\lambda)$ attains its least value for the greatest value of $|\lambda|$ and the least values of the first $m-1$ coordinates of the point $\lambda$, that is, for $\lambda = k\, e_m$. Hence $L(\lambda) \geqq L(k\, e_m)$. On the other hand, $L(k\, e_m) > L(K e_m) \geqq L(i + K e_m)$, since $L(i) \leqq 0$ for arbitrary $i \geqq 0$. We have thus proved the inequality (19) and together with it we have established $(3'_k)$.

Now suppose $k \geqq K$. Taking account of formula (16) for the functional $F$ in (11), we arrive at the equation

$$\mathscr{G}_i^k \eta^\lambda - \mathscr{G}_i^{k-1} \eta^\lambda = \sum_j \delta_{i-j} Q_k \sum_{i'} R_j^{i'} \mathscr{D}_{i'}^{k-1} \eta^\lambda. \tag{20.4}$$

By analogy with Eq. (17), it follows that the values of the indices $i, i', j, \lambda$, for which the terms on the righthand side of (20) are different from zero, satisfy the first two inequalities of (18). In place of the third, we have now the inequality $L(j + K e_m) \geqq L(i + K e_m)$. This follows from the relation $i \geqq j$. Thus, we obtain the chain of inequalities

$$L(\lambda) \geqq L(i') \geqq L(j + K e_m) \geqq L(i + K e_m).$$

Comparing the left and right hand side, we arrive at the conclusion that the difference $\mathscr{G}_i^k \eta^\lambda - \mathscr{G}_i^{k-1} \eta^\lambda$ differs from zero only for $L(\lambda) \geqq L(i + K e_m)$. This fact, together with the assertion $(3'_k)$ which has been proved for all $k < K$, allows us to establish $(3'_k)$ by induction for all $k$.

**5°. Proof of the fourth property.** We refine the construction of the functionals $F$. Let us suppose, to begin with, that $k < K$. Let $M_k$ be the minor of the matrix $\Delta_k$ appearing in the theorem of § 3, and let $\sigma_1, \ldots, \sigma_\rho$, $(\rho = \rho_k)$ be the indices of the columns containing this minor. The column $F$ will be defined as follows: its components with indices different from $\sigma_1, \ldots, \sigma_\rho$, will be set equal to zero. Then for the contracted column $\hat{F} = (F \sigma_1, \ldots, F \sigma_\rho)$, the system (8) can be rewritten as:

$$M_k \hat{F} = \delta_{\mathscr{I}_k} \mathscr{D}^{k-1}.$$

The solution of this system we write in the form

$$\hat{F} = M_k^{-1} \delta_{\mathscr{I}_k} \mathscr{D}^{k-1}. \tag{21.4}$$

We note that by the theorem of § 3, the minor $M_k$ is non-singular at every point and that on every set of the form $N_\nu \backslash N_{\nu+1}$ the number of columns and rows containing it is constant, and its elements are polynomials in $z$.

In the case $k \geqq K$ the column $F = \{F_j\}$ was defined by means of the system (15), by applying the fundamental lemma of § 2. In this lemma we put $M = M_{K'}$ where $M_K$ is the minor developed in the theorem of § 3. In accordance with the Fundamental Lemma, the elements of the matrix $R_j^{i'}$ appearing in (16) are polynomials in the elements of the matrices $\Delta_K$ and $M_K^{-1}$ and, accordingly, are rational regular functions of $z$ on every set $N_\nu \backslash N_{\nu+1}$.

Thus, we have proved that for arbitrary $k$ the elements of the column $F$ are linear combinations of the functionals $\mathscr{D}_i^{k-1}$ with coefficients which are rational (matrix) functions on all of the sets $N_\nu \backslash N_{\nu+1}$. On the other hand, the functionals $\mathscr{D}_i^k$ and $\mathscr{G}_i^k$ are linear combinations of the functionals $F, \mathscr{D}_i^{k-1}, \mathscr{G}_i^{k-1}$ with coefficients (equal to $\pm \delta_i P_k$, $\pm \delta_i Q_k$, or 1) which are polynomials in $z$. Thus, for arbitrary $k$, the functionals $\mathscr{D}_i^k$ and $\mathscr{G}_i^k$ are constructed from the functionals $\mathscr{D}_i^{k-1}$ and $\mathscr{G}_i^{k-1}$ by means of the operations of addition and of multiplication by functions of the type described above. Hence by (6) there follows the first part of the fourth assertion of the theorem.

We now turn to the proof of the inequality (3). We establish inductively the inequality

$$|(\mathscr{D}_i^k - \delta_i)\eta^\lambda| \leqq B^{|L(\lambda) - L(i) + l_k^i| + 1}, \qquad |i| \geqq k, \tag{$4_k$}$$

where $B$ is defined by the formulae

$$B = 2^{m+1} b \prod_0^K B_k, \qquad B_k = 2(|M_k^{-1}| + 1)\Big[\sum_{|i|=k} (|\delta_i P_k| + |\delta_i Q_k|) + 1\Big], \tag{22.4}$$

and $b$ is a constant derived in the Fundamental Lemma of § 2, where we set $M = M_K$. By ($3_k$), we can assume that $L(\lambda) - L(i) + l_k^i \geqq 0$, since otherwise the inequality ($4_k$) is trivially satisfied.

We suppose to begin with that $k < K$. Then by (21)

$$|F\eta^\lambda| \leqq |M_k^{-1}| \cdot |\delta_k \mathscr{D}^{k-1} \eta^\lambda|.$$

It then follows from Eq. (9) and formula (22) that

$$\begin{aligned} |\mathscr{D}_i^k \eta^\lambda| &\leqq |\mathscr{D}_i^{k-1} \eta^\lambda| + |\delta_i P_k F \eta^\lambda| \\ &\leqq |\mathscr{D}_i^{k-1} \eta^\lambda| + \left(\frac{B_k}{2} - 1\right) |\delta_k \mathscr{D}^{k-1} \eta^\lambda| \leqq \frac{B_k}{2} \max_{i'} |\mathscr{D}_{i'}^{k-1} \eta^\lambda|. \end{aligned}$$

Applying this inequality repeatedly and taking into account the fact that $|\mathscr{D}_i^{-1}\eta^\lambda|=|\delta_i\eta^\lambda|\leqq 1$, we obtain

$$|\mathscr{D}_i^k\eta^\lambda|\leqq 2^{-(k+1)}B_0\cdot B_1\cdot\dots\cdot B_k. \tag{23.4}$$

Since by hypothesis the quantity $|L(\lambda)-L(i)+l_k^i|$ is nonnegative and $B\geqq 2$, it follows from (23) that

$$|(\mathscr{D}_i^k-\delta_i)\eta^\lambda|\leqq|\mathscr{D}_i^k\eta^\lambda|+|\delta_i\eta^\lambda|\leqq\frac{B}{2}+1\leqq B\leqq B^{|L(\lambda)-L(i)+l_k^i|+1}.$$

But this proves the inequality $(4_k)$ for all $k<K$.

Now suppose $k\geqq K$. We have from Eq. (17) and the inequality (11.2)

$$\begin{aligned}|(\mathscr{D}_i^k-\delta_i)\eta^\lambda|&\leqq|(\mathscr{D}_i^{k-1}-\delta_i)\eta^\lambda|+\sum_{|i'|=k}\Big(\sum_{j\leqq i}|\delta_{i-j}P_K|\cdot|R_j^{i'}|\Big)|\mathscr{D}_{i'}^{k-1}\eta^\lambda|\\ &\leqq|(\mathscr{D}_i^{k-1}-\delta_i)\eta^\lambda|+\sum_{|i'|=k}\left(\frac{B_K}{2}\max_{j\leqq i}b^{|L(i')-L(j+Ke_m)|+1}\right)\cdot|\mathscr{D}_{i'}^{k-1}\eta^\lambda|.\end{aligned} \tag{24.4}$$

It follows from the inequalities (18) that the exponent on the righthand side does not exceed $|L(i')-L(i)+l_k^i|+1$. Therefore, the quantity in the curved bracket is not greater than

$$\frac{B_K}{2}b^{|L(i')-L(i)+l_k^i|+1}\leqq[2^{-(m+2)}B]^{|L(i')-L(i)+l_k^i|+1}.$$

We assume that the inequality $(4_{k-1})$ is satisfied. Then, for $|i'|=k$ we have $l_{k-1}^{i'}=-e_m$ and, accordingly,

$$\begin{aligned}|\mathscr{D}_{i'}^{k-1}\eta^\lambda|&\leqq|(\mathscr{D}_{i'}^{k-1}-\delta_{i'})\eta^\lambda|+|\delta_{i'}\eta^\lambda|\\ &\leqq B^{|L(\lambda)-L(i')|}+1\leqq 2B^{|L(\lambda)-L(i')|},\end{aligned} \tag{25.4}$$

since $|L(\lambda)-L(i')|\geqq 0$ by (18). Then from (24) there follows the inequality

$$\begin{aligned}&|(\mathscr{D}_i^k-\delta_i)\eta^\lambda|\\ &\quad\leqq|(\mathscr{D}_i^{k-1}-\delta_i)\eta^\lambda|+\sum_{|i'|=k}[2^{-(m+2)}B]^{|L(i')-L(i)+l_k^i|+1}\,2B^{|L(\lambda)-L(i')|}\\ &\quad\leqq B^{|L(\lambda)-L(i)+l_{k-1}^i|+1}\\ &\qquad+\Big[\sum_{i'}2^{-(m+2)(|L(i')-L(i)+l_k^i|+1)+1}\Big]B^{|L(\lambda)-L(i)+l_k^i|+1}.\end{aligned} \tag{26.4}$$

We note that the sum with respect to $i'$ on the righthand side does not exceed the sum $\sum 2^{-(m+2)(|j|+1)+1}$, taken over all $j\in Z^m$. But this latter sum can be easily shown to be equal to $2^{-(m+1)}[1-2^{-(m+2)}]^{-m}$. We note that $|l_{k-1}^i|<|l_k^i|$ for all $i$, $|i|\geqq k$. Therefore the righthand side of (26) is bounded by

$$\{B^{-1}+2^{-(m+1)}[1-2^{-(m+2)}]^{-m}\}\, B^{|L(\lambda)-L(i)+l_k^i|+1}. \tag{27.4}$$

It is clear from formula (22) that $B \geqq 2^{m+1}$, whence $B^{-1} \leqq 2^{-(m+1)}$. After a straightforward calculation, we can conclude that the first factor in (27) does not exceed unity, which leads us finally to the inequality $(4_k)$.

The inequality $(4_k)$ is now completely proved. We obtain from it, for $k=|i|$

$$|\mathscr{D}_i \eta^\lambda| = |\mathscr{D}_i^{|i|} \eta^\lambda| \leqq B^{|L(\lambda)-L(i)+l_k^i|+1}+1 \leqq (B+1)^{|L(\lambda)-L(i)+l_k^i|+1},$$

which implies the first of the inequalities (3) with $\mathscr{B}=B+1$.

Let us estimate the constant $\mathscr{B}$. Since $P_k$ and $Q_k$ are polynomials in $z$ and $\eta$, all the functions of the form $\delta_i P_k$ and $\delta_i Q_k$ and, in particular, all elements of the matrix $\Delta_k$ are polynomials in z. Therefore, they are all bounded above in absolute value by a quantity of the form $C(|z|+1)^q$. On the other hand, by the theorem of § 3, the determinants of the minors $M_k$, $k \leqq K$, are bounded in absolute value from below, by functions of the form $c\,\theta^q(z, \mathscr{N})$. Therefore, taking account of the inequality (1.1), we obtain

$$|M_k^{-1}| \leqq |\mathscr{I}_k| \frac{|\Delta_k|^{|\mathscr{I}_k|-1}}{|\det M_k|} \leqq C\,\theta^q(z, \mathscr{N}), \qquad k \leqq K,$$

with suitable values of $C$ and $q$. From formula (22), it is clear that a similar process provides an upper bound for the quantities $B_k$ and, from formula (12.2), the quantity $b$ also. Taking account of formula (22) for $B$, we arrive at (4). Thus, the fourth assertion of the theorem for the operator $\mathscr{D}$ has been proved.

We now establish the inequality (3) for the quantities $\mathscr{G}_i \eta^\lambda$. In view of $(3'_k)$, we may assume that the inequality $L(\lambda) \geqq L(i+K e_m)$. Suppose to begin with $k<K$. Then from (11) and (23) there follows the inequality

$$\begin{aligned} |\mathscr{G}_i^k \eta^\lambda - \mathscr{G}_i^{k-1} \eta^\lambda| &\leqq \frac{B_k}{2} \max_{i'} |\mathscr{D}_{i'}^{k-1} \eta^\lambda| \leqq 2^{-(k+1)} B_0 \ldots B_k \\ &\leqq 2^{-(k+2+m)} B \leqq [2^{-(k+2+m)} B]^{|L(\lambda)-L(i+K e_m)|+1}, \end{aligned} \tag{28.4}$$

since by hypothesis $|L(\lambda)-L(i+K e_m)| \geqq 0$. If $k \geqq K$, we have from (20) and (11.2)

$$|\mathscr{G}_i^k \eta^\lambda - \mathscr{G}_i^{k-1} \eta^\lambda| \leqq \sum_{|i'|=k} \left( \frac{B_K}{2} \max_{j \leqq i} b^{|L(i')-L(j+K e_m)|+1} \right) |\mathscr{D}_{i'}^{k-1} \eta^\lambda|.$$

Since $j \leqq i$, we have $L(j+K e_m) \geqq L(i+K e_m)$. Therefore, the exponent on the righthand side of the inequality is not greater than $|L(i')-L(i+$

$K e_m)|+1$. Therefore, each term on the right side does not exceed the quantity

$$[2^{-(m+2)}B]^{|L(i')-L(i+K e_m)|+1}\,|\mathscr{D}_{i'}^{k-1}\eta^\lambda|.$$

But this inequality, together with (25), yields

$$|\mathscr{G}_i^k\eta^\lambda-\mathscr{G}_i^{k-1}\eta^\lambda|\leqq\Big(\sum_{i'}2^{-(m+2)(|L(i')-L(i+K e_m)|+1)+1}\Big)B^{|L(\lambda)-L(i+K e_m)|+1}.$$

Summing these inequalities, together with (28), we obtain

$$\begin{aligned}|\mathscr{G}_i\eta^\lambda|&=\sum_{k=0}^{K-1}|\mathscr{G}_i^k\eta^\lambda-\mathscr{G}_i^{k-1}\eta^\lambda|+\sum_{k=K}^{|i|+K}|\mathscr{G}_i^k\eta^\lambda-\mathscr{G}_i^{k-1}\eta^\lambda|\\&\leqq\left(\sum_{k=0}^{K-1}2^{-(k+2+m)}+\sum_{j\in Z_+^m}2^{-(m+2)(|j|+1)+1}\right)B^{|L(\lambda)-L(i+K e_m)|+1}.\end{aligned}$$

The first factor on the righthand side is equal to

$$2^{-(m+1)}[1-2^{-(K+1)}]+2^{-(m+1)}[1-2^{-(m+2)}]^{-m}\leqq 1.$$

From this finally we have

$$|\mathscr{G}_i\eta^\lambda|\leqq B^{|L(\lambda)-L(i+K e_m)|+1},$$

which completes the proof of the inequality (3), and also the proof of the theorem. ▯

### 6°. Remarks and Corollaries

*Remark* 1. A $p$-decomposition defines a representation of the space $\mathscr{S}^t$ as a direct sum of subspaces $\mathscr{S}_{\mathscr{I}}^t\oplus p\,\mathscr{S}^s$. In fact the images of the operators $\mathscr{D}$ and $p\mathscr{G}$ belong respectively to the subspaces $\mathscr{S}_{\mathscr{I}}^t$ and $p\,\mathscr{S}^s$, and by the properties of 1 and 2, we have

$$\mathscr{D}=\mathscr{D}E=\mathscr{D}\mathscr{D},\qquad p\mathscr{G}=p\mathscr{G}E=p\mathscr{G}p\mathscr{G},$$

that is, $\mathscr{D}$ and $p\mathscr{G}$ are projection operators.

We note that in the case $p(z)\equiv 0$, we have $p(z)\,\mathscr{S}^s\equiv 0$ and, accordingly, $\mathscr{D}(z)\equiv E$.

*Remark* 2. Since $\mathscr{D}$ is an identity operator on the subspace $\mathscr{S}_{\mathscr{I}}^t$, its coefficients $\mathscr{D}_{\tau,i}$ with indices $(\tau,i)\in t\,Z_+^m\setminus\mathscr{I}$ are linearly independent. Since the operator $\mathscr{D}$ acts from $\mathscr{S}^t$ to $\mathscr{S}_{\mathscr{I}}^t$, the rest of its coefficients vanish.

We note that for arbitrary $i\in Z_+^m$

$$\deg\mathscr{D}_i\leqq|i|,\qquad \deg\mathscr{G}_i\leqq|i|+K. \tag{29.4}$$

In fact, by the third property, the quantity $\mathscr{D}_i \eta^\lambda$ is non-zero only for $L_m(\lambda) \geqq L_m(i) + L_m(K e_m - K e_1)$, that is, only for $|\lambda| \leqq |i|$. From this, the first inequality of (29) follows. The second inequality is verified in a similar fashion.

*Remark* 3. Property 2 of the operator $\mathscr{G}(z)$ can be written in the following strengthened form:

$$\mathscr{G}_i(z, D)\,\phi = 0 \qquad \text{for all } i \text{ and for arbitrary } \phi \in \mathscr{S}^t_{\mathscr{I}} \tag{30.4}$$

(where $\mathscr{G}_i(z, D)$ is the differential operator corresponding to the functional $\mathscr{G}_i(z, \delta)$, see § 1). In fact, the quantity $D^{\tau, \lambda} \eta^{\sigma, j}$ is non-zero only when $\sigma = \tau$, and $j \geqq \lambda$. On the other hand, the operator $\mathscr{G}_i(z, D)$ contains the differentiated $D^{\tau, \lambda}$ only if $(\tau, \lambda) \in \mathscr{I}$. Since the set $\mathscr{I}$ is monotone, every pair $(\sigma, j)$ satisfying the conditions $\sigma = \tau$, and $j \geqq \lambda$, also belongs to this set, and this implies (30).

**Corollary 1.** *There exists a power function $r_z \leqq 1$ on the stratification $\mathscr{N}$ such that for arbitrary $Z \in C^n$ and $\varepsilon$, $0 < \varepsilon \leqq 1$, the functions $\phi \in \mathscr{C}^t_\varepsilon$ can be represented in the form*

$$\phi = \mathscr{D}(z)\,\phi + p(z)\,\mathscr{G}(z)\,\phi,$$

*where $\mathscr{D}(z)\,\phi$ and $\mathscr{G}(z)\,\phi$ are convergent for $|\eta| \leqq \varepsilon\, r_z$ and satisfy the inequalities*

$$\|\mathscr{D}(z)\,\phi\|_{\varepsilon r_z} \leqq \frac{1}{r_z}\,\|\phi\|_\varepsilon, \qquad \|\mathscr{G}(z)\,\phi\|_{\varepsilon r_z} \leqq \frac{1}{\varepsilon^K\, r_z}\,\|\phi\|_\varepsilon. \tag{31.4}$$

*Proof.* We estimate the coefficients $\mathscr{D}_. \phi$ of the power series $\mathscr{D}(z)\,\phi$. They can be written in the form

$$\mathscr{D}_i \phi = \sum_\lambda \mathscr{D}_i \eta^\lambda \cdot \delta_\lambda \phi.$$

Since the order of the functional $\mathscr{D}_i$ does not exceed $|i|$ (see (29)), the summation on the righthand side runs only over those $\lambda$ that satisfy $|\lambda| \leqq |i|$. Therefore the number of terms in this sum does not exceed $(|i|+1)^m$. Further, since $L(i)$ is a linear operator in $R^m$, the inequality $|-L(i)| \leqq l\,|i|$ holds on the set $Z^m_+$ for sufficiently large $l$. Assuming that the constant $l$ is not smaller than $|-L(K e_m - K e_1)| + 1$, we obtain from the inequality (3)

$$|\mathscr{D}_i \eta^\lambda| \leqq \mathscr{B}^{l(|i|+1)},$$

since $|L(\lambda)| \leqq 0$. Writing $\mathscr{A} = \sqrt{m}\,\mathscr{B}^l$ and taking account of Proposition 2 of § 1, we find

$$|\mathscr{D}_i \phi| \leqq \sum_\lambda |\mathscr{D}_i \eta^\lambda| \cdot |\delta_\lambda \phi| \leqq (|i|+1)^m \mathscr{A}^{(|i|+1)}\, \varepsilon^{-|i|}\, \|\phi\|_\varepsilon.$$

Hence for $|\eta| \leqq \frac{\varepsilon}{2\mathscr{A}}$, we have

$$|\sum \eta^i \mathscr{D}_i \phi| \leqq \sum |\eta|^{|i|} |\mathscr{D}_i \phi| \leqq C\mathscr{A} \|\phi\|_\varepsilon.$$

Choosing the power function $r_z$ not to exceed $(2\mathscr{A})^{-1}$ and $(C\mathscr{A})^{-1}$, we arrive at the first of the inequalities (31). The second of these inequalities follows in a similar way from (3) and (29).

**7°. *p*-decompositions for analytic matrices *p*.** We note that all the arguments of §§ 1–4 can be carried over, with the obvious modifications, to the case when $p$ is a matrix formed of arbitrary functions that are holomorphic in some open set $\Omega \subset C^n$. In order to accomplish this generalization, it is sufficient to replace the words "polynomial in $z$" in the foregoing argument by the words "function holomorphic in $\Omega$," and to replace the concept of algebraic stratification of $C^n$ by the similar concept of analytic stratification of $\Omega$, which means an arbitrary strongly-decreasing sequence $\mathscr{N}$:

$$\Omega = N_0 \supset N_1 \supset \cdots \supset N_\nu \supset N_{\nu+1} \supset \cdots$$

of analytic subsets of the domain $\Omega$. It is well known that every such sequence is finite on every compact, that is, for an arbitrary compact $K \subset \Omega$ the intersections $K \cap N_\nu$ are empty from some $\nu$ onward.

The function $\theta(z, N)$, corresponding to such a stratification is constructed exactly in the same way as for an algebraic stratification. The power function on an analytic stratification $\mathscr{N}$ is defined as an arbitrary function of the form $C(z)\,\theta^{q(z)}(z, \mathscr{N})$ where $C(z) > 0$ and $q(z)$ are some real continuous functions in $\Omega$. We now formulate the general result, which is the analog of the Fundamental Theorem.

**Theorem 2.** *There exists an analytic stratification $\mathscr{N} = \{N_\nu\}$ of the set $\Omega$ such that on every stratum $N_\nu \backslash N_{\nu+1}$ the basis set $\mathscr{I}$ is constant. At every point $z \in C^n$, the decomposition* (1) *is defined, and the operators $\mathscr{D}(z)$ and $\mathscr{G}(z)$ satisfy the conditions* 1–3 *of the Fundamental Theorem, and also the condition:*

4′) *On every stratum $N_\nu \backslash N_{\nu+1}$, the quantities* (2) *are analytic functions of $z$ and satisfy the inequality* (3) *in which $\mathscr{B}$ is a power function on the stratification $\mathscr{N}$.*

Chapter III

# Cohomologies of Analytic Functions of Bounded Growth

In this chapter we consider a complex of analytic cochains in $C^n$ (or in some region $C^n$), satisfying the inequality

$$|\phi_{i_0,\ldots,i_\nu}(z)| \leqq C M_\alpha(z), \qquad \exists\, C,\ \alpha,\ \forall\, i_0,\ldots,i_\nu,\ z \in C^n,$$

where $M_\alpha(z)$, $\alpha=1,2,\ldots$, is a fixed increasing sequence of positive functions. We call this sequence a family of majorants; when it satisfies certain special conditions, we shall show that the homologies of the complex are trivial. This result will be established in § 4 on the basis of a theorem on the solubility of an inhomogeneous equation with the operator $\partial/\partial\bar{z}$ which we obtained in § 2. We have collected all definitions in § 1 and 3. In § 5, we consider the operator which acts by multiplication by a polynomial matrix, on complexes of the type we are now considering. We shall show that the kernel and the cokernel of this operator are homologically trivial.

## § 1. The space of holomorphic functions

### 1°. Families of majorants and spaces

**Definition.** Suppose that in the complex space $C^n$ there is given a nonincreasing sequence of open sets

$$C^n \supset \Omega_1 \supset \Omega_2 \supset \cdots \supset \Omega_\alpha \supset \cdots.$$

Suppose further that for every $\alpha$, there is given a function $M_\alpha(z)$ in $C^n$, which is finite and positive in $\Omega_\alpha$, and equal to $\infty$ outside of $\Omega_\alpha$. The sequence of functions $M_\alpha(z)$, $\alpha=1,2,\ldots$, will be called a family of majorants in $C^n$, if for every $\alpha$ the inequalities

a) $(|z|+1)\, M_\alpha(z) \leqq C_\alpha M_{\alpha+1}(z)$,

b) $\sup\limits_{|z'-z|\leqq\varepsilon_\alpha} M_\alpha(z') \leqq C_\alpha M_{\alpha+1}(z)$,

hold, where $C_\alpha$ and $\varepsilon_\alpha$ are positive numbers. The inequality b) implies, in particular, that the $\varepsilon_\alpha$-neighborhood of the set $\Omega_{\alpha+1}$ belongs to $\Omega_\alpha$.

We now fix a family of majorants $\mathcal{M}=\{M_\alpha\}$. For every integer $\alpha>0$ and $k\geqq 0$ we consider the norm

$$\|\phi\|_\alpha^k = \max_{|i|+|j|\leqq k}\ \sup_{\Omega_\alpha} \frac{|D^{i,\bar{j}}\phi(z)|}{M_\alpha(z)}, \tag{1.1}$$

defined for functions given in $\Omega_\alpha$, and having derivatives up to the order $k$. Here

$$i=(i_1, \ldots, i_n), \qquad j=(j_1, \ldots, j_n)\in Z_+^n,$$

and

$$D^{i,\bar{j}}=D_z^{i,\bar{j}}=\frac{\partial^{|i|}}{\partial z_1^{i_1}\ldots\partial z_n^{i_n}}\,\frac{\partial^{|j|}}{\partial \bar{z}_1^{j_1}\ldots\partial \bar{z}_n^{j_n}}.$$

The space of functions for which the norm (1) is finite will be denoted by $\mathscr{H}_\alpha^{0,k}$. This space endowed with the norm (1) is a normed space. An arbitrary function defined in a wider set $\Omega\supset\Omega_\alpha$ will also be taken to be an element of the space $\mathscr{H}_\alpha^{0,k}$, if its norm (1) is finite. We extend the upper bound in (1) to the whole set $\Omega$, setting $\frac{1}{M_\alpha(z)}=0$ outside $\Omega_\alpha$. We shall say that the function $\phi$ is equal to zero as an element of the space $\mathscr{H}_\alpha^{0,k}$, if $\phi=0$ on $\Omega_\alpha$.

Let $m$ be an integer lying between 0 and $n$. The subspace of $\mathscr{H}_\alpha^{0,k}$ formed by functions that are holomorphic in $z_1, \ldots, z_m$, will be denoted by $\mathscr{H}_\alpha^{m,k}$. For fixed $m$ and $\alpha$ the spaces $\mathscr{H}_\alpha^{m,k}$ form a decreasing sequence of subspaces of the space $\mathscr{H}_\alpha^{m,0}$ $\mathscr{H}_\alpha^{m,k+1}\to\mathscr{H}_\alpha^{m,k}$. We write[1]

$$\mathscr{H}_\alpha^m=\varprojlim_k \mathscr{H}_\alpha^{m,k}.$$

The operation of restricting functions defined in $\Omega_\alpha$ to the subregion $\Omega_{\alpha+1}$ defines a continuous mapping $\mathscr{H}_\alpha^{m,k}\to\mathscr{H}_{\alpha+1}^{m,k}$. Then the diagram

$$\begin{array}{ccc} \mathscr{H}_\alpha^{m,k} & \longrightarrow & \mathscr{H}_{\alpha+1}^{m,k} \\ \uparrow & & \uparrow \\ \mathscr{H}_\alpha^{m,k+1} & \longrightarrow & \mathscr{H}_{\alpha+1}^{m,k+1} \end{array}$$

is obviously commutative. Accordingly, we may define the mapping of the limit spaces $\mathscr{H}_\alpha^m\to\mathscr{H}_{\alpha+1}^m$. Thus, the spaces $\mathscr{H}_\alpha^m$, $-\infty<\alpha<\infty$ form a family of linear topological spaces. This family will be denoted by $\mathscr{H}_{\mathscr{M}}^m$.

The identity mappings of the spaces $\mathscr{H}_\alpha^m\to\mathscr{H}_\alpha^{m-1}$ define mappings of their family $\mathscr{H}_{\mathscr{M}}^m\to\mathscr{H}_{\mathscr{M}}^{m-1}$. The spaces $\mathscr{H}_\alpha^{m,k}$ provide the means for studying the spaces $\mathscr{H}_\alpha^{n,0}$, which from now on we will denote more concisely by $\mathscr{H}_\alpha^{n,0}=\mathscr{H}_\alpha$. The family of these spaces will be denoted by $\mathscr{H}_{\mathscr{M}}$. We write $\overline{\mathscr{H}}_{\mathscr{M}}=\varinjlim \mathscr{H}_{\mathscr{M}}$. The space $\overline{\mathscr{H}}_{\mathscr{M}}$ consists of functions each defined and analytic in some region $\Omega_\alpha$, and equal to $0(M_\alpha(z))$.

**2°. Bounds for the derivatives in the spaces $\mathscr{H}_\alpha^{m,0}$.** The result we now intend to establish is a particular case of a well-known property of elliptic differential operators.

---

1 For the definition of limit spaces, we refer to Ch. V, §1, 2°.

Let us consider the operator $''d$, which maps the function $\phi(z)$ to the vector formed from the functions

$$\frac{\partial \phi}{\partial \bar{z}_m}, \qquad m=1, \ldots, n.$$

**Proposition 1.** *Let $\omega$ be an $r$-neighborhood of the point $z \in C^n$, where $0<r \leqq 1$. Then for an arbitrary function $\phi \in L_2(\omega)$ we have the inequality*

$$|D^{i,\bar{j}}\phi(z)| \leqq C r^{-n-|i|-|j|} \|\phi\|_{L_2(\omega)} + C \sup_{\omega} |D^{i,\bar{j}}\, '' d\phi| \tag{2.1}$$

*under the condition that the vector function $''d\phi$ has the necessary derivatives.*

*Proof.* We shall suppose that $z=0$. Initially, let $r=1$. Let $E$ denote the fundamental solution for the Laplace operator $\Delta = \sum_1^n \frac{\partial^2}{\partial z_m \partial \bar{z}_m}$ and let us consider the vector function $\mathscr{E} = \left\{\frac{\partial E}{\partial z_m}, m=1, \ldots, n\right\}$. This function satisfies the equation $''d\mathscr{E} = \delta$ ($\delta$ being the delta-function), and its components are locally summable. Let $h$ be an infinitely differentiable function, with support in $\omega$ being equal to unity in the neighborhood of zero. We have

$$''dh\mathscr{E} + ''d(1-h)\mathscr{E} = \delta,$$

where $h\mathscr{E}$ is a summable function, and $(1-h)\mathscr{E}$ is infinitely differentiable. Hence

$$D^{i,\bar{j}}\phi(0) = (\overline{D^{i,\bar{j}}}\delta, \phi) = (\bar{h}\bar{\mathscr{E}}, D^{i,\bar{j}}\, ''d\phi) + (\overline{D^{i,\bar{j}}\, ''d(1-h)\mathscr{E}}, \phi).$$

The first term on the right side does not exceed $C \sup_{\omega} |D^{i,\bar{j}}\, ''d\phi|$, and the second term does not exceed $C\|\phi\|_{L_2(\omega)}$. Thus, the inequality (2) is proved for the case $r=1$.

Now suppose $r<1$. We apply the inequality just proved to the function $\phi(rz)$, and we obtain

$$r^{|i|+|j|}|D^{i,\bar{j}}\phi(0)| \leqq C r^{-n}\|\phi\|_{L_2(\omega)} + C r^{|i|+|j|+1} \sup_{\omega} |D^{i,\bar{j}}\, ''d\phi|.$$

Dividing both sides by $r^{|i|+|j|}$, we arrive at (2). □

**Corollary 1.** *In the space $\mathscr{H}_\alpha^{0,0}$ we have the inequality*

$$\|\phi\|_{\alpha+1}^k \leqq C_\alpha(\|\phi\|_\alpha^0 + \|''d\phi\|_\alpha^k), \qquad k=0, 1, 2, \ldots. \tag{3.1}$$

*Proof.* In Proposition 1, we set $r=\varepsilon_\alpha$. Since $\phi \in \mathscr{H}_\alpha^{0,0}$ is bounded in the $r$-neighborhood of an arbitrary point $z \in \Omega_{\alpha+1}$, we have $\|\phi\|_{L_2(\omega)} \leqq$

$C\,r^n \sup\limits_{\omega} |\phi|$. Hence

$$|D^{i,\bar{j}}\,\phi(z)| \leqq \frac{C}{r^{|i|+|j|}} \sup_{|\zeta|\leqq\varepsilon_\alpha} |\phi(z+\zeta)| + C \sup_{|\zeta|\leqq\varepsilon_\alpha} |D^{i,\bar{j}\,\prime\prime} d\phi(z+\zeta)|. \tag{4.1}$$

Dividing the left side by $M_{\alpha+1}(z)$, and the right side by $M_\alpha(z+\zeta)\leqq M_{\alpha+1}(z)$ and taking the upper bound with respect to $z$ and $(i,j)$ with $|i|+|j|\leqq k$, we obtain (3). □

**Corollary 2.** *For arbitrary* $\alpha$, *we have* $\mathscr{H}_\alpha^{n,0}\subset\mathscr{H}_{\alpha+1}^n$, *and the identity imbedding* $\mathscr{H}_\alpha^{n,0}\to\mathscr{H}_{\alpha+1}^n$ *is continuous. The families* $\mathscr{H}_{\mathscr{M}}$ *and* $\mathscr{H}_{\mathscr{M}}^n$ *are equivalent.*

### 3°. Logarithmically convex functions

**Definition.** A function $\mathscr{I}(y)$, defined in $R^n$, is said to be logarithmically convex (l.c.), if it is positive, that is, if it takes only positive values and the value $+\infty$, and if the set in $R^1\times R^n$, consisting of the points $(t,y)$ at which $t\geqq \ln \mathscr{I}(y)$, is convex and closed.

Let $\mathscr{I}$ be an l.c. function. By $g$ we denote the set in $R^n$ where $\mathscr{I}(y)<\infty$. This coincides with the set on which $\ln\mathscr{I}(y)<\infty$, and is therefore the projection in $R^n$ of the set in $R^1\times R^n$ defined by the inequality $t\geqq\ln\mathscr{I}(y)$. It follows that the set $g$ is convex. We shall suppose that it has inner points; then it is the closure of some open set $\omega$. Thus, the function $\ln\mathscr{I}$ is finite and convex in the open set $\omega$, and is therefore continuous on it.

The logarithmic reciprocal of $\mathscr{I}$ is the function $I$ defined on the dual Euclidian space by the formula

$$I(\sigma)=\sup_y \frac{\exp((y,\sigma))}{\mathscr{I}(y)}. \tag{5.1}$$

We note that $\mathscr{I}(y)\exp(-(y,\sigma))$ is an l.c. function for any $\sigma$. Hence for an arbitrary $c\in R$ the set $g_c=\{y\colon \mathscr{I}(y)\exp(-(y,\sigma))\leqq c\}$ is convex and closed. If for all $c$ the set $g_c$ is bounded, the supremum (5) is attained on some compact set. We denote by $y(\sigma)$ some point of this set. In the contrary case there exists $c$ such that $g_c$ is unbounded. Since this set is convex it contains at least one ray. The unit vector of this ray we shall consider as a non-proper point of $R^n$ and denote by $y(\sigma)$ also.

**Proposition 2.** *Let* $\mathscr{I}$ *be an l.c. function. Then for arbitrary* $\sigma$, *when the point* $y$ *moves along a straight line to the point* $y(\sigma)$ *(along a ray parallel to the vector* $y(\sigma)$*), the function* $\mathscr{I}(y)\exp(-(y,\sigma))$ *does not increase.*

*Proof.* We suppose that the point $y(\sigma)$ is proper. By definition it is contained in any set $g_c\neq\varnothing$. Since this set is convex its intersection with the segment $[y,y(\sigma)]$ is connected for any $y\in R^n$. The assertion follows.

Let $y(\sigma)$ be a non-proper point. It is sufficient to prove that for any $c$ and $y \in g_c$ the set $g_c$ contains the whole ray $\{y+r\,y(\sigma), r>0\}$. By definition of the vector $y(\sigma)$ there exists a point $y_0 \in R^n$ such that the function $\mathscr{I}'(y)=\mathscr{I}(y)\exp(-(y,\sigma))$ is bounded by some constant $c_0$ on the ray $\{y_0+s\,y(\sigma), s>0\}$. The function $\ln \mathscr{I}'$ being convex, its value at the point $y_s=y+r/s(y_0+s\,y(\sigma)-y)$ belonging to the segment $[y, y_0+s\,y(\sigma)]$ does not exceed $\ln c + r/s \ln c_0$. If $r$ is fixed and $s\to\infty$, then $y_s \to y+r\,y(\sigma)$. Since the set $\{t \geqq \ln \mathscr{I}'(y)\}$ is closed, it contains the limit point $(t=\ln c, y+r\,y(\sigma))$; this means that $\mathscr{I}'(y+r\,y(\sigma)) \leqq c$ for any $r>0$. □

By definition of the logarithmic reciprocal function we have the inequality

$$\mathscr{I}(y)\,I(\sigma) \geqq \exp((y,\sigma)), \tag{6.1}$$

for any $y$ and $\sigma$.

**Proposition 3.** *The function $I$, logarithmically reciprocal to $\mathscr{I}$, is logarithmically convex. If $\mathscr{I}$ itself is logarithmically convex, then it coincides with the logarithmic reciprocal of $I$.*

*Proof.* From the definition of $I$

$$\ln I(\sigma) = \sup_y [(y,\sigma) - \ln \mathscr{I}(y)].$$

Therefore, the set $\{t \geqq \ln \mathscr{I}(y)\}$ is the intersection of the closed half-spaces $\{t \geqq (y,\sigma) - \ln \mathscr{I}(y)\}$ and is, therefore, closed and convex.

Let us prove the second assertion. Let $\mathscr{I}$ be an l.c. function. We shall show that it coincides with the function $\mathscr{I}'$, which is logarithmically reciprocal to $I$. The set $\{t \geqq \ln \mathscr{I}(y)\}$, which is convex and closed, coincides with the intersection of the half-spaces $\{t \geqq (y,\sigma) - t_\sigma\}$, which contain it. Fixing $\sigma \in R^n$, we determine the smallest $t_\sigma$ for which the corresponding half-space contains the set $\{t \geqq \ln \mathscr{I}(y)\}$:

$$\min t_\sigma = \sup_y [(y,\sigma) - \ln \mathscr{I}(y)] = \ln I(\sigma).$$

Therefore, the set $\{t \geqq \ln \mathscr{I}(y)\}$ coincides with the intersection of the half-spaces

$$t \geqq (y,\sigma) - \ln I(\sigma).$$

Hence

$$\ln \mathscr{I}(y) = \sup_\sigma [(y,\sigma) - \ln I(\sigma)] = \ln \mathscr{I}'(y),$$

q.e.d. □

**4°. Families of majorants of type $\mathscr{I}$.** Let $z=x+iy$. A family of majorants $\{M_\alpha\}$ in $C^n$ will be called a family of type $\mathscr{I}$, if for arbitrary $\alpha$ the function $M_\alpha$ has the form $M_\alpha(z)=\mathscr{R}_\alpha(z)\,\mathscr{I}_\alpha(y)$, where $\mathscr{R}_\alpha$ is an everywhere finite positive function in $C^n_z$, and $\mathscr{I}_\alpha$ is logarithmically convex in $R^n_y$. We also require that the functions $\mathscr{R}_\alpha$, $-\infty<\alpha<\infty$, form a family of majorants in $C^n$, and that the functions $\mathscr{I}_\alpha$, $-\infty<\alpha<\infty$, form a family of majorants in $R_y^n$, that is, that for arbitrary $\alpha$ we have the inequalities

a′) $$(|y|+1)\mathscr{I}_\alpha(y) \leqq C_\alpha \mathscr{I}_{\alpha+1}(y),$$

b′) $$\sup_{|y'-y|\leqq \varepsilon_\alpha} \mathscr{I}_\alpha(y') \leqq C_\alpha \mathscr{I}_{\alpha+1}(y).$$

The most important condition is as follows: for arbitrary $\alpha$ there exists an entire function $e_\alpha(z) \not\equiv 0$ in $C^n$, which is non-negative, square-summable in $R_x^n$, and subject to the inequality

$$|e_\alpha(z)| \leqq r_\alpha(z)\, i_\alpha(y),$$

where the functions $r_\alpha$ and $i_\alpha$ satisfy

$$r_\alpha(\pm(z-\lambda))\, \mathscr{R}_\alpha(\lambda) \leqq C_\alpha \mathscr{R}_{\alpha+1}(z), \qquad z, \lambda \in C^n;$$

$$i_\alpha(y)\, \mathscr{I}_\alpha(y) \leqq C_\alpha \mathscr{I}_{\alpha+1}(y), \qquad y \in R^n.$$

We remark that since by hypothesis the function $\mathscr{R}_\alpha$ is everywhere finite, the domain $\Omega_\alpha$ in which the function $M_\alpha$ is finite has the form $R_x^n \times \omega_\alpha$, where $\omega_\alpha$ is the open set in which $\mathscr{I}_\alpha$ is finite. The region $\omega_\alpha$ is not empty, since by inequality b′) it contains the $\varepsilon_\alpha$-neighborhood of the set on which $\mathscr{I}_{\alpha+1} < \infty$ (we do not consider the case when $\mathscr{I}_\alpha \equiv \infty$ for all $\alpha$).

We choose an integer $m$ arbitrarily, in the range 1 to $n$, and we divide the variable $z$ into two groups: $v=(z_1, \ldots, z_m)$ and $w=(z_{m+1}, \ldots, z_n)$. The restriction of $f(z)$, defined in $C_z^n$ on the subspace $w=W$, will be denoted by $f_W(v)$. Then we have the self-evident

**Proposition 4.** *Let $\{M_\alpha\}$ be a family of majorants of type $\mathscr{I}$ and $C^n$, satisfying the conditions* a), b), a′), b′) *for some constants $\varepsilon_\alpha$, $C_\alpha$ and some functions $e_\alpha$. Then for arbitrary $W$ the sequence of functions $M_{\alpha,W}(v)$ is a family of majorants of type $\mathscr{I}$ in $C_v^n$, satisfying the same conditions with the same constants, and with the functions $e_{\alpha,W}$.*

*Remark.* Let $\mathscr{M} = \{M_\alpha\}$ be an arbitrary family of majorants in $C^n$. Multiplying the functions $M_\alpha$ by a suitable positive constant, we can always arrange matters so that the inequalities a) and b) are satisfied, with the constants $C_\alpha$ equal to unity. Clearly, if $\mathscr{M}$ is a family of type $\mathscr{I}$, such a transformation does not change the type, and we can also arrange matters so that the constants $C_\alpha$ in the inequalities a′) and b′) are also equal to unity. From now on, we shall suppose that such a transformation has been carried out, and, accordingly, all the constants $C_\alpha$ that we encounter will be equal to unity. Without loss of generality, we shall also suppose that

$$\varepsilon_{\alpha+1} \leqq \frac{\varepsilon_\alpha}{2} \quad \text{and} \quad \varepsilon_\alpha \leqq 1.$$

## § 2. The operator $D_{\bar{z}}$ in spaces of type $\mathscr{I}$

We consider a sequence of mappings of the spaces

$$0 \to \mathscr{H}_\alpha^m \to \mathscr{H}_\alpha^{m-1} \xrightarrow{\frac{\partial}{\partial \bar{z}_m}} \mathscr{H}_\alpha^{m-1} \to 0 \qquad (0<m \leqq n).$$

Since these mappings commute with the identity mappings of the form $\mathscr{H}_\alpha^m \to \mathscr{H}_{\alpha+1}^m$, we can go over to the mappings of the families

$$0 \to \mathscr{H}_{\mathscr{M}}^m \to \mathscr{H}_{\mathscr{M}}^{m-1} \xrightarrow{\frac{\partial}{\partial \bar{z}_m}} \mathscr{H}_{\mathscr{M}}^{m-1} \to 0. \tag{1.2}$$

This sequence is obviously exact in the first two terms. In this section we shall show that if $\mathscr{M}$ is a family of type $\mathscr{I}$, the sequence is exact in the third term also.

**Theorem 1.** *Let $\mathscr{M}$ be a family of majorants of type $\mathscr{I}$. Then for arbitrary $m$, $1 \leqq m \leqq n$, and $\alpha>0$, there exists a continuous operator $\mathfrak{R}_\alpha$: $\mathscr{H}_\alpha^{m-1} \to \mathscr{H}_{\alpha'}^{m-1}$, $\alpha'=\alpha+2n+12$, such that the composition $\frac{\partial}{\partial \bar{z}_m}\mathfrak{R}_\alpha$: $\mathscr{H}_\alpha^{m-1} \to \mathscr{H}_{\alpha'}^{m-1}$ is an identity mapping and for all $k \geqq 0$, the inequalities*

$$\|\mathfrak{R}_\alpha \phi\|_{\alpha'}^k \leqq C_k \|\phi\|_\alpha^{k+\kappa}, \qquad \kappa=\text{const.} \tag{2.2}$$

*are satisfied.*

The exactness of (1) evidently follows from this theorem.

**1°. Auxiliary norms and spaces.** We fix our attention on an arbitrary family of majorants $\mathscr{M}=\{M_\alpha(z)\}$ of type $\mathscr{I}$. For arbitrary integers $\alpha>0$ and $k \geqq 0$, we consider the norm

$$\|\phi\|_{(\alpha)}^k = \max_{|i|+|j| \leqq k} \sup_{\Omega_\alpha} \frac{|D^{i,\bar{j}}\phi(z)|}{\mathscr{I}_\alpha(y)}, \qquad (M_\alpha(z)=\mathscr{R}_\alpha(z)\mathscr{I}_\alpha(y)),$$

defined for functions $\phi$ given in the regions $\Omega_\alpha=R_x^n \times \omega_\alpha$ ($\omega_\alpha$ being the open set in which the function $\mathscr{I}_\alpha$ is finite). The space of functions for which this norm is finite will be denoted by $J_\alpha^{0,k}$. The subspace consisting of functions holomorphic in $\Omega_\alpha$ in the variables $z_1, \ldots, z_m$, will be denoted by $J_\alpha^{m,k}$. In $J_\alpha^{m,k}$ we consider a subspace $\check{J}_\alpha^{m,k}$, consisting of functions for which the norm

$${}^{\vee}\|\phi\|_{(\alpha)}^k = \max_{|i|+|j| \leqq k} \sup_{\Omega_\alpha} (|x|+1)^{n+1} \frac{|D^{i,\bar{j}}\phi(z)|}{\mathscr{I}_\alpha(y)}, \qquad (z=x+iy).$$

is finite. We set

$$J_\alpha^m = \varprojlim_k J_\alpha^{m,k}, \qquad \check{J}_\alpha^m = \varprojlim_k \check{J}_\alpha^{m,k}.$$

For functions defined in the region $R_x^n \times \omega_\alpha$, we define the operator $F$ as the Fourier transform with respect to $x$:

$$F\colon \phi(x+iy) \to \psi(\sigma+iy) = \int_{R_x^n} e^{i(x,\sigma)} \phi(x+iy)\,dx.$$

If the function $\phi$ is summable in $x$ for every $y$, then its transform is continuous and bounded with respect to $\sigma$ for all $y$. Then for arbitrary $\alpha$

$$\sup_{\sigma,y} \frac{|\psi(\sigma+iy)|}{\mathscr{I}_\alpha(y)} \leqq C \sup_z (|x|+1)^{n+1} \frac{|\phi(x+iy)|}{\mathscr{I}_\alpha(y)} = C\ {}^{\vee}\|\phi\|_{(\alpha)}^0,$$

where the constant $C$ depends only on $n$.

By a known property of Fourier transforms, if the norm ${}^{\vee}\|\phi\|_{(\alpha)}^k$ is finite, then the operator $F$ carries $D_x^j \phi$, $|j| \leqq k$, into $i^{|j|} \sigma^j F\phi$. Hence for arbitrary $k \geqq 0$

$$\sup_{\sigma,y} \frac{(|\sigma|+1)^k\,|\psi(\sigma+iy)|}{\mathscr{I}_\alpha(y)} \leqq C_{n,k}\ {}^{\vee}\|\phi\|_{(\alpha)}^k. \tag{3.2}$$

The inverse Fourier transform (with respect to $\sigma$)

$$F^{-1}\colon \psi(\sigma+iy) \to \phi(x+iy) = (2\pi)^{-n} \int_{R_\rho^n} e^{-i(\sigma,x)} \psi(\sigma+iy)\,d\sigma$$

is the inverse of the operator $F$. Since it differs from $F$ only inessentially, we can set $k=0$ in (3) and interchange the roles of $\phi$ and $\psi$, interchanging the roles of $x$ and $\sigma$. We then arrive at the inequality

$$\|\phi\|_{(\alpha)}^0 \leqq C_n \sup_{y,\sigma} (|\sigma|+1)^{n+1} \frac{|\psi(\sigma+iy)|}{\mathscr{I}_\alpha(y)}. \tag{4.2}$$

**Lemma 1.** *For arbitrary $\phi \in \check{J}_\alpha^{n-1}$ and for an arbitrary point $\sigma$ the integral*

$$e^{-(y,\sigma)} \psi(\sigma+iy) = \int_{y=\text{const}} e^{i(z,\sigma)} \phi(z)\,dx, \tag{5.2}$$

*where $y=(y_1,\dots,y_n) \in \omega_\alpha$, does not depend on $y_1,\dots,y_{n-1}$.*

*Proof.* Let $y'$ and $y''$ be two arbitrary values of $y$, belonging to the region $\omega_\alpha$, and projecting into the same point on the $y_n$-axis. We join these points by the straight line segment $l$ and we consider the manifold $L = R_x^n \times l$. In this manifold we construct a cylinder having as base the sphere in $R_x^n$ with center at the origin of coordinates, a radius $r$, and having as generator $l$. By Stokes' theorem, the integral

$$\int_S e^{i(z,\sigma)} \phi(z)\,dz,$$

taken over the surface of this cylinder is equal to the integral, taken over its interior, of the differential form

$$d(e^{i(z,\sigma)}\phi(z)\,dz)=e^{i(z,\sigma)}\frac{\partial\phi(z)}{\partial\bar{z}_n}\,d\bar{z}_n\,dz.$$

But since our cylinder lies in the space $z_n=\text{const}$, we have $d\bar{z}_n=0$, whence

$$\int_S e^{i(z,\sigma)}\phi(z)\,dz=0.$$

This integral is the sum of integrals over the bases of the cylinder, which as $r\to\infty$ tend to the integrals (5) with $y=y'$ and $y=y''$, plus the integral over the lateral surface, which tends to zero as $r\to\infty$, since $e^{i(z,\sigma)}\phi(z)$ is $O((|x|+1)^{-(n+1)})$. □

**2°. Lemma 2.** *For arbitrary $\alpha$ and $m$ there exists a continuous operator $R_\alpha\colon \check{J}_\alpha^{m-1}\to J_{\alpha+5}^{m-1}$ such that the composition*

$$\frac{\partial}{\partial\bar{z}_m}R_\alpha\colon \check{J}_\alpha^{m-1}\to J_{\alpha+5}^{m-1}$$

*is an identity mapping and for arbitrary $k\geqq n+2$ the inequality*

$$\|R_\alpha\phi\|_{(\alpha+5)}^k\leqq C_k\,{}^{\vee}\|\phi\|_{(\alpha)}^k. \tag{6.2}$$

*is satisfied.*

*Proof.* We suppose that $m=n$. Let $y(\sigma)$ be the function corresponding to the logarithmically convex function $\mathscr{I}_{\alpha+2}(y)$. We write

$$\chi(\sigma+iy)=\int_{y(\sigma)}^{y} e^{(y,\sigma)-(\eta,\sigma)}\psi(\sigma+i\eta)\,d\eta_n,\qquad \psi=F\phi, \tag{7.2}$$

where the integration is carried out over the segment joining the points $y$ and $y(\sigma)$; if $y(\sigma)$ is improper, the integration is carried out over the ray $\{y+r\,y(\sigma),\,r>0\}$. We estimate this integral for the function $\phi\in\check{J}_\alpha^{m-1}$.

By Proposition 2 of § 1 we have the inequality

$$\frac{\mathscr{I}_{\alpha+2}(\eta)}{e^{(\eta,\sigma)}}\leqq\frac{\mathscr{I}_{\alpha+2}(y)}{e^{(y,\sigma)}},$$

along the segment (ray) of integration and, therefore

$$\begin{aligned}|\chi(\sigma+iy)|&\leqq\mathscr{I}_{\alpha+2}(y)\int_{y(\sigma)}^{y}\frac{|\psi(\sigma+i\eta)|}{\mathscr{I}_{\alpha+2}(\eta)}|d\eta_n|\\&\leqq\mathscr{I}_{\alpha+2}(y)\int_{y(\sigma)}^{y}\frac{|\psi(\sigma+i\eta)|}{\mathscr{I}_\alpha(\eta)(|\eta|+1)^2}|d\eta_n|\leqq C\mathscr{I}_{\alpha+2}(y)\sup_\eta\frac{|\psi(\sigma+i\eta)|}{\mathscr{I}_\alpha(\eta)}.\end{aligned}$$

Hence

$$\sup_{\sigma, y} \frac{(|\sigma|+1)^{n+2} |\chi(\sigma+i y)|}{\mathcal{I}_{\alpha+2}(y)} \leqq C \sup_{\sigma, y} \frac{(|\sigma|+1)^{n+2} |\psi(\sigma+i y)|}{\mathcal{I}_{\alpha}(y)}. \tag{8.2}$$

In view of the inequality (4), the lefthand side of (8) is not less than $C_n' \|F^{-1}\chi\|^0_{(\alpha+2)}$. Because of (3), the righthand side of (8) does not exceed $C_n'' \,{}^{\vee}\|\phi\|^{n+2}_{(\alpha)}$. Therefore,

$$\|R_\alpha \phi\|^0_{(\alpha+2)} \leqq b_n \,{}^{\vee}\|\phi\|^{n+2}_{(\alpha)}, \qquad \text{where } R_\alpha \phi = -2i F^{-1}\chi, \tag{9.2}$$

for some constant $b_n$, depending only on $n$.

We evaluate the derivative of the function $\chi$ with respect to $y_k$, $k=1, \ldots, n$. We write, omitting the integrand $e^{-(\eta, \sigma)} \psi(\sigma+i\eta)\, d\eta_n$

$$\begin{aligned} &\chi(\sigma+i(y+\Delta y)) - \chi(\sigma+i y) \\ &\qquad = (e^{(\Delta y, \sigma)}-1)\, e^{(y, \sigma)} \int_{y(\sigma)}^{y+\Delta y} + e^{(y, \sigma)} \left( \int_{y(\sigma)}^{y+\Delta y} - \int_{y(\sigma)}^{y} \right). \end{aligned} \tag{10.2}$$

Since the function $e^{-(\eta, \sigma)} \psi(\sigma+i\eta)$ depends only on $\eta_n$ the second term on the right side is equal to

$$e^{(y, \sigma)} \int_{y}^{y+\Delta y_n}$$

where $\Delta y_n$ is the $n$-th coordinate of the vector $\Delta y$. This transformation is permissible if the point $y(\sigma)$ is proper or if $y(\sigma)$ is non-proper but the $n$-th coordinate of the vector $y(\sigma)$ does not vanish. In these cases (10) implies

$$\frac{\partial}{\partial y_k} \chi(\sigma+i y) = \sigma_k \chi(\sigma+i y) + \delta_n^k \psi(\sigma+i y) \tag{11.2}$$

where $\delta_n^k$ is the Kronecker delta.

Now we consider the exceptional case when $y(\sigma)$ is a vector orthogonal to the $y_n$-axis. Then by (7) $\chi(\sigma+i y)=0$ for all $y$. In virtue of Proposition 2 § 1 the function $\mathcal{I}_{\alpha+2}(y) \cdot \exp(-(y, \sigma))$ does not increase along any ray which is parallel to the vector $y(\sigma)$. Therefore $\mathcal{I}_\alpha(y) = o(\exp(y, \sigma))$ along such a ray. Since the right side of (8) is finite, the function $\exp(-(y, \sigma))\, \psi(\sigma+i y)$ tends to zero on every ray of this type. On the other hand this function is constant on any of these rays. Hence $\psi(\sigma+i y) \equiv 0$ for given $\sigma$. Therefore the relation (11) is valid again. Thus this relation is proved for all $\sigma$.

By virtue of the inequality (8), the function $(|\sigma|+1)\, \chi(\sigma+i y)$ is absolutely summable with respect to $\sigma$, and, therefore, we can carry out the following transformation:

$$F^{-1}\left\{\left[\frac{\partial}{\partial y_k} - \sigma_k\right] \chi\right\} = \left[\frac{\partial}{\partial y_k} - i \frac{\partial}{\partial x_k}\right] F^{-1}\chi = \frac{\partial}{\partial \bar{z}_k} R_\alpha \phi.$$

Hence by (11)

$$\frac{\partial}{\partial \bar{z}_k} R_\alpha \phi = \delta_n^k \phi, \qquad k=1, \ldots, n. \tag{12.2}$$

The Eqs. (12), for $k=1, \ldots, n-1$, show that the operator $R_\alpha$ carries the functions of the space $\check{J}_\alpha^{n-1}$ into functions that are holomorphic with respect to the variables $z_1, \ldots, z_{n-1}$, and Eq. (12) with $k=n$, shows that

$$\frac{\partial}{\partial \bar{z}_n} R_\alpha \phi = \phi.$$

Now by Corollary 1 of § 1 and the inequality (9), we have

$$\|R_\alpha \phi\|_{(\alpha+3)}^k \leqq C(\|R_\alpha \phi\|_{(\alpha+2)}^0 + \|\phi\|_{(\alpha+2)}^k) \leqq C\ {}^\vee\|\phi\|_{(\alpha)}^k \qquad (k \geqq n+2).$$

Thus, $R_\alpha$ is the operator we were seeking. □

**3°. The case when $m$ is arbitrary.** We write $v=(z_1, \ldots, z_m)$ and $w=(z_{m+1}, \ldots, z_n)$. By $C_v$ and $C_w$ we denote the coordinate subspaces of the variables $v$ and $w$. For arbitrary $\alpha$ and $w' \in C_w$ we denote by $M_{\alpha, w'}$, $R_{\alpha, w'}$, and $\mathscr{I}_{\alpha, w'}$ the restrictions of the functions $M_\alpha$, $\mathscr{R}_\alpha$, and $\mathscr{I}_\alpha$ on the subspace $w=w'$. By Proposition 4 of § 1 the sequence of functions $M_{\alpha, w}(v) = \mathscr{R}_{\alpha, w}(v) \mathscr{I}_{\alpha, w}(v'')$, $v=v'+iv''$ is a family of majorants in $C_v$ of type $\mathscr{I}$, satisfying the inequalities b) and b′) with a constant $\varepsilon_\alpha$, not depending on $w$. Starting with this family, we define a series of norms like the norms introduced in 1°:

$$\|\phi\|_{(\alpha), w}^k = \max_{|i|+|j| \leqq k} \sup_{C_v} \frac{|D^{i, \bar{j}} \phi(v)|}{\mathscr{I}_{\alpha, w}(v'')},$$

$${}^\vee\|\phi\|_{(\alpha), w}^k = \max_{|i|+|j| \leqq k} (|v'|+1)^{m+1} \frac{|D^{i, \bar{j}} \phi(v)|}{\mathscr{I}_{\alpha, w}(v'')}.$$

We shall denote by $J_{\alpha, w}^{l, k}$ and $\check{J}_{\alpha, w}^{l, k}$ the spaces of functions in $C_v$ which are holomorphic in the variables $z_1, \ldots, z_l$ $(0 \leqq l \leqq m)$, for which the norms $\|\cdot\|_{(\alpha), w}^k$ and ${}^\vee\|\cdot\|_{(\alpha), w}^k$, $k=0, 1, 2, \ldots$, respectively are finite.

We now choose $w$ arbitrarily and we apply the results of 2° to the family of majorants $\{M_{\alpha, w}\}$. Also, for every $\alpha$ we construct the continous operator $R_{\alpha, w}: \check{J}_{\alpha, w}^{m-1} \to J_{\alpha+3, w}^{m-1}$ such that the composition $\frac{\partial}{\partial \bar{z}_m} R_\alpha$ is an identity mapping, which for arbitrary $k \geqq 0$ satisfies the inequality

$$\|R_{\alpha, w} \phi\|_{(\alpha+3), w}^k \leqq C_k\ {}^\vee\|\phi\|_{(\alpha), w}^{k+\kappa}, \qquad \kappa = m+2. \tag{13.2}$$

Here it is essential that the constant $C_k$ in these inequalities depends only on $n$, $k$ and on the constant $\varepsilon_{\alpha+2}$, corresponding to the family of majorants $\{M_{\alpha,w}\}$. However, as we noted earlier, the quantity $\varepsilon_{\alpha+2}$ does not depend on $w$, and therefore, the constants $C_k$ are also independent of $w$. In the next step we need to collate the operators $R_{\alpha,w}$.

**4°. Construction of the partition of the identity in $C_w$.** Let $\delta$ be an arbitrary positive number. We cover the space $C_w$ with a sequence of cubes $K_\tau$, $\tau=1,2,\ldots$, with edges equal to $\delta/\sqrt{2n}$, such that they intersect each other only along their boundaries. We denote the center of the cube $K_\tau$ by $w_\tau$.

Suppose now that $h(w)$ is an non-negative, infinitely differentiable function equal to unity in the cube $K_1-w_1$ and zero outside the concentric twice as big cube. We consider the functions

$$h_\tau(w)=\frac{h(w-w_\tau)}{\sum_\sigma h(w-w_\sigma)}.$$

Since at every point $w$, at least one of the functions $h(w-w_\tau)$ is equal to unity and the rest are non-negative, we have

$$\sum_\sigma h(w-w_\sigma)\geqq 1.$$

It follows that the functions $h_\tau$ are infinitely differentiable. It is clear that the function $h_\tau(w+w_\tau)$ does not depend on $\tau$. Therefore, for arbitrary $i$ and $j$

$$|D^{i,\bar{j}}h_\tau(w)|\leqq C_{i,j},$$

where the constant $C_{i,j}$ does not depend on $w$ and $\tau$. We note that the support of the function $h_\tau$ lies in the $2\delta$-neighborhood $U_\tau$ of the point $w_\tau$.

**5°. The collage of the operators $R_{\alpha,w}$.** In the construction of Subsection 4° we set $\delta=\frac{1}{2}\min(\varepsilon_\alpha,\varepsilon_{\alpha+4})$. Then for arbitrary $\tau$ and $w'\in U_\tau$ we have the inequality $|w'-w_\tau|\leqq\varepsilon_\alpha$. It therefore follows from inequality b′) of § 1 that $\mathscr{I}_{\alpha,w'}\leqq\mathscr{I}_{\alpha+1,w_\tau}$. Hence, for arbitrary $w'\in U_\tau$ and for an arbitrary function $\phi\in\mathring{J}_\alpha^{m-1}$ its restriction $\phi_{w'}$ on the subspace $w=w'$ belongs to the space $\mathring{J}^{m-1}_{\alpha+1,w_\tau}$, and

$$\|\phi_{w'}\|^k_{(\alpha+1),w_\tau}\leqq\|\phi\|^k_{(\alpha)}$$

for arbitrary $k\geqq 0$. An arbitrary derivative of the function $\phi$ of the form $(D^{i,\bar{j}}\phi)_{w'}$ also belongs to $\mathring{J}^{m-1}_{\alpha+1,w_\tau}$. Therefore the operator $R_{\alpha+1,w_\tau}$ may be applied to it. We write $\psi_{\tau,w}=\mathrm{R}_{\alpha+1,w_\tau}\phi_w$. Since the operator $R_{\alpha+1,w_\tau}$ is linear, it commutes with the differentiation operator with respect to the variable $w$. Therefore,

$$D_w^{i,\bar{j}}\psi_{\tau,w}=R_{\alpha+1,w_\tau}(D_w^{i,\bar{j}}\phi)_w\in J^{m-1}_{\alpha+4,w_\tau}.$$

It follows that for arbitrary $i$ and $j$, the function

$$D_w^{i,\bar{j}}\{h_\tau(w)\,\psi_{\tau,w}\} \tag{14.2}$$

also belongs to the space $J^{m-1}_{\alpha+4,w}$.

Since $2\delta \leqq \varepsilon_{\alpha+4}$, we have for arbitrary $w \in U_\tau$ the inequality $\|\cdot\|_{(\alpha+5),w} \leqq \|\cdot\|_{(\alpha+4),w_\tau}$. Therefore the function (14) also belongs to the space $J^{m-1}_{\alpha+5,w}$ and

$$\begin{aligned} \|D_w^{i,\bar{j}}\{h_\tau(w)\,\psi_{\tau,w}\}\|^k_{(\alpha+5),w} &\leqq C_{i,j} \max_{|i'|\leqq|i|,\,|j'|\leqq|j|} \|D_w^{i',\bar{j}'}\,\psi_{\tau,w}\|^k_{(\alpha+4),w_\tau} \\ &\leqq C_{i,j}\,C_k \max_{|i'|\leqq|i|,\,|j'|\leqq|j|} \|D_w^{i',\bar{j}'}\,\phi\|^{k+\kappa}_{(\alpha),w}, \end{aligned} \tag{15.2}$$

where $C_k$ are the constants in (13). Since the function $h_\tau$ vanishes outside of $U_\tau$, this inequality holds for all points $w$.

We set $\psi_w = \sum_\tau h_\tau\,\psi_{\tau,w}$. Since the $C_k$ are independent of $w$, and the number of the $h_\tau$ that do not vanish simultaneously is bounded, (15) implies the inequality

$$\|D_w^{i,\bar{j}}\,\psi_w\|^k_{(\alpha+5),w} \leqq C_{i,j,k} \max_{|i'|\leqq|i|,\,|j'|\leqq|j|} \|D_w^{i',\bar{j}'}\,\phi\|^{k+\kappa}_{(\alpha),w},$$

in which the constants $C_{i,j,k}$ are independent of $w$. Passing to the upper bound with respect to $w$, we arrive at the inequality

$$\|\psi\|^k_{(\alpha+5)} \leqq C_k\,\|\phi\|^{k+\kappa}_\alpha,$$

where $\psi = \psi_w(v)$ is a function belonging to the space $J^{m-1}_{\alpha+5}$. It follows that the mapping $R_\alpha\colon \phi \to \psi$ is a continuous operator from $J^{m-1}_\alpha$ to $\check{J}^{m-1}_{\alpha+5}$. Because of the properties of the operator $R_{\alpha+1,w_\tau}$, we have

$$\frac{\partial}{\partial \bar{z}_m} R_\alpha\,\phi = \frac{\partial}{\partial \bar{z}_m} \sum h_\tau\,\psi_{\tau,w} = \sum h_\tau \frac{\partial}{\partial \bar{z}_m} R_{\alpha+1,w_\tau}\,\phi = \sum h_\tau\,\phi = \phi.$$

Accordingly, $R_\alpha$ is the operator we were seeking. □

**6°. Proof of the theorem.** We fix $m$ and $\alpha$. Since by hypothesis $\mathscr{M}$ is a family of majorants of type $\mathscr{I}$, there exists an entire function $e(z) \not\equiv 0$, non-negative, square-summable on $R^n_x$, and satisfying the inequality

$$|e(z)| \leqq r(z)\,\frac{\mathscr{I}_{\beta+2}(y)}{\mathscr{I}_{\beta+1}(y)}, \qquad \beta = \alpha + n + 1,$$

where

$$r(z-\lambda)\,\mathscr{R}_{\beta+1}(z) \leqq \mathscr{R}_{\beta+2}(\lambda)$$

for arbitrary $z$, $\lambda \in C^n$. We set

$$e_\lambda(z) = \frac{1}{\mathscr{R}_{\beta+2}(\lambda)} e(z-\lambda), \qquad \lambda \in R^n.$$

and we find a bound for the derivatives of this function. Let $\varepsilon = \min(\varepsilon_\beta, \varepsilon_{\beta+2})$. Then

$$\begin{aligned} |D_z^i e_\lambda(z)| &\leqq \frac{C_i}{\mathscr{R}_{\beta+2}(\lambda)} \sup_{|z-z'|\leqq\varepsilon} r(z'-\lambda) \frac{\mathscr{I}_{\beta+2}(y')}{\mathscr{I}_{\beta+1}(y')} \\ &\leqq \frac{C_i}{\mathscr{R}_{\beta+2}(\lambda)} \sup_{|z-z'|\leqq\varepsilon} \frac{\mathscr{R}_{\beta+2}(\lambda)}{\mathscr{R}_{\beta+1}(z')} \frac{\mathscr{I}_{\beta+3}(y')}{\mathscr{I}_{\beta+1}(y')} \leqq \frac{C_i}{\mathscr{R}_\beta(z)} \frac{\mathscr{I}_{\beta+3}(y)}{\mathscr{I}_\beta(y)} \qquad (16.2) \\ &\qquad (y' = \operatorname{Im} z'). \end{aligned}$$

Hence for arbitrary $\phi \in \mathscr{H}_\alpha^0$

$$\begin{aligned} &\sup (|x|+1)^{n+1} \frac{|D^{i,\bar{j}}\{e_\lambda(z)\,\phi(z)\}|}{\mathscr{I}_{\beta+3}(y)} \\ &\quad \leqq C_{i,j} \sup (|x|+1)^{n+1} \frac{\mathscr{R}_\alpha(z)}{\mathscr{R}_\beta(z)} \max_{|i'|\leqq|i|} \sup \frac{|D^{i',\bar{j}}\phi|}{\mathscr{R}_\alpha \mathscr{I}_\alpha}. \end{aligned}$$

Since the functions $\mathscr{R}_\alpha(z)$ form a family of majorants by hypothesis, we have the inequality $(|x|+1)^{n+1} \mathscr{R}_\alpha(z) \leqq \mathscr{R}_\beta(z)$. It follows that the second factor on the right side does not exceed unity. The third factor does not exceed $\|\phi\|_\alpha^k$, where $k = |i| + |j|$, and the left side is equal to $^\vee\|D^{i,\bar{j}}\{e_\lambda \phi\}\|_{(\beta+3)}^0$. Taking the maximum with respect to $i$ and $j$ in this relation, with $|i|+|j| \leqq k$, we obtain the inequality

$$^\vee\|e_\lambda \phi\|_{(\beta+3)}^k \leqq C_k \|\phi\|_\alpha^k, \qquad (17.2)$$

in which $k \geqq 0$ is arbitrary, and the constants $C_k$ are independent of $\lambda$.

We shall suppose that the function $\phi$ belongs to the space $\mathscr{H}_\alpha^{m-1}$, i.e. that it is holomorphic in the variables $z_1, \ldots, z_{m-1}$. Since the functions $e_\lambda(z)$ are holomorphic, the product $e_\lambda \phi$ is also holomorphic in these variables. Because the left side of (17) is finite, it follows that $e_\lambda \phi$ belongs to $\check{J}_{\beta+3}^{m-1}$. Therefore, we may apply to these functions the operator $R_{\beta+3}$ constructed in Lemma 2. We set $\psi_\lambda = R_{\beta+3}\, e_\lambda \phi$. From the inequalities (17) and (6), we derive the inequality

$$\|\psi_k\|_{(\beta+8)}^k \leqq C_k' \|\phi\|_\alpha^{k+\kappa}, \qquad (18.2)$$

Again making use of the property of $\mathscr{M}$, we find an entire function $e'(z) \not\equiv 0$, non-negative square summable in $R_x^n$, and satisfying the inequality

$$|e'(z)| \leqq r'(z) \frac{\mathscr{I}_{\gamma+2}(y)}{\mathscr{I}_{\gamma+1}(y)}, \qquad \gamma = \beta + n + 8,$$

where

$$r'(z-\lambda)\,\mathscr{R}_{\gamma+1}(\lambda) \leqq \mathscr{R}_{\gamma+2}(z), \qquad z, \lambda \in C^n.$$

Since both $e$ and $e'$ are positive and square summable in $R^n_x$, their product is non-negative, not identically zero and summable. Therefore

$$a = \int e(x)\, e'(x)\, dx > 0.$$

We set

$$e'_\lambda(z) = \frac{\mathscr{R}_{\beta+2}(\lambda)}{a}\, e'(z-\lambda).$$

By analogy with (16) we establish the inequality

$$|D^j_z\, e'_\lambda(z)| \leqq C'_i\, \mathscr{R}_{\beta+2}(\lambda) \frac{\mathscr{R}_{\gamma+3}(z)}{\mathscr{R}_\gamma(\lambda)} \frac{\mathscr{I}_{\gamma+3}(y)}{\mathscr{I}_\gamma(y)}.$$

And making use of this bound, we obtain

$$\|D^{i,\bar{j}}\{e'_\lambda\, \psi_\lambda\}\|^0_{\gamma+3} \leqq \frac{C'_i}{(|\lambda|+1)^{n+1}} \sup(|\lambda|+1)^{n+1} \frac{\mathscr{R}_{\beta+2}(\lambda)}{\mathscr{R}_\gamma(\lambda)} \max_{|i'| \leqq |i|} \sup \frac{|D^{i',\bar{j}}\, \psi_\lambda|}{\mathscr{I}_\gamma(y)}. \tag{19.2}$$

Since the functions $\mathscr{R}_\alpha(z)$ form a family of majorants, the second factor on the right side does not exceed unity. Passing to the maximum with respect to $i$ and $j$, with $|i|+|j| \leqq k$ in (19), and taking account of the inequality (18), we find the inequality

$$\|e'_\lambda\, \psi_\lambda\|^k_{\gamma+3} \leqq \frac{C_k}{(|\lambda|+1)^{n+1}} \|\psi_\lambda\|^k_{(\gamma)} \leqq \frac{C_k}{(|\lambda|+1)^{n+1}} \|\phi\|^{k+\kappa}_\alpha.$$

which holds for arbitrary $k \geqq 0$. Hence, it follows that the integral $\psi = \int_{R^n} e'_\lambda\, \psi_\lambda\, d\lambda$ is absolutely convergent in the norm $\|\cdot\|^k_{\gamma+3}$, $k \geqq 0$, and

$$\|\psi\|^k_{\gamma+3} = \|\int e'_\lambda\, \psi_\lambda\, d\lambda\|^k_{\gamma+3} \leqq C'_k\, \|\phi\|^{k+\kappa}_\alpha, \tag{20.2}$$

Accordingly, $\psi \in \mathscr{H}^0_{\alpha'}$, $\alpha' = \gamma + 3 = \alpha + 2n + 12$. Since $e'_\lambda$ are entire functions and the functions $\psi_\lambda$ are holomorphic in the variables $z_1, \ldots, z_{m-1}$, the function $\psi$ is also holomorphic in these variables, that is, $\psi \in \mathscr{H}^{m-1}_{\alpha'}$. Thus, we have constructed a continuous operator $\mathfrak{R}_\alpha$: $\mathscr{H}^{m-1}_\alpha \ni \phi \to \psi \in \mathscr{H}^{m-1}_{\alpha'}$, satisfying the inequality (2) in accordance with (20).

It remains to verify that the composition $\frac{\partial}{\partial \bar{z}_m} \mathfrak{R}_\alpha$ is an identity mapping, that is $\frac{\partial}{\partial \bar{z}_m} \psi = \phi$ for arbitrary $\phi \in \mathscr{H}_\alpha^{m-1}$. We have

$$\begin{aligned} \frac{\partial}{\partial \bar{z}_m} \psi &= \int e'_\lambda \frac{\partial}{\partial \bar{z}_m} \psi_\lambda \, d\lambda = \int e'_\lambda \, e_\lambda \, \phi \, d\lambda \\ &= \frac{1}{a} \int e'(z-\lambda) \, e(z-\lambda) \, \phi(z) \, d\lambda. \end{aligned} \tag{21.2}$$

Since the right side converges as a function belonging to $\mathscr{H}_{\alpha'}^{m-1}$, it converges for arbitrary $z \in \Omega_{\alpha'}$ and is equal to

$$\phi(z) \frac{1}{a} \int_{R^n} e'(z-\lambda) \, e(z-\lambda) \, d\lambda.$$

By Cauchy's theorem this integral is equal to

$$\int_{R^n} e'(x-\lambda) \, e(x-\lambda) \, d\lambda = a,$$

and, accordingly, the integral in (21) converges to $\phi$, whence

$$\frac{\partial}{\partial \bar{z}_m} \psi = \phi. \quad \square$$

**7°. The operator $D_{\bar{z}}$ in a bounded region.** We shall now construct an operator which acts in the space of functions defined in a bounded convex region of $C^n$ and which has the properties of the operator $\mathfrak{R}_\alpha$ defined in Theorem 1. We shall obtain this result as a consequence of the known fact that the $''d$-cohomology is trivial in a convex region.

Let $\Omega$ be a bounded open set in $C^n$. For every integer $k \geqq 0$ we consider the norm

$$\|\phi\|_\Omega^k = \max_{|i|+|j| \leqq k} \sup_\Omega |D^{i,\bar{j}} \phi|,$$

which is defined for functions having derivatives up to order $k$ in $\Omega$.

**Theorem 2.** *Let $\Omega$ be a convex open set in $C^n$ of diameter not exceeding unity, $0 < r \leqq 1$, and let $\Omega_r$ be an $r$-neighborhood of $\Omega$. For every integer $m$, $0 < m \leqq n$, and for every function $\phi$ which is infinitely differentiable and holomorphic in the variables $z_1, \dots, z_{m-1}$ in $\Omega_r$, there exists a function $\psi$ which is infinitely differentiable and holomorphic with respect to the variables $z_1, \dots, z_{m-1}$ in $\Omega$ and such that $\frac{\partial \psi}{\partial \bar{z}_m} = \phi$ and*

$$\|\psi\|_\Omega^k \leqq C_k \|\phi\|_{\Omega_r}^k, \qquad k = 0, 1, 2, \dots, \tag{22.2}$$

*where the constant $C_k$ is independent of $\Omega$.*

*Proof.* A $''d$-form of order $k$ is a form of the type $f=\sum f^{j_1,\dots,j_k}\, d\bar{z}_{j_1}\wedge\cdots\wedge d\bar{z}_{j_k}$. The operator $''d$ operates in accordance with the formula

$$''d\colon\ f\to\sum_j\sum\frac{\partial f^{j_1,\dots,j_k}}{\partial\bar{z}_j}\,d\bar{z}_j\wedge d\bar{z}_{j_1}\wedge\cdots\wedge d\bar{z}_{j_k}.$$

Let $k\geqq 0$ be an integer, and let $\omega$ be an open set in $C^n$. We consider the space $L^k(\omega)$, consisting of $''d$-forms of order $k$ in $\omega$ with square-summable coefficients, and with the norm

$$\|f\|_\omega=\left|\sum\|f^{j_1,\dots,j_k}\|^2_{L_2(\omega)}\right|^{\frac{1}{2}}.$$

Let $\omega$ be a bounded, pseudo-convex open set of diameter $\delta$. By a theorem of Hörmander[2] for every form $f\in L^{k+1}(\omega)$, satisfying the equation $''df=0$, there exists a form $g\in L^k(\omega)$ such that $''dg=f$ and

$$\|g\|_\omega\leqq e\,\delta^2\,\|f\|_\omega.$$

We shall now prove our theorem in the case $m=n$. We set $\omega=\Omega_r$ and $f=\phi\,d\bar{z}_n$. The region $\Omega_r$ is bounded and convex and its diameter does not exceed 2. Since the function $\phi$ is analytic with respect to $z_1,\dots,z_{n-1}$, the form $f\in L^1(\omega)$ satisfies the equation $''df=0$. Applying the results formulated above, we determine a form $g\in L^0(\omega)$. In the Hilbert space $L^0(\omega)$ we project the function $g$ onto the subspace orthogonal to the kernel of the operator $''d$; its projection will be denoted by $\psi$. The operator $R\colon\phi\to\psi$ is linear; it is continuous as an operator from $L^1(\omega)$ to $L^0(\omega)$ and it satisfies the equation $\frac{\partial}{\partial\bar{z}_i}R=\delta^i_n$, since $''d\psi=\phi$. From Proposition 1 of § 1 it follows that the function $\psi$ is infinitely differentiable in $\Omega$ and satisfies the inequalities (22). The constants $C$ in these inequalities depend only on $r$ and the constant $e\delta^2$ in (23), which do not exceed 12. Thus, in the case $m=n$, our theorem is proved.

Now suppose $m<n$. We return to the arguments of 4°–5°. For every $W\in C_w$ the intersection $\Omega_W$ of the set $\Omega$ with the subspace $w=W$ is convex. According to what we have already proved, for every $w$ there exists an operator $R_w$, satisfying the relation

$$\frac{\partial}{\partial\bar{z}_i}R_w=\delta^i_m,\qquad i\leqq m,$$

and the inequalities

$$\|R_w\,\phi_w\|^k_{\Omega_W}\leqq C_k\,\|\phi_w\|^k_{\Omega'_W},\qquad \Omega'=\Omega_{r/2},$$

in which the constants $C_k$ are independent of $\Omega$ and $w$. Let $\{h_\tau\}$ be a decomposition of unity in $C_w$ with $\delta=r/2$ (see 4°). Then, it is easy to see that the operator $\phi\to\sum h_\tau\,R_{w_\tau}\,\phi$ is the one that we are looking for. □

2 See Hörmander [2], Theorem 2.2.3.

## § 3. $\mathcal{M}$-cohomologies

**1°. Cochains.** Let $\Omega$ be an open set in $C^n$. The set of domains $U=\{U_i, i\in I\}$ in $C^n$ is called a covering of $\Omega$, if $\bigcup U_i=\Omega$. Let $U$ be a covering of $\Omega$, and let $L$ be a linear space over the field $C$. We suppose that for every finite set of indices $i_0, \dots, i_\nu \in I$ we have a linear topological space $\Phi(U_{i_0}\cap\cdots\cap U_{i_\nu})$, consisting of some functions defined in the region $U_{i_0}\cap\cdots\cap U_{i_\nu}$ with values in $L$. We suppose, further, that the operation of restricting of the functions of this space on the subregions $U_{i_0}\cap\cdots\cap U_{i_{\nu+1}}$ (for arbitrary $i_0, \dots, i_{\nu+1}\in I$) defines a continuous mapping of $\Phi(U_{i_0}\cap\cdots\cap U_{i_\nu})$ into $\Phi(U_{i_0}\cap\cdots\cap U_{i_{\nu+1}})$. In such a situation, we can speak of spaces of cochains on the covering $U$ with values in these spaces. For convenience, we shall resort to the following formalism.

A cochain of order $\nu$ ($\nu=0, 1, 2, \dots$) on the covering $U$ with values in the spaces $\Phi(U_{i_0}\cap\cdots\cap U_{i_\nu})$ will be represented in the following form:

$$\phi=\sum_{i_0,\dots,i_\nu}\phi_{i_0,\dots,i_\nu}U_{i_0}\wedge\cdots\wedge U_{i_\nu}, \qquad \phi_{i_0,\dots,i_\nu}\in\Phi(U_{i_0}\cap\cdots\cap U_{i_\nu}),$$

where the sum extends over all possible ordered sets of $\nu$ elements of the set $I$. The functions $\phi_{i_0,\dots,i_\nu}$ will be called components of the cochain $\phi$. The operation denoted by the symbol $\wedge$ has the customary properties of skew multiplication: associativity and anti-commutativity. Thus an interchange of positions of two neighboring factors in one of the terms in $\phi$ leads to a change of sign before that term.

This formalism allows us to operate with cochains on coverings by analogy with differential forms. In particular, an arbitrary cochain of order $\nu$ can be represented in skew-symmetric form, that is, in the form of the sum

$$\phi=\sum_{i_0,\dots,i_\nu}\phi'_{i_0,\dots,i_\nu}U_{i_0}\wedge\cdots\wedge U_{i_\nu},$$

in which the functions $\phi'_{i_0,\dots,i_\nu}$ depend in a skew-symmetric way on the indices $i_0, \dots, i_\nu$. From now on we shall consider only cochains written in skew-symmetric form.

The ordinary functions defined in the region $\Omega=\bigcup U_i$, will also be called cochains of order $-1$ on the covering $U$. We suppose that the space $L$ is finite-dimensional. We shall say that the cochain $\phi$ on the covering $U$ is defined, differentiable, holomorphic, or equal to zero, in some region $\omega$, if every one of its components $\phi_{i_0,\dots,i_\nu}$ is respectively defined, differentiable, holomorphic, or equal to zero in the region $U_{i_0}\cap\cdots\cap U_{i_\nu}\cap\omega$.

A covering of the region $\Omega\cap\omega$ consisting of regions $U_i\cap\omega$ will be denoted by $U\cap\omega$. Thus, a cochain on the covering $U$ which is defined in $\omega$ is the same thing as a cochain on the covering $U\cap\omega$.

**2°. The coboundary operator.** Let $U=\{U_i, i\in I\}$ be some such covering of the region $\Omega$. We consider the following operator, defined on the cochains on this covering:

$$\partial:\quad \phi\to\sum_{i\in I} U_i\wedge\phi.$$

In extensive form the action of this operator can be expressed as follows:

$$\begin{aligned}\partial\phi&=\sum_i U_i\wedge\sum_{i_0,\ldots,i_\nu}\phi_{i_0,\ldots,i_\nu}U_{i_0}\wedge\cdots\wedge U_{i_\nu}\\&=\sum_{i,i_0,\ldots,i_\nu}\phi^{(i)}_{i_0,\ldots,i_\nu}U_i\wedge U_{i_0}\wedge\cdots\wedge U_{i_\nu}\end{aligned}\tag{1.3}$$

where $\phi^{(i)}_{i_0,\ldots,i_\nu}$ is the restriction of the function $\phi_{i_0,\ldots,i_\nu}$ on the region $U_i\cap U_{i_0}\cap\cdots\cap U_{i_\nu}$. In particular, if $\phi$ is a cochain of order $-1$, then $\partial\phi=\sum\phi^{(i)}U_i$. In order to write the right side of (1) in skew-symmetric form, we rewrite every term of it as:

$$\begin{aligned}&\phi^{(i)}_{i_0,\ldots,i_\nu}U_i\wedge U_{i_0}\wedge\cdots\wedge U_{i_\nu}\\&\quad=\frac{1}{\nu+2}\sum_{\mu=0}^{\nu+1}(-1)^\mu\phi^{(i)}_{i_0,\ldots,i_\nu}U_{i_0}\wedge\cdots\wedge U_{\mu-1}\wedge U_i\wedge U_\mu\wedge\cdots\wedge U_{i_\nu}.\end{aligned}\tag{2.3}$$

Hence

$$\partial\phi=\frac{1}{\nu+2}\sum_{i_0,\ldots,i_{\nu+1}}\left(\sum_{\mu=0}^{\nu+1}(-1)^\mu\phi^{(i_\mu)}_{i_0,\ldots,\hat{i}_\mu,\ldots,i_{\nu+1}}\right)U_{i_0}\wedge\cdots\wedge U_{i_{\nu+1}}.$$

In this formula the components of the cochain $\partial\phi$ are skew-symmetric in their indices[3].

The equations $\sum U_i\wedge\sum U_i=-\sum U_i\wedge\sum U_i=0$ imply that $\partial\partial=0$.

Suppose we have two coverings $U=\{U_i, i\in I\}$ and $V=\{V_j, j\in J\}$. We shall say that the covering $V$ is inscribed in the covering $U$ if we are given a mapping $\theta\colon J\to I$, such that $V_j\subset U_{\theta(j)}$ for arbitrary $j\in J$. We can then define the conjugate mapping $\theta^*$, which acts on cochains:

$$\theta^*:\quad \phi=\sum_{i_0,\ldots,i_\nu}\phi_{i_0,\ldots,i_\nu}U_{i_0}\wedge\cdots\wedge U_{i_\nu}\to\phi|_V=\sum\hat{\phi}_{\theta(j_0),\ldots,\theta(j_\nu)}V_{j_0}\wedge\cdots\wedge V_{j_\nu},$$

where $\hat{\phi}_{\theta(j_0),\ldots,\theta(j_\nu)}$ is the restriction of the function $\phi_{\theta(j_0),\ldots,\theta(j_\nu)}$ on the region $V_{j_0}\cap\cdots\cap V_{j_\nu}$.

**3°. Elementary coverings.** An elementary covering $U$ of the space $C^n$, is a covering consisting of spheres

$$U_z=\{\zeta\colon|\zeta-z|<r(|z|+1)^{-\rho}\},\qquad z\in C^n.$$

3 Thus the operator $\partial$ differs from the usual coboundary operator only in the constant coefficient $1/(\nu+2)$.

The quantities $\rho=\rho(U)$ and $r=r(U)$ will be called respectively the parameter and the radius of the elementary covering $U$. We shall consider elementary coverings only for $\rho(U)\geqq 0$ and $r(U)>0$. If $\rho(V)\geqq\rho(U)$ and $r(V)\leqq r(U)$, the elementary covering $V$ is inscribed in the elementary covering $U$, provided we choose as the mapping $\theta$ the identity mapping $C^n\to C^n$.

Let $S$ be some centrally symmetric set in $C^n$, and let $\lambda$ be a positive number. We denote by $\lambda S$ the set obtained from $S$ by a dilation with coefficient $\lambda$ and center at the center of $S$. If $U=\{U_z\}$ is an elementary covering we denote by $\lambda U$ the covering consisting of spheres $\lambda U_z$, that is, the elementary covering with the parameter $\rho(U)$ and radius $\lambda r(U)$.

**Proposition 1.** *Let $\mathcal{N}$ be some algebraic stratification of the space $C^n$, and let $C>0$, $\varepsilon>0$, and $q$ be real numbers. In the covering $S$, consisting of the spheres*

$$S_z=\{\zeta: |\zeta-z|<C\,\varepsilon\,\theta^q(z,\mathcal{N})\}^4, \qquad z\in C^n,$$

*we can inscribe the elementary covering with parameter $\rho$ and radius $r=c\,\varepsilon^Q$, where the quantities $\rho$, $c$ and $Q$ are independent of $\varepsilon$.*

*Proof.* Let $\mathcal{N}=\{C^n\supset N_1\supset N_2\supset\cdots\supset N_m\supset\varnothing\}$. We shall suppose that for some number $\nu\leqq m$ we have constructed a covering $S^\nu$, inscribed in $S$, and consisting of the spheres $S_z^\nu$, $z\in N_\nu$, and having the following properties: the center of the sphere $S_z^\nu$ is at the point $z$, and the radius $r_z^\nu$ is not less than $r_\nu(|z|+1)^{-\rho_\nu}$ for some $r_\nu$, $1\geqq r_\nu>0$ and $\rho_\nu\geqq 0$. We remark that if $\nu=m$, such a covering can be constructed by setting $S_z^m=S_z$, $z\in N_m$. The necessary lower bound for $r_z^m$ is satisfied, since by definition $\theta(z,N)=(|z|^2+1)^{-1}$ on the variety $N_m$ and, accordingly $r_z^m=C\,\varepsilon\,\theta^q(z,\mathcal{N})=C\varepsilon(|z|+1)^{-2q}$. We shall now suppose that we have constructed the covering $S^\nu$, and we construct the covering $S^{\nu-1}$.

We assume that for a given point $z\in N_{\nu-1}$ the inequality

$$\rho(z,N_\nu)\leqq r_z'\overset{\text{def}}{=}\frac{r_\nu}{2}\left(|z|+1+\frac{r_\nu}{2}\right)^{-\rho_\nu}$$

is satisfied. We choose the point $z'\in N_\nu$ so that $\rho(z,N_\nu)=\rho(z,z')$. Since $\rho(z,z')\leqq\frac{r_\nu}{2}$, we have

$$r_{z'}^\nu\geqq r_\nu(|z'|+1)^{-\rho_\nu}\geqq r_\nu\left(|z|+1+\frac{r_\nu}{2}\right)^{-\rho_\nu}=2r_z'\geqq\rho(z,N_\nu)+r_z'.$$

It follows that the sphere $S_{z'}^\nu$ contains the open sphere $S_z^{\nu-1}$ with center $z$ and radius $r_z'\geqq r_\nu\,2^{-1-\rho_\nu}(|z|+1)^{-\rho_\nu}$.

---

4 The definition of the function $\theta$ is given in Chapter II, § 3.

We shall now suppose that the converse inequality $\rho(z, N_\nu) > r'_z$ is satisfied. Then

$$C\,\varepsilon\,\theta^q(z, \mathcal{N}) = C\,\varepsilon \left(\frac{\rho(z, N_\nu)}{|z|^2+1}\right)^q > C\,\varepsilon \left(\frac{r'_z}{|z|^2+1}\right)^{|q|},$$

whence it follows that the sphere $S_z^{\nu-1}$ with center at the point $z$ and radius $C\,\varepsilon \left(\frac{r'_z}{|r|^2+1}\right)^q$ lies within the sphere $S_z$. Accordingly, for arbitrary $z \in N_{\nu-1}$ the radius of the sphere $S_z^{\nu-1}$ is not less than $r_{\nu-1}(|z|+1)^{-\rho_{\nu-1}}$, where

$$\begin{aligned} r_{\nu-1} &= \min(r_\nu 2^{-1-\rho_\nu}, C\,\varepsilon[r_\nu 2^{-1-\rho_\nu}]^q), \\ \rho_{\nu-1} &= \max(\rho_\nu, |q|(\rho_\nu+2)). \end{aligned} \tag{3.3}$$

Thus, the desired covering $S^{\nu-1} = \{S_z^{\nu-1}\}$ has been constructed. If we carry out the induction, we construct the covering $S^0 = \{S_z^0\}$. If we decrease the radius of each of the spheres $S_z^0$ to $r_0(|z|+1)^{-\rho_0}$, we obtain a covering satisfying the hypotheses of the proposition. Since $r_m = C\,\varepsilon$, and $\rho_m = 2q$, we find from the recurrence formula (3) that $r_0 \geqq c\,\varepsilon^Q$, where the constants $c$, $Q$ and $\rho_0$ are independent of $\varepsilon$. □

**4°. The space of cochains on elementary coverings.** Let $\mathcal{M} = \{M_\alpha(z)\}$ be a family of majorants in $C^n$, let $U = \{U_z\}$ be an elementary covering with parameter zero, and let $\nu \geqq 0$ be an integer. For arbitrary integers $\alpha > 0$ and $k \geqq 0$, we consider the following norm defined on the cochains of order $\nu$ on the covering $U \cap \Omega_\alpha$

$$\|\phi\|_{\alpha, U}^k = \max_{|i|+|j| \leqq k} \sup_{z_0, \ldots, z_\nu} \sup_{U_{z_0} \cap \cdots \cap U_{z_\nu} \cap \Omega_\alpha} \frac{|D^{i,\bar{j}} \phi_{z_0, \ldots, z_\nu}(z)|}{M_\alpha(z)}. \tag{4.3}$$

The space of these cochains for which the norm is finite will be denoted by ${}^\nu H_\alpha^{0,k}(\mathrm{U})$. If $\phi$ is some cochain of order $\nu$ on a wider covering $V$, we shall say that $\phi$ is an element of the space ${}^\nu H_\alpha^{0,k}(U)$, if its norm (4) is finite. As earlier, we shall set $\frac{1}{M_\alpha(z)} = 0$ outside $\Omega_\alpha$. In particular, if $\phi \equiv 0$ in $\Omega_\alpha$, the cochain $\phi$ is equal to zero as an element of the space $\mathcal{H}_\alpha^{0,k}(U)$.

The subspace in ${}^\nu\mathcal{H}_\alpha^{0,k}(U)$ consisting of cochains that are holomorphic in $z_1, \ldots, z_m$ in $\Omega_\alpha$, will be denoted by ${}^\nu\mathcal{H}_\alpha^{m,k}(U)$. We set ${}^\nu\mathcal{H}_\alpha^m(U) = \varprojlim_{k \to \infty} {}^\nu\mathcal{H}_\alpha^{m,k}(U)$. It is convenient to introduce the notation

$${}^{-1}\mathcal{H}_\alpha^{m,k}(U) = \mathcal{H}_\alpha^{m,k}, \qquad {}^{-1}\mathcal{H}_\alpha^m(U) = \mathcal{H}_\alpha^m.$$

It is clear from formula (2) that for arbitrary $\alpha$, $m$, $k$ and $\nu \geqq -1$ the operator $\partial$ defines a continuous mapping $\partial\colon {}^\nu\mathcal{H}_\alpha^{m,k}(U) \to {}^{\nu+1}\mathcal{H}_\alpha^{m,k}(U)$ with norm not exceeding unity, and also the limit mapping $\partial\colon {}^\nu\mathcal{H}_\alpha^m(U) \to$

${}^{\nu+1}\mathcal{H}_\alpha^m(U)$. We denote by ${}^\nu\mathcal{Z}_\alpha^{m,k}$ and ${}^\nu\mathcal{Z}_\alpha^m$ the kernels, and by ${}^{\nu+1}\mathcal{B}_\alpha^{m,k}$ and ${}^{\nu+1}\mathcal{B}_\alpha^m$ the images, of these mappings.

If $\alpha' \geqq \alpha$, and the covering $V$ is inscribed in $U$, then there is defined also a continuous mapping of the restriction ${}^\nu\mathcal{H}_\alpha^{m,k}(U) \to {}^\nu\mathcal{H}_{\alpha'}^{m,k}(V)$, the norm of which does not exceed unity. Such mappings, and the corresponding limit mapping, will be called identity mappings.

By $U_\alpha$, $0<\alpha<\infty$, we denote an elementary covering with parameter zero and radius $\varepsilon_\alpha$. For arbitrary $\alpha$ the covering $U_{\alpha+1}$ is inscribed in $U_\alpha$ and therefore the identity mapping ${}^\nu\mathcal{H}_\alpha^m(U_\alpha) \to {}^\nu\mathcal{H}_{\alpha+1}^m(U_{\alpha+1})$ is defined.

Thus the spaces ${}^\nu\mathcal{H}_\alpha^m(U_\alpha)$, with respect to the identity mappings, form a family of linear topological spaces. This family will be denoted by ${}^\nu\mathcal{H}_\mathcal{M}^m$. Since the identity mappings commute with the operators $\partial$, the sequence of family mappings

$$0 \to \mathcal{H}_\mathcal{M}^m \xrightarrow{\partial} {}^0\mathcal{H}_\mathcal{M}^m \xrightarrow{\partial} {}^1\mathcal{H}_\mathcal{M}^m \to \cdots \to {}^\nu\mathcal{H}_\mathcal{M}^m \xrightarrow{\partial} {}^{\nu+1}\mathcal{H}_\mathcal{M}^m \to \cdots$$

is defined.

Among the spaces that we are considering, we shall most frequently make use of the space ${}^\nu\mathcal{H}_\alpha(U) = {}^\nu\mathcal{H}_\alpha^{n,0}(U)$. We denote by ${}^\nu\mathcal{Z}_\alpha(U)$ and ${}^{\nu+1}\mathcal{B}_\alpha(U)$ the kernel and the image of the mapping $\partial\colon {}^\nu\mathcal{H}_\alpha(U) \to {}^{\nu+1}\mathcal{H}_\alpha(U)$. The spaces ${}^\nu\mathcal{H}_\alpha(U_\alpha)$, $\alpha = 1, 2, \ldots$, form a family which we denote by ${}^\nu\mathcal{H}_\mathcal{M}$. The coboundary operator defines the sequence of mappings

$$0 \to \mathcal{H}_\mathcal{M} \xrightarrow{\partial} {}^0\mathcal{H}_\mathcal{M} \xrightarrow{\partial} {}^1\mathcal{H}_\mathcal{M} \xrightarrow{\partial} \cdots \to {}^\nu\mathcal{H}_\mathcal{M} \xrightarrow{\partial} {}^{\nu+1}\mathcal{H}_\mathcal{M} \to \cdots. \tag{5.3}$$

This sequence is semi-exact, and therefore we may consider the factor-families

$$\operatorname{Ker}\{\partial\colon {}^\nu\mathcal{H}_\mathcal{M} \to {}^{\nu+1}\mathcal{H}_\mathcal{M}\}/\partial^{\nu-1}\mathcal{H}_\mathcal{M}.$$

These factor-families will be called $\mathcal{M}$-cohomologies. We shall say that the $\mathcal{M}$-cohomologies are trivial, if the sequence (5) is exact.

**5°. Proposition 2.** *Given an arbitrary elementary covering $U$ of parameter zero, given a number $\theta$, $0<\theta<1$, and given integers $\alpha>0$, $\nu \geqq 0$, the operation of restricting the functions of the space ${}^\nu\mathcal{H}_\alpha(U)$ on the inscribed covering $\theta U$ maps this space continuously on ${}^\nu\mathcal{H}_{\alpha+1}^n(\theta U)$.*

*Proof.* We set $U = \{U_z\}$ and $V_z = \theta U_z$. For arbitrary $z_0, \ldots, z_\nu$, the region $V_{z_0} \cap \cdots \cap V_{z_\nu}$ is contained in the region $U_{z_0} \cap \cdots \cap U_{z_\nu}$ together with its neighborhood, of radius $(1-\theta)\, r(U)$. Let $\phi$ be an arbitrary cochain belonging to the space ${}^\nu\mathcal{H}_\alpha(U)$. To an arbitrary component $\phi_{z_0, \ldots, z_\nu}$ of this cochain, we apply Proposition 1 of § 1 (see p. 90) with $r = \min\{(1-\theta)\, r(U), \varepsilon_\alpha\}$ at any point $z \in V_{z_0} \cap \cdots \cap V_{z_\nu} \cap \Omega_{\alpha+1}$:

$$|D^{i,\bar{j}} \phi_{z_0, \ldots, z_\nu}(z)| \leqq C_{ij}\, M_{\alpha+1}(z) \sup_{U_{c_0} \cap \cdots \cap U_{c_\nu} \cap \Omega_\alpha} \frac{|\phi_{z_0, \ldots, z_\nu}(\zeta)|}{M_\alpha(\zeta)}.$$

Hence, for arbitrary $k \geqq 0$

$$\|\phi\|^k_{\alpha+1,\,\theta U} \leqq C_k \| \, \|\phi\|^0_{\alpha,\,U}. \tag{6.3}$$

But this inequality yields the assertion. □

It follows in particular from Proposition 2 that the family ${}^{\nu}\mathscr{H}_{\mathscr{M}}$ is a subfamily of the family ${}^{\nu}\mathscr{H}^n_{\mathscr{M}}$, and is equivalent to this family.

## § 4. The theorem on the triviality of $\mathscr{M}$-cohomologies

**1°. Theorem 1.** *Let $\mathscr{M}$ be a family of majorants of type $\mathscr{I}$. Then for arbitrary $m$, $0 \leqq m \leqq n$, the sequence of families*

$$0 \to \mathscr{H}^m_{\mathscr{M}} \xrightarrow{\partial} {}^0\mathscr{H}^m_{\mathscr{M}} \xrightarrow{\partial} {}^1\mathscr{H}^m_{\mathscr{M}} \to \cdots \to {}^{\nu}\mathscr{H}^m_{\mathscr{M}} \xrightarrow{\partial} {}^{\nu+1}\mathscr{H}_{\mathscr{M}} \to \cdots \tag{1.4}$$

*is exact.*

The proof is based on two lemmas. In Lemma 1, we state that for $m=0$, the sequence (1) is exact for an arbitrary family of majorants. With the help of Lemma 2, the exactness of the sequence (1) will be proved by induction on $m$.

**Lemma 1.** *For an arbitrary family of majorants $\mathscr{M}$, the sequence* (1) *with $m=0$ is exact.*

*Proof.* Let $r$ be an arbitrary number lying between 0 and 1, and let $U=\{U_z\}$ be an elementary covering of parameter zero and radius $r$. We construct a partition of the identity subordinate to it. Let $h(\zeta)$ be a non-negative, infinitely differentiable function in $C^n$ with support in the sphere $|\zeta|<1$, whose integral over $C^n$ is equal to unity. We write $h_z(\zeta)= r^{-2n} h[(z-\zeta)/r]$, $z \in C^n$. The functions $h_z$ form a partition of the identity subordinate to the covering $U$ and

$$\int h_z(\zeta)\, dz\, d\bar{z} = 1.$$

It is clear that the derivatives of the functions $h_z$ satisfy the inequality

$$|D^{i,\bar{j}}_{\zeta} h_z(\zeta)| \leqq C_{ij}, \qquad z \in C^n,$$

where $C_{i,j}$ are constants, independent of $r$. Hence

$$\int |D^{i,\bar{j}}_{\zeta} h_z(\zeta)\, dz\, d\bar{z}| \leqq C'_{i,j}. \tag{2.4}$$

We now prove the lemma. We fix an index $\alpha$. Let

$$\phi = \sum_{z_0,\ldots,z_\nu} \phi_{z_0,\ldots,z_\nu} U_{z_0} \wedge \cdots \wedge U_{z_\nu}$$

be an arbitrary cochain belonging to the space ${}^{\nu}\mathscr{Z}_{\alpha}^{0}(U)$. We need to construct the cochain $\psi \in {}^{\nu-1}\mathscr{H}_{\alpha}^{0}(U)$ such that $\partial\psi=\phi$. For arbitrary points $z, z_1, \ldots, z_\nu$ in the region $\Omega_\alpha \cap U_{z_1} \cap \cdots \cap U_{z_\nu}$ we consider the function $(h_z\,\phi_{z,z_1,\ldots,z_\nu})'$, equal to $h_z\,\phi_{z,z_1,\ldots,z_\nu}$ in $\Omega_\alpha \cap U_{z_1} \cap \cdots \cap U_{z_\nu} \cap U_z$ and zero in $\Omega_\alpha \cap U_{z_1} \cap \cdots \cap U_{z_\nu} \setminus U_z$. We set

$$\psi_{z_1,\ldots,z_\nu} = (\nu+1)\int (h_z\,\phi_{z,z_1,\ldots,z_\nu})'\,dz\,d\bar{z}. \tag{3.4}$$

This inequality (2) implies that this integral converges in $\Omega_\alpha \cap U_{z_1} \cap \cdots \cap U_{z_\nu}$. We shall show that the cochain $\psi = \sum_{z_1,\ldots,z_\nu} \psi_{z_1,\ldots,z_\nu}\, U_{z_1} \wedge \cdots \wedge U_{z_\nu}$ — is the one we are looking for. We have

$$\begin{aligned}\partial\psi &= \frac{1}{\nu+1} \sum_{z_0,\ldots,z_\nu} \sum_{\mu=0}^{\nu} (-1)^\mu\, \psi_{z_0,\ldots,z_{\mu-1},z_{\mu+1},\ldots,z_\nu}\, U_{z_0} \wedge \cdots \wedge U_{z_\nu} \\ &= \sum_{z_0,\ldots,z_\nu} \int h_z \sum_{\mu=0}^{\nu} (-1)^\mu\, \phi_{z,z_0,\ldots,z_{\mu-1},z_{\mu+1},\ldots,z_\nu}\, dz\,d\bar{z}\; U_{z_0} \wedge \cdots \wedge U_{z_\nu}.\end{aligned} \tag{4.4}$$

By definition, the component of the cochain $\partial\phi$ at the product $U_z \wedge U_{z_0} \wedge \cdots \wedge U_{z_\nu}$ is equal to

$$\frac{1}{\nu+2}\left[\sum_{\mu=0}^{\nu} (-1)^{\mu+1}\, \phi_{z,z_0,\ldots,z_{\mu-1},z_{\mu+1},\ldots,z_\nu} + \phi_{z_0,\ldots,z_\nu}\right].$$

when $z \neq z_\mu$, $\mu=0,\ldots,\nu$. But, by hypothesis, $\partial\phi=0$, and, accordingly,

$$\phi_{z_0,\ldots,z_\nu} = \sum_{\mu=0}^{\nu} (-1)^\mu\, \phi_{z,z_0,\ldots,z_{\mu-1},z_{\mu+1},\ldots,z_\nu}.$$

Thus, when $z \neq z_\mu$, $\mu=0,\ldots,\nu$, the inner sum on the right side of (4) is equal to $\phi_{z_0,\ldots,z_\nu}$. In the excluded cases $z=z_\mu$, $\mu=0,\ldots,\nu$, only one term is present in this sum, and this is equal to $\phi_{z_0,\ldots,z_\nu}$. Therefore, the right side of (7) can be rewritten as:

$$\sum_{z_0,\ldots,z_\nu} \int h_z\,dz\,d\bar{z}\,\phi_{z_0,\ldots z_\nu}\, U_{z_0} \wedge \cdots \wedge U_{z_\nu} = \phi.$$

Whence $\partial\psi=\phi$.

Let us find bounds for the derivatives of the cochain $\psi$. From (3) and (2)

$$\begin{aligned}|D_\zeta^{i,\bar{j}}\,\psi_{z_1,\ldots,z_\nu}(\zeta)| &\leqq (\nu+1)\int |D_\zeta^{i,\bar{j}}\,(h_z(\zeta)\,\phi_{z,z_1,\ldots,z_\nu}(\zeta))'\,dz\,d\bar{z}| \\ &\leqq C_{i,j} \max_{i'\leqq i,\, j'\leqq j} \int |D_\zeta^{i-i',\overline{j-j'}}\, h_z(\zeta)\,dz\,d\bar{z}| \sup_z |D_\zeta^{i',\bar{j}'}\,\phi_{z,z_1,\ldots,z_\nu}(\zeta)| \\ &\leqq C'_{i,j} \max_{i'\leqq i,\, j'\leqq j} \sup_z |D^{i',\bar{j}}\,\phi_{z,z_1,\ldots,z_\nu}(\zeta)|.\end{aligned}$$

Dividing both sides of this inequality by $M_\alpha(\zeta)$ and passing to the upper bound with respect to $\zeta, z_1, \ldots, z_\nu$, we arrive at the inequality

$$\|\psi\|_{\alpha, U}^k \leqq C_k \|\phi\|_{\alpha, U}^k, \qquad k=0, 1, 2, \ldots. \tag{5.4}$$

Thus, we have constructed the operator $S_U: {}^\nu\mathscr{Z}_\alpha^0(U) \ni \phi \to \psi \in {}^{\nu-1}\mathscr{H}_\alpha^0(U)$ which is such that the composition $\partial S_U: {}^\nu\mathscr{Z}_\alpha^0(U) \to {}^\nu\mathscr{Z}_\alpha^0(U)$ is an identity mapping. Hence follows the exactness of the sequence (1) with $m=0$. □

**2°. Lemma 2.** *Let $M$ be an arbitrary family of majorants and let $\alpha>0$, $\nu \geqq 0$, and $m$, $0<m\leqq n$, be integers. There exists a continuous operator $T: {}^\nu\mathscr{H}_\alpha^{m-1}(U_\alpha) \to {}^\nu\mathscr{H}_{\alpha+1}^{m-1}(U_{\alpha+1})$, satisfying the equation*

$$\frac{\partial}{\partial \bar{z}_m} T\phi = \phi.$$

*Proof.* We set $U_\alpha = \{U_z\}$, $U_{\alpha+1} = \{V_z\}$ and we choose the points $z_0, \ldots, z_\nu$ arbitrarily in such a way that the region $\omega = V_{z_0} \cap \cdots \cap V_{z_\nu} \cap \Omega_{\alpha+1}$ is not empty. The region $\omega' = U_{z_0} \cap \cdots \cap U_{z_\nu} \cap \Omega_\alpha$ contains a neighborhood of $\omega$ of radius $r = \varepsilon_\alpha - \varepsilon_{\alpha+1}$. We choose an arbitrary cochain $\phi \in {}^\nu\mathscr{H}_\alpha^{m-1}(U_\alpha)$ and we apply to the component $\phi_{z_0, \ldots, z}$ Theorem 2 of § 2, setting $\Omega = \omega$. We thus determine a function $\psi_{z_0, \ldots, z_\nu}$, which is infinitely differentiable in $\omega$, is holomorphic with respect to $z_1, \ldots, z_{m-1}$ satisfies

$$\frac{\partial \psi_{z_0, \ldots, z_\nu}}{\partial \bar{z}_m} = \phi_{z_0, \ldots, z_\nu}$$

and

$$\|\psi_{z_0, \ldots, z_\nu}\|_\omega^k \leqq C_k \|\phi_{z_0, \ldots, z_\nu}\|_{\omega'}^k,$$

where the constants $C_k$ are independent of $z_0, \ldots, z_\nu$ and $\alpha$. Since the diameter of the region $\omega'$ does not exceed $\varepsilon_\alpha$, we obtain the inequality

$$\max_{|i|+|j|\leqq k} \sup_\omega \frac{|D^{i,\bar{j}} \psi_{z_0, \ldots, z_\nu}|}{M_{\alpha+1}} \leqq C_k \max_{|i|+|j|\leqq k} \sup_{\omega'} \frac{|D^{i,\bar{j}} \phi_{z_0, \ldots, z_\nu}|}{M_\alpha}. \tag{6.4}$$

Since the function $\phi_{z_0, \ldots, z_\nu}$ depends anti-symmetrically on the indices $z_0, \ldots, z_\nu$, we can choose the function $\psi_{z_0, \ldots, z_\nu}$ so that it has the same property. We form the cochain $\psi = \sum \psi_{z_0, \ldots, z_\nu} \times V_{z_0} \wedge \cdots \wedge V_{z_\nu}$. The inequality (6) implies that it belongs to ${}^\nu\mathscr{H}_{\alpha+1}^{m-1}(U_{\alpha+1})$ and depends continuously on $\phi$. By construction, $\dfrac{\partial \psi}{\partial \bar{z}_m} = \phi$. Thus, the operator $T: \phi \to \psi$ is the desired one. □

**Corollary 1.** *For an arbitrary family of majorants $\mathscr{M}$, and integers $\nu \geqq 0$ and $m$, the sequence*

$$0 \to {}^\nu\mathscr{H}_{\mathscr{M}}^m \to {}^\nu\mathscr{H}_{\mathscr{M}}^{m-1} \xrightarrow{\frac{\partial}{\partial \bar{z}_m}} {}^\nu\mathscr{H}_{\mathscr{M}}^{m-1} \to 0$$

*is exact.*

*We now prove the theorem.* We shall assume that the sequence obtained from (1) by the replacement of $m$ by $m-1$ is exact. We shall prove the exactness of the sequence (1) itself. We consider the following commutative diagram of families:

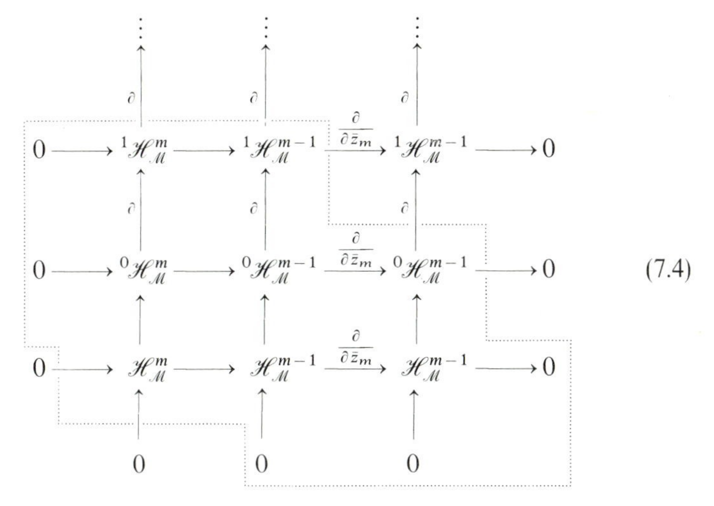

(7.4)

Corollary 1 guarantees the exactness of all rows of this diagram except the lowest. The exactness of the lowest follows from Theorem 1 of § 2. The induction hypothesis implies the exactness of the second and third columns. The portion of this diagram enclosed in the frame has a form symmetric to ($\beta$) of § 2, Chapter I. Applying Theorem 1 of § 2 of Chapter I to this portion, we can establish the exactness of the first column at the term ${}^0\mathcal{H}^m_{\mathcal{M}}$. Further, lifting the frame one step higher and applying the same argument, we find that the first column is exact at the term ${}^1\mathcal{H}^m_{\mathcal{M}}$ and so on. Thus, the exactness of the whole first column is proved.

To complete the proof, it remains to show that for $m=0$ this sequence is exact in view of Lemma 1. □

**Corollary 2.** *Let $\mathcal{M}$ be a family of majorants of type $\mathcal{I}$. Then the sequence*

$$0 \to \mathcal{H}_{\mathcal{M}} \xrightarrow{\partial} {}^0\mathcal{H}_{\mathcal{M}} \xrightarrow{\partial} {}^1\mathcal{H}_{\mathcal{M}} \to \cdots \to {}^\nu\mathcal{H}_{\mathcal{M}} \xrightarrow{\partial} {}^{\nu+1}\mathcal{H}_{\mathcal{M}} \to \cdots$$

*is exact.*

*Proof.* The exactness of the sequence follows from the exactness of (1) with $m=n$, since by what we proved in § 3, ${}^\nu\mathcal{H}_{\mathcal{M}}$ is a subfamily of ${}^\nu\mathcal{H}^n_{\mathcal{M}}$ and equivalent to it.

**3°. Remark.** By Corollary 2, for arbitrary $\nu \geqq -1$ the mapping

$$R^\nu\colon \quad \operatorname{Ker}\{\partial\colon {}^{\nu+1}\mathcal{H}_{\mathcal{M}} \to {}^{\nu+2}\mathcal{H}_{\mathcal{M}}\} \to {}^\nu\mathcal{H}_{\mathcal{M}}/\partial^{\nu-1}\mathcal{H}_{\mathcal{M}},$$

inverse to the mapping $\partial$, is defined. We shall show that the mapping $R^\nu$ can always be so chosen that its order (see Definition 2 of § 1, Chapter I) is some fixed function $\alpha \to \beta(\alpha)$, not depending on the family of majorants $\mathcal{M}$. To this end we begin by proving a similar assertion for the mappings

$$R_m^\nu: \quad \operatorname{Ker}\{\partial: {}^{\nu+1}\mathcal{H}_{\mathcal{M}}^m \to {}^{\nu+2}\mathcal{H}_{\mathcal{M}}^m\} \to {}^\nu\mathcal{H}_{\mathcal{M}}^m / \partial\, {}^{\nu-1}\mathcal{H}_{\mathcal{M}}^m,$$

which are inverse to the mappings $\partial$.

According to Lemma 1, given an arbitrary family of majorants $\mathcal{M}$, there exists a mapping $R_0^\nu$, $\nu = -1, 0, 1, \ldots$, inverse to $\partial$, and of order equal to the function $\beta(\alpha) \equiv \alpha$. We shall now suppose that there exists a mapping $R_{m-1}^\nu$, inverse to $\partial$, with an order not depending on the family $\mathcal{M}$. We consider the diagram (7). By Theorem 1 of § 2 and 2 of this section, given arbitrary $\nu \geqq -1$, there exists a mapping $T^\nu$: ${}^\nu\mathcal{H}_{\mathcal{M}}^{m-1} \to {}^\nu\mathcal{H}_{\mathcal{M}}^{m-1} / {}^\nu\mathcal{H}_{\mathcal{M}}^m$, inverse to the mapping $\partial/\partial \bar{z}_m$, with an order independent of $\mathcal{M}$. Therefore in diagram (7), we can apply the remark of § 2, Chapter I, according to which, corresponding to an arbitrary $\nu \geqq -1$, there exists a mapping $R_m^\nu$, inverse to $\partial$, with an order dependent only on the order of $R_{m-1}^\nu$ and $T^\nu$, and, therefore, not dependent on $\mathcal{M}$. Thus, we have carried out the induction with respect to $m$ and we can say that there exists a mapping $R_n^\nu$, $\nu \geqq 1$, inverse to $\partial$, with an order not dependent on $\mathcal{M}$. Finally, from Proposition 2 of § 3, it follows that the isomorphisms ${}^\nu\mathcal{H}_{\mathcal{M}}^n \cong {}^\nu\mathcal{H}_{\mathcal{M}}$ can be determined with the help of mappings whose order does not depend on $\mathcal{M}$. These mappings (more exactly the mappings associated with them) when suitably compounded with the mapping $R_n^\nu$ yield the desired mapping $R^\nu$. Clearly, the order of the mapping $R^\nu$ constructed in this fashion, does not depend on the family $\mathcal{M}$. Our assertion is thus proved.

**4°. Cohomologies of analytic cochains in bounded regions.** It is well known that in an arbitrary pseudo-convex open set, the cohomologies of analytic cochains are trivial. We now establish a variant of this classic result by turning our attention to an estimate of the norms of the corresponding operators.

Let $\Omega$ be an open set in $C^n$, and let $U$ be an elementary covering. We consider the norm

$$\|\phi\|_{\Omega,U}^0 = \sup_{z_0,\ldots,z_\nu} \sup_{U_{z_0,\ldots,z_\nu}} |\phi|, \qquad U_{z_0,\ldots,z_\nu} = U_{z_0} \cap \cdots \cap U_{z_\nu} \cap \Omega,$$

defined on cochains over the covering $U \cap \Omega$.

**Theorem 2.** *Let $\omega$ be an open set of diameter not greater than* 1, *suppose $0 < r \leqq 1$, let $\Omega$ be an $r$-neighborhood of $\omega$, and let $U$ and $V$ be elementary coverings of parameter zero and radii $2r$ and $r$. For every holomorphic cochain $\phi$ on the covering $U \cap \Omega$ with a finite norm $\|\phi\|_{\Omega,U}^0$, such that*

$\partial\phi=0$, *there exists a holomorphic cochain* $\psi$ *on* $V\cap\omega$, *such that* $\partial\psi=\phi$ *and*

$$\|\psi\|^0_{\omega,V}\leqq\frac{C}{r^n}\|\phi\|^0_{\Omega,U}, \tag{8.4}$$

*where the constant* $C$ *is independent of* $\omega$ *and* $r$.

*Proof.* For arbitrary integers $v\geqq 0$ and $k\geqq 0$ we consider the space ${}^vL^k$, consisting of cochains $F$ of order $v$, on the covering $U\cap\Omega$, the components of which are $''d$-forms of order $k$ (see 7°, §2) with square-summable coefficients:

$$F=\sum_{z_0,\ldots,z_v}\Big(\sum_{j_1,\ldots,j_k} f^{j_1,\ldots,j_k}_{z_0,\ldots,z_v}\,d\bar z_{j_1}\wedge\cdots\wedge d\bar z_{j_k}\Big)\,U_{z_0}\wedge\cdots\wedge U_{z_v}.$$

We define the topology in ${}^vL^k$ by the norm

$$\|F\|_{\Omega,U}=\sup_{z_0,\ldots,z_v}\Big|\sum_{j_1,\ldots,j_k}\|f^{j_1,\ldots,j_k}_{z_0,\ldots,z_v}\|^2_{L_2(U_{z_0,\ldots,z_v})}\Big|^{\frac12}.$$

We consider the commutative diagram:

$$\begin{array}{ccccccccc}
 & & \vdots & & \vdots & & \vdots & & \\
 & & \uparrow & & \uparrow & & \uparrow & & \\
0 & \longrightarrow & {}^1H(U) & \longrightarrow & {}^1L^0 & \xrightarrow{''d} & {}^1L^1 & \xrightarrow{''d} & \cdots \\
 & & \uparrow{\scriptstyle\partial} & & \uparrow{\scriptstyle\partial} & & \uparrow{\scriptstyle\partial} & & \\
0 & \longrightarrow & {}^0H(U) & \longrightarrow & {}^0L^0 & \xrightarrow{''d} & {}^0L^1 & \xrightarrow{''d} & \cdots \\
 & & \uparrow & & \uparrow & & \uparrow & & \\
0 & \longrightarrow & H(\Omega) & \longrightarrow & L^0 & \xrightarrow{''d} & L^1 & \xrightarrow{''d} & \cdots \\
 & & \uparrow & & \uparrow & & \uparrow & & \\
 & & 0 & & 0 & & 0 & &
\end{array} \tag{9.4}$$

Here $L^k$, $k=0,1,2,\ldots$, is the kernel of the mapping $\partial\colon {}^0L^k\to{}^1L^k$, that is, the space of $''d$-forms in $\Omega$ of order $k$ with square-summable coefficients, and ${}^vH(U)$ is a space of holomorphic cochains of order $v$ on the covering $U\cap\Omega$ with the norm

$$\|\phi\|_{\Omega,U}=\sup_{z_0,\ldots,z_v}\|\phi^{\boldsymbol\cdot}_{z_0,\ldots,z_v}\|_{L_2(U_{z_0,\ldots,z_v})}.$$

$H(\Omega)$ is the space of square-summable functions holomorphic in $\Omega$.

We shall prove the exactness of the rows of this diagram. Let $F\in{}^vL^{k+1}$ and $''dF=0$. We choose $z_0,\ldots,z_v$ arbitrarily and we consider the $''d$-form

$$f=f_{z_0,\ldots,z_v}=\sum_{j_1,\ldots,j_k} f^{j_1,\ldots,j_k}_{z_0,\ldots,z_v}\,d\bar z_{j_1}\wedge\cdots\wedge d\bar z_{j_k}.$$

It has a finite norm $\|f\|_{U_{z_0,\ldots,z_\nu}}$ and satisfies the equation $''dF=0$. According to the theorem of Hörmander, formulated in 7° of § 2, there exists a $''d$-form $g$ such that

$$dg=f, \qquad \|g\|_{U_{z_0,\ldots,z_\nu}} \leqq 12\,\|f\|_{U_{z_0,\ldots,z_\nu}}. \tag{10.4}$$

Assuming that the form $g$ depends in an anti-symmetric fashion on $z_0,\ldots,z_\nu$, we make up the cochain $G=\sum g\,U_{z_0}\wedge\cdots\wedge U_{z_\nu}$. It follows from (10) that

$$''dG=F, \qquad \|G\|_{\Omega,U}\leqq 12\|F\|_{\Omega,U}.$$

Thus we have established the exactness of all rows of the diagram (9), except the lowest. The exactness of the lowest row is proved in an entirely similar fashion if we remember that the norm $\|\cdot\|_{\Omega,U}$ in the space $L^k\subset{}^0L^k$ is equivalent to the norm $\|\cdot\|_{\Omega}$.

We now turn our attention to the columns. Making use of the notation of Lemma 1, we write for every cochain $F\in{}^\nu L^k$ such that $\partial F=0$,

$$g_{z_1,\ldots,z_\nu}=(\nu+1)\int (h_z\,f_{z,z_1,\ldots,z_\nu})'\,dz\,d\bar{z}.$$

We note that $0\leqq h_z\leqq 1$ and that for arbitrary fixed $z_1,\ldots,z_\nu$ the integrand differs from zero only when $|z-z_1|\leqq 2$. Therefore the square of the right side does not exceed

$$C\int |f_{z,z_1,\ldots,z_\nu}|^2\,|dz\,d\bar{z}|,$$

where by $|f\ldots|^2$ we mean $\sum |f_{\ldots}^{j_1,\ldots,j_k}|^2$, and $C$ depends only on $n$ and $\nu$. Hence

$$\int |g_{z_1,\ldots,z_\nu}(\zeta)|^2\,|d\zeta\,d\bar{\zeta}|\leqq C\int |f_{z,z_1,\ldots,z_\nu}(\zeta)|^2\,|d\zeta\,d\bar{\zeta}\,dz\,d\bar{z}|.$$

Therefore the operator

$$F\to G=\sum g_{z_1,\ldots,z_\nu}\,U_{z_1}\wedge\cdots\wedge U_{z_\nu}\in{}^{\nu-1}L^k$$

is defined and continuous. It follows from the calculations made in Lemma 1 that $\partial G=F$. Thus we have established the exactness of all the columns of diagram (9), except the leftmost.

The exactness of the leftmost column follows from Theorem 1 of § 2, Chapter I. We may therefore assert that for an arbitrary cochain $\phi\in{}^{\nu+1}H(U)$ satisfying $\partial\phi=0$, there exists a cochain $\psi\in{}^\nu H(U)$ such that $\partial\psi=\phi$, and

$$\|\psi\|_{\Omega,U}\leqq C\,\|\phi\|_{\Omega,U}. \tag{11.4}$$

We remark that we are justified in supposing that the constant $C$ in this inequality is independent of $\omega$ and $r$, since the constants which appear in the preceding argument have this property (see 5° of § 2, Chapter I).

The right side of (11), clearly, does not exceed the norm $C\|\phi\|^0_{\Omega,U}$. On the other hand, every region $V_{z_0,\ldots,z_\nu}=V_{z_0}\cap\cdots\cap V_{z_\nu}\cap\omega$ belongs to $U_{z_0,\ldots,z_\nu}$ together with its $r$-neighborhood. Since the components of the cochain $\psi$ are holomorphic, it follows from Proposition 1 of § 1 that we have the inequality

$$\sup_{V_{z_0,\cdots,z_\nu}} |\psi_{z_0,\ldots,z_\nu}|\leqq\frac{C}{r^n}\|\psi_{z_0,\ldots,z_\nu}\|_{L_2(U_{z_0,\ldots,z_\nu})}$$

where $C$ does not depend on $\omega$ and $r$. Hence $\|\psi\|^0_{\omega,V}\leqq C\|\psi\|_{\Omega,U}$. Combining this inequality with (11), we arrive at (8). ▯

**5°. Examples of non-trivial $\mathcal{M}$-cohomologies.** In this subsection, we shall say something about the general problem: to write down all families of majorants $\mathcal{M}$ which correspond to trivial $\mathcal{M}$-cohomologies. Certainly, the families of type $\mathcal{I}$ are far from exhausting the set of all families for which $\mathcal{M}$-cohomologies are trivial. For example, a well-known theorem of Oka-Cartan-Serre provides another example of such families. A more general class of families of majorants of this type has been given by Hörmander [2].

We now produce two examples of families of majorants with non-trivial $\mathcal{M}$-cohomologies. In the first example, the family $\mathcal{M}$ will be so chosen that the sequence (5.3) is not algebraically exact at the term ${}^1\mathcal{H}_{\mathcal{M}}$. In the second example, the mapping $\partial$: ${}^0\mathcal{H}_{\mathcal{M}}\to{}^1\mathcal{H}_{\mathcal{M}}$ is not a homomorphism.

**Example 1.** Let $n=1$, and

$$M_\alpha(z)=(|z|+1)^\alpha\exp\left(-\frac{1}{\alpha}|\operatorname{Im} z|^{1+\varepsilon}\right),\qquad \alpha=1,2,\ldots,$$

where $\varepsilon>0$ is an arbitrary fixed quantity. It is not difficult to see that the sequence of functions $M_\alpha(z)$ forms a family of majorants. Suppose that $U=\{U_z\}$ is an elementary covering of parameter zero, and that $R_+$ is the open half-plane $\operatorname{Im} z>0$, and $R_-$ is the closed half-plane $\operatorname{Im} z\leqq 0$. We consider the following holomorphic first order cochain on the covering $U$:

$$\phi=\sum\phi_{z_1,z_2}U_{z_1}\wedge U_{z_2},$$

where

$$\phi_{z_1,z_2}(z)=\begin{cases}0, & z_1,z_2\in R_\pm;\\ 1, & z_1\in R_+,\quad z_2\in R_-,\qquad z\in U_{z_1}\cap U_{z_2};\\ -1, & z_1\in R_-,\quad z_2\in R_+.\end{cases}$$

It is clear that the components of the cochain $\phi$ differ from zero only in the strip $|\operatorname{Im} z|\leqq r(U)$ and are bounded in the ensemble. It follows that the function $\phi$ belongs to the space ${}^1\mathscr{H}_1(U)$. It is easy to verify that $\partial\phi=0$, and therefore, $\phi\in{}^1\mathscr{Z}_1(U)$.

We shall show that for any covering $V$ and integer $\alpha$, there exists no cochain $\psi\in{}^0\mathscr{H}_\alpha(V)$ such that $\partial\psi=\phi$. It will then follow that the sequence (5.4) is not algebraically exact at the term ${}^1\mathscr{H}_{\mathscr{M}}$. Let us suppose the contrary. Then suppose that on some covering $V=\{V_z\}$, there exists a cochain $\psi=\sum\psi_z V_z\in{}^0\mathscr{H}_\alpha(V)$ such that $\partial\psi=\phi$. This equation implies that for arbitrary $z_1, z_2\in R_+$ or $z_1, z_2\in R_-$, $\psi_{z_1}\equiv\psi_{z_2}$ in $V_{z_1}\cap V_{z_2}$. Accordingly, the functions $\psi_z$ with $z\in R_\pm$ are restrictions of some holomorphic function $\psi_\pm$, defined in $R_\pm$. But the inclusion relation $\psi\in{}^0\mathscr{H}_\alpha(V)$ implies the inequality

$$|\psi_\pm(z)|\leqq C(|z|+1)^\alpha \exp\left(-\frac{1}{\alpha}|\mathrm{Im}\, z|^{1+\varepsilon}\right).$$

Hence it follows that $\psi_\pm\equiv 0$, that is $\psi=0$. Therefore, the equation $\partial\psi=\phi$ cannot hold.

**Example 2.** We again suppose $n=1$ and we let $h(\sigma)$ be a non-negative function in $R^1$ which is not identically zero, which is infinitely differentiable and has a support contained in the segment $[-1,1]$. As is known, the Fourier transform of such a function satisfies the inequality

$$|\tilde{h}(z)|\leqq\mu(x)\exp(|y|), \qquad z=x+iy.$$

where $\mu(x)$ is a positive function decreasing at infinity faster than an arbitrary power of $|x|$. We set

$$\mu_\alpha(x)=\sup\{\mu(x'), |x-x'|\leqq 1-2^{-\alpha}\}$$

and

$$M_\alpha(z)=\begin{cases}(|z|+1)^{\alpha-1}\mu_\alpha(x)\exp(|y|), & |x|\geqq|y|, \qquad \varepsilon>0,\ \alpha=1,2,\ldots,\\ (|z|+1)^{\alpha-1}\mu_\alpha(x)\exp[|y|+(1+\varepsilon)|y-x|], & |y|\geqq|x|.\end{cases}$$

We shall show that the mapping $\partial\colon {}^0\mathscr{H}_{\mathscr{M}}\to{}^1\mathscr{H}_{\mathscr{M}}$ is not a homomorphism. We assume the contrary: $\partial$ is a homomorphism. Then, in particular, if $U=\{U_z\}$ is an elementary covering of parameter zero and radius $\frac{1}{2}$, there must exist a covering $V$and an integer $\alpha$ and a continuous operator

$$R\colon {}^1\mathscr{B}_1(U)\to{}^0\mathscr{H}_\alpha(V)/{}^0\mathscr{Z}_\alpha(V)$$

such that the composition $\partial R$ is an identity mapping. From the continuity of the operator $R$, there follows the inequality

$$\inf\{\|\phi-\partial\chi\|^0_{\alpha,V}, \chi\in\mathscr{H}_\alpha\}\leqq C\,\|\partial\phi\|^0_{1,U}, \phi\in{}^0\mathscr{H}_1(U). \tag{12.4}$$

We shall show that this inequality cannot hold, by constructing a sequence of functions for which the right side of (12) is infinitely small compared to the left.

For every integer $\lambda \geqq 2$ we consider the cochain

$$\psi^\lambda = \sum \psi_{z_1, z_2} U_{z_1} \wedge U_{z_2},$$

the components of which are defined in the following fashion. Let $S_\lambda$ be a closed sphere of radius 1 with center at the point $i\lambda$. Then

$$\psi_{z_1, z_2}(z) = \begin{cases} \dfrac{\tilde{h}(z)}{z - i\lambda}, & z_1 \bar{\in} S_\lambda, \quad z_2 \in S_\lambda, \\ -\dfrac{\tilde{h}(z)}{z - i\lambda}, & z_1 \in S_\lambda, \quad z_2 \bar{\in} S_\lambda; \quad z \in U_{z_1} \cap U_{z_2}. \end{cases}$$

$$\psi_{z_1, z_2}(z) = 0 \begin{cases} z_1, z_2 \in S_\lambda, \\ z_1, z_2 \bar{\in} S_\lambda. \end{cases}$$

It is clear that for arbitrary $z_1$ and $z_2$

$$|\psi_{z_1, z_2}(z)| \leqq 2|\tilde{h}(z)| \leqq 2\mu(x)\exp(|y|).$$

It follows that the cochain $\psi^\lambda$ belongs to the space ${}^1\mathcal{H}_1(U)$. We shall a derive a bound for its norm. Since an arbitrary component of it is different from zero only in the $\frac{3}{2}$-neighborhood of the point $i\lambda$, we have

$$\begin{aligned} \|\psi^\lambda\|_{1,U}^0 &= \sup_{z_1, z_2} \sup_{U_{z_1} \cap U_{z_2}} \frac{|\psi_{z_1, z_2}(z)|}{M_1(z)} \\ &\leqq \sup_{\frac{1}{2} \leqq |z - i\lambda| \leqq \frac{3}{2}} \frac{|\tilde{h}(z)|}{|z - i\lambda|\, \mu'(x)\exp(|y| + (1+\varepsilon)|y - x|)} \qquad (13.4) \\ &\leqq 2\sup_{\frac{1}{2} \leqq |z - i\lambda| \leqq \frac{3}{2}} \frac{\exp(|y|)}{\exp(|y| + (1+\varepsilon)|y - x|)} \\ &\leqq 2\exp(-(1+\varepsilon)|\lambda - \tfrac{3}{2}|). \end{aligned}$$

On the other hand, $\psi^\lambda = \partial\phi$, where

$$\phi^\lambda = \sum \phi_\zeta U_\zeta, \qquad \phi_\zeta = \begin{cases} \dfrac{\tilde{h}(z)}{z - i\lambda} & \text{in } U_\zeta, \ \zeta \bar{\in} S_\lambda, \\ 0 & \text{in } U_\zeta, \ \zeta \in S_\lambda. \end{cases}$$

The cochain $\phi^\lambda$ belongs to the space ${}^0\mathcal{H}_1(U)$, since

$$|\phi_\zeta(z)| \leqq 2|\tilde{h}(z)| \leqq 2\mu(x)\exp(|y|) \leqq 2M_1(z).$$

We bound the left side of (12) with $\phi=\phi^\lambda$. Parseval's equation implies that for an arbitrary function $\chi\in\mathscr{H}_\alpha$ we have the inequality

$$\begin{aligned}\|\phi^\lambda-\partial\chi\|_\alpha^0 &\geqq \sup_{|z-i\lambda|\geqq\frac{1}{2}}\left|\frac{\tilde{h}(z)}{z-i\lambda}-\chi(z)\right|\frac{1}{M_\alpha(z)}\\ &\geqq C\sup_x(|x|+1)\left|\frac{\tilde{h}(x)}{x-i\lambda}-\chi(x)\right|\geqq C'\left\|\frac{\tilde{h}(x)}{x-i\lambda}-\chi(x)\right\|_{L_2} \qquad (14.4)\\ &=C''\left\|F^{-1}\left[\frac{\tilde{h}(x)}{x-i\lambda}\right]-F^{-1}[\chi(x)]\right\|_{L_2},\end{aligned}$$

where $F^{-1}$ is the operator of the inverse Fourier transform. The Phragmen-Lindelöf principle implies that an arbitrary function $\chi$ belonging to the space $\mathscr{H}_\alpha$, that is, satisfying the inequality

$$|\chi(z)|\leqq CM_\alpha(z), \qquad (15.4)$$

also satisfies the inequality

$$|\chi(z)|\leqq C'\exp(|y|). \qquad (16.4)$$

Accordingly, the inverse Fourier transform of the function $\chi$ is an infinitely differentiable function, with support belonging to the segment $[-1,1]$. Hence, for an arbitrary function $\chi\in\mathscr{H}_\alpha$ the right side of (14) is not smaller than

$$\left\|F^{-1}\left[\frac{\tilde{h}(x)}{x-i\lambda}\right]\right\|_{L_2(R_+)}$$

where $R_+$ is the pair of rays $|\sigma|>1$. The inverse Fourier transform of the product $\frac{\tilde{h}(x)}{x-i\lambda}$ is the convolution

$$h(\sigma)*F^{-1}\left[\frac{1}{x-i\lambda}\right]=h(\sigma)*\theta_\lambda(\sigma), \qquad \theta_\lambda(\sigma)=\begin{cases}0, & \sigma>0,\\ ie^{\lambda\sigma}, & \sigma<0.\end{cases}$$

Since the function $h(\sigma)$ is non-negative, the convolution $-i\,h(\sigma)*\theta_\lambda(\sigma)$ is also non-negative and $-i\,h(\sigma)*\theta_\lambda(\sigma)=b\,e^{\lambda\sigma}$, $\sigma\leqq-1$ for some $b>0$.

Therefore

$$\|h(\sigma)*\theta_\lambda(\sigma)\|_{\alpha_2(R_+)}\geqq\frac{b}{\sqrt{2\lambda}}e^{-\lambda}.$$

If we take account of this inequality and use (14), we arrive finally at

$$\inf\{\|\phi^\lambda-\partial\chi\|_{\alpha,V}^0,\ \chi\in\mathscr{H}_\alpha\}\geqq\frac{b}{\sqrt{2\lambda}}e^{-\lambda}$$

for arbitrary $\lambda$. Combining this inequality with the inequality (13), we arrive at a contradiction with the inequality (12).

The examples 1 and 2 can of course be generalized in various directions. We shall take note of only one such generalisation, and that without proving it. Let $i(\phi)$ be some continuous periodic function in the segment $[0, 2\pi]$. Then the family of majorants

$$M_\alpha(z) = \exp\left[\left(i(\arg z) - \frac{1}{\alpha}\right)|z|\right]; \qquad \alpha = 1, 2, \ldots, \tag{17.4}$$

defines trivial $\mathscr{M}$-cohomologies, if and only if the function $i(\phi)$ is trigonometrically convex.

This statement illustrates rather clearly the idea of Examples 1 and 2, that non-trivial $\mathscr{M}$-cohomologies arise when the family of majorants $\mathscr{M}$ in a certain sense does not correspond to the inventory of functions in the space $\bar{\mathscr{H}}_{\mathscr{M}} = \lim \mathscr{H}_{\mathscr{M}}$. Thus, in Example 1, the space $\bar{\mathscr{H}}_{\mathscr{M}}$ consists of the single function that vanishes everywhere. In Example 2, this lack of correspondence consists in the fact that the functions of the space $\bar{\mathscr{H}}_{\mathscr{M}}$ satisfy the inequality (16), which improves the inequality (15) in the region $|y| \geqq |x|$. Our final assertion shows that the $\mathscr{M}$-cohomologies defined by the family of majorants (17) are trivial if and only if there exists an entire function of first-order growth, whose indicatrix is equal to $i(\phi)$. These observations lead us to the following condition as necessary in order that the family $\mathscr{M}$ should have trivial $\mathscr{M}$-cohomologies: the functions $M_\alpha$ cannot be essentially decreased without changing the inventory of functions of the space $\bar{\mathscr{H}}_{\mathscr{M}}$, or, more exactly: there exists no family of majorants $\mathscr{M}'$ such that $\bar{\mathscr{H}}_{\mathscr{M}'} = \bar{\mathscr{H}}_{\mathscr{M}}$, while $\bar{\mathscr{H}}^0_{\mathscr{M}'} \subset \bar{\mathscr{H}}^0_{\mathscr{M}}$ and $\bar{\mathscr{H}}^0_{\mathscr{M}'} \neq \bar{\mathscr{H}}^0_{\mathscr{M}}$. Probably the sufficient condition for the triviality of $\mathscr{M}$-cohomologies should be close to this necessary condition.

## § 5. Cohomologies connected with $\mathscr{P}$-matrices

**1°. Formulation of the theorem.** Let $\mathscr{M}$ be a family of majorants and let ${}^\nu\mathscr{H}_{\mathscr{M}} = \{{}^\nu\mathscr{H}_\alpha(U_\alpha), \alpha = 1, 2, \ldots\}$, $\nu = -1, 0, 1, \ldots$, be the corresponding family of spaces. Further, let $p\colon \mathscr{P}^s \to \mathscr{P}^t$ be a $\mathscr{P}$-matrix, that is, a matrix of size $t \times s$, consisting of polynomials in $C^n$. For arbitrary $\nu$, $\alpha$ and arbitrary covering $U$, the multiplication of holomorphic functions on the covering $U$ by the matrix $p$ defines the continuous operator

$$p\colon [{}^\nu\mathscr{H}_\alpha(U)]^s \to [{}^\nu\mathscr{H}_{\alpha+\mu}(U)]^t, \tag{1.5}$$

where $\mu$ is the highest order of the elements of the matrix $p$. This mapping clearly commutes with the identity mapping of the space ${}^\nu\mathscr{H}_\alpha(U)$. In particular, for arbitrary $\alpha$, the following diagram is commutative:

$$\begin{array}{ccc} [{}^{\nu}\mathscr{H}_{\alpha+1}(U_{\alpha+1})]^s & \xrightarrow{p} & [{}^{\nu}\mathscr{H}_{\alpha+1+\mu}(U_{\alpha+1+\mu})]^t \\ \uparrow & & \uparrow \\ [{}^{\nu}\mathscr{H}_{\alpha}(U_{\alpha})]^s & \xrightarrow{p} & [{}^{\nu}\mathscr{H}_{\alpha+\mu}(U_{\alpha+\mu})]^t. \end{array}$$

It follows that the horizontal mappings in this diagram are components of the family mapping

$$p\colon [{}^{\nu}\mathscr{H}_{\mathscr{M}}]^s \to [{}^{\nu}\mathscr{H}_{\mathscr{M}}]^t. \tag{2.5}$$

Let $\mathscr{D}$ be the $p$-operator constructed in Theorem 1 of § 4, Chapter II, with $m=n$. By $[{}^{\nu}\mathscr{H}_{\alpha}(U)]^t \cap \operatorname{Ker}\mathscr{D}$ we denote the subspace in $[{}^{\nu}\mathscr{H}_{\alpha}(U)]^t$, consisting of cochains $\phi$ such that $\mathscr{D}(z)\,\phi_{z_0', \dots, z_\nu}(z)\equiv 0$ for all $z_0, \dots, z_\nu$. By $[{}^{\nu}\mathscr{H}_{\mathscr{M}}]^t \cap \operatorname{Ker}\mathscr{D}$ we denote the subfamily in the family $[{}^{\nu}\mathscr{H}_{\mathscr{M}}]^t$, consisting of the subspaces $[{}^{\nu}\mathscr{H}_{\alpha}(U_{\alpha})]^t \cap \operatorname{Ker}\mathscr{D}$, $\alpha=1, 2, \dots$. It follows from the properties of the $p$-operator $\mathscr{D}$ that the image of the mapping (1) belongs to the subspace $[{}^{\nu}\mathscr{H}_{\alpha+\mu}(U)]^t \cap \operatorname{Ker}\mathscr{D}$. Therefore the image of the mapping (2) belongs to the subfamily $[{}^{\nu}\mathscr{H}_{\mathscr{M}}]^t \cap \operatorname{Ker}\mathscr{D}$.

**Theorem.** *For an arbitrary family of majorants $\mathscr{M}$ of type $\mathscr{I}$, and for an arbitrary $\mathscr{P}$-matrix p, the sequence*

$$[\mathscr{H}_{\mathscr{M}}]^s \xrightarrow{p} [\mathscr{H}_{\mathscr{M}}]^t \cap \operatorname{Ker}\mathscr{D} \to 0 \tag{3.5}$$

*is exact.*

The proof is postponed for the following construction.

**2°. Auxiliary norms and spaces.** Let $Z=(Z_1, \dots, Z_n)$ be an arbitrary point of the space $C^n$, and let $R=(R_1, \dots, R_n)$ be an arbitrary point of the cube $K=\{R\colon 0<R_i \leqq \frac{1}{2},\ i=1, \dots, n\}$. We shall denote by $e^{Z,R}$ the open ellipsoid in $C^n$, defined by the inequality

$$\sum \frac{|z-Z_i|^{\frac{1}{2}}}{R_i^2} < 1.$$

The $R_1, \dots, R_n$ will be called the semi-axes of the ellipsoid $e^{Z,R}$. For an arbitrary integer $\alpha$, we write

$$e_{\alpha}^{Z,R} = \left(1+\frac{1}{\alpha}\right) e^{Z,R},$$

and we denote by $U_{\alpha}^{Z,R}$ the elementary covering of $C^n$ with parameter zero and radius

$$\left(\frac{R}{|Z|+1}\right)^{\alpha} = \left(\frac{R_1 \dots R_n}{|Z|+1}\right)^{\alpha}.$$

For arbitrary $\nu$ and $\alpha$ we consider the norm

$$\|\phi\|_{\alpha}^{Z,R} = \sup_{z_0,\ldots,z_\nu} \sup_{U_{z_0}\cap\cdots\cap U_{z_\nu}\cap\Omega_\alpha} |\phi_{z_0,\ldots,z_\nu}(\zeta)|; \quad \{U_z\} = U_{\alpha}^{Z,R},$$

which is defined on the holomorphic cochains, of order $\nu$, on the covering $U_{\alpha}^{Z,R}\cap e_{\alpha}^{Z,R}$. We denote by ${}^{\nu}E_{\alpha}^{Z,R}$ the space of holomorphic cochains of order $\nu$ on the covering $U_{\alpha}^{Z,R}\cap e_{\alpha}^{Z,R}$, for which the norm is finite. In particular, ${}^{-1}E_{\alpha}^{Z,R}$ is the space of ordinary functions, holomorphic and bounded on $e_{\alpha}^{Z,R}$. We shall often use the more concise notation $E_{\alpha}^{Z,R}$ for this space.

For arbitrary $\nu \geqq -1$ we consider the space $(C^n\times K,\ {}^{\nu}E_{\alpha}^{Z,R})$[5] which consists of all functions $\Phi = \phi^{Z,R}$, defined on $C^n\times K$ and such that $\phi^{Z,R}\subset {}^{\nu}E_{\alpha}^{Z,R}$ for arbitrary $Z\in C^n$ and $R\in K$. The subspace consisting of functions for which the norm

$${}^{\pi}\|\Phi\|_{\alpha} = \sup_{Z,R} \left(\frac{R}{|Z|+1}\right)^{\alpha-1} \|\phi^{Z,R}\|_{\alpha}^{Z,R},$$

is finite will be denoted by ${}^{\nu}\Pi_{\alpha}$.

For a fixed $Z$ and $R$ the spaces ${}^{\nu}E_{\alpha}^{Z,R}$ form a family of relatively identical mappings. In fact, if $\alpha'>\alpha$ then $e_{\alpha'}^{Z,R}\subset e_{\alpha}^{Z,R}$ and the covering $U_{\alpha'}^{Z,R}$ is inscribed in $U_{\alpha}^{Z,R}$. The norm $\|\cdot\|_{\alpha}^{Z,R}$ is a nonincreasing function of $\alpha$. Accordingly, the spaces ${}^{\nu}\Pi_{\alpha}$ also form a family, which we denote by ${}^{\nu}\Pi$, and the norm ${}^{\pi}\|\cdot\|_{\alpha}$ is a non-increasing function of $\alpha$.

The formalism we have just developed will be applied in the following situation. Suppose that for arbitrary $Z$ and $R$ there are given certain subspaces $F^{Z,R}\subset {}^{\nu}E_{\alpha}^{Z,R}$ and $G^{Z,R}\subset {}^{\mu}E_{\alpha'}^{Z,R}$ and that there is defined an operator

$$L_{Z,R}\colon\ F^{Z,R}\to {}^{\mu}E_{\alpha'}^{Z,R}/G^{Z,R},$$

which is continuous in the natural topology. This series of operators yields the mapping

$$L^{\pi}\colon\ \Phi = \phi^{Z,R}\to \Psi = L_{Z,R}\,\phi^{Z,R},$$

which carries the functions defined on $C^n\times K$ with values in the spaces $F^{Z,R}$ into functions defined on the same set with values in the factor-spaces ${}^{\mu}E_{\alpha'}^{Z,R}/G^{Z,R}$. It is clear that the functions with values in the spaces ${}^{\mu}E_{\alpha'}^{Z,R}/G^{Z,R}$ can be looked on as elements of the factor-space $(C^n\times K,\ {}^{\mu}E_{\alpha'}^{Z,R})$ relative to the subspace $(C^n\times K,\ G^{Z,R})$. This correspondence is linear and biunique. Therefore, we may formulate the following statement.

**Proposition 1.** *Let $\mathscr{F}$ and $\mathscr{G}$ be subspaces in ${}^{\nu}\Pi_{\alpha}$ and ${}^{\mu}\Pi_{\alpha'}$ consisting of functions on $C^n\times K$, whose values belong to the corresponding subspaces*

5 $K$ is the cube introduced earlier.

*$F^{Z,R}$ and $G^{Z,R}$. The series of operators $L_{Z,R}$ defines the mapping*

$$L^\pi: \mathscr{F} \to {}^\mu\Pi_{\alpha'}/\mathscr{G}$$

*if and only if the inequality*

$$\|L_{Z,R}\| \leqq C\left(\frac{|Z|+1}{R}\right)^{\alpha'-\alpha}. \tag{4.5}$$

*is satisfied.*

*Proof.* The norm in the spaces ${}^\mu E_{\alpha'}^{Z,R}/G^{Z,R}$ and ${}^\mu\Pi_{\alpha'}/\mathscr{G}$ can be written by definition in the following form:

$$*\|\phi\|_{\alpha'}^{Z,R} = \inf\{\|\phi-\psi\|_{\alpha'}^{Z,R}, \psi \in G^{Z,R}\},$$
$$*\|\Phi\|_{\alpha'} = \inf\{{}^\pi\|\Phi-\Psi\|_{\alpha'}, \Psi \in \mathscr{G}\}^{-}.$$

Therefore the continuity of the operator $L^\pi$ implies that the following inequality is satisfied:

$$\begin{aligned} *\|L^\pi\Phi\|_{\alpha'} &= \inf_{\Psi\in\mathscr{G}} \sup_{Z,R}\left(\frac{R}{|Z|+1}\right)^{\alpha'-1} \|L_{Z,R}\,\phi^{Z,R}-\psi^{Z,R}\|_{\alpha'}^{Z,R} \\ &\leqq \|L^\pi\| \sup_{Z,R}\left(\frac{R}{|Z|+1}\right)^{\alpha-1} \|\phi^{Z,R}\|_{\alpha}^{Z,R}. \end{aligned} \tag{5.5}$$

where $\Phi = \phi^{Z,R}$, and $\Psi = \phi^{Z,R}$. Given arbitrary functions $\phi \in F^{Z,R}$ and $\psi \in G^{Z,R}$ we may consider the functions $\Phi \in \mathscr{F}$ and $\Psi \in \mathscr{G}$, which are equal to the corresponding $\phi$ and $\psi$ at the point $(Z, R)$ and zero at all other points. Inserting functions of this type in the inequality (5), we arrive at the inequality

$$\begin{aligned} &\inf\left\{\left(\frac{R}{|Z|+1}\right)^{\alpha'-1} \|L_{Z,R}\,\phi-\psi\|_{\alpha'}^{Z,R}, \psi \in G^{Z,R}\right\} \\ &\qquad \leqq \|L^\pi\|\left(\frac{R}{|Z|+1}\right)^{\alpha-1} \|\phi\|_{\alpha}^{Z,R}, \end{aligned} \tag{6.5}$$

whence

$$*\|L_{Z,R}\,\phi\|_{\alpha'}^{Z,R} \leqq \|L^\pi\|\left(\frac{|Z|+1}{R}\right)^{\alpha'-\alpha} \|\phi\|_{\alpha}^{Z,R},$$

which is equivalent to (4). We remark that the set of inequalities (6) with $Z \in C^n$, $R \in K$ is equivalent to the inequality (5). □

**3°. The coboundary operator.** It is easily seen from the formulae of § 3 that the norm of the operator $\partial$: ${}^\nu E_\alpha^{Z,R} \to {}^{\nu+1}E_\alpha^{Z,R}$, does not exceed unity. Accordingly, we may define the operator $\partial$: ${}^\nu\Pi_\alpha \to {}^{\nu+1}\Pi_\alpha$, whose norm also does not exceed unity. The set of all such operators defines a mapping of the families $\partial = \partial^\nu$: ${}^\nu\Pi \to {}^{\nu+1}\Pi$. The kernels of the mapping we consider in this subsection will be denoted respectively by ${}^\nu\mathscr{Z}_\alpha^{Z,R}$, ${}^\nu\mathscr{Z}_\alpha$ and ${}^\nu\mathscr{Z}$.

**Lemma 1.** *The sequence*

$$0 \to {}^{-1}\Pi \xrightarrow{\partial^{-1}} {}^0\Pi \xrightarrow{\partial^0} {}^1\Pi \to \cdots \to {}^\nu\Pi \xrightarrow{\partial^\nu} {}^{\nu+1}\Pi \to \cdots \tag{7.5}$$

*is exact.*

*Proof.* We arbitrarily fix $\alpha$, $\nu \geqq 0$, $Z$, and $R$. We choose an arbitrary cochain $\phi \in {}^{\nu+1}E_\alpha^{Z,R}$ such that $\partial\phi = 0$. By definition, the cochain $\phi$ is defined on $U_\alpha^{Z,R} \cap e_\alpha^{Z,R}$, where $U_\alpha^{Z,R}$ is an elementary covering of parameter zero and radius

$$\left(\frac{R}{|Z|+1}\right)^\alpha.$$

We apply Theorem 2 of § 4, in which we set

$$\omega = e_{\alpha+1}^{Z,R}, \qquad V = U_{\alpha+1}^{Z,R}, \qquad r = \left(\frac{R}{|Z|+1}\right)^{\alpha+1}.$$

The region $e_\alpha^{Z,R}$ contains the $r$-neighborhood of $\omega$, since

$$r \leqq 2^{-\alpha-1} \leqq \frac{1}{\alpha(\alpha+1)},$$

and the radius of the covering $U_\alpha^{Z,R}$ is not less than $2r$. According to this theorem, there exists a holomorphic cochain $\psi$ on the covering $U_{\alpha+1}^{Z,R} \cap e_{\alpha+1}^{Z,R}$ such that $\partial\psi = \phi$, and

$$\|\psi\|_{\alpha+1}^{Z,R} \leqq \frac{c}{r^n} \|\phi\|_\alpha^{Z,R} = C\left(\frac{|Z|+1}{R}\right)^n \|\phi\|_\alpha^{Z,R}.$$

In view of Proposition 1, this implies that there is defined a continuous operator

$$L\colon {}^{\nu+1}\mathscr{Z}_\alpha \to {}^\nu\Pi_{\alpha+n}/{}^\nu\mathscr{Z}_{\alpha+n},$$

inverse to the operator $\partial$. □

**4°. The operator *p* in the family ${}^\nu\Pi$.** For simplicity of notation in the rest of this paragraph we will omit the square brackets and the index $k$ in expressions of the form $[{}^\nu E_\alpha^{Z,R}]^k$, $[{}^\nu\Pi_\alpha]^k$ and so on.

Multiplication by the matrix $p(z)$ is clearly a continuous operation in the space ${}^\nu E_\alpha^{Z,R}$ (in the complete notation we are, of course, talking about the operator $p$: $[{}^\nu E_\alpha^{Z,R}]^s \to [{}^\nu E_\alpha^{Z,R}]^t$). Here

$$\|p\phi\|_\alpha^{Z,R} \leqq \sup_{e_\alpha^{Z,R}} |p(z)| \, \|\phi\|_\alpha^{Z,R} \leqq C(|Z|+1)^\mu \|\phi\|_\alpha^{Z,R},$$

where the constant $C$ does not depend on $Z$ and $R$. Therefore, the operation of multiplying functions on $C^n \times K$ with values in the spaces ${}^vE_\alpha^{Z,R}$ by the matrix $p$ defines a continuous operator $p: {}^v\Pi_\alpha \to {}^v\Pi_{\alpha+\mu}$. The set of all such operators defines a family mapping $p: {}^v\Pi \to {}^v\Pi$.

We shall use ${}^vE_\alpha^{Z,R} \cap \operatorname{Ker} \mathscr{D}$, ${}^v\Pi_\alpha \cap \operatorname{Ker} \mathscr{D}$ to denote the subspace of the respective spaces $[{}^vE_\alpha^{Z,R}]'$, $[{}^v\Pi_\alpha]'$, consisting of the functions $\phi$, $\Phi = \phi^{Z,R}$ such that $\mathscr{D}\phi = 0$, $\mathscr{D}\phi^{Z,R} = 0$ for all $Z \in C^n$, $R \in K$. We use ${}^v\Pi \cap \operatorname{Ker} \mathscr{D}$ to denote the subfamily in $[{}^v\Pi]'$, consisting of the subspaces ${}^v\Pi_\alpha \cap \operatorname{Ker} \mathscr{D}$. We use the notation ${}^v\Pi \cap \operatorname{Ker} p\,\partial$, ${}^v\mathscr{Z} \cap \operatorname{Ker} \mathscr{D}$ and so forth, with a similar meaning.

**Lemma 2.** *The sequence*

$$ {}^0\Pi \cap \operatorname{Ker} p\,\partial \xrightarrow{p} {}^0\mathscr{Z} \cap \operatorname{Ker} \mathscr{D} \to 0 \tag{8.5} $$

*is exact.*

*Proof.* At first we describe the idea of the proof. Every element of the space ${}^0Z_\alpha \cap \operatorname{Ker} \mathscr{D}$ is by definition a function on $C^n \times K$, whose values are functions $\phi \in E_\alpha^{Z,R}$ such that $\mathscr{D}\phi \equiv 0$. We can write any such function $\phi$ in the form $p\,\mathscr{G}(z)\,\phi$, where $\mathscr{G}(z)\,\phi$ is an analytic function in the neighborhood of an arbitrary point $z \in e_{\alpha+1}^{Z,R}$. Thus, we have $\phi = p\,\psi$, where $\psi$ is a cochain of order zero, on some covering of $e_{\alpha+1}^{Z,R}$. Taking the restriction of $\psi$ on a suitable inscribed covering, we obtain $\psi \in {}^0E_{\alpha'}^{Z,R} \cap \operatorname{Ker} p\,\partial$, which leads us to our goal after we have estimated $|\psi|$.

In the proof itself, we shall use a rather more general construction, which we shall also use elsewhere.

We fix our attention on an arbitrary integer $\alpha$, arbitrary point $Z \in C^n$ and $R \in K$. Let $\phi$ be an arbitrary function in the space $E_\alpha^{Z,R}$, that is, a function that is bounded and holomorphic in $e_\alpha^{Z,R}$. Further, let $\mathscr{D}$ and $\mathscr{G}$ be $p$-operators, as constructed in Theorem 1, § 4, Chapter II for the case $m = n$. If we apply the decomposition (1.4) of this theorem to the function $\phi$ at an arbitrary point $z \in e_\alpha^{Z,R}$ we obtain the equation

$$ \phi(\zeta) = \chi_z(\zeta) + p(\zeta)\,\psi_z(\zeta), \qquad \chi_z = \mathscr{D}(z)\,\phi, \qquad \psi_z = \mathscr{G}(z)\,\phi. \tag{9.5} $$

We find bounds for the functions $\chi_z$, $\psi_z$ at the point $z \in e_{\alpha+1}^{Z,R}$. Since

$$ \rho(e_{\alpha+1}^{Z,R}, \complement\, e_\alpha^{Z,R}) > \varepsilon = \frac{R^1}{(\alpha+1)(\alpha+2)}, $$

every such point belongs to $e_\alpha^{Z,R}$ together with its $\varepsilon$-neighborhood, therefore the function $\phi$ is holomorphic and is bounded in the $\varepsilon$-neighborhood of the point $z$. According to Corollary I, § 4, Chapter II, the operators $\mathscr{D}(z)$ and $\mathscr{G}(z)$ carry functions that are holomorphic in the $\varepsilon$-neighborhood of $z$ into functions holomorphic in the $\varepsilon\, r_z$-neighborhood of that point, and their norms do not exceed $(\varepsilon^K r_z)^{-1}$, where $r_z = C\,\theta^q(z, \mathscr{N})$, and $\mathscr{N}$ is an algebraic stratification of $C^n$. It follows that for arbitrary $z \in e^{Z,R}_{\alpha+1}$ the functions $\chi_z$ and $\psi_z$ are holomorphic in the neighborhood $S_z$ of the point $z$ of radius $\varepsilon r_z$ and

$$\max\left[\sup_{S_z}|\chi_z|, \sup_{S_z}|\psi_z|\right] \leqq \frac{1}{\varepsilon^K r_z} \sup_{e^{Z,R}_{\alpha}} |\phi| = \frac{1}{\varepsilon^K r_z} \|\phi\|^{Z,R}_{\alpha}. \tag{10.5}$$

Now at every point $z \in C^n$, we construct the sphere $S_z$ of radius $\varepsilon r_z$. According to Proposition 1, § 3, we can inscribe in the covering $S = \{S_z, z \in C^n\}$ an elementary covering $V = \{V_z\}$ with some parameter $\rho$ and radius $b\,\varepsilon^q$, where the quantities $\rho$, $b$ and $q$ are independent of $\varepsilon$. Suppose that some sphere $V_z$ intersects the ellipsoid $e^{Z,R}_{\alpha+2}$. Since the covering $V$ is inscribed in the covering $S$, this sphere lies inside some sphere $S_{z'}$. The radius of the latter is equal to $\varepsilon r_{z'} \leqq \varepsilon \leqq \rho(e^{Z,R}_{\alpha+2}, \complement\, e^{Z,R}_{\alpha+1})$, hence the points $z$ and $z'$ belong to $e^{Z,R}_{\alpha+1}$. It follows that the covering $V \cap e^{Z,R}_{\alpha+2}$ is inscribed in the covering $\{S_z, z \in e^{Z,R}_{\alpha+1}\}$, and the radii of the sphere $V_z$, intersecting with $e^{Z,R}_{e+2}$, are not less than

$$b\,\varepsilon^q(|Z|+2)^{-\rho} \geqq b' \left(\frac{\varepsilon}{|Z|+1}\right)^{q'},$$

where $b' = 2^{-\rho} b$, and $q' = \max(q, \rho)$. Since $R^1 \leqq \frac{1}{2}$, we can find a sufficiently large constant $Q$, independent of $Z$ and $R$, such that

$$b' \left(\frac{\varepsilon}{|Z|+1}\right)^{q'} \geqq \left(\frac{R}{|Z|+1}\right)^{Q}.$$

But this inequality implies that the covering $U^{Z,R}_Q \cap e^{Z,R}_{\alpha+2}$ is inscribed in $V \cap e^{Z,R}_{\alpha+2}$ and *a fortiori* the covering $U^{Z,R}_{\alpha'} \cap e^{Z,R}_{\alpha'}$, where $\alpha' = \max(\alpha+2, Q)$, is inscribed in $V \cap e^{Z,R}_{\alpha+2}$, and, therefore, in $\{S_z, z \in e^{Z,R}_{\alpha+1}\}$. Accordingly, we have defined the restrictions $\chi'$, $\psi'$ of the cochains $\sum \chi_z S_z$ and $\sum \psi_z S_z$ on the covering $U^{Z,R}_{\alpha'} \cap e^{Z,R}_{\alpha'}$. Eq. (9) implies that

$$\partial\phi = \chi' + p\,\psi'. \tag{11.5}$$

We set $\chi' = \sum \chi'_z U_z$ and $\psi' = \sum \psi'_z U_z$, where $\{U_z\} = U^{Z,R}_{\alpha'} \cap e^{Z,R}_{\alpha'}$. If some region $U_z$ of this covering lies within the sphere $S_{z'}$, then the corresponding sphere of the covering $U^{Z,R}_Q$ also lies within $S_{z'}$ and, therefore,

$$\varepsilon r_{z'} \geqq \left(\frac{R}{|Z|+1}\right)^{Q}, \quad \text{whence} \quad r_{z'} \geqq \left(\frac{R}{|Z|+1}\right)^{Q},$$

since $\varepsilon \leqq 1$. So

$$\sup_{U_z}|\chi'_z| \leqq \sup_{S_{z'}}|\chi_{z'}| \leqq \left(\frac{|Z|+1}{R}\right)^{Q} r_{z'} \sup_{S_{z'}}|\mathscr{D}(z')\,\phi|. \tag{12.5}$$

Combining this inequality with (10), we obtain

$$\|\chi'\|_{\alpha'}^{Z,R} \leqq C' \left(\frac{|Z|+1}{R}\right)^{Q'} \|\phi\|_{\alpha}^{Z,R}.$$

We derive the inequality

$$\|\psi'\|_{\alpha'}^{Z,R} \leqq C'' \left(\frac{|Z|+1}{R}\right)^{Q'} \|\phi\|_{\alpha}^{Z,R}. \tag{13.5}$$

in a similar fashion. We must emphasize the fact that the constants $C'$, $C''$ and $Q'$ are independent of $Z$ and $R$.

We now prove the lemma. Let $\phi'$ be an arbitrary element of the space ${}^0Z_{\alpha}^{Z,R} \cap \operatorname{Ker} \mathscr{D}$. This can be written in the form $\partial\phi$, $\phi \in E_{\alpha}^{Z,R}$, where $\|\phi\|_{\alpha}^{Z,R} = \|\phi'\|_{\alpha}^{Z,R}$. If we apply Eq. (11) to the function $\phi$ we obtain $\phi = p\psi'$, since by hypothesis $\mathscr{D}\phi' \equiv 0$. Thus, we have constructed the continuous operator

$$L_{Z,R}\colon {}^0\mathscr{Z}_{\alpha}^{Z,R} \cap \operatorname{Ker} \mathscr{D} \ni \phi' \to \psi' \in {}^0E_{\alpha'}^{Z,R}$$

such that the composition $p\,L_{Z,R}$ is an identity mapping. It follows from the inequality (13) that the norm of this operator does not exceed

$$C'' \left(\frac{|Z|+1}{R}\right)^{Q'}.$$

Therefore, according to Proposition 1, the set of these operators defines a continuous operator $L^{\pi}\colon {}^0\mathscr{Z}_{\alpha} \cap \operatorname{Ker} \mathscr{D} \to {}^0\Pi_{\alpha'}$, such that the composition $p\,L^{\pi}$ is an identity operator. It follows from the equations $\partial p\, L^{\pi} = p\,\partial L^{\pi} = 0$ that the image of $L^{\pi}$ belongs to the kernel of the operator $p\,\partial_0$, that is, the subspace ${}^0\Pi_{\alpha'} \cap \operatorname{Ker} p\,\partial$. Since the operator $L^{\pi}$ was constructed for an arbitrary integer $\alpha$, the exactness of the sequence (8) has been proved. □

**5°. Lemma 3.** *For arbitrary* $v \geqq -1$ *the sequence*

$${}^v\Pi \xrightarrow{p} {}^v\Pi \cap \operatorname{Ker} \mathscr{D} \to 0 \tag{14.5}$$

*is exact.*

*The proof* will be carried through by induction on the cohomological dimension $\delta(\operatorname{Ker} p)$ of the $\mathscr{P}$-module $\operatorname{Ker}\{p\colon \mathscr{P}^s \to \mathscr{P}^t\}$[6]. We set $\delta = \delta(\operatorname{Ker} p)$ and we assume that the lemma has been proved for all matrices $q$ with $\delta(\operatorname{Ker} q) < \delta$ (if $\delta = -1$, that is $\operatorname{Ker} p = 0$, then the hypothesis is not required). The definition of the number $\delta$ implies the existence of an exact sequence of $\mathscr{P}$-mappings of the form

$$0 \to \mathscr{P}^{s_\delta} \xrightarrow{q_\delta} \mathscr{P}^{s_{\delta-1}} \xrightarrow{q_{\delta-1}} \cdots \xrightarrow{q_1} \mathscr{P}^{s_0} \xrightarrow{q_0} \mathscr{P}^{s} \xrightarrow{p} \mathscr{P}^{t}. \tag{15.5}$$

6 The function $\delta(M)$ was defined in Chapter I, § 3.

We remark that by the induction hypothesis, the lemma holds for an arbitrary matrix $q_i$, $i=0, \ldots, \delta$, since $\delta(\operatorname{Ker} q_i)<\delta$.

We now consider the commutative diagram (16.5) below. We are at the moment interested only in a selected portion of this diagram.

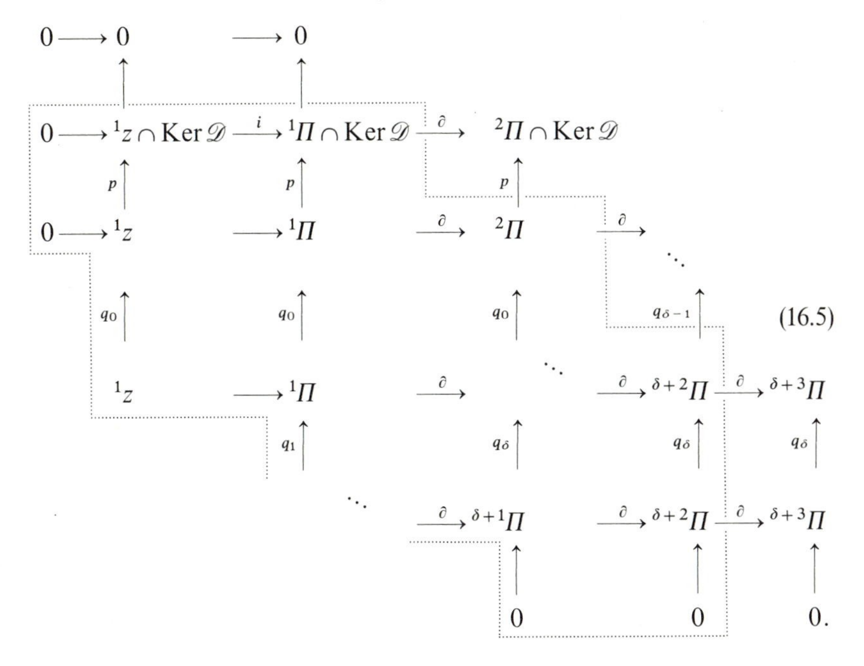

(16.5)

We are to prove the exactness of the left column of this portion. Let $\mathscr{D}^j$ be a $q_j$-operator, where $j=0, \ldots, \delta+1$, and suppose that $q_{\delta+1}=0$. By Corollary 2, § 4, Chapter II, we have for an arbitrary point $Z \in C^n$ the equation $\mathcal{O}_z^{s_{j-1}} \cap \operatorname{Ker} \mathscr{D}^j = \mathcal{O}_z^{s_{j-1}} \cap \operatorname{Ker} q_{j-1}$[7] where the notations are self-evident. It follows that for an arbitrary $v$ ${}^v\Pi \cap \operatorname{Ker} \mathscr{D}^j = {}^v\Pi \cap \operatorname{Ker} q_{j-1}$. Since by the induction hypothesis, our lemma is proved for all matrices $q_j$, it follows that all the sequences

$$ {}^v\Pi \xrightarrow{q_j} {}^v\Pi \cap \operatorname{Ker} q_{j-1} \to 0, \qquad j=0, \ldots, \delta+1 \;\; (q_{-1}=p) \qquad (17.5)$$

are exact.

So we have proved that all the columns of diagram (16), beginning with the second, are algebraically exact, and all the mappings $q_j$, $j=0, \ldots, \delta$, of these columns are homomorphisms. The exactness of the rows in the diagram follows from Lemma 1. Applying Theorem 1 of § 2, Chapter I, we find that the left column is algebraically exact, and the mapping $q_0$, which occurs in it, is a homomorphism.

---

7 $\mathcal{O}_z$ is the space of analytic functions in the neighborhood of $z$ (see Chapter II, § 1).

We shall now consider a new commutative diagram:

$$
\begin{array}{ccccccccc}
0 \longrightarrow & 0 & \longrightarrow & 0 & & & & & \\
 & \uparrow & & \uparrow & & & & & \\
0 \longrightarrow & {}^{-1}\Pi \cap \operatorname{Ker} \mathscr{D} & \xrightarrow{\partial^{-1}} & {}^{0}z \cap \operatorname{Ker} \mathscr{D} & \xrightarrow{\partial^{0}} & 0 & & & \\
 & p \uparrow & & p \uparrow & & \uparrow & & & \\
 & {}^{-1}\Pi & \xrightarrow{\partial^{-1}} & {}^{0}\Pi \cap \operatorname{Ker} p\,\partial & \xrightarrow{\partial^{0}} & {}^{1}z \cap \operatorname{Ker} p & \longrightarrow & 0 & \\
 & & & q_0 \uparrow & & q_0 \uparrow & & & \\
 & & & {}^{0}\Pi & \xrightarrow{\partial^{0}} & {}^{1}z & \longrightarrow & 0. &
\end{array}
\tag{18.5}
$$

By establishing the exactness of the left column, we establish our lemma for the case $v=-1$. Lemma 1 implies that the second and fourth rows are exact. We shall prove the exactness of the third row. The mapping $\partial^{-1}$ in this row coincides with the mapping $\partial^{-1}$ of the sequence (7) and is therefore a homomorphism. The kernel of the mapping $\partial^0$ coincides with the subfamily ${}^0Z$. Therefore the algebraic exactness of the third row, at the term ${}^0\Pi \cap \operatorname{Ker} p\,\partial$ follows from the exactness of (7). The exactness of (7) also implies that the mapping $\partial^0$ is an epimorphism. It remains to show that this mapping is a homomorphism. To this end we consider another commutative diagram:

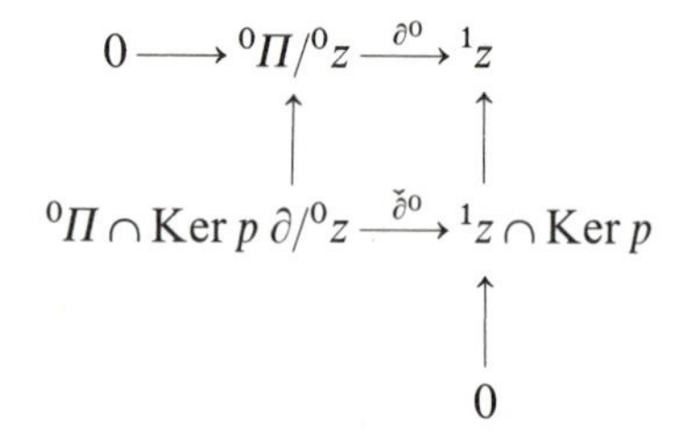

All the mappings in this diagram, with the possible exception of the mapping $\breve{\partial}^0$ are homomorphisms. If we apply Statement B) of Lemma 1, § 2, Chapter I, we see that the mapping $\breve{\partial}^0$ is also a homomorphism. Thus, the exactness of the third row of diagram (18) has been completely proved.

Let us consider the columns of diagram (18). The exactness of the third column follows from the properties of the left column of diagram (16). The exactness of the second column at the term ${}^0Z \cap \operatorname{Ker} \mathscr{D}$ follows from Lemma 2, and its exactness at the term ${}^0\Pi \cap \operatorname{Ker} p\,\partial$ follows from the exactness of the sequence (17) with $j=0$. Thus, we again have the right to apply Theorem 1, § 2, Chapter I, from which follows the exactness of the first column. With this, we have proved the exactness of (14) with $v=-1$.

Now, we establish the exactness of the sequence (14) for arbitrary $v \geqq 0$. The exactness of (14) with $v=-1$ implies that there exists a continuous operator

$$B\colon {}^{-1}\Pi_1 \cap \operatorname{Ker} \mathscr{D} \to {}^{-1}\Pi_a / {}^{-1}\Pi_a \cap \operatorname{Ker} p, \qquad a \geqq 1,$$

such that the composition $pB\colon {}^{-1}\Pi_1 \cap \operatorname{Ker} \mathscr{D} \to {}^{-1}\Pi_{a+\mu} \cap \operatorname{Ker} \mathscr{D}$ is an identity mapping. In accordance with Proposition 1, the operator $B$ generates a series of operators

$$B_{Z,R}\colon E_1^{Z,R} \cap \operatorname{Ker} \mathscr{D} \to E_a^{Z,R} / E_a^{Z,R} \cap \operatorname{Ker} p$$

such that the composition $p B_{Z,R}$ is an identity mapping for arbitrary $Z$ and $R$, and

$$\|B_{Z,R}\| \leqq C \left( \frac{|Z|+1}{R} \right)^{a-1}. \tag{19.5}$$

We arbitrarily choose $v \geqq 0$, $\alpha$, $Z$ and $R$. Let the integer $b$ be so chosen so that

$$2^{-b} \leqq \frac{b}{3\alpha(\alpha+b)}.$$

Then

$$\left( \frac{R}{|Z|+1} \right)^{\alpha+b} \leqq \tfrac{1}{3} \rho(e_{\alpha+b}^{Z,R}, \complement\, e_\alpha^{Z,R}),$$

since $R^1 \leqq \frac{1}{2}$. Therefore, if the sphere $V_z$ of the covering $U_{\alpha+b}^{Z,R}$ intersects the ellipsoid $e_{\alpha+b}^{Z,R}$, the sphere $2V_z$ lies within $e_\alpha^{Z,R}$. Since $b \geqq 2$, we have $V_z \subset \frac{1}{4} U_z$, where $U_z$ is a sphere of the covering $U_\alpha^{Z,R}$. Let $z_0, \ldots, z_v$ be an arbitrary point such that the intersection $V_{z_0} \cap \cdots \cap V_{z_v} \cap e_{\alpha+b}^{Z,R}$ is not empty. From what we have said earlier, we can derive the inclusion

$$V_{z_0} \cap \cdots \cap V_{z_v} \cap e_{\alpha+b}^{Z,R} \subset V_{z_0} \subset e_a^{z_0,r} \subset e_1^{z_0,r} = 2V_{z_0} \subset U_{z_0} \cap \cdots \cap U_{z_v} \cap e_\alpha^{Z,R},$$

where

$$r = r(U_{\alpha+b}^{Z,R}) = \left( \frac{R}{|Z|+1} \right)^{\alpha+b}.$$

Let $\phi = \sum \phi_{z_0,\ldots,z_v} U_{z_0} \wedge \cdots \wedge U_{z_v}$ be an arbitrary element of the space ${}^vE_\alpha^{Z,R} \cap \operatorname{Ker} \mathscr{D}$. The restriction of the function $\phi_{z_0,\ldots,z_v}$ on the sphere $e_1^{z_0,r}$ is an element of the space $E_1^{z_0,r} \cap \operatorname{Ker} \mathscr{D}$. If we apply to this element the operator $B_{z_0,r}$, we obtain a coset belonging to $E_a^{z_0,r}/E_a^{z_0,r} \cap \operatorname{Ker} p$. Let $\psi_{z_0,\ldots,z_v}$ be an arbitrary element of this coset. From (19) we have

$$\begin{aligned} &\inf \left\{ \sup_{e_a^{z_0,r}} |\psi_{z_0,\ldots,z_v} - \chi|,\ \chi \in E_a^{z_0,r} \cap \operatorname{Ker} p \right\} \\ &\quad \leqq C \left( \frac{|z_0|+1}{r} \right)^{a-1} \sup_{e_1^{z_0,r}} |\phi_{z_0,\ldots,z_v}|. \end{aligned} \tag{20.5}$$

Restricting each of the functions $\psi_{z_0,\ldots,z_\nu}$ to the region $V_{z_0}\cap\cdots\cap V_{z_\nu}\cap e^{Z,R}_{\alpha+b}$, we construct the cochain $\psi=\sum\psi_{z_0,\ldots,z_\nu}V_{z_0}\wedge\cdots\wedge V_{z_\nu}$ on the covering $U^{Z,R}_{\alpha+b}\cap e^{Z,R}_{\alpha+b}$. (We can, of course, suppose that the function $\psi_{z_0,\ldots,z_\nu}$ is skew-symmetric in its indices.) By the properties of the operator $B_{z_0,r}$

$$p\,\psi=\phi. \tag{21.5}$$

The inequality (20) implies that

$$\inf\{\|\psi-\chi\|^{Z,R}_{\alpha+b},\ \chi\in{}^{\nu}E^{Z,R}_{\alpha+b}\cap\operatorname{Ker}p\}\leqq C\sup_{e^{Z,R}_{\alpha}}\left(\frac{|z_0|+1}{r}\right)^{a-1}\|\phi\|^{Z,R}_{\alpha}. \tag{22.5}$$

We remark that on the left side of this inequality, we have the norm of the cochain $\psi$ as an element of the factor-space ${}^{\nu}E^{Z,R}_{\alpha+b}/{}^{\nu}E^{Z,R}_{\alpha+b}\cap\operatorname{Ker}p$. We estimate the second factor on the right side. Since $|z_0-Z|\leqq 1$, we have $|z_0|+1\leqq 2(|Z|+1)$. Using this in (22) and substituting the expression for $r$, we arrive at the inequality

$$^{*}\|\psi\|^{Z,R}_{\alpha+b}\leqq C'\left(\frac{|Z|+1}{R}\right)^{q}\|\phi\|^{Z,R}_{\alpha}.$$

Hence, in view of Proposition 1, it follows that the series of operators $A_{Z,R}\colon\phi\to\psi$ defines a continuous operator

$$A^{\pi}\colon\ {}^{\nu}\Pi_{\alpha}\cap\operatorname{Ker}\mathscr{D}\to{}^{\nu}\Pi_{\alpha'},\qquad \alpha'=\alpha+b+q,$$

and it follows from (21) that the composition $p\,A^{\pi}$ is an identity mapping. With this the exactness of the sequence (14) is proved for arbitrary $\nu$. □

**6°. Corollaries of lemma 3.** We mention two corollaries which we shall use in the next chapter.

**Corollary 1.** *Let $r(z)$ be a power function on some algebraic stratification, not exceeding the function $r_z$, which appears in Corollary 1, § 4, Chapter II. Then for arbitrary $Z$ and $R$ and $\phi$, $\phi$ being holomorphic and bounded in $2e^{Z,R}$, we have the equation $\phi=\phi_{\mathscr{D}}+p\,\psi$, where $\phi_{\mathscr{D}}$ and $\psi$ are functions holomorphic in $e^{Z,R}$, and*

$$\begin{aligned}\sup_{e^{Z,R}}|\phi_{\mathscr{D}}|\leqq C\left(\frac{|Z|+1}{R}\right)^{q}\sup\Bigg\{r(z)\sup_{|\xi|\leqq R^{1}r(z)}|\mathscr{D}(z,\xi)\,\phi(z)|,\ z\in 2e^{Z,R},\ r(z)\\ \geqq c\left(\frac{R}{|Z|+1}\right)^{q}\Bigg\}\end{aligned} \tag{23.5}$$

*for some constants $C$, $c$ and $q$, independent of $Z$ and $R$.*

*Proof.* We consider diagram (16) as a whole. In view of Lemma 3, all of its columns, beginning with the second, are exact. It follows from

Lemma 1 that all of the rows are exact, beginning with the third. It is clear that the second row is algebraically exact, and the mapping $i$ is a homomorphism. Applying Theorem 1, § 2, Chapter I, we can establish the exactness of the left column.

We consider a new commutative diagram:

$$\begin{array}{ccccccc}
0 & & 0 & & & & \\
\uparrow & & \uparrow & & & & \\
{}^0z & \longrightarrow & {}^0\Pi \cap \operatorname{Ker} \mathscr{D}\,\partial / {}^0\Pi \cap \operatorname{Ker} \mathscr{D} & \longrightarrow & 0 & & \\
\uparrow & & \uparrow & & \uparrow & & \\
{}^0z & \longrightarrow & {}^0\Pi \cap \operatorname{Ker} \mathscr{D}\,\partial & \xrightarrow{\partial} & {}^1z \cap \operatorname{Ker} \mathscr{D} & \longrightarrow & 0 \\
 & & \uparrow{\scriptstyle p} & & \uparrow{\scriptstyle p} & & \\
 & & {}^0\Pi & \xrightarrow{\partial} & {}^1z & \longrightarrow & 0
\end{array}$$

The exactness of the third column of this diagram follows from the exactness of the left column of diagram (16). The exactness of the second column follows from Lemma 3. The exactness of the third row follows from Lemma 1. We consider the second row. Its algebraic exactness at the term ${}^0\Pi$ Ker $\mathscr{D}\,\partial$ is obvious. The mapping $\partial$ in the second row is an epimorphism since by Lemma 1, $\partial({}^0\Pi \cap \operatorname{Ker} \mathscr{D}\,\partial) = {}^1\mathscr{Z} \cap \operatorname{Ker} \mathscr{D}$. The mapping $\partial$ is a homomorphism since the mapping $\partial^0$ in the sequence (7) is a homomorphism. Thus, we have established the exactness of the second row. Applying Theorem 1, § 2, Chapter I, we conclude that the first row is also exact. From the exactness of the first row, it follows that for arbitrary $\alpha$ there is a continuous operator

$$I\colon\ {}^0\Pi_\alpha \cap \mathscr{D}\,\partial / {}^0\Pi_\alpha \cap \operatorname{Ker} \mathscr{D} \to {}^0\mathscr{Z}_\beta / {}^0\mathscr{Z}_\beta \cap \operatorname{Ker} \mathscr{D}, \qquad \beta = \beta(\alpha),$$

inverse to the identity mapping. Arguments like those of Proposition 1 show that the operator I generates a series of operators

$$I_{Z,R}\colon\ {}^0E_\alpha^{Z,R} \cap \operatorname{Ker} \mathscr{D}\,\partial / {}^0E_\alpha^{Z,R} \cap \operatorname{Ker} \mathscr{D} \to {}^0\mathscr{Z}_\beta^{Z,R} / {}^0\mathscr{Z}_\beta^{Z,R} \cap \operatorname{Ker} \mathscr{D}, \tag{24.5}$$

whose norms are bounded by functions of the type

$$C\left(\frac{|Z|+1}{R}\right)^q.$$

Let $\phi$ be an arbitrary function belonging to $E_1^{Z,R}$. We apply to it the arguments of Lemma 2. From the decomposition (11), we find $\partial\chi' = -p\,\partial\psi'$. Therefore the cochain $\chi'$ belongs to the space ${}^0E_{\alpha'}^{Z,R} \cap \operatorname{Ker} \mathscr{D}\,\partial$. Now in (24), we replace $\alpha$ by $\alpha'$ and we consider the cochain $\chi'$ as an element of the factor-space in the left side. Let $\phi' \in {}^0\mathscr{Z}_{\beta'}^{Z,R}$, $\beta' = \beta(\alpha')$ be a cochain running over the image $\chi'$ under the mapping $I_{Z,R}$. From the bound on the norm of the operator $I_{Z,R}$ we derive the inequality

$$\inf \|\phi'\|_{\beta'}^{Z,R} \leqq C \left(\frac{|Z|+1}{R}\right)^q \|\chi'\|_{\alpha}^{Z,R}. \tag{25.5}$$

The properties of the operator $I_{Z,R}$ imply the equation

$$\mathscr{D}(\partial\phi - \phi') = 0. \tag{26.5}$$

It follows from Eq. (25) that for arbitrary $\varepsilon > 0$, we can find a cochain $\phi' \in I_{Z,R}\chi'$ such that

$$\|\phi'\|_{\beta'}^{Z,R} \leqq C \left(\frac{|Z|+1}{R}\right)^q \|\chi'\|_{\alpha'}^{Z,R} + \varepsilon. \tag{27.5}$$

If the first term on the right side is different from zero, we can suppose that $\varepsilon$ is equal to it. Then we obtain the inequality

$$\|\phi'\|_{\beta'}^{Z,R} \leqq 2C \left(\frac{|Z|+1}{R}\right)^q \|\chi'\|_{\alpha'}^{Z,R}. \tag{28.5}$$

Let us suppose that the first term on the right side of (27) is equal to zero. Then from (11), we have $\partial\phi = p\psi'$, and, accordingly, $\mathscr{D}(z)\,\phi(z) \equiv 0$ in $e_{\alpha'}^{Z,R}$. Then we may set $\phi' \equiv 0$, which again leads to (26) and (28). We denote by $\phi_{\mathscr{D}} \in E_{\beta'}^{Z,R}$ a function such that $\phi' = \partial\phi_{\mathscr{D}}$.

We find a bound for $\|\chi'\|_{\alpha'}^{Z,R}$ with the aid of the inequality (12). We remark that in the construction of the decomposition, we can replace the function $r_z$ by $r(z)$, since by hypothesis $r(z)$ has those properties of $r_z$ which we used in 4°. Therefore, we can derive from (12) the inequality

$$\|\chi'\|_{\alpha'}^{Z,R} \leqq C \left(\frac{|Z|+1}{R}\right)^Q$$
$$\times \sup\left\{r(z) \sup_{|\xi| \leqq \varepsilon r(z)} |\mathscr{D}(z,\xi)\,\phi(z)|,\ z \in e_1^{Z,R},\ r(z) \geqq \left(\frac{R}{|Z|+1}\right)^{Q}\right\}.$$

Together with (28) this inequality yields (23). It follows from (26) that $\phi - \phi_{\mathscr{D}} \in E_{\beta'}^{Z,R} \cap \operatorname{Ker}\mathscr{D}$. Therefore, in view of Lemma 3, $\phi - \phi_{\mathscr{D}} = p\psi$, where $\psi \in E_{\beta''}^{Z,R}$. □

**Corollary 2.** *Let $V$ be an elementary covering of parameter zero, and let $\phi$ be a cochain of order zero on this covering, and let it be holomorphic and bounded on some open sphere $S$, and let $\mathscr{D}\,\partial\phi = 0$. Then on the sphere $\frac{1}{2}S$ there exists a bounded holomorphic function $\psi$ such that $\mathscr{D}(\phi - \partial\psi) = 0$.*

*Proof.* The cochain $\phi$ belongs by hypothesis to the space ${}^0E_1^{z,r}\cap\operatorname{Ker}\mathscr{D}\,\partial$, where $z$ is the center of the sphere $S$, and $2r$ is its radius. Let $\psi'\in{}^0\mathscr{Z}_{\beta(1)}^{Z,R}$ be an arbitrary element of the coset $I_{z,r}\,\phi$. It is easy to see that the function $\psi\in E_{\beta(1)}^{z,r}$ such that $\partial\psi=\psi'$ is the function we are looking for. ▯

**7°. Lemma 4.** *For an arbitrary family of majorants $\mathscr{M}$ and for arbitrary integer $v\geqq 0$ the sequence*

$$[{}^v\mathscr{H}_{\mathscr{M}}]^s \xrightarrow{p} [{}^v\mathscr{H}_{\mathscr{M}}]^t\cap\operatorname{Ker}\mathscr{D}\to 0. \tag{29.5}$$

*is exact.*

*The proof* of this lemma is similar to the reasoning of 6°. We arbitrarily choose the integers $v\geqq 0$ and $\alpha$. We set $U_\alpha=\{U_z\}$ and $U_{\alpha+2}=\{V_z\}$. Let the points $z_0,\dots,z_v$ be such that the intersection $V_{z_0}\cap\dots\cap V_{z_v}\cap\Omega_{\alpha+2}$ is not empty. Since

$$r(U_{\alpha+2})\leqq\frac{1}{4}\,r(U_\alpha)=\frac{\varepsilon_\alpha}{4},$$

we have the inclusions

$$V_{z_0}\cap\dots\cap V_{z_v}\cap\Omega_{\alpha+2}\subset V_{z_0}\subset e_a^{z_0,r}\subset e_1^{z_0,r}=2\,V_{z_0}\subset U_{z_0}\cap\dots\cap U_{z_v}\cap\Omega_\alpha,$$
$$r=r(U_{\alpha+2}).$$

Let $\phi$ be an arbitrary cochain in the space ${}^v\mathscr{H}_\alpha(U_\alpha)\cap\operatorname{Ker}\mathscr{D}$ and let $\phi_{z_0,\dots,z_v}$ be the component of $\phi$ corresponding to the regions $U_{z_0},\dots,U_{z_v}$. We apply to its restriction on $e_1^{z_0,r}$ the operator $B_{z_0,r}$, constructed in subsection 5°. Let $\psi_{z_0,\dots,z_v}$ be a element of the coset $B_{z_0,r}\,\phi_{z_0,\dots,z_v}$. From Eq. (19), we have

$$\inf\left\{\sup_{e^{z_0,r}}|\psi_{z_0,\dots,z_v}-\chi|,\ \chi\in E_a^{z_0,r}\right\}\leqq C(|z_0|+1)^{a-1}\sup_{e_1^{z_0,r}}|\phi_{z_0,\dots,z_v}|.$$

Hence, since $r<\varepsilon_{\alpha+1}<\varepsilon_\alpha$,

$$\inf_\chi\sup\frac{|\psi_{z_0,\dots,z_v}-\chi|}{(|z|+1)^{a-1}M_{\alpha+2}(z)}\leqq C\sup\frac{|\phi_{z_0,\dots,z_v}|}{M_\alpha(z)}.$$

Setting $\psi=\sum\psi_{z_0,\dots,z_v}V_{z_0}\wedge\dots\wedge V_{z_v}$ (we choose $\psi_{z_0,\dots,z_v}$ so that this function is skew-symmetric in $z_0,\dots,z_v$), we arrive at the inequality

$$\inf\|\psi-\chi\|^0_{\alpha',U_\alpha},\qquad \alpha'=\alpha+a+1, \tag{30.5}$$

where the lower bound is taken over the cochain $\chi\in{}^v\mathscr{H}_{\alpha'}(U_{\alpha'})$ where $p\chi=0$. It follows from the properties of the operator $B_{z_0,r}$ that $p\psi=\phi$. This relation together with the inequality (30) proves the exactness of the sequence (29). ▯

**8°. Proof of the theorem.** Let $\mathscr{M}$ be a family of majorants of type $\mathscr{I}$. We consider the following commutative diagram

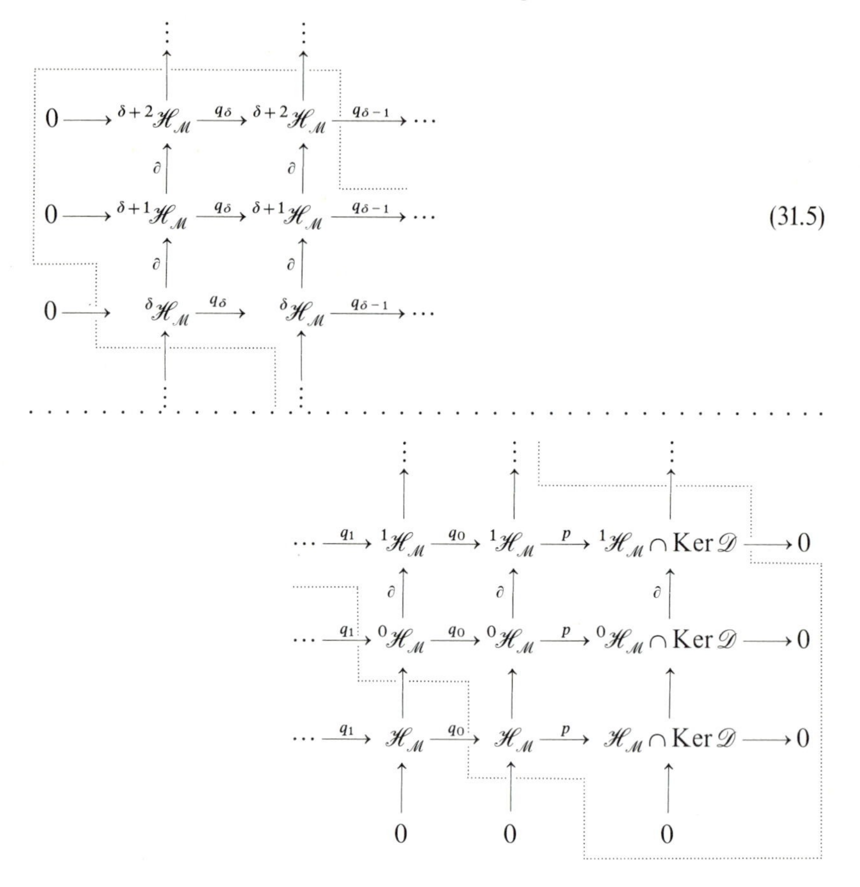

(31.5)

where the $q_j$, $j=0, \ldots, \delta$, are the $\mathscr{P}$-matrices from (15). Since for arbitrary $j=0, \ldots, \delta$ ${}^{v}\mathscr{H}_{\mathscr{M}} \cap \operatorname{Ker} \mathscr{D}^j = {}^{v}\mathscr{H}_{\mathscr{M}} \cap \operatorname{Ker} q_{j-1}$, the exactness of all the rows except the lowest follows from Lemma 4. The exactness of all columns, except the right column, follows from Corollary 2, § 4. It is clear that the right column is exact in the two lowest terms. Applying Theorem 1, § 2, Chapter I, to the selected portion, we establish the exactness of the lowest row in the last term. ▯

**9°. A remark and a corollary.** The mapping

$$\mathscr{H}_{\mathscr{M}} \cap \operatorname{Ker} \mathscr{D} \to \mathscr{H}_{\mathscr{M}} / \mathscr{H}_{\mathscr{M}} \cap \operatorname{Ker} p, \tag{32.5}$$

which is inverse to the mapping $p$ constructed in this theorem, has an order independent of the family of majorants $\mathscr{M}$. In fact, for arbitrary

$v \geqq 0$, the mapping $p^{-1}$: ${}^{v}\mathscr{H}_{\mathscr{M}} \cap \operatorname{Ker} \mathscr{D} \to {}^{v}\mathscr{H}_{\mathscr{M}}/{}^{v}\mathscr{H}_{\mathscr{M}} \cap \operatorname{Ker} p$, constructed in Lemma 4, has order $\alpha \to \alpha + a + 1$, which is independent of $\mathscr{M}$. Therefore, the analogous mappings $q_j^{-1}$ also have orders independent of $\mathscr{M}$. It follows from the remark of § 4 that the mappings inverse to $\partial$ in the columns of diagram (31) (with the exception of the last column), have orders independent of $\mathscr{M}$. Therefore, on the basis of the remark given in § 2, Chapter I, we conclude that the order of the mappings (32) is also independent of $\mathscr{M}$.

**Corollary 3.** *The right column of diagram* (31) *is exact.*

In fact, as we have already proved, all the remaining columns and all the remaining rows of this diagram are exact. In order to establish the exactness of the right column, it is sufficient to apply Theorem 1, § 2, Chapter I. □

Chapter IV

# The Fundamental Theorem

This chapter centers on the study of Noetherian operators: §§ 3 and 4. The role of these operators is defined by the theorem of § 3; the proof of the theorem is founded on the construction given in § 2. The results of these sections, together with the theorem on the triviality of cohomologies obtained in the preceding chapter, lead to Theorem 2 of § 5, which contains the description of the factor-space $\mathscr{H}_{\mathscr{M}}^{t}/p\,\mathscr{H}_{\mathscr{M}}^{s}$ in terms of Noetherian operators localized on the variety $N$. The fundamental theorem, given in § 5, represents the result of Theorem 2 in invariant form.

## § 1. Some properties of finite $\mathscr{P}$-modules

**1°. Primary decompositions.** In this subsection, we recall certain known definitions and theorems[1]. We continue the notation $\mathscr{P}$ for the ring of all polynomials with complex coefficients in the variables $z \in C^n$. Let $\mathscr{I}$ be an ideal in the ring $\mathscr{P}$. $N(\mathscr{I})$ denotes the algebraic variety in $C^n$, consisting of the common roots of the polynomials belonging to $\mathscr{I}$. If $N$ is an algebraic variety in $C^n$, then $\mathscr{I}(N)$ will denote the ideal in $\mathscr{P}$, consisting of all polynomials which vanish on $\mathscr{N}$. The ideal $\mathscr{I} \subset \mathscr{P}$

1 See, for example, Zarisky and Samuel [1].

is said to be prime, if $fg \in \mathscr{I}$, where $f$ and $g$ are polynomials, implies that at least one of the two polynomials belongs to $\mathscr{I}$. The variety $\mathscr{N}$ is irreducible if and only if the ideal $\mathscr{I}(N)$ is prime.

Let us now fix some finite $\mathscr{P}$-module $E$. Let $\mathfrak{p}$ be a submodule of $E$. The radical of the submodule $\mathfrak{p}$ is the ideal $\mathfrak{r}_E(\mathfrak{p})$ in $\mathscr{P}$, consisting of the polynomials $f$ which have the following property: there exists an integer $\rho$, such that $f^\rho F \in \mathfrak{p}$ for arbitrary $F \in E$. The algebraic variety $N(\mathfrak{p}) \overset{\text{def}}{=} N(\mathfrak{r}_E(\mathfrak{p}))$ is said to be associated with the submodule $\mathfrak{p}$. The submodule $\mathfrak{p}$ is said to be primary if the inclusion $fF \in \mathfrak{p}$, where $f \in \mathscr{P}$, and $F \in E$, implies that either $F \in \mathfrak{p}$, or $f \in \mathfrak{r}_E(\mathfrak{p})$. The radical of a primary submodule $\mathfrak{p}$ is a prime ideal, and accordingly, the variety $N(\mathfrak{p})$ associated with $\mathfrak{p}$ is irreducible. Every ideal $\mathscr{I}$ in the ring $\mathscr{P}$ can be looked on as a submodule of $\mathscr{P}$. The associated variety coincides with $N(\mathscr{I})$, introduced above.

Every submodule $\mathfrak{p} \subset E$ has at least one reduced primary decomposition, that is, it can be represented in the form of an intersection

$$\mathfrak{p} = \mathfrak{p}_0 \cap \cdots \cap \mathfrak{p}_l \tag{1.1}$$

of primary submodules $\mathfrak{p}_\lambda \subset E$, $\lambda = 0, \ldots, l$, which are such that none of them contains the intersections of the others and all the ideals $\mathfrak{r}(\mathfrak{p}_\lambda)$ are distinct. The ideals $\mathfrak{r}(\mathfrak{p}_\lambda)$ are defined uniquely by $\mathfrak{p}$. The set of varieties $N^\lambda \overset{\text{def}}{=} N(\mathfrak{p}_\lambda)$, $\lambda = 0, \ldots, l$, will be said to be associated with the submodule $\mathfrak{p} \subset E$. If the decomposition (1) is fixed, then the modules $\mathfrak{p}_\lambda$ are called the primary components of the submodule $\mathfrak{p}$.

Let $N$ be an algebraic variety in $C^n$. We write down the reduced primary decomposition of the ideal $\mathscr{I}(N)$, which we shall look on as a submodule of $\mathscr{P}$. Then $N = \bigcup N^\lambda$ is a decomposition of this variety into irreducible components.

Let $M$ be an arbitrary finite $\mathscr{P}$-module. The variety associated with the module $M$ is defined as the variety $N(M) \overset{\text{def}}{=} N(\mathfrak{r}_M(0))$, which is associated with the zero submodule of $M$. Let

$$0 = M_0 \cap \cdots \cap M_l$$

be a reduced primary decomposition of this submodule. From what we have said above, it follows that the set of varieties $N^\lambda = N(\mathfrak{r}_M(M_\lambda))$ associated with the submodules $M_\lambda$ of $M$, depends only on $M$. This ensemble will be said to be associated with $M$. The notions of varieties associated with submodules and with modules, are connected in the following way.

**Proposition 1.** *Let $M \cong E/\mathfrak{p}$, where $\mathfrak{p}$ is a submodule of $E$. Then the ensemble of varieties $N^\lambda$, associated with $M$, coincides with the ensemble of varieties associated with the submodule $\mathfrak{p} \subset E$.*

*Proof.* Let (1) be a reduced primary decomposition of the submodule $\mathfrak{p}$. Factoring this equation modulo $\mathfrak{p}$, we obtain:

$$0=\mathfrak{p}_0/\mathfrak{p}\cap\cdots\cap\mathfrak{p}_l/\mathfrak{p}$$

for the submodules of $M$. We show that this equation is a reduced primary decomposition of the zero submodule of $M$. We first establish the fact that every submodule $\mathfrak{p}_\lambda/\mathfrak{p}\subset M$ is primary. Let $f\,F\in\mathfrak{p}_\lambda/\mathfrak{p}$, where $f\in\mathscr{P}$, and $F\in M\setminus\mathfrak{p}_\lambda/\mathfrak{p}$. We choose a representative $F'\in F$. Since $f\,F'\in\mathfrak{p}_\lambda$, but $F'\bar{\in}\mathfrak{p}_\lambda$, we have $f\in\mathfrak{r}_E(\mathfrak{p}_\lambda)$. Conversely, every element of the ideal $\mathfrak{r}_E(\mathfrak{p}_\lambda)$, obviously, belongs to $\mathfrak{r}_M(\mathfrak{p}_\lambda/\mathfrak{p})$. Hence it follows that every submodule $\mathfrak{p}_\lambda/\mathfrak{p}$ is primary, and $\mathfrak{r}_M(\mathfrak{p}_\lambda/\mathfrak{p})=\mathfrak{r}_E(\mathfrak{p}_\lambda)$. It remains to prove that none of the modules $\mathfrak{p}_\lambda/\mathfrak{p}$ contains the intersection of any other. This follows from the fact that none of the modules $\mathfrak{p}_\lambda$ contains the intersection of others. □

**Proposition 2.** *Let $p\colon \mathscr{P}^s\to\mathscr{P}^t$ be an arbitrary $\mathscr{P}$-matrix, and let $M=\mathscr{P}^t/p\,\mathscr{P}^s$. Then the following four sets in $C^n$ coincide:*

1) *The variety $N(M)$, associated with $M$,*
2) *the set of points $z$, at which $M\otimes\mathcal{O}_z\neq 0$,*
3) *the set of points $z$, at which rank $p(z)<t$,*
4) *the set on which the p-operator $\mathscr{D}(z)$ $(m=n)$ differs from zero.*

*Proof.* These sets will be denoted respectively by $N_1$, $N_2$, $N_3$, $N_4$. We shall establish in turn the inclusions

$$N_1\supset N_2\supset N_3\supset N_4\supset N_1.$$

Before we begin the proof, we take note of the following isomorphisms:

$$M\otimes\mathcal{O}_z\cong\mathscr{P}^t\otimes\mathcal{O}_z/p\,\mathscr{P}^s\otimes\mathcal{O}_z\cong\mathcal{O}_z^t/p\,\mathcal{O}_z^s. \tag{2.1}$$

These follow from the results of 2°, § 3, Chapter I, since $\mathcal{O}_z$ is a flat $\mathscr{P}$-module (see Proposition 3, § 1, Chapter II).

We establish the first of the inclusions: $N_1\supset N_2$. Suppose $z\bar{\in}N_1$; then there exists a polynomial $f\in\mathscr{P}$, which is different from zero at the point $z$, and such that $f\,\mathscr{P}^t\subset p\,\mathscr{P}^s$. Since $f(z)\neq 0$ the function $f^{-1}$ belongs to $\mathcal{O}_z$. Therefore, for arbitrary $\phi\in\mathcal{O}_z^t$ we have

$$\phi=\phi f^{-1}\cdot f\subset p\,\mathcal{O}_z^s.$$

Hence, taking account of (2), we conclude that $M\otimes\mathcal{O}_z=0$, that is, $z\bar{\in}N_2$, which proves the first inclusion.

We now consider the second. Let $z\in N_3$. Then rank $p(z)<t$, and, accordingly, there exists a vector $\zeta\in C^t$, which is not representable in the form $p(z)\,\lambda$, where $\lambda\in C^s$. Considering this vector as an element of

the space $\mathcal{O}_z^t$, we conclude that it does not belong to the subspace $p\,\mathcal{O}_z^s$. Hence $\mathcal{O}_z^t \neq p\,\mathcal{O}_z^s$ and, accordingly, $M \otimes \mathcal{O}_z \neq 0$, that is $z \in N_2$, q.e.d.

We now prove the third inclusion: $N_3 \supset N_4$. Let $z \in N_4$; then $\mathcal{D}(z) \neq 0$ and, accordingly, the basis set of $\mathcal{I}(z)$ does not coincide with $Z_+^n$ (see Remark 2, § 4, Chapter II). Therefore the zero coefficient $\mathcal{D}_0(z)$ of the operator $\mathcal{D}(z)$ is different from zero. From Eq. (29.4), Chapter II, it follows that the functional $\mathcal{D}_0(z)$ has order zero, that is, it is equal to $v'\,\delta_0$, where $v \in C^t$. From the properties of the $p$-operator, it follows that $\mathcal{D}_0(z)\,p(z)\,\phi = 0$ for arbitrary $\phi \in \mathcal{O}_z^s$, whence $v'\,p(z) = 0$, which implies the inequality rank $p(z) < t$.

We now prove the last inclusion: $N_4 \supset N_1$. Let $z \,\bar{\in}\, N_4$, that is, $\mathcal{D}(z) = 0$. Suppose, further, that $\{e_\tau\}$ is a canonical basis of the module $\mathcal{P}^t$ (that is, the set of columns of the unit matrix). For arbitrary $\tau$ we have $\mathcal{D}(z)\,e_\tau = 0$, whence by Corollary 1, § 4, Chapter II, $e_\tau \in p\,\mathcal{O}_z^s$. It then follows from Proposition 4, § 1, Chapter II, that the columns $e_\tau$ belong to $p\,R_z^s$, where $R_z$ is a ring of rational functions in $C^n$ with denominator different from zero at the point $z$. Accordingly, we can choose the polynomial $f$, different from zero at $z$, such that $f\,e_\tau \in p\,\mathcal{P}^s$ for arbitrary $\tau$. Hence $f\,\mathcal{P}^t \subset p\,\mathcal{P}^s$, and accordingly, $f \in \mathrm{r}(M)$. Since $f(z) \neq 0$, the point $z$ does not belong to the variety $N(M)$, q.e.d. □

The set characterized in Proposition 2 will also be called the variety associated with the matrix $p$, and we denote it by $N(p)$. Thus, for an arbitrary $\mathcal{P}$-matrix $p$, the terms: variety associated with $p$, variety associated with the submodule $p\,\mathcal{P}^s \subset \mathcal{P}^t$, and variety associated with the module $M = \mathcal{P}^t / p\,\mathcal{P}^s$, all have the same meaning.

**Proposition 3** [2]**.** *Let $M$ and $L$ be arbitrary finite $\mathcal{P}$-modules. Then*

$$N(M \otimes L) = N(M) \cap N(L).$$

### 2°. The decomposition of a $p$-operator corresponding to the decomposition of a submodule $p\,\mathcal{P}^s$

**Proposition 4.** *Let the $\mathcal{P}$-matrices $p$, $p_1$ and $p_2$ of sizes $t \times s$, $t \times s_1$ and $t \times s_2$ be such that*

$$p\,\mathcal{P}^s = p_1\,\mathcal{P}^{s_1} \cap p_2\,\mathcal{P}^{s_2}. \tag{3.1}$$

*Further, let*

$$\mathcal{D},\ \mathcal{G};\ \mathcal{D}_1,\ \mathcal{G}_1;\ \mathcal{D}_2,\ \mathcal{G}_2;\ \mathcal{D}_{12},\ \mathcal{G}_{12}$$

*be operators, corresponding respectively to the $\mathcal{P}$-matrices $p$, $p_1$, $p_2$ and $p_1 \oplus p_2$ (with $m = n$). Then we have the equation*

$$\mathcal{D} = \Lambda_1\,\mathcal{D}_1 + \Lambda_2\,\mathcal{D}_2,$$

2 The proof follows from Serre [4], Proposition 10, Chapter I.

*where $\Lambda_1$ and $\Lambda_2$ are polynomials, with integer coefficients, in the operators $\mathscr{D}$, $\mathscr{D}_2$, $\mathscr{G}_{12}$, $p_1$, $p_2$ and the projection operator $\pi_1: \mathscr{S}^{s_1+s_2} \to \mathscr{S}^{s_1}$.*

*Proof.* We fix an arbitrary point $z \in C^n$ and we set $\mathscr{D} = \mathscr{D}(z)$, $\mathscr{G} = \mathscr{G}(z)$, and so on. Let $\phi$ be an arbitrary element of the space $\mathscr{S}^t$. From the equation

$$\phi - \mathscr{D}_1 \phi = p_1 \mathscr{G}_1 \phi = \mathscr{D}_2(\phi - \mathscr{D}_1 \phi) + p_2 \mathscr{G}_2(\phi - \mathscr{D}_1 \phi) \tag{4.1}$$

it is clear that the series $\mathscr{D}_2(\phi - \mathscr{D}_1 \phi)$ belongs to the arithmetic sum of the subspaces

$$p_1 \mathscr{S}^{s_1} + p_2 \mathscr{S}^{s_2} = (p_1 \oplus p_2) \mathscr{S}^{s_1+s_2},$$

and accordingly

$$\begin{aligned}\mathscr{D}_2(\phi - \mathscr{D}_1 \phi) &= (p_1 \oplus p_2) \mathscr{G}_{12} \mathscr{D}_2(\phi - \mathscr{D}_1 \phi)\\ &= p_1 \pi_1 \mathscr{G}_{12} \mathscr{D}_2(\phi - \mathscr{D}_1 \phi) + p_2 \pi_2 \mathscr{G}_{12} \mathscr{D}_2(\phi - \mathscr{D}_1 \phi).\end{aligned}$$

We substitute this expression in the right side of (4) and we move the term $p_1 \pi_1 \mathscr{G}_{12} \mathscr{D}_2(\phi - \mathscr{D}_1 \phi)$ to the left. As a result, we obtain the equation

$$\begin{aligned}\phi' &\overset{\text{def}}{=} \phi - \mathscr{D}_1 \phi - p_1 \pi_1 \mathscr{G}_{12} \mathscr{D}_2(\phi - \mathscr{D}_1 \phi)\\ &= p_1 \mathscr{G}_1 \phi - p_1 \pi_1 \mathscr{G}_{12} \mathscr{D}_2(\phi - \mathscr{D}_1 \phi)\\ &= p_2 \pi_2 \mathscr{G}_{12} \mathscr{D}_2(\phi - \mathscr{D}_1 \phi) + p_2 \mathscr{G}_2(\phi - \mathscr{D}_1 \phi),\end{aligned} \tag{5.1}$$

from which it follows that the series $\phi'$ belongs to the intersection $p_1 \mathscr{S}^{s_1} \cap p_2 \mathscr{S}^{s_2}$. Since the $\mathscr{P}$-module $\mathscr{S}$ is flat, it follows that $p \mathscr{S}^s = p \mathscr{P}^s \otimes \mathscr{S}$ for an arbitrary $\mathscr{P}$-matrix $p$ (see 2°, § 3, Chapter I). Therefore Eq. (3) and Proposition 4, § 3, Chapter I imply that the intersection $p_1 \mathscr{S}^{s_1} \cap p_2 \mathscr{S}^{s_2}$ coincides with $p \mathscr{S}^s$. Hence $\mathscr{D} \phi' = 0$. Therefore, applying the operator $\mathscr{D}$ to the left side of (5), we obtain the relation

$$\mathscr{D} \phi = (\mathscr{D} - \mathscr{D} p_1 \pi_1 \mathscr{G}_{12} \mathscr{D}_2) \mathscr{D}_1 \phi + (\mathscr{D} p_1 \pi_1 \mathscr{G}_{12}) \mathscr{D}_2 \phi,$$

which is the desired decomposition. ▯

**Corollary 1.** *Let the $\mathscr{P}$-matrices $p, p_0, \ldots, p_l$ be such that*

$$p \mathscr{P}^s = p_0 \mathscr{P}^{s_0} \cap \cdots \cap p_l \mathscr{P}^{s_l}.$$

*Then the p-operator $\mathscr{D}$ admits the following decomposition:*

$$\mathscr{D} = \sum \Lambda_\lambda \mathscr{D}^\lambda, \tag{6.1}$$

*where $\mathscr{D}^\lambda$, $\lambda = 0, \ldots, l$, is a $p_\lambda$-operator, and $\Lambda_\lambda$ is an operator in $\mathscr{S}^t$, having the following property: for an arbitrary $z \in C^n$ and $\delta$, $0 < \delta \leqq 1$, $\Lambda_\lambda(z)$ is a continuous operator from $\mathscr{O}_\delta^t$ to $\mathscr{O}_{\delta\rho}^t$ with a norm not exceeding $\delta^{-q} \rho^{-1}$,* *where $\rho=\rho(z)$ is some power function on an algebraic stratification, and $q\geqq 1$ is a constant.*

*Proof.* Repeated application of Proposition 4 leads to the decomposition (6), in which the $\Lambda_\lambda$ are certain polynomials in the operators corresponding to the matrices $p_\lambda, p_0\oplus\cdots\oplus p_\lambda$, $\lambda=0,\ldots,l$, in the matrices themselves and in the projection operators defined by the decomposition of $\mathscr{S}^{s_0+\cdots+s_l}$ into the direct sum $\mathscr{S}^{s_0}\oplus\cdots\oplus\mathscr{S}^{s_l}$. Each of these operators satisfies the condition: for arbitrary $\delta$, $0<\delta\leqq 1$, it acts from $\mathscr{O}^\tau_\delta$ to $\mathscr{O}^\sigma_{\delta\rho'}$, where $\rho'$ is a power function on some algebraic stratification, with norm not exceeding $\delta^{-q'}\rho'^{-1}$, where $q'$ is a constant. An arbitrary polynomial in such operators has a similar property in view of Proposition 3, § 3, Chapter II. ▯

**3°. The dimension of $\mathscr{P}$-modules and algebraic varieties.** Let $N\subset C^n$ be an algebraic variety. The number $d\geqq 0$ is called the dimension of $N$, if any $d+1$ polynomials $f_1,\ldots,f_{d+1}$ are algebraically dependent in the ring $\mathscr{P}/\mathscr{I}(N)$ (that is, if there exists a polynomial $F\neq 0$ in $d+1$ variables, such that $F(f_1,\ldots,f_{d+1})\equiv 0$) and if there exist $d$ polynomials which are not algebraically dependent in $\mathscr{P}/\mathscr{I}(N)$. If $\mathfrak{p}$ is a submodule of $E$, its dimension is that of the associated variety $N(\mathfrak{p})$. The submodule $\mathfrak{p}$ is said to be unmixed, if all its primary components have the same dimension. This definition does not depend on the choice of the decomposition (1), since there appear in it only the radicals of the primary components. The dimension of a finite $\mathscr{P}$-module $M$ is that of the associated variety $N(M)$.

We choose some integer $m$, lying between 1 and $n$, and we divide the variables $z$ into two groups $v=(z_1,\ldots,z_m)$ and $w=(z_{m+1},\ldots,z_n)$. The space $C^n$ will be represented as the direct product of the coordinate subspaces $C_v=C^m$ and $C_w=C^{n-m}$.

**Proposition 5.** *Let $N\subset C^n$ be an algebraic variety of dimension $d$. Then, either*

I) *$N$ belongs to a variety of the form $C_v\times\mu$, where $\mu$ is a proper subvariety of $C_w$, or*

II) *the variables $z$ can be divided into two groups $v'$ and $w'\in C^d$ such that the group $w'$ contains the group $w$, and there exists a proper algebraic subvariety $\delta\subset C_{w'}$ such that the set $N\setminus N_\delta$, where $N_\delta=N\cap(C_{v'}\times\delta)$, in the neighborhood of each of its points is the graph of a holomorphic function $v'=v'(w')$, and*

$$|\operatorname{grad} v'(w')|\leqq c\,\theta^q(w',\delta)^{3} \tag{7.1}$$

*for some $c$ and $q$ not depending on the point in question.*

3 For an arbitrary set $\Delta\subset C^n$, we set $\theta(\zeta,\Delta)=\rho(\zeta,\Delta)(|\zeta|+1)^{-2}$.

*Proof.* Let us suppose that the monomials $z_{m+1}, \ldots, z_n$ are algebraically dependent as elements of the ring $\mathscr{P}/\mathscr{I}(N)$. This means that there exists a polynomial $F(z_{m+1}, \ldots, z_n) \in \mathscr{I}(N)$, which is not identically zero. Then case I) holds.

Now let us suppose that the monomials $z_{m+1}, \ldots, z_n$ are algebraically independent in the ring $\mathscr{P}/\mathscr{I}(N)$. We add some of the monomials $z_1, \ldots, z_m$ (suppose for simplicity that these are the monomials $z_{h+1}, \ldots, z_m$) such that we obtain the maximal system of algebraically independent monomials. We write $v' = (z_1, \ldots, z_h)$ and $w' = (z_{h+1}, \ldots, z_n)$. Since the system $w'$ is maximal, $n - h = d$ and an arbitrary monomial $z_i$ with $i \leqq h$ depends algebraically on the monomials $w'$ in the ring $\mathscr{P}/\mathscr{I}(N)$, that is, there exists a polynomial $f_i(z_i, w')$, belonging to the ideal $\mathscr{I}(N)$, that is, vanishing on $N$. Therefore, the variety $N$ lies within the variety $N'$ defined by the system of equations

$$f_1(z_1, w') = \cdots = f_h(z_h, w') = 0, \qquad h = n - d. \tag{8.1}$$

Let $d_i(w')$ be the discriminant of the polynomial $f_i$ with respect to $z_i$. We may suppose that the polynomials $f_i$ have no multiple divisors, and therefore $d_i \not\equiv 0$. Let $\delta_i$ be the variety of the roots of $d_i$. We write $\delta' = \cup \delta_i$. We choose an arbitrary point $z$, belonging to the variety $N'$, and not lying in $C_{v'} \times \delta'$. The derivative of $f_i$ with respect to $z_i$ at this point does not vanish for any $i$. Therefore the vectors grad $f_i$, $i = 1, \ldots, h$, are linearly independent in the neighborhood of $z$. Therefore, by the theorem on implicit functions, the variety $N'$ is the graph of a holomorphic function $v' = v'(w') \overset{\text{def}}{=} (z_1(w'), \ldots, z_h(w'))$, in the neighborhood of the point $z$, and

$$\frac{\partial z_i(w')}{\partial z_j} = \frac{\partial f_i(z)}{\partial z_j} : \frac{\partial f_i(z)}{\partial z_i}, \qquad j = h+1, \ldots, n.$$

Since the denominator is different from zero on $N' \setminus (C_{v'} \times \delta)$, we can make use of Corollary 2, § 3, Chapter I, to find a bound for the right side of the form $C\,\theta^q(z, C_{v'} \times \delta')$. The identity of $f_i(z_i(w'), w') \equiv 0$ implies that the function $|z_i(w')|$ is bounded by a quantity of the form $C\,\theta^q(w', \delta')$. Hence

$$C\,\theta^q(z, C_{v'} \times \delta')|_{N'} \leqq C\,\theta^q(w', \delta'),$$

which leads to the inequality (12), if $\delta \supset \delta'$.

We now decompose the variety $N$ into irreducible components. As is known, an arbitrary irreducible variety of dimension $d$ is a regular surface of dimension $d$ at an arbitrary one of its points, with the exception of a proper subvariety. Let $N_i$ be a $d$-dimensional irreducible component of $N$, and $N_i^*$ be a subvariety consisting of the non-regular points of this component. From what we have said, it follows that the set $N_i \setminus N_i^*$ coincides at each of its points with the variety $N' \setminus (C_{v'} \times \delta')$. The pairwise intersection of the sets of the form $N_i \setminus N_i^*$ belongs to $C_{v'} \times \delta'$, since all the points of the variety $N' \setminus (C_{v'} \times \delta')$ are regular. If $N_i$ is an irreducible component of $N$ of dimension less than $d$, we write $N_i^* = N_i$. Let $\gamma_i$ be the projection of the variety $N_i^*$, on $C_{w'}$. We set $\gamma = \cup \gamma_i$ and $\delta = \delta' \cup \gamma$.

There remains to show that $\delta$ is a proper subvariety of $C_{w'}$. It will be sufficient to prove that the dimension of each of the varieties $N_i^*$ is less than $d$. In fact, if this is so, the monomials of $w'$ are algebraically dependent in the ring $\mathscr{P}/\mathscr{I}(N_i^*)$, and, accordingly, there exists a polynomial $g(w') \in \mathscr{I}(N_i^*)$, which is not identically zero, and therefore $\gamma_i \neq C_{w'}$. We shall show that the dimension of each of the varieties $N_i^*$ is less than $d$. If the dimension of the component $N_i$ is less than $d$, then by definition $N_i^* = N_i$ and, accordingly, our statement is proved. If the dimension of $N_i$ is equal to $d$, our assertion is implied by the following lemma.

**Lemma 1**[4]**.** *Let $L$ be an irreducible variety of dimension $d$. An arbitrary proper submanifold $K \subset L$ has dimension less than $d$.*

The variety $N_\delta$ will be called the discriminant with respect to the variables $v'$.

### 4°. Normally placed varieties and modules

**Definition.** We shall say that an algebraic variety $N \subset C^n$ is normally placed in a given system of coordinates in $C^n$, if it belongs to the variety of the roots of the system of equations

$$f_1(z_1, w') = \cdots = f_h(z_h, w') = 0, \qquad w' = (z_{h+1}, \ldots, z_n), \tag{9.1}$$

where $f_i$, $i = 1, \ldots, h$, is a polynomial, of which the coefficient of the term in the highest power of $z_i$, is equal to unity, and $d = n - h$ is the dimension of $N$. We shall say that the submodule $\mathfrak{p} \subset E$ is normally placed, if the variety $N(\mathfrak{p})$ is normally distributed.

In other words, the variety $N$ of dimension $d$ is normally placed in $C^n$, if the monomials $z_1, \ldots, z_h$ are algebraic integers relative to the $w'$ in the ring $\mathscr{P}/\mathscr{I}(N)$. We know that this situation can always be brought about if we choose a suitable system of coordinates, and therefore, for an arbitrary variety, we can find a system of coordinates in which it is normally placed. Moreover, the systems of coordinates that do not have this property form a nowhere dense subset[5].

**Proposition 6.** Let $\mathfrak{p} \subset E$ be a submodule, and let the variety $N(\mathfrak{p})$ belong to the variety (9). In order that the module $\mathfrak{p}$ be unmixed and of

4 See, for example, van der Waerden [1].

5 See, for example, Zarisky and Samuel [1].

dimension $d$, it is necessary and sufficient that for an arbitrary polynomial $q(w')\not\equiv 0$ the conditions $qF\in\mathfrak{p}$, $F\in E$ should imply $F\in\mathfrak{p}$.

*Proof. Sufficiency*. We shall suppose that some primary component $\mathfrak{p}_\lambda$ of the module $\mathfrak{p}$ has dimension less than $d$. Then there exists a polynomial $q(w')\in\mathfrak{r}(\mathfrak{p}_\lambda)$, which is not identically zero. Since the decomposition (1) is reducible, the set $\bigcap_{\nu\neq\lambda}\mathfrak{p}_\nu\setminus\mathfrak{p}_\lambda$ is not empty. Let $F$ be an element of this set. It follows from the definition of the ideal $\mathfrak{r}(\mathfrak{p}_\lambda)$ that $q^\rho F\in\mathfrak{p}$ for some integer $\rho$. Hence $q^\rho F\in\mathfrak{p}$, while $F\bar{\in}\mathfrak{p}$. The sufficiency is thereby established.

*Necessity*. We shall suppose that there exists a polynomial $q(w')\not\equiv 0$ and an element $F\in E\setminus\mathfrak{p}$ such, that $qF\in\mathfrak{p}$. Suppose for example that $F\bar{\in}\mathfrak{p}_l$. Then $q\in\mathfrak{r}(\mathfrak{p}_l)$, since the module $\mathfrak{p}_l$ is primary. This means that the monomials $w'$ are algebraically dependent in the ring $\mathscr{P}/\mathfrak{r}(\mathfrak{p}_l)$. On the other hand, by hypothesis, the monomials $z_i$, $i\leqq h$, depend algebraically on the monomials $w'$ in this ring. Therefore, the dimension of the ideal $\mathfrak{r}(\mathfrak{p}_l)$ is less than $d$. ☐

**5°. The restriction of modules on subspaces in $C^n$.** We shall denote by $\mathscr{P}_v$, $\mathscr{P}_{v'}$ and so on, the rings of polynomials with complex coefficients in the variables $v$, $v'$, and so forth. In particular $\mathscr{P}_z=\mathscr{P}$. For an arbitrary point $W\in C_w$ we may define the mapping

$$\mathscr{P}^t\ni F\to F_W\in\mathscr{P}_v^t, \tag{10.1}$$

which relates the vector polynomial $F$ to its restriction on the subspace $w=W$. This is a mapping of rings when $t=1$ and of modules on these rings in the general case. If $\mathfrak{p}$ is a submodule of $\mathscr{P}^t$, we denote by $\mathfrak{p}_W$ its image under the mapping (10). It is clear that $\mathfrak{p}_W$ is a submodule of $\mathscr{P}_v^t$. If $p$: $\mathscr{P}^s\to\mathscr{P}^t$ is a $\mathscr{P}$-matrix, then we shall denote by $p_W$ the $\mathscr{P}_v$-matrix which is the restriction of $p$ on the subspace $w=W$. We shall suppose that $\mathfrak{p}=p\,\mathscr{P}^s$, and then $\mathfrak{p}_w=p_w\,\mathscr{P}_v^s$.

**Lemma 2.** *For arbitrary $W$, the variety $N(\mathfrak{p}_W)$ coincides with the intersection of the variety $N(\mathfrak{p})$ and the subspace $w=W$.*

*Proof.* For $t=1$, the kernel of the mapping (10) is the ideal $\mathscr{I}$, consisting of the polynomials that vanish on the subspace $w=W$, and therefore, this mapping is an isomorphism of the rings $\mathscr{P}/\mathscr{I}$ and $\mathscr{P}_v$. For arbitrary $t$ the mapping (10) establishes an isomorphism of the modules $\mathscr{P}^t/\mathscr{I}^t$ and $\mathscr{P}_v^t$ over these rings. Therefore the image of the natural mapping $\mathfrak{p}\to\mathscr{P}^t/\mathscr{I}^t$ is isomorphic to the module $\mathfrak{p}_W$. On the other hand, this image is equal to $[\mathfrak{p}+\mathscr{I}^t]/\mathscr{I}^t$. Accordingly, the variety associated with this submodule is equal to $N(\mathfrak{p}_W)$. We write $E=\mathscr{P}^t/\mathscr{I}^t$ in Proposition 1, and we obtain the equation $N(\mathfrak{p}_W)=N(M)$, where

$$M=[\mathscr{P}^t/\mathscr{I}^t]/[\mathfrak{p}+\mathscr{I}^t/\mathscr{I}^t]\cong\mathscr{P}^t/[\mathfrak{p}+\mathscr{I}^t].$$

By Proposition 3, § 3, Chapter I, this module is equal to $\mathscr{P}^t/\mathfrak{p}\otimes\mathscr{P}/\mathscr{I}$. It follows from Proposition 3 that $N(M)$ is the intersection of the variety $N(\mathscr{P}^t/\mathfrak{p})=N(\mathfrak{p})$ and the variety associated with the ideal $\mathscr{I}$. The latter clearly coincides with the subspace $w=W$. □

**Theorem 1.** *Let $\mathfrak{p}$ be an unmixed submodule of the module $\mathscr{P}^t$, of dimension $d=n-h$, $h\leqq m$, and let it be normally placed in a given system of coordinates. In the space $C_w$ there exists a set $\mathscr{M}$ of first category, such that if $w\bar{\in}\mathscr{M}$, the submodule $\mathfrak{p}_W\subset\mathscr{P}_v^t$ is unmixed and of dimension $m-h$, and is normally placed in $C_v$.*

*Proof.* Since by hypothesis the variety $N(\mathfrak{p})$ belongs to a variety (8), it follows from Lemma 2 that the manifold $N(\mathfrak{p}_W)$ belongs to the variety of solutions of the system

$$f_1(z_1, w'', w)=\cdots=f_h(z_h, w'', w)=0, \qquad w=\text{const},$$

where $w''=(z_{h+1}, \ldots, z_m)$. Therefore, by Proposition 6, we can prove that the submodule $\mathfrak{p}_W$ is unmixed and of dimension $m-h$, if we show that for an arbitrary polynomial $q(w'')\not\equiv 0$, the condition $q\,f\in\mathfrak{p}_W$ implies $f\in\mathfrak{p}_W$. If we establish this fact, we will also prove that the submodule $\mathfrak{p}_W$ is normally placed.

We denote by $\Omega$ the algebraic closure of the field $R_w$ of rational functions in the variables $w$. Extending the underlying field $C$ to the field $\Omega$, we extend the ring $\mathscr{P}_v$ to the ring $\Omega[v]$, the ring $\mathscr{P}_{w'}$ to the ring $\Omega[w'']$, and the submodule $\mathfrak{p}\subset\mathscr{P}^t$ to the submodule $\pi\subset\Omega^t[v]$. Let $\mathfrak{p}=p\,\mathscr{P}^s$ for some $\mathscr{P}$-matrix $p$. Then $\pi=p\,\Omega^s[v]$. We shall prove the following statement.

(H). Given an arbitrary non-zero polynomial $q\in\Omega[w'']$ and element $f\in\Omega^t[v]$, such that $q\,f=p\,g$, where $g\in\Omega^s[v]$, there exists an element $g'\in\Omega^s[v]$ such that $f=p\,g'$.

Since the field $\Omega$ is algebraic over $R_w$ and, therefore, over $\mathscr{P}_w$, and since the monomials $w''$ are obviously algebraic over $\mathscr{P}_{w'}$, it follows that the polynomial $q\in\Omega[w'']$ is algebraic over $\mathscr{P}_{w'}$, that is, it satisfies an equation of the form

$$\lambda_k\,q^k+\cdots+\lambda_1\,q+\lambda_0=0,$$

where $\lambda_k, \ldots, \lambda_0\in\mathscr{P}_{w'}$. If $\lambda_0=0$, then we can divide the lefthand side by $q\neq 0$. Therefore we may suppose that $\lambda_0\neq 0$. It follows that

$$\lambda_0\,f=p\,\gamma \tag{11.1}$$

for some $\gamma \in \Omega^s[v]$. We now look on the field $\Omega$ as a vector space over the field $R_w$. In the subspace spanning the coefficients of the polynomials which form the vectors $f$ and $\gamma$, we choose a basis $\{\omega_\alpha \in \Omega\}$. Expanding these coefficients in this basis, we rewrite (11) in the form

$$\frac{1}{\Delta} \sum \omega_\alpha (\lambda_0 f_\alpha - p \gamma_\alpha) = 0, \tag{12.1}$$

where $f_\alpha \in \mathscr{P}^t$, $\gamma_\alpha \in \mathscr{P}^s$, and $\Delta \in \mathscr{P}_w$. Since the $\omega_\alpha$ are linearly independent over $R_w$, the equation (12) implies that $\lambda_0 f_\alpha = p \gamma_\alpha$, that is $\lambda_0 f_\alpha \in \mathfrak{p}$ for all $\alpha$. By hypothesis the module $\mathfrak{p}$ is unmixed of dimension $d$ and is normally placed. Therefore, Proposition 6 implies that $f_\alpha \in \mathfrak{p}$ for all $\alpha$ and, therefore $f = \frac{1}{\Delta} \sum \omega_\alpha f_\alpha \in \pi$. With this, statement (H) is proved.

For an arbitrary vector $j \in \Omega^k[v]$ or $j \in \mathscr{P}_v^k$, we denote by deg $j$ the highest order of any components of this vector. Let $\sigma$ be an arbitrary integer. We denote by $(\mathrm{H}_\sigma)$ the particular cases of statement (H), for which $\deg g \leqq \sigma$. We shall show that there exists an integer $\tau \geqq \sigma$, such that the statement $(\mathrm{H}_\sigma)$ can be reformulated as follows.

$(\mathrm{H}_\sigma)$. The conditions $q f = p g, q \neq 0, \deg g \leqq \sigma$ imply $f = p g', \deg g' \leqq \tau$. In fact, let $\mu$ be the highest order of the elements of the matrix $p$. Then $\deg p g \leqq \sigma + \mu$ and, accordingly, $\deg q \leqq \sigma + \mu$ and $\deg f \leqq \sigma + \mu$. For every integer $k \geqq 0$, we shall consider the subspace $\Omega_k[v]$ in $\Omega[v]$, consisting of the elements $j$ with $\deg j \leqq k$. It is finite dimensional (as a linear space over the field $\Omega$). From what we said earlier, it follows that the vector $f$ belongs to the intersection $\Omega^t_{\sigma+\mu}[v] \cap p\, \Omega^s[v]$. This intersection is finite dimensional and is the union of an increasing sequence of subspaces $\Omega^t_{\sigma+\mu}[v] \cap p\, \Omega^s_\tau[v]$, $\tau = 0, 1, 2, \ldots$. Therefore it coincides with one of the spaces $\Omega^t_{\sigma+\mu}[v] \cap p\, \Omega^s_\tau[v]$, q.e.d.

In the next step we write the statement $(\mathrm{H}_\sigma)$ in analytic form. For this purpose, we first dissect the following condition:

$$f = p g, \qquad g \in \Omega^s[v], \qquad \deg g \leqq \sigma. \tag{13.1}$$

Since $\deg f \leqq \sigma + \mu$, we can rewrite the condition in an equivalent form, by comparing coefficients on both sides of Eq. (13) for powers of the unknowns up to order $\sigma + \mu$. By this means we obtain a system of linear equations over the field $\Omega$:

$$F = P G. \tag{14.1}$$

Here $F$ and $G$ are vectors, consisting of the coefficients of the polynomials which appear in $f$ and $g$, and $P$ is a matrix whose elements are linear functions, with integer coefficients, of the polynomials of the matrix $p$.

We remark that the form of the system (14) does not depend on the underlying field.

Let $r$ be the rank of the matrix $P$. By the Kronecker-Capelli theorem, the solubility conditions of the system (14) amount to saying that the extended matrix is also of rank $r$. This condition can be written in the form of a system of equations

$$M_i^\sigma(f, p)=0,$$

where the $M_i$ are all possible minors of order $r+1$ of the extended matrix. We remark that the minors $M_i(f, p)$ are polynomials in the elements of the vector $F$ and the matrix $P$. Thus, the statement $(\mathrm{H}_\sigma)$ can be rewritten as:

$$M_i^\sigma(q\, f, p)=0, \quad \forall i,\ q\neq 0 \ \Rightarrow\ M_j^\tau(f, p)=0,\ \forall j. \tag{15.1}$$

The condition $q\neq 0$ means that not all the coefficients $q_i$ of the polynomial $q$ are equal to zero. Since $\deg q \leqq \sigma+\mu$, it is sufficient to deal with the coefficients $q_i$ for which $|i| \leqq \sigma+\mu$. Therefore, the assertion (15) is equivalent to:

$$M_k^\sigma(q\, f, p)=0, \qquad \forall k \ \Rightarrow\ q_i\, M_j^\tau(f, p)=0,\ \forall i,\ j, \quad |i| \leqq \sigma+\mu. \tag{16.1}$$

We denote by $Q$ the vector consisting of the coefficients $q_i$ with $|i| \leqq \sigma+\mu$. The quantities $M_k^\sigma(q\, f, p)$ and $q_i\, M_j^\tau(f, p)$ are polynomials in the variables $F$ and $Q$, whose values belong to the field $\Omega$. Assertion (16) means that if the polynomials $M_k^\sigma(q\, f, p)$ vanish, then so do the polynomials $q_i\, M_j^\tau(f, p)$. Since the field $\Omega$ is algebraically closed, the Hilbert Theorem on roots[6] implies that for arbitrary $i$ and $j$ there exist an integer $\rho$ and polynomials $a_k \in \Omega[v]$ such that

$$[q_i\, M_j^\tau(f, p)]^\rho = \sum_k a_k\, M_k^\sigma(q\, f, p). \tag{17.1}$$

We note that in this equation, the coefficients of all the polynomials (with the possible exception of the $a_k$) belong to $\mathscr{P}_w$. In the space $\Omega$ we choose a finite number of elements $\omega_\alpha$, which are linearly independent over $R_w$ and such that one of them is unity, and the space that spans them contains all the coefficients of the polynomials $a_k$. We expand the coefficients of the polynomials $a_k$ with respect to these elements, and we eliminate all terms not belonging to $R_w$. Clearly, Eq. (17) is not disturbed by this process. Thus, we can suppose that the coefficients of the polynomials $a_k$ belong to $R_w$ for arbitrary $i, j, k$. We denote by $\mu'_\sigma \subset C_w$ the variety of the roots of the least common denominator of all these coefficients.

6 See, for example, van der Waerden [1].

We now fix an arbitrary $w \in C_w \setminus \mu'_\sigma$ and we consider the following condition:

$$f = p_w g, \quad g \in \mathscr{P}_v^s, \quad \deg g \leqq \sigma, \tag{18.1}$$

imposed on the vectors $f \in \mathscr{P}_v^t$. Using an analogy with the condition (13) we conclude that condition (18) is equivalent to the solubility of the following system of linear equations over the field $C$

$$F = P_w G, \tag{19.1}$$

where $F$ and $G$ are vectors, consisting of the coefficients of $f$ and $g$, and $P_w$ is the value of the matrix $P$ for a fixed $w$. Since the rank of the matrix $P$ is equal to $r$, the rank of the matrix $P_w$ is also $r$, if the point $w$ does not belong to some proper algebraic subvariety $\mu''_\sigma \subset C_w$.

We shall suppose that $w \in \mu''_\sigma$. Then the solubility of the system (19) is equivalent to the condition that the system of equations $M_k(f, p_w) = 0$, $\forall k$ be satisfied. We write $\mathscr{M} = \bigcup_\sigma (\mu'_\sigma \cup \mu''_\sigma)$ and we suppose that $w \in \mathscr{M}$. Then we are allowed to evaluate the coefficients in Eq. (17) at the point $w$. Since this equation holds for all $i$ and $j$, we arrive at the following assertion:

$$M_k^\sigma(q\,f, p_w) = 0, \quad \forall k \Rightarrow q_i M_j^\tau(f, p_w) = 0, \quad \forall i, j.$$

This implication means that for arbitrary $q \in \mathscr{P}_{w''}$ and $f \in \mathscr{P}_v^t$, the condition $q\,f = p_w g$, where $g \in \mathscr{P}_v^s$, $\deg g \leqq \sigma$, implies that $f = p\,g'$, where $g' \in \mathscr{P}_v^s$, $\deg g' \leqq \tau$. Since $\sigma$ is arbitrary, we conclude that the module $\mathfrak{p}_w = p_w \mathscr{P}_v^s$ is unmixed and of dimension $m-h$. It remains only to note that the set $\mathscr{M}$ is of first category since it is the union of a countable number of nowhere dense sets $\mu'_\sigma$ and $\mu''_\sigma$. □

Let $\mathscr{S}$ be the space of power series in the variables $\eta = (\eta_1, \ldots, \eta_m)$. If $L$ is a subspace in $\mathscr{S}^t$, then we denote by $\eta_m L$ the subspace in $L$, consisting of series of the form $\eta_m \phi$, where $\phi \in L$.

**Theorem 2.** *Let $p: \mathscr{P}^s \to \mathscr{P}^t$ be a $\mathscr{P}$-matrix such that $p\,\mathscr{P}^s$ is an unmixed submodule of $\mathscr{P}^t$ of dimension $d > n-m$, and is normally placed in $C^n$. Then there exists in the space $C_w$ a proper algebraic variety $\mu$ such that for an arbitrary point $Z \in C^n \setminus (C_v \times \mu)$ we have the equation*

$$p(Z)\,\mathscr{S}^s \cap \eta_m \mathscr{S}^t = \eta_m\, p(Z)\,\mathscr{S}^s. \tag{20.1}$$

*Proof.* The preceding theorem implies the existence of a subset $\mathscr{M} \subset C$ of first category, such that for an arbitrary point $w \bar{\in} \mathscr{M}$ the submodule $\mathfrak{p}_w = p_w \mathscr{P}_v^s$ of $\mathscr{P}_v^t$ is unmixed and of dimension $m-h$, and normally placed in $C_v$. By Proposition 6 it follows that for arbitrary $Z_m \in C^1$ the condition $(z_m - Z_m)\,f \in \mathfrak{p}_w$ implies $f \in \mathfrak{p}_w$ $(m > h = n-d)$. Thus we have established the equation

$$\mathfrak{p}_w \cap (z_m - Z_m)\,\mathscr{P}_v^t = (z_m - Z_m)\,\mathfrak{p}_w, \qquad w \,\bar{\in}\, \mathscr{M}. \tag{21.1}$$

We imbed the space $C^n$ in the space $C^{n+1} = C^n \times C^1$, the points of which will be denoted by $(z, Z_m)$. Any polynomial in $C^n$ can be looked on as a polynomial in $C^{n+1}$, which is constant in $Z_m$. We thus obtain a mapping of the ring $\mathscr{P} = \mathscr{P}_z$ into the ring $\mathscr{P}' = \mathscr{P}_{z, Z_m}$ consisting of polynomials in the variables $z, Z_m$. We denote by $\check{p}$ the $\mathscr{P}'$-matrix $(z_m - Z_m)\,p$, and we denote by $e$ the unit matrix of order $t$. Applying Theorem 1, § 4, Chapter II to the $\mathscr{P}'$-matrices $p, \check{p}(z_m - Z_m)\,e$, $p \oplus (z_m - Z_m)\,e$, we construct the operators

$$d,\ g;\quad \check{d},\ \check{g};\quad d^0,\ g^0;\quad d',\ g', \tag{22.1}$$

which act from the space $\mathscr{S}^t$ to the space $\mathscr{S}^{t'}$, where $t'$ is equal to $t, s, t+s$, and which have the following properties: there exists an algebraic stratification $\mathscr{N} = \{N_\nu\}$ of the space $C^{n+1}$, such that on every set $N_\nu \backslash N_{\nu+1}$ the coefficients of all these operators are rational, regular functions (in fact, Theorem 1, § 4, Chapter II guarantees the existence of such a stratification for each of the matrices enumerated above – in order to construct the common stratification, it is sufficient to apply Proposition 3, § 3, Chapter II).

From (21) and Proposition 4 we derive the equation

$$\check{d}(z, Z_m) = \lambda(z, Z_m)\,d(z, Z_m) + \lambda^0(z, Z_m)\,d^0(z, Z_m), \qquad w \,\bar{\in}\, \mathscr{M}, \tag{23.1}$$

where $\lambda$ and $\lambda^0$ are polynomials in the operators (22), the matrix $p$, $(z_m - Z_m)\,e$ and the projection operator $\mathscr{S}^{s+t} \to \mathscr{S}^s$. It follows that the coefficients of the operators $\lambda$ and $\lambda^0$ are rational, regular functions on each of the sets $N_\nu \backslash N_{\nu+1}$.

We fix an arbitrary index $\nu \geqq 0$ and we resolve the variety $N_\nu$ into irreducible components. Let $N'$ be one of its components not belonging to $N_{\nu+1}$. We divide the coordinates in the space $C^{n+1}$ into two groups $\check{v} = (v, Z_m)$ and $w$ and we apply Proposition 5 to the variety $N'$. We assume that for this variety we have case II. Then the variety $N'$, with the exception of its discriminant subvariety $N_\delta$, is projected into an open subset in $C_{w'}$, and accordingly into an open subset in $C_w$. Therefore the set $C_w \backslash \mathscr{M}$ is everywhere dense in the projection of $N' \backslash N_\delta$ on $C_w$. Therefore, the set $C^{n+1} \backslash (C_v \times \mathscr{M} \times C^1)$ is everywhere dense in $N' \backslash N_\delta$. We remark that the subvariety $N_\delta$, together with the subvariety $N' \cap N_{\nu+1}$, is nowhere dense in $N'$. This follows from Lemma 1, since the variety $N'$ is irreducible. Hence it follows that the set $C^{n+1} \backslash (C_v \times \mathscr{M} \times C^1)$ is everywhere dense in $N' \backslash N_\delta$. Since Eq. (23) holds on this set, and both sides of the equation have as coefficients the rational functions in $N' \backslash N_{\nu+1}$, and these are non-infinite, this equation is satisfied on the whole set $N' \backslash N_{\nu+1}$.

We denote by $\mu_\nu$ the projection on $C_w$ of the union of all the irreducible components of the variety $N_\nu$, for which case I of Proposition 5 holds. From what we have proved, Eq. (23) is valid for all points $w \in \mu = \cup \mu_\nu$. We show that the variety $\mu$ satisfies the conditions of the theorem. In fact, the left side of (21) always contains the righthand side. Let $w \in \mu$. Since the operators $d$ and $d^0$ vanish on the lefthand side of (21), the operator $\check{d}$ also vanishes on the left side in view of (23). By the properties of the operator $\check{d}$, this means that the left side lies within the right. □

## § 2. Local $p$-operators

From here on, until the end of the chapter, we fix an arbitrary $\mathscr{P}$-matrix $p\colon \mathscr{P}^s \to \mathscr{P}^t$. We recall that the ring $\mathscr{P}$ is interpreted as a ring of polynomials with complex coefficients defined in $C^n$. The points of the space $C^n$ will be denoted by $z = (z_1, \ldots, z_n)$ or $\xi = (\xi_1, \ldots, \xi_n)$. The letter $z$ will often be used to denote a fixed point, at which some $p$-decomposition is carried out. The letter $\xi$ will always denote an active variable – active in the operators that occur in these decompositions.

The content of this section consists in the following: beginning with a $p$-decomposition, corresponding to some value of the parameter $m$, we construct a $p$-decomposition, corresponding to a larger value of $m$, having some local property, which we shall discuss later.

**1°. The initial $p$-decomposition.** We shall now partially reproduce the fundamental theorem of Chapter II. We fix an arbitrary integer $m$, lying between 0 and $n-1$. We set $n = (\xi_1, \ldots, \xi_m)$; we shall also regard $\eta$ as a point in $C^n$, lying in the corresponding coordinate subspace. According to the fundamental theorem, as applied to the matrix $p$ and the variables $\eta$, the identity operator $E_\eta$ in the space $\mathscr{S}^t[\eta]$ for arbitrary $z \in C^n$ admits the following decomposition:

$$E_\eta = d^m(z) + p(z+\eta)\, g^m(z), \tag{1.2}$$

where the $p$-operators $d^m$ and $g^m$, in particular, have the following properties:

1. The operator $d^m(z)$ vanishes on the subspace $p(z)\, \mathscr{S}^s[\eta]$.

2. The coefficients $d_i^j = \delta_i\, d^m\, \eta^j$ and $g_i^j = \delta_i\, g^m\, \eta^j$ of these operators differ from zero only if

$$L(j) \geqq L(i) - l, \qquad l = -L(K\, e_m) \geqq 0,$$

where $K$ is a constant, and $L\colon Z^m \to Z^m$ is the linear operator constructed in Chapter II.

3. We have the inequality

$$\max\{|d_i^j|, |g_i^j|\} \leqq a^{|L(j)-L(i)+l|+1},$$

where $a \geqq 1$ is some power function on a certain algebraic stratification $\mathcal{N}^m$ of the space $C^n$. As a consequence of this property, we have

4. For arbitrary $\varepsilon$, $0<\varepsilon\leqq 1$, $d^m$ and $g^m$ are continuous operators from $\mathcal{O}_\varepsilon^t$ to $\mathcal{O}_{\varepsilon\rho}^t$ and $\mathcal{O}_{\varepsilon\rho}^s$, respectively, and

$$\|d^m \phi\|_{\varepsilon\rho} \leqq \frac{1}{\rho} \|\phi\|_\varepsilon, \qquad \|g^m \phi\|_{\varepsilon\rho} \leqq \frac{1}{\varepsilon^K \rho} \|\phi\|_\varepsilon, \tag{2.2}$$

where $\rho=\rho_m(z)$ is a power function on the stratification $\mathcal{N}^m$.

**2°. The tensor product of operators on spaces of power series.** We now fix an arbitrary integer $m^*$, satisfying the inequality $m<m^*\leqq n$. We consider the groups of variables $\omega=(\xi_{m+1}, \ldots, \xi_{m^*})$ and $\eta^*=(\eta, \omega)=(\xi_1, \ldots, \xi_{m^*})$ and the corresponding spaces of power series $\mathcal{S}[\omega]$ and $\mathcal{S}[\eta^*]$; $\mathcal{S}[\eta]$ and $\mathcal{S}[\omega]$ are subspaces of the space $\mathcal{S}[\eta^*]$, which we will denote, for brevity, by the single character $\mathcal{S}$.

We first define the operation of tensor multiplication of functions on the spaces $\mathcal{S}[\eta]$ and $\mathcal{S}[\omega]$. We denote by $\delta_i^\eta$, $i\in Z_+^m$ and $\delta_j^\omega$, $j\in Z_+^\mu$, $\mu=m^*-m$, the basis functionals on these spaces. We set

$$\delta_i^\eta \otimes \delta_j^\omega \overset{\text{def}}{=} \delta_{(i,j)}, \qquad (i,j)\in Z_+^{m^*},$$

where $\delta_{(i,j)}$ is a basis functional on $\mathcal{S}$. Extending this definition by linearity, we write

$$f\otimes g=\sum f^i \delta_i^\eta \otimes \sum g^j \delta_j^\omega = \sum f^i \otimes g^j \delta_{(i,j)},$$

for the functional $f$ on $\mathcal{S}^a[\eta]$ with values in $C^b$ and for the functional $g$ on $\mathcal{S}^\alpha[\omega]$ with values in $C^\beta$. It is clear that $f\otimes g$ is a functional on $\mathcal{S}^{a\alpha}$ with values in $C^{b\beta}$.

Now suppose that $A\colon \mathcal{S}^a[\eta]\to\mathcal{S}^b$ and $B\colon \mathcal{S}^\alpha[\omega]\to\mathcal{S}^\beta$ are arbitrary operators. Their tensor product is defined as:

$$A\otimes B=\sum \eta^{*i} A_i(\delta^\eta)\otimes \sum \eta^{*j} B_j(\delta^\omega)=\sum \eta^{*i+j} A_i(\delta^\eta)\otimes B_j(\delta^\omega).$$

The operator so obtained acts from $\mathcal{S}^{a\alpha}$ to $\mathcal{S}^{b\beta}$.

Let us dwell for a moment on a particular case of tensor multiplication of operators. First of all, we note that every series $\phi\in\mathcal{O}$, $\mathcal{O}=\mathcal{O}[\eta^*]$ can be looked on as an element of the space $\mathcal{S}[\eta]$, depending analytically on the parameters $\omega$ in some neighborhood of zero; in this case we write $\phi=\phi(\omega)$. We further consider the operator

$$\delta_i^\eta \otimes E_\omega = \sum_{j\in Z_+^\mu} \omega^j \delta_{(i,j)}\colon \mathcal{S}\to\mathcal{S}[\omega].$$

This operator acts on the series $\phi \in \mathcal{O}$ by the formula

$$(\delta_i^\eta \otimes E_\omega)\,\phi = \delta_i^\eta\,\phi(\omega). \tag{3.2}$$

Let $A$: $\mathcal{S}[\eta] \to \mathcal{S}$ be an arbitrary operator which carries the space $\mathcal{O}[\eta]$ into $\mathcal{O}$. We write it in the form $\sum \eta^i A_i^j(\omega)\,\delta_j^\eta$, where $A_i^j(\omega) \in \mathcal{O}[\omega]$ and we assume in addition that all the series $A_i^j(\omega)$ converge in some common neighborhood of zero. In this case, the sum $\sum \eta^i A_i^j(\omega)\,\delta_j^\eta$ can be regarded as an operator in $\mathcal{S}[\eta]$, which depends analytically on $\omega$ in this neighborhood; to accord with this, we write the operator $A$ as $A(\omega)$. Then we can derive from (3) the following important formula

$$(A \otimes E_\omega)\,\phi = A(\omega)\,\phi(\omega), \qquad \phi \in \mathcal{O}.$$

Let us make one further remark. Let $A$ be an operator in $\mathcal{S}[\eta]$. Admitting to a certain lack of precision in the notation, we shall sometimes regard it as an operator in $\mathcal{S}$. Then its action will consist in eliminating from the series $\phi \in \mathcal{S}$ those terms containing $\omega$, and then applying the operator $A$. The resulting series belonging to $\mathcal{S}[\eta]$ will be regarded as an element of the wider space $\mathcal{S}$.

**3°. Formulation of the theorem.** We denote by $\omega\,\mathcal{S}^k$ the space in $\mathcal{S}^k$, consisting of series of the form

$$\sum_1^\mu \omega_i\,\phi_i,$$

where all the $\phi_i \in \mathcal{S}^k$, and we write $\omega_i = \xi_{m+i}$, $i = 1, \ldots, \mu$. We denote by $p(z)\,\omega\,\mathcal{S}^s$ the image of the mapping $p(z)$: $\omega\,\mathcal{S}^s \to \mathcal{S}^t$.

**Theorem.** *Suppose that the equation*

$$p(z)\,\mathcal{S}^s \cap \omega\,\mathcal{S}^t = p(z)\,\omega\,\mathcal{S}^s,$$

*is satisfied at a given point $z \in C^n$. Then the identity operator in $\mathcal{S}^t$ admits the decomposition*

$$E = \mathcal{D}^*(z) + p(z+\eta^*)\,\mathcal{G}^*(z), \tag{4.2}$$

*in which*

$$\mathcal{D}^*(z) = d^*(z) \otimes E_\omega, \qquad \mathcal{G}^*(z) = g^*(z) \otimes E_\omega, \tag{5.2}$$

*and $d^*(z)$ and $g^*(z)$ are operators acting from $\mathcal{S}^t[\eta]$ to $\mathcal{S}^t$ and $\mathcal{S}^s$, respectively, while*

I. *The operator $\mathcal{D}^*(z)$ vanishes on the subspace $p(z)\,\mathcal{S}^s$.*

II. *For arbitrary $\varepsilon$, $0 < \varepsilon \leqq 1$, $\mathcal{D}^*(z)$ is a continuous operator acting from $\mathcal{O}_\varepsilon^t$ to $\mathcal{O}_{\varepsilon^q \rho^*}^t$, and satisfies the inequality*

$$\|\mathcal{D}^*(z)\,\phi\|_{\varepsilon^q \rho^*} \leqq \frac{1}{\rho^*}\,\|\phi\|_\varepsilon,$$

*where $q$ is a constant and $\rho^* = \rho^*(z)$ is a power function on the stratification $\mathcal{N}^m$.*

*Remark.* From what we said in 2°, it follows that the operator $\mathcal{D}^*$ acts in accordance with the formula

$$\mathcal{D}^* \phi = d^*(\omega)\, \phi(\omega), \qquad \phi \in \mathcal{O}^t.$$

In other words the function $\mathcal{D}^* \phi$, for an arbitrary fixed value of the parameter $\omega = \omega_0$, is the result of applying the operator $d^*$, in which we have also fixed $\omega = \omega_0$, to the restriction of the function $\phi$ on the subspace $\omega = \omega_0$. Hence, in particular, it follows that the value of the function $\mathcal{D}^* \phi$ on the subspace $\omega = \omega_0$ depends on the value of the function $\phi$ on this subspace only. This property of localization of the operator $\mathcal{D}^*$ lies at the basis of our subsequent reasoning in Chapter IV.

Let us note two further properties of the decomposition (4), which are important in other applications of it. Let $\mathcal{I}$ be a basis set corresponding to the decomposition (1).

III. *The image of the operator $\mathcal{D}^*$ belongs to the space $\mathcal{S}^t_{\mathcal{I} \times Z^\mu_+}$, where $\mathcal{I} \times Z^\mu_+$ is a subset of $t\, Z^m_+{}^*$, consisting of the couples $(\tau, (i, j))$, where $(\tau, i) \in \mathcal{I}$, and $j \in Z^\mu_+$.*

IV. *The operator $\mathcal{G}^*$ vanishes on $\mathcal{S}^t_{\mathcal{I} \times Z^\mu_+}$.*

For the proof of this assertion, it is sufficient to note that by the formulae of 7° $\mathcal{D}^* = (d^m \otimes E_\omega) A$, and $\mathcal{G}^* = B(g^m \otimes E_\omega)$, where $A$ and $B$ are certain operators.

**4°. Proof of the theorem.** We replace $p(z+\eta)$ in the decomposition (1) by $p(z+\eta^*)$, and obtain

$$E_\eta = d(z) + p(z+\eta^*)\, g(z) + R(z) \qquad (d = d^m,\ g = g^m), \tag{6.2}$$

where

$$R(z) = [p(z+\eta) - p(z+\eta^*)]\, g(z).$$

The operator $R(z)$ acts from $\mathcal{S}^t[\eta]$ to $\mathcal{S}^t$ and, clearly, does not contain terms independent of $\omega$. The following most important step in the proof consists in applying the method of sequential approximations to Eq. (6), in order to eliminate the term $R(z)$. Analytically, the method consists in multiplying both sides of (6) on the right by a Neumann series of the operator $R(z)$. In this method, it is essential that we prove the convergence of the Neumann series as an operator in the spaces of convergent series.

For every integer $k$, we write

$$R^k(z) = \overbrace{[R(z) \otimes E_\omega] \ldots [R(z) \otimes E_\omega]}^{k-1 \text{ brackets}}\, R(z), \qquad R^0(z) = E_\eta.$$

It is also evident that for arbitrary $k \geqq 1$, the operator $R^k(z)$ contains no term with $\omega^i$, $|i|<k$. Therefore the series

$$N(z)=R^0(z)+R^1(z)+\cdots+R^k(z)+\cdots \tag{7.2}$$

converges as an operator from $\mathscr{S}^t[\eta]$ to $\mathscr{S}^t$. This series will be called the Neumann series for the operator $R$. It is clear that the series

$$N\otimes E_\omega=E+R\otimes E_\omega+[R\otimes E_\omega]^2+\cdots+[R\otimes E_\omega]^k+\cdots$$

satisfies the relation

$$[(E_\eta-R)\otimes E_\omega]\,[N\otimes E_\omega]=E, \quad [N\otimes E_\omega]\,[E_\eta-R]=E_\eta. \tag{8.2}$$

**5°. Convergence of the Neumann series in the space $\mathcal{O}^t$**

**Lemma.** *For arbitrary $\varepsilon$, $0<\varepsilon\leqq 1$, the Neumann series (7) converges as a continuous operator from $\mathcal{O}_\varepsilon^t[\eta]$ to $\mathcal{O}_{\varepsilon^q\rho_0}^t$, and*

$$\|N(z)\,\phi\|_{\varepsilon^q\rho_0}\leqq C\,\|\phi\|_\varepsilon,$$

*where $C$ and $q$ are constants not depending on $z$ and $\varepsilon$, and $\rho_0=\rho_0(z)$ is a power function on the stratification $\mathcal{N}^m$.*

*Proof of the lemma.* For every pair $i$ and $j\in Z_+^m$, we write $R_i^{k,j}=\delta_i^\eta R^k(\omega)\,\eta^j$. Every one of the coefficients $R_i^{k,j}$ is a polynomial in $\omega$, whose coefficients are matrices of size $t\times t$, with elements in $C$. We estimate these coefficients supposing that $|\omega|\leqq 1$.

**Inductive proposition**

$$R_i^{k,j}\neq 0, \quad \textit{only if} \quad L(j)\geqq L(i)-k\,l; \tag{$a_k$}$$

$$|R_i^{k,j}(\omega)|\leqq \frac{|\omega|^k}{(k!)^m}\,[L(j)-L(i)+k(l+e)]^k A^{|L(j)-L(i)+k(l+e)|}, \tag{$b_k$}$$

*where $e=(1,\ldots,1)\in Z^m$, and $A\geqq 2\sqrt{m}$ is some power function on the stratification $\mathcal{N}^m$. Here, and later, we denote by $[i]^k$, where $i\in R^m$, the expression $(i_1,\ldots,i_m)^k$.*

*Proof of the inductive proposition.* For $k=0$ we have, by definition, $R_i^{0,j}=\delta_i^j$, where $\delta_i^j$ is the Kronecker symbol. Accordingly, the assertions $(a_0)$ and $(b_0)$ are true. Assuming the truth of $(a_k)$ and $(b_k)$, we prove $(a_{k+1})$ and $(b_{k+1})$.

We consider the polynomial

$$r(r,\eta^*)=p(z+\eta)-p(z+\eta^*).$$

For every $i\in Z_+^m$ the polynomial $r_i(z,\omega)=\delta_i^\eta\, r(z,\eta,\omega)$ contains no term without $\omega$. Therefore, taking into account the fact that $|\omega|\leqq 1$, we are allowed to write the inequality

$$\sum_i |r_i(z,\omega)| \leqq |\omega|\, M(z), \tag{9.2}$$

where $M(z)$ is some function of the form $C(|z|+1)^q$. The power function $A$ is defined by the condition $A \geqq \max\{a(M+1), 2\sqrt{m}\}$ (where $a$ is the function in Condition 3, 1°).

The coefficients of the operator $(g\otimes E_\omega)R^k$ will be denoted by

$$(g\,R^k)_i^j \overset{\text{def}}{=} \delta_i^\eta (g\otimes E_\omega)\, R^k\, \eta^j.$$

It is clear that

$$(g\,R^k)_i^j = \sum_\alpha g_i^\alpha R_\alpha^{k,j}.$$

Hence, and also from property 3, 1°,

$$\begin{aligned} |(g\,R^k)_i^j| &\leqq \sum_\alpha |g_i^\alpha| \cdot |R_\alpha^{k,j}| \\ &\leqq \frac{|\omega|^k}{(k!)^m}\, a\cdot A^{|L(j)-L(i)+k(l+e)+l|} \cdot \sum_{\alpha\in\pi_{ij}} |L(j)-L(\alpha)+k(l+e)]^k, \end{aligned} \tag{10.2}$$

where $\pi_{ij}$ is the set of all $\alpha$, for which we have simultaneously $g_i \neq 0$ and $R^{k,j} \neq 0$. From property 2, 1°, and from $(a_k)$, it follows that the set $\pi_{ij}$ belongs to the set of points $\alpha$, for which the inequality $L(j)+k\,l \geqq L(\alpha) \geqq L(i)-l$, holds. Hence, in particular, it follows that

$$(g\,R^k)_i^j \neq 0, \quad \text{only if} \quad L(j) \geqq L(i)-(k+1)\,l. \tag{11.2}$$

Writing $\lambda_0 = L(j)+k(l+e)$ and substituting $\lambda = L(\alpha)$ in the sum on the right in (10), we arrive at the following larger sum:

$$\sum \{[\lambda_0-\lambda]^k, \lambda\in Z_+^m, \lambda_0 \geqq \lambda \geqq L(i)-l\}. \tag{12.2}$$

We will estimate this sum in turn by a suitable integral of the function $[\lambda_0-\lambda]^k$. This function does not decrease when the coordinates of the point $\lambda$ decrease. Therefore, its value at an arbitrary point $\lambda \leqq \lambda_0$ does not exceed its integral over the unit cube with vertices at the points $\lambda$ and $\lambda - e$. Therefore, the sum (12) is not greater than the integral

$$\int [\lambda_0-\lambda]^k\, d\lambda_1 \dots d\lambda_m,$$

taken in the range $\lambda_0 \geqq \lambda \geqq L(i)-l-e$. This integral, as we may calculate without difficulty, is equal to

$$\frac{1}{(k+1)^m}[\lambda_0-L(i)+l+e]^{k+1} = \frac{1}{(k+1)^m}[L(j)-L(i)+(k+1)(l+e)]^{k+1}.$$

Summing up, we arrive at the inequality

$$|(g\,R^k)_i^j| \leqq \frac{|\omega|^k}{((k+1)!)^m}\,[L(j)-L(i)+(k+1)(l+e)]^{k+1} \cdot a \cdot A^{|L(j)-L(i)+k(l+e)+l|}. \tag{13.2}$$

To obtain the operator $R^{k+1}$, we must multiply $(g \otimes E_\omega)\,R^k$ by the polynomial $r(z, \eta^*)$. Therefore

$$R_i^{k+1,j} = \sum_\alpha r_\alpha (g\,R^k)_{i-\alpha}^j. \tag{14.2}$$

Hence, in particular, it follows that $R_i^{k+1,j} \neq 0$, only if $(g\,R^k)_{i-\alpha}^j \neq 0$ for some $\alpha \geqq 0$. According to (11) this last inequality can be satisfied only if

$$L(j) \geqq L(i-\alpha)-(k+1)\,l = L(i)-L(\alpha)-(k+1)\,l.$$

Since $-L(\alpha) \geqq 0$ in accordance with Proposition 1, § 2, Chapter II, it follows that $L(j) \geqq L(i)-(k+1)\,l$. With this, the assertion $(a_{k+1})$ is proved.

Further, we have from (14), (9) and (13)

$$\begin{aligned} |R_i^{k+1,j}| &\leqq \sum_{\alpha \geqq 0} |r_\alpha| \cdot \max_{\alpha \geqq 0} |(g\,R^k)_{i-\alpha}^j| \\ &\leqq \frac{|\omega|^{k+1}}{[(k+1)!]^m}\,[L(j)-L(i)+(k+1)(l+e)]^{k+1} \\ &\quad \times a\,M\,A^{|L(j)-L(i)+k(l+e)+l|}. \end{aligned}$$

Here, we have made use of the fact that the coordinates of the point $-L(i-\alpha)$ reach their largest values on the set $\alpha \geqq 0$ for $\alpha = 0$. Taking account of the fact that $a\,M \leqq A \leqq A^{|e|}$, we obtain finally

$$|R_i^{k+1,j}| \leqq \frac{|\omega|^{k+1}}{[(k+1)!]^m}\,[L(j)-L(i)+(k+1)(l+e)]^{k+1} \times A^{|L(j)-L(i)+(k+1)(l+e)|}.$$

With this the proof of the inductive proposition is completed. □

**6°. Proof of the lemma.** We approximate the results of the action of the operator $R^k$ on an arbitrary series $\phi \in \mathscr{C}_\varepsilon^t$, where $0 < \varepsilon \leqq 1$. We write the series in the form $\phi = \phi(\omega) = \sum \eta^i\,\phi_i(\omega)$, where $\phi_i(\omega) \in \mathcal{O}^t[\omega]$. For $|\omega| \leqq \varepsilon/2$ this series converges uniformly in the sphere $|\eta| \leqq \varepsilon/2$ and is bounded in modulus by the quantity $\|\phi\|_\varepsilon$. Therefore, we obtain from Proposition 2, § 1, Chapter II, the inequality

$$\sup_{|\omega| \leqq \frac{\varepsilon}{2}} |\phi_i(\omega)| \leqq \left(\frac{2\sqrt{m}}{\varepsilon}\right)^{|i|} \|\phi\|_\varepsilon.$$

Further, we have

$$R^k \phi = \sum_i \eta^i \sum_j R_i^{k,j} \phi_j.$$

We estimate each term of the inner sum, supposing that $|\eta| \leqq \delta$ where $\delta > 0$ is a parameter not yet defined:

$$\begin{aligned} |\eta^i \sum_j R_i^{k,j} \phi_j| \leqq \|\phi\|_\varepsilon \cdot \frac{|\omega|^k}{(k!)^m} \sum_{L(j) \geqq L(i) - kl} [L(j) - L(i) + k(l+e)]^k \\ \times \delta^{|i|} A^{|L(j) - L(i) + k(l+e)|} \cdot \left(\frac{2\sqrt{m}}{\varepsilon}\right)^{|j|}. \end{aligned} \tag{15.2}$$

We now approximate the sum on the left side. We take note of the inequalities

$$|i| \leqq |-L(i)| \leqq \|L\| \cdot |i|, \qquad i \geqq 0.$$

The left inequality follows from the fact that the $m$-th coordinate of the point $-L(i)$ is equal to $|i|$, and the remaining coordinates are non-negative in accordance with the Proposition of § 2, Chapter II. The right inequality follows from the fact that $L$ is a bounded operator in the cone $Z_+^m$, under the norm $i \to |i|$. We now set

$$\delta = \tfrac{1}{2} e^{-m} \cdot \varepsilon \cdot A^{-\|L\|}.$$

Then the product of the last three factors on the right side of (15) does not exceed

$$2^{-|i|} \cdot e^{-m|i|} \cdot \varepsilon^{|i| - |j|} \cdot \left(\frac{2\sqrt{m}}{A}\right)^{|j|} \cdot A^{|k(l+e)|}. \tag{16.2}$$

Since $A \geqq 2\sqrt{m}$ in accordance with the inductive proposition, the factor

$$\left(\frac{2\sqrt{m}}{A}\right)^{|j|}$$

can be eliminated without decreasing the expression itself.

The inequality $L(j) \geqq L(i) - kl$, which determines the range of the summation on the right side of (15), implies that $-|j| \geqq |i| - k|l|$ and, accordingly $|i| - |j| \geqq -k|l|$. Therefore, the quantity (16) does not exceed the product $2^{-|i|} \cdot e^{-m|i|} \cdot \varepsilon^{-k|l|} \cdot A^{|k(l+e)|}$. The inequality $|i| + k|l| \geqq |j|$ implies also that the number of terms of the sum in (15) does not exceed $(|i| + k|l| + 1)^m$. Furthermore, the quantity $[L(j) - L(i) + k(l+e)]^k$ attains its greatest value in the range $j \geqq 0$ for $j = 0$. Therefore, this greatest value is equal to $[\Lambda]^k$ where $\Lambda = -L(i) + k(l+e)$. Since all the coordinates of the point $\Lambda$ are non-negative, $[\Lambda]^k \leqq (|\Lambda|)^{km}$. To sum up, we have established that the sum on the right side of (15) does not exceed the expression

$$\begin{aligned}&(|i|+k|l|+1)^m|-L(i)+k(l+e)|^{km}\cdot 2^{-|i|}\cdot e^{-m|i|}\\&\times\varepsilon^{-k|l|}A^{k|l+e|}\leqq 2^{-|i|}(B(|i|+k+1))^{(k+1)m}e^{-m|i|}\left(\frac{A}{\varepsilon}\right)^{kB},\end{aligned}\tag{17.2}$$

where $B$ is some sufficiently large constant, depending only on $m$.

We now determine the largest value of the product of the second and third factors on the right side for $i\geqq 0$. Considering $|i|$ as a continuous parameter, we differentiate the product with respect to $|i|$. This derivative is equal to

$$\left(\frac{(k+1)m}{|i|+k+1}-m\right)(B(|i|+k+1))^{(k+1)m}e^{-m|i|}.$$

It is clear that the first parenthesis is negative, and therefore, the whole expression is negative. It follows from this that the product of the second and third factors on the right side of (17) decreases for $|i|\geqq 0$ and, therefore, cannot exceed its value at $|i|=0$, that is, cannot exceed $(B(k+1))^{(k+1)m}$. Therefore, the right side of (15) is not greater than

$$2^{-|i|}\|\phi\|_\varepsilon B^m\left(\frac{|\omega|B^m A^B}{\varepsilon^B}\right)^k\left(\frac{(k+1)^{(k+1)}}{k!}\right)^m.\tag{18.2}$$

It follows from Sterling's formula that the fraction $(k+1)^{(k+1)}/k!$ does not exceed $C3^k$, where $C$ is an absolute constant, and therefore the expression (18) is not greater than

$$2^{-|i|}C\|\phi\|_\varepsilon\left(\frac{|\omega|\mathscr{A}}{\varepsilon^B}\right)^k,\quad\text{where}\quad \mathscr{A}=B^m A^B,$$

and the constant $C$ depends only on $m$. Hence, finally

$$\begin{aligned}\|R^k(\omega)\,\phi(\omega)\|_\delta&\leqq\sup_{|\eta|\leqq\delta}\sum_i|\eta^i\sum_j R_i^{k,j}\phi_j|\\&\leqq\sum_i 2^{-|i|}C\|\phi\|_\varepsilon\left(\frac{|\omega|\mathscr{A}}{\varepsilon^B}\right)^k=C'\|\phi\|_\varepsilon\left(\frac{|\omega|\mathscr{A}}{\varepsilon^B}\right)^k\end{aligned}$$

Assuming that

$$\frac{|\omega|\mathscr{A}}{\varepsilon^B}\leqq\frac{1}{2},\quad\text{that is,}\quad|\omega|\leqq\delta'=\frac{\varepsilon^B}{2\mathscr{A}},$$

we obtain

$$\sup_{|\omega|\leqq\delta'}\|\sum_k R^k(\omega)\,\phi(\omega)\|^\delta\leqq\sup_{|\omega|\leqq\delta'}\sum_k\|R^k(\omega)\,\phi(\omega)\|_\delta\leqq C''\|\phi\|_\varepsilon.\tag{19.2}$$

Thus the series $\sum R^k \phi$ converges absolutely in the space $\mathcal{O}_\delta^t[\eta]$ and depends analytically on the parameters $\omega$ in the sphere $|\omega| \leqq \delta'$. Accordingly, the sum of the series $\sum R^k \phi$ is an analytic function of the variables $\eta^*$ in the product of the spheres $\{|\eta| \leqq \delta\}$ and $\{|\omega| \leqq \delta'\}$. The product of these spheres contains a sphere of the form $\{\eta^*: |\eta^*| \leqq \varepsilon^q \rho_0\}$, where $q$ is a suitable constant and $\rho_0$ is a power function on the stratification $\mathcal{N}^m$. In view of the inequality (19) the proof of the lemma is complete. □

**7°. The construction of a local *p*-decomposition.** We set

$$d^*(z) = [d(z) \otimes E_\omega]\, N(z), \qquad g^*(z) = [g(z) \otimes E_\omega]\, N(z),$$

and we define the operators $\mathcal{D}^*$ and $\mathcal{G}^*$ by formula (5). By property 3 of the operator $d$ (see 1°) and from the lemma, it follows that the operator $\mathcal{D}^*$ satisfies Condition II of the theorem.

Furthermore, we have from formulae (6) and (8)

$$\begin{aligned}\mathcal{D}^*(z) + p(z+\eta^*)\, \mathcal{G}^*(z) &= [d^*(z) + p(z+\eta^*)\, g^*(z)] \otimes E_\omega \\ &= [(d(z) + p(z+\eta^*)\, g(z)) \otimes E_\omega][N(z) \otimes E_\omega] \\ &= [(E_\eta - R(z)) \otimes E_\omega][N(z) \otimes E_\omega] = E,\end{aligned}$$

whence (4) follows.

Thus it remains to show that the operator $\mathcal{D}^*(z)$ vanishes on the subspace $p(z)\mathcal{S}^s$. Let $\phi$ be an arbitrary element of this subspace. It is clear that the series $E_\eta \phi$ (obtained by eliminating all terms containing $\omega$) belongs to the subspace $p(z)\mathcal{S}^s[\eta]$. Therefore

$$d(z)\,\phi = 0. \tag{20.2}$$

We set $\psi = [E_\eta - R(z)]\,\phi$. Since the operator $R(z)$ does not contain terms of order zero with respect to $\omega$, the series $\phi' = \phi - \psi = R(z)\,\phi$ likewise contains no such terms. From (8) and (20), we have

$$\begin{aligned}\mathcal{D}^*(z)\,\psi &= [d(z) \otimes E_\omega][N(z) \otimes E_\omega][E_\eta - R(z)]\,\phi \\ &= [d(z) \otimes E_\omega]\, E_\eta\,\phi = d(z)\,\phi = 0.\end{aligned}$$

Applying the decomposition (4) to $\psi$, we obtain

$$\psi = p(z+\eta^*)\, \mathcal{G}^*(z)\,\psi \in p(z)\,\mathcal{S}^s.$$

Thus the series $\phi' = \phi - \psi$ belongs to the subspace $p(z)\mathcal{S}^s$, does not contain terms in which $\omega$ is missing, and satisfies the equation

$$\mathcal{D}^*(z)\,\phi' = \mathcal{D}^*(z)\,\phi.$$

Since the series $\phi'$ belongs to the intersection $p(z)\mathscr{S}^s \cap \omega \mathscr{S}^t$, according to the hypothesis of the theorem, it can be written in the form

$$\phi' = \sum \omega_i \phi_i, \qquad \phi_i \in p(z)\mathscr{S}^s.$$

On the other hand, the operator $d^*(z)$ contains no functional $\delta_j^\omega$, and accordingly, the operator $\mathscr{D}^*(z)$ commutes with multiplication by $\omega_i$, $i=1, \ldots, \mu$, whence

$$\mathscr{D}^*(z)\,\phi' = \sum \omega_i\, \mathscr{D}^*(z)\,\phi_i.$$

Applying the preceding arguments to each of the series $\phi_i$, $i=1, \ldots, \mu$, we construct the series $\phi_i'$, which contain no terms in which $\omega$ is missing, which belong to the subspace $p(z)\mathscr{S}^s$, and are such that

$$\mathscr{D}^*(z)\,\phi_i' = \mathscr{D}^*(z)\,\phi_i.$$

We write $\phi'' = \sum \omega_i \phi_i'$. Then

$$\mathscr{D}^*(z)\,\phi'' = \sum \omega_i\, \mathscr{D}^*(z)\,\phi_i' = \sum \omega_i\, \mathscr{D}^*(z)\,\phi_i = \mathscr{D}^*(z)\,\phi,$$

and the series $\phi''$ contains no terms with $\omega^i$, where $|i|<2$, and so forth. Continuing this process, we construct for arbitrary $k>0$ the series $\phi^{(k)} \in p(z)\mathscr{S}^s$, containing no terms with $\omega^i$, $|i|<k$ and such that

$$\mathscr{D}^*(z)\,\phi^{(k)} = \mathscr{D}^*(z)\,\phi.$$

Since $\mathscr{D}^*(z)$ is an operator in $\mathscr{S}^t$, each of its coefficients $\delta_i\, \mathscr{D}^*(z)$ is a functional on $\mathscr{S}^t$ of finite order and, therefore, vanishes on $\phi^{(k)}$ for sufficiently large $k$. Hence $\mathscr{D}^*(z)\,\phi = 0$, q.e.d. □

## § 3. The fundamental inequality for the operator $\mathscr{D}$

**1°. The Noetherian operator.** We shall assume that the matrix $p$ is such that the submodule $p\mathscr{P}^s \subset \mathscr{P}^t$ is primary. We choose some system of coordinates in $C^n$, $z=(z_1, \ldots, z_n)$, in which the associated variety $N = N(p)$ is normally placed. We write

$$\dim N = d = n - h, \quad v' = (z_1, \ldots, z_h), \quad w' = (z_{h+1}, \ldots, z_n), \quad \eta = (\xi_1, \ldots, \xi_h).$$

Because of the choice of the coordinates, the variety $N$ belongs to a variety of the type (9.1). Therefore, its intersection $N_{W'}$ with an arbitrary subspace of the form $w' = W'$ consists of a finite number of points. On the other hand, by Lemma 2, § 1, this intersection $N_{W'}$ is the variety associated with the submodule $p_{W'}\mathscr{P}_{v'}^s$ of the module $\mathscr{P}_{v'}^t$. The submodule is therefore zero dimensional. Accordingly, we can apply the known theorem of Noether [7] which asserts that a necessary and sufficient condition that the polynomial $f \in \mathscr{P}^t_{v'}$ belong to the submodule $p_{W'} \mathscr{P}^s_{v'}$ is that at every point $V' \in N_{W'}$ the Taylor series $f(V'+\eta')$ belongs to the subspace $p(Z)\mathscr{S}^s[\eta']$; here we write $Z=(V', W')$. This is called the Noether condition.

On the other hand, by the properties of the operator $d^h$ (see § 2), the condition

$$d^h(Z) f = \sum \eta'^{\tau, i} d^h_{\tau, i}(Z) f = 0, \qquad d^h_{\tau, i} = \delta_{\tau, i} d^h \tag{1.3}$$

is necessary and sufficient in order that the series $f(V'+\eta')$ belong to $p(Z)\mathscr{S}^s[\eta']$, that is, the condition (1) coincides with the Noether condition. It is known that the Noether condition can be written in the following form:

$$f \in p(Z)\mathscr{S}^s[\eta'] + m^t_{\kappa+1}$$

for some $\kappa \geqq 0$. In particular, every polynomial which vanishes at the point $V'$ with a multiplicity not less than $\kappa+1$, satisfies the Noether condition at that point. Accordingly, the orders of the functions $d^h_{\tau, i}$ in (1) do not exceed $\kappa$. On the other hand, from Remark 1, § 4, Chapter II, $d^h_{\tau, i} \eta'^{\tau, i} = 1$ for arbitrary $(\tau, i) \in t Z^h_+ \setminus \mathscr{I}$. Therefore, the order of each of the non-zero functionals $d^h_{\tau, i}$ is exactly equal to $|i|$. Consequently in the sum (1) the only terms present are those for which $|i| \leqq \kappa$. Hence, the Noether condition at the point $V'$ can be written as: $d^h_{\tau, i}(Z) f = 0$, $|i| \leqq \kappa$.

Let $\mathscr{N}^h = \{N^h_\nu\}$ be an algebraic stratification corresponding to the $p$-operator $d^h$ (see § 2). The functionals $d^h_{\tau, i}(z)$ have rational, everywhere finite coefficients on every set $N^h_\nu \setminus N^h_{\nu+1}$. Since the Noether condition is meaningful only at the points $V' \in N_{W'}$, the functionals $d^h_{\tau, i}$ do not all vanish except on the variety $N$. We find $\nu$ such that $N^h_\nu \setminus N$, and $N^h_{\nu+1} \not\supset N$. The coefficients of the functionals $d^h_{\tau, i}$ are rational regular functions on the set $N \setminus N^h_{\nu+1}$. Since at every point $z \in N$, only a finite number of these functionals are different from zero, only a finite number $e$ of these functionals do not vanish identically on $N \setminus N^h_{\nu+1}$. Let $\Delta(z)$ be the product of the denominators of the coefficients of these functions. We denote by $N^*$ the union of the varieties $N^h_{\nu+1} \cap N$ and the intersection of $N$ with the variety of roots $\Delta(z)$. $N^*$ is a proper subvariety of the irreducible variety $N$ and therefore has a lower dimension in accordance with Lemma 1, § 1. On the set $N \setminus N^*$ the polynomial $\Delta(z)$ has no zero. We set

$$d_{\tau, i}(z, D) = \Delta(z)\, d^h_{\tau, i}(z, D).$$

Here $d^h_{\tau, i}(z, D)$ is the differential operator obtained from the functional $d^h_{\tau, i}(z)$ by replacing in its expansion the functionals $\delta^{\eta'}_i$ by the operators $\frac{1}{i!} D^i_{v'}$. Placing the (vector) operators $d_{\tau, i}(z, D)$ into columns, we obtain

7 See, for example, van der Waerden [1].

a matrix of size $e \times t$. This matrix will be denoted by $d(z, D)$ and we call it the Noetherian operator associated with the matrix $p$.

**2°. The fundamental inequality.** We denote by $U_\varepsilon$, where $\varepsilon > 0$, the closed sphere in $C^n$ with radius $\varepsilon$ and center at the origin.

**Theorem.** *If the submodule $p\,\mathscr{P}^s \subset \mathscr{P}^t$ is primary, then the p-operator $\mathscr{D} = d^n$ satisfies the following inequality:*

$$\begin{aligned} &\varepsilon^q r(Z) \sup\{|\mathscr{D}(Z,\xi)\,\phi(Z)|,\ |\xi| \leqq \varepsilon^q r(Z)\} \\ &\quad \leqq \sup\{|d(z,D)\,\phi(z)|,\ z \in N \cap (Z+U_\varepsilon),\ \rho(z,L) \geqq \varepsilon^q r(Z)\}. \end{aligned} \tag{2.3}$$

*The point $Z \in C^n$, the number $\varepsilon$, $0 < \varepsilon \leqq 1$, and the function $\phi$, which is holomorphic in $Z + U_\varepsilon$, are all arbitrary. L is an arbitrary proper algebraic subvariety of N, $r(Z)$ is some power function on some algebraic stratification, independent of L; and $q \geqq 1$.*

*Proof.* We shall say that an algebraic stratification $\mathscr{M}$ is more fine than the algebraic stratification $\mathscr{L}$, if we have the inequality $c\,\theta^q(z,\mathscr{M}) \leqq \theta(z,\mathscr{L})$ for some positive $c$ and $q$.

**The inductive proposition.** *Let m be an integer lying between h and n. We write $v = (z_1, \dots, z_m)$, $w = (z_{m+1}, \dots, z_n)$, $\eta = (\xi_1, \dots, \xi_m)$. We have the inequality*

$$\begin{aligned} &\varepsilon^{q_m} r_m(Z) \sup\{|d^m(Z,\eta)\,\phi(Z)|,\ |\eta| \leqq \varepsilon^{q_m} r_m(Z)\} \\ &\quad \leqq \sup\{|d(z,D)\,\phi(z)|,\ z \in N \cap (Z+U_\varepsilon),\ \rho(z,L) \geqq \varepsilon^{q_m} r_m(z)\}. \end{aligned} \tag{3.3}$$

*Here Z is an arbitrary point not belonging to the variety $C_v \times \mu_m$, where $\mu_m$ is some proper subvariety of $C_w$; the function $\phi$ is holomorphic in $Z + U_\varepsilon$, $0 < \varepsilon \leqq 1$; $r_m(Z) \leqq \rho_m(Z)$ is some power function on an algebraic stratification $\hat{\mathscr{N}}^m$, more fine than $\mathscr{N}^m$; and $q_m$ is a positive constant.*

For $m = n$ the theorem follows from this inequality. We shall prove the inductive proposition. We consider first the case $h = n$. Then the variety $N$, being irreducible and zero dimensional, reduces to a single point. Accordingly, the subvariety $L$ is empty, and $\Delta(z) \equiv 1$. Therefore the inequality (3) is obvious.

Now let us suppose that $h < n$. We fix some integer $m$, $h \leqq m < n$, and we suppose that the inductive proposition is true for this number $m$. We shall prove it for $m+1$.

**3°. Two lemmas**

**Lemma 1.** *There exists an algebraic stratification $\hat{\mathscr{N}} = \{\hat{N}_v\}$ and an integer $v_0$, satisfying the following conditions.*

1. *The stratification $\hat{\mathscr{N}}$ is more fine than $\hat{\mathscr{N}}^m$.*

2. *If* $\nu<\nu_0$, *then Assertion II of Proposition* 5, § 1 *holds for the variety* $\hat{\mathcal{N}}_\nu$ *(relative to the decomposition* $(v, w)$*), and the discriminant subvariety belongs to* $\hat{\mathcal{N}}_{\nu+1}$.

3. *The inclusion relations*

$$C_v\times\mu_m\subset\hat{N}_{\nu_0}\subset C_v\times\lambda_m,$$

*hold, where* $\lambda_m$ *is some proper subvariety of* $C_w$.

*Proof.* Let $\hat{\mathcal{N}}^m=\{\hat{N}_\nu^m\}$. We write $\hat{N}_0=C^n$. We suppose that for some integer $k\geqq 0$ we have constructed a decreasing sequence of varieties $\hat{N}_0,\ldots,\hat{N}_k$, containing $C_v\times\mu_m$, and satisfying Condition 2 with $\nu=0,\ldots, k-1$. We construct the variety $\hat{N}_{k+1}$ in such a way that it satisfies Conditions 2 or 3 with $\nu=k$. We apply Proposition 5, § 1, to the variety $\hat{N}_k$. We suppose that Assertion I holds for it. Then we can set $\nu_0=k$.

We now suppose that for the variety $\hat{N}_k$ Assertion II of Proposition 5, § 1, is valid. Let $N'$ be some irreducible component of $\hat{N}_k$, for which Assertion II also holds, and suppose that $N_\delta$ is its discriminant subvariety. We denote by $\hat{N}_{k+1}$ the union of the remaining irreducible components of $\hat{N}_k$, the largest of the varieties $\hat{N}_k\cap N_\nu^m$ not containing $N'$, and finally the varieties $N_\delta$ and $C_v\times\mu_m$. It is clear that in constructing the variety $\hat{N}_{k+1}$, we have satisfied Condition 2 with $\nu=k$, supposing that $k<\nu_0$.

Since the variety $\hat{N}_{k+1}$ is strictly smaller than $\hat{N}_k$, if we continue this construction, we arrive at a variety $\hat{N}_k$ for which Assertion 1 of Proposition 5, § 1, is valid. Setting $\nu_0=k$, we satisfy Condition 3. Further, we denote by $\hat{N}_{\nu_0+1}$ the largest of the varieties $\hat{N}_{\nu_0}\cap\hat{N}_\nu^m$ not containing $\hat{N}_{\nu_0}$, we denote by $\hat{N}_{\nu_0+2}$ the largest of the varieties $\hat{N}_{\nu_0+1}\cap\hat{N}_\nu^m$ not containing $\hat{N}_{\nu_0+1}$, etc. The desired stratification $\hat{\mathcal{N}}=\{\hat{N}_k\}$ has now been constructed. There remains to prove that it satisfies Condition 1. It is obvious from the construction that for an arbitrary $\nu$ we can find a $\nu'$ such that $\hat{N}_\nu\setminus\hat{N}_{\nu+1}\subset\hat{N}_{\nu'}^m\setminus\hat{\ }^m_{\nu'+1}$. This inclusion relation, plus Proposition 2, § 3, of Chapter II, implies that on the set $\hat{N}_\nu\setminus\hat{N}_{\nu+1}$ we have the inequality

$$\rho(z,\hat{N}_{\nu+1})\leqq\rho(z,\hat{N}_\nu\cap\hat{N}_{\nu'+1}^m)\leqq C\left(\frac{\rho(z,\hat{N}_{\nu'+1}^m)}{|z|^2+1}\right)^q$$

for some positive $C$ and $q$. From this follows the inequality $c\,\theta^q(z,\hat{\mathcal{N}})\leqq\theta(z,\hat{\mathcal{N}}^m)$, where $c>0$ and $q>0$, and this means that the stratification $\hat{\mathcal{N}}$ is more fine than $\hat{\mathcal{N}}^m$. □

**Lemma 2.** *At every point* $Z\in C^n\setminus\hat{N}_{\nu_0}$ *the identity operator* $E$ *in the space* $\mathcal{O}^t[\xi]$ *admits the following decomposition:*

$$E=\mathscr{D}^m(Z)+p(Z)\,\mathscr{G}^m(Z), \tag{4.3}$$

*where for arbitrary $\beta$, $0<\beta\leqq 1$, $\mathscr{D}^m(Z)$ and $\mathscr{G}^m(Z)$ are operators from $\mathcal{O}^t_\beta$ to $\mathcal{O}^t_{\beta\hat{r}}$ and $\mathcal{O}^s_{\beta\hat{r}}$, and the operator $\mathscr{D}^m$ satisfies the inequality*

$$\begin{aligned}\hat{r}\sup\{|\mathscr{D}^m(Z,\xi)\,\phi(Z)|\,|\xi|\leqq\beta\,\hat{r}\}\leqq\sup\{r_m(z)\,|d^m(z,\eta)\,\phi(z)|,\\ |z-Z|\leqq\beta,\hat{r}\leqq r_m(z),|\eta|\leqq\beta\,r_m(z)\},\end{aligned}\tag{5.3}$$

*where $\phi$ is an arbitrary function holomorphic in $Z+U_\beta$, and $\hat{r}=\hat{r}(Z)$ is a power function on the stratification $\hat{\mathscr{N}}$, and $\hat{r}(Z)\leqq r_m(Z)$.*

*Proof.* The set $C^n\backslash\hat{N}_{\nu_0}$ is the union of the sets $\hat{N}_\nu\backslash\hat{N}_{\nu+1}$ with $\nu<\nu_0$. We fix an arbitrary $\nu<\nu_0$ and we prove the lemma for points of the set $\hat{N}_\nu\backslash\hat{N}_{\nu+1}$. It follows from Lemma 1 that there exists a division of the variables $z$ into two groups $\hat{v}$ and $\hat{w}$ such that the group $\hat{w}$ contains the variables of the group $w$, and the set $\hat{N}_\nu\backslash\hat{N}_{\nu+1}$ in the neighborhood of any of its points is the graph of some holomorphic function $\hat{v}=\hat{v}(\hat{w})$ which satisfies the inequality

$$|\mathrm{grad}\,\hat{v}(\hat{w})|\leqq c\,\theta^{-q}(\hat{w},\hat{\mu}),\tag{6.3}$$

where $\hat{\mu}$ is the projection on $C_{\hat{w}}$ of the discriminant subvariety of $\hat{N}_\nu$.

Let $Z$ be an arbitrary point of the set $\hat{N}_\nu\backslash\hat{N}_{\nu+1}$, and let $(\hat{V},\hat{W})$ be its $(\hat{v},\hat{w})$-coordinates. In $C_{\hat{w}}$, we consider the sphere $\hat{U}$, with center at the point $\hat{W}$, of radius $\frac{1}{2}\rho(\hat{W},\hat{\mu})$. Clearly, this does not intersect $\hat{\mu}$, and therefore, at every one of its points the set $\hat{N}_\nu\backslash\hat{N}_{\nu+1}$ consists of a finite number of pieces each of which is the locus of a holomorphic function of the form $\hat{v}=\hat{v}(\hat{w})$. Any such function, defining a set $\hat{N}_\nu\backslash\hat{N}_{\nu+1}$ in the neighborhood of the point $Z$, may be extended analytically to the whole sphere $\hat{U}$. We thus obtain a holomorphic surface of the form $\hat{v}=\hat{v}(\hat{w})$, $\hat{w}\in\hat{U}$ which in its neighborhood coincides with that portion of the set $\hat{N}_\nu\backslash\hat{N}_{\nu+1}$ which lies over $U$.

For every point $z\in C_{\hat{v}}\times\hat{U}$ we denote by $z_0(z)$ its projection onto this surface parallel to the axis of $\hat{v}$, that is, the point $(\hat{v}(\hat{w}),\hat{w})$, where $z=(\hat{v},\hat{w})$. We construct the power function $\hat{r}=\hat{r}(Z)=\hat{c}\,\theta^{\hat{q}}(Z,\hat{\mathscr{N}})$ in such a way that for an arbitrary $\beta$, $0<\beta\leqq 1$, and for an arbitrary point $z\in Z+U_{\beta\hat{r}}$ the inequalities

$$|z-Z|\leqq\tfrac{1}{2}\,\theta(\hat{W},\hat{\mu}),\tag{7.3}$$

hold, whence $z\in C_{\hat{v}}\times\hat{U}$,

$$\theta\big(z_0(z),\hat{\mathscr{N}}\big)\geqq\tfrac{1}{2}\,\theta(Z,\hat{\mathscr{N}}),\tag{8.3}$$

$$|z_0(z)-Z|\leqq\frac{\beta}{8}\,r_m\big(z_0(z)\big)\tag{9.3}$$

and

$$|z-z_0(z)|\leqq\frac{\beta}{4}\,r_m\big(z_0(z)\big).\tag{10.3}$$

We have $\rho(\hat{w}, \hat{\mu}) = \rho(z, C_{\hat{v}} \times \hat{\mu})$. The inclusion relations $\hat{N}_{\nu+1} \supset (C_{\hat{v}} \times \hat{\mu}) \cap \hat{N}_\nu$ and Proposition 2, § 3, Chapter II, imply the inequality

$$c\left(\frac{\rho(z, \hat{N}_{\nu+1})}{|z|^2+1}\right)^q \leqq \rho(\hat{w}, \hat{\mu}).$$

Therefore, choosing $\hat{c}$ and $\hat{q}$ in a suitable way, we obtain the inequality $\hat{r} \leqq \frac{1}{4}\theta(\hat{W}, \hat{\mu})$. With this the inequality (7) is established. If the number $\hat{c}$ is sufficiently small, and $\hat{q} \geqq 0$, then $\hat{r} \leqq \frac{1}{4}$. Hence

$$\theta(\hat{w}, \hat{\mu}) = \frac{\rho(\hat{w}, \hat{\mu})}{|\hat{w}|^2+1} \geqq \frac{\frac{3}{4}\rho(\hat{W}, \hat{\mu})}{\frac{5}{4}|\hat{W}|^2+\frac{3}{2}} \geqq \frac{2}{3} \cdot \frac{3}{4} \cdot \frac{\rho(\hat{W}, \hat{\mu})}{|\hat{W}|^2+1} = \frac{1}{2}\theta(\hat{W}, \hat{\mu}) \tag{11.3}$$

on the condition that $|w - W| \leqq \hat{r}$. It follows from the inequalities (6) and (11) that, under this condition

$$|\operatorname{grad} \hat{v}(\hat{w})| \leqq c' \theta^{-q'}(\hat{W}, \hat{\mu}).$$

Since $\hat{V} = \hat{v}(\hat{W})$, we have

$$\begin{aligned} |\hat{v}(\hat{w}) - \hat{V}| &= |\hat{v}(\hat{w}) - \hat{v}(\hat{W})| \\ &\leqq |\hat{w} - \hat{W}| \sup\{|\operatorname{grad} \hat{v}(\hat{w})|, |\hat{w} - \hat{W}| \leqq \hat{r}\} \leqq |\hat{w} - \hat{W}|\, c' \theta^{-q'}(\hat{W}, \hat{\mu}), \end{aligned}$$

whence

$$|z_0(z) - Z| = |(\hat{v}(\hat{w}) - \hat{V}, \hat{w} - \hat{W})| \leqq |z - Z| \left(c' \theta^{-q'}(\hat{W}, \hat{\mu}) + 1\right).$$

Thus, if the inequality

$$\hat{r}\left(c' \theta^{-q'}(\hat{W}, \hat{\mu}) + 1\right) \leqq \min\left(\tfrac{1}{4}, \tfrac{1}{4}\theta(Z, \hat{\mathcal{N}})\right),$$

is satisfied, the inequality (8) also holds. Choosing the constants and $\hat{c}$ and $\hat{q}$ in a suitable way, we satisfy this inequality also.

To obtain the inequality (9), it is sufficient to subject the function $\hat{r}$ to the condition

$$\beta\, \hat{r}\left(c' \theta^{-q'}(\hat{W}, \hat{\mu}) + 1\right) \leqq \frac{\beta}{8} r_m(z_0(z)). \tag{12.3}$$

Since the stratification $\hat{\mathcal{N}}$ is more fine than $\hat{\mathcal{N}}^m$, and since the inequality (8) holds on the set $Z + U_{\beta\hat{r}}$, we have $r_m(z_0(z)) \geqq c''\, \theta^{q''}(Z, \hat{\mathcal{N}})$ for some $c'' > 0$ and $q''$. Therefore, in order that the inequality (12) be satisfied, it is sufficient that

$$\hat{r}\left(c'\, \theta^{-q'}(\hat{W}, \hat{\mu}) + 1\right) \leqq \tfrac{1}{8} c''\, \theta^{q''}(Z, \hat{\mathcal{N}}).$$

This inequality can be satisfied also by a suitable choice of the constants $\hat{c}$ and $\hat{q}$. We remark that the inequality (12) implies that in the region $Z + U_{\beta\hat{r}}$ we have

$$|z - Z| \leqq \beta\, \hat{r} \leqq \frac{\beta}{8} r_m(z_0(z)),$$

which together with (9) implies the inequality (10). Thus the desired function $\hat{r}$ has been constructed.

Let $\phi$ be an arbitrary function which is holomorphic in $Z+U_\beta$, and let $z$ be an arbitrary point of the sphere $Z+U_{\beta\hat{r}}$. From the inequality (9) we find

$$|z_0(z)-Z|\leqq\frac{\beta}{8}$$

(since $r_m\leqq 1$) and, accordingly, the function $\phi$ is holomorphic in a $\beta/2$-neighborhood of the point $z_0(z)$. We write

$$\begin{aligned}\mathscr{D}^m(Z,z-Z)\,\phi(Z)&=d^m\big(z_0(z),\hat{v}-\hat{v}(\hat{w})\big)\,\phi\big(z_0(z)\big),\\ \mathscr{G}^m(Z,z-Z)\,\phi(Z)&=g^m\big(z_0(z),\hat{v}-\hat{v}(\hat{w})\big)\,\phi\big(z_0(z)\big).\end{aligned}\tag{13.3}$$

Here $\hat{v}-\hat{v}(\hat{w})$ is a vector in the space $C_{\hat{v}}$ regarded as a coordinate subspace of $C_v$. The inequality (2.2) implies that for arbitrary $\hat{w}$ the righthand side of (13) is holomorphic in $\hat{v}$ in the sphere

$$|\hat{v}-\hat{v}(\hat{w})|\leqq\beta\, r_m\big(z_0(z)\big),\tag{14.3}$$

since $r_m\big(z_0(z)\big)\leqq\rho_m\big(z_0(z)\big)$. We remark that it follows from (10) that this inequality is automatically satisfied, if $z\in Z+U_{\beta\hat{r}}$.

The functions (13) are holomorphic in $\hat{w}$ also, provided $z\in Z+U_{\beta\hat{r}}$. In fact, these functions are the sums of series that converge absolutely in the region (14)

$$\begin{aligned}&\sum\big(\hat{v}-\hat{v}(\hat{w})\big)^i\,d_i^m\big(z_0(z)\big)\,\phi\big(z_0(z)\big),\\ &\sum\big(\hat{v}-\hat{v}(\hat{w})\big)^i\,g_i^m\big(z_0(z)\big)\,\phi\big(z_0(z)\big),\end{aligned}$$

in which the function $\phi\big(z_0(z)\big)$ and the coefficients of the functionals $d_i^m\big(z_0(z)\big)$ and $g_i^m\big(z_0(z)\big)$ are holomorphic functions of $z_0(z)$ and, consequently, are holomorphic functions of $z$.

Thus, we have shown that the functions (13) are holomorphic in $Z+U_{\beta\hat{r}}$. Moreover, we have established the fact that $\mathscr{G}^m$ is a continuous operator from $\mathscr{O}_\beta^t$ to $\mathscr{O}_{\beta\hat{r}}^s$. The inequality (5) follows from (13), (12) and (9). $\square$

**4°. Proof of the inductive proposition.** We write

$$v^*=(z_1,\ldots,z_{m+1})=(v,z_{m+1}),\qquad w^*=(z_{m+2},\ldots,z_n),\qquad \eta^*=(\xi_1,\ldots,\xi_{m+1}),$$

and we denote by $C_{v^*}$ and $C_{w^*}$ the corresponding coordinate subspaces in $C^n$.

Let $q(w)$ be some polynomial which is not identically zero and which vanishes on the variety $\lambda_m$ constructed in Lemma 1. Let $\lambda$ be the variety of the roots of $q$, let $\nu$ be its order relative to $z_{m+1}$ and let $q_0(w^*)$ be the highest order coefficient. If $m=h$, we subject the polynomial $q$ to an additional condition: it vanishes on $L$ and $q_0(w^*)\equiv\text{const}\neq 0$. The latter condition can be satisfied if we choose the system of coordinates in the space $C_w$ in a suitable fashion. We note that a transformation of coordinates in this space has no influence on the preceding reasoning for $m=h$.

We apply Theorem 2, § 1 to the matrix $p$ and the division of the variables $z$ into the groups $v^*$ and $w^*$. Let $\mu\subset C_{w^*}$ be the variety developed in this theorem. We denote by $\mu_{m+1}$ the union of $\mu$ and the variety of the roots of $q_0$. Let $\check{\mathcal{N}}=\{\check{N}_\nu\}$ be the product of the stratifications $\hat{\mathcal{N}}$, $\mathcal{N}^{m+1}$, $\mathcal{N}^n$ and the stratification $C^n\supset C_{v^*}\times\mu_{m+1}\supset\phi$. We note that if $m=h$, $q_0\equiv\text{const}\neq 0$ and therefore the variety $\mu_{m+1}$, and consequently the decomposition $\check{\mathcal{N}}$, do not depend on the variety $L$.

Let $Z=(V, Z_{m+1}, W^*)$ be an arbitrary point of the set $C^n\setminus(C_{v^*}\times\mu_{m+1})$, and let $\varepsilon$ be an arbitrary positive number not exceeding unity. We consider the complex line $l$, parallel to the axis $z_{m+1}$, defined by the equations $v=V$, $w^*=W^*$. On this line there are $\nu$ roots of the polynomial $q(w)$. We consider $\nu+1$ circles

$$|z_{m+1}-Z_{m+1}|=\gamma\frac{k}{\nu+1}, \qquad k=1,\dots,\nu+1,$$

placed on the line $l$. Here $\gamma$ is a quantity, not yet defined, not exceeding $\varepsilon/4$. At least one of the circles $S$ lies at a distance not less than $\gamma\dfrac{1}{2(\nu+1)}$ from the roots of the polynomial $q$ that are placed along $l$. On this line, $|q(w)|$ is equal to the product of the distances from the point $w$ to these roots, multiplied by $|q_0(W^*)|$. Accordingly, on $S$

$$|q(w)|\geqq|q_0(W^*)|\left(\frac{\gamma}{2(\nu+1)}\right)^{\nu}.$$

Since the roots of the polynomial $q_0$ lie in $\mu_{m+1}$, it follows from the rightmost inequality in (8.3), Chapter II, that

$$|q_0(W^*)|\geqq C\,\theta^q(W^*,\mu_{m+1})\geqq C\,\theta^q(Z, C_{v^*}\times\mu_{m+1})\geqq c_1\,\theta^{q_1}(Z,\check{\mathcal{N}}).$$

Therefore $|q(w)|\geqq c_2\big(\gamma\,\theta(Z,\check{\mathcal{N}})\big)^{q_2}$ for some positive $c_2$ and $q_2$. Hence, in view of the leftmost inequality of (8.3), Chapter II, the distance from $S$ to the variety of the roots of $q$ is not less than $\delta=c_3\big(\gamma\,\theta(Z,\check{\mathcal{N}})\big)^{q_3}$, where $c_3$ and $q_3$ are some positive constants. We shall suppose that these constants are so chosen that $\delta\leqq\varepsilon$. Since the polynomial $q$ depends only on the variables $w$, the distance from the cylinder $C_v\times S$ to the variety of the roots of $q$ is also not less than $\delta$.

Let $z^0$ be an arbitrary point of the circle $S$, and (in accordance with the notation of § 5, Chapter III) let $e^{z^0, R}$ denote the complex ellipsoid with center at the point $z^0$, and semi-axes parallel to the axes of coordinates and equal to $R_1 = \cdots = R_m = \varepsilon/8$, $R_{m+1} = \cdots = R_n = \delta/8$. Since the ellipsoid $2e^{z^0, R}$ lies in the $\frac{\delta}{2}$-neighborhood of the cylinder $C_v \times S$, its distance from the variety of the roots of $q$ is not less than $\delta/2$. Since the radius of $S$ does not exceed $\varepsilon/4$, and the diameter of $2e^{z^0, R}$ is not greater than $\varepsilon/4$, the ellipsoid $2e^{z^0, R}$ together with some neighborhood of it, lies within the sphere $Z + U_\varepsilon$.

Let $\phi$ be some holomorphic function in $Z + U_\varepsilon$. From what we have said above, it follows that it is holomorphic in the neighborhood of $2e^{z^0, R}$ and, accordingly, lies in the space $E_1^{z^0, R}$ (see § 5, Chapter III). We consider the function $r_z = \hat{r}\, \rho_n$, where we have set $\hat{r} = \hat{r}(z)$ and $\rho_n = \rho_n(z)$. This is a power function on the product of the stratifications $\hat{\mathcal{N}}$ and $\mathcal{N}^n$ and it does not exceed $\rho_n$ (since $\hat{r} \leqq 1$). Therefore, we can substitute this function in Corollary 1, § 5, Chapter III. Applying this corollary to the function $\phi$, we obtain the decomposition

$$\phi = \phi_{\mathscr{D}} + p\,\psi, \tag{15.3}$$

in which $\phi_{\mathscr{D}}$ and $\psi$ are functions holomorphic in $e^{z^0, R}$, and

$$\begin{aligned} &\sup\{|\phi_{\mathscr{D}}(z)|, \quad z \in e^{z^0, R}\} \\ &\quad \leqq C\left(\frac{|z^0|+1}{R}\right)^q \sup\left\{\hat{r}\,\rho_n\,|d^n(z,\xi)\,\phi(z)|, \quad z \in 2e^{z^0, R},\right. \\ &\quad\quad \left.\hat{r}\,\rho_n \geqq c\left(\frac{R}{|z^0|+1}\right)^q, \ |\xi| \leqq R^1\,\hat{r}\,\rho_n\right\}. \end{aligned}$$

The right side of this inequality can be bounded with the aid of the function $\mathscr{D}(z)\,\phi(z)$. For this purpose, we apply the decomposition (4) to the function $\phi$ at an arbitrary point $z \in 2e^{z^0, R}$, and then we act on it with the operator $d^n$. As a result we obtain the equation

$$d^n(z,\xi)\,\phi(z) = d^n(z,\xi)\,\mathscr{D}^m(z)\,\phi(z).$$

Setting $m = n$ and $\varepsilon = R^1\,\hat{r}$ in (2.2), we arrive at the inequality

$$\begin{aligned} &\sup\{|d^n(z,\xi)\,\phi(z)|, |\xi| \leqq R^1\,\hat{r}\,\rho_n\} \\ &\qquad\qquad \leqq \frac{1}{\rho_n} \sup\{|\mathscr{D}^m(z,\xi)\,\phi(z)|, |\xi| \leqq R^1\,\hat{r}\}. \end{aligned} \tag{16.3}$$

Setting $\beta = R^1$ in (5), we obtain a bound for the right side of (16):

$$\frac{1}{\rho_n} \sup \{r_m(\zeta)\, |d^m(\zeta, \eta)\, \phi(\zeta)|,\ |\zeta - z| \leqq \beta,\ \hat{r} \leqq r_m(\zeta),\ |\eta| \leqq \beta\, r_m(\zeta)\}.$$

Thus, finally

$$\sup\{|\phi_{\mathscr{D}}(z)|,\ z \in e^{z^0, R}\} \leqq H(\phi), \tag{17.3}$$

where

$$H(\phi) = \frac{1}{\sigma} \sup \{r_m(\zeta)\, |d^m(\zeta, \eta)\, \phi(\zeta)|,\ z \in 2\, e^{z^0, R},\ \hat{r}\, \rho_n \geqq \sigma,\ |\zeta - z| \leqq \beta,\ \hat{r} \leqq r_m(\zeta),\ |\eta| \leqq \beta\, r_m(\zeta)\},$$

and

$$\sigma = \sigma(Z) = c \left( \frac{R}{|Z|+1} \right)^q$$

for some positive $c$ and $q$.

We now make use of the theorem of § 2, writing $m^* = m+1$. This is possible, since by hypothesis the point $W^*$ does not belong to $\mu$, and therefore by Theorem 2 of § 1

$$p(Z)\, \mathscr{S}^s \cap \xi_{m+1}\, \mathscr{S}^t = p(Z)\, \xi_{m+1}\, \mathscr{S}^s, \qquad \mathscr{S} = \mathscr{S}[\eta^*].$$

By the theorem of § 2, we have the decomposition

$$E_{\eta^*} = \mathscr{D}^*(Z) + p(Z)\, \mathscr{G}^*(Z), \tag{18.3}$$

where $\mathscr{D}^*(Z)$ and $\mathscr{G}^*(Z)$ are operators from $\mathscr{S}^t$ to $\mathscr{S}^t$ and $\mathscr{S}^s$ respectively, and the operator $\mathscr{D}^*$ has the following properties.

1. *For arbitrary* $\alpha$, $0 < \alpha \leqq 1$, *the operator* $\mathscr{D}^*(Z)$ *acts from* $\mathscr{O}^t_\alpha$ *to* $\mathscr{O}^t_{\alpha^q \rho^*}$ *with a norm not exceeding* $1/\rho^*$, *where* $\rho^* = \rho^*(Z)$ *is some power function on the stratification* $\mathscr{N}^m$, *and* $q \geqq 1$ *is a constant.*

2. *The operator* $\mathscr{D}^*(Z)$ *vanishes on* $p(Z)\, \mathscr{S}^s$.

3. *For arbitrary* $t$, $|t| \leqq \alpha^q \rho^*$, *the value of the function* $\mathscr{D}^*(Z\, \eta^*)\, \phi$, *where* $\phi \in \mathscr{O}^t_\alpha$, *evaluated on the subspace* $\xi_{m+1} = t$, *depends only on the values of the function* $\phi$ *on the same subspace.*

The following step, which is the central step in the proof, consists in bounding the effect of the operator $\mathscr{D}^*(Z)$ on $\phi$ by the upper bound of the quantity $H(\phi)$ in the disc $S$, all this in the form of the inequality (23). Further, expressing the operator $d^{m+1}$ in terms of $\mathscr{D}^*$ and substituting in the inequality mentioned, we shall after some simplification make the passage from $m$ to $m+1$ in the inductive proposition. Let $z^0_{m+1}$ be the

$z_{m+1}$-th coordinate of the point $z^0$. In Eq. (15) we write $w=(z^0_{m+1}, W^*)$:

$$\phi(v, z^0_{m+1}, W^*)=\phi_{\mathscr{D}}(v, z^0_{m+1}, W^*)+p(v, z^0_{m+1}, W^*)\,\psi(v, z^0_{m+1}, W^*). \quad (19.3)$$

The functions $\phi_{\mathscr{D}}(v, z^0_{m+1}, W^*)$, and $\psi(v, z^0_{m+1}, W^*)$, regarded as functions of the variables $v^*$, are constant with respect to $z_{m+1}$, and will be denoted respectively by $\hat{\phi}_{\mathscr{D}}(v^*)$ and $\hat{\psi}(v^*)$. From (19),

$$\phi(v^*, W^*)|_{z_{m+1}=z^0_{m+1}}=\{\hat{\phi}_{\mathscr{D}}(v^*)+p(v^*, W^*)\,\hat{\psi}(v^*)\}|_{z_{m+1}=z^0_{m+1}}. \quad (20.3)$$

Since the function $\phi$ is holomorphic in $Z+U_\varepsilon$, and the functions $\phi_{\mathscr{D}}$ and $\psi$ are holomorphic in $e^{z^0, R}$, the functions (20) are holomorphic in $v^*$ in the $\varepsilon/8$-neighborhood of the point $V^*=(V, Z_{m+1})$. We apply the operator $\mathscr{D}^*(Z)$ to these functions at the point $V^*$. By its Property 1 we obtain the functions

$$\mathscr{D}^*(Z, v^*-V^*)\,\phi(Z),\ \mathscr{D}^*(Z, v^*-V^*)\,\{\hat{\phi}_{\mathscr{D}}(V^*)+p(Z)\,\hat{\psi}(V^*)\}, \quad (21.3)$$

which are holomorphic in the sphere

$$|v^*-V^*|\leqq\left(\frac{\varepsilon}{8}\right)^q \rho^*. \quad (22.3)$$

The constant $\gamma$, which appears in the construction of the circle $S$, will now be given the fixed value $\gamma=\varepsilon^q\rho'$, $\rho'=8^{-q-1}\rho^*$. We shall assume that $\rho^*\leqq 1$, $q\geqq 1$, and accordingly, $\gamma\leqq\varepsilon/4$, as we supposed earlier. It is clear that an arbitrary point of the circle $S$, together with the $\gamma$-neighborhood of it, lies within the sphere (22). Therefore the functions (21) are holomorphic in the $\gamma$-neighborhood of $S$.

The functions (21), generally speaking, do not coincide. However, by Property 3 of the operator $\mathscr{D}^*(Z)$, the values of these functions for $z_{m+1}=z^0_{m+1}$ depend only on the values of the functions (20) for $z_{m+1}=z^0_{m+1}$. Therefore, by Eq. (20) and Property (2)

$$\mathscr{D}^*(Z, v^*-V^*)\,\phi(Z)\,|_{z_{m+1}=z^0_{m+1}}=\mathscr{D}^*(Z, v^*-V^*)\,\phi_{\mathscr{D}}(Z)\,|_{z_{m+1}=z^0_{m+1}}.$$

Applying Property 1 with $\alpha=\frac{\varepsilon}{8}$, and the inequality (17), we obtain

$$\sup\{|\mathscr{D}^*(Z, v^*-V^*)\,\phi(Z)\,|_{z_{m+1}=z^0_{m+1}}|,\ |v-V|\leqq\varepsilon^q\rho'\}$$
$$\leqq\frac{1}{\rho^*(Z)}\sup\left\{|\hat{\phi}_{\mathscr{D}}(v^*)|,\ |v-V|\leqq\frac{\varepsilon}{8}\right\}\leqq\frac{1}{\rho^*(Z)}H(\phi).$$

In this inequality we pass to the upper bound over the circle $S$. On the left we obtain an upper bound of the function $\mathscr{D}^*(Z, v^*-V^*)\,\phi(Z)$ on the set $u\times S$, where $u$ is a sphere in $C_v$ with radius $\varepsilon^q\rho'$ and center at the point $V$. Since this function is holomorphic in the sphere (22), which contains the set $u \times S$, this upper bound is not less than the upper bound over the set $u \times U$, where $U$ is the disc bounded by the circle $S$. The set $u \times U$, obviously, contains the sphere $|v^* - V^*| \leqq \gamma^* = \dfrac{\gamma}{2(\nu+1)}$. Thus,

$$\begin{aligned}&\sup\{|\mathscr{D}^*(Z,\eta^*)\,\phi(Z)|,\ |\eta^*|\leqq\gamma^*\}\\&\qquad\leqq\frac{1}{\rho^*}\sup\{H(\phi),\ z^0\in S\}\qquad(\eta^*=v^*-V^*).\end{aligned}\tag{23.3}$$

Applying the decomposition (18) to the function $\phi(Z+\eta^*)$, and then applying the operator $d^{m+1}(Z)$, we obtain

$$d^{m+1}(Z,\eta^*)\,\phi(Z) = d^{m+1}(Z,\eta^*)\,\mathscr{D}^*(Z)\,\phi(Z).$$

In view of the inequality (2.2), both sides of this inequality converge absolutely for $|\eta^*| \leqq \varepsilon^q \rho''$, where $\rho'' = \rho''(Z)$ is some power function on $\check{\mathscr{N}}$. Then from (23)

$$\begin{aligned}&\sup\{|d^{m+1}(Z,\eta^*)\,\phi(Z)|,\ |\eta^*|\leqq\varepsilon^q\rho''\}\\&\qquad\leqq\frac{1}{\rho''}\sup\{|\mathscr{D}^*(Z,\eta^*)\,\phi(Z)|,\ |\eta^*|\leqq\gamma^*\}\\&\qquad\leqq\frac{1}{\rho''\rho^*}\sup\{H(\phi),\ z^0\in S\}\\&\qquad=\frac{1}{\rho''\rho^*\sigma}\sup\{r_m(\zeta)\,|d^m(\zeta,\eta)\,\phi(\zeta)|,\ z^0\in S,\ z\in 2e^{z^0,R},\\&\hat{r}(z)\,\rho_n(z)\geqq\sigma,\ |\zeta-z|\leqq\beta,\ \hat{r}(z)\leqq r_m(\zeta),\ |\eta|\leqq\beta\, r_m(\zeta)\}.\end{aligned}\tag{24.3}$$

On the right side of this inequality

$$|\zeta - Z| \leqq |\zeta - z| + |z - z^0| + |z^0 - Z| \leqq \beta + \frac{\varepsilon}{4} + \gamma \leqq R^1 + \frac{\varepsilon}{4} + \frac{\varepsilon}{4} \leqq \varepsilon,$$

and

$$r_m(\zeta) \geqq \hat{r}(z) \geqq \hat{r}(z)\,\rho_n(z) \geqq \sigma(Z) = c\left(\frac{R}{|Z|+1}\right)^q \geqq \varepsilon^{q'}\, r^*(Z),$$

where $r^*(Z)$ is some suitable power function on $\check{\mathscr{N}}$. Accordingly, we have from (24) the inequality

$$\begin{aligned}&\rho''\rho^*\sigma\sup\{|d^{m+1}(Z,\eta^*)\,\phi(Z)|,\ |\eta^*|\leqq\varepsilon^q\rho''\}\\&\qquad\leqq\sup\{r_m(\zeta)\,|d^m(\zeta,\eta)\,\phi(\zeta)|,\ |\zeta-Z|\\&\qquad\leqq\varepsilon,\ r_m(\zeta)\geqq\varepsilon^{q'}\,r^*(Z),\ |\eta|\leqq\varepsilon\, r_m(\zeta)\}.\end{aligned}\tag{25.3}$$

We now assume that $m>h$. Then, applying the inductive proposition to the bound standing on the right side of (25), we obtain the inequality

$$\begin{aligned}
&\rho''\,\rho^*\,\sigma \sup\{|d^{m+1}(Z,\eta^*)\,\phi(Z)|,\ |\eta^*|\leqq \varepsilon^q \rho''\}\\
&\quad\leqq \sup\{|d(\zeta',D)\,\phi(\zeta')|,\ \zeta'\in N\cap(\zeta+U_\varepsilon),\ \rho(\zeta',L)\\
&\quad\geqq \varepsilon^{q_m} r_m(\zeta),\ |\zeta-Z|\leqq\varepsilon,\ r_m(\zeta)\geqq \varepsilon^{q'} r^*(Z)\}\\
&\quad\leqq \sup\{|d(\zeta',D)\,\phi(\zeta')|,\ \zeta'\in N\cap(Z+U_{2\varepsilon}),\ \rho(\zeta',L)\geqq \varepsilon^{q'+q_m} r^*(Z)\}.
\end{aligned}\tag{26.3}$$

Here replacing $2\varepsilon$ by $\varepsilon$, we arrive at the inequality (3) with $m+1$ instead of $m$, with some suitable power function $r_{m+1}$ over the stratification $\check{\mathscr{N}}$, and with the constant $q_{m+1}\geqq 1$.

Now suppose $m=h$. Then the inequality (25) can be sharpened in the following way. Since the polynomial $q$ vanishes on $L$, we have $\rho(2e^{z^0,R}L)\geqq\delta/2$. Since on the right side of (24) $|\zeta-z|\leqq\beta=R^1\leqq\delta/4$, we have $\rho(\zeta,L)\geqq\delta/4$. Therefore, on the right side of (25) the upper bound can be extended only over those $\zeta$ for which

$$|\zeta-Z|\leqq\varepsilon,\qquad \rho(\zeta,L)\geqq\varepsilon^{q^*} r^*(Z).$$

Here the constant $q^*$ and the power function $r^*$ on the stratification $\check{\mathscr{N}}$ will be chosen so that $\varepsilon^{q^*} r^*(Z)\leqq\delta/4$. Further we have

$$d^h(\zeta,\eta)\,\phi(\zeta)=\sum_{|i|\leqq\kappa}\eta^{\tau,i}\,d^h_{\tau,i}(\zeta,D)\,\phi(\zeta),\tag{27.3}$$

whence

$$\sup\{|d^h(\zeta,\eta)\,\phi(\zeta)|,\ |\eta|\leqq\varepsilon\}\leqq e\max_{\tau,i}|d^h_{\tau,i}(\zeta,D)\,\phi(\zeta)|,$$

since $\varepsilon\leqq 1$. Accordingly, from (25)

$$\begin{aligned}
&\rho''\,\rho^*\,\sigma\sup\{|d^{h+1}(Z,\eta^*)\,\phi(Z)|,\ |\eta^*|\leqq\varepsilon^q\rho''\}\\
&\quad\leqq e\max_{\tau,i}\sup\{r_h(\zeta)\,|d^h_{\tau,i}(\zeta,D)\,\phi(\zeta)|,\ |\zeta-Z|\leqq\varepsilon,\ \rho(\zeta,L)\geqq\varepsilon^{q^*} r^*(Z)\}.
\end{aligned}\tag{28.3}$$

We now recall that the choice of the stratification $\hat{\mathscr{N}}^h$ and the function $r_h$ was in no way constrained. We set $\hat{\mathscr{N}}^h=\mathscr{N}^h$, and we subject the function $r_h$ to the condition $r_h(z)\leqq\Delta(z)$, $z\in N\setminus N^*$. Without loss of generality, we may suppose that the variety $L$ contains $N^*$. Then the right side of (28) does not exceed the quantity

$$\sup\{|d(\zeta,D)\,\phi(\zeta)|,\ |\zeta-Z|\leqq\varepsilon,\ \rho(\zeta,L)\geqq\varepsilon^{q^*} r^*(Z)\}.$$

We have now established the inductive proposition for $m+1=h+1$. Thus, the inductive proposition has been proved for all $m$, and therefore the proof of the theorem is complete. □

**5°. Noetherian operators for arbitrary *P*-matrices.** Let $p: \mathscr{P}^s \to \mathscr{P}^t$ be an arbitrary $\mathscr{P}$-matrix. We write down the primary decomposition for the submodule $p\,\mathscr{P}^s \subset \mathscr{P}^t$:

$$p\,\mathscr{P}^s = \mathfrak{p}_0 \cap \dots \cap \mathfrak{p}_l.$$

For arbitrary $\lambda$ we can find a $\mathscr{P}$-matrix $p: \mathscr{P}^{s_\lambda} \to \mathscr{P}^t$ such that $p_\lambda \mathscr{P}^{s_\lambda} = \mathfrak{p}_\lambda$. Since the module $\mathfrak{p}_\lambda$ is primary, the construction of 1° is applicable to matrix $p_\lambda$. Let $N^\lambda$ and $d^\lambda(z, D)$ be the manifold and the Noetherian operator associated with $p_\lambda$. The set of varieties $N^\lambda$ and the operators $d^\lambda(z, D)$ $\lambda = 0, \dots, l$ will be said to be associated with the matrix $p$.

**Corollary 1.** *The p-operator D satisfies the following inequality:*

$$\begin{aligned} &\varepsilon^q\, r(Z) \sup\{|\mathscr{D}(Z, \xi)\,\phi(Z)|, |\xi| \leqq \varepsilon^q\, r(Z)\} \\ &\quad \leqq \max_\lambda \sup\{|d^\lambda(z, D)\,\phi(z)|, z \in N^\lambda(Z + U_\varepsilon), \rho(z, L^\lambda) \geqq \varepsilon^q\, r(Z)\}. \end{aligned} \tag{29.3}$$

*Here the point $Z \in C^n$, the number $\varepsilon$, $0 < \varepsilon \leqq 1$, and the function $\phi$, which is holomorphic on $Z + U_\varepsilon$, are arbitrary. The $L^\lambda$ are arbitrary proper subvarieties of the $N^\lambda$, $\lambda = 0, \dots, l$; $r(Z)$ is a power function on some algebraic subvariety not depending on $L^\lambda$, and $q \geqq 1$ is a constant.*

*Proof.* Suppose that $\mathscr{D}^\lambda$ is a $p_\lambda$-operator, $\lambda = 0, \dots, l$. Corollary 1, § 1, implies the equation

$$\mathscr{D}(Z) = \sum \Lambda_\lambda(Z)\,\mathscr{D}^\lambda(Z),$$

where $\Lambda_\lambda(Z)$ is an operator in $\mathscr{O}^t$, which for arbitrary $\delta$, $0 < \delta \leqq 1$, acts from $\mathscr{O}^t_\delta$ to $\mathscr{O}^t_{\delta\rho}$ with a norm not exceeding $\delta^{-q} \rho^{-1}$, where $\rho = \rho(Z)$ is a power function on some stratification $\mathscr{N}'$. Hence for an arbitrary function $\rho$, holomorphic in $Z + U_\delta$, we have the inequality

$$\begin{aligned} &\delta^q\, \rho \sup\{|\mathscr{D}(Z, \xi)\,\phi(Z)|, |\xi| \leqq \delta\rho\} \\ &\quad \leqq \max_\lambda \sup\{|\mathscr{D}^\lambda(Z, \xi)\,\phi(Z)|, |\xi| \leqq \delta\}. \end{aligned} \tag{30.3}$$

Applying the theorem to the matrices $p_\lambda$, we obtain for arbitrary $\lambda$ the inequality

$$\begin{aligned} &\varepsilon^{q_\lambda}\, r_\lambda(Z) \sup\{|\mathscr{D}^\lambda(Z, \xi)\,\phi(Z)|, |\xi| \leqq \varepsilon^{q_\lambda}\, r_\lambda(Z)\} \\ &\quad \leqq \sup\{|d^\lambda(z, D)\,\phi(z)|, z \in N^\lambda \cap (Z + U_\varepsilon), \rho(z, L^\lambda) \geqq \varepsilon^{q_\lambda}\, r_\lambda(Z)\} \end{aligned} \tag{31.3}$$

governing the functions $\phi$, which are holomorphic in $Z + U_\varepsilon$. Here $r_\lambda(Z)$ is a power function on some stratification $N^\lambda$. We write $q = \max q_\lambda$, and we denote by $\mathscr{N}$ the product of all the stratifications $\mathscr{N}^\lambda$ and $\mathscr{N}'$. We choose the power function $r$ on $\mathscr{N}$ in such a way that $r \leqq \min r_\lambda$. We substitute $\delta = \varepsilon^q r(Z)$ in (30) and we combine this inequality with the inequalities (31). We are then led to the inequality (29), in which $r$ is a power function on the stratification $\mathcal{N}$. □

## § 4. Noetherian operators

The construction introduced in § 3 for the Noetherian operators admits arbitrariness. For instance, a change of coordinate system in $C^n$ leads to other Noetherian operators. It is clear that there are also other methods for constructing the operators $d(z, D)$, satisfying the theorem of § 3. We now describe some of the classes of differential operators having this property.

**1°. Noetherian operators in the wide sense of the word.** We fix a $\mathscr{P}$-matrix $p: \mathscr{P}^s \to \mathscr{P}^t$.

**Definition 1.** We shall suppose that the submodule $p\mathscr{P}^s \subset \mathscr{P}^t$ is primary. The matrix $\partial(z, D)$ of size $E \times t$ will be called the Noetherian operator (in the wide sense) associated with the matrix $p$, if

I. This matrix consists of differential operators with polynomial coefficients.

II. There exists a proper algebraic subvariety $N_*$ of the variety $N = N(p)$ such that for an arbitrary point $Z \in N \setminus N_*$, the condition

$$\partial(z, D)\, \phi(z)|_N = 0$$

is necessary and sufficient in order that the function $\phi \in \mathcal{O}_Z^t$ belong to $p\,\mathcal{O}_Z^s$.

The manifold $N_*$ will be called exceptional.

It follows from the theorem of § 3 that the operator $d(z, D)$, constructed in 1°, is a Noetherian operator in the wide sense for an arbitrary choice of the coordinate system in $C^n$, if the submodule $p\mathscr{P}^s$ is normally placed and $N_* = \varnothing$. In fact, the theorem of § 3 remains valid when the operator $d(z, D)$ is replaced by an arbitrary Noetherian operator in the wide sense[8]. We shall establish a partial result, which is sufficient for our purposes.

**Theorem.** *Let $p\mathscr{P}^s \subset \mathscr{P}^t$ be a primary submodule of dimension $n-h$, normally placed in the coordinate system $z = (v, w)$, where $v = (z_1, \ldots, z_h)$ and $w = (z_{h+1}, \ldots, z_n)$. Further, let $\partial(z, D)$ be a Noetherian operator in the wide sense, containing differentiations with respect to the variable $v$ only. Then the theorem of* § 3 *remains valid when the operator $d(z, D)$ is replaced by the operator $\partial(z, D)$.*

8 See V. P. Palamodov [14].

*Proof.* We shall show that there exists a matrix $\Lambda$, consisting of rational functions, such that

$$d(z, D) = \Lambda(z)\,\partial(z, D). \tag{1.4}$$

In every element of the matrix $\partial(z, D)$ we replace the operators $(1/i!)\,D_v^i$ by the functionals $(\delta_i^\eta, \eta=(\xi_1, \dots, \xi_h)$. We obtain the matrix $\partial(z, \delta)$, in which the rows are the linear functionals $\partial_k(z, \delta)\colon \mathscr{S}^t[\eta] \to C, k=1, \dots, E$. Let $\gamma$ be the highest order of these functionals. We consider the matrix

$$\{\partial_k(z, \delta)\,\eta^{\tau, i}, 1 \leqq k \leqq E, |i| \leqq \gamma, 1 \leqq \tau \leqq t\} \tag{2.4}$$

of size $E \times t \sum_\gamma$. Let $M$ be a minor of this matrix of maximal order $\rho$, and not identically singular. Since the matrix (2) consists of polynomials, the minor $M$ is not singular on the set $N \setminus N'$, where $N'$ is some proper subvariety of $N$. Let $\alpha_\sigma = (\tau_\sigma, i_\sigma) \in t\,Z_+^h$, $\sigma = 1, \dots, \rho$ be the indices of the columns, and $1, \dots, \rho$ the indices of the rows making up the minor $M$. We consider the system of linear equations

$$\sum_1^\rho \lambda_k(z)\,\partial_k(z, \delta)\,\eta^{\alpha_\sigma} = d(z, \delta)\,\eta^{\alpha_\sigma}, \qquad \sigma = 1, \dots, \rho, \tag{3.4}$$

in the unknown vectors $\lambda_k(z)$ of order $e$ ($e$ is the number of components of the vector $d(z, \delta)\,\eta^{\alpha_\sigma}$). This system has a solution consisting of rational functions, in which the denominators are $\det M$. We write $\lambda_k(z) \equiv 0$ for $k = \rho+1, \dots, E$. We shall show that

$$\sum_1^E \lambda_k(z)\,\partial_k(z, \delta) = d(z, \delta). \tag{4.4}$$

on the set $N \setminus N''$, where $N'' = N' \cup N_* \cup N_\delta$, and $N_\delta$ is the discriminant subvariety of $N$ relative to $v$.

We fix an arbitrary point $Z \in N \setminus N''$. In the neighborhood of $Z$ the variety $N$ is given by the equation $v = v(w)$, where $v(w)$ is a holomorphic function. Since the operator $\partial(z, D)$ contains differentiations only with respect to $v$, we have in the neighborhood of $Z$

$$\partial(z, D)\,(v - v(w))^{\tau, i}|_N = \partial(z, \delta)\,\eta^{\tau, i}, \qquad (\tau, i) \in t\,Z_+^h,$$

where $(v - v(w))^{\tau, i} = (v - v(w))^i\, e_\tau$, and $e_\tau$ is the column with index $\tau$ of the unit matrix of order $t$. It is clear from the construction of the operator $d(z, D)$ that it contains differentiations only with respect to $v$, and therefore, has a similar property.

We now fix an arbitrary couple $(\tau, i)$ and we consider the system of equations

$$\sum_\sigma \mu_\sigma(z)\,\partial_k(z, \delta)\,\eta^{\alpha_\sigma} = \partial_k(z, \delta)\,\eta^{\tau, i}, \qquad k = 1, \dots, \rho,$$

with respect to the functions $\mu_\sigma(w)$, which are defined in the neighborhood of $W$, where $Z=(V, W)$. This system is soluble, and its solution is holomorphic in the neighborhood of $W$, since its matrix is the minor $M$ which is non-singular on $N \setminus N'$. Therefore,

$$\partial_k(z, \delta)\left(\eta^{\tau, i} - \sum_\sigma \mu_\sigma(w)\,\eta^{\alpha_\sigma}\right) = 0 \tag{5.4}$$

for all $k=1, \ldots, \rho$, and also for $k=\rho+1, \ldots, E$, since the rows $\partial_k(z, \delta)$ with $k=1, \ldots, \rho$ form a basis for all the rows of the matrix $\partial(z, \delta)$. Hence $\partial(z, D) f(z)|_N = 0$, where

$$f(z) = \left(v - v(w)\right)^{\tau, i} - \sum_\sigma \mu_\sigma(w)\left(v - v(w)\right)^{\alpha_\sigma} \in \mathcal{O}^t_Z.$$

Since $\partial(z, D)$ is a Noetherian operator in the wide sense, it follows that $f \in p\,\mathcal{O}^s_Z$. Therefore $d(z, D) f(z)|_N = 0$. Since this equation is satisfied in particular at the point $Z$, we have

$$d(Z, \delta)\left(\eta^{\tau, i} - \sum_\sigma \mu_\sigma(W)\,\eta^{\alpha_\sigma}\right) = 0.$$

Combining this equation with (3) and (5), we obtain

$$\left[\sum_1^E \lambda_k(Z)\,\partial_k(Z, \delta) - \mathrm{d}(Z, \delta)\right] \eta^{\tau, i} = 0.$$

The couple $(\tau, i)$ being arbitrary, the quantity in square brackets is equal to zero. Since the point $Z \in N \setminus N''$ is arbitrary, Eq. (4) follows.

Eq. (4) implies Eq. (1), in which $\Lambda(z)$ is a matrix consisting of rational functions which are regular on $N \setminus N''$. In the theorem of § 3, we now choose the variety $L$ so that $N'' \subset L$. Proposition 2, § 3, Chapter II, implies that for arbitrary $Z \in N$ and $\varepsilon$, $0 < \varepsilon \leqq 1$, we have on the set $N_\varepsilon = [N \cap (Z + U_\varepsilon)] \setminus [L + U_{\varepsilon^q r}]$ the inequality

$$|\Lambda(z)| \leqq c\,\theta^{-q}(z, L) \leqq \frac{1}{\varepsilon^{q'}\, r^{q'}}, \qquad r = r(Z).$$

Employing this inequality to find a bound for the righthand side of (2.3), we obtain

$$\sup\{|d(z, D)\,\phi(z)|,\, z \in N_\varepsilon\} \leqq \frac{1}{\varepsilon^{q'}\, r^{q'}} \sup\{|\partial(z, D)\,\phi(z)|,\, z \in N_\varepsilon\}. \tag{6.4}$$

Taking account of this inequality in the estimate (2.3) and multiplying both sides by $\varepsilon^{q'} r^{q'}$, we arrive at a similar estimate, in which the role of the operator $d(z, D)$ is played by the operator $\partial(z, D)$. □

Noetherian operators, satisfying the conditions of this theorem, will be said to be normal.

**2°. Properties of normal Noetherian operators.** We note that the class of Noetherian operators associated with a given matrix $p$, depends only on the submodules $p\mathscr{P}^s \subset \mathscr{P}^t$. In fact, since the $\mathscr{P}$-module $\mathcal{O}_Z$ is flat, we have the equation

$$p\,\mathcal{O}_Z^s = p\,\mathscr{P}^s \otimes \mathcal{O}_Z.$$

that is, the space $p\mathcal{O}_Z^s$ depends only on $p\mathscr{P}^s$. Therefore, we may speak of Noetherian operators associated with a given primary submodule $\mathfrak{p} \subset \mathscr{P}^t$.

Let $\mathfrak{p} = \mathrm{p}\mathscr{P}^s$ be an arbitrary submodule of $\mathscr{P}^t$ and let

$$\mathfrak{p} = \mathfrak{p}_0 \cap \cdots \cap \mathfrak{p}_l \tag{7.4}$$

be a reduced primary decomposition of this submodule. Then, further, let $\partial^\lambda(z, D)$, $\lambda = 0, \ldots, l$, be Noetherian operators associated with the submodules $\mathfrak{p}_\lambda$. We shall say that their ensemble is associated with the submodule $\mathfrak{p} \subset \mathscr{P}^t$ and the matrix $p$. Until the end of this subsection, we will fix the arbitrary matrix $p$ and a selection of normal Noetherian operators associated with it. Let $\{N^\lambda\}$ be the set of associated varieties.

**Corollary 1.** *Corollary* 1 *of* § 3 *is valid when the operators* $d^\lambda(z, D)$ *are replaced by the operators* $\partial^\lambda(z, D)$.

*Proof.* According to the theorem of this section, we may replace the operators $d^\lambda(z, D)$ by the operators $\partial^\lambda(z, D)$ in the inequality (31.3). This altered inequality leads to the inequality (29.3), where a similar replacement has been made. ▯

**Corollary 2.** *For an arbitrary point* $Z \in N(p)$ *the function* $\phi \in \mathcal{O}_Z^t$ *belongs to the subspace* $p\,\mathcal{O}_Z^s$ *if and only if the equations*

$$\partial^\lambda(z, D)\,\phi(z)|_{N^\lambda} = 0, \qquad \lambda = 0, \ldots, l, \tag{8.4}$$

*hold.*

*Proof.* If the Eq. (8) is valid, then by Corollary 1, $\mathscr{D}(Z)\,\phi(Z) = 0$, whence $\phi \in p\,\mathcal{O}_Z^s$. Conversely suppose that $\phi \in p\,\mathcal{O}_Z^s$. Then the properties of Noetherian operators imply that for arbitrary $\lambda$. Eq. (8) holds on the set $N^\lambda \setminus N_*^\lambda$, where $N_*^\lambda$ is the exceptional variety of $N^\lambda$. Since this subvariety is a proper one, and the variety $N^\lambda$ is irreducible, $N_*^\lambda$ is nowhere dense in $N^\lambda$. Therefore the continuity of the coefficients of the operator $\partial^\lambda(z, D)$ implies that Eq. (8) is satisfied on the whole variety $N^\lambda$. ▯

**Corollary 3.** *Let* $B$ *be a set of points in* $C^n$, *such that in every variety* $N^\lambda$, $\lambda = 0, \ldots, l$, *there lies at least one of the points of this set. Further, suppose that* $\mathscr{M}$ *is a family of majorants of type* $\mathscr{I}$, *such that* $M_\alpha(z) < \infty$ *for all* $z \in C^n$ *and* $\alpha$. *Then, in order that the functions* $\phi \in [\overline{\mathscr{H}}_{\mathscr{M}}]^t$ *belong to the subspace* $p[\overline{\mathscr{H}}_{\mathscr{M}}]^s$, *it is necessary and sufficient that one of the two following equivalent conditions be fulfilled:*

I. *For every* $Z \in B$

$$\mathscr{D}(Z)\,\phi(Z)=0.$$

II. *In the neighborhood of every point of* $B$, *the Eq.* (8) *holds.*

*Either of these conditions is also necessary and sufficient for the polynomial* $\phi \in \mathscr{P}^t$ *to belong to* $p\mathscr{P}^s$.

*Proof.* The equivalence of the conditions I and II follows from Corollary 2. Their necessity is self-evident. We shall prove the sufficiency of Condition II. Since $\Omega_\alpha = C^n$ for arbitrary $\alpha$, all the functions of the space $[\bar{\mathscr{H}}_{\mathscr{M}}]^t$ are entire functions. For arbitrary $\lambda$ the coefficients of the operator $\partial^\lambda(z, D)$ are polynomials and therefore, for an arbitrary function $\phi \in [\bar{\mathscr{H}}_{\mathscr{M}}]^t$ the function $\partial^\lambda(z, D)\,\phi(z)$ is holomorphic on the regular part of the variety $N^\lambda$. By hypothesis, Eq. (8) is satisfied in the neighborhood of some point $Z \in N^\lambda$. Since the regular points of an algebraic variety are everywhere dense, Eq. (8) holds in the neighborhood of some regular point $Z'$ of the variety $N^\lambda$. By the properties of analytic functions, this equation continues to be satisfied at all points of the variety which can be linked to $Z'$ by an arc lying in the regular part of $N^\lambda$. But, as is known, the regular part of an irreducible variety is connected[9]. Therefore, Eq. (8) is satisfied on the whole of the regular part and therefore on the whole of the variety $N^\lambda$.

Therefore, $\partial^\lambda(z, D)\,\phi(z)=0$ identically on $N^\lambda$, $\lambda=0, \ldots, l$. It follows from Corollary 2 that $\mathscr{D}(z)\,\phi(z)\equiv 0$, that is, $\phi \in [\bar{\mathscr{H}}_{\mathscr{M}}]^t \cap \operatorname{Ker} \mathscr{D}$. In view of the theorem of § 5, Chapter III, the subspace $[\bar{\mathscr{H}}_{\mathscr{M}}]^t \cap \operatorname{Ker} \mathscr{D}$ coincides with the subspace $p[\bar{\mathscr{H}}_{\mathscr{M}}]^s$, and therefore $\phi \in p[\bar{\mathscr{H}}_{\mathscr{M}}]^s$, q.e.d. If $\phi \in \mathscr{P}^t$, then the equation $\mathscr{D}(z)\,\phi(z)\equiv 0$ implies that $\phi \in p\mathscr{C}^s_z$ for arbitrary $z \in C^n$. Hence $\phi \in p\mathscr{P}^s$ by Proposition 4, § 1, Chapter II. □

**3°. Normal Noetherian operators in certain particular cases.** We consider the decomposition (7). We shall suppose that it contains an $n$-dimensional primary component. The variety associated with this $n$-dimensional primary component being $C^n$, the stratifications contain no more than one $n$-dimensional component, since by hypothesis the radicals of the modules $p_\lambda$ and therefore the varieties $N(p_\lambda)$ are all distinct. The $n$-dimensional primary component in (7) will be denoted by $\mathfrak{p}_0$. If there is no such component, we write $\mathfrak{p}_0 = \mathscr{P}^t$.

**Proposition 1.** *Let* $p_1$ *be a* $\mathscr{P}$*-matrix, such that the sequence*

$$\mathscr{P}^{t_2} \xrightarrow{p_1'} \mathscr{P}^t \xrightarrow{p'} \mathscr{P}^s \tag{9.4}$$

*is exact* ($p_1'$ *and* $p'$ *are the transported matrices*). *Then the matrix* $p_1(z)$ *is a normal Noetherian operator associated with the submodule* $\mathfrak{p}_0 \subset \mathscr{P}^t$.

9 See, for example, Hervé [1].

*Proof.* We establish the exactness of the sequence

$$0 \longrightarrow \mathfrak{p}_0 \longrightarrow \mathscr{P}^t \xrightarrow{p_1} \mathscr{P}^{t_2}. \tag{10.4}$$

The submodule $\mathfrak{p}_0 \subset \mathscr{P}^t$ is characterized by the fact that for any element $F$, we can find a polynomial $f \neq 0$ such that $f F \in p \mathscr{P}^s$. From the exactness of (9), we derive $p_1 p = 0$. Hence $p_1 f F = 0$, and accordingly, $p_1 F = 0$ and $f \neq 0$. Therefore, the sequence (10) is semi-exact.

Let $F \in \mathscr{P}^t$ be an arbitrary element, which is annihilated by the matrix $p_1$. In order to show that $F \in \mathfrak{p}_0$, it is sufficient to find a polynomial $f \neq 0$, such that $f F \in p \mathscr{P}^s$. Let $\hat{p}$ be some minor of the matrix $p$, not identically singular, and of maximal rank $\rho$, and let $p_1, \ldots, p_\rho$ be the rows of this matrix, in which the minor lies. We shall suppose for simplicity that the minor $\hat{p}$ lies in the first $\rho$ columns. We consider the system of linear equations

$$\begin{gathered} p_\tau G = \det \hat{p} \cdot F_\tau, \qquad F = (F_1, \ldots, F_t), \\ G = (G_1, \ldots, G_\rho, 0, \ldots, 0), \qquad \tau = 1, \ldots, \rho. \end{gathered} \tag{11.4}$$

This system, obviously, has a solution $G \in \mathscr{P}^s$. We shall show that the solution satisfies the system

$$p G = \det \hat{p} F. \tag{12.4}$$

also.

Since the rows $p_1, \ldots, p_\rho$ contain a minor of the matrix $p$ of maximal rank, any other row $p_\tau$, $\tau = \rho + 1, \ldots, t$, of the matrix $p$ can be expressed as follows: $a p_\tau = a_1 p_1 + \cdots + a_\rho p_\rho$, where $a$ and $a_1, \ldots, a_\rho$ are polynomials and $a \neq 0$. This equation can be written in the form $A' p = 0$ with some $A \in \mathscr{P}^t$. We shall show that $A' F = 0$. The equation $A' p = 0$ implies that $p' A = 0$. Therefore, because (9) is exact $A \in p_1' \mathscr{P}^{t_2}$, that is $A = p_1' B$, where $B \in \mathscr{P}^{t_2}$. Hence $A' = B' p_1$ and, accordingly $A' F = B' p_1 F = 0$, which is what we set out to prove. This fact, taken together with (11), leads to the equations $a p_\tau G = a \det \hat{p} F_\tau$. Since $a \neq 0$, it follows that $p_\tau G = \det \hat{p} F_\tau$. Since $\tau$ is arbitrary, Eq. (12) is established. And with this the exactness of (10) is established.

Tensoring the sequence (10) by the flat $\mathscr{P}$-module $\mathscr{O}_z$, we obtain the exact sequence

$$0 \to \mathfrak{p}_0 \otimes \mathscr{O}_Z \to \mathscr{O}_Z^t \xrightarrow{p_1} \mathscr{O}_Z^{t_2}.$$

From the exactness of this sequence, it follows that the matrix $p_1(z)$ is a Noetherian operator associated with $\mathfrak{p}_0$. Its normality is obvious. ▯

**Proposition 2.** *Let*

$$\mathscr{P}^s \xrightarrow{p} \mathscr{P}^t \xrightarrow{r} \mathscr{P}^v \tag{13.4}$$

*be an exact sequence of $\mathscr{P}$-mappings. Then the submodule $p \mathscr{P}^s \subset \mathscr{P}^t$ is primary and of dimension $n$, and $r$ is a normal Noetherian operator associated with this module.*

*Proof.* Let $fF \in p\mathscr{P}^s$, where $f \in \mathscr{P}$, and $F \in \mathscr{P}^t$. The exactness of (13) implies that $rfF=0$, whence either $f=0$, or $rF=0$, that is $F \in p\mathscr{P}^s$. Therefore the submodule $p\mathscr{P}^s$ is primary, and its radical is a null ideal, that is, the dimension of the submodule $p\mathscr{P}^s$ is equal to $n$. Since $\mathcal{O}_z$ is a flat $\mathscr{P}$-module, the exactness of (13) implies the exactness of the sequence

$$\mathcal{O}_z^s \xrightarrow{p} \mathcal{O}_z^t \xrightarrow{r} \mathcal{O}_z^v.$$

This means that the matrix $r$ is a Noetherian operator associated with $p\mathscr{P}^s$. The normality of this operator is self-evident. □

**Proposition 3.** *Let $s=t=1$ and let the coefficients of the polynomial $p$, corresponding to the highest power of $z_1$, be constant. Suppose that*

$$p = p_1^{\alpha_1} \dots p_l^{\alpha_l}$$

*is a decomposition of the polynomial $p$ into the product of powers of irreducible polynomials, and that $N^\lambda$ is the variety of the roots $p_\lambda$, $\lambda=1, \dots, l$. Then the set of varieties $N^\lambda$ and normal Noetherian operators*

$$d^\lambda(z, D) = \left\{ \frac{\partial^i}{\partial z_1^i},\ i=0, \dots, \alpha_\lambda - 1 \right\}, \qquad \lambda = 1, \dots, l,$$

*is associated with $p$.*

*Proof.* We shall establish the equation

$$p\mathscr{P} = p_1^{\alpha_1}\mathscr{P} \cap \dots \cap p_l^{\alpha_l}\mathscr{P} \tag{14.4}$$

and we shall show that it is a reduced primary decomposition of the ideal $p\mathscr{P}$. The left side, clearly, lies within the right side. Conversely, if the polynomial $f$ belongs to the right side, it is divisible by every one of the polynomials $p_\lambda^{\alpha_\lambda}$ and, accordingly, is divisible by their product, since they are relatively prime, that is, $f$ belongs to the left side. This establishes Eq. (14). We note that for arbitrary $\lambda$ the ideal $p_\lambda^{\alpha_\lambda}\mathscr{P}$ is primary, and its radical is equal to $p_\lambda\mathscr{P}$. In fact, if the product of the polynomials $f$ and $g$ is divisible by $p_\lambda^{\alpha_\lambda}$, but $g$ is not divisible by $p_\lambda$, then the polynomial $f$ must necessarily be divisible by $p_\lambda^{\alpha_\lambda}$. Conversely, if $g$ is divisible by $p_\lambda$, then $g^{\alpha_\lambda} \in p_\lambda^{\alpha_\lambda}\mathscr{P}$. All the radicals $p_\lambda\mathscr{P}$, $\lambda=1, \dots, l$, are distinct, since the polynomials $p_\lambda$ are distinct, and none of the ideals can be contained in the intersection of any others, else the corresponding irreducible polynomial $p_\lambda$ would divide the product of two other irreducible polynomials. Thus, (14) is a reduced primary decomposition of $p\mathscr{P}$.

The variety associated with the ideal $p_\lambda\mathscr{P}$, is $N^\lambda$; this variety is normally placed, since the highest coefficient of the polynomial $p$ with respect to $z_1$ is by hypothesis a constant. Let $N_\delta^\lambda$ be the discriminant subvariety of $N^\lambda$ with respect to $z_1$, that is, the subset where the derivative $p'_\lambda$ with respect to $z_1$ vanishes. We fix an arbitrary point $Z \in N^\lambda \setminus N_\delta^\lambda$. In the neighborhood of this point, the set $N^\lambda \setminus N_\delta^\lambda$ is defined by the equation $z_1 = \zeta(z')$, where $\zeta(z')$ is a holomorphic function of the variables $z' = (z_2, \dots, z_n)$. The condition

$$d^\lambda(z, D)\,\phi(z)|_{N^\lambda} = 0, \qquad \phi \in \mathcal{O}_Z,$$

means that for every fixed $z'$, $\phi$ is a function of $z_1$ and vanishes at the point $z_1 = \zeta(z')$ with a multiplicity not less than $\alpha_\lambda$. Since $p'_\lambda(Z) \neq 0$, this condition is necessary and sufficient in order that the function $\phi$ should be expressible in the form $p_\lambda^{\alpha_\lambda}\psi$, where $\psi \in \mathscr{C}_Z$. We have thus proved that Definition 1 is satisfied for the matrix $p_\lambda^{\alpha_\lambda}$ and the operator $d^\lambda(z, D)$ with $N_* = N_\delta^\lambda$, that is, $d^\lambda(z, D)$ is a Noetherian operator. Its normality is self-evident. □

**Proposition 4.** *Let the matrix p be such that the module $M = \mathscr{P}^t / p\,\mathscr{P}^s$ is zero dimensional. Then the following assertions are true.*

1. *The set of points $z^\lambda$, $\lambda = 1, \dots, l$, of the set $N(p)$ forms a set of algebraic varieties associated with the matrix p.*

2. *Let $\mathscr{D}(z)$ be a p-operator and let $\mathscr{I}(z)$ be the associated basis set. For every $\lambda$ we denote by $d^\lambda(D)$ the column consisting of the differential operators $\mathscr{D}_{\tau,i}(z^\lambda, D)$ with $(\tau, i) \in t\,Z_+^n \setminus \mathscr{I}(z^\lambda)$[10]. The set of differential operators $d^\lambda(D)$ forms a set of normal Noetherian operators associated with p.*

3. *Let $l_\lambda = |t\,Z_+^n \setminus \mathscr{I}(z^\lambda)|$ be the number of elements in the column $d^\lambda$. The sum $l = \sum l_\lambda$ is equal to the dimension of the module M as a linear space over the field C.*

*Proof.* From the construction of the operator $d^\lambda(D)$ and from Corollary 1, § 4, Chapter II, it follows that for arbitrary $\lambda$ the following assertion is true:

$$d^\lambda(D)\,\phi(z^\lambda) = 0, \qquad \phi \in \mathcal{O}_{z^\lambda}^t \Leftrightarrow \phi \in \mathcal{O}_{z^\lambda}^s. \tag{15.4}$$

For every $\lambda$ we set $\mathfrak{p}_\lambda = \mathscr{P}^t \cap \mathfrak{p}\,\mathcal{O}_{z^\lambda}^s$. It is clear that $\mathfrak{p}_\lambda$ is a submodule of $\mathscr{P}^t$. We shall show that this submodule is primary and that its radical coincides with the ideal $I_\lambda$ consisting of all the polynomials vanishing at the point $z^\lambda$.

Let $\kappa_\lambda$ be the order of the differential operator $d^\lambda$. It is clear that for an arbitrary polynomial $f \in I_\lambda$ and an arbitrary column $F \in \mathscr{P}^t$, the column $\phi = f^{\kappa_\lambda + 1} F$ satisfies the condition (15) and, therefore, belongs to $\mathfrak{p}_\lambda$. Hence, we conclude that $I_\lambda \subset \mathfrak{r}(\mathfrak{p}_\lambda)$. Since the ideal $I_\lambda$ is maximal and the radical $\mathfrak{r}(\mathfrak{p}_\lambda)$ is distinct from the whole ring $\mathscr{P}$ (since $\mathfrak{p}_\lambda \neq \mathscr{P}^t$), we have $I_\lambda = \mathfrak{r}(\mathfrak{p}_\lambda)$, q.e.d.

---

10 The set $t\,Z_+^n \setminus \mathscr{I}(z^\lambda)$ is finite, since the module $M$ is zero dimensional (see the reasoning of 1°, § 3).

It remains to show that the submodule $\mathfrak{p}_\lambda \subset \mathscr{P}^t$ is primary. Suppose that $f F \in \mathfrak{p}_\lambda$, where $f \in \mathscr{P}$, and $F \in \mathscr{P}^t$. If the polynomial $f$ does not belong to $\mathfrak{r}(\mathfrak{p}_\lambda) = I_\lambda$, then it can be inverted in the ring $\mathcal{O}_{z^\lambda}$, and accordingly, $F = (1/f) f F \in p\, \mathcal{O}_{z^\lambda}$, that is $F \in \mathfrak{p}_\lambda$. From this it follows that the submodule $\mathfrak{p}_\lambda$ is primary. From everything we have said, it follows that the point $z^\lambda$ is the algebraic variety and $d^\lambda(D)$ is a normal Noetherian operator associated with this submodule.

We now establish the equation

$$\mathfrak{p}_1 \cap \cdots \cap \mathfrak{p}_l = p\mathscr{P}^s. \tag{16.4}$$

That the left side contains the right side is obvious. Conversely, if the column $F \in \mathscr{P}^t$ belongs to the left side, then it belongs to $p\,\mathcal{O}_z^s$ for an arbitrary point $z \in C^n$. Hence, in accordance with Proposition 1, § 4, Chapter II, it follows that $F \in p\mathscr{P}^s$. This establishes Eq. (16). We remark that all the modules on the left side are primary submodules in $\mathscr{P}^t$ and that none of them contains the intersection of the others. Accordingly, (16) is a reduced primary decomposition of the submodule $p\mathscr{P}^s \subset \mathscr{P}^t$; this completes the proof of the assertions 1 and 2.

We shall prove the third assertion. In accordance with Remark 2, § 4, Chapter II, for arbitrary $\lambda$, all the differential operators $\mathscr{D}_{\tau,i}(z^\lambda, D)$ with $(\tau, i) \in t\, Z_+^n \setminus \mathscr{I}(z^\lambda)$ are linearly independent. Therefore, all the $l = \sum l_\lambda$ linear conditions

$$\mathscr{D}_{\tau,i}(z^\lambda, D)\, F(z^\lambda) = 0, \qquad (\tau, i) \in t\, Z_+^n \setminus \mathscr{I}(z^\lambda), \qquad \lambda = 1, \ldots, l.$$

are linearly independent. According to what we have proved, they distinguish the subspace $p\mathscr{P}^s$ in the space $\mathscr{P}^t$. Therefore, the number of these conditions is equal to the dimension of the factor-space $\mathscr{P}^t / p\mathscr{P}^s$. ▯

**4°. An example of a $\mathscr{P}$-matrix for which there exists no Noetherian operator with constant coefficients.** We shall suppose that the submodule $p\mathscr{P}^s$ is primary. In the applications it is important to know whether there is at least one Noetherian operator with constant coefficients corresponding to this submodule. We established earlier, with the help of Proposition 3, that there always is such an operator in the case $s = t = 1$. In the general case this is not so, even for $t = 1$. As an example we consider the $\mathscr{P}$-matrix

$$p = (z_1^2, z_2^2, z_2 - z_1 z_3), \qquad n = 3, s = 3, t = 1.$$

The corresponding variety $N = N(p)$ is, obviously, the $z_3$-axis.

We shall show that for this matrix there exists no Noetherian operator with constant coefficients (a little later we shall show that the submodule $p\mathscr{P}^s$ is primary). Let us suppose the contrary. Let $d = (d_1(D), \ldots,$

$d_e(D)$) be a Noetherian operator with constant coefficients. We write each of the operators $d_\alpha(D)$ in the form

$$d_\alpha(D) = \sum_0^\kappa \frac{\partial^i}{\partial z_3^i} d_\alpha^i(D'), \qquad D' = \left(\frac{\partial}{\partial z_1}, \frac{\partial}{\partial z_2}\right).$$

Since all the polynomials of the ideal $p\mathscr{P}^3$ vanish on the $z_3$-axis, all the operators

$$\frac{\partial^i}{\partial z_3^i} d_\alpha^i(D'), \qquad i > 0,$$

vanish on this ideal. Therefore, all the operators $d_\alpha^0(D')$ also vanish on the ideal $p\mathscr{P}^s$. We note that all the polynomials of the form $z_1^{i_1} \cdot z_2^{i_2}$, $i_1 + i_2 \geqq 2$, belong to the ideal $p\mathscr{P}^s$, since this ideal contains the polynomials $z_1^2$, $z_2^2$ and $z_1 z_2 = z_1(z_2 - z_1 z_3) + z_3 z_1^2$. Therefore, all the operators $d_\alpha^0(D')$ have orders no higher than the first.

Eliminating terms of order zero, we obtain homogeneous differential operators $\hat{d}_\alpha$ of first order which also vanish on $p\mathscr{P}^3$. Let $\lambda$ be the number of linearly independent operators $\hat{d}_\alpha$, $\alpha = 1, \ldots, e$. We suppose that $\lambda = 0$, that is, that all the operators $d_\alpha^0$ are of order zero. Then the operators $d_\alpha(D)$ vanish on the monomial $z_1$, which does not belong to the subspace $p\mathcal{O}_Z^3$ for any $Z \in N$. This contradicts the hypothesis that the operator $d$ is Noetherian.

Thus we have arrived at the conclusion that $\lambda > 0$. Therefore, one of the operators $\hat{d}_\alpha$ has the form

$$a_1 \frac{\partial}{\partial z_1} + a_2 \frac{\partial}{\partial z_2},$$

where either $a_1$ or $a_2$ is different from zero. But in this case

$$\left(a_1 \frac{\partial}{\partial z_1} + a_2 \frac{\partial}{\partial z_2}\right)(z_2 - z_1 z_3) = a_2 - a_1 z_3.$$

The function $a_2 - a_1 z_3$ is obviously not identically zero on $N$, and this contradicts the hypothesis that $d(D)$ is Noetherian. The contradiction that we have obtained proves that the matrix $p$ has no Noetherian operator with constant coefficients.

It is very easy, however, to construct a Noetherian operator with variable coefficients for this matrix. For example, the operator

$$\partial(z, D) = \begin{pmatrix} 1 \\ \dfrac{\partial}{\partial z_1} + z_3 \dfrac{\partial}{\partial z_2} \end{pmatrix}$$

is Noetherian. We shall prove this. Since the result of applying the operator to the polynomials $z_1^2$, $z_2^2$ and $z_2 - z_1 z_3$ is equal to zero on $N$, the operator vanishes on $p\mathscr{C}_Z^3$ for arbitrary $Z \in N$. Conversely, suppose it is known that

$$\partial(z, D)\,\phi(z)|_N = 0, \qquad \phi \in \mathcal{O}_Z. \tag{17.4}$$

We shall show that the function $\phi$ belongs to $p\mathscr{C}_Z^3$. To this end, we expand the function in the neighborhood of the point $Z = (0, 0, Z_3) \in N$ in a series of powers of $z_1$ and $z_2$

$$\phi(z) = \sum_{i,j} \phi_{ij}(z_3)\, z_1^i\, z_2^j.$$

Here $\phi_{ij}(z_3)$ is a function that is holomorphic in the neighborhood of $Z_3$. From (17) we have the equations

$$\phi_{00}(z_3) \equiv 0, \qquad \phi_{10}(z_3) + z_3\, \phi_{01}(z_3) \equiv 0,$$

whence

$$\phi(z) = (z_2 - z_1 z_3)\, \phi_{01}(z_3) + \sum_{i+j \geqq 2} \phi_{ij}(z_3)\, z_1^i\, z_2^j.$$

The second term can be represented in the form

$$z_1^2\, \psi_1(z) + z_1 z_2\, \psi_2(z) + z_2^2\, \psi_3(z),$$

where $\psi_1$, $\psi_2$ and $\psi_3$ are functions holomorphic in the neighborhood of $Z$. Therefore, we have established that the function $\phi$ belongs to the space $p\mathscr{P}^3 \otimes \mathcal{O}_Z = p\,\mathcal{O}_Z^3$. Consequently, the condition (17) is, in fact, necessary and sufficient, in order that $\phi \in p\mathscr{C}_Z^3$.

We still must show that the submodule $p\mathscr{P}^3$ is primary. We note that the line of reasoning given above shows also that the condition (17), when applied to the polynomials $\phi \in \mathscr{P}$, is necessary and sufficient in order that $\phi \in p\mathscr{P}^3$. We now find the radical of the ideal $p\mathscr{P}^3$. If the polynomial $f$ vanishes on the $z_3$-axis, then $f^2 g \in p\mathscr{P}^3$ for an arbitrary polynomial $g \in \mathscr{P}$. Conversely, if the polynomial $f$ belongs to $\mathfrak{r}(p\mathscr{P}^3)$, then some power of it belongs to $p\mathscr{P}^3$. Accordingly, it vanishes on the $z_3$-axis. Thus the ideal $\mathfrak{r}(p\mathscr{P}^3)$ consists of polynomials vanishing on the $z_3$-axis. Suppose the product $fg$ belongs to $p\mathscr{P}^3$, and that the polynomial $f$ does not vanish identically on the $z_3$-axis. Then the equation $\partial(z, D)\, f g|_N = 0$ implies that $\partial(z, D)\, g|_N = 0$, whence $g \in p\mathscr{P}^3$. Therefore, the ideal $p\mathscr{P}^3$ is in fact primary.

### 5°. Noetherian operators associated with an arbitrary finite *P*-module

**Definition 2.** Let $M$ be a finite $\mathscr{P}$-module. We choose a representation in the form

$$M \cong \mathscr{P}^t/\mathfrak{p}. \tag{18.4}$$

The set of Noetherian operators associated with $M$ is defined as the collection of Noetherian operators associated with the submodule $\mathfrak{p} \subset \mathscr{P}^t$.

The set of Noetherian operators associated with $M$ depends, of course, on the choice of the representation (18) (and also on the choice of the primary decomposition of the submodule $\mathfrak{p}$). Therefore, in speaking of this set, we shall indicate the corresponding representation of the module $M$. In the next section, we show that the choice of the representation (18) does not essentially influence the form of the normal Noetherian operators associated with $M$. We remark that Proposition 1, § 1, implies that the set of varieties $N^\lambda$, associated with the submodule $\mathfrak{p}$, does not depend on the representation (18), since the module $M$ itself characterizes them.

**Proposition 5.** *Let z and w be two groups of complex variables. Suppose, moreover, that*

$$M \cong \mathscr{P}_z^t / p\, \mathscr{P}_z^s, \qquad L \cong \mathscr{P}_w^\tau / q\, \mathscr{P}_w^\sigma$$

*are arbitrary* $\mathscr{P}_z$*- and* $\mathscr{P}_w$*-modules, and* $\{N^\lambda \subset C^n, \partial^\lambda(z, D),\ \lambda = 0, \ldots, l\}$ *and* $\{K^\mu \subset C^\nu, g^\mu(w, D)\ \mu = 0, \ldots, m\}$ *are the collections of varieties and Noetherian operators associated with these modules. Regarding the matrices p and q as* $\mathscr{P}$*-matrices, where* $\mathscr{P} = \mathscr{P}_{z,w}$*, we construct the* $\mathscr{P}$*-modules* $\mathscr{M} = \mathscr{P}^t / p\, \mathscr{P}^s$ *and* $\mathscr{L} = \mathscr{P}^\tau / q\, \mathscr{P}^\sigma$*. Then*

$$\operatorname{Tor}_i(\mathscr{M}, \mathscr{L}) = 0, \quad i \geqq 1,$$

*and* $\{N^\lambda \times K^\mu \subset C^{n+\nu}, \partial^\lambda \otimes g^\mu, \lambda = 0, \ldots, l; \mu = 0, \ldots, m\}$ *is the set of varieties and Noetherian operators associated with the module* $\mathscr{M} \otimes \mathscr{L}$*, represented in the form*

$$\mathscr{M} \otimes \mathscr{L} \cong \mathscr{P}^{t\tau} / [p\, \mathscr{P}^{s\tau} + q\, \mathscr{P}^{t\sigma}]$$

(see Proposition 3, § 3, Chapter I). *If the operators* $\partial^\lambda$ *and* $g^\mu$ *are normal, then the operators* $\partial^\lambda \otimes g^\mu$ *are also normal.*

We leave the proof of this theorem to the reader.

## § 5. The fundamental theorem

### 1°. Holomorphic $p$-functions

**Definition.** Let $p\colon \mathscr{P}^s \to \mathscr{P}^t$ be a $\mathscr{P}$-matrix, and let

$$p\, \mathscr{P}^s = \mathfrak{p}_0 \cap \cdots \cap \mathfrak{p}_l$$

be a reduced primary decomposition of the submodule $p\, \mathscr{P}^s \subset \mathscr{P}^t$, and let $d = \{d^\lambda(z, D),\ \lambda = 0, \ldots, l\}$ be the associated set of normal Noetherian operators. Further let $\Omega$ be an open set in $C^n$. A $p$-function holomorphic in $\Omega$ is defined as an arbitrary set of functions $f=\{f^\lambda, \lambda=0, \ldots, l\}$, defined on the sets $N^\lambda \cap \Omega$, where $N^\lambda = N(\mathfrak{p}_\lambda)$, and satisfying the following condition: For an arbitrary point $Z$ we can find a holomorphic function $F_Z \in \mathcal{O}_Z^t$, such that in the neighborhood of $Z$

$$f^\lambda(z) = d^\lambda(z, D)\, F_Z(z)\,|_{N^\lambda}, \qquad \lambda=0, \ldots, l.$$

The functions $f^\lambda$, $\lambda=0, \ldots, l$, will be called the components of the $p$-function $f$.

Let $\mathcal{M}=\{M_\alpha(z)\}$ be a family of majorants in $C^n$ and $\Omega_\alpha$, $\alpha=1, 2, \ldots$, open sets in which the functions $M_\alpha$ are finite. For every $\alpha$ we consider the normed space $\mathcal{H}_\alpha\{p, d\}$, consisting of $p$-functions holomorphic in $\Omega_\alpha$, for which the norm:

$$\|f\|_\alpha = \max_\lambda \sup_{N^\lambda} \frac{|f^\lambda(z)|}{M_\alpha(z)}.$$

is finite. If $\alpha' \geqq \alpha$, then the identity mapping $\mathcal{H}_\alpha\{p, d\} \to \mathcal{H}_{\alpha'}\{p, d\}$, is defined and continuous. Therefore, the spaces $\mathcal{H}_\alpha\{p, d\}$, $\alpha=1, 2, \ldots$, form a family of spaces. This family will be denoted by $\mathcal{H}_\mathcal{M}\{p, d\}$. We shall sometimes abbreviate, writing $\mathcal{H}_\alpha\{p\}$ and $\mathcal{H}_\mathcal{M}\{p\}$ instead of $\mathcal{H}_\alpha\{p, d\}$ and $\mathcal{H}_\mathcal{M}\{p, d\}$.

**2°. Isomorphic expressions of the family $\mathcal{H}_\mathcal{M}\{p, d\}$.** We recall that in § 3, Chapter III, we considered the family ${}^\nu\mathcal{H}_\mathcal{M}$, $\nu=0, 1, \ldots$. For arbitrary $\nu \geqq 0$, ${}^\nu\mathcal{H}_\mathcal{M}$ is the family of spaces ${}^\nu\mathcal{H}_\alpha(U_\alpha)$, $\alpha=1, 2, \ldots$, where $U_\alpha$ is the elementary covering of parameter zero and radius $\varepsilon_\alpha$, and ${}^\nu\mathcal{H}_\alpha(U_\alpha)$ is the space of holomorphic cochains of order $\nu$ on the covering $U_\alpha=\{U_z\}$, for which the norm

$$\|\phi\|_{\alpha, U_\alpha} = \sup_{z_0, \ldots, z_\nu} \sup \left\{ \frac{|\phi_{z_0, \ldots, z_\nu}|}{M_\alpha}, z \in U_{z_0} \cap \cdots \cap U_{z_\nu} \cap \Omega_\alpha \right\}$$

is finite. Let $\mathcal{D}$ be a $p$-operator. We denoted by $[{}^\nu\mathcal{H}_\mathcal{M}]^t \cap \operatorname{Ker} \mathcal{D}$ the subfamily $[{}^\nu\mathcal{H}_\mathcal{M}]^t$, consisting of the subspaces $[{}^\nu\mathcal{H}_\alpha(U_\alpha)]^t \cap \operatorname{Ker} \mathcal{D}$. We recall that $[{}^\nu\mathcal{H}_\alpha(U_\alpha)]^t \cap \operatorname{Ker} \mathcal{D}$ is a subspace of $[{}^\nu\mathcal{H}_\alpha(U_\alpha)]^t$, consisting of the cochains $\phi$ such that $\mathcal{D}\,\phi_{z_0, \ldots, z_\nu} \equiv 0$ for arbitrary $z_0, \ldots, z_\nu$.

It is clear that the co-boundary operator $\partial$: $[{}^0\mathcal{H}_\alpha(U_\alpha)]^t \to [{}^1\mathcal{H}_\alpha(U_\alpha)]^t$ carries the subspace $[{}^0\mathcal{H}_\alpha(U_\alpha)]^t \cap \operatorname{Ker} \mathcal{D}$ into $[{}^1\mathcal{H}_\alpha(U_\alpha)]^t \cap \operatorname{Ker} \mathcal{D}$ and accordingly, generates an operator which acts in the corresponding factor-spaces. We write $\mathcal{Z}_\alpha(p)$ to denote the kernel of the following mapping:

$$\mathcal{Z}_\alpha(p) = \operatorname{Ker} \{[{}^0\mathcal{H}_\alpha(U_\alpha)]^t / [{}^0\mathcal{H}_\alpha(U_\alpha)]^t \cap \operatorname{Ker} \mathcal{D} \xrightarrow{\partial} [{}^1\mathcal{H}_\alpha(U_\alpha)]^t / [{}^1\mathcal{H}_\alpha(U_\alpha)]^t \cap \operatorname{Ker} \mathcal{D}\}.$$

We write $\mathscr{Z}_{\mathscr{M}}(p)$ to denote the family consisting of the spaces $\mathscr{Z}_\alpha(p)$ and their identity mappings. Clearly, $\mathscr{Z}_{\mathscr{M}}(p)$ is the kernel of the mapping

$$\partial\colon\ [{}^0\mathscr{H}_{\mathscr{M}}]^t/[{}^0\mathscr{H}_{\mathscr{M}}]^t \cap \operatorname{Ker}\mathscr{D} \to [{}^1\mathscr{H}_{\mathscr{M}}]^t/[{}^1\mathscr{H}_{\mathscr{M}}]^t \cap \operatorname{Ker}\mathscr{D}. \tag{1.5}$$

We fix an arbitrary integer $\alpha$. Let $\Phi$ be an element of the space $\mathscr{Z}_\alpha(p)$. By the definition of this space, $\Phi$ is a coset belonging to the factor-space $[{}^0\mathscr{H}_\alpha(U_\alpha)]^t/[{}^0\mathscr{H}_\alpha(U_\alpha)]^t \cap \operatorname{Ker}\mathscr{D}$, consisting of the cochains $\phi=\sum \phi_Z U_Z$ such that $\partial\phi \in [{}^1\mathscr{H}_\alpha(U_\alpha)]^t \cap \operatorname{Ker}\mathscr{D}$. This inclusion relation implies that the functions

$$f^\lambda(z)=d^\lambda(z,D)\,\phi_Z(z)|_{N_\lambda}, \qquad \lambda=0,\dots,l,$$

do not depend on $Z\in\Omega_\alpha$ and are therefore components of some $p$-function $f$ holomorphic in $\Omega_\alpha$. If the cochain $\phi$ itself belongs to the subspace $[{}^0\mathscr{H}_\alpha(U_\alpha)]^t \cap \operatorname{Ker}\mathscr{D}$, then $f^\lambda\equiv 0$, and accordingly, the function $f$ depends only on the coset $\Phi$. Thus, we have constructed an operator $d\colon \Phi\to f$, which maps $p$-functions that are holomorphic in the region $\Omega_\alpha$ into elements of the space $\mathscr{Z}_\alpha(p)$. We shall call this operator a Noetherian operator.

**Theorem 1.** *Let $\mathscr{M}$ be an arbitrary family of majorants. Then a Noetherian operator defines a family mapping*

$$d\colon\ \mathscr{Z}_{\mathscr{M}}(p)\to\mathscr{H}_{\mathscr{M}}\{p,d\}, \tag{2.5}$$

*which is an isomorphism.*

*Proof.* We fix $\alpha$ arbitrarily. Let $\Phi$ be an arbitrary element of the space $\mathscr{Z}_\alpha(p)$. We estimate the function $f=d\Phi$ in the region $\Omega_{\alpha+1}$. By the definition of the norm in the space $\mathscr{Z}_\alpha(p)$, corresponding to arbitrary positive $\varepsilon$ we can find a representative $\phi\in\Phi$ such that $\|\phi\|_\alpha \leqq \|\Phi\|_\alpha+\varepsilon$. Suppose that $\phi=\sum\phi_Z U_Z$. For an arbitrary point $Z\in\Omega_{\alpha+1}$ the function $\phi_Z$ is bounded and holomorphic in the $\varepsilon_\alpha$-neighborhood of that point. Let $\kappa^\lambda$ be the order of the operator $d^\lambda(z,D)$ as a polynomial in $z$ and $D$. From Cauchy's theorem, we derive the inequality

$$\begin{aligned}\max_\lambda |f^\lambda(Z)| &\leqq \max_\lambda \sum_{|i|\leqq\kappa_\lambda} |d^\lambda(Z,\delta)\,\xi^i|\,|\delta_i\,\phi_Z(Z)| \\ &\leqq C(|Z|+1)^\kappa \sup\{\phi_Z(z)|,\ |z-Z|\leqq\varepsilon_\alpha\} \\ &\leqq C(|Z|+1)^\kappa M_{\alpha+1}(Z)\,\|\phi\|_\alpha \leqq C M_{\alpha'}(Z)\,\|\phi\|_\alpha, \\ \kappa&=\max\kappa^\lambda, \qquad \alpha'=\alpha+1+\kappa.\end{aligned}$$

Hence

$$\|f\|_{\alpha'}\leqq C\,\|\phi\|_\alpha\leqq C(\|\Phi\|_\alpha+\varepsilon).$$

Since the function $f$ depends only on the coset $\Phi$, this equation holds for arbitrary $\varepsilon>0$, and therefore, the term $\varepsilon$ on the right side can be discarded.

Thus, we have established that the mapping

$$d: \quad \mathscr{Z}_\alpha(p) \ni \Phi \to d\Phi \in \mathscr{H}_{\alpha'}\{p, d\} \tag{3.5}$$

is defined and continuous. The set of all such mappings defines the family mapping (2).

We shall show that the mapping (2) is a monomorphism. Suppose that the element $\Phi \in \mathscr{Z}_\alpha(p)$ lies in the kernel of the mapping (3), that is, $d\phi = 0$ for an arbitrary representative $\phi \in \Phi$. Corollary 2, § 4, implies that $\mathscr{D}\phi_Z = 0$, that is $\phi \in [{}^0\mathscr{H}_\alpha(U_\alpha)]^t \cap \operatorname{Ker} \mathscr{D}$, whence $\Phi = 0$, q.e.d.

We construct the mapping inverse to (2). We fix $\alpha$. Let $f = \{f^\lambda\}$ be an arbitrary holomorphic $p$-function, lying in the space $\mathscr{H}_\alpha\{p, d\}$. We set $U_{\alpha+3} = \{U_Z\}$. By definition, $U_Z$ is a neighborhood of the point $Z$ of radius $\varepsilon_{\alpha+3} \leqq \frac{1}{8}\varepsilon_\alpha$. Let the point $Z$ belong to $\Omega_{\alpha+1}$. Then $A U_Z \subset\subset \Omega_\alpha$, and therefore, the definition of the holomorphic $p$-function implies that we can find an elementary covering $V$ of diameter zero and sufficiently small radius and a holomorphic cochain $F$, of zero order, on the covering $V \cap 4U_Z$, such that $dF = f$.

Employing Corollary 2, § 5, Chapter III, we find a function $\phi$ holomorphic on $2U_Z$ such that $\mathscr{D}(\partial\phi - F) = 0$. We apply to the function $\phi$ Corollary 1 of the same section. We have then represented the function $\phi$ in the form of the sum $\phi_Z + p\psi$, where $\phi_Z$ and $\psi$ are functions holomorphic in $U_Z$, and

$$\begin{aligned} &\sup\{|\phi_Z(z)|, z \in U_Z\} \\ &\quad \leqq C(|Z|+1)^q \sup\{r(z)|\mathscr{D}(z,\xi)\,\phi(z)|, z \in 2U_Z, |\xi| \leqq \varepsilon r(z)\}, \end{aligned} \tag{4.5}$$

where $\varepsilon = \varepsilon_{\alpha+3}$ and $r(z)$ is a power function on some algebraic stratification appearing in inequality (2.3). (The use of the function $r(z)$ in this inequality is legitimate, since we may, of course, suppose that this function does not exceed the function $r_z$, appearing in Corollary 1, § 4, Chapter II.) In view of the inequality (2.3) the right side of (4) does not exceed the expression

$$C(|Z|+1)^q \max_\lambda \sup\{|d^\lambda(z, D)\,\phi(z)|, z \in N^\lambda \cap 4U_Z\}.$$

We remark that

$$d^\lambda(z, D)\,\phi(z) = d^\lambda(z, D)\,F = f^\lambda(z), \qquad \lambda = 0, \ldots, l. \tag{5.5}$$

Therefore we arrive at the inequality

$$\sup\{|\phi_Z(z)|, z \in U_Z\} \leqq C(|Z|+1)^q \max_\lambda \sup\{|f^\lambda(z)|, z \in N^\lambda \cap 4U_Z\}.$$

Since the diameter of the sphere $4U_Z$ is less than $\varepsilon_\alpha$, we are allowed to write the inequality

$$\sup_z \frac{|\phi_Z(z)|}{(|z|+1)^q M_{\alpha+1}(z)} \leqq C \max_\lambda \sup_z \frac{|f^\lambda(z)|}{M_\alpha(z)}.$$

Setting $\phi = \sum \phi_Z U_Z$, we obtain a cochain of zero order on the covering $U_{\alpha+3}$, satisfying the inequality

$$\|\phi\|_{\alpha', U_{\alpha'}} \leqq C \|f\|_{\alpha}, \quad \alpha' = \alpha + 3 + [q]. \tag{6.5}$$

Therefore, the cochain belongs to the space ${}^0\mathscr{H}_{\alpha'}(U_{\alpha'})$, and it follows from (5) that $d\phi = f$. Hence, in particular, it follows that $\mathscr{D}\,\partial\phi = 0$, and accordingly, the cochain $\phi$ can be regarded as an element of the space $\mathscr{Z}_{\alpha'}(p)$. Thus, we have constructed the correspondence

$$\mathscr{H}_{\alpha}\{p, d\} \ni f \to \phi \in \mathscr{Z}_{\alpha'}(p),$$

which has the property that $d\phi = f$. Since the operator $d$ is biunique, this correspondence is single-valued and therefore, linear. It follows from the inequality (6) that the operator is continuous. The set of all operators defines a family mapping inverse to (2). □

**Theorem 2.** *Let* $p: \mathscr{P}^s \to \mathscr{P}^t$ *be an arbitrary* $\mathscr{P}$*-matrix, and let* $\mathscr{M}$ *be a family of majorants of type* $\mathscr{I}$. *Then the sequence of families*

$$[\mathscr{H}_{\mathscr{M}}]^s \xrightarrow{p} [\mathscr{H}_{\mathscr{M}}]^t \xrightarrow{d} \mathscr{H}_{\mathscr{M}}\{p, d\} \to 0 \tag{7.5}$$

*is exact. The mapping d, which appears in this sequence, is the composition of the identity mappings*

$$[\mathscr{H}_{\mathscr{M}}]^t \to \mathscr{Z}_{\mathscr{M}}(p) \tag{8.5}$$

*and the mapping* (2).

*Proof.* From what we said at the beginning of this subsection it follows that the coboundary operator in the sequence (5.3), Chapter III, acts from the subfamily $[{}^{\nu}\mathscr{H}_{\mathscr{M}}]^t \cap \operatorname{Ker}\mathscr{D}$ to the subfamily $[{}^{\nu+1}\mathscr{H}_{\mathscr{M}}]^t \cap \operatorname{Ker}\mathscr{D}$ and, therefore, defines a mapping of the corresponding factor-families. All these mappings are represented in the commutative diagram:

$$\begin{array}{ccccccccc}
 & & \vdots & & \vdots & & \vdots & & \\
 & & \uparrow & & \uparrow & & \uparrow & & \\
0 & \longrightarrow & [{}^1\mathscr{H}_{\mathscr{M}}]^t \cap \operatorname{Ker}\mathscr{D} & \longrightarrow & [{}^1\mathscr{H}_{\mathscr{M}}] & \longrightarrow & [{}^1\mathscr{H}_{\mathscr{M}}]^t/[{}^1\mathscr{H}_{\mathscr{M}}]^t \cap \operatorname{Ker}\mathscr{D} & \longrightarrow & 0 \\
 & & \partial\uparrow & & \partial\uparrow & & \partial\uparrow & & \\
0 & \longrightarrow & [{}^0\mathscr{H}_{\mathscr{M}}]^t \cap \operatorname{Ker}\mathscr{D} & \longrightarrow & [{}^0\mathscr{H}_{\mathscr{M}}]^t & \longrightarrow & [{}^0\mathscr{H}_{\mathscr{M}}]^t/[{}^0\mathscr{H}_{\mathscr{M}}]^t \cap \operatorname{Ker}\mathscr{D} & \longrightarrow & 0 \\
 & & \partial\uparrow & & \partial\uparrow & & \partial\uparrow & & \\
0 & \longrightarrow & [\mathscr{H}_{\mathscr{M}}]^t \cap \operatorname{Ker}\mathscr{D} & \longrightarrow & [\mathscr{H}_{\mathscr{M}}]^t & \longrightarrow & [\mathscr{H}_{\mathscr{M}}]^t/[\mathscr{H}_{\mathscr{M}}]^t \cap \operatorname{Ker}\mathscr{D} & \longrightarrow & 0 \\
 & & \uparrow & & \uparrow & & \uparrow & & \\
 & & 0 & & 0 & & 0 & &
\end{array} \tag{9.5}$$

The exactness of all the rows of this diagram is self-evident. The exactness of the second column follows from Corollary 2, § 4, Chapter III. The exactness of the first column is the content of Corollary 3, § 5, Chapter III. On the basis of Theorem 1, § 2, Chapter II, we conclude that the righthand column is also exact.

The exactness of this column in the second term from the bottom implies that the family $[\mathscr{H}_{\mathscr{M}}]^t/[\mathscr{H}_{\mathscr{M}}]^t \cap \operatorname{Ker} \mathscr{D}$ is isomorphic to the kernel of (1), that is, to the family $\mathscr{Z}_{\mathscr{M}}(p)$, and the isomorphism is realized through the mapping (8). Taking account of the isomorphism of Theorem 1, we arrive at the isomorphism

$$[\mathscr{H}_{\mathscr{M}}]^t/[\mathscr{H}_{\mathscr{M}}]^t \cap \operatorname{Ker} \mathscr{D} \overset{d}{\cong} \mathscr{H}_{\mathscr{M}}\{p, d\}. \tag{10.5}$$

Hence it follows that the mapping $d$ in (7) is a homomorphism and an epimorphism and its kernel is the subfamily $[\mathscr{H}_{\mathscr{M}}]^t \cap \operatorname{Ker} \mathscr{D}$. In view of the theorem of § 5, Chapter III, this subfamily is the image of the mapping $p$ in (7), and this mapping is also a homomorphism. Thus, the exactness of the sequence (7) is proved. □

*Remark.* We shall show that the mappings $d$ and $d^{-1}$, which establish the isomorphism (10), have orders independent of the family $\mathscr{M}$. For this we must return to the beginning of the section.

From the proof of Theorem 1, it is evident that the mappings which establish the isomorphism (2), have orders that are independent of $\mathscr{M}$. We now return to the diagram (9). What we have proved implies that the mappings of the form $\operatorname{Ker} \partial \to \operatorname{Coim} \partial$, which are inverse to the mapping denoted by the symbol $\partial$, have orders not depending on $\mathscr{M}$. For the mapping $\partial$ in the second column, this was established in 3°, § 4, Chapter III, and for the mapping in the first column, it follows from 9°, § 5, Chapter III. The remark in § 2, Chapter I, implies that the mapping which establishes the isomorphism

$$\mathscr{Z}_{\mathscr{M}}(p) \cong [\mathscr{H}_{\mathscr{M}}]^t/[\mathscr{H}_{\mathscr{M}}]^t \cap \operatorname{Ker} \mathscr{D} \tag{11.5}$$

has a similar property. We remark that the order of $\partial$ also does not depend on $\mathscr{M}$.

The mapping that establishes the isomorphism (10) is the composition of the mappings that establish the isomorphisms (2) and (11). With this the remark formulated above is proved.

**3°. Supplement to Theorem 2.** Our remark allows us to formulate Theorem 2 in a stronger version. We introduce the following notation. The doubly infinite sequence of functions $\mathscr{M} = \{M_\alpha(z), \alpha = \cdots -1, 0, 1, \ldots\}$, defined in $C^n$, will be called a complete family of majorants, if for an arbitrary integer $\beta$, the sequence $\mathscr{M}_\beta = \{M_{\beta+\alpha}(z), \alpha = 1, 2, \ldots\}$ is a family of majorants. We shall say that $\mathscr{M}$ is a complete family of majorants of type $\mathscr{I}$, if for an arbitrary integer $\beta$, $\mathscr{M}_\beta$ is a family of majorants of type $\mathscr{I}$.

The construction of § 1, Chapter III, and also subsections 1° and 2° of this section, can be carried over in an obvious fashion to complete families of majorants $\mathscr{M}$. In particular, if $\mathscr{M}$ is a complete family of majorants, $\mathscr{H}_\mathscr{M}$ is a doubly infinite family of spaces $\mathscr{H}_\alpha$, $\alpha = \cdots, -1, 0, 1, \ldots$, where $\mathscr{H}_\alpha$ is the space of functions holomorphic in $\Omega_\alpha$ and having finite norms $\|\cdot\|_\alpha^0$ (see (1.1), Chapter III). The symbol $\mathscr{H}_\mathscr{M}\{p, d\}$ has an analogous meaning. The families $\mathscr{H}_\mathscr{M}$, $\mathscr{H}_\mathscr{M}\{p, d\}$ and so on, corresponding to the complete family of majorants $\mathscr{M}$, will be called complete families. The matrix $p$ and the associated Noetherian operator $d$ define the mapping (7), where $\mathscr{M}$ is an arbitrary, complete family of majorants. We remark that the order of each of these mappings is a function of the form $\alpha \to \alpha + a$, where $a$ is a constant depending only on $p$ and $d$. In fact, for the mapping $p$, the constant $a$ is equal to $\mu = \deg p$, and for the mapping $d$ according to (6), it is equal to $3 + [q]$. We shall now formulate the assertion in which we are interested.

*Supplement.* Let $\mathscr{M}$ be a complete family of majorants of type $\mathscr{I}$. Then the sequence (7) is exact and, moreover, there are defined the mappings of complete families

$$d^{-1}\colon\ \mathscr{H}_\mathscr{M}\{p, d\} \to \mathscr{H}_\mathscr{M}^t/\operatorname{Ker} d,$$
$$p^{-1}\colon\ \operatorname{Ker} d \to \mathscr{H}_\mathscr{M}^s/\operatorname{Ker} p,$$

inverse to the mappings $d$ and $p$ in (7), the order of which has the form $\alpha \to \alpha + A$, where the constant $A$ depends only on $p$ and $d$.

*Proof.* We fix an arbitrary integer $\beta$ and we consider the family of majorants $\mathscr{M}_\beta$, consisting of the functions $\mathscr{M}_{\beta+\alpha}$, $\alpha = 1, 2, \ldots$. By hypothesis, this family is of type $\mathscr{I}$. Therefore, by Theorem 2, there is defined the mapping

$$d^{-1}\colon\ \mathscr{H}_{\mathscr{M}_\beta}\{p, d\} \to \mathscr{H}_{\mathscr{M}_\beta}^t/\operatorname{Ker} d,$$

inverse to $d$. In view of the remark made in connection with Theorem 2, the function $\alpha \to \gamma(\alpha)$, which is the order of the mapping $d^{-1}$, does not depend on the family $\mathscr{M}_\beta$, that is, depends only on $p$ and $d$.

We denote by $\mathscr{H}_\alpha'$ and $\mathscr{H}_\alpha'\{p, d\}$ the spaces forming the families $\mathscr{H}_{\mathscr{M}_\beta}$ and $\mathscr{H}_{\mathscr{M}_\beta}\{p, d\}$. Let

$$(d^{-1})_1'\colon\ \mathscr{H}_1'\{p, d\} \to [\mathscr{H}_\gamma']^t/(\operatorname{Ker} d)_\gamma',$$
$$(\operatorname{Ker} d)_\gamma' = \operatorname{Ker}\{[\mathscr{H}_\gamma']^t \xrightarrow{d} \mathscr{H}_{\gamma+a}'\{p, d\}\}, \qquad \gamma = \gamma(1)$$

be the component of the mapping $d^{-1}$. We remark that the spaces $\mathscr{H}_1'\{p, d\}$ and $\mathscr{H}_\gamma'$ are also components of the families $\mathscr{H}_\mathscr{M}\{p, d\}$ and

$\mathscr{H}_{\mathscr{M}}$ respectively, with the indices $\beta+1$ and $\beta+\gamma$. Therefore, the mapping $(d^{-1})'_1$ can be rewritten as:

$$\mathscr{H}_{\beta+1}\{p, d\} \to \mathscr{H}^t_{\beta+\gamma}/(\operatorname{Ker} d)_{\beta+\gamma}.$$

We denote this mapping by $(d^{-1})_{\beta+1}$. Since the composition of the mappings $(d^{-1})'_1$ and $d$ is the identity mapping, acting from $\mathscr{H}'_1\{p, d\}$ to $\mathscr{H}'_{\gamma+a}\{p, d\}$, and since the mapping $d$, defined on $[\mathscr{H}'_\gamma]^t/(\operatorname{Ker} d)'_\gamma$, is biunique, the mappings $(d^{-1})_\beta$ form a mapping of families $d^{-1}$. This is the one we are looking for, since its order is equal to the function $\alpha \to \alpha + \gamma - 1$, where $\gamma = \gamma(1)$ is a quantity depending only on $p$ and $d$.

In a similar fashion, we construct the desired mapping $p^{-1}$. □

**4°. The invariant notion of holomorphic *p*-function.** By Definition 1, the concept of holomorphic $p$-function depends not only on the matrix $p$ itself, but also on the choice of the associated set of normal Noetherian operators. We shall now show that in fact the content of this concept depends only on the $\mathscr{P}$-module $M \cong \operatorname{Coker} p = \mathscr{P}^t/p\mathscr{P}^s$. We remark that in view of Theorem 1, the family $\mathscr{H}_{\mathscr{M}}\{p, d\}$ of the spaces of holomorphic $p$-functions is isomorphic to the family $\mathscr{Z}_{\mathscr{M}}(p)$, which does not depend on the construction of Noetherian operators. We shall show that for an arbitrary finite $\mathscr{P}$-module $M$ and for an arbitrary family of majorants $\mathscr{M}$, the families $\mathscr{Z}_{\mathscr{M}}(p)$ with the matrices $p$ such that $\operatorname{Coker} p \cong M$ are connected among themselves by natural isomorphisms.

**Proposition 1**

I. *Let $\mathscr{M}$ be a family of majorants and let $f\colon M \to M'$ be a mapping of finite $\mathscr{P}$-modules with $M \cong \mathscr{P}^t/p\mathscr{P}^s$ and $M' \cong \mathscr{P}^{t'}/p'\mathscr{P}^{s'}$ for some $\mathscr{P}$-matrices $p$ and $p'$. Then there is defined a mapping of families $f_p^{p'}\colon \mathscr{Z}_{\mathscr{M}}(p) \to \mathscr{Z}_{\mathscr{M}}(p')$, depending only on $f$, $p$ and $p'$, and satisfying the following condition.*

II. *If $f$ is an isomorphism and $p = p'$, then $f_p^{p'}$ is also an isomorphism.*

III. *If*

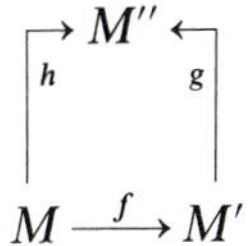

*is a commutative diagram of mappings of finite $\mathscr{P}$-modules, then for arbitrary $\mathscr{P}$-matrices $p$, $p'$, $p''$, the cokernels of which coincide respectively with $M$, $M'$, and $M''$, the diagram*

$$\begin{array}{ccc} & \mathscr{Z}_{\mathscr{M}}(p'') & \\ {\scriptstyle h_p^{p''}} \nearrow & & \nwarrow {\scriptstyle g_{p'}^{p''}} \\ \mathscr{Z}_{\mathscr{M}}(p) & \xrightarrow{\;f_p^{p'}\;} & \mathscr{Z}_{\mathscr{M}}(p') \end{array}$$

*is also commutative.*

*Proof.* Suppose given the mapping $f: M \to M'$. We construct the mapping $f_p^{p'}$. In view of Proposition 1, § 3, Chapter I, there exist $\mathscr{P}$-matrices $f_0$ and $f_1$, such that the diagram

$$\begin{array}{ccccccc} \mathscr{P}^{s'} & \xrightarrow{p'} & \mathscr{P}^{t'} & \longrightarrow & M' & \longrightarrow & 0 \\ {\scriptstyle f_1}\uparrow & & {\scriptstyle f_0}\uparrow & & {\scriptstyle f}\uparrow & & \\ \mathscr{P}^{s} & \xrightarrow{p} & \mathscr{P}^{t} & \longrightarrow & M & \longrightarrow & 0 \end{array} \tag{12.5}$$

is commutative. But its commutativity implies the relation $f_0\, p = p' f_1$. Hence, it is easy to see that the operation of multiplication by the matrix $f_0$ in the space $\mathcal{O}_Z^t$ (the point $Z \in C^n$ is arbitrary) carries the subspace $p\,\mathcal{O}_Z^s$ into the subspace $p'\,\mathcal{O}_Z^{s'} \subset \mathcal{O}_Z^{t'}$. For an arbitrary integer $\alpha$ the operation of multiplication by the matrix $f_0$ defines a continuous mapping

$$[{}^{\nu}\mathscr{H}_\alpha(U_\alpha)]^t \to [{}^{\nu}\mathscr{H}_{\alpha+m}(U_{\alpha+m})]^{t'}, \tag{13.5}$$

where $m$ is the order of this matrix. Let $\mathscr{D}$ and $\mathscr{D}'$ be $p$- and $p'$-operators respectively. If the cochain $\phi$ belongs to the subspace $[{}^{\nu}\mathscr{H}_\alpha(U_\alpha)]^t \cap \operatorname{Ker}\mathscr{D}$, then by the property of the operator $\mathscr{D}$, for arbitrary $Z \in C^n$ the cochain $\phi$ belongs to $p\,\mathcal{O}_Z^s$ (that is, each of its components belongs to $p\,\mathcal{O}_Z^s$). Therefore, the cochain $f_0\,\phi$ belongs to $[{}^{\nu}\mathscr{H}_{\alpha+m}(U_{\alpha+m})]^{t'} \cap \operatorname{Ker}\mathscr{D}'$. Thus we have established the fact that the mapping (13) carries the subspace $[{}^{\nu}\mathscr{H}_\alpha(U_\alpha)]^t \cap \operatorname{Ker}\mathscr{D}$ into $[{}^{\nu}\mathscr{H}_{\alpha+m}(U_{\alpha+m})]^{t'} \cap \operatorname{Ker}\mathscr{D}'$ and, therefore, defines a mapping of the corresponding factor-spaces. The set of all such mappings forms the family mapping

$$\check{f}_0:\quad {}^{\nu}\mathscr{H}_{\mathscr{M}}^t / {}^{\nu}\mathscr{H}_{\mathscr{M}}^t \cap \operatorname{Ker}\mathscr{D} \to {}^{\nu}\mathscr{H}_{\mathscr{M}}^{t'} / {}^{\nu}\mathscr{H}_{\mathscr{M}}^{t'} \cap \operatorname{Ker}\mathscr{D}'. \tag{14.5}$$

Clearly, mappings of this kind commute with the coboundary mappings of the form (1). Therefore, the mapping (14) carries $\mathscr{Z}_{\mathscr{M}}(p)$ into $\mathscr{Z}_{\mathscr{M}}(p')$, that is, it defines the desired mapping

$$f_p^{p'}:\quad \mathscr{Z}_{\mathscr{M}}(p) \to \mathscr{Z}_{\mathscr{M}}(p').$$

We shall show that this mapping depends only on $f$, $p$ and $p'$ and does not depend on the matrices $f_0$ and $f_1$. An arbitrary pair of matrices $(f_0, f_1)$, which render the diagram (12) commutative, will be said to correspond to the mapping $f$. Let $(f_0, f_1)$ and $(g_0, g_1)$ be two pairs of matrices corresponding to the mapping $f$. In accordance with Proposition 1, § 3, Chapter I, we have the equation $f_0 = g_0 + p'\mu$, where $\mu$ is some $\mathscr{P}$-matrix of size $s' \times t$. Since the image of the mapping $p'\mu$: ${}^{\nu}\mathscr{H}_{\mathscr{M}}^t \to {}^{\nu}\mathscr{H}_{\mathscr{M}}^{t'}$ belongs to the subfamily $p'\,{}^{\nu}\mathscr{H}_{\mathscr{M}}^{s'}$, in going to a mapping of the type (14) we obtain a null mapping. Hence, $\check{f}_0 = \check{g}_0$, q.e.d.

We shall establish Property II. Let the mapping $f$ be an isomorphism and let $p=p'$. Then we may choose for $f_0$ and $f_1$ the unit matrices. Clearly the associated mapping $\check{f}_0$ is an identity mapping.

We shall establish the third property. Let $(f_0, f_1)$ and $(g_0, g_1)$ be pairs corresponding to the mappings $f$ and $g$. From the commutativity of the diagram

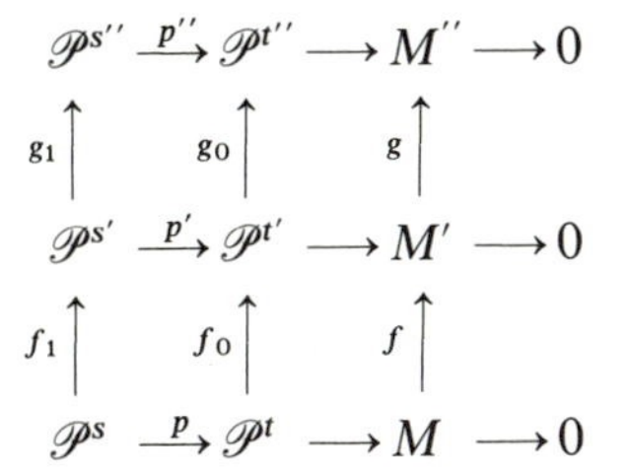

it is obvious that the pair $(g_0 f_0, g_1 f_1)$ corresponds to the mapping $gf=h$. By definition, the mapping $h_p^{p''}$ is associated with the matrix $g_0 f_0$ and, accordingly, is the composition of the mappings $f_p^{p'}$ and $g_{p'}^{p''}$, associated with the matrices $f_0$ and $g_0$. □

Suppose that the mapping $f\colon M \to M'$ is an isomorphism. Then our proposition implies the existence of the mapping

$$\mathscr{L}_{\mathscr{M}}(p) \underset{(f^{-1})_{p'}^{p}}{\overset{f_p^{p'}}{\rightleftarrows}} \mathscr{L}_{\mathscr{M}}(p'), \tag{15.5}$$

which establishes the isomorphism of the families $\mathscr{L}_{\mathscr{M}}(p)$ and $\mathscr{L}_{\mathscr{M}}(p')$. Thus for arbitrary pairs of $\mathscr{P}$-matrices $p$ and $p'$ such that Coker $p \cong$ Coker $p' \cong M$, we have established an isomorphism of the families $\mathscr{L}_{\mathscr{M}}(p)$, and $\mathscr{L}_{\mathscr{M}}(p')$. These isomorphisms agree in the sense that for an arbitrary three matrices $p$, $p'$ and $p''$, the cokernels of which are isomorphic to $M$, the diagram

$$\begin{array}{ccc} & \mathscr{L}_{\mathscr{M}}(p'') & \\ \updownarrow & & \updownarrow \\ \mathscr{L}_{\mathscr{M}}(p) & \longleftrightarrow & \mathscr{L}_{\mathscr{M}}(p') \end{array}$$

which is formed by these isomorphisms is commutative. (Its commutativity follows from Property III.)

For an arbitrary finite $\mathscr{P}$-module $M$, we regard all the families $\mathscr{L}_{\mathscr{M}}(p)$ for which Coker $p \cong M$, and isomorphisms of type (15), as a new object, which we will denote by the symbol $\mathscr{L}_{\mathscr{M}}(M)$. Isomorphisms of the type (15) will be referred to as inner isomorphisms.

Let $f\colon M \to M'$ be a mapping of finite $\mathscr{P}$-modules. By Proposition 1, for any matrices $p$, $q$ and $p'$, $q'$, whose cokernels are isomorphic to $M$

and $M'$, the diagram

$$\begin{array}{ccc} \mathscr{L}_{\mathcal{M}}(p) & \xrightarrow{f_p^{p'}} & \mathscr{L}_{\mathcal{M}}(p') \\ \updownarrow & & \updownarrow \\ \mathscr{L}_{\mathcal{M}}(q) & \xrightarrow{f_q^{q'}} & \mathscr{L}_{\mathcal{M}}(q') \end{array}$$

in which the columns contain inner isomorphisms, is commutative. The set of mappings $f_p^{p'}$ will be regarded as mappings of the corresponding objects.

We now take the final step. For every finite $\mathscr{P}$-module $M$, we shall consider the new object $\mathscr{H}_{\mathcal{M}}\{M\}$, formed by the family $\mathscr{H}_{\mathcal{M}}\{p, d\}$, where Coker $p \cong M$. These families are interconnected by isomorphisms, which are compositions of isomorphisms of the type (2) and (15). If we are given a mapping of modules $f: M \to M'$, selected matrices $p$ and $p'$, the cokernels of which are isomorphic to $M$ and $M'$, and if we have constructed the corresponding Noetherian operators $d$ and $d'$, then we can construct the mapping $f_{p,d}^{p',d'}: \mathscr{H}_{\mathcal{M}}\{p, d\} \to \mathscr{H}_{\mathcal{M}}\{p', d'\}$, which is defined by the condition that the diagram

$$\begin{array}{ccc} \mathscr{H}_{\mathcal{M}}\{p, d\} & \xrightarrow{f_{p,d}^{p',d'}} & \mathscr{H}_{\mathcal{M}}\{p', d'\} \\ {\scriptstyle d}\uparrow & & {\scriptstyle d'}\uparrow \\ \mathscr{L}_{\mathcal{M}}(p) & \xrightarrow{f_p^{p'}} & \mathscr{L}_{\mathcal{M}}(p') \end{array}$$

be commutative. It is clear from the construction that such mappings commute with the inner isomorphisms of the objects $\mathscr{H}_{\mathcal{M}}\{M\}$ and $\mathscr{H}_{\mathcal{M}}\{M'\}$. The set of all such mappings will be denoted by: $f_{\mathcal{M}}: \mathscr{H}_{\mathcal{M}}\{M\} \to \mathscr{H}_{\mathcal{M}}\{M'\}$. Thus we have constructed the functor

$$M \sim \to \mathscr{H}_{\mathcal{M}}\{M\}, \tag{16.5}$$

which relates to every finite $\mathscr{P}$-module $M$ the object $\mathscr{H}_{\mathcal{M}}\{M\}$, and to every mapping $f$ of such modules the mapping $f_{\mathcal{M}}$ of the corresponding object.

We characterize the object $\mathscr{H}_{\mathcal{M}}\{M\}$ in the case $M = \mathscr{P}^t$, $t \geqq 0$. To do this, we shall describe the simplest family belonging to this object. We choose the family $\mathscr{H}_{\mathcal{M}}\{p\}$, where $p$ is a matrix of size $t \times 1$, consisting of zeroes. Clearly, the cokernel of such a matrix is equal to $\mathscr{P}^t$. From the exactness of the sequence

$$\mathscr{P} \xrightarrow{p} \mathscr{P}^t \xrightarrow{e} \mathscr{P}^t,$$

where $e$ is the unit matrix, it follows, in view of Proposition 2, § 4, that the matrix $e$ is a normal Noetherian operator, associated with $p$, and $N(p) = C^n$. Since Ker $e = 0$, the isomorphism (10) can be written as follows:

$$\mathscr{H}_{\mathcal{M}}^t \xrightarrow{e} \mathscr{H}_{\mathcal{M}}\{p, e\}.$$

Thus the family $\mathscr{H}_{\mathscr{M}}\{p, e\}$ coincides with $\mathscr{H}_{\mathscr{M}}^{t}$. Let $f\colon \mathscr{P}^{t} \to \mathscr{P}^{t'}$ be a $\mathscr{P}$-matrix. As we have just determined, the objects $\mathscr{H}_{\mathscr{M}}\{\mathscr{P}^{t}\}$, $\mathscr{H}_{\mathscr{M}}\{\mathscr{P}^{t'}\}$ contain the families $\mathscr{H}_{\mathscr{M}}^{t}$, $\mathscr{H}_{\mathscr{M}}^{t'}$, corresponding to the null matrices $p$ and $p'$. It is not difficult to see that the corresponding mapping $f_{p}^{p'}\colon \mathscr{H}_{\mathscr{M}}^{t} \to \mathscr{H}_{\mathscr{M}}^{t'}$ coincides with a multiplication by the matrix $f$.

To every sequence

$$M' \xrightarrow{f} M \xrightarrow{g} M'' \tag{17.5}$$

of mappings of finite $\mathscr{P}$-modules corresponds the sequence

$$\mathscr{H}_{\mathscr{M}}\{M'\} \xrightarrow{f_{\mathscr{M}}} \mathscr{H}_{\mathscr{M}}\{M\} \xrightarrow{g_{\mathscr{M}}} \mathscr{H}_{\mathscr{M}}\{M''\}. \tag{18.5}$$

We shall say that this sequence is exact if every sequence of the form

$$\mathscr{H}_{\mathscr{M}}\{p'\} \xrightarrow{f_{p'}^{p}} \mathscr{H}_{\mathscr{M}}\{p\} \xrightarrow{g_{p}^{p''}} \mathscr{H}_{\mathscr{M}}\{p''\},$$

is exact, where the matrices $p, p', p''$ are so chosen that their cokernels are isomorphic to the modules $M$, $M'$, $M''$. From what we said earlier, it follows that if such a sequence is exact, for some special choice of the matrices $p, p', p''$, it is exact for any other arbitrary choice of these matrices, subject only to the above condition. We may now formulate

**5°. The fundamental theorem.** *For an arbitrary family of majorants $\mathscr{M}$ of type I, the functor* (16) *is exact, that is, an arbitrary exact sequence* (17) *corresponds to an exact sequence* (18).

*Proof.* We suppose at first that the sequence

$$0 \to M' \xrightarrow{f} M \xrightarrow{g} M'' \to 0.$$

is exact. By the lemma of § 3, Chapter I, such a sequence can be imbedded in a commutative diagram of the form

$$\begin{array}{ccccccc}
0 & & 0 & & 0 & & \\
\uparrow & & \uparrow & & \uparrow & & \\
\mathscr{P}^{s''} & \xrightarrow{p''} & \mathscr{P}^{t''} & \longrightarrow & M'' & \longrightarrow & 0 \\
\uparrow & & \uparrow & & \uparrow & & \\
\mathscr{P}^{s} & \xrightarrow{p} & \mathscr{P}^{t} & \longrightarrow & M & \longrightarrow & 0 \\
\uparrow & & \uparrow & & \uparrow & & \\
\mathscr{P}^{s'} & \xrightarrow{p'} & \mathscr{P}^{t'} & \longrightarrow & M' & \longrightarrow & 0 \\
\uparrow & & \uparrow & & \uparrow & & \\
0 & & 0 & & 0 & &
\end{array} \tag{19.5}$$

in which all the rows and columns are exact. We consider another diagram

$$\begin{array}{ccccccc}
0 & & 0 & & 0 & & \\
\uparrow & & \uparrow & & \uparrow & & \\
\mathscr{H}_{\mathscr{M}}^{s''} & \xrightarrow{p''} & \mathscr{H}_{\mathscr{M}}^{t''} & \xrightarrow{d''} & \mathscr{H}_{\mathscr{M}}\{p'', d''\} & \longrightarrow & 0 \\
\uparrow & & \uparrow & & \uparrow & & \\
\mathscr{H}_{\mathscr{M}}^{s} & \xrightarrow{p} & \mathscr{H}_{\mathscr{M}}^{t} & \xrightarrow{d} & \mathscr{H}_{\mathscr{M}}\{p, d\} & \longrightarrow & 0 \\
\uparrow & & \uparrow & & \uparrow & & \\
\mathscr{H}_{\mathscr{M}}^{s'} & \xrightarrow{p'} & \mathscr{H}_{\mathscr{M}}^{t'} & \xrightarrow{d'} & \mathscr{H}_{\mathscr{M}}\{p', d'\} & \longrightarrow & 0 \\
\uparrow & & \uparrow & & \uparrow & & \\
0 & & 0 & & 0 & & .
\end{array} \tag{20.5}$$

From the exactness of the columns of diagram (19) and from the theorem of § 5, Chapter III, there follows the exactness of the columns of diagram (20), except the rightmost. The exactness of the rows follows from Theorem 2 of this section. Applying Theorem 1, § 2, Chapter I, we establish the exactness of the right column. In view of the remark made at the end of 4°, we may conclude that the sequence

$$0 \to \mathscr{H}_{\mathscr{M}}\{M'\} \xrightarrow{f_{\mathscr{M}}} \mathscr{H}_{\mathscr{M}}\{M\} \xrightarrow{g_{\mathscr{M}}} \mathscr{H}_{\mathscr{M}}\{M''\} \to 0. \tag{21.5}$$

is exact.

Now suppose we have an arbitrary exact sequence (17). In view of the exactness of the sequences

$$0 \to \operatorname{Coim} f \xrightarrow{\check{f}} M \xrightarrow{\check{g}} \operatorname{Im} g \to 0,$$

$$0 \to \operatorname{Ker} f \longrightarrow M' \longrightarrow \operatorname{Coim} f \to 0,$$

$$0 \to \operatorname{Im} g \to M'' \to \operatorname{Coker} g \to 0$$

the sequences

$$0 \to \mathscr{H}_{\mathscr{M}}\{\operatorname{Coim} f\} \to \mathscr{H}_{\mathscr{M}}\{M\} \to \mathscr{H}_{\mathscr{M}}\{\operatorname{Im} g\} \to 0,$$

$$\mathscr{H}_{\mathscr{M}}\{M'\} \to \mathscr{H}_{\mathscr{M}}\{\operatorname{Coim} f\} \to 0 \to \mathscr{H}_{\mathscr{M}}\{\operatorname{Im} g\} \to \mathscr{H}_{\mathscr{M}}\{M''\}.$$

are also exact. Hence, it follows that the mappings

$$f_{\mathscr{M}}\colon\ \mathscr{H}_{\mathscr{M}}\{M'\} \to \mathscr{H}_{\mathscr{M}}\{\operatorname{Coim} f\} \to \mathscr{H}_{\mathscr{M}}\{M\},$$

$$g_{\mathscr{M}}\colon\ \mathscr{H}_{\mathscr{M}}\{M\} \to \mathscr{H}_{\mathscr{M}}\{\operatorname{Im} g\} \to \mathscr{H}_{\mathscr{M}}\{M''\}$$

are homomorphisms and $\operatorname{Im} f_{\mathscr{M}} \sim \operatorname{Ker} g_{\mathscr{M}}$ (that is, these properties are enjoyed by the corresponding representatives of the sets $f_{\mathscr{M}}$ and $g_{\mathscr{M}}$). □

### 6°. Remarks

*Remark* 1. Theorem 2 is a particular case of the Fundamental Theorem, corresponding to the exact sequence

$$\mathscr{P}^s \xrightarrow{p} \mathscr{P}^t \to M \to 0.$$

*Remark* 2. We shall formulate a sharpening of the Fundamental Theorem. Let $f: M \to M'$ be a mapping of finite modules, and let $p$ and $p'$ be $\mathscr{P}$-matrices, whose cokernels are isomorphic to $M$ and $M'$. Then for an arbitrary family of majorants $\mathscr{M}$ of type $\mathscr{I}$, the mapping $f_{p,d}^{p',d'}: \mathscr{H}_{\mathscr{M}}\{p,d\} \to \mathscr{H}_{\mathscr{M}}\{p',d'\}$ is a homomorphism and the order of the mapping and its inverse does not depend on $\mathscr{M}$.

The proof of this assertion is left to the reader.

PART TWO

# Differential Equations with Constant Coefficients

Chapter V

## Linear Spaces and Distributions

In § 1 we continue the study of families of linear topological spaces that was begun in § 1, Chapter I. We consider two special types of spaces: Frechet spaces and Schwartz spaces. Our attention is mainly focussed on inductive and projective limiting processes. In §§ 2 and 3 we set forth the theory of distributions and Fourier transforms in a form suited to our needs.

### § 1. Limiting processes in families of linear spaces

By a linear space, or more simply, a space, we shall mean a linear topological locally convex space – we shall use the abbreviation l.t.s. – over the field $C$ of complex numbers. We remark that this concept is a special case of the notion of a topological module in the sense of § 1 Chapter I. We now introduce families of l.t.s. of a more general type than those in § 1 Chapter I.

**1°. Increasing families of l.t.s.** The set $\mathscr{A}$ is said to be a directed set if there is an order relation among its elements, $<(\leqq)$, such that for arbitrary $\alpha$ and $\alpha'$ belonging to $\mathscr{A}$ there exists an $\alpha''$ such that $\alpha, \alpha' \leqq \alpha''$.

**Definition 1.** Let $\mathscr{A}$ be a directed set. An increasing family of l.t.s., defined on $\mathscr{A}$, is a system $X = \{X_\alpha, i_\alpha^{\alpha'}\}$ consisting of functions $X_\alpha$ defined on $\mathscr{A}$ and having as values l.t.s., together with a set of continuous mappings $i_\alpha^{\alpha'}\colon X_\alpha \to X_{\alpha'}$ defined for arbitrary $\alpha$ and $\alpha' \geqq \alpha$, and satisfying the conditions

(i) for arbitrary $\alpha$ the mapping $i_\alpha^\alpha$ is an identity mapping, and

(ii) for arbitrary $\alpha < \alpha' < \alpha''$ we have $i_{\alpha'}^{\alpha''}\, i_\alpha^{\alpha'} = i_\alpha^{\alpha''}$.

To every increasing family $X=\{X_\alpha, i_\alpha^{\alpha'}\}$ there corresponds an inductive limit, which is defined as follows: We consider the union $\bigcup X_\alpha$ of all the spaces $X_\alpha$ and we introduce an equivalence relation in it. The elements $x_\alpha \in X_\alpha$ and $x_{\alpha'} \in X_{\alpha'}$ will be said to be equivalent if there exists an $\alpha'' \geqq \alpha, \alpha'$ such that $i_\alpha^{\alpha''} x_\alpha = i_{\alpha'}^{\alpha''} x_{\alpha'}$. It is easily seen that this is indeed an equivalence relation. The set of equivalence classes of the elements of $\bigcup X_\alpha$ will be denoted by $\vec{X}$. We now introduce linear operations on these equivalence classes. Multiplication of an equivalence class by a scalar is defined as the multiplication of each element of the class by the given scalar. To define addition in $\vec{X}$ we begin by letting $x_\alpha \in X_\alpha$ and $x_{\alpha'} \in X_{\alpha'}$ be arbitrary elements of $\bigcup X_\alpha$. We choose an element $\alpha'' \in \mathscr{A}$ such that $\alpha, \alpha' \leqq \alpha''$ and we consider the sum $i_\alpha^{\alpha''} x_\alpha + i_{\alpha'}^{\alpha''} x_{\alpha'}$. The equivalence class containing this sum is uniquely defined by the classes containing $x_\alpha$ and $x_{\alpha'}$.

The linear operations thus defined in $\vec{X}$ convert the set into a linear space.

For every $\alpha$ the correspondence that maps the element $x_\alpha \in X_\alpha$ into its equivalence class will be denoted by $i_\alpha$. This correspondence defines a linear mapping $i_\alpha$: $X_\alpha \to \vec{X}$, which we call a canonical mapping. Note that for arbitrary $\alpha$ and $\alpha' > \alpha$ we have $i_{\alpha'} i_\alpha^{\alpha'} = i_\alpha$.

We introduce a topology in $\vec{X}$, the finest of all the locally convex topologies in $\vec{X}$ for which all the mappings $i_\alpha$ are continuous. A neighborhood of zero in $\vec{X}$ is defined as an arbitrary set containing a convex set $U$ whose pre-image $i_\alpha^{-1}(U)$ is for arbitrary $\alpha$ a neighborhood of zero in $X_\alpha$. The l.t.s. constructed in this way will be denoted by

$$\vec{X} = \varinjlim X = \varinjlim \{X_\alpha, i_\alpha^{\alpha'}\}.$$

We observe that the union of all the subspaces in $i_\alpha(X_\alpha)$ in $\vec{X}$ coincides with the whole of $\vec{X}$.

**2°. Decreasing families of l.t.s.**

**Definition 2.** A decreasing family of l.t.s. defined on a directed set $\mathscr{A}$ is a system $X=\{X_\alpha, i_\alpha^{\alpha'}\}$ consisting of functions $X_\alpha$ defined on $\mathscr{A}$ and having l.t.s. as values, together with a set of linear continuous mappings $i_\alpha^{\alpha'}: X_\alpha \to X_{\alpha'}$, defined for arbitrary $\alpha'$ and $\alpha \geqq \alpha'$, and satisfying conditions like (i) and (ii) as given in Definition 1.

Every decreasing family of l.t.s. has a projective limit, defined as follows. A thread is an arbitrary set of the form

$$x = \{x_\alpha \in X_\alpha, \alpha \in \mathscr{A}\}.$$

in which the elements $x_\alpha$ satisfy the relations $i_\alpha^{\alpha'} x_\alpha = x_{\alpha'}$ for arbitrary $\alpha'$ and $\alpha \geqq \alpha'$. The set of all threads will be denoted by $\overleftarrow{X}$. Threads may be multiplied termwise by scalars, and added termwise; the set $\overleftarrow{X}$ is therefore a linear space.

We denote by $i^\alpha: \overleftarrow{X} \to X_\alpha$ the linear operation which maps the thread $\{x_\alpha\}$ into the element $x_\alpha$. This will be called a canonical mapping. We now introduce a topology into $\overleftarrow{X}$, the weakest of all topologies for which the mappings $i^\alpha$ are all continuous. A neighborhood of zero in $\overleftarrow{X}$ is defined as an arbitrary set containing a set of the form $(i^\alpha)^{-1}(U_\alpha)$, where $U_\alpha$ is a neighborhood of zero in $X_\alpha$. The l.t.s. so constructed will be written in the form

$$\overleftarrow{X} = \varprojlim X = \varprojlim \{X_\alpha, i_\alpha^{\alpha'}\}.$$

**3°. Cofinal subfamilies.** Let $\mathscr{A}$ be a directed set. An arbitrary subset $\mathscr{A}'$ will be said to be cofinal if for arbitrary $\alpha \in \mathscr{A}$ there exists an element $\alpha' \in \mathscr{A}'$ such that $\alpha \leqq \alpha'$. A cofinal subset $\mathscr{A}'$, with the induced order relation, is obviously a directed set.

Let $X = \{X_\alpha, i_\alpha^{\alpha'}\}$ be a family of l.t.s. (either increasing or decreasing) defined on $\mathscr{A}$, and let $\mathscr{A}'$ be a cofinal subset of $\mathscr{A}$. A cofinal subfamily in $X$ is defined as the family (increasing or decreasing) formed by the spaces $X_\alpha$ and the mappings $i_\alpha^{\alpha'}$ with $\alpha, \alpha' \in \mathscr{A}'$.

**Proposition 1.** *Let $\mathscr{A}$ be a directed set and let $\mathscr{A}'$ be a cofinal subset. Let $X = \{X_\alpha, i_\alpha^{\alpha'}\}$ be a family defined on $\mathscr{A}$ and let $X'$ be the corresponding cofinal family. Then there exist natural isomorphisms*

$$\vec{X} \cong \vec{X}', \qquad \overleftarrow{X} \cong \overleftarrow{X}'$$

*if the family $X$ is either increasing or decreasing.*

*Proof.* Let $X$ be an increasing family. The inclusion relation $\bigcup_{\mathscr{A}'} X_\alpha \subset \bigcup_{\mathscr{A}} X_\alpha$ conserves the equivalence relations, and therefore defines a linear mapping $\vec{X}' \to \vec{X}$, which is obviously continuous. Let us construct its inverse. Let $x_\alpha \in X_\alpha$ be an arbitrary element of $\bigcup_{\mathscr{A}} X_\alpha$. We choose an $\alpha' \in \mathscr{A}'$ such that $\alpha \leqq \alpha'$. The element $i_\alpha^{\alpha'} x_\alpha$ belongs to $\bigcup_{\mathscr{A}'} X_\alpha$ and is equivalent to $x_\alpha$. We have thus constructed the inverse mapping $\vec{X} \to \vec{X}'$; its continuity is self-evident.

Now let $X$ be a decreasing family. Corresponding to every thread $\{x_\alpha, \alpha \in \mathscr{A}\} \in \overleftarrow{X}$ we select the thread $\{x_\alpha, \alpha \in \mathscr{A}'\} \in \overleftarrow{X}'$ which is a subset of the original. This correspondence sets up a linear mapping from $\overleftarrow{X}$ to $\overleftarrow{X}'$. Conversely, let $\{x_\alpha, \alpha \in \mathscr{A}'\}$ be a thread belonging to $\overleftarrow{X}'$. Corresponding to every $\alpha \in \mathscr{A}$ we select an $\alpha' \in \mathscr{A}'$ such that $\alpha \leqq \alpha'$ and we write $x_\alpha = i_{\alpha'}^{\alpha} x_{\alpha'}$. The element $x_\alpha \in X_\alpha$ is clearly independent of the choice of $\alpha'$, and the set of all such $x_\alpha$ forms a thread belonging to $\overleftarrow{X}$. But this means that we have constructed the inverse linear mapping $\overleftarrow{X}' \to \overleftarrow{X}$; both are obviously continuous. ∎

**4°. Countable families.** We shall often encounter increasing and decreasing families of l.t.s. defined on the set $Z_+$ of natural numbers, ordered by increase in value. Another particular case that will be of importance to us is the family in the sense of § 1, Chapter I, an increasing family defined on the set $Z$ of all integers. The set $Z$ has two order relations – by increasing value and by decreasing value. In both cases, $Z$ is a directed set. Thus, every family of l.t.s. in the sense of § 1 Chapter I is both an increasing and a decreasing family, and therefore has both an inductive and a projective limit.

We remark that every increasing or decreasing family of l.t.s. defined on the set $Z_+$ can be looked on as a family in the sense of § 1 Chapter I. For, if $\{X_\alpha, i_\alpha^{\alpha'}\}$ is an increasing family of l.t.s. defined on $Z_+$, we may supplement it with null spaces $X_\alpha$ and null mappings $i_\alpha^{\alpha'}$ for all integer $\alpha \leqq 0$; we obtain an increasing family defined on $Z$. If $\{X_\alpha, i_\alpha^{\alpha'}\}$ is a decreasing family defined on $Z_+$, we may set

$$Y_\alpha = \begin{cases} X_{-\alpha}, & \alpha < 0, \\ 0, & \alpha \geqq 0, \end{cases} \qquad j_\alpha^{\alpha'} = \begin{cases} i_{-\alpha}^{-\alpha'}, & \alpha \leqq \alpha' < 0, \\ 0, & \alpha \geqq 0, \end{cases}$$

and obtain an increasing family defined on $Z$. This means that we may on occasion limit ourselves to the consideration of families of l.t.s. without being compelled to make parallel investigations of increasing and decreasing families defined on $Z_+$.

From now on we shall refer to families defined on $Z_+$ as decreasing or increasing families, and we shall refer to increasing families defined on $Z$ merely as families.

**5°. General properties of limits.** We recall that an l.t.s. $E$ is a Frechet space, or $\mathscr{F}$-space, if it is a complete metric space.

**Proposition 2.** *Let $X = \{X_\alpha, i_\alpha^{\alpha'}\}$ be a decreasing family in which all the $X_\alpha$ are Frechet spaces. Then the projective limit $\overleftarrow{X}$ is also a Frechet space. If for every $\alpha$ the image of the mapping $i_{\alpha+1}^{\alpha}: X_{\alpha+1} \to X_\alpha$ is dense in $X_\alpha$, then for arbitrary $\alpha$ the image of the mapping $i^\alpha: \overleftarrow{X} \to X_\alpha$ is also dense in $X_\alpha$.*

*Proof.* We shall first show that the space $\overleftarrow{X}$ is separable. Let $x$ be an arbitrary non-zero element of it. For some $\alpha$ the element $x_\alpha = i^\alpha x \in X_\alpha$ is also different from zero. Since $X_\alpha$ is a Frechet space, it is separable, and we can therefore find in $X_\alpha$ a neighborhood $U_\alpha$ of zero not containing $x_\alpha$. Its pre-image under the mapping $i^\alpha$ is a neighborhood of zero in $\overleftarrow{X}$ and does not contain $x$. Hence $X$ is separable.

Since $X_\alpha$ is a metric space for arbitrary $\alpha$, it has a countable fundamental system of neighborhoods of zero, $U_\alpha^i$, $i = 1, 2, \ldots$. The pre-images $(i^\alpha)^{-1}(U_\alpha^1)$, $\alpha, i = 1, 2, \ldots$ of all these neighborhoods form a countable fundamental system of neighborhoods of zero in $\overleftarrow{X}$. Since $\overleftarrow{X}$ is separable, it is also metric.

We must show that $\overleftarrow{X}$ is complete. Let $\{x^i\}$ be a fundamental system in $\overleftarrow{X}$. For arbitrary $\alpha$ the sequence $i^\alpha x^i$ is fundamental in $X_\alpha$ and therefore has a limit $x_\alpha \in X_\alpha$ since $X_\alpha$ is complete. In the equation $i^\alpha_{\alpha+1} i^{\alpha+1} x^i = i^\alpha x^i$ we pass to the limit with respect to $i$ and we obtain $i^\alpha_{\alpha+1} x_{\alpha+1} = x_\alpha$. Therefore the elements $x_\alpha$ form a thread which is an element $x \in \overleftarrow{X}$. Since for arbitrary $\alpha$ the sequence $i^\alpha x^i$ converges to $x_\alpha$ in $X_\alpha$ the sequence $x^i$ converges to $x$ in $\overleftarrow{X}$. Thus $\overleftarrow{X}$ is complete.

We now take up the second assertion. Let $\rho_\alpha(.,.)$ be the metric in $X_\alpha$. For convenience we shall suppose that $\rho_\alpha$ is a non-decreasing function of $\alpha$, that is, we suppose that for arbitrary $\alpha$ we have the inequality

$$\rho_\alpha(i^\alpha_{\alpha+1} x, i^\alpha_{\alpha+1} y) \leqq \rho_{\alpha+1}(x,y), \qquad x,y \in X_{\alpha+1}.$$

If this is not so, we may replace $\rho_\alpha$ by the equivalent metric

$$\rho'_\alpha(x,y) = \sum_{\beta \leqq \alpha} \rho_\beta(i^\beta_\alpha x, i^\beta_\alpha y), \qquad x,y \in X_\alpha.$$

Let us show, for example, that $i^1(\overleftarrow{X})$ is dense in $X_1$. We choose an arbitrary element $x_1 \in X_1$, and a number $\varepsilon > 0$. Using our assumption, we inductively construct the sequence of elements $x_\alpha \in X_\alpha$, $\alpha = 2, 3, \ldots$ such that

$$\rho_\alpha(a_\alpha, i^\alpha_{\alpha+1} x_{\alpha+1}) \leqq \frac{\varepsilon}{2^\alpha}, \qquad \alpha = 1, 2, \ldots. \tag{1.1}$$

We fix $\beta \geqq 1$ and we consider the sequence of elements $x^\beta_\alpha = i^\beta_\alpha x_\alpha$ with $\alpha \geqq \beta$. From (1), using the fact that $\rho_\alpha$ is an increasing function of $\alpha$, we obtain

$$\rho_\beta(x^\beta_\alpha, x^\beta_{\alpha+1}) \leqq \rho_\alpha(x_\alpha, i^\alpha_{\alpha+1} x_{\alpha+1}) \leqq \frac{\varepsilon}{2^\alpha}, \tag{2.1}$$

and therefore the sequence $\{x^\beta_\alpha\}$ is fundamental in $X_\beta$. Let $x^\beta$ be its limit. In the equation $i^\beta_{\beta+1} x^{\beta+1}_\alpha = x^\beta_\alpha$ we pass to the limit with respect to $\alpha$ and obtain $i^\beta_{\beta+1} x^{\beta+1} = x^\beta$, i.e. the elements $x^\beta$ form a thread which we denote by $x \in \overleftarrow{X}$. Summing the inequalities (2) with $\beta = 1$, we obtain the inequality

$$\rho_1(x_1, i^1 x) = \rho_1(x_1, x^1) \leqq \sum_1^\infty \rho_1(x^1_\alpha, x^1_{\alpha+1}) \leqq \varepsilon.$$

**6°. Duality in families of l.t.s.** Let $E$ be an l.t.s. Its conjugate space $E^*$ will always be assigned the strong topology, i.e. the topology in which the polars[1] of the bounded sets in $E$ form a fundamental system of neighborhoods of zero. If $\phi: E \to F$ is a continuous mapping of l.t.s.

1 The polar of a set $G \subset E$ is the set of all functionals $f \in E^*$ satisfying the condition $|(f,x)| \leqq 1 \ \forall\ x \in G$.

the conjugate mapping $\phi^*: F^* \to E^*$ defined by the formula $(\phi^* f', e) = (f', \phi e)$ is also continuous.

Let $X = \{X_\alpha, i_\alpha^{\alpha'}\}$ be an increasing family of l.t.s. We consider the ensemble of conjugate spaces $X_\alpha^*$ and conjugate mappings

$$j_\alpha^{\alpha'} = (i_{\alpha'}^\alpha)^*: \; X_\alpha^* \to X_{\alpha'}^*.$$

These spaces and mappings clearly satisfy the conditions given in Definition 2 and therefore form a decreasing family, which we call the conjugate family and denote by $X^*$. Similarly, for every decreasing family $X$ we define the conjugate increasing family $X^*$.

Suppose that $X = \{X_\alpha, i_\alpha^{\alpha'}\}$ is an increasing family, and that $B_\alpha$ is a bounded subset of one of the spaces $X_\alpha$. Its image $i_\alpha(B_\alpha) \subset \vec{X}$ is obviously absorbed by an arbitrary neighborhood of zero in $\vec{X}$ and is therefore bounded in $\vec{X}$. We shall say that the family $X$ is regular if the converse assertion is true: every set bounded in $\vec{X}$ is for some $\alpha$ equal to $i_\alpha(B_\alpha)$, where $B_\alpha$ is a bounded subset of $X_\alpha$.

**Proposition 3.** *Let $X$ be a regular increasing family. Then there exists a natural isomorphism*

$$(\vec{X})^* \cong \overleftarrow{X}^* = \varprojlim \{X_\alpha^*, j_\alpha^{\alpha'} = (i_{\alpha'}^\alpha)^*\}. \tag{3.1}$$

*Proof.* Let $f$ be an arbitrary continuous functional on $\vec{X}$. Since the canonical mapping $i_\alpha: X_\alpha \to \vec{X}$ is continuous, we may consider the continuous functional $f_\alpha = i_\alpha^* f \in X_\alpha^*$. The ensemble of these functionals is obviously a thread in the family $X^*$ and therefore is an element of the limit $\overleftarrow{X}^*$. Conversely, let $\{f_\alpha \in X_\alpha^*\}$ be a thread which is an element of $\overleftarrow{X}^*$. We consider the functional defined on $\vec{X}$ by the formula

$$(f, i_\alpha x_\alpha) = (f_\alpha, x_\alpha), \qquad x_\alpha \in X_\alpha.$$

It is clearly linear and continuous, and is therefore an element of $(\vec{X})^*$. But this establishes the algebraic isomorphism (3).

We show that this isomorphism is also topological. By definition, the polars of the bounded sets $B \subset \vec{X}$ form a fundamental system of neighborhoods of zero in $(\vec{X})^*$. On the other hand, in each of the spaces $X_\alpha^*$ a fundamental system of neighborhoods of zero is formed by the polars of the bounded sets $B_\alpha \subset X_\alpha$. Therefore, the sets of the form $(j^\alpha)^{-1}(B_\alpha^0)$ form a fundamental system of neighborhoods of zero in $\overleftarrow{X}^*$ ($j^\alpha$ is the canonical mapping in the family $X^*$). We now note that $(j^\alpha)^{-1}(B_\alpha^0) = (i_\alpha(B_\alpha))^0$. There remains to be added the fact that by hypothesis the classes of sets of the forms $B$ and $i_\alpha(B_\alpha)$ coincide. $\square$

**Proposition 4.** *Let $X$ be an increasing family consisting of reflexive Banach spaces. Then the limit $\vec{X}$ is also reflexive and the isomorphism* (3) *is valid.*

*If $X$ is a decreasing family consisting of reflexive Banach spaces, the limit $\overleftarrow{X}$ is also reflexive and there exists an isomorphism $(\overleftarrow{X})^* \cong \overrightarrow{X}^*$.*

*Proof.* As we know, an increasing family consisting of reflexive Banach spaces is regular and its limit is reflexive[2]. The isomorphism (3) follows from Proposition (3). Suppose that $X$ is a decreasing family of reflexive Banach spaces. The conjugate family $X^*$ is an increasing family and also consists of reflexive Banach spaces. By what we have already shown, the limit $\overrightarrow{X}^*$ is reflexive and $(\overrightarrow{X}^*)^* = \overleftarrow{X}$. Taking the conjugates of both sides of this isomorphism, we obtain the isomorphism $(\overleftarrow{X})^* = \overrightarrow{X}^*$, q.e.d. □

**7°. Schwartz spaces.** We recall that the mapping $\phi: E \to F$ of l.t.s. is said to be compact, or completely continuous, if it carries a neighborhood of zero into a relatively compact set[3]. If $E$ and $F$ are Hilbert spaces, the conjugate mapping $\phi^*: F^* \to E^*$ is compact if and only if $\phi$ itself is compact.

The l.t.s. $E$ is a Schwartz space if, corresponding to an arbitrary neighborhood of zero $U$ in $E$, we can find a neighborhood $V$ of zero which for arbitrary $\varepsilon > 0$ admits a finite $\varepsilon U$-net, that is, can be covered by a finite number of translations of the set $\varepsilon U$.

Our main concern is with Schwartz spaces that are also Frechet spaces, which we refer to as $\mathscr{F}\mathscr{S}$-spaces; we shall encounter them as projective limits.

**Proposition 5.** *Let $X = \{X_\alpha, i_\alpha^{\alpha'}\}$ be a decreasing family having the following properties: for arbitrary $\alpha$, $X_\alpha$ is an $\mathscr{F}$-space, the mapping $i_{\alpha+1}^\alpha$ is compact, and its image is dense in $X_\alpha$. Then $\overleftarrow{X}$ is an $\mathscr{F}\mathscr{S}$-space.*

*Proof.* By Proposition 2, $\overleftarrow{X}$ is an $\mathscr{F}$-space and for arbitrary $\alpha$ the image of the mapping $i^\alpha: \overleftarrow{X} \to X_\alpha$ is dense in $X_\alpha$. Let $U$ be an arbitrary neighborhood of zero in $\overleftarrow{X}$. It contains a neighborhood of the form $(i^\alpha)^{-1}(U_\alpha)$, where $U_\alpha$ is a neighborhood of zero in $X_\alpha$. Let $U_{\alpha+1}$ be a neighborhood of zero in $X_{\alpha+1}$ such that the set $i_{\alpha+1}^\alpha(U_{\alpha+1})$ is relatively compact in $X_\alpha$. We set $V = (i^{\alpha+1})^{-1}(U_{\alpha+1})$ and we show that the neighborhood $V$, for arbitrary $\varepsilon > 0$, admits a finite $\varepsilon U$-net.

Since the set $i_{\alpha+1}^\alpha(U_{\alpha+1})$ is relatively compact in $X_\alpha$, it belongs to a finite union of the form

$$\bigcup_\lambda \left(x_\alpha^\lambda + \frac{\varepsilon}{2} U_\alpha\right),$$

where all the $x_\alpha^\lambda \in X_\alpha$. We choose the elements $x^\lambda \in \overleftarrow{X}$ so that for all $\lambda$ we have

$$i^\alpha x^\lambda - x_\alpha^\lambda \in \frac{\varepsilon}{2} U_\alpha.$$

2 See Kantorovič and Akilov [1].

3 The set $K$ in the l.t.s. $E$ is said to be relatively compact if its closure is compact.

Then

$$i^{\alpha}_{\alpha+1}(U_{\alpha+1}) \subset \bigcup_{\lambda} (i^{\alpha} x^{\lambda} + \varepsilon U_{\alpha}).$$

Applying to both sides the operation $(i^{\alpha})^{-1}$ we obtain the inclusion $V \subset \bigcup (x^{\lambda} + \varepsilon U)$. But this constructs the finite $\varepsilon U$-net for the set $V$. □

We note that every $\mathscr{F}\mathscr{S}$-space is reflexive. For, by a general criterion[4], for an $\mathscr{F}$-space $X$ to be reflexive it is sufficient (and necessary) that every closed bounded set $B$ be weakly compact. If $X$ is an $\mathscr{F}\mathscr{S}$-space, then for every neighborhood of zero $U$ there exists a neighborhood of zero $V$ which admits a finite $\varepsilon U$-net for any $\varepsilon > 0$. Since $B$ is bounded it is contained in $\lambda V$ for some $\lambda > 0$ and consequently has a finite $U$-net. The space being complete and $B$ being closed, $B$ is compact and accordingly weakly compact q.e.d.

Let $E$ be an l.t.s. and let $F$ be a subspace of it. The polar of the subspace $F$ is a subspace $F^0 \subset E^*$ consisting of those functionals that vanish on $F$. If $E$ is reflexive, the second polar $F^{00}$ is a subspace in $E$, and $F \subset F^{00}$. We will have $F = F^{00}$ if and only if $F$ is closed.

Let $F$ be a closed subspace of $E$. We shall establish the following algebraic isomorphisms[5]:

$$F^* \cong E^*/F^0, \qquad F^0 \cong (E/F)^*. \tag{4.1}$$

By the Hahn-Banach theorem, every continuous functional on the subspace $F$ can be extended to a continuous functional on the whole space $E$. The extension is defined up to a functional vanishing on $F$. But this defines a linear mapping $F^* \to E^*/F^0$. Conversely, every element of the factor-space $E^*/F^0$ can be looked on as a continuous functional on $F$. This establishes the first of the isomorphisms (4).

Let us construct the second. To every functional $f \in F^0$ we may relate a functional $\check{f}$ on the factor-space $E/F$, which we shall call the functional associated to $f$. Conversely, every functional on $E/F$ may be looked on as an element of $E^*$ that vanishes on $F$.

The significance of the $\mathscr{F}\mathscr{S}$-spaces is brought out in the following proposition.

**Proposition 6**[6]**.** *Let $E$ be an $\mathscr{F}\mathscr{S}$-space, and let $F$ be a closed subspace of it. Then $F$ and $E/F$ are $\mathscr{F}\mathscr{S}$-spaces, and the algebraic isomorphisms* (4) *are topological isomorphisms.*

We note that under the hypotheses of this proposition all the spaces in (4) are reflexive. Therefore if we pass to the conjugate isomorphisms

4 See, for example, Bourbaki [1], Ch. IV, § 3, No. 3.

5 We recall that unless we explicitly state the contrary case, we endow every subspace $F \subset E$ with the induced topology, and every factor space $E/F$ with the canonical topology.

6 For the proof see Grothendieck [1].

we obtain $F \cong (E^*/F^0)^*$, $(F^0)^* \cong E/F$. These isomorphisms show that Proposition 6 holds also when $E^*$ is the conjugate of a Schwartzian $\mathscr{F}\mathscr{S}$-space and $F^0$ is an arbitrary closed subspace of it.

**8°. Duality in exact sequences of l.t.s.** We recall some definitions from § 1, Chapter I. A continuous mapping $\phi: E \to F$ of l.t.s. is said to be a homomorphism if the associated mapping $\phi$: Coim $\phi = E/\mathrm{Ker}\,\phi \to \mathrm{Im}\,\phi$ is a topological isomorphism. The mapping $\phi$ will be a homomorphism if and only if the image of an arbitrary neighborhood of zero $U \subset E$ is a neighborhood of zero in Im $\phi$.

The sequence of mappings of l.t.s.

$$E \xrightarrow{\phi} F \xrightarrow{\psi} G \tag{5.1}$$

is algebraically exact if $\mathrm{Im}\,\phi = \mathrm{Ker}\,\psi$. This sequence is said to be exact if it is algebraically exact and the mappings $\phi$ and $\psi$ are homomorphisms. The mapping $\psi$ in this sequence is called an (algebraic) monomorphism if the sequence (5) is (algebraically) exact and $\phi = 0$. We call $\phi$ an (algebraic) epimorphism if (5) is (algebraically) exact and $\psi = 0$. In other words, to say that a mapping $\phi: E \to F$ is an algebraic monomorphism means that $\mathrm{Ker}\,\phi = 0$, and to say that it is an algebraic epimorphism means that $\mathrm{Coker}\,\phi = 0$. A mapping $\phi$ is a monomorphism (epimorphism) if and only if it is an algebraic monomorphism (algebraic epimorphism) and also a homomorphism.

**Proposition 7**

I. *Let E and F be $\mathscr{F}$-spaces, and let $\phi: E \to F$ be a continuous mapping. Then the three following assertions are equivalent:*

a) *$\phi$ is a homomorphism;*

b) *the subspace* Im $\phi$ *is closed in F;*

b*) *the subspace* Im $\phi^*$ *is closed in $E^*$;*

II. *If E and F are $\mathscr{F}\mathscr{S}$-spaces, these assertions are also equivalent to the following:*

a*) *$\phi^*$ is a homomorphism.*

*Proof.* The equivalence of the first three statements is known [7]. Let us prove the assertion II. We suppose that the conditions a), b), and b*) are satisfied. Condition a) means that the mapping $\check{\phi}$: $E/\mathrm{Ker}\,\phi \to \mathrm{Im}\,\phi$ associated with $\phi$ is an isomorphism. Therefore the conjugate mapping $(\check{\phi})^*: (\mathrm{Im}\,\phi)^* \to (E/\mathrm{Ker}\,\phi)^*$ is an isomorphism. Proposition 6 implies the two isomorphisms $(\mathrm{Im}\,\phi)^* \cong F^*/(\mathrm{Im}\,\phi)^0$ (since $\mathrm{Im}\,\phi^*$ is closed) and $(E/\mathrm{Ker}\,\phi)^* \cong (\mathrm{Ker}\,\phi)^0$. Taking account of these two isomorphisms and the equations $(\mathrm{Im}\,\phi)^0 = \mathrm{Ker}\,\phi^*$ and $(\mathrm{Ker}\,\phi)^0 = \mathrm{Im}\,\phi^*$ (since $\mathrm{Im}\,\phi^*$ is

7 See Dieudonné and Schwartz [1].

closed), we may rewrite the mapping $(\check{\phi})^*$ in the form $F^*/\text{Ker}\,\phi^* \to \text{Im}\,\phi^*$. We easily see that this mapping is associated with $\phi^*$; this establishes that $\phi^*$ is a homomorphism.

Conversely, suppose that $\phi^*$ is a homomorphism, i.e. that the associated mapping $\check{\phi}^*$: $F^*/\text{Ker}\,\phi^* \to \text{Im}\,\phi^*$ is an isomorphism. By Proposition 6 the space conjugate to $F^*/\text{Ker}\,\phi^*$ is isomorphic to a closed subspace $(\text{Ker}\,\phi^*)^0 \subset F$ and therefore it is an $\mathscr{F}\mathscr{S}$-space. Since it is reflexive, its conjugate is $F^*/\text{Ker}\,\phi^*$, which is therefore conjugate to an $\mathscr{F}$-space and consequently complete. Since $\text{Im}\,\phi^*$ is isomorphic to $F^*/\text{Ker}\,\phi^*$, it is also complete and is therefore a closed subspace in $E^*$. This proves b*). ▯

**Proposition 8**

I. *Suppose that*

$$E \xrightarrow{\phi} F \xrightarrow{\psi} G \tag{6.1}$$

*is a sequence of continuous mappings of l.t.s. If it is algebraically exact, and if $\psi$ is a homomorphism, the conjugate sequence*

$$E^* \xleftarrow{\phi^*} F^* \xleftarrow{\psi^*} G^* \tag{7.1}$$

*is algebraically exact.*

II. *Let E, F, and G be $\mathscr{F}\mathscr{S}$-spaces. Then the following four conditions are equivalent:*

a) *the sequence* (6) *is exact;*

b) *the sequence* (6) *is algebraically exact and the subspace* $\text{Im}\,\psi$ *is closed;*

a*) *the sequence* (7) *is exact;*

b*) *the sequence* (7) *is algebraically exact and the subspace* $\text{Im}\,\phi^*$ *is closed.*

*Proof.* We shall establish the first assertion. The equations $\phi^*\psi^* = (\psi\phi)^* = 0$ imply the inclusion $\text{Im}\,\psi^* \subset \text{Ker}\,\phi^*$. We prove the inverse inclusion. Let $f'$ be an arbitrary element of $\text{Ker}\,\phi^*$; that is, let $f'$ be a continuous functional on $F$ which vanishes on $\text{Im}\,\phi$. Since (6) is algebraically exact, $\text{Im}\,\phi = \text{Ker}\,\psi$. Therefore to the functional $f'$ there corresponds an associated functional $\check{f}$ on $F/\text{Ker}\,\psi$. By hypothesis the mapping $\check{\psi}$: $F/\text{Ker}\,\psi \to \text{Im}\,\psi$ associated to $\psi$ is an isomorphism. We write $g' = (\check{\psi}^{-1})^* f$; $g'$ is a continuous functional on $G$ and

$$(\psi^* g', f) = (g', \psi f) = (\check{f}, \check{\psi}^{-1}\psi f) = (f', f), \qquad f \in F,$$

whence $\psi^* g' = f' \in \text{Im}\,\psi^*$. This proves the inclusion $\text{Ker}\,\phi^* \subset \text{Im}\,\psi^*$.

The second assertion follows from the first and from the preceding proposition. ▯

**9°. Passage to the limit in mappings of families.** Let us recall certain definitions from § 1 Chapter I. Suppose that $X=\{X_\alpha, i_\alpha^{\alpha'}\}$ and $Y=\{Y_\alpha, j_\alpha^{\alpha'}\}$ are two families of l.t.s. The mapping $\phi\colon X\to Y$ is an ensemble of continuous linear operators $\phi_\alpha\colon X_\alpha\to Y_{\beta(\alpha)}$, defined for all integer $\alpha$, and such that for arbitrary $\alpha$ and $\alpha'>\alpha$ the commutative relations $\phi_{\alpha'} i_\alpha^{\alpha'}= j_{\beta(\alpha)}^{\beta(\alpha')}\phi_\alpha$ are satisfied. The function $\alpha\to\beta(\alpha)$, which we call the order of the mapping $\phi$, must be monotonely increasing and must tend to $\pm\infty$ with $\alpha$. If $\phi\colon X\to Y$ and $\psi\colon Y\to Z$ are two family mappings with the respective orders $\beta(\alpha)$ and $\gamma(\alpha)$, their compositions is the mapping $\psi\,\phi\colon X\to Z$ with the order $\gamma(\beta(\alpha))$ and the components $\psi_{\beta(\alpha)}\phi_\alpha$($\psi_\beta$ and $\phi_\alpha$ are the components of $\psi$ and $\phi$). The identity mapping $I\colon X\to X$ is a mapping with components of the form $i_\alpha^{\alpha'}$.

**Proposition 9**

I. *Let $\phi\colon X\to Y$ be a family mapping. Then there exist uniquely defined mappings $\vec{\phi}\colon \vec{X}\to\vec{Y}$ and $\overleftarrow{\phi}\colon \overleftarrow{X}\to\overleftarrow{Y}$ such that for arbitrary $\alpha$ the following diagrams are commutative:*

$$\begin{array}{ccc} \vec{X} & \xrightarrow{\vec{\phi}} & \vec{Y} \\ {\scriptstyle i_\alpha}\uparrow & & \uparrow{\scriptstyle j_\beta} \\ X_\alpha & \xrightarrow{\phi_\alpha} & Y_\beta \end{array} \qquad \begin{array}{ccc} \overleftarrow{X} & \xrightarrow{\overleftarrow{\phi}} & \overleftarrow{Y} \\ {\scriptstyle i^\alpha}\downarrow & & \downarrow{\scriptstyle j^\beta} \\ X_\alpha & \xrightarrow{\phi_\alpha} & Y_\beta \end{array} \qquad \beta=\beta(\alpha), \tag{8.1}$$

*where $i_\alpha$, $j_\beta$, $i^\alpha$ and $j^\beta$ are canonical mappings. The mappings $\vec{\phi}$ and $\overleftarrow{\phi}$ are called the limits of $\phi$.*

II. *If $\phi$, $\phi'\colon X\to Y$ are equivalent mappings* (see § 1 Chapter I) *their limits coincide.*

III. *For any two mappings $\phi\colon X\to Y$ and $\psi\colon Y\to Z$ we have the relations $\overrightarrow{\psi\,\phi}=\vec{\psi}\,\vec{\phi}$ and $\overleftarrow{\psi\,\phi}=\overleftarrow{\psi}\,\overleftarrow{\phi}$.*

*Proof.* Let us establish the first assertion. The function $\beta\colon\alpha\to\beta(\alpha)$, i.e. the order of $\phi$, is by hypothesis monotone and tends to $\pm\infty$ as $\alpha\to\pm\infty$. Therefore the image of the mapping $\beta\colon Z\to Z$ is cofinal in $Z$ for both the ordering relations in $Z$ discussed in 4°. Let us consider the corresponding cofinal subfamily $Y'=\{Y'_\alpha=Y_{\beta(\alpha)}\}$. The mapping $\phi$ can be considered a mapping from $X$ to $Y'$, and its order is the identity function $\alpha\to\alpha$. On the other hand, by Proposition I, both the limits of the subfamily $Y'$ are isomorphic to the limits of the family $Y$. Therefore we may suppose from the outset that $\beta(\alpha)\equiv\alpha$.

Let us construct the mapping $\vec{\phi}$. The diagram (8) will be commutative if and only if it carries an arbitrary element of the form $i_\alpha x_\alpha$ into $j_\alpha\phi_\alpha x_\alpha$. We shall show that this condition is correctly formulated, i.e. show that the element $j_\alpha\phi_\alpha x_\alpha$ does not depend on $x_\alpha$, but only on $i_\alpha x_\alpha$. If

$i_\alpha x_\alpha = i_{\alpha'} x_{\alpha'}$ there exists by definition an $\alpha'' \geqq \alpha, \alpha'$ such that $i_\alpha^{\alpha''} x_\alpha = i_{\alpha'}^{\alpha''} x_{\alpha'}$, whence

$$j_\alpha^{\alpha''} \phi_\alpha x_\alpha = \phi_{\alpha''} i_\alpha^{\alpha''} x_\alpha = \phi_{\alpha''} i_{\alpha'}^{\alpha''} x_{\alpha'} = j_{\alpha'}^{\alpha''} \phi_{\alpha'} x_{\alpha'},$$

that is, the elements $\phi_\alpha x_\alpha$ and $\phi_{\alpha'} x_{\alpha'}$ are equivalent, and therefore $j_\alpha \phi_\alpha x_\alpha = j_{\alpha'} \phi_{\alpha'} x_{\alpha'}$, q.e.d. We have thus proved that there exists a unique mapping $\vec{\phi}$ which makes the diagram (8) commutative. The continuity of $\vec{\phi}$ follows from the definitions.

Now let us construct the mapping $\overleftarrow{\phi}$. Let $x = \{x_\alpha\}$ be an arbitrary thread belonging to $X$. The elements $\phi_\alpha x_\alpha \in Y_\alpha$ also form a thread. This follows from the computation

$$j_\alpha^{\alpha'} \phi_\alpha x_\alpha = \phi_{\alpha'} i_\alpha^{\alpha'} x_\alpha = \phi_{\alpha'} x_{\alpha'}, \qquad \alpha' > \alpha.$$

We write $\overleftarrow{\phi} x = \{\phi_\alpha x_\alpha\}$. This definition of $\overleftarrow{\phi}$ is necessary and sufficient for the commutativity of the diagram (8). It is easy to prove that the mapping is continuous. Then the second assertion is proved.

The third assertion is obvious. It is also obvious that the limits of an arbitrary identity mapping are identity operators. Now suppose that $\phi, \phi'$: $X \to Y$ are equivalent mappings. By definition, this means that there exist identity operators $J, J'$: $Y \to Y$ such that $J\phi = J'\phi'$. It follows from what we have said that $\overrightarrow{J\phi} = \vec{J}\vec{\phi} = \vec{\phi}$, whence $\vec{\phi} = \vec{\phi}'$. Similarly, $\overleftarrow{\phi} = \overleftarrow{\phi}'$. $\square$

We note some consequences of Proposition 9. Let $I$: $X \to X$ be a unit mapping, i.e. a mapping whose components are the operators $i_\alpha^{\alpha'}$. We apply our proposition to it. Since (8) is commutative, the limit mappings $\vec{I}$ and $\overleftarrow{I}$ are identity mappings. Now let $I$ be an identity mapping of the family $X$. It is equivalent to a unit mapping and therefore Assertion II of our proposition implies that the limits $\vec{I}$ and $\overleftarrow{I}$ are also identity operators.

Let us suppose that the mappings $\phi$: $X \to Y$ and $\psi$: $Y \to X$ establish an isomorphism of these families. We shall show that the limits

$$\vec{X} \underset{\vec{\psi}}{\overset{\vec{\phi}}{\rightleftarrows}} \vec{Y}, \qquad \overleftarrow{X} \underset{\overleftarrow{\psi}}{\overset{\overleftarrow{\phi}}{\rightleftarrows}} \overleftarrow{Y}$$

are also isomorphisms. By hypothesis, $\psi\phi = I$ and $\phi\psi = J$, where $I$ and $J$ are identity mappings. Using Assertion III, we find that $\overrightarrow{\psi\phi} = \vec{\psi}\vec{\phi}$ and $\overrightarrow{\phi\psi} = \vec{\phi}\vec{\psi}$. On the other hand, it follows from what we have said above that the mappings $\overrightarrow{\psi\phi} = \vec{I}$ and $\overrightarrow{\phi\psi} = \vec{J}$ are identity operators. Hence it follows that the operators $\vec{\phi}$ and $\vec{\psi}$ are mutually inverse. Similar arguments will show that the operators $\overleftarrow{\phi}$ and $\overleftarrow{\psi}$ are also mutually inverse.

Proposition 9 proves that the operation of passing to either the inductive or projective limit is a functor acting from the category of classes of equivalent families of l.t.s. and the classes of equivalent mappings of these families to the category of linear topological spaces. We now prove two propositions characterizing the exactness of these functors.

**10°. Passage to the inductive limit in exact sequences**

**Proposition 10.** *Let*

$$\begin{gathered} 0 \to X \xrightarrow{\phi} Y \xrightarrow{\psi} Z \to 0, \\ X = \{X_\alpha, i_\alpha^{\alpha'}\}, \qquad Y = \{Y_\alpha, j_\alpha^{\alpha'}\}, \qquad Z = \{Z_\alpha, k_\alpha^{\alpha'}\} \end{gathered} \tag{9.1}$$

*be a sequence of increasing families and let*

$$0 \to \vec{X} \xrightarrow{\vec{\phi}} \vec{Y} \xrightarrow{\vec{\psi}} \vec{Z} \to 0 \tag{10.1}$$

*be the sequence of their inductive limits.*

A) *If the sequence* (9) *is algebraically exact in the term Y, the sequence* (10) *is algebraically exact in the term* $\vec{Y}$.

B) *If the sequence* (9) *is exact in the term Z, the sequence* (10) *is exact in the term* $\vec{Z}$.

*Remark.* Assertion A) implies that a passage to the inductive limit conserves algebraical exactness in an arbitrary sequence of increasing families.

We precede the proof of the proposition by a few general remarks. We shall suppose that (9) is an algebraically exact sequence of families.

I. The exactness of (9) implies that $\psi\phi \sim 0$, i.e. there exists an identity mapping $K$ of the family $Z$ such that $K\psi\phi = 0$. Replacing the mapping $\psi$ by the equivalent mapping $\psi' = K\phi$ we obtain the relation $\psi'\phi = 0$ and still maintain the exactness of (9) and do not change the limit mappings $\vec{\psi}$ and $\overleftarrow{\psi}$. Therefore we may suppose that the original mappings satisfy $\psi\phi = 0$.

II. Suppose that

$$\phi_\alpha\colon X_\alpha \to Y_{\beta(\alpha)}, \qquad \psi_\alpha\colon Y_\alpha \to Z_{\gamma(\alpha)}$$

are the components of the mappings $\phi$ and $\psi$. By hypothesis $\beta(\alpha)$ and $\gamma(\alpha)$, the orders of these mappings, are monotone increasing and tend to $\pm\infty$ with $\alpha$. The composite function $\gamma(\beta(\alpha))$ has the same property. It follows that the subfamilies

$$Y' = \{Y_{\beta(\alpha)}, j_{\beta(\alpha)}^{\beta(\alpha')}\}, \qquad Z' = \{Z_{\gamma(\beta(\alpha))}, k_{\gamma(\beta(\alpha))}^{\gamma(\beta(\alpha'))}\}$$

are cofinal respectively in the families $Y$ and $Z$. If we replace $Y$ and $Z$ in (9) by $Y'$ and $Z'$, the orders of the mappings $\phi$ and $\psi$ become the identity mapping $\alpha \to \alpha$. We may therefore suppose from the outset that $\beta(\alpha) \equiv \gamma(\alpha) \equiv \alpha$.

III. By definition, the exactness of (9) implies that there exist mappings

$$\phi^{-1}\colon \operatorname{Ker}\psi \to X, \qquad \psi^{-1}\colon Z \to Y/\operatorname{Ker}\psi \tag{11.1}$$

and identity mappings $I, J, K$ of the respective families $X, Y, Z$ such that

$$\operatorname{Ker}\phi \subset \operatorname{Ker} I, \qquad \phi\phi^{-1} = J, \qquad \psi\psi^{-1} = K. \tag{12.1}$$

Let $a(\alpha)$, $b(\alpha)$, $c(\alpha)$ be the orders of the mappings $I, J, K$. We note that the orders of $\phi^{-1}$ and $\psi^{-1}$ are $b(\alpha)$ and $c(\alpha)$ respectively. We choose some monotone increasing function $\lambda(\alpha)$, tending to $\pm\infty$ with $\alpha$, which for arbitrary $\alpha$ satisfies the inequalities

$$a(\lambda(\alpha)), b(\lambda(\alpha)), c(\lambda(\alpha)) \leqq \lambda(\alpha+1). \tag{13.1}$$

If we replace the mappings $I, J, K, \phi^{-1}, \psi^{-1}$ by their compositions with suitable identity mappings, we may so increase their orders $a(\alpha)$, $b(\alpha)$, and $c(\alpha)$ that the inequalities (13) become equations. Relation (12) is not disturbed. Next we replace $X$, $Y$, and $Z$ by the cofinal families $\{X_{\lambda(\alpha)}\}$, $\{Y_{\lambda(\alpha)}\}$, $\{Z_{\lambda(\alpha)}\}$. The result is that the orders of all the mappings $I, J, K, \phi^{-1}, \psi^{-1}$ are now equal to the function $\alpha \to \alpha+1$. We may therefore suppose that the mappings $I, J, K, \phi^{-1}, \psi^{-1}$ constructed for the initial sequence (9) have the same property.

*Proof of Proposition* 10. We shall establish Assertion A). The equation $\psi\phi = 0$ implies (see Remark I) that $\vec{\psi}\vec{\phi} = 0$, whence $\operatorname{Im}\vec{\phi} \subset \operatorname{Ker}\vec{\psi}$. Let us prove the converse inclusion. Let $y$ be an arbitrary element of the space $\operatorname{Ker}\vec{\psi}$. Since

$$\vec{Y} = \bigcup_\alpha j_\alpha(Y_\alpha),$$

where $j_\alpha$ is a canonical mapping, we have $y = j_\alpha y_\alpha$ for some $y_\alpha \in Y_\alpha$. The equation $\vec{\psi} y = 0$ and the relation $\vec{\psi} j_\alpha = k_\alpha \psi_\alpha$ (see Remark II) imply that $k_\alpha \psi_\alpha y_\alpha = 0$. This means that $k_\alpha^\beta \psi_\alpha y_\alpha = 0$ for some $\beta \geqq \alpha$. Since $k_\alpha^\beta \psi_\alpha y_\alpha = \psi_\beta j_\alpha^\beta y_\alpha$ the element $j_\alpha^\beta y_\alpha$ belongs to the kernel of $\psi_\beta$. The relations (11) and (12) imply that the identity mapping $J$ carries $\operatorname{Ker}\psi$ into $\operatorname{Im}\phi$, and therefore its component $j_\beta^{\beta+1}$ carries $\operatorname{Ker}\psi_\beta$ into $\operatorname{Im}\phi_{\beta+1}$. Hence $j_\alpha^{\beta+1} y_\alpha = \phi_{\beta+1} x_{\beta+1}$ for some $x_{\beta+1} \in X_{\beta+1}$. Therefore $y = \vec{\phi} x$, where $x = i_{\beta+1} x_{\beta+1}$, q.e.d.

Let us now prove Assertion B). By hypothesis the mapping associated to $\psi$, namely $\hat{\psi}\colon Y/\operatorname{Ker}\psi \to Z$, is an isomorphism. The limit mapping

$$\vec{\hat{\psi}}\colon \overrightarrow{Y/\operatorname{Ker}\psi} \to \vec{Z}$$

is an isomorphism, by Proposition 9. Let $\pi$: $Y \to Y/\operatorname{Ker}\psi$ be a canonical mapping of the family on its factor-family, and let

$$\vec{\pi}\colon \vec{Y} \to \overrightarrow{Y/\operatorname{Ker}\psi}$$

be its limit. Since $\psi = \psi\,\pi$ we have $\vec{\psi} = \vec{\psi}\,\vec{\pi}$. Thus, we have only to show that $\vec{\pi}$ is a homomorphism. This means that the image of an arbitrary convex neighborhood of zero $V$ in $\vec{Y}$ is a neighborhood of zero in $\overrightarrow{Y/\operatorname{Ker}\psi}$.

Suppose that for every $\alpha$ the mapping $\pi_\alpha$: $Y_\alpha \to Y_\alpha/\operatorname{Ker}\psi_\alpha$ is a component of the mapping $\pi$, and that $j'_\alpha$: $Y_\alpha/\operatorname{Ker}\psi_\alpha \to \overrightarrow{Y/\operatorname{Ker}\psi}$ is the canonical mapping. The equation

$$j'_\alpha\,\pi_\alpha\, j_\alpha^{-1}(V) = \vec{\pi}\, j_\alpha\, j_\alpha^{-1}(V) \subset \vec{\pi}(V)$$

implies the relation

$$\pi_\alpha\big(j_\alpha^{-1}(V)\big) \subset (j'_\alpha)^{-1}\big(\pi(V)\big). \tag{14.1}$$

Since by hypothesis we have for every $\alpha$ that $j_\alpha^{-1}(V)$ is a neighborhood of zero in $Y_\alpha$, and the mapping $\pi_\alpha$ is a homomorphism, the left side of (14) is a neighborhood of zero in $Y_\alpha/\operatorname{Ker}\psi_\alpha$. Thus, for arbitrary $\alpha$ the set $(j'_\alpha)^{-1}\big(\pi(V)\big)$ is a neighborhood of zero in $Y_\alpha/\operatorname{Ker}\psi_\alpha$. Therefore the set $\pi(V)$, being convex, is a neighborhood of zero in $\overrightarrow{Y/\operatorname{Ker}\psi}$. □

### 11°. Passage to the projective limit in exact sequences of families

**Proposition 11.** *Let* (9) *be a sequence of decreasing families and mappings, and let*

$$0 \to \overleftarrow{X} \xrightarrow{\overleftarrow{\Phi}} \overleftarrow{Y} \xrightarrow{\overleftarrow{\Psi}} \overleftarrow{Z} \to 0 \tag{15.1}$$

*be the sequence of their projective limits.*

I. *If the sequence* (9) *is algebraically exact in the terms $X$ and $Y$, and if the mapping $\phi$ is a homomorphism, then the sequence* (17) *is algebraically exact in the terms $\overleftarrow{X}$ and $\overleftarrow{Y}$, and the mapping $\overleftarrow{\phi}$ is a homomorphism.*

II. *Suppose that the sequence* (9) *is exact and that it satisfies the following conditions:*

a) *for arbitrary $\alpha$, $X_\alpha$ is an $\mathscr{F}$-space;*

b) *for arbitrary $\alpha$ the image of the mapping $i^\alpha_{\alpha+1}$: $X_{\alpha+1} \to X_\alpha$ is dense in $X_\alpha$;*

*Then the sequence* (15) *is also exact.*

*Proof.* Since a decreasing family is also a family, we may adopt the stipulations I, II, and III in 10°, replacing $\alpha+1$ by $\alpha-1$ in III. We must note that these stipulations do not violate the conditions a) and b), since to adopt them we must replace the families in (9) by cofinal families.

Such a substitution obviously does not violate condition a). Since an arbitrary mapping $i_\alpha^{\alpha'}$ is a composition of the mappings $i_\beta^{\beta-1}$, the condition b) implies that the image of any $i_\alpha^{\alpha'}$ is dense in $X_{\alpha'}$. Therefore the condition b) is also conserved under a passage to a cofinal family.

Let us prove the first assertion of our proposition. We first show that $\overleftarrow{\phi}$ is biunique. Let $x$ be any element of the space $\operatorname{Ker}\overleftarrow{\phi}$, and let $\{x_\alpha\}$ be the corresponding thread. By hypothesis $\phi_\alpha x_\alpha = 0$ for all $\alpha$. Because (9) is algebraically exact in the term X, we conclude from the first relation in (12) that $i_\alpha^{\alpha-1} x_\alpha = 0$ for all $\alpha$. Since $\{x_\alpha\}$ is a thread, we have $i_\alpha^{\alpha-1} x_\alpha = x_{\alpha-1}$, and therefore the thread consists of zeroes, i.e. $x=0$, q.e.d. But this proves that (15) is exact in the term $\overleftarrow{X}$.

Let us now show that $\overleftarrow{\phi}$ is a homomorphism. Since $\phi$ is a homomorphism, and since (9) is algebraically exact in $X$, the associated mapping $\breve{\phi}\colon X \to \operatorname{Im}\phi$ is a family isomorphism. By Proposition 9 the limit mapping $\overleftarrow{\breve{\phi}}\colon \overleftarrow{X} \to \overleftarrow{\operatorname{Im}\phi}$ is an isomorphism of spaces. But $\overleftarrow{\phi}$ is the composition of the isomorphism $\overleftarrow{\breve{\phi}}$ and the natural mapping $e\colon \overleftarrow{\operatorname{Im}\phi} \to \operatorname{Im}\overleftarrow{\phi}$. Thus there remains only to show that $e$ is also an isomorphism.

The spaces $\overleftarrow{\operatorname{Im}\phi}$ and $\operatorname{Im}\overleftarrow{\phi}$ coincide algebraically and $e$ is an identity mapping. Therefore we need only prove that the topologies on these spaces coincide. Let $v$ be a neighborhood of zero in $\overleftarrow{\operatorname{Im}\phi}$. By definition it contains a set of the form $(j^\alpha)^{-1}(v_\alpha)$, where $v_\alpha$ is a neighborhood of zero in $\operatorname{Im}\phi_\alpha$, i.e. a set of the form $V_\alpha \cap \operatorname{Im}\phi_\alpha$, where $V_\alpha$ is a neighborhood of zero in $Y_\alpha$. If we think of $j^\alpha$ as a mapping from $\overleftarrow{\operatorname{Im}\phi}$ to $\operatorname{Im}\phi_\alpha$ we obtain the relations

$$v \supset (j^\alpha)^{-1}(v_\alpha) = (j^\alpha)^{-1}(V_\alpha \cap \operatorname{Im}\phi_\alpha) = (j^\alpha)^{-1}(V_\alpha).$$

The set $(j^\alpha)^{-1}(V_\alpha)$ is a neighborhood of zero in $\operatorname{Im}\overleftarrow{\phi}$. Conversely, every neighborhood of zero in $\operatorname{Im}\overleftarrow{\phi}$ contains a set of this form, and therefore contains a set of the form $(j^\alpha)^{-1}(v_\alpha)$, which is a neighborhood of zero in $\overleftarrow{\operatorname{Im}\phi}$. This proves that the topolgies in $\overleftarrow{\operatorname{Im}\phi}$ and $\operatorname{Im}\overleftarrow{\phi}$ coincide, and this, in turn, completes the proof that $\overleftarrow{\phi}$ is a homomorphism.

To complete the proof of the first assertion there remains to be proved the equation $\operatorname{Im}\overleftarrow{\phi} = \operatorname{Ker}\overleftarrow{\psi}$. The equation $\psi\phi = 0$ implies that $\overleftarrow{\psi}\,\overleftarrow{\phi} = 0$, i.e. $\operatorname{Im}\overleftarrow{\phi} \subset \operatorname{Ker}\overleftarrow{\psi}$. We must prove the converse inclusion. Let $y = \{y_\alpha\}$ be an arbitrary element of $\operatorname{Ker}\overleftarrow{\psi}$. From (11) and from the second relation in (12) it follows that the mapping $J$, which is of order $\alpha \to \alpha - 1$, carries the subfamily $\operatorname{Ker}\psi$ into $\operatorname{Im}\phi$. Therefore, for arbitrary $\alpha$ the element $y_\alpha = j_{\alpha+1}^\alpha y_{\alpha+1}$ belongs to $\operatorname{Im}\phi_\alpha$, that is, it is equal to $\phi_\alpha x_\alpha$, where $x_\alpha \in X_\alpha$. We shall show that the elements $x'_\alpha = i_{\alpha+1}^\alpha x_{\alpha+1}$ form a thread. For any $\alpha$

$$\phi_\alpha(x_\alpha - i_{\alpha+1}^\alpha x_{\alpha+1}) = \phi_\alpha x_\alpha - j_{\alpha+1}^\alpha \phi_{\alpha+1} x_{\alpha+1} = y_\alpha - j_{\alpha+1}^\alpha y_{\alpha+1} = 0,$$

since the elements $y_\alpha$ form a thread. Thus the difference $x_\alpha - i^\alpha_{\alpha+1} x_{\alpha+1}$ belongs to the kernel of $\phi_\alpha$ and so to the kernel of $i^{\alpha-1}_\alpha$ in accordance with the first of the inclusions in (12). Hence

$$x'_{\alpha-1} - i^{\alpha-1}_\alpha x'_\alpha = i^{\alpha-1}_\alpha (x_\alpha - i^\alpha_{\alpha+1} x_{\alpha+1}) = 0,$$

i.e. the elements $x'_\alpha$ form a thread, which we shall denote by $x'$. The relations

$$\phi_{\alpha-1} x'_{\alpha-1} = j^{\alpha-1}_\alpha \phi_\alpha x_\alpha = j^{\alpha-1}_\alpha y_\alpha = y_{\alpha-1}$$

imply that $\overleftarrow{\phi} x' = y$, i.e. the element $y$ belongs to $\operatorname{Im} \overleftarrow{\phi}$, q.e.d. This completes the proof of the first assertion.

Now we pass to the second assertion. We suppose that the metric $\rho_\alpha$ in $X_\alpha$ is a nondecreasing function of $\alpha$ (cf. Proposition 2). We fix $\varepsilon > 0$ and an element $z = \{z_\alpha\} \in \overleftarrow{Z}$. For any $\alpha$ we choose $y_\alpha \in \psi^{-1}_{\alpha+1} z_{\alpha+1}$, where $\psi^{-1}_{\alpha+1}$ is the component of the mapping $\psi^{-1}$. By the third relation (12) we have $\psi_\alpha y_\alpha = z_\alpha$. Hence

$$\psi_\alpha (y_\alpha - j^\alpha_{\alpha+1} y_{\alpha+1}) = \psi_\alpha y_\alpha - k^\alpha_{\alpha+1} \psi_{\alpha+1} y_{\alpha+1} = z_\alpha - k^\alpha_{\alpha+1} z_{\alpha+1} = 0$$

i.e. $y_\alpha - j^\alpha_{\alpha+1} y_{\alpha+1} \in \operatorname{Ker} \psi_\alpha$. By the second relation (12) $j^{\alpha-1}_\alpha$ maps $\operatorname{Ker} \psi_\alpha$ in $\operatorname{Im} \phi_{\alpha-1}$. Therefore for any $\alpha > 1$ we obtain the inclusion

$$j^{\alpha-1}_\alpha y'_\alpha - y'_{\alpha-1} \in \operatorname{Im} \phi_{\alpha-1} \tag{16.1}$$

where $y'_\alpha = j^\alpha_{\alpha+1} y_{\alpha+1}$.

We shall construct a sequence of elements $y^*_\alpha \in Y_\alpha$, satisfying for arbitrary $\beta$ the conditions

(i) $$y^*_\beta - y'_\beta \in \operatorname{Im} \phi_\beta$$

(ii) $$j^{\beta-1}_\beta y^*_\beta - y^*_{\beta-1} = \phi_{\beta-1}(x_{\beta-1}), \qquad x_{\beta-1} \in X_{\beta-1},$$

$$\rho_{\beta-1}(x_{\beta-1}, 0) \leqq \frac{\varepsilon}{2^\beta}.$$

with $y^*_1 = y'_1$. Let us suppose that we have constructed the elements $y^*_\beta$ with $\beta \leqq \alpha$ and we construct the element $y^*_{\alpha+1}$. The inclusions (16) and (i) imply that

$$j^\alpha_{\alpha+1} y'_{\alpha+1} - y^*_\alpha = \phi_\alpha(\xi_\alpha), \qquad \xi_\alpha \in X_\alpha.$$

By condition b) there exists an element $\xi_{\alpha+1} \in X_{\alpha+1}$ such that

$$\rho_\alpha(x_\alpha) \leqq \frac{\varepsilon}{2^{\alpha+1}}$$

where $x_\alpha = \xi_\alpha - i^\alpha_{\alpha+1} \xi_{\alpha+1}$. We set $y^*_{\alpha+1} = y'_{\alpha+1} - \phi_{\alpha+1}(\xi_{\alpha+1})$ and verify (ii):

$$j^\alpha_{\alpha+1} y^*_{\alpha+1} - y^*_\alpha = j^\alpha_{\alpha+1} y'_{\alpha+1} - y^*_\alpha - j^\alpha_{\alpha+1} \phi_{\alpha+1}(\xi_{\alpha+1}) = \phi_\alpha(x_\alpha).$$

So, the sequence $\{y^*_\alpha\}$ has been found.

In view of (ii) the series

$$x_\alpha + i^\alpha_{\alpha+1} x_{\alpha+1} + \cdots + i^\alpha_\beta x_\beta + \cdots$$

converges in $X_\alpha$ for every $\alpha$ since $X_\alpha$ is complete. Let $x^\alpha$ be the sum of this series. Since $X_\alpha$ is a separable space we have $x^\alpha - i^\alpha_{\alpha+1} x^{\alpha+1} = x_\alpha$. Therefore the elements $y^0_\alpha = y^*_\alpha + \phi_\alpha(x^\alpha) \in Y_\alpha$ $\alpha = 1, 2, \ldots$ form a thread $y^0 \in \overleftarrow{Y}$. Since the operator $\psi_\alpha$ annihilates the subspace Im $\phi_\alpha$ we have the equations

$$\psi_\alpha y^0_\alpha = \psi_\alpha y^*_\alpha = \psi_\alpha y'_\alpha = z_\alpha,$$

whence $\overleftarrow{\psi} y^0 = z$. This proves that $\overleftarrow{\psi}$ is an algebraic epimorphism.

It remains to show that $\overleftarrow{\psi}$ is a homomorphism. For an arbitrary neighborhood of zero $V$ in $\overleftarrow{Y}$ we must find a neighborhood of zero $W$ in $\overleftarrow{Z}$ such that $\overleftarrow{\psi}(V) \supset W$. By the definition of the topology in $\overleftarrow{Y}$, for some $\gamma$ there exists a neighborhood $V_\gamma$ in $Y_\gamma$ such that $V$ contains the pre-image of $2V_\gamma$. For simplicity we suppose that $\gamma = 1$. Since the mapping $\psi_3^{-1}$: $Z_3 \to Y_2$ is continuous, there exists a neighborhood of zero $W_3$ in $Z_3$ such that for arbitrary $z_3 \in W_3$ we can find an element $y_2 \in \psi_3^{-1} z_3$ such that $y'_1 = j^1_2 y_2 \in V_1$. So if $W$ is the pre-image of $W_3$, $z \in W$ and $z_3$ is the element of the thread $z$, then the element $y'_1$ which appears in the preceding argument can be chosen in $V_1$. From what we have proved it follows that

$$y^0_1 = y'_1 + \phi_1(x^1), \qquad \rho_1(x^1, 0) \leqq \varepsilon.$$

If $\varepsilon$ is sufficiently small, than $\phi_1(x^1) \in V_1$, consequently $y^0_1 \in 2V_1$ and $y^0 \in V$. □

## § 2. Functional spaces

**1°. Spaces of functions of finite smoothness.** Let $f$ be a function measurable in $R^n$. Its support is the smallest set supp $f \subset R^n$ such that for an arbitrary point $\xi \bar{\in}$ supp $f$ we can find a neighborhood in which $f = 0$ almost everywhere. It is obvious that a support is always closed. Let $\Omega$ be a region[8] in $R^n$. We denote by $\mathscr{D}(\Omega)$ the space of all infinitely differentiable functions defined on $R^n$ and having compact supports lying in $\Omega$. We shall soon endow this space with a topology.

For every non-negative integer $q$ we consider in the space $\mathscr{D}(R^n)$ the following scalar product:

$$\langle \phi, \psi \rangle^q = \sum_{|j| \leqq q} \int_{R^n} \overline{D^j \phi}\, D^j \psi \, d\xi, \qquad D^j = \frac{\partial^{|j|}}{\partial \xi_1^{j_1} \ldots \partial \xi_n^{j_n}}. \tag{1.2}$$

8 By a region we shall mean an arbitrary open set, not necessarily connected.

We denote the corresponding norm by

$$\|\phi\|^q = |\langle\phi, \phi\rangle^q|^{\frac{1}{2}} = \Big|\sum_{|j|\leqq q} \|D^j\phi\|^2_{L_2(R^n)}\Big|^{\frac{1}{2}}.$$

The completion of $\mathscr{D}(R^n)$ in the norm $\|\cdot\|^q$ will be denoted by $\mathscr{E}^q$. The norm $\|\cdot\|^q$ and the corresponding scalar product $\langle\cdot,\cdot\rangle^q$ can be extended to this completion. This makes $\mathscr{E}^q$ a Hilbert space.

The most important of the scalar products (2) is $\langle\cdot,\cdot\rangle^0$, which we will denote by $(\cdot,\cdot)$. Let us fix the integer $q \geqq 0$. For any function $\phi \in \mathscr{D}(R^n)$ we define a linear continuous functional on $\mathscr{E}^q$ by the formula

$$\psi \to (\phi, \psi) = \int_{R^n} \bar{\phi}\,\psi\, d\xi, \qquad \psi \in \mathscr{E}^q.$$

This functional is obviously continuous. We have therefore constructed a mapping of $\mathscr{D}(R^n)$ into the conjugate of $\mathscr{E}^q$. This mapping is biunique, since if $(\phi, \psi) = 0$ for all $\psi \in \mathscr{E}^q$ it follows that $\phi \equiv 0$.

We shall show that the image of the mapping $\mathscr{D}(R^n) \to (\mathscr{E}^q)^*$ is dense in $(\mathscr{E}^q)^*$. In fact, if this were not so, the Hahn-Banach theorem would imply the existence of a non-zero element $\chi$ in the second conjugate space $(\mathscr{E}^q)^{**}$ such that $(\phi, \chi) = 0$ for all the functions $\phi \in \mathscr{D}(R^n)$. Since $\mathscr{E}^q$ is a Hilbert space, it is reflexive, and therefore the element $\chi$ can be identified with some function $\chi \in \mathscr{E}^q$. But if $(\phi, \chi) = 0$ for all $\phi \in \mathscr{D}(R^n)$, we have $\chi \equiv 0$ almost everywhere. Therefore the element $\chi \in (\mathscr{E}^q)^{**}$ is the zero element, and the contradiction we have obtained shows that $\mathscr{D}(R^n)$ is dense in $(\mathscr{E}^q)^*$.

When we look on $\mathscr{D}(R^n)$ as a subspace of $(\mathscr{E}^q)^*$ there is an induced topology for it, which is generated by the norm

$$\|\phi\|^{-q} = \sup\left\{\frac{|(\phi, \psi)|}{\|\psi\|^q},\ \psi \in \mathscr{E}^q,\ \|\psi\|^q \neq 0\right\}.$$

We denote by $\mathscr{E}^{-q}$ the completion of $\mathscr{D}(R^n)$ in this norm. Our earlier remarks imply that the space $\mathscr{E}^{-q}$ is isomorphic to the conjugate of $\mathscr{E}^q$ and is therefore a Hilbert space. Thus the spaces $\mathscr{E}^q$ and $\mathscr{E}^{-q}$ are mutually conjugate.

The space $\mathscr{E}^0$ obviously coincides, as an l.t.s., with $L_2(R^n)$, and the space $\mathscr{E}^{-0}$, being its conjugate, is also isomorphic with $L_2(R^n)$. This implies the isomorphism $\mathscr{E}^0 \cong \mathscr{E}^{-0}$. These two spaces will be identified with each other.

We have thus defined a Hilbert space $\mathscr{E}^q$ for every integer $q$; for all $q$ the space $\mathscr{E}^{q+1}$ is a subspace of $\mathscr{E}^q$ (in the algebraic sense), and the identity mapping $\mathscr{E}^{q+1} \to \mathscr{E}^q$ is a continuous biunique operation.

The support of a function $\phi\in\mathscr{E}^q$ where $-\infty<q<\infty$, is defined as the smallest closed set supp $\phi$ such that $(\phi,\psi)=0$ for an arbitrary function $\psi\in\mathscr{D}(R^n\setminus\operatorname{supp}\phi)$. This definition obviously agrees with the definition at the beginning of the subsection. Let $F$ be a closed set in $R^n$. We denote by $\mathscr{D}_F^q$ the subspace of $\mathscr{E}^q$ consisting of functions whose supports belong to $F$. This subspace is closed, since every relation of the form $(\phi,\psi)=0$, $\psi\in\mathscr{D}(R^n)$ is conserved in the limiting process with respect to $\phi\in\mathscr{E}^q$. Since $\mathscr{D}_F^q$ is a closed subspace of a Hilbert space, it is itself a Hilbert space. Let us consider the factor-space

$$\mathscr{E}_F^q=\mathscr{E}^q/\mathscr{D}_G^q, \qquad G=\overline{R^n\setminus F}. \tag{2.2}$$

Since the subspace $\mathscr{D}_G^q\subset\mathscr{E}^q$ is closed, the factor-space $\mathscr{E}_F^q$ is also a Hilbert space. The norm in the factor-space will be denoted by $\|\cdot\|_F^q$. The elements of the space $\mathscr{E}_F^q$ will be looked on as functions defined on the set $F$ and admitting an extension in $R^n$ to functions belonging to $\mathscr{E}^q$.

If the function $f$ is defined and has continuous derivatives, up to order $q\geqq 0$, that belong to $L_2$ in an $\varepsilon$-neighborhood of $F$, its restriction on $F$ is an element of $\mathscr{E}_F^q$. On the other hand, if $q\geqq\nu$ where $\nu=[n/2]+1$, every element of $\mathscr{E}_F^q$ is a function defined on $F$ and having in int $F$ continuous derivatives up to order $q-\nu$. This follows from Proposition 3, § 3.

In the spaces $\mathscr{D}_F^q$ and $\mathscr{E}_F^q$ we may introduce an operation of multiplication by an arbitrary function $f$ defined in the $\varepsilon$-neighborhood of $F$ and having derivatives up to order $|q|$ that are bounded in this neighborhood. To do this we construct a function $f$ having bounded derivatives up to order $|q|$ in the whole of $R^n$ and coinciding with $f$ in the neighborhood of $F$. Clearly, there is a defined and continuous multiplication in $\mathscr{E}^q$ by such a function $f$ and this multiplication carries $\mathscr{D}_F^q$ and $\mathscr{D}_G^q$ into themselves. It follows that this operation can be extended over the factor-space (2). In the spaces $\mathscr{D}_F^q$ and $\mathscr{E}_F^q$ the multiplication depends only on the original function $f$.

Now let us suppose that $F$ is the closure of some region $\Omega$. The region $\Omega$ and the set $F$ will be said to be admissible if corresponding to every point $\xi\in\partial\Omega$ we can find a neighborhood $U$ and a vector $\eta\in R^n$ such that under an arbitrary, but sufficiently small, translation in the direction of $\eta$ the region $\Omega\cap U$ does not go outside $\Omega$. If the boundary of $\Omega$ has a continuously changing normal, the region itself is admissible. Clearly $\Omega$ is admissible or not simultaneously with $R^n\setminus F$.

**Proposition 1**[9]**.** *If the region $\Omega$ is admissible, then for an arbitrary integer $q$ the subspace $\mathscr{D}(\Omega)$ is dense in $\mathscr{D}_F^q$.*

---

9 The proof of this assertion is given in a paper by Volevič and Paneyah [1], § 3, 2°. They use the notation $H_{\bar\Omega}^{\mu}$ and $\overset{0}{H}{}^{\mu}(\Omega)$ for the space $\mathscr{D}_F^q$ and the closure of $\mathscr{D}(\Omega)$ in it.

**Proposition 2.** *Let $F=\bar{\Omega}$ be an admissible set. Then the duality relation between the spaces $\mathscr{D}_F^q$ and $\mathscr{E}_F^{-q}$ created by the extension of the scalar product $(\cdot,\cdot)$ makes these spaces mutually conjugate.*

*Proof.* By what we have already proved, the Hilbert spaces $\mathscr{E}^q$ and $\mathscr{E}^{-q}$ are mutually conjugate with respect to the scalar product $(\cdot,\cdot)$. If we show that the subspaces $\mathscr{D}_F^q$ and $\mathscr{D}_G^{-q}$, where $G=R^n\setminus\Omega$, are orthogonal complements of one another we will be able to reach the desired duality relation between $\mathscr{D}_F^q$ and $\mathscr{E}_F^{-q}$. It is obvious that the orthogonal complement of $\mathscr{D}_F^q$ consists of functions whose supports belong to $R^n\setminus F$. Therefore it belongs to $\mathscr{D}_G^{-q}$. We must show that $(\phi,\psi)=0$ for arbitrary $\phi\in\mathscr{D}_F^q$ and $\psi\in\mathscr{D}_G^{-q}$. But this is obvious if $\phi\in\mathscr{D}(\Omega)$. The preceding proposition implies that the functions $\phi\in\mathscr{D}(\Omega)$ are dense in $\mathscr{D}_F^q$. Therefore the equation $(\phi,\psi)=0$ holds for arbitrary $\phi\in\mathscr{D}_F^q$ and $\psi\in\mathscr{D}_G^{-q}$, q.e.d. □

**2°. Infinitely differentiable functions and distributions on compacts.** Let $F$ be the closure of some admissible region $\Omega$. We fix an integer $q$. The imbedding $\mathscr{E}^{q+1}\to\mathscr{E}^q$ defines a continuous mapping of the subspaces

$$d_{q+1}\colon\ \mathscr{D}_F^{q+1}\to\mathscr{D}_F^q,$$

and $\mathscr{D}_G^{q+1}$ is carried into $\mathscr{D}_G^q$. Therefore this imbedding $\mathscr{E}^{q+1}\to\mathscr{E}^q$ can be continuously extended to a mapping of the factor-spaces

$$e_{q+1}\colon\ \mathscr{E}_F^{q+1}=\mathscr{E}^{q+1}/\mathscr{D}_G^{q+1}\to\mathscr{E}_F^q=\mathscr{E}^q/\mathscr{D}_G^q.$$

We note that the mapping $e_{q+1}$ is biunique. In fact, if some function $\phi\in\mathscr{E}^{q+1}$ when imbedded in $\mathscr{E}^q$ falls in the subspace $\mathscr{D}_G^q$, its support belongs to $G$ and therefore the function itself belongs to $\mathscr{D}_G^{q+1}$.

We remark that since by Proposition 2 the spaces $\mathscr{D}_F^q$ and $\mathscr{E}_F^{-q}$ are mutually conjugate, the mappings $d_{q+1}$ and $e_{-q}$ are also mutually conjugate.

The ensemble of the spaces $\mathscr{D}_F^q$ and the mappings $d_q$ forms a decreasing family of spaces. The spaces $\mathscr{E}_F^q$ and the mappings $e_q$ also form a decreasing family. We write

$$\mathscr{D}_F=\varprojlim_{q\to\infty}\{\mathscr{D}_F^q,d_q\},\qquad \mathscr{E}_F=\varprojlim_{q\to\infty}\{\mathscr{E}_F^q,e_q\}. \tag{3.2}$$

Since $\mathscr{D}_F^q$ and $\mathscr{E}_F^q$ are Hilbert spaces and are therefore Frechet spaces, the projective limits (3) are also $\mathscr{F}$-spaces, by Proposition 2, § 1. Let us consider the space $\mathscr{D}_F$ in more detail. Since all the mappings $d_q$ are biunique, the mappings $d^q\colon\ \mathscr{D}_F\to\mathscr{D}_F^q$ are also biunique. Therefore the projective limit $\mathscr{D}_F$ may be identified with the intersection $\bigcap_q\mathscr{D}_F^q$. The functions belonging to $\mathscr{D}_F^q$, $q\geqq\nu$ have continuous derivatives up to order $q-\nu$ and vanish outside $F$. Hence it follows that the elements of $\mathscr{D}_F$ are infinitely differentiable functions in $R^n$, with supports lying in $F$. Conversely, if a function $\phi$ is infinitely differentiable in $R^n$, if supp $\phi \subset F$, and if all its derivatives $D^j \phi$ belong to $L_2(R^n)$, then $\phi \in \mathscr{D}_F$. If $F$ is compact, the conditions $D^j \phi \in L_2(R^n)$ may be omitted, since they follow from the continuity of $D^j \phi$.

By the definition of the topology in the projective limit, the neighborhoods of zero in the space $\mathscr{D}_F$ are sets of the form $\mathscr{D}_F \cap U$, where $U$ is a neighborhood of zero in one of the spaces $\mathscr{D}_F^q$. It follows that the sets of the form $\{\phi\colon \|\phi\|^q \leqq \varepsilon\}$ constitute a fundamental system of neighborhoods of zero in $\mathscr{D}_F$.

Let us now describe the space $\mathscr{E}_F$. It is clear that the space $\mathscr{E}_{R^n}$ consists of all infinitely differentiable functions in $R^n$ whose derivatives all belong to $L_2(R^n)$. For every integer $q$ we consider the exact sequence of Hilbert spaces

$$0 \to \mathscr{D}_G^q \to \mathscr{E}^q \to \mathscr{E}_F^q \to 0.$$

Since the set $F$ is admissible, the region $R^n \setminus F$ is also admissible, and therefore the space $\mathscr{D}(R^n \setminus F)$ is dense in each of the spaces $\mathscr{D}_G^q$. Therefore, for arbitrary $q$ the subspace $\mathscr{D}_G^{q+1}$ is dense in $\mathscr{D}_G^q$. Then Proposition 11, § 1 implies that the sequence of projective limits

$$0 \to \mathscr{D}_G \to \mathscr{E} \to \mathscr{E}_F \to 0$$

is also exact. Hence $\mathscr{E}_F \cong \mathscr{E}/\mathscr{D}_G$.

We shall show that the spaces conjugate to the spaces (3) admit the following representation:

$$\begin{aligned} \mathscr{D}_F^* &\cong \underrightarrow{\lim} \{(\mathscr{D}_F^q)^*, d_{q+1}^*\} \cong \underrightarrow{\lim} \{\mathscr{E}_F^{-q}, e_{-q}\}; \\ \mathscr{E}_F^* &\cong \underrightarrow{\lim} \{(\mathscr{E}_F^q)^*, e_{q+1}^*\} \cong \underrightarrow{\lim} \{\mathscr{D}_F^{-q}, d_{-q}\}. \end{aligned} \tag{4.2}$$

Since all the spaces $\mathscr{D}_F^q$ and $\mathscr{E}_F^q$ are Hilbert spaces, they are reflexive Banach spaces. Therefore the formulae (4) follow from Proposition 4, § 1, which also implies that the spaces $\mathscr{D}_F$ and $\mathscr{E}_F$ are reflexive. We note that the spaces $\mathscr{D}_F^*$ and $\mathscr{E}_F^*$ are complete, since they are conjugates of $\mathscr{F}$-spaces.

Since all the mappings $e_{-q}$ are biunique, the inductive limit $\underrightarrow{\lim} \{\mathscr{E}_F^{-q}, e_{-q}\}$ is identical to the union $\bigcup_q \mathscr{E}_F^{-q}$. When $F$ is compact, the elements of $\mathscr{D}_F^* = \bigcup \mathscr{E}_F^{-q}$ are called distributions on it. For every distribution $f \in \mathscr{D}_F^*$ the quantity

$$\deg_F f = \inf\{q\colon f \in \mathscr{E}_F^{-q}, \ -\infty < q < \infty\} < \infty$$

is called its order on the compact $F$.

**Proposition 3**[10]. *Let $F$ be compact. Then for arbitrary integer $q$ the mappings $d_q$ and $e_q$ are compact.*

**3°. The space $\mathscr{E}(\Omega)$.** Let $F$ and $F' \supset F$ be closed sets. The restriction operation defined over $F'$ on the subset $F$ defines a continuous mapping $\mathscr{E}^q_{F'} \to \mathscr{E}^q_F$ for arbitrary integer $q$. It commutes with the operators $e_q$ and therefore can be extended to a mapping of the limit spaces $\mathscr{E}_{F'} \to \mathscr{E}_F$ and $\mathscr{D}^*_{F'} \to \mathscr{D}^*_F$.

Let $\Omega$ be an arbitrary region in $R^n$. We choose a sequence of admissible compacts $K_\alpha$, $1 \leqq \alpha < \infty$, satisfying the conditions

$$\dots K_\alpha \subset\subset K_{\alpha+1} \subset\subset \dots \Omega^{11}, \qquad \bigcup K_\alpha = \Omega. \tag{5.2}$$

Such a sequence will be said to be a strictly increasing sequence of compacts tending to $\Omega$. We remark that (5) implies that an arbitrary compact $K \subset \Omega$ is contained in one of the compacts of the sequence.

For every natural $\alpha$ we consider the operator $i_{\alpha+1}: \mathscr{E}^{\alpha+1}_{K_{\alpha+1}} \to \mathscr{E}^\alpha_{K_\alpha}$, which is the composition of the identity imbedding $e_{\alpha+1}: \mathscr{E}^{\alpha+1}_{K_{\alpha+1}} \to \mathscr{E}^\alpha_{K_{\alpha+1}}$, and the restriction operation $\mathscr{E}^\alpha_{K_{\alpha+1}} \to \mathscr{E}^\alpha_{K_\alpha}$. By Proposition 3 the imbedding $e_{\alpha+1}$ is a compact operation and therefore the operator $i_{\alpha+1}$ is compact. We shall show that its image is dense in $\mathscr{E}^\alpha_{K_\alpha}$. In fact, by construction the space $\mathscr{D}(R^n)$ is dense in $\mathscr{E}^\alpha$, and therefore the image of the composition of the mappings

$$\mathscr{D}(R^n) \to \mathscr{E}^\alpha \to \mathscr{E}^\alpha / \mathscr{D}^\alpha_{CK_\alpha} = \mathscr{E}^\alpha_{K_\alpha}$$

is dense in $\mathscr{E}^\alpha_{K_\alpha}$. But this image is contained in the image of $i_{\alpha+1}$ and therefore the image of the operator $i_{\alpha+1}$ is also dense in $\mathscr{E}^\alpha_{K_\alpha}$.

Let us consider the decreasing family consisting of the spaces $\mathscr{E}^\alpha_{K_\alpha}$ and the operators $i_\alpha$, and let us write

$$\mathscr{E}(\Omega) = \varprojlim \{\mathscr{E}^\alpha_{K_\alpha}, i_\alpha\}. \tag{6.2}$$

We characterize the elements of this space. By definition, each element is a thread consisting of functions $f_\alpha \in \mathscr{E}^\alpha_{K_\alpha}$, that satisfy the condition: for arbitrary $\alpha$ the restriction of $f_{\alpha+1}$ on $K_\alpha$ coincides with $f_\alpha$. Therefore the functions $f_\alpha$ represent the restrictions of some global function $f$ defined on the region $\Omega$. Since an arbitrary compact $K \subset \Omega$ is contained in all the compacts $K_\alpha$ from some $\alpha$ onward, the restriction of the function $f$ on $K$ belongs to all the spaces $\mathscr{E}^\alpha_K$ and therefore is an infinitely differentiable function within $K$, and so in the whole region $\Omega$. Conversely, every infinitely differentiable function in $\Omega$ can be looked on as an element of the space $\mathscr{E}(\Omega)$.

---

10 See for example the paper of Volevič and Paneyah [1], Theorem 8.1, § 8.

11 The symbol $A \subset\subset B$ means that a neighborhood of $A$ is contained in $B$.

The topology in $\mathscr{E}(\Omega)$ can be characterized as follows: the sets

$$\{f\in\mathscr{E}(\Omega),\ \|f\|_K^q<\varepsilon,\ 0<q<\infty,\ K\subset\Omega,\ \varepsilon>0\}$$

form a fundamental system of neighborhoods of zero. The topology of that space does not depend on the choice of the sequence $\{K_\alpha\}$.

We now list some topological properties of the space. By Proposition 5, § 1, what we have said about the operators $i_\alpha$ implies that $\mathscr{E}(\Omega)$ is an $\mathscr{F}\mathscr{S}$-space. Hence, in particular, $\mathscr{E}(\Omega)$ is reflexive and an arbitrary bounded set in it is relatively compact. Since all the $\mathscr{E}_{K_\alpha}^\alpha$ are Hilbert spaces, we may apply Proposition 4, § 1 to the decreasing family in (6) and we obtain

$$\mathscr{E}^*(\Omega)\cong\varinjlim\{(\mathscr{E}_{K_\alpha}^\alpha)^*,\ i_\alpha^*\}\cong\varinjlim\{\mathscr{D}_{K_\alpha}^{-\alpha},\ i_\alpha^*\}. \tag{7.2}$$

We note that the mappings $i_\alpha^*\colon \mathscr{D}_{K_\alpha}^{-\alpha}\to\mathscr{D}_{K_{\alpha+1}}^{-\alpha-1}$ appearing on the right side are biunique. Therefore we will identify (algebraically) the space $\mathscr{E}_{-\alpha}^*(\Omega)$ with the union $\bigcup\mathscr{D}_{K_\alpha}^{-\alpha}$. This means that $\mathscr{E}^*(\Omega)$ is the space of all distributions on $R^n$ with compact supports belonging to $\Omega$.

If $\Omega'$ is a subregion of $\Omega$ the restriction of functions belonging to $\mathscr{E}(\Omega)$ on $\Omega$ is a continuous mapping from $\mathscr{E}(\Omega)$ to $\mathscr{E}(\Omega')$. The conjugate mapping $\mathscr{E}^*(\Omega')\to\mathscr{E}^*(\Omega)$, which acts in spaces of distributions with compact supports, is an imbedding.

### 4°. The connection between the spaces $\mathscr{E}_K^*$ and $\mathscr{E}^*(\Omega)$

**Proposition 4**

I. *For an arbitrary region $\Omega$ and an admissible compact $K\subset\Omega$ the subspace of $\mathscr{E}^*(\Omega)$ consisting of distributions with supports lying in $K$ coincides with $\mathscr{E}_K^*$.*

II. *Let $\{K_\alpha\}$ be a strictly increasing sequence of compacts tending to $\Omega$. Then*

$$\varinjlim\mathscr{E}_{K_\alpha}^*\cong\mathscr{E}^*(\Omega).$$

*Proof.* We establish the first assertion. Write $G=R^n\setminus K$. The mapping $\mathscr{E}\to\mathscr{E}(\Omega)$ is obviously continuous and it carries the space $\mathscr{D}_G$ into the space $\mathscr{E}(\Omega, G)$ consisting of the functions in $\mathscr{E}(\Omega)$ that have carriers in $G$. We may therefore define a continuous mapping of the factor-spaces

$$\mathscr{E}/\mathscr{D}_G\to\mathscr{E}(\Omega)/\mathscr{E}(\Omega, G). \tag{8.2}$$

Let us construct the inverse mapping. We choose a function $h\in\mathscr{D}(\Omega)$ which is equal to unity in some neighborhood of the compact $K$. For an arbitrary function $f\in\mathscr{E}(\Omega)$ the product $hf$ has a compact support and therefore can be looked on as an element of $\mathscr{E}$. We thus obtain a continuous mapping $\mathscr{E}(\Omega)\to\mathscr{E}$ which does not enlarge the support and therefore maps the subspace $\mathscr{E}(\Omega, G)$ into $\mathscr{D}_G$. The associated mapping of the factor-spaces is the inverse of the mapping (8). This proves that (8) is an isomorphism.

Since the compact $K$ is admissible, the left side of (8) is isomorphic to $\mathscr{E}_K$, by what we proved in 2°. On the other hand, $\mathscr{E}(\Omega)$ is an $\mathscr{F}\mathscr{S}$-space and $\mathscr{E}(\Omega, G)$ is a closed subspace of it. Therefore, taking the conjugates of both sides of (8) we obtain on the left space $\mathscr{E}_K^*$, and on the right the subspace $(\mathscr{E}(\Omega, G))^0\subset\mathscr{E}^*(\Omega)$, by Proposition 6 § 1. This subspace consists of all those, and only those, $\phi\in\mathscr{E}^*(\Omega)$ having supports in $K$. This completes the proof.

Let us prove the second assertion. Since $\mathscr{E}^*(\Omega)$ is the space of all distributions with compact supports belonging to $\Omega$, the set of its elements coincides with the set of elements in the inductive limit $\varinjlim \mathscr{E}_{K_\alpha}^*$. We need only show that the corresponding topologies coincide. By our first assertion, an arbitrary neighborhood of zero in $\mathscr{E}^*(\Omega)$ contains a neighborhood of zero in an arbitrary one of the spaces $\mathscr{E}_{K_\alpha}^*$ and therefore in their inductive limit.

Conversely, let $U$ be a convex neighborhood of zero in $\varinjlim \mathscr{E}_{K_\alpha}^*$. Then for arbitrary $\alpha$ the set $U_\alpha=U\cap\mathscr{E}_{K_\alpha}^*$ is a neighborhood of zero in $\mathscr{E}_{K_\alpha}^*$ and therefore contains the polar $B_\alpha^0$ of some bounded set $B_\alpha$ in $\mathscr{E}_{K_\alpha}$. To complete the proof we must find a bounded set $B$ in $\mathscr{E}(\Omega)$ such that $B^0\subset U$.

We choose a sequence of functions $h_\alpha\subset\mathscr{D}(\Omega)$, which constitute a partition of the unity in $\Omega$ and for which:

$$\operatorname{supp} h_\alpha\subset K_{\alpha-1}\setminus K_{\alpha-3}.$$

We set

$$B=\{2^\alpha h_\alpha f, f\in B_\alpha, \alpha=1,2,\ldots\}.$$

It follows from the properties of $h_\alpha$ and $B_\alpha$ that this set is bounded in $\mathscr{E}(\Omega)$. Let $\phi$ be an arbitrary element in the polar of $B$. Since the functions $\bar{h}_\alpha$ form a partition of the unity, we have $\phi=\sum\bar{h}_\alpha\phi$, and this sum is finite. For arbitrary $\alpha$

$$|(\bar{h}_\alpha\phi, f)|=|(\phi, h_\alpha f)|\leqq\frac{1}{2^\alpha},\qquad f\in B_\alpha.$$

This inequality shows that the function $\bar{h}_\alpha\phi$ belongs to the set $2^{-\alpha}B_\alpha^0\subset 2^{-\alpha}U$. Therefore $\phi$ belongs to the set

$$\bigcup_{\beta=1}^{\infty}\sum_{1}^{\beta}2^{-\alpha}U\subset U,\qquad \text{q.e.d.}\quad \square$$

**5°. The space $\mathscr{D}(\Omega)$.** Again let $F$ and $F'\supset F$ be closed sets in $R^n$. For arbitrary integer $q$ the space $\mathscr{D}_{F'}^q$ contains $\mathscr{D}_F^q$ and the identity imbedding

$\mathscr{D}_F^q \to \mathscr{D}_{F'}^q$ is continuous. Since this operation commutes with the operators $d_q$, it can be extended to the limit spaces $\mathscr{D}_F \to \mathscr{D}_{F'}$ and $\mathscr{E}_F^* \to \mathscr{E}_{F'}^*$.

Let $\Omega$ be a region in $R^n$ and let $\{K_\alpha\}$ be a strictly increasing sequence of compacts tending to $\Omega$. The spaces $\mathscr{D}_{K_\alpha}$ and the imbeddings $\mathscr{D}_{K_\alpha} \to \mathscr{D}_{K_{\alpha+1}}$ form an increasing family. We consider its inductive limit

$$\mathscr{D}(\Omega) = \varinjlim \mathscr{D}_{K_\alpha}. \tag{9.2}$$

The space $\mathscr{D}(\Omega)$ has the same elements as the union $\bigcup \mathscr{D}_{K_\alpha}$, that is, it is the space of all infinitely differentiable functions in $R^n$ having compact supports in $\Omega$. Thus the space $\mathscr{D}(\Omega)$ defined by (9) coincides with the space introduced in 1° and denoted by the same symbol. Formula (9) defines a topology in this space; a neighborhood of zero in $\mathscr{D}(\Omega)$ is an arbitrary set containing a convex set $U$ whose intersection with an arbitrary one of the spaces $\mathscr{D}_{K_\alpha}$ is a neighborhood of zero in the latter. In other words there is a fundamental system of neighborhoods of zero in $\mathscr{D}(\Omega)$ consisting of convex sets $U$, each of which contains for arbitrary $\alpha$ a set of the form

$$\{\phi \in \mathscr{D}_{K_\alpha} : \|\phi\|^q \leqq \varepsilon, \ 0 < q < \infty, \ \varepsilon > 0\}. \tag{10.2}$$

The following assertion characterizes the topology in $\mathscr{D}(\Omega)$.

**Proposition 5.** *The increasing family* (9) *is regular.*

*Proof.* We must show that an arbitrary bounded set $B$ in $\mathscr{D}(\Omega)$ is contained in, and bounded in, one of the spaces $\mathscr{D}_{K_\alpha}$. We first show that $B$ belongs to one of the $\mathscr{D}_{K_\alpha}$. Let us suppose the contrary. Then for an arbitrary $\alpha$ we may find a function $\phi_\alpha \in B$ and not contained in $\mathscr{D}_{K_\alpha}$, i.e. such that $\operatorname{supp} \phi_\alpha \not\subset K_\alpha$. We choose the point $\xi_\alpha \in R^n \setminus K_\alpha$ so that $\phi_\alpha(\xi_\alpha) \neq 0$, and we consider the set $U$ in $\mathscr{D}(\Omega)$ consisting of all the functions $\phi$ for which

$$|\phi(\xi_\alpha)| \leqq \frac{1}{\alpha} |\phi_\alpha(\xi_\alpha)|, \quad \alpha = 1, 2, \ldots.$$

The set $U$ is a neighborhood of zero in $\mathscr{D}(\Omega)$ since it is convex and for arbitrary $\alpha$ the intersection $U \cap \mathscr{D}_{K_\alpha}$ is a neighborhood of zero in $\mathscr{D}_{K_\alpha}$. On the other hand, for no $\lambda > 0$ does the set $\lambda U$ contain $B$, i.e. $U$ does not absorb $B$, which contradicts the assumption that $B$ is a bounded set.

Thus $B \subset \mathscr{D}_{K_\alpha}$ for some $\alpha$. The sets of the form

$$V = \{\phi : \|\phi\|^q \leqq \varepsilon\}$$

constitute a fundamental system of neighborhoods of zero in $\mathscr{D}_{K_\alpha}$ and are also neighborhoods of zero in $\mathscr{D}(\Omega)$. The set $B$, being bounded in $\mathscr{D}(\Omega)$ is absorbed by each of them and is therefore bounded in $\mathscr{D}_{K_\alpha}$. □

We obtain the following formula from Proposition 3 § 1.

$$\mathscr{D}^*(\Omega) = \varprojlim \mathscr{D}^*_{K_\alpha}.$$

Using it, we can in particular describe the elements of $\mathscr{D}^*(\Omega)$: every continuous functional on $\mathscr{D}(\Omega)$ is characterized by the fact that its restriction on the subspace $\mathscr{D}_K$, where $K \subset \Omega$ is an arbitrary compact, is a continuous function on $\mathscr{D}_K$; that is, it is a distribution on $K$. The quantity $\deg_K f$ may depend on $K$ and is a non-decreasing function of $K$. The elements of $\mathscr{D}^*(\Omega)$ are called distributions on the region $\Omega$.

If $\Omega'$ is a subregion of $\Omega$, the identity imbedding $\mathscr{D}(\Omega') \to \mathscr{D}(\Omega)$ is defined and continuous. The conjugate operation, from $\mathscr{D}^*(\Omega)$ to $\mathscr{D}^*(\Omega')$ will be called the restriction of the distributions in $\Omega$ to the subregion $\Omega'$.

Every infinitely differentiable function $f \in \mathscr{E}(\Omega)$ and, more generally, every locally summable function in $\Omega$, can be looked on as a distribution in $\Omega$ defined by the formula

$$\phi \to (f, \phi) = \int_\Omega \bar{f} \phi \, d\xi.$$

The support of the distribution $f \in \mathscr{D}^*(\Omega)$ is the smallest relatively closed subset $\operatorname{supp} f \subset \Omega$ such that $(f, \phi) = 0$ for all $\phi \in \mathscr{D}(\Omega \setminus \operatorname{supp} f)$. The support of the singularities of the function $f \in \mathscr{D}^*(\Omega)$ is the smallest relatively closed subset $\operatorname{sing\,supp} f \subset \Omega$ such that the restriction of $f$ on the subregion $\Omega \setminus \operatorname{sing\,supp} f$ coincides with some function belonging to $\mathscr{E}(\Omega \setminus \operatorname{sing\,supp} f)$. Thus the subspace $\mathscr{E}(\Omega) \subset \mathscr{D}^*(\Omega)$ is characterized by the condition $\operatorname{sing\,supp} f = \emptyset$.

Since there exists a continuous imbedding $\mathscr{D}(\Omega) \to \mathscr{E}(\Omega)$, every continuous functional on $\mathscr{E}(\Omega)$ can be looked on as an element of the space $\mathscr{D}^*(\Omega)$, that is, as a distribution on $\Omega$.

In each of the spaces $\mathscr{D}^q_{K_\alpha}$ and $\mathscr{E}^q_{K_\alpha}$ there exists, in accordance with 1°, a continuous multiplication by an arbitrary function belonging to $\mathscr{E}(\Omega)$. These multiplications commute with the mappings $e_q$ and $d_q$, and therefore may be extended to the limit spaces $\mathscr{D}(\Omega)$, $\mathscr{E}(\Omega)$, $\mathscr{D}^*(\Omega)$, and $\mathscr{E}^*(\Omega)$. It should be noted that the resulting multiplication by a function $f$ in $\mathscr{E}^*(\Omega)$ and $\mathscr{D}^*(\Omega)$ is conjugate to multiplication by $\bar{f}$ in the spaces $\mathscr{E}(\Omega)$ and $\mathscr{D}(\Omega)$.

We introduce one more concept related to distributions on $\Omega$:

$$\deg_\Omega f = \sup \{\deg_K f, K \subset \Omega\}.$$

The quantity $\deg_\Omega f$ is generally infinite. In $\mathscr{D}^*(\Omega)$ we single out the subspace consisting of distributions of finite order, i.e. functions for which $\deg_\Omega f$ is finite. This subspace, with the topology induced by $\mathscr{D}^*(\Omega)$, will be denoted by $\mathscr{D}^{*F}(\Omega)$.

**6°. Bounded sets in $\mathscr{D}(\Omega)$.** Let $b=b(\eta)$ be an arbitrary positive function of one variable, defined on the ray $\eta \geqq 0$. Suppose further that $F$ is an arbitrary closed set in $R^n$. In the space $\mathscr{D}_F$ we single out the subspace $\mathscr{D}_F^b$ consisting of functions for which the norm

$$\|\phi\|^b=\sup_j \frac{1}{b(|j|)} \|D^j \phi\|^0$$

is finite. The role of the spaces $\mathscr{D}_F^b$ is characterized by the following proposition.

**Proposition 6.** *Let $\Omega$ be an arbitrary region in $R^n$. Then*

I. *We have*

$$\mathscr{D}(\Omega)=\bigcup \mathscr{D}_K^b, \tag{11.2}$$

*where the union is taken over all compacts $K \subset \Omega$ and all functions $b$ of the type defined above.*

II. *The sets of the form*

$$B=\{\phi \in \mathscr{D}_K^b, \|\phi\|^b \leqq 1\} \tag{12.2}$$

*constitute a fundamental system of bounded sets in $\mathscr{D}(\Omega)$.*

*Proof.* We first prove Eq. (11). The inclusion $\supset$ is obvious. Let us prove the converse. Let $\phi$ be an arbitrary function in $\mathscr{D}(\Omega)$. By definition, the set $K=\operatorname{supp}\phi$ is compact and belongs to $\Omega$. Therefore $\phi$ belongs to the space $\mathscr{D}_K^b$, where

$$b(\eta)=\max_{|j| \leqq \eta} \|D^j \phi\|^0.$$

The function $b$ is obviously positive if $\phi \not\equiv 0$.

Let us prove the second assertion. We first show that every set of the form (12) is bounded. As we noted in 5°, every neighborhood of zero in $\mathscr{D}(\Omega)$ contains a set of the form (10) for arbitrary $\alpha$. If $\alpha$ is large enough, $K \subset K_\alpha$ and therefore the set (10) absorbs the set (12), since the functions in (12) belong to $\mathscr{D}_K^b$ and are uniformly bounded, together with all their derivatives. The set (12) is therefore bounded.

Conversely, let $B$ be an arbitrary bounded set in $\mathscr{D}(\Omega)$. Proposition 5 implies that $B$ is contained in one of the spaces $\mathscr{D}_{K_\alpha}$ and is bounded in it. Therefore $B$ is contained in an arbitrary set of the form (10), i.e. the functions in $B$ are uniformly bounded, together with all their derivatives. Therefore the function

$$b_0(\eta)=\max_{|j| \leqq \eta} \sup \{\|D^j \phi\|^0, \phi \in B\}$$

is finite for all $\eta \geqq 0$. Hence it is clear that the set $B$ is contained in the set (12) with $K=K_\alpha$ and $b=b_0$. $\square$

We now note the following well-known fact.

**Proposition 7.** *For arbitrary numbers $\varepsilon>0$ and $\beta>1$ there exists an infinitely differentiable function $e$ in $R^n$, whose support is contained in the sphere $|\xi|\leqq\varepsilon$, such that $\int e(\xi)\,d\xi=1$, and*

$$\sup|D^j e(\xi)|\leqq C B^{|j|}\,|j|^{|j|\beta} \tag{13.2}$$

*for some $B>0$.*

**Proposition 8.** *Let $F$ be the closure of some admissible region $\Omega$, and let the function $b$ satisfy the inequality*

$$b(\eta)\geqq C B_0^\eta\,\eta^{\eta\beta}, \qquad \eta\geqq 0,\ \beta>1, \tag{14.2}$$

*for some $B_0>0$. Then the space $\mathscr{D}_F^b$ is dense in $\mathscr{D}_F^q$ for any integer $q$.*

*Proof.* By Proposition 1 the space $\mathscr{D}(\Omega)$ is dense in $\mathscr{D}_F^q$. Therefore it is sufficient to show that an arbitrary function belonging to $\mathscr{D}(\Omega)$ can be approximated in the norm $\|\cdot\|^q$ by functions belonging to $\mathscr{D}_F^b$.

Let $\phi$ be an arbitrary function belonging to $\mathscr{D}(\Omega)$ and let $e$ be a function satisfying the conditions of Proposition 7, where the constant $\beta$ is taken to be the same as that appearing in the inequality (14) and $\varepsilon=\rho(\operatorname{supp}\phi,\ \complement\,\Omega)$. We consider the sequence of functions $e_\alpha(\xi)=\alpha^n e(\alpha\,\xi)$, $\alpha=1, 2, \ldots$. The support of each of these is contained in the $\varepsilon$-neighborhood of zero, and therefore the support of each of the convolutions $e_\alpha*\phi$ is contained in the set $F$. We estimate the derivatives of these convolutions:

$$|D^i(e_\alpha*\phi)|=|D^i e_\alpha*\phi|\leqq C_\alpha(\alpha B)^{|i|}\,|i|^{|i|\beta},$$

where $B$ is the constant introduced in Proposition 7. Since this constant may chosen as near to zero as we please, the constant $\alpha B$ can be made less than the constant $B_0$ in the inequality (14). Hence it follows that the function $e_\alpha*\phi$ belongs to $\mathscr{D}_F^b$ for arbitrary $\alpha$.

Since the integral over $R^n$ of the function $e_\alpha$ is equal to unity, and its support tends to zero as $\alpha\to\infty$, the function $e_\alpha*\phi$ tends uniformly to $\phi$. The relation $D^i(e_\alpha*\phi)=e_\alpha*D^i\phi$ implies that an arbitrary derivative $D^i(e_\alpha*\phi)$ of this function tends uniformly to the derivative $D^i\phi$. Hence it follows that $e_\alpha*\phi\to\phi$ in the norm $\|\cdot\|^q$. □

By what we have proved, the image of the continuous mapping $\mathscr{D}_F^b\to\mathscr{D}_F^q$ is dense in $\mathscr{D}_F^q$ and therefore the conjugate mapping $\mathscr{E}_F^{-q}\to(\mathscr{D}_F^b)^*$ is biunique. Hence the space $\mathscr{D}_F^*=\bigcup_q\mathscr{E}_F^q$ can be looked on as a subspace in $(\mathscr{D}_F^b)^*$. When $F$ is compact, the elements of $(\mathscr{D}_F^b)^*$ not belonging to $\mathscr{D}_F^*$ are called ultra-distributions.

**7°. Sheaves of spaces of distributions.** We recall a known definition.

**Definition 1.** A presheaf of linear topological spaces in $R^n$ is a correspondence $\Phi$: $\Omega \sim \to \Phi(\Omega)$, which to every region $\Omega \subset R^n$ correlates some l.t.s. $\Phi(\Omega)$ and to every pair of regions $\Omega' \subset \Omega$ a continuous linear mapping $\rho_\Omega^{\Omega'}: \Phi(\Omega) \to \Phi(\Omega')$ which satisfies the following two conditions: (i) for an arbitrary region $\Omega$, $\phi_\Omega^\Omega$ is an identity mapping; (ii) for any three regions $\Omega'' \subset \Omega' \subset \Omega$ we have the equation $\rho_{\Omega'}^{\Omega''} \rho_\Omega^{\Omega'} = \rho_\Omega^{\Omega''}$.

The presheaf $\Phi$ is called a sheaf if it has the following property: Let $U = \{U_\alpha\}$ be an arbitrary covering of the region $\Omega$. Then

I. If $f$ is an element of $\Phi(\Omega)$ such that $\rho_\Omega^U \alpha f = 0$ for arbitrary $\alpha$, we have $f = 0$;

II. If there exist elements $f_\alpha \subset \Phi(U_\alpha)$ such that for arbitrary $\alpha$ and $\beta$ the equation $\rho_{U_\alpha}^{U_\alpha \cap U_\beta} f_\alpha = \rho_{U_\beta}^{U_\alpha \cap U_\beta} f_\beta$ holds, then there exists an element $f \in \Phi(\Omega)$ such that $\rho_\Omega^{U_\alpha} f = f_\alpha$ for arbitrary $\alpha$.

We note that given any covering $U$ of the region $\Omega$ we can always choose a locally finite subcovering, that is, a subcovering $U' = \{U'_\alpha\}$ with the following property: an arbitrary compact $K \subset \Omega$ intersects only a finite number of the regions $U'_\alpha$. The conditions I and II need be verified only for the covering $U'$, and we may therefore always suppose that the covering $U$ that appears in the definition of a sheaf is itself locally finite.

We consider the correspondence

$$\mathscr{E}: \Omega \sim \to \mathscr{E}(\Omega). \tag{15.2}$$

For every pair of regions $\Omega' \subset \Omega$ we choose the mapping $\rho_\Omega^{\Omega'}$ to be the restriction on the subregion $\Omega'$ of functions defined in $\Omega$. The relations (i) and (ii) are obviously satisfied and therefore these mappings, together with the spaces (15), form a presheaf. It is easy to verify that this presheaf is a sheaf. Similarly, the correspondence

$$\mathscr{D}^*: \Omega \sim \to \mathscr{D}^*(\Omega)$$

together with the set of restriction mappings $\rho_\Omega^{\Omega'}: \mathscr{D}^*(\Omega) \to \mathscr{D}^*(\Omega')$ forms a presheaf. The assertion that we are about to prove shows that the presheaf is a sheaf.

Let $U = \{U_i\}$ be some locally finite covering of a region $\Omega$. We shall say that $\Phi = \{\Phi(\omega), \rho_\omega^{\omega'}\}$ is a presheaf on the covering $U$ if the spaces $\Phi(\omega)$ and the mappings $\rho_\omega^{\omega'}$, satisfy the conditions (i) and (ii) and are defined for regions $\omega$, $\omega'$ having the form

$$\omega = \Omega, \qquad U_{i_0, \dots, i_\nu} = U_{i_0} \cap \dots \cap U_{i_\nu}, \qquad \nu = 0, 1, 2, \dots. \tag{16.2}$$

Let $\Phi$ be an arbitrary presheaf defined on the covering $U$. For every integer $\nu \geqq 0$ we consider the l.t.s. ${}^\nu\Phi(U)$ consisting of cochains of order $\nu$

on the covering $U$ with coefficients in the spaces $\Phi(U_{i_0,\dots,i_\nu})$; that is expressions of the form

$$\phi=\sum \phi_{i_0,\dots,i_\nu} U_{i_0}\cap\cdots\cap U_{i_\nu}, \qquad \text{where } \phi_{i_0,\dots,i_\nu}\in\Phi(U_{i_0,\dots,i_\nu}), \quad (17.2)$$

and where the element $\phi_{i_0,\dots,i_\nu}$ is skew-symmetric with respect to its indices (see the analogous definition in § 3, Chapter III). The topology in the space ${}^\nu\Phi(U)$ is defined via the isomorphism

$${}^\nu\Phi(U)\cong \prod_{i_0<\cdots<i_\nu} \Phi(U_{i_0,\dots,i_\nu}),$$

in which the right side is the topological direct product of l.t.s. In this topology a fundamental system of neighborhoods of zero is formed by sets of the form

$$\{\phi\colon \phi_\sigma\in V_\sigma,\ \sigma\in\Sigma\},$$

where $\Sigma$ is a finite set of vectors $\sigma=(i_0,\dots,i_\nu)$, and the $V_\sigma$ are neighborhoods of zero in the spaces $\Phi(U_\sigma)$.

We also recall the definition of the coboundary operators $\partial^\nu$. For arbitrary $\nu\geqq 0$ the effect of the operator $\partial^\nu$ is defined by the formula

$$\partial^\nu\phi=\frac{1}{\nu+2}\sum_{i_0,\dots,i_{\nu+1}}\sum_{j=0}^{\nu+1}(-1)^j\hat{\phi}_{i_0,\dots,\hat{i}_j,\dots,i_{\nu+1}} U_{i_0}\cap\cdots\cap U_{i_{\nu+1}}, \quad (18.2)$$

where $\hat{\phi}_{i_0,\dots}$ is the restriction of the element $\phi_{i_0,\dots}$ on $U_{i_0,\dots,i_{\nu+1}}$ (i.e. the result of applying the mapping $\rho^{U\dots}_{U\dots}$). It is easy to see that these operators define continuous mappings $\partial^\nu\colon {}^\nu\Phi(U)\to{}^{\nu+1}\Phi(U)$, $\nu=0, 1, 2, \dots$. This collection of mappings is completed by the continuous mapping $\partial^{-1}$: $\Phi(\Omega)\to{}^0\Phi(U)$, which carries the element $\phi$ into the cochain $\Sigma\hat{\phi}\,U_i$. Sometimes, for generality of our notation, we shall write ${}^{-1}\Phi(U)=\Phi(\Omega)$. We arrange the so-constructed mappings in the sequence

$$0\to\Phi(\Omega)\xrightarrow{\partial^{-1}}{}^0\Phi(U)\xrightarrow{\partial^0}\cdots\longrightarrow{}^\nu\Phi(U)\xrightarrow{\partial^\nu}{}^{\nu+1}\Phi(U)\to\cdots. \quad (19.2)$$

It is semi-exact, since $\partial^\nu\partial^{\nu-1}=0$ for $\nu\geqq 0$.

**Proposition 9.** *We suppose that the presheaf $\Phi$, defined on the locally finite covering $U$, satisfies the following conditions:*

A) *For an arbitrary region $\omega$ (of the form* (16)*) $\Phi(\omega)$ is a subspace (in the algebraic sense) of the space $\mathscr{D}^*(\omega)$, and for arbitrary regions $\omega'\subset\omega$ the mapping $\phi^{\omega'}_\omega$ is a restriction on the subregion $\omega'$.*

B) *Let $\{\omega_i\}$ be a locally finite covering of the region $\omega$. Then the series $\sum f_i$, where $f_i\in\Phi(\omega)$ and supp $f_i\subset\omega_i$, converges in $\Phi(\omega)$, and the* *mapping*

$$\prod_i \Phi(\omega) \ni \{f_i\} \to \sum f_i \in \Phi(\omega) \tag{20.2}$$

*is continuous.*

C) *These is defined in $\Phi(\omega)$ a continuous operation of multiplication by an arbitrary function belonging to $\mathscr{E}(\omega)$.*

D) *For arbitrary regions $\omega$ and $\upsilon$ the following condition is satisfied: if $f$ is a distribution belonging to $\mathscr{D}^*(\omega)$ supp $f \subset \upsilon$ and its restriction on $\omega \cap \upsilon$ is an element of $\Phi(\omega \cap \upsilon)$, then $f \in \Phi(\omega)$.*

*Under these conditions, the sequence* (19) *is exact.*

Let us verify that the conditions A)–D) are satisfied by the presheaves $\mathscr{E}$ and $\mathscr{D}^*$. For the sheaf $\mathscr{E}$ they are obvious. For the presheaf $\mathscr{D}^*$ we need only B) in the proof. By hypothesis the covering $\{\omega_i\}$ which consists of regions of the form (16), is locally finite. Therefore an arbitrary compact $\kappa \subset \omega$ intersects only a finite number of the regions $\omega_i$; for definiteness, say these are $\omega_1, \ldots, \omega_\nu$. Therefore the series

$$(f, \phi) = \sum (f_i, \phi) \tag{21.2}$$

converges for an arbitrary function $\phi \in \mathscr{D}_\kappa$ and is a continuous function on $\mathscr{D}_\kappa$, since only the first $k$ terms of the series can be different from zero.

Since the compact $\kappa \subset \omega$ was chosen arbitrarily, the functional $f$ is continuous on all the spaces $\mathscr{D}_\kappa$, and therefore on their inductive limit, which is the whole space $\mathscr{D}(\omega)$. Therefore $f \in \mathscr{D}^*(\omega)$. Further, let $U$ be a neighborhood of zero in $\mathscr{D}^*(\omega)$. By the definition of the topology in the conjugate space $\mathscr{D}^*(\omega)$ the neighborhood $U$ contains the polar of some set $B$ bounded in $\mathscr{D}(\omega)$. By Proposition 5, the set $B$ is contained in one of the spaces $\mathscr{D}_\kappa$ and is bounded in it. Therefore, on the right side of (21) only the first $k$ terms can be different from zero on the set $B$. We shall suppose that the functionals $f_1, \ldots, f_k$ belong to the polar of the set $kB$ (this set is a neighborhood of zero in $\mathscr{D}^*(\omega)$). Then the functional $f$ belongs to the polar of $B$ and therefore belongs to $U$. But this proves the continuity of the mapping (20) when $\Phi = \mathscr{D}^*$, q.e.d.

The exactness of (19) implies in particular that the presheaf $\mathscr{D}^*$ is a sheaf.

Let us turn to the proof of Proposition 9. We construct the set of functions $\alpha_i \in \mathscr{E}(\Omega)$ satisfying the following conditions: a) for arbitrary $i$, supp $\alpha_i \subset U_i$; b) $\sum \alpha_i \equiv 1$ in $\Omega$. Such a set of functions will be called an infinitely differentiable partition of the unity in $\Omega$ subject to the covering $U$.

Let $\phi$ be an arbitrary element of the kernel of the mapping $\partial^\nu : {}^\nu\Phi(U) \to {}^{\nu+1}\Phi(U)$, $\nu \geqq 0$. We fix the indices $i_1, \ldots, i_\nu$ in an arbitrary way and we choose an index $i_0$ at will. Let $\phi_{i_0, \ldots, i_\nu}$ be the corresponding coefficient of the cochain $\phi$. Since the function $\alpha_{i_0}$ is infinitely differentiable in $\Omega$, Condition C) implies that the product $\alpha_{i_0}\phi_{i_0,\ldots,i_v}\in\Phi(U_{i_0,\ldots,i_v})$ is defined. This product vanishes near $\partial U_{i_0}\cap U_{i_1,\ldots,i_v}$, and therefore it can be extended as a distribution in the region $U_{i_1,\ldots,i_v}$ with support belonging to $U_{i_0,\ldots,i_v}$. This distribution will be denoted by $(\alpha_{i_0}\phi_{i_0,\ldots,i_v})'$. Condition D) implies that it belongs to $\Phi(U_{i_1,\ldots,i_v})$. We now consider the sum

$$\psi_{i_1,\ldots,i_v}=(v+1)\sum_{i_0}(\alpha_{i_0}\phi_{i_0,\ldots,i_v})'. \tag{22.2}$$

Since the distribution $\psi_{i_1,\ldots,i_v}$ is obviously skew-symmetric in its indices, we may consider the cochain

$$\psi=\sum\psi_{i_1,\ldots,i_v}U_{i_1,\ldots,i_v}.$$

We shall show that the series on the right in (22) converges in $\Phi(\omega)$, where $\omega=U_{i_1,\ldots,i_v}$. Since $\operatorname{supp}(\alpha_{i_0}\phi_{i_0,\ldots,i_v})'\subset U_{i_0}\cap\omega$, and the regions $U_{i_0}\cap\omega$ form a locally finite covering of the region $\omega$, this follows from Condition B), which also implies that the mapping

$$\prod_{i_0}\Phi(U_{i_0}\cap\omega)\ni\{\phi_{i_0,\ldots,i_v}\}\to\psi_{i_1,\ldots,i_v}\in\Phi(\omega)$$

is continuous for arbitrary fixed indices $i_1,\ldots,i_v$. Therefore the product

$${}^v\Phi(U)\cong\prod\Phi(U_{i_0,\ldots,i_v})\ni\phi\to\psi\in\prod\Phi(U_{i_1,\ldots,i_v})\cong{}^{v-1}\Phi(U) \tag{23.2}$$

of these mappings is also continuous.

We must show that the mapping (23) is the inverse of the operator $\partial^{v-1}$, i.e. that $\partial^{v-1}\psi=\phi$. We have

$$\partial^{v-1}\psi=\sum_{i_0,\ldots,i_v}\sum_i\alpha_i\sum_{j=0}^{v}(-1)^j\phi_{i,i_0,\ldots,\hat{\imath}_j,\ldots,i_v}U_{i_0}\cap\cdots\cap U_{i_v}. \tag{24.2}$$

It is clear from (18) that the coefficient of the cochain $\partial^v\phi$ corresponding to the region $U_{i,i_0,\ldots,i_v}$ is equal to

$$\frac{1}{v+2}\left[-\sum_{j=0}^{v}(-1)^j\phi_{i,i_0,\ldots,\hat{\imath}_j,\ldots,i_v}+\phi_{i_0,\ldots,i_v}\right]$$

when $i\neq i_j$, $j=0,\ldots,v$. Since by hypothesis $\partial^v\phi=0$ this expression vanishes, and therefore the inner sum on the right side of (24) is equal to $\phi_{i_0,\ldots,i_v}$. If however $i=i_j$ for some $j$, $0\leqq j\leqq v$, this inner sum is also equal to $\phi_{i_0,\ldots,i_v}$. Accordingly, the right side of (24) is equal to $\phi$, q.e.d.

We have established the exactness of (19) in all terms except the first. We now establish its exactness in the first term, by showing that the operator $\partial^{-1}$ is biunique. Suppose that $\partial^{-1}\phi=0$, where $\phi\in\Phi(\Omega)$. For every $i$ we write $\phi_i = \rho_\Omega^{U_i} \phi$, and we extend the distribution $\alpha_i \phi_i$ in $\Omega$, setting it equal to zero on $\Omega \setminus U_i$. The extended distribution $(\alpha_i \phi_i)$ belongs to $\Phi(\Omega)$, by Condition D), and the series $\Sigma(\alpha_i \phi_i)$ converges in $\Phi(\Omega)$ by Condition B). Its sum in the space $\mathscr{D}^*(\Omega)$ is obviously $\phi$. We have only to note that all the terms of this series vanish, since by hypothesis $\phi_i = 0$ for all $i$. It follows that $\phi = 0$. □

**8°. Cosheaves of spaces of distributions.** We now employ a concept dual to the concept of the sheaf.

**Definition 2.** A precosheaf of l.t.s., defined in $R^n$, is a mapping $\Psi$: $\Omega \sim\to \Psi(\Omega)$, that to every region $\Omega \subset R^n$ pairs some l.t.s. $\Psi(\Omega)$, and to every pair of regions $\Omega \subset \Omega'$ some continuous mapping $e_\Omega^{\Omega'}: \Psi(\Omega) \to \Psi(\Omega')$, satisfying the conditions (i) $e_\Omega^\Omega = I$ and (ii) $e_{\Omega'}^{\Omega''} e_\Omega^{\Omega'} = e_\Omega^{\Omega'}$ for an arbitrary triplet of regions $\Omega \subset \Omega' \subset \Omega''$.

The precosheaf $\Psi$ is a cosheaf if the mapping $\Phi$: $\Omega \sim\to \Psi(C\Omega)$ is a sheaf.

The deciphering of this latter definition is left to the reader. We use only the notion of the precosheaf. We note the following fact: if $\Phi = \{\Phi(\Omega), \rho_\Omega^{\Omega'}\}$ is a presheaf, the set of conjugate spaces and mappings $\Phi^* = \{\Phi^*(\Omega), (\rho_\Omega^{\Omega'})^*\}$, is a precosheaf, which we call the conjugate of the presheaf $\Phi$. Conversely, if $\Psi = \{\Psi(\Omega), e_\Omega^{\Omega'}\}$ is a precosheaf,

$$\Psi^* = \{\Psi^*(\Omega), (e_\Omega^{\Omega'})^*\}$$

is a presheaf.

In particular, we may consider the precosheaves

$$\mathscr{D}: \Omega \sim\to \mathscr{D}(\Omega), \qquad \mathscr{E}^*: \Omega \sim\to \mathscr{E}^*(\Omega), \tag{25.2}$$

conjugate to the presheaves $\mathscr{D}^*$ and $\mathscr{E}$. By definition, for an arbitrary pair of regions $\Omega \subset \Omega'$ the mapping $e_\Omega^{\Omega'}$ needed for the construction of the precosheaves (25) is the conjugate of the restriction mapping $\rho_{\Omega'}^\Omega$ in the presheaves $\mathscr{D}^*$ and $\mathscr{E}$ respectively. Therefore $e_\Omega^{\Omega'}$ is precisely the imbedding $\mathscr{D}(\Omega) \to \mathscr{D}(\Omega')$, or $\mathscr{E}^*(\Omega) \to \mathscr{E}^*(\Omega')$.

Now let $U$ be a locally finite covering of some region $\Omega$. We shall say that on this covering there is defined a precosheaf $\Psi$ if the spaces $\Psi(\omega)$ and the mappings $e_\omega^{\omega'}$ (satisfying the conditions (i) and (ii)) are defined for regions of the form (16). Let $\Psi$ be a precosheaf on the covering $U$. A chain of order $v$ on $U$ with coefficients in this precosheaf is an arbitrary sum of the form (17) in which the coefficients $\phi_{i_0, \ldots, i_v}$ belong to the spaces $\Psi(U_{i_0, \ldots, i_v})$, are skew-symmetric in their indices, and have only a finite number different from zero. The $v$-th order chains with coefficients in $\Psi$ obviously form a linear space; we denote it by ${}_v\Psi(U)$. We may assign it a topology originating in the fact that ${}_v\Psi(U) \cong \sum\limits_{i_0 < \cdots < i_v} \Psi(U_{i_0, \ldots, i_v})$.

For every $\nu \geqq 0$ we consider the boundary operator $\partial_\nu$: ${}_\nu\Psi(U) \to {}_{\nu-1}\Psi(U)$, defined by the formula

$$\partial_\nu: \quad \phi \to \psi = \sum_{i_1, \ldots, i_\nu} \left(\sum_{i_0} \breve{\phi}_{i_0, \ldots, i_\nu}\right) U_{i_1} \wedge \cdots \wedge U_{i_\nu}, \tag{26.2}$$

where $\breve{\phi}_{i_0, \ldots, i_\nu}$ is the extension of the coefficient $\phi_{i_0, \ldots, i_\nu}$ to the region $U_{i_1, \ldots, i_\nu}$ (i.e. the result of applying the operator $e_{U\ldots}^{U\ldots}$). The operators $\partial_\nu$ are obviously continuous and form a sequence

$$0 \longleftarrow \Psi(\Omega) \xleftarrow{\partial_0} {}_0\Psi(U) \xleftarrow{\partial_1} \cdots \longleftarrow {}_{\nu-1}\Psi(U) \xleftarrow{\partial_\nu} {}_\nu\Psi(U) \longleftarrow \cdots. \tag{27.2}$$

Let us verify that this sequence is semi-exact. The application of the operator $\partial_{\nu-1}\,\partial_\nu$ to the chain $\phi$ results in a chain formed by the coefficients $\sum_{i_0, i_1} \breve{\phi}_{i_0, i_1, \ldots, i_\nu}$. Since the elements $\phi_{i_0, \ldots, i_\nu}$ are skew-symmetric with respect to their indices, every such coefficient vanishes. This implies that the sequence (27) is semi-exact.

Let $\Phi = \{\Phi(\omega)\}$ be some presheaf on the covering $U$, and let $\Psi = \{\Psi(\omega)\}$ be the conjugate precosheaf on the same covering. For arbitrary $\nu \geqq 0$ the spaces ${}^\nu\Phi(U)$ and ${}_\nu\Psi(U)$ admit the duality relation

$$(\psi, \phi) = \sum_{i_0, \ldots, i_\nu} (\psi_{i_0, \ldots, i_\nu}, \phi_{i_0, \ldots, i_\nu}), \quad \psi \in {}_\nu\Psi(U), \quad \phi \in {}^\nu\Phi(U).$$

We show that the operator $\partial_\nu$ in (27) is the conjugate of the operator $\partial^{\nu-1}$ in (19). For arbitrary $\phi \in {}^{\nu-1}\Phi(U)$ and $\psi \in {}_\nu\Psi(U)$ we have

$$\begin{aligned}(\psi, \partial^{\nu-1}\phi) &= \sum_{i_0, \ldots, i_\nu} (\psi_{i_0, \ldots, i_\nu}, \hat{\phi}_{i_1, \ldots, i_\nu}) \\ &= \sum_{i_1, \ldots, i_\nu} \left(\sum_{i_0} \breve{\psi}_{i_0, \ldots, i_\nu}, \phi_{i_1, \ldots, i_\nu}\right) = (\partial_\nu \psi, \phi),\end{aligned}$$

i.e. $\partial_\nu = (\partial^{\nu-1})^*$, q.e.d. Conversely, if $\Psi$ is a precosheaf on the covering $U$, and $\Phi$ is the conjugate presheaf, then for arbitrary $\nu \geqq 0$ we have $\partial^{\nu-1} = (\partial_\nu)^*$.

**9°. The complex connected with a family of supports.** In this subsection we consider a more complicated construction. Let $U$ be a locally finite covering of the region $\Omega$, and let $\Psi$ be a precosheaf defined on the given covering and satisfying the condition: A) for every region $\omega$ of the form (16), $\Psi(\omega)$ is a subspace (in the algebraic sense) of the space $\mathscr{D}^*(\Omega)$ and the support of an arbitrary function $\psi \in \Psi(\omega)$ belongs to $\omega$: if $\omega \subset \omega'$, then $\Psi(\omega)$ is a subspace in $\Psi(\omega')$ and $e_\omega^{\omega'}$ is the restriction of the identity mapping of $\mathscr{D}^*(\Omega)$ imbedding $\Psi(\omega)$ in $\Psi(\omega')$.

A cochain with coefficients in the precosheaf $\Psi$ is an arbitrary sum of the form (17) in which the coefficients $\phi_{i_0, \ldots, i_\nu}$, are skew-symmetric in their indices, and belong to the spaces $\Psi(U_{i_0, \ldots, i_\nu})$. It follows from A) that every coefficient of such a cochain $\phi$ is a distribution with a support belonging to $U_{i_0,\dots,i_\nu}$. The union of the supports of all the functions $\phi_{i_0,\dots,i_\nu}$ is called the support of the cochain $\phi$.

Let $S=\{s_\alpha\}$ be an ensemble of subsets of the region $\Omega$ having the following properties: 1) an arbitrary set $s$ contained in one of the sets $s_\alpha \in S$ itself belongs to $S$, and 2) the union of an arbitrary finite collection of the sets in $S$ belongs to $S$. An ensemble $S$ having these properties is called a family of supports in $\Omega$.

We now fix some family of supports $S$ in $\Omega$. For every integer $\nu \geqq 0$ we denote by ${}^\nu\Psi_S(U)$ the set of all cochains with coefficients in the pre-cosheaf $\Psi$ whose supports belong to $S$. This set is a linear space. Similarly we denote by $\Psi_S(\omega)$ the subspace in $\Psi(\omega)$ consisting of functions whose supports are in $S$.

To define the boundary operator $\partial_\nu\, {}^\nu\Psi_S(U) \to {}^{\nu-1}\Psi_S(U)$ in accordance with formula (26) we must impose another condition on the pre-cosheaf $\Psi$: B) let $\{\omega_i\}$ be a locally finite covering of the region $\omega$ (all these regions are of the form (16)); then the series $\sum \Psi_i$, in which $\psi_i \in \Psi(\omega_i)$, converges in $\Psi(\omega)$ if $\bigcup \operatorname{supp} \psi_i \in S$. The condition B) implies that the inner sum on the right side of (26) converges in $\Psi(U_{i_1,\dots,i_\nu})$ for arbitrary $i_1, \dots, i_\nu$. In fact, let us write $\omega = U_{i_1,\dots,i_\nu}$. Since the regions $\omega_{i_0} = U_{i_0,\dots,i_1}$ form a locally finite covering of $\omega$, and $\phi_{i_0,\dots,i_1} \in \Psi(\omega_{i_0})$ and

$$\bigcup_{i_0} \operatorname{supp} \phi_{i_0,\dots,i_1} \in S,$$

(since $\operatorname{supp} \phi \in S$), the sum $\sum_{i_0} \phi_{i_0,\dots,i_\nu}$ converges in $\Psi(\omega)$. It follows that $\partial_\nu \phi \in {}^{\nu-1}\Psi(U)$, and since obviously $\operatorname{supp} \partial_\nu \phi \subset \operatorname{supp} \phi \in S$, it also follows that $\partial_\nu \phi \in {}^{\nu-1}\Psi_S(U)$. This completes the construction of the operator $\partial_\nu$. These operators form a sequence

$$0 \longleftarrow \Psi_S(\Omega) \xleftarrow{\partial_0} {}^0\Psi_S(U) \xleftarrow{\partial_1} \cdots \longleftarrow {}^{\nu-1}\Psi_S(U) \xleftarrow{\partial_\nu} {}^\nu\Psi_S(U) \longleftarrow \cdots \tag{28.2}$$

**Proposition 10.** *Let $S$ be a family of supports in $\Omega$, and let $\Psi$ be a pre-cosheaf defined on some locally finite covering $U$ of the region $\Omega$, satisfying the conditions* A) *and* B), *and also the following conditions:*

C) *For an arbitrary region $\omega$ there is defined in the space $\Psi(\omega)$ an operation of multiplication by an arbitrary function in $\mathscr{E}(\omega)$.*

D) *For any two arbitrary regions $\omega$ and $\upsilon$ the inclusions $\psi \in \Psi(\omega)$ and* $\operatorname{supp} \psi \subset \omega \cap \upsilon$ *imply that $\psi \in \Psi(\omega \cap \upsilon)$.*

*Under these conditions the sequence* (28) *is algebraically exact.*

It is easy to see that the conditions A) – D) are fulfilled for an arbitrary region $\Omega$, locally finite covering $U$, family of supports $S$ in $\Omega$, and pre-cosheaves $\mathscr{D}$ and $\mathscr{E}^*$.

*Proof of the proposition.* Let $\{\alpha_i\}$ be an infinitely differentiable partition of the unity in $\Omega$ subject to the covering $U$. Further, let $\phi$ be an arbitrary element of the kernel of the mapping $\partial_{\nu-1}$ in (28), where $\nu \geqq 0$ (and $\partial_{-1}=0$). For arbitrary indices $i_0, \ldots, i_\nu$ we consider the product $\alpha_{i_0}\phi_{i_1,\ldots,i_\nu}$ where $\phi_{i_1,\ldots,i_\nu}$ is a coefficient in the cochain $\phi$. By Condition B) this product belongs to $\Psi(U_{i_1,\ldots,i_\nu})$, and its support is contained in the region $U_{i_0,\ldots,i_\nu}$. Therefore Condition D) implies that $\alpha_{i_0}\phi_{i_1,\ldots,i_\nu} \in \Psi(U_{i_0,\ldots,i_\nu})$. It follows that the sum

$$\psi_{i_0,\ldots,i_\nu} = \sum_{j=0}^{\nu} (-1)^j \alpha_{i_j} \phi_{i_0,\ldots,\hat{i}_j,\ldots,i_\nu}$$

also belongs to $\Psi(U_{i_0,\ldots,i_\nu})$. The element $\psi_{i_0,\ldots,i_\nu}$ is skew-symmetric in its indices and therefore we may consider the cochain $\psi = \sum \psi_{i_0,\ldots,i_\nu} U_{i_0} \wedge \cdots \wedge U_{i_\nu} \in {}^\nu\Psi(U)$. It is clear that the support of this cochain belongs to $\operatorname{supp}\phi \in S$ and that it is therefore an element of the family $S$. Hence $\psi \in {}^\nu\Psi_S(U)$.

Let us establish the relation $\partial_\nu \psi = \phi$. We have

$$\begin{aligned}\partial_\nu \psi &= \sum_{i_1,\ldots,i_\nu} \Big(\sum_{i_0} \psi_{i_0,\ldots,i_\nu}\Big) U_{i_1} \wedge \cdots \wedge U_{i_\nu} \\ &= \sum_{i_1,\ldots,i_\nu} \left[\sum_{j=1}^{\nu} (-1)^j \alpha_{i_j} \sum_{i_0} \phi_{i_0,\ldots,\hat{i}_j,\ldots,i_\nu} + \sum_{i_0} \alpha_{i_0} \phi_{i_1,\ldots,i_\nu}\right] U_{i_1} \wedge \cdots \wedge U_{i_\nu}.\end{aligned}$$

The condition $\partial_{\nu-1}\phi = 0$ implies that the first sum over $i_0$ on the right is equal to zero. Thus the right side can be written in the form

$$\sum_{i_1,\ldots,i_\nu} \sum_{i_0} \alpha_{i_0} \phi_{i_1,\ldots,i_\nu} U_{i_1} \wedge \cdots \wedge U_{i_\nu} = \phi,$$

whence $\partial_\nu \psi = \phi$. □

## § 3. Fourier transforms

**1°. Special classes of functional spaces.** We now consider certain classes of spaces of smooth functions and of distributions characterized by restrictions on their growth or by conditions of decreasing at infinity.

The function $I$, defined in $R^n$, is said to be admissible if it is constructed in the following way. In some region $\Omega_I$ it is positive, continuous, and satisfies the inequality $I(\xi) \geqq 1$; in the complement of $\Omega_I$ it is infinite, $I(\xi) = +\infty$. Let $I$ be an admissible function and let $q$ be an arbitrary non-negative integer. In the space $\mathscr{D}(\Omega_I)$ we consider the norm

$$\|\phi\|_I^q = \sum_{|j| \leqq q} \|I(\xi) D^j \phi(\xi)\|_{L_2}. \tag{1.3}$$

The continuity of the function $I$ implies that it is bounded on an arbitrary compact $K \subset \Omega_I$. Therefore the topology defined by the norm (1) is weaker than the topology of the space $\mathscr{D}(\Omega_I)$. We consider the completion $S_I^q$ of the space $\mathscr{D}(\Omega_I)$ in this topology. We assign to the space $S_I^q$ the norm $\|\cdot\|_I^q$, the extension of the norm (1), and obtain a Banach space. The inequality $I(\xi) \geqq 1$ makes the norm $\|\cdot\|_I^q$ stronger than $\|\cdot\|^q$. Therefore the space $S_I^q$ is a subspace of $\mathscr{D}_{\bar{\Omega}_I}^q$.

We note that the norm (1) in $\mathscr{D}(\Omega_I)$ is generated by the scalar product

$$\langle \phi, \psi \rangle_I^q = \sum_{|j| \leqq q} (I D^j \phi, I D^j \psi).$$

Extending this product to the whole of $S_I^q$ we obtain the scalar product generating the norm $\|\cdot\|_I^q$. Thus $S_I^q$ is a Hilbert space.

The function $I^{-1}$ is defined and continuous in $\Omega_I$; we extend it with the value zero in $C\Omega_I$. In the space $\mathscr{D}(R^n)$ we consider the semi-norm

$$\|\phi\|_{I^{-1}}^q = \left| \sum_{|j| \leqq q} \left\| \frac{D^j \phi(\xi)}{I(\xi)} \right\|_{L_2}^2 \right|^{\frac{1}{2}}. \tag{2.3}$$

It is clearly continuous in the topology of $\mathscr{D}(R^n)$. Since $I^{-1}$ is positive in $\Omega_I$, the subspace on which the semi-norm vanishes coincides with $\mathscr{D}_{C\Omega_I}$. Therefore the semi-norm induces a norm on the factor-space $\mathscr{D}(R^n)/\mathscr{D}_{C\Omega_I}$. The completion of the factor-space in the induced norm will be denoted by $\mathscr{E}_I^q$. With the norm resulting from the extension of the semi-norm (2), this space becomes a Hilbert space, since the semi-norm $\|\cdot\|_{I^{-1}}^q$ corresponds to the scalar product $\langle \cdot, \cdot \rangle_{I^{-1}}^q$.

Since the function $I^{-1}$ is bounded from below by a positive number on an arbitrary compact $K \subset \Omega_I$, the norm $\|\cdot\|_{I^{-1}}^q$ majorizes the norm $\|\cdot\|_K^q$. Therefore the elements of $\mathscr{E}_I^q$ on an arbitrary compact $K \subset \Omega_I$ belong to the space $\mathscr{E}_K^q$, i.e. they are distributions in $\Omega_I$ of order not exceeding $-q$.

To every function $\phi$ in the space $\mathscr{D}(\Omega_I)$ there corresponds a continuous functional on the space $\mathscr{E}_I^q$, as defined by the formula $\psi \to (\phi, \psi)$. This mapping defines an imbedding of the space $\mathscr{D}(\Omega_I)$ in the strong conjugate of $\mathscr{E}_I^q$. The norm in this conjugate space can be written as:

$$\|\phi\|_I^{-q} = \sup \left\{ \frac{|(\phi, \psi)|}{\|\psi\|_{I^{-1}}^q}, \ \psi \in \mathscr{D}(R^n), \ \|\psi\|_{I^{-1}}^q \neq 0 \right\}. \tag{3.3}$$

The completion of $\mathscr{D}(\Omega_I)$ in the topology defined by this norm will be denoted by $S_I^{-q}$; we endow $S_I^{-q}$ with the norm resulting from the extension of (3).

With this, $S_I^{-q}$ is a subspace of the strong conjugate of $\mathscr{E}_I^q$. Let us show that in fact $S_I^{-q}$ coincides with $(\mathscr{E}_I^q)^*$. To do this we need only show that $\mathscr{D}(\Omega_I)$ is dense in $(\mathscr{E}_I^q)^*$. Let us suppose the contrary. Then there exists an element $\psi \neq 0$ in the second conjugate of $\mathscr{E}_I^q$ that vanishes on all the elements of $\mathscr{D}(\Omega_I)$. But since $\mathscr{E}_I^q$ is a Hilbert space it is reflexive, and therefore $\psi \in \mathscr{E}_I^q$; then the equation $(\phi, \psi)=0$ for all $\phi \in \mathscr{D}(\Omega_I)$ implies that $\psi=0$. This contradiction proves that $S_I^{-q} \cong (\mathscr{E}_I^q)^*$. Hence, in particular, it follows that the space $S_I^{-q}$ is also a Hilbert space and $\mathscr{E}_I^q = (S_I^{-q})^*$.

Since $I(\xi) \geqq 1$, the topology in $\mathscr{E}_I^q$ is weaker than the topology in $\mathscr{E}_{\bar{\Omega}_I}^q$. Therefore the norm $\|\cdot\|_I^{-q}$ is stronger than the norm $\|\cdot\|^{-q}$, and so $S_I^{-q}$ is a subspace (in the algebraic sense) of the space $\mathscr{D}_{\bar{\Omega}_I}^{-q}$. Therefore it consists of distributions in $R^n$ of order not exceeding $q$ with supports lying in $\bar{\Omega}_I$ (but the converse is not true).

We remark that as a linear topological space $\mathscr{E}_I^0$ coincides with the space of functions defined in $\Omega_I$ and belonging to $L_2$ with weight $I^{-1}$. Therefore the conjugate space $S_I^{-0}$ coincides, as an l.t.s., with the space of functions defined in $\Omega_I$ and belonging to $L_2$ with weight $I$. Thus we have the isomorphism $S_I^{-0} \cong S_I^0$.

We have constructed the Hilbert space $S_I^q$ for arbitrary $q$, and the imbedding $S_I^{q+1} \to S_I^q$ is defined and continuous. We remark that for an arbitrary vector $j \in Z_+^n$ the differential operator $D^j$ acts continuously from $S_I^q$ to $S_I^{q-|j|}$.

Let us again suppose that $q$ is an arbitrary non-negative integer. In the space $\mathscr{D}(R^n)$ we introduce the topology of the strong conjugate of $S_I^q$. This topology is defined by the semi-norm

$$\|\phi\|_I^{-q} = \sup \left\{ \frac{|(\phi, \psi)|}{\|\psi\|_I^q}, \ \psi \in \mathscr{D}(\Omega_I), \ \|\psi\|_I^q \neq 0 \right\}. \tag{4.3}$$

The subspace on which this semi-norm vanishes is $\mathscr{D}_{C\Omega_I}$. We complete the factor-space $\mathscr{D}(R^n)/\mathscr{D}_{C\Omega_I}$ in the norm induced by our semi-norm; we denote the completion by $\mathscr{E}_I^{-q}$, and we assign it the norm extending (4). Arguments like those we have just made lead us to the conclusion that $\mathscr{E}_I^{-q}$ is a Hilbert space and is mutually conjugate to $S_I^q$, and that $\mathscr{E}_I^{-0} \cong \mathscr{E}_I^0$. The elements of $\mathscr{E}_I^{-q}$ are distributions in $\Omega_I$ of order not exceeding $q$.

Thus, for arbitrary integer $q$ we have constructed the Hilbert space $\mathscr{E}_I^q$ mutually conjugate to $S_I^{-q}$. Also for arbitrary $q$ the space $\mathscr{E}_I^{q+1}$ is continuously imbedded in $\mathscr{E}_I^q$, and the operator $D^j$ acts continuously from $\mathscr{E}_I^q$ to $\mathscr{E}_I^{q-|j|}$.

Let $F$ be the closure of an admissible region $\Omega$, and let $I_F$ be a function equal to unity on $\Omega$ and $\infty$ outside $\Omega$. It is clearly an admissible function, and $\Omega_I = \Omega$. We remark that the norm $\|\cdot\|_{I_F}^q$ coincides on $\mathscr{D}(\Omega)$ with

$\|\cdot\|^q$, and the norm $\|\cdot\|^q_{I_F^{-1}}$ is equal to $\|\cdot\|^q_F$. Hence we derive the isomorphisms

$$\mathscr{E}^q_{I_F} \cong \mathscr{E}^q_F, \quad S^q_{I_F} \cong \mathscr{D}^q_F, \quad -\infty < q < \infty.$$

**2°. Families of logarithmically convex functions.** We recall that the function $I$, defined in $R^n$ and taking on finite or infinite values, is said to be logarithmically convex (l.c.) if it is positive and the set in $R^{n+1}$ consisting of the points $(t, \xi)$ at which $t \geqq \ln I(\xi)$, is convex and closed. The logarithmic reciprocal of $I$ is the function

$$\mathscr{I}(y) = \sup_\xi \frac{\exp(y, \xi)}{I(\xi)}.$$

The function $\mathscr{I}$ is always logarithmically convex. If $I$ itself is l.c., it is the logarithmic reciprocal of $\mathscr{I}$.

In our further work we shall most often be considering the spaces $S^q_I$ and $\mathscr{E}^q_I$ for logarithmically convex $I$. We shall suppose that the function $I$ satisfies the inequality

$$I(\xi) \geqq C \exp(\varepsilon |\xi|), \tag{5.3}$$

and that the set on which $I(\xi) < \infty$ has inner points. Then this set, being convex, is the closure of some convex region, which we shall denote by $\Omega_I$. So the function $I$ is admissible. Its logarithmic reciprocal $\mathscr{I}$ is finite in the region $|y| \leqq \varepsilon$, since the function $I$ itself satisfies the inequality (5). Therefore the set on which $\mathscr{I}(y) < \infty$ is also the closure of some nonempty convex region.

**Definition.** A sequence $I_\alpha$, $-\infty < \alpha < \infty$, infinite in both directions, composed of functions defined in $R^n$, is said to be an increasing family of l.c. functions if each of the functions is l.c. and if for every $\alpha$ the inequalities

$$\text{I.} \quad \exp(\varepsilon_\alpha |\xi|)\, I_\alpha(\xi) \leqq C_\alpha I_{\alpha+1}(\xi),$$

$$\text{II.} \quad \sup_{|\xi' - \xi| \leqq \varepsilon_\alpha} I_\alpha(\xi') \leqq C_\alpha I_{\alpha+1}(\xi)$$

hold for some positive $\varepsilon_\alpha$ and $C_\alpha$.

The sequence $I_\alpha$, $-\infty < \alpha < \infty$, is said to be a decreasing family of l.c. functions if the sequence $\{I_{-\alpha}\}$ is an increasing family of l.c. functions; that is, if all the functions $I_\alpha$ are logarithmically convex and for arbitrary $\alpha$

$$\text{I}'. \quad \exp(\varepsilon_\alpha |\xi|)\, I_{\alpha+1}(\xi) \leqq C_\alpha I_\alpha(\xi),$$

$$\text{II}'. \quad \sup_{|\xi' - \xi| \leqq \varepsilon_\alpha} I_{\alpha+1}(\xi') \leqq C_\alpha I_\alpha(\xi)$$

for some positive $\varepsilon_\alpha$ and $C_\alpha$.

**Proposition 1.** *Let $\{I_\alpha\}$ be a decreasing family of l.c. functions. Then the sequence of logarithmic reciprocals $\mathscr{I}_\alpha$ forms an increasing family of l.c. functions satisfying the conditions I and II with the same constants $\varepsilon_\alpha$.*

*Conversely, if $\{I_\alpha\}$ is an increasing family of l.c. functions, the functions $\mathscr{I}_\alpha$ form a decreasing family of l.c. functions satisfying the conditions I′ and II′ with the same constants $\varepsilon_\alpha$.*

*Proof.* Let us prove the first assertion. The inequality $I'$ yields

$$\sup_{|y'-y|\leqq\varepsilon_\alpha} \mathscr{I}_\alpha(y') = \sup_{|y'-y|\leqq\varepsilon_\alpha} \sup_\xi \frac{\exp(y',\xi)}{I_\alpha(\xi)}$$

$$= \sup_\xi \frac{\exp(\varepsilon_\alpha|\xi|)\exp(y,\xi)}{I_\alpha(\xi)} \leqq C_\alpha \sup_\xi \frac{\exp(y,\xi)}{I_{\alpha+1}(\xi)} = C_\alpha \mathscr{I}_{\alpha+1}(y).$$

This proves the inequality II. Making use of II′, we obtain

$$\exp(\varepsilon_\alpha|y|)\mathscr{I}_\alpha(y) = \exp(\varepsilon_\alpha|y|)\sup_\xi \frac{\exp(y,\xi)}{I_\alpha(\xi)}$$

$$= \sup_{|\xi-\xi'|\leqq\varepsilon_\alpha} \frac{\exp(y,\xi')}{I_\alpha(\xi)} \leqq C_\alpha \sup_{\xi'} \frac{\exp(y,\xi')}{I_{\alpha+1}(\xi')} = C_\alpha \mathscr{I}_{\alpha+1}(y).$$

This establishes I. The second assertion is proved in the same way. □

Let $\Omega$ be a non-empty convex bounded region, and let $K$ be its closure. Then the function $I_K$ which is equal to unity on $K$ and $\infty$ outside $K$ is logarithmically convex and admissible, and satisfies (5). Its logarithmic reciprocal will be denoted by $\mathscr{I}_K$. Let $K_\alpha, -\infty<\alpha<\infty$, be a strictly increasing sequence of convex compacts of which none is empty. Since for arbitrary $\alpha$ the compact $K_\alpha$ contains a neighborhood of the non-empty compact $K_{\alpha-1}$ it is the closure of some convex region. It follows that the functions $I_{K_\alpha}$ are admissible and form a decreasing family of l.c. functions, since they satisfy the conditions I and II with

$$\varepsilon_\alpha = \rho(K_\alpha, CK_{\alpha+1}).$$

**3°. Fourier transforms in the spaces $S_I^q$.** We now describe the Fourier transforms of the spaces $S_I^q$ when the function $I$ is logarithmically convex. We recall that the direct and inverse Fourier transforms are defined as follows:

$$F\phi = \tilde{\phi}(x) = \int \exp(i(x,\xi))\,\phi(\xi)\,d\xi,$$
$$F^{-1}\psi = \tilde{\psi}(\xi) = \frac{1}{(2\pi)^n}\int \exp(-i(x,\xi))\,\psi(x)\,dx. \tag{6.3}$$

The operators $F$ and $F^{-1}$ establish an isomorphism between the spaces $L_2(R^n)$ and $L_2(R^n_x)$ if the improper integral (6) is interpreted as its principal value. For arbitrary functions $\phi, \psi \in L_2$ we have Parseval's equation $(2\pi)^n(\phi,\psi)=(F\phi, F\psi)$.

The Fourier transform maps the differential operator $(iD_\xi)^j$ into a multiplication by $x^j$. More precisely, for arbitrary $\phi \in \mathscr{E}^q_{R^n}$, where $q$ is a non-negative integer, we have the relations

$$F[(iD_\xi)^j\phi]=x^j F[\phi], \qquad |j| \leqq q. \tag{7.3}$$

To verify them it is only necessary to remark that on the subspace $\mathscr{D}(R^n)$, which is dense in $\mathscr{E}^q_{R^n}$, they follow from the formula for integration by parts.

Let us suppose that for a given function $\phi$ there exists in $R^n_\xi$ a region $\Omega \subset R^n_y$ containing the origin of coordinates and such that for arbitrary $y \in \Omega$ the function $\exp(-y,\xi)\,\phi(\xi)$ belongs to some bounded set in $L_2$. In this case for arbitrary $y \in \Omega$ the space $L_2$ also contains the function $\exp(-y,\xi)\exp(\rho\,|\xi|)\,\phi(\xi)$, where $\rho=\rho(y, C\Omega)$. Accordingly, for arbitrary $z \in R^n_x \times \Omega$ we may consider the integral

$$\tilde{\phi}(z)=\int \exp\big(i(z,\xi)\big)\,\phi(\xi)\,d\xi,$$

which converges uniformly in the neighborhood of every point of the region $R^n_x \times \Omega$. Since the kernel of this integral is an entire function, it follows that the function $\tilde{\phi}(z)$ is analytic in the region $R^n_x \times \Omega$ and is therefore the analytic extension of the Fourier transform $\tilde{\phi}(x)$ defined on the real axis. We note that if $\phi \in S^0_I$, where $I$ is an admissible l.c. function, then we may set $\Omega=\Omega_\mathscr{I}$ in these arguments, since by (5) the region $\Omega_\mathscr{I}$ contains an $\varepsilon$-neighborhood of the origin of coordinates.

We now consider a space of analytic functions which we shall use to characterize the Fourier transform of the space $S^q_I$.

Let $R(z)$ be a continuous positive function on $C^n$ and let $\mathscr{I}(y)$ be an admissible function in $R^n_y$ (we do not now require that $\mathscr{I}(y) \geqq 1$ be satisfied). We denote by $S^\mathscr{I}_R$ the space of functions analytic in $R^n_x \times \Omega_\mathscr{I}$ for which the norm

$$\|\psi\|^\mathscr{I}_R = \sup \frac{|\psi(z)|}{R(z)\,\mathscr{I}(y)}.$$

is finite. We use this norm to introduce a topology in $S^\mathscr{I}_R$. This space is then obviously complete. When $R(z)=(|z|+1)^{-q}$, where $q$ is some integer, the space $S^\mathscr{I}_R$ and the norm $\|\cdot\|^\mathscr{I}$ will be denoted by $S^\mathscr{I}_q$ and $\|\cdot\|^\mathscr{I}_q$. When $\mathscr{I}=\mathscr{I}_K$, these symbols become, respectively, $S^K_R$, $\|\cdot\|^K_R$, $S^K_q$, and $\|\cdot\|^K_q$.

The symbol $*$ denotes an operation, defined for functions that are given in some region of the space $C^n$, which replaces the function $\phi(z)$

by $\phi^*(z)=\phi(\bar{z})$. This is an anti-linear operation, since it maps $\lambda\phi$ into $\bar{\lambda}\phi^*$. If the function $\phi$ is holomorphic in some region $\Omega$, then $\phi^*$ is analytic in the region $\Omega^*$ which is symmetric to $\Omega$ with respect to the real subspace, and the Taylor coefficients of the functions $\phi$ and $\phi^*$ at complex conjugate points are themselves conjugate to each other. It is clear that for an arbitrary function $\phi$ we have $\phi=\phi^{**}$.

In particular, for arbitrary $\mathscr{I}(y)$ in $R_y^n$ we denote by $\mathscr{I}^*(y)$ the function $\mathscr{I}(-y)$. If $\mathscr{I}=\mathscr{I}_K$, then $\mathscr{I}^*=\mathscr{I}_{K^*}$ where $K^*$ is the compact that is centrally symmetric to $K$. For arbitrary functions $R$ and $\mathscr{I}$ the operation $*$ establishes an anti-isomorphism between the spaces $S_R^{\mathscr{I}}$ and $S_{R^*}^{\mathscr{I}}$.

**Proposition 2.** *Let $\{I_\alpha, -\infty<\alpha<\infty\}$ be a decreasing family of logarithmically convex functions and $\{\mathscr{I}_\alpha\}$ be the family of logarithmic reciprocal functions. For arbitrary integers $\alpha$ and $q$ there exist continuous and mutually inverse operators*

$$F\colon S_{I_\alpha}^q \to S_q^{\mathscr{I}_{\alpha+1}^*}, \qquad F^{-1}\colon S_q^{\mathscr{I}_\alpha^*} \to S_{I_{\alpha+2}}^{q-\nu}, \qquad \nu=\left[\frac{n}{2}\right]+1,$$

*and the operator $F$ on the subspace $\mathscr{D}(\Omega_{I_\alpha})$ coincides with the Fourier transform.*

Since the subspace $\mathscr{D}(\Omega_{I_\alpha})$ is dense in $S_{I_\alpha}^q$, the operator $F$ coincides with the Fourier transform also for an arbitrary function in its domain of definition belonging to $L_2$. Therefore the operator $F^{-1}$ coincides, for functions belonging to $L_2$, with the inverse Fourier transform. By the same reason, the operator $F$ does not depend on $q$ and $\alpha$; that is, the operator $F$, constructed on a given space $S_{I_\alpha}^q$, is the restriction of operators $F$ constructed for any wider space $S_{I_{\alpha'}}^{q'}$. The operator $F^{-1}$ has a similar property.

*Proof of Proposition* 2. Let us fix $\alpha$ and $q$, and construct the operator $F$. We suppose at first that $q\geqq 0$. Let $\phi$ be an arbitrary element of $\mathscr{D}(\Omega_{I_\alpha})$. Since the support of $\phi$ is compact, its Fourier transform can be extended in $C^n$ to an entire function $\tilde{\phi}(z)$. Let us estimate it at an arbitrary point $z=x+iy$, where

$$y\in\Omega_{\mathscr{I}_{\alpha+1}^*}\colon\ |\tilde{\phi}(z)|=|(\phi, \exp(i\,z,\xi))|\leqq \|\exp(i\,z,\xi)\|_{I_\alpha^{-1}}^0\,\|\phi\|_{I_\alpha}^0.$$

We find a bound for the first factor on the right by using the inequality $I'$:

$$\|\exp(i\,z,\xi)\|_{I_\alpha^{-1}}^0\leqq C\sup_\xi \frac{\exp(\varepsilon_\alpha|\xi|)\exp(-y,\xi)}{I_\alpha(\xi)}$$

$$\leqq C'\sup_\xi \frac{\exp(-y,\xi)}{I_{\alpha+1}(\xi)}=C'\,\mathscr{I}_{\alpha+1}(-y).$$

Hence

$$|\tilde{\phi}(z)| \leqq C' \mathscr{I}_{\alpha+1}(-y) \, \|\phi\|_{I_\alpha}^0 .$$

Applying this inequality to the arbitrary function $\phi$ of order $q$ and taking account of (7), we obtain

$$\|\tilde{\phi}\|_q^{\mathscr{I}_\alpha^*+1} \leqq C' \, \|\phi\|_{I_\alpha}^q . \tag{8.3}$$

Now suppose $q<0$. We have

$$|\tilde{\phi}(z)| \leqq \|\exp(i\,z,\xi)\|_{I_\alpha^{-1}}^{-q} \, \|\phi\|_{I_\alpha}^q ,$$

where

$$\|\exp(i\,z,\xi)\|_{I_\alpha^{-1}}^{-q} = \sum_{|j| \leqq -q} |z^j| \, \|\exp(i\,z,\xi)\|_{I_\alpha^{-1}}^0 \leqq C(|z|+1)^{-q} \mathscr{I}_{\alpha+1}(-y),$$

and accordingly

$$|\tilde{\phi}(z)| \leqq C(|z|+1)^{-q} \mathscr{I}_{\alpha+1}(-y) \, \|\phi\|_{I_\alpha}^q ,$$

which again leads to (8).

From (8) it follows that $\tilde{\phi} \in S_q^{\mathscr{I}_\alpha^*+1}$ and the operator $F\colon S_{I_\alpha}^q \to S_q^{\mathscr{I}_\alpha^*+1}$, which is defined momentarily only on the subspace $\mathscr{D}(\Omega_{I_\alpha})$, is continuous. Since this subspace is dense in $S_{I_\alpha}^q$, and the space $S_q^{\mathscr{I}_\alpha^*+1}$ is complete, we may extend the operator $F$ by continuity to the whole space $S_{I_\alpha}^q$. This extension is the desired operator.

We now construct the operator $F^{-1}$. We suppose that $q \leqq v$ and we introduce the notation $r=q-v$ and $\varepsilon=\min(\varepsilon_\alpha, \varepsilon_{\alpha+1})$. For $\phi \in S_q^{\mathscr{I}_\alpha^*}$ we denote by $F^{-1}\phi$ the functional on $\mathscr{D}(R^n)$ defined by the formula

$$(F^{-1}\phi, \psi) = \frac{1}{(2\pi)^n} (\phi, F\psi).$$

Fixing an arbitrary point $\xi_0 \in R^n$, we estimate the value of the functional $F^{-1}$ on the functions $\psi$ whose supports are contained in the $\varepsilon$-neighborhood of the point $\xi_0$. Since the functions $\phi$ and $F\psi$ are analytic for $y \in \Omega_{\mathscr{I}_\alpha^*}$, and as $|x| \to \infty$ $F\psi$ decreases faster than an arbitrary power, the quantity $(\phi, F\psi)$ can be written as an integral of $\phi^* \tilde{\psi}$ over an arbitrary subspace of the form $y=\eta \in \Omega_{\mathscr{I}_\alpha^*}$. Therefore

$$\begin{aligned} (2\pi)^n |(F^{-1}\phi, \psi)| &= |(\phi, F\psi)| = \Big| \int_{y=\eta} \phi^*(z) \tilde{\psi}(z) \, dz \Big| \\ &\leqq \left\| \frac{\phi^*(z)}{(|z|+1)^{-r}} \bigg|_{y=\eta} \right\|^0 \|(|z|+1)^{-r} \tilde{\psi}(z)|_{y=\eta}\|^0 \qquad (9.3) \\ &\leqq C \sup_{y=\eta} \frac{|\phi^*(z)|}{(|z|+1)^{-q}} \|(|z|+1)^{-r} \tilde{\psi}|_{y=\eta}\|^0 . \end{aligned}$$

The second factor on the right side does not exceed $\mathscr{I}_\alpha(\eta)\,\|\phi\|_q^{\mathscr{I}_\alpha^*}$ and the third does not exceed

$$c\sum_{|j|\leqq -r}\|\widetilde{D^j\psi}|_{y=\eta}\|^0.$$

We remark that the value of the function $\widetilde{D^j\psi}$ at the point $x+i\eta$ is the same as the value of the function

$$\overline{D^j\psi\exp(-\eta,\xi)}$$

at the point $x$. Therefore the third factor does not exceed

$$\begin{aligned}c\sum_{|j|\leqq -r}\|\overline{D^j\psi\exp(-\eta,\xi)}\|^0 &= c'\sum_{|j|\leqq -r}\|D^j\psi\exp(-\eta,\xi)\|^0\\ &\leqq c'\sup_{|\xi-\xi_0|\leqq\varepsilon}\exp(-\eta,\xi)\sum_{|j|\leqq -r}\|D^j\psi\|^0\\ &\leqq c'\exp(\varepsilon|\eta|)\exp(-\eta,\xi_0)\,\|\psi\|^{-r}.\end{aligned}$$

Returning to (9), we obtain the inequality

$$|(F^{-1}\phi,\psi)|\leqq c\frac{\mathscr{I}_{\alpha+1}(\eta)}{\exp(\eta,\xi_0)}\,\|\phi\|_q^{\mathscr{I}_\alpha^*}\,\|\psi\|^{-r}. \tag{10.3}$$

It is clear from this inequality that $F^{-1}\phi$ is a continuous functional on $\mathscr{D}(R^n)$. Suppose that $\xi_0\in\complement\,\Omega_{I_{\alpha+1}}$. Then, as we saw in 3° § 1, Chapter III, we can find a sequence of points $\eta$ such that the corresponding values of the second factor on the right tend to zero. Since the remaining terms in the inequality (10) do not depend on $\eta$, we conclude that $(F^{-1}\phi,\psi)=0$ for an arbitrary function $\psi$ with a support contained in the $\varepsilon$-neighborhood of the point $\xi_0$. It follows that the support of the functional $F^{-1}\phi$ is contained in $\Omega_{I_{\alpha+1}}$.

Now suppose that $\xi_0\in\Omega_{I_{\alpha+1}}$. In this case we can find a point $\eta=\eta(\xi_0)$, at which the value of the second factor on the right side of (10) is equal to $I_{\alpha+1}^{-1}(\xi_0)$. By the inequality II′ this quantity does not exceed the lower bound $C\inf I_{\alpha+2}^{-1}(\xi)$ on the $\varepsilon$-neighborhood of $\xi_0$. Hence $I_{\alpha+1}^{-1}(\xi_0)\,\|\psi\|^{-r}\leqq c\,\|\psi\|_{I_{\alpha+2}^{-1}}^{-r}$

$$|(F^{-1}\phi,\psi)|\leqq c\,\|\phi\|_q^{\mathscr{I}_\alpha^*}\,\|\psi\|_{I_{\alpha+2}^{-1}}^{-r}. \tag{11.3}$$

This establishes the fact that the functional $F^{-1}\phi$ is defined and continuous on the subspace of $\mathscr{D}(R^n)$ spanned over the functions whose supports are contained in the $\varepsilon$-neighborhood of some point. This subspace is obviously dense in $\mathscr{D}(R^n)$ and therefore in $\mathscr{E}_{I_{\alpha+2}}^{-r}$. Thus the functional $F^{-1}\phi$ can be extended by continuity to the whole space $\mathscr{E}^{-r}_{I_\alpha+2}$. The space of continuous functionals on $\mathscr{E}^{-r}_{I_\alpha+2}$ will be identified with $S^r_{I_\alpha+2}$, and therefore $F^{-1}\phi\in S^r_{I_\alpha+2}$. We have from (11) that

$$\|F^{-1}\phi\|^r_{I_\alpha+2}\leqq c\,\|\phi\|^{\mathscr{I}^*_\alpha}_q.$$

Thus $F^{-1}$ is the desired operator.

We note that for arbitrary functions $\phi\in S^{\mathscr{I}^*_\alpha}_q$ and vectors $j\in Z^n_+$ we have the equation

$$F^{-1}[z^j\phi]=(i\,D)^j F^{-1}[\phi]. \tag{12.3}$$

This is a consequence of the following computation:

$$\begin{aligned}(2\pi)^n(F^{-1}[z^j\phi],\psi)&=(z^j\phi,\tilde{\psi})=(\phi,x^j\tilde{\psi})\\&=(\phi,\widetilde{(i\,D)^j\psi})=(2\pi)^n((i\,D)^j F^{-1}\phi,\psi).\end{aligned}$$

Let us now look at the case $q>v$. Suppose that $\phi\in S^{\mathscr{I}^*_\alpha}_q$. We note that all the functions $z^j\phi$ with $|j|\leqq r$ belong to $S^{\mathscr{I}^*_\alpha}_v$. Since for the space $S^{\mathscr{I}^*_\alpha}_v$ the desired operator $F^{-1}$ has been constructed and satisfies (12), we have

$$F^{-1}[z^j\phi]=(i\,D)^j F^{-1}[\phi]\in S^0_{I_\alpha+2}$$

for arbitrary $j$, $|j|\leqq r$. Hence $F^{-1}\phi\in S^r_{I_\alpha+2}$ and the mapping

$$S^{\mathscr{I}^*_\alpha}_q\ni\phi\to F^{-1}\phi\in S^r_{I_\alpha+2}$$

is continuous, i.e. the operator $F^{-1}$ is the desired one for $q>v$.

We must show that the operators $F$ and $F^{-1}$ are mutually inverse. We first show that the composition of the operators

$$S^q_{I_\alpha}\xrightarrow{F}S^{\mathscr{I}^*_\alpha+1}_q\xrightarrow{F^{-1}}S^{q-v}_{I_\alpha+3} \tag{13.3}$$

is an identity mapping. By the construction hypothesis, the operator $F$ is a Fourier transform on the subspace $\mathscr{D}(\Omega_{I_\alpha})$ and therefore is defined on $L_2$. It is clear from the construction of the operator $F^{-1}$ that it is an inverse Fourier transform on $L_2$. Thus the composition of operators (12) is an identity operator on the subspace $\mathscr{D}(\Omega_{I_\alpha})$ which is dense in $S^q_{I_\alpha}$. Therefore this composition is an identity operator on the whole subspace $S^q\,I_\alpha$.

We now show that the composition of operators

$$S^{\mathscr{I}^*_\alpha}_q\xrightarrow{F^{-1}}S^{q-v}_{I_\alpha+2}\xrightarrow{F}S^{\mathscr{I}^*_\alpha+3}_{q-v}$$

is also an identity operator. Suppose that $h \in \mathscr{D}(R^n)$ is a function equal to unity at the origin of coordinates. We write

$$h_k(\xi) = h\left(\frac{\xi}{k}\right), \qquad k = 1, 2, \ldots.$$

We choose an arbitrary point $z_0 \in R_x^n \times \Omega_{\mathscr{I}_\alpha^* + 3}$. The sequence of functions $h_k(\xi) \exp(-i\bar{z}_0, \xi)$ belongs to $\mathscr{D}(R^n)$ and tends to $\exp(-i\bar{z}_0, \xi)$ in the topology of the space $\mathscr{E}_{I_{\alpha+2}}^{\nu-q}$, since the function $\exp(-i\bar{z}_0, \xi) I_{\alpha+2}^{-1}(\xi)$ is $O(\exp(-\varepsilon_{\alpha+2}|\xi|))$ as $|\xi| \to \infty$. It follows that for an arbitrary function $\phi \in S_q^{\mathscr{I}_\alpha^*}$ we can carry out the following calculation:

$$\begin{aligned} FF^{-1}\phi|_{z=z_0} &= \overline{(F^{-1}\phi, \exp(i z_0, \xi))} = \overline{(F^{-1}\phi, \exp(-i\bar{z}_0, \xi))} \\ &= \lim_{k\to\infty} \overline{(F^{-1}\phi, h_k(\xi)\exp(-i\bar{z}_0, \xi))} \qquad (14.3) \\ &= \frac{1}{(2\pi)^n} \lim \overline{(\phi, F[k_k \exp(-i z_0, \xi)])} = \frac{1}{(2\pi)^n} \lim \overline{(\phi, \tilde{h}_k(z - \bar{z}_0))}. \end{aligned}$$

Since every function $h_k$ belongs to $\mathscr{D}(R^n)$ its Fourier transform is an entire function which as $|x| \to \infty$ decreases faster than any power. It follows that the scalar product on the right side of (14) can be rewritten as

$$\int \phi(x)\, \tilde{h}_k^*(x - z_0)\, dx = \int \phi(x + z_0)\, \tilde{h}_k^*(x)\, dx. \qquad (15.3)$$

Since $\tilde{h}_k^*(x) = k^n \tilde{h}^*(kx)$, we find that outside of an arbitrary neighborhood of the origin of coordinates the function $\tilde{h}_k^*(x)$ tends to zero in the norm of $L_1$. It follows that the right side of (15) tends to the product of $\phi(z_0)$ by the integral of $\tilde{h}^*$ over $R_x^n$. The inversion formula for the Fourier transform implies that this integral is equal to $(2\pi)^n h(0) = (2\pi)^n$. This proves that the right side is equal to $\phi(z_0)$, □

**4°. Local properties of the functions belonging to the spaces $\mathscr{E}_I^q$**

**Proposition 3.** *Let $I$ be an admissible function and let $q \geqq \nu = [n/2] + 1$ be an integer. Then*

I. *The elements of the space $\mathscr{E}_I^q$ are functions defined in $\Omega_I$, having continuous derivatives up to order $q - \nu$, and satisfying the inequalities*

$$|D^j \phi(\xi)| \leqq C\, I_\varepsilon(\xi)\, \|\phi\|_{I^{-1}}^q, \qquad |j| \leqq q - \nu,$$

*where*

$$I_\varepsilon(\xi) = \sup_{|\xi' - \xi| \leqq \varepsilon} I(\xi'),$$

*and the number $\varepsilon > 0$ is arbitrary.*

II. *The space $S_I^{-q}$ contains all functionals of the form*

$$\delta^j(\xi-\xi_0)\colon\ \phi \to D^j\phi|_{\xi=\xi_0}, \qquad |j| \leqq q-\nu, \qquad \xi_0 \in \Omega_I,$$

*and*

$$\|\delta^j(\xi-\xi_0)\|_I^{-q} \leqq C\, I_\varepsilon(\xi).$$

*Proof.* Let us establish the first assertion. We arbitrarily fix $\varepsilon > 0$ and we choose an infinitely differentiable function $h(\xi)$ which is equal to unity in the neighborhood of zero and has a support lying in the sphere $S = \{\xi\colon |\xi| \leqq \varepsilon\}$. Let $\xi_0$ be an arbitrary point contained, together with its $\varepsilon$-neighborhood, in $\Omega_I$. We consider the operator

$$\mathscr{D}(R^n) \ni \phi(\xi) \to h(\xi)\,\phi(\xi+\xi_0) \in \mathscr{D}_S. \tag{16.3}$$

It is obviously continuous if we assign to $\mathscr{D}(R^n)$ the semi-norm $\|\cdot\|_{I^{-1}}^q$, and to $\mathscr{D}_S$ the norm $\|\cdot\|^q$, and the norm of the operator itself does not exceed $I_\varepsilon(\xi_0)$. If we complete the spaces (16) in the semi-norm $\|\cdot\|_{I^{-1}}^q$ and the norm $\|\cdot\|^q$, respectively, we extend our operator to a continuous operator acting from $\mathscr{E}_I^q$ to $\mathscr{D}_S^q$ with the same norm.

Parseval's equation implies of the function $\psi \in \mathscr{D}_S^q$ that

$$\sum_{|j| \leqq \nu} \|x^j \tilde{\psi}\|_{L_2} = C \sum_{|j| \leqq \nu} \|D^j \psi\|^0 = C\,\|\psi\|^\nu.$$

Since $\nu > n/2$, the finiteness of the left side implies that the function $\tilde{\psi}(x)$ belongs to $L_1$. If we apply the inverse Fourier transform to $\tilde{\psi}$ we arrive at the inequality

$$|\psi(\xi)| \leqq \|\tilde{\psi}\|_{L_1} \leqq C \sum_{|j| \leqq \nu} \|x^j \tilde{\psi}\|_{L_2} = C\,\|\psi\|^\nu.$$

Combining this with the bound for the norm of the operator (16), we finally obtain

$$|D^j \phi(\xi_0)| = |D^j h(\xi)\,\phi(\xi+\xi_0)|_{\xi=0}| \leqq C\,\|h(\xi)\,\phi(\xi+\xi_0)\|^q \leqq C\, I_\varepsilon(\xi_0)\,\|\phi\|_{I^{-1}}^q,$$

where $|j| \leqq q-\nu$.

This establishes the first assertion. The second is an immediate consequence of the first, since $S_I^{-q} = (\mathscr{E}_I^q)^*$. □

*Remark.* Let us suppose that the function I that appears in the hypothesis of this proposition is logarithmically convex. Then by Proposition 2 the Fourier transform operator $F$ is defined on the space $S_I^{-q}$. Let us compute the Fourier transform of the functional $\delta^j(\xi-\xi_0)$. We have

$$\widetilde{\delta^j(\xi-\xi_0)} = \overline{(\overline{\delta^j(\xi-\xi_0)}, \exp(z, i\,\xi))} = \overline{(\delta^j(\xi-\xi_0), \exp(\bar{z}, -i\,\xi))}$$

$$= \overline{D_\xi^j \exp(\bar{z}, -i\,\xi)}|_{\xi=\xi_0} = (i\,z)^j \exp(z, i\,\xi_0).$$

### 5°. Sequences of spaces $S_I^q$ and $\mathscr{E}_I^q$

**Proposition 4.** *Let $I$ and $J$ be two admissible functions in $R^n$ such that $\Omega_I = \Omega_J = R^n$ and $I = o(J)$ as $|\xi| \to \infty$. Then for arbitrary integer $q$ the imbeddings*

$$S_J^{q+1} \to S_I^q, \quad \mathscr{E}_I^{q+1} \to \mathscr{E}_J^q \tag{17.3}$$

*are compact mappings.*

*Proof.* Let us first establish the compactness of the imbedding $S_J^{q+1} \to S_I^q$ for $q \geqq 0$. Since these are Banach spaces it suffices to show that we may select from an arbitrary sequence $\{\phi_\alpha \in S_J^{q+S}\}$, bounded in the norm

$$\|\phi_\alpha\|_J^{q+1} = \sum_{|j| \leqq q+1} \int |J D^j \phi_\alpha|^2 \, d\xi, \tag{18.3}$$

a subsequence that converges in the norm $\|\cdot\|_I^q$. By Proposition 3 § 2, for an arbitrary compact $K$ the imbedding $\mathscr{E}_K^{q+1} \to \mathscr{E}_K^q$ is compact. Since the norm (18) majorizes the norm $\|\cdot\|_K^{q+1}$ we can select from the sequence $\{\phi_\alpha\}$ a subsequence converging in the norm $\|\cdot\|_K^q$. If we choose some increasing sequence of compacts $K$ tending to $R^n$, we may carry out this operation for each of these compacts. Then, selecting the diagonal subsequence, we obtain a subsequence $\{\phi'_\alpha\}$ which converges in any norm $\|\cdot\|_K^q$, where $K$ is compact.

For an arbitrary compact $K$ we may write

$$\|\phi\|_I^q = \sum_{|j| \leqq q} \int_K |I D^j \phi|^2 \, d\xi + \sum_{|j| \leqq q} \int_{CK} |I D^j \phi|^2 \, d\xi. \tag{19.3}$$

Since $I$ is continuous in $R^n$, the first term on the right does not exceed $c\|\phi\|_K^q$. Since the sequence $\{\phi'_\alpha\}$ converges in the semi-norm $\|\cdot\|_K^q$, it converges in the semi-norm defined by the first term in (19). Since the functions $\phi'_\alpha$ are bounded in the norm (18) and $I = o(J)$ as $|\xi| \to \infty$, the second term on the right in (19), with $\phi = \phi'_\alpha$, converges uniformly to zero as $K \to R^n$. It follows that the sequence $\{\phi'_\alpha\}$ converges in the norm (19), q.e.d.

The mapping $\mathscr{E}_I^{q+1} \to \mathscr{E}_J^q$ is also compact for $q \geqq 0$, since we may apply the foregoing arguments to it, with $J$ being replaced by $I^{-1}$ and $I$ by $J^{-1}$.

When $q < 0$ an arbitrary mapping of the type (17) is conjugate to a mapping of the same type for some $q \geqq 0$. The latter is compact, by what we have just proved, and therefore the former is also, since all the spaces (17) are Hilbert spaces. □

Suppose that $I = \{I_\alpha, -\infty < \alpha < \infty\}$ is a decreasing family of logarithmically convex functions. For arbitrary $\alpha$ we have the continuous mappings

$$S_{I_\alpha}^{-\alpha} \to S_{I_{\alpha+1}}^{-\alpha-1}, \quad \mathscr{E}_{I_{\alpha+1}}^{\alpha+1} \to \mathscr{E}_{I_\alpha}^{\alpha}. \tag{20.3}$$

The first is an imbedding, and the second is the composition of the imbedding $\mathscr{E}_{I_{\alpha+1}}^{\alpha+1} \to \mathscr{E}_{I_{\alpha+1}}^{\alpha}$ with the restriction mapping $\mathscr{E}_{I_{\alpha+1}}^{\alpha} \to \mathscr{E}_{I_\alpha}^{\alpha}$ on the subregion $\Omega_{I_\alpha} \subset \Omega_{I_{\alpha+1}}$. The spaces and mappings that appear in (20) form an increasing family $\{S_{I_\alpha}^{-q}\}$ and a decreasing family $\{\mathscr{E}_{I_\alpha}^{\alpha}\}$. We write

$$S_I = \varprojlim_{\alpha\to-\infty} S_{I_\alpha}^{-\alpha}, \qquad \mathscr{E}_I = \varprojlim_{\alpha\to\infty} \mathscr{E}_{I_\alpha}^{\alpha}. \tag{21.3}$$

We shall establish the following topological isomorphisms:

$$S_I^* \cong \varinjlim (S_{I_\alpha}^{-\alpha})^* = \varinjlim \mathscr{E}_{I_\alpha}^{\alpha}, \qquad \mathscr{E}_I^* \cong \varinjlim (\mathscr{E}_{I_\alpha}^{\alpha})^* = \varinjlim S_{I_\alpha}^{-\alpha}.$$

Since they are similar to one another, we need prove only the first. All the spaces $S_{I_\alpha}^{-\alpha}$ are Hilbert spaces. Therefore it follows from Proposition 4, § 1 that the inductive limit $\varinjlim (S_{I_\alpha}^{-\alpha})^*$ is reflexive and mutually conjugate to the space $S_I = \varprojlim S_{I_\alpha}^{-\alpha}$, q.e.d.

The spaces (21) will be used only when for arbitrary $\alpha$ we have $\Omega_{I_\alpha} = R^n$. Then Proposition 3 implies that every element of $\mathscr{E}_I$ is infinitely differentiable in $R^n$, with all of its derivatives of order $O(I_\alpha(\xi))$ for arbitrary $\alpha$. The converse is obvious: every infinitely differentiable function in $R^n$ satisfying these conditions at infinity belongs to $\mathscr{E}_I$. Applying Proposition 3 to the functions $I = I_\alpha^{-1}$, we obtain a similar description of the space $S_I$: its elements are infinitely differentiable in $R^n$ and such that $D^j \phi(\xi) = O(I_\alpha^{-1})$ for arbitrary $j$ and $\alpha$.

We also note the following fact: the spaces (21) are Schwarzian $\mathscr{F}\mathscr{S}$-spaces. This follows from Proposition 5, § 1, since all the spaces $S_{I_\alpha}^{-\alpha}, \mathscr{E}_{I_\alpha}^{\alpha}$, being Hilbert spaces, are $\mathscr{F}$-spaces, $\mathscr{D}(R^n)$ is dense in them, and all the mappings (20) are compact by Proposition 4, since $I_{\alpha+1} = o(I_\alpha)$ because of the inequality $I'$.

**6°. The spaces $S_I^b$.** We denote by $b$ an arbitrary non-decreasing function in $R^1$, defined and finite for $\eta \geqq 0$, and equal to unity at zero. Suppose further that $I$ is some admissible function in $R^n$. In the space $\bigcap_q S_I^q$ we consider the subspace $S_I^b$ consisting of functions for which the norm

$$\|\phi\|_I^b = \sup_j \frac{1}{b(|j|)} \|D^j \phi\|_I^0.$$

is finite. It is obvious that with this norm the space is complete, i.e. is a Banach space.

If $F$ is the closure of some region, the space $S_{I_F}^b$ coincides as an l.t.s. with $\mathscr{D}_F^b$.

We shall suppose throughout that the function $b$ satisfies an inequality of the type (14.2). Then the space $S_{I_F}^b$ is not empty, since it contains the non-empty spaces $\mathscr{D}_K^b$ (see Proposition 8, § 2).

**Proposition 5.** *Let $\{I_\alpha\}$ be a decreasing family of logarithmically convex functions in $R^n$, and let $\{\mathcal{I}_\alpha\}$ be the family of their logarithmic reciprocals. Further let $\{b_\alpha\}$ be an increasing family of logarithmically convex functions in $R^1$, for which the inequalities* I *and* II *are satisfied with $\varepsilon_\alpha = 1$, and let each of the functions $b_\alpha$ satisfy the conditions imposed in this subsection. We denote by $B_\alpha$ the logarithmic reciprocal of $b_\alpha$ (we write $b_\alpha(\eta) = \infty$ for $\eta < 0$), and we set*

$$R_\alpha(z) = \frac{1}{B_\alpha(\ln|z|)}, \qquad -\infty < \alpha < \infty.$$

*Then the direct and inverse Fourier transforms define the continuous operators*

$$F\colon\ S^{b_\alpha}_{I_\alpha} \to S^{\mathcal{I}^*_{\alpha+1}}_{R_{\alpha+v}}, \qquad F^{-1}\colon\ S^{\mathcal{I}^*_\alpha}_{R_\alpha} \to S^{b_{\alpha+v}}_{I_{\alpha+2}}.$$

We first establish the fact that we may legitimately consider these spaces. We need to show that every function $R_\alpha(z)$ is defined and continuous in $C^n$. Since by hypothesis all the $b_\alpha$ satisfy an inequality of the type of (14.2), they increase as $\eta \to \infty$ faster than any function of the form $\exp(a_\eta)$. It follows that the function $B_\alpha(t)$ is defined and continuous on all the axes. We have $B_\alpha(t) = 1$ for $t < 0$, since $b_\alpha(\eta) = \infty$ for $\eta < 0$, $b(0) = 1$ and $b(\eta) \geqq 1$ for $\eta > 0$. Therefore the function $R_\alpha(z)$ is defined and continuous in $C^n$ since $R_\alpha(z) \equiv 1$ in the sphere $|z| \leqq 1$.

*We now take up the proof.* Since $S^b_I$ is a subspace in any of the spaces $S^q_I$, we know that for an arbitrary function $\phi \in S^b_I$ every one of the derivatives $D^j \phi$ also belongs to all the $S^q_I$. Therefore the continuity of the operator $F$ in Proposition 2 implies that

$$\frac{|z^j \tilde{\phi}(z)|}{\mathcal{I}_{\alpha-1}(-y)} \leqq \|z^j \tilde{\phi}\|_0^{\mathcal{I}^*_\alpha+1} \leqq C\,\|\phi\|^{|j|}_{I_\alpha} \leqq C\, b_\alpha(|j|)\, \|\phi\|^{b_\alpha}_{I_\alpha}.$$

We divide both sides of the inequality by $b_\alpha(|j|)$ and take the upper bound with respect to $j$:

$$\sup_j \frac{|z^j|}{b_\alpha(|j|)} \frac{|\tilde{\phi}(z)|}{\mathcal{I}_{\alpha+1}(-y)} \leqq C\,\|\phi\|^{b_\alpha}_{I_\alpha}. \tag{22.3}$$

Let us approximate the upper bound on the left, from below. The inequality $\sqrt{n}\max_i |z_i| \geqq |z|$ yields

$$n^{k/2} \max_{|j| \leqq k} |z^j| \geqq |z|^k, \qquad k = 0, 1, 2, \ldots.$$

Hence

$$\sup_j \frac{|z^j|}{b_\alpha(|j|)} \geqq \sup_{k\leqq 0} \frac{|z|^k}{n^{k/2}\, b_\alpha(k)} \geqq \sup_{k\geqq 0} \frac{|z|^k}{\exp\left(\frac{\ln n}{2} k\right) b_\alpha(k)}$$
$$\geqq c \sup_{k\geqq 0} \frac{|z|^k}{b_{\alpha+\nu-1}(k)}, \tag{23.3}$$

since the inequality II, with $\varepsilon_\alpha \equiv 1$, applied $\nu-1$ times, implies that

$$\exp\left(\frac{\ln n}{2} k\right) b_\alpha(k) \leqq \exp\left(\left[\frac{n}{2}\right] k\right) b_\alpha(k) \leqq C\, b_{\alpha+\nu-1}(k).$$

We observe that for arbitrary $z$ and $\eta$, where $k \leqq \eta \leqq k+1$, we have

$$\max(|z|^{k+1}, |z|^k) \geqq |z|^\eta, \qquad b_{\alpha+\nu-1}(k) \leqq b_{\alpha+\nu-1}(k+1) \leqq C\, b_{\alpha+\nu}(\eta).$$

Hence

$$c \max\left\{\frac{|z|^k}{b_{\alpha+\nu-1}(k)}, \frac{|z|^{k+1}}{b_{\alpha+\nu-1}(k+1)}\right\} \geqq \frac{|z|^\eta}{b_{\alpha+\nu}(\eta)}.$$

Therefore the right side of (23) is not less than

$$c \sup_{\eta\geqq 0} \frac{|z|^\eta}{b_{\alpha+\nu}(\eta)} = c \sup_{\eta\geqq 0} \frac{\exp(\ln|z|\,\eta)}{b_{\alpha+\nu}(\eta)} = B_{\alpha+\nu}(\ln|z|) = \frac{1}{R_{\alpha+\nu}(z)}.$$

Using this in (22), we obtain

$$\|\tilde{\phi}\|_{R_{\alpha+\nu}}^{\mathscr{I}_\alpha^*+1} \leqq C\, \|\phi\|_{I_\alpha}^{b_\alpha}.$$

We have now proved that the Fourier transform is a continuous mapping from $S_{I_\alpha}^{b_\alpha}$ to $S_{R_{\alpha+\nu}}^{\mathscr{I}_\alpha^*+1}$; that is, we have constructed the operator $F$.

Let us now construct $F^{-1}$. Since the space $S_{R_\alpha}^{\mathscr{I}_\alpha^*}$ is contained in $S_q^{\mathscr{I}_\alpha^*}$ for arbitrary $q$, we may apply the operator $F^{-1}$ constructed in Proposition II to an arbitrary function of the form $z^j\psi$ where $\psi \in S_{R_\alpha}^{\mathscr{I}_\alpha^*}$. Since the operator is continuous, we arrive at the inequality

$$\|D^j\tilde{\psi}\|_{I_{\alpha+2}}^0 \leqq C\, \|z^j\psi\|_\nu^{\mathscr{I}_\alpha^*} \leqq C \sup R_\alpha(z)(|z|+1)^\nu\, |z|^{|j|}\, \|\psi\|_{R_\alpha}^{\mathscr{I}_\alpha^*}, \tag{24.3}$$

where $\tilde{\psi} = F^{-1}(\psi)$. We now estimate, from above, the upper bound on the right. Using the fact that $R_\alpha(z) \equiv 1$ for $|z| \leqq 1$, we have

$$\sup R_\alpha(z)(|z|+1)^\nu\, |z|^{|j|} = \sup_{|z|\geqq 1} R_\alpha(z)(|z|+1)^\nu\, |z|^{|j|} \leqq C \sup R_\alpha(z)\, |z|^{\nu+|j|}$$
$$= C \sup \frac{\exp(\ln|z|\,(\nu+|j|))}{B_\alpha(\ln|z|)}$$
$$= C\, b_\alpha(\nu+|j|) \leqq C'\, b_{\alpha+\nu}(|j|).$$

Substituting this in (24), we arrive at the inequality

$$\|\tilde{\psi}\|_{I_\alpha+2}^{b_\alpha+\nu} \leqq C\,\|\psi\|_{R_\alpha}^{\mathscr{I}_\alpha^*},$$

which proves the continuity of the operator $F^{-1}$ as a mapping from $S_{R_\alpha}^{\mathscr{I}_\alpha^*}$ to $S_{I_\alpha+2}^{b_\alpha+\nu}$. □

*Remark.* We observe that under the hypotheses of the proposition we have just proved, the differential operator $D^j$ defines a continuous mapping $D^j\colon S_{I_\alpha}^{b_\alpha} \to S_{I_\alpha}^{b_\alpha+|j|}$. In fact, using the inequality II $|j|$ times with $\varepsilon_\alpha = 1$, we obtain

$$\|D^{i+j}\phi\|_{I_\alpha}^0 \leqq b_\alpha(|i|+|j|)\,\|\phi\|_{I_\alpha}^{b_\alpha} \leqq C\,b_{\alpha+|j|}(|i|)\,\|\phi\|_{I_\alpha}^{b_\alpha}$$

for an arbitrary function $\phi \in S_{I_\alpha}^{b_\alpha}$. Hence

$$\|D^j\phi\|_{I_\alpha}^{b_\alpha+|j|} \leqq C\,\|\phi\|_{I_\alpha}^{b_\alpha},$$

which is what we were to prove.

**7°. Connection with the families $\mathscr{H}_{\mathscr{M}}$**

**Proposition 6.** *Let $\{I_\alpha\}$ $\{\mathscr{I}_\alpha\}$, $\{b_\alpha\}$ and $\{R_\alpha\}$ be sequences of functions satisfying the conditions of Proposition 5. Then any of the sequences $\{S_{-\alpha}^{\mathscr{I}_\alpha}\}$, $\{S_{R_\alpha}^{\mathscr{I}_\alpha^*}\}$ is a complete family $\mathscr{H}_{\mathscr{M}}$ corresponding to the family of majorants $\mathscr{M}$ of type $\mathscr{I}$.*

*Proof.* According to Proposition I, the functions $\mathscr{I}_\alpha$ form an increasing family of logarithmically convex functions. It follows that these functions satisfy the conditions a) and b) of § 1, Chapter III, and that they also satisfy the inequalities

$$\mathscr{I}_\alpha(y)\exp(\varepsilon_\alpha|y|) \leqq C_\alpha\,\mathscr{I}_{\alpha+1}(y) \tag{25.3}$$

for some positive $\varepsilon_\alpha$ and $C_\alpha$. The functions $\mathscr{R}_\alpha(z) = (|z|+1)^\alpha$, $-\infty < \alpha < \infty$, form a complete family of majorants in $C^n$ in the sense of 3° § 5, Chapter IV. Therefore the sequence of functions

$$\mathscr{M} = \{M_\alpha(z) = \mathscr{R}_\alpha(z)\,\mathscr{I}_\alpha(y),\ -\infty < \alpha < \infty\} \tag{26.3}$$

also forms a complete family of majorants in $C^n$. It is clear that the corresponding complete family of spaces $\mathscr{H}_{\mathscr{M}}$ coincides with the sequence $\{S_{-\alpha}^{\mathscr{I}_\alpha}\}$.

To prove that $\mathscr{M}$ is a family of type $\mathscr{I}$ we must construct the functions $e_\alpha$. By definition, $e_\alpha(z)$ is an entire function for every $\alpha$, is not identically zero, is non-negative, is square summable in $R_x^n$, and satisfies the inequality

$$|e_\alpha(z)| \leqq r_\alpha(z)\,i_\alpha(y), \tag{27.3}$$

where the functions $r_\alpha$ and $i_\alpha$ are such that

$$\mathscr{I}_\alpha(y)\, i_\alpha(y) \leqq c_\alpha \mathscr{I}_{\alpha+1}(y) \tag{28.3}$$

and

$$r_\alpha(\pm(z-\lambda))\, \mathscr{R}_\alpha(\lambda) \leqq c_\alpha \mathscr{R}_{\alpha+1}(z), \qquad z, \lambda \in C^n. \tag{29.3}$$

Let $h$ be an infinitely differentiable function in $R^n$ whose support is contained in the sphere $|\xi| < \varepsilon_\alpha/2$ and which satisfies $\int h(\xi)\, d\xi = 1$. Proposition 2 implies that its Fourier transform is an entire function satisfying the inequality

$$|\tilde{h}(z)| \leqq C(|z|+1)^{-q} \exp\left(\frac{\varepsilon_\alpha}{2}|y|\right), \qquad q = 1, 2, \ldots. \tag{30.3}$$

We have $\tilde{h}(0) = \int h(\xi)\, d\xi = 1$, and therefore the function is not identically zero. We now write $e(z) = \tilde{h}(z)\, \tilde{h}^*(z)$. The inequality (30) implies (27) with $i_\alpha(y) = \exp(\varepsilon_\alpha |y|)$, and we have $r_\alpha(z) = O(|z|^{-q})$ as $|z| \to \infty$ for arbitrary $q$. Then (28) follows from (25).

We must prove (29). The obvious inequality

$$(|\lambda|+1)(|z-\lambda|+1) \geqq |z|+1$$

implies that

$$\frac{1}{|z-\lambda|+1} \leqq \frac{|\lambda|+1}{|z|+1}$$

for arbitrary $z$ and $\lambda$. Hence for arbitrary integer $\alpha$ we have

$$r_\alpha(\pm(z-\lambda)) \leqq \frac{1}{(|z-\lambda|+1)^{|\alpha|}} \leqq \left(\frac{|\lambda|+1}{|z|+1}\right)^{-\alpha} \leqq \frac{\mathscr{R}_{\alpha+1}(z)}{\mathscr{R}_\alpha(\lambda)},$$

which is what we were to prove.

Let us now turn to the sequence of spaces $S_{R_\alpha}^{\mathscr{I}_\alpha}$. The sequence is a complete family $\mathscr{H}_{\mathscr{M}}$ where $\mathscr{M}$ is a family of majorants of the form (26), and $\mathscr{R}_\alpha(z) = R_\alpha(z)$. We shall show that the functions $R_\alpha(z)$ form a complete family of majorants (whence it will follow that the functions (26) also form a complete family of majorants). Since the functions $b_\alpha$ satisfy the inequality II with $\varepsilon_\alpha = 1$, Proposition 1 implies that the functions $B_\alpha$ satisfy the inequality I with $\varepsilon_\alpha = 1$. Hence follows the estimate

$$(|z|+1)\, R_\alpha(z) = |z|\, R_\alpha(z) + R_\alpha(z) \leqq C R_{\alpha+1}(z) + R_\alpha(z) \leqq C' R_{\alpha+1}(z).$$

This proves that Condition a), § 1, Chapter III is satisfied. Condition b) follows from the stronger inequality

$$R_\alpha(z-\lambda)\, R_\alpha(\lambda) \leqq C R_{\alpha+1}(z), \tag{31.3}$$

which we shall now prove.

By definition

$$\frac{1}{R_\alpha(z)} = B_\alpha(\ln|z|) = \sup_{\eta \geqq 0} \frac{\exp(\ln|z|\,\eta)}{b_\alpha(\eta)} = \sup_{\eta \geqq 0} \frac{|z|^\eta}{b_\alpha(\eta)},$$

and we may therefore rewrite (31) in the form

$$\sup \frac{|\lambda|^\eta}{b_\alpha(\eta)} \cdot \sup \frac{|z-\lambda|^\eta}{b_\alpha(\eta)} \geqq C \sup \frac{|z|^\eta}{b_{\alpha+1}(\eta)}. \tag{32.3}$$

For arbitrary points $z, \lambda \in C^n$ at least one of the quantities $|z-\lambda|$ and $|\lambda|$ is not less than $\frac{1}{2}|z|$. Suppose for example that $|\lambda| \geqq \frac{1}{2}|z|$. Then the first factor on the left side of (32) is not less than

$$\sup \frac{|z|^\eta}{2^\eta b_\alpha(\eta)}. \tag{33.3}$$

Since the functions $b_\alpha$ satisfy Condition I with $\varepsilon_\alpha = 1$, we have

$$2^\eta b_\alpha(\eta) \leqq \exp(\eta)\, b_\alpha(\eta) \leqq C\, b_{\alpha+1}(\eta),$$

and therefore the quantity (33) is not less than the right side of (32). The second factor on the left in (32) is not less than the value of the function

$$\frac{|z-\lambda|^\eta}{b_\alpha(\eta)} \qquad \text{for } \eta = 0,$$

i.e. is not less than unity. This proves (32) and with it (31) as well.

Let us now construct the function $e_\alpha$. Let $h$ be an arbitrary function belonging to the space $D_S^{b_\alpha - \nu}$, where $S$ is the sphere $|\xi| \leqq \varepsilon_\alpha/3$, and let $h$ satisfy the condition $\int h(\xi)\, d\xi = 1$. By Proposition 5 its Fourier transform is an entire function satisfying the inequality

$$|\tilde{h}(z)| \leqq C R_\alpha(z) \exp\left(\frac{\varepsilon_\alpha}{2}|y|\right).$$

The entire function $e = h\, h^*$ is equal to unity at the origin of coordinates and satisfies the inequality (27) with $r_\alpha = C R_\alpha$, and $i_\alpha = \exp(\varepsilon_\alpha |y|)$. The inequality (28) follows from (25), and (29) follows from (31). □

*Remark.* For an arbitrary function $b(\eta)$ defined and finite for $\eta \geqq 0$, we can construct a sequence $\{b_\alpha\}$ satisfying the conditions of Proposition 5 and such that

$$b(\eta) \leqq c\, b_\alpha(\eta), \qquad -\infty < \alpha < \infty. \tag{34.3}$$

To do this, we proceed as follows: We choose some logarithmically convex function $b'(\eta)$, non-decreasing for $\eta \geqq 0$, satisfying an inequality of the form (14.2), equal to unity at zero, and such that $b(\eta) \leqq c\, b'(\eta)$. We further construct some strictly increasing sequence of numbers $a_\alpha$, $-\infty < \alpha < \infty$, all not less than unity, and we write $b_\alpha(\eta) = b'(a_\alpha \eta)$. The inequality (34) is obviously fulfilled.

Let us verify that the sequence of functions $\{b_\alpha\}$ satisfies the conditions of Proposition 5. From their construction we see at once that none of them decrease, that they are all equal to unity at zero, that they are logarithmically convex, and satisfy an inequality of the form (14.2). The function $b'$, being logarithmically convex, satisfies the inequality

$$b'(\eta)\, b'(\xi) \leqq b'(\eta + \xi)$$

for arbitrary points $\eta > 0$ and $\xi > 0$. Using this inequality, and (14.2) with $b = b'$, we obtain

$$b_\alpha(\eta) \exp(\eta) \leqq C\, b'(a_\alpha \eta)\, b'((a_{\alpha+1} - a_\alpha)\eta) \leqq C\, b'(a_{\alpha+1} \eta) = C\, b_{\alpha+1}(\eta)$$

and

$$\sup_{|\eta' - \eta| \leqq 1} b_\alpha(\eta') = b'(a_\alpha(\eta + 1)) \leqq b'(a_\alpha)\, b'(a_\alpha \eta) = C\, b_\alpha(\eta) \leqq C'\, b_{\alpha+1}(\eta).$$

This completes our argument.

**8°. Examples.** Let us look at two examples of logarithmically convex functions and let us compute their logarithmic reciprocals.

**Example 1.** The function $I(\xi) = \exp(A\,|\xi|^{1/a})$ for arbitrary $a$, $0 < a \leqq 1$, and for $A > 0$, is logarithmically convex, since the function $A\,|\xi|^{1/a}$ is convex. Let us find its logarithmic reciprocal

$$\mathscr{I}(y) = \sup_\xi \frac{\exp(y, \xi)}{\exp(A\,|\xi|^{1/a})}. \tag{35.3}$$

Suppose first that $a < 1$. Since the function $I$ is spherically symmetric, we may suppose that $n = 1$, and we consider the upper bound (35) only for $y > 0$ and $\xi > 0$. To find the upper bound, we take the logarithm of the fraction in (35) and differentiate it with respect to $\xi$. Equating the derivative to zero we find that the upper bound is attained when

$$y - \frac{1}{a} A\, \xi^{\frac{1}{a} - 1} = 0, \quad \text{i.e.} \quad \xi = \left(\frac{a\,y}{A}\right)^{\frac{a}{1-a}},$$

and is therefore equal to

$$\mathscr{I}(y) = \exp\left(A'\,|y|^{\frac{a}{1-a}}\right), \quad \text{where} \quad A' = \left(\frac{1}{A}\right)^{\frac{a}{1-a}} \left(a^{\frac{a}{1-a}} - a^{\frac{1}{1-a}}\right).$$

When $a=1$ we find immediately from (35) that

$$\mathscr{I}(y)=\begin{cases}1, & |y|\leqq A\\ \infty, & |y|>A.\end{cases}$$

**Example 2.** We consider the function $b(\eta)=B^\eta \eta^{\eta\beta}$ for $\eta\geqq 0$, where $\beta$ and $B$ are positive and $b(\eta)=\infty$ for $\eta<0$. It is logarithmically convex, since the function $\ln b(\eta)=\eta(\ln B+\beta\ln\eta)$ is convex for $\eta\geqq 0$. Let us calculate the logarithmic reciprocal

$$B(t)=\sup_{\eta\geqq 0}\frac{\exp(\eta\, t)}{B^\eta\,\eta^{\eta\beta}}.$$

Taking the logarithm and differentiating with respect to $\eta$, we find that the upper bound is attained when $t=\ln B+\beta\ln\eta+\beta$ and is therefore equal to

$$B(t)=\exp[\eta(\ln B+\beta\ln\eta+\beta)-\eta(\ln B+\beta\ln\eta)]=\exp\left(\frac{\beta}{e}\exp\left(\frac{t-\ln B}{\beta}\right)\right)$$

for $t\geqq 0$ and has the value $B(t)=1$ for $t<0$. The corresponding function $R(z)$ (cf. Proposition 5) is equal to

$$R(z)=\frac{1}{B(\ln|z|)}=\exp\left(-\frac{\beta}{e}\left(\frac{|z|}{B}\right)^{1/\beta}\right).$$

We remark that since the function $b(\eta)$ is logarithmically convex, the logarithmic reciprocal of $B(t)$ coincides with $b(\eta)$. Hence follows an equation that we shall need later:

$$\sup_{\tau\geqq 0}\tau^\eta\exp\left(-\frac{\beta}{e}\left(\frac{\tau}{B}\right)^{1/\beta}\right)\underset{(t=\ln\tau)}{=}\sup\frac{\exp(t\,\eta)}{B(t)}=b(\eta)=B^\eta\,\eta^{\eta\beta}.$$

**9°. The structure of a $\mathscr{P}$-module in the "base" space and in the "conjugate" space.** Let $\mathscr{P}$ be the ring of all polynomials in $n$ variables with complex coefficients. To each element $p$ of this ring we correlate an operator in each of the spaces contemplated in §§ 2 and 3 (generally speaking, not confined to the space in question). This operator is defined individually for each of the several types of space. In the "base" spaces $\mathscr{D}_K^q$, $\mathscr{D}_K$, $\mathscr{D}(\Omega)$, $\mathscr{E}^*(\Omega)$, $S_I^q$, $S_I^b$ the operator $p$ acts as a differential operator $p(i\,D)$ with constant coefficients, where

$$i\,D=\left(i\frac{\partial}{\partial\xi_1},\ldots,i\frac{\partial}{\partial\xi_n}\right).$$

In the spaces $S_q^{\mathscr{I}}$, $S_R^{\mathscr{I}}$, which arise from the application of the Fourier transform to the "base" spaces, the action of $p$ is a multiplication by the polynomial $p(z)$. We note that this operation commutes with the Fourier transform. In fact, the operator $p(iD)$ becomes the operator $p(z)$ after Fourier transformation.

Corresponding to every polynomial $p \in \mathscr{P}$ we denote by $p^*$ the polynomial obtained from $p$ by replacing each coefficient with its complex conjugate. This notion agrees with the operation $*$ introduced in 3°. The operator $p$ in the "conjugate" spaces $\mathscr{E}_K^q$, $\mathscr{E}_K$, $\mathscr{E}(\Omega)$, $\mathscr{D}^*(\Omega)$, $\mathscr{E}_I^q$ is defined as the conjugate of the operator $p^*$ in the fundamental spaces, i.e. it is defined by

$$(p\,f, \phi) = (f, p^*\,\phi).$$

We remark that this definition agrees with the mapping $\mathscr{E}(\Omega) \to \mathscr{D}^*(\Omega)$, which we have defined by the formula

$$f \to (f, \phi) = \int \bar{f}\,\phi\,d\xi, \qquad \phi \in \mathscr{D}(\Omega),$$

that is, the operator $p(iD)$ acts on a function $f \in \mathscr{E}(\Omega)$ just as the operator $p$ acts on the corresponding functional $f \in \mathscr{D}^*(\Omega)$. This agreement of the operators follows from the computation

$$(p\,f, \phi) = \int \overline{p(iD)\,f}\,\phi\,d\xi = \int \bar{f}\,\overline{p(-iD)}\,\phi\,d\xi = \int \bar{f}\,p^*(iD)\,\phi\,d\xi = (f, p^*\,\phi).$$

We note also that the operator $p$ acts in the spaces $\mathscr{D}(\Omega)$, $\mathscr{E}(\Omega)$, $\mathscr{D}_K$, $\mathscr{E}_K$ and their conjugates without going outside of these spaces, and is continuous. This converts the spaces into topological $\mathscr{P}$-modules.

Let $\Phi$ be one of the functional spaces considered in §§ 2 and 3, and let $k$ be a natural number. We denote by $\Phi^k$ the direct sum of $k$ copies of the spaces, and we interpret the elements of $\Phi^k$ as columns of height $k$ made up of functions belonging to $\Phi$. In accordance with the definition given in § 3, Chapter I, a $\mathscr{P}$-matrix of size $t \times s$ is an arbitrary rectangular matrix with $t$ rows and $s$ columns, consisting of elements of the ring $\mathscr{P}$. When the space $\Phi$ is a $\mathscr{P}$-module, every such $\mathscr{P}$-matrix can be correlated with an operator $p: \Phi^s \to \Phi^t$, which acts by ordinary matrix multiplication, and the elements of $p$ and the elements of the column $\phi \in \Phi^s$ are multiplied by the rule given above.

Corresponding to the $\mathscr{P}$-matrix $p$ we denote by $p^*$ the matrix obtained from the transpose $p'$ by replacing each element $p_{ij}$ by $p_{ij}^*$. If $p$: $[\Phi^*]^s \to [\Phi^*]^t$ is the operator corresponding to the matrix $p$ and acting in some one of the "conjugate spaces," the conjugate operator has the form $p^*: \Phi^t \to \Phi^s$.

Chapter VI

# Homogeneous Systems of Equations

In § 4, we obtain an exponential representation for the solution of homogeneous systems of equations, of general form, as discussed in the introduction. This representation is the foundation of the greater part of the results in this and succeeding chapters. In § 5, we apply the exponential representation to the study of hypoelliptic and partially hypoelliptic operators. In § 6 we establish for the partially hypoelliptic operators a uniqueness theorem for the solution of the generalized Cauchy problem.

## § 4. The exponential representation of solutions of homogeneous systems of equations

The ring $\mathscr{P}$ will now be interpreted as the ring of all polynomials in a vector $iD$, consisting of $n$ differential operators

$$i\frac{\partial}{\partial \xi_1}, \ldots, i\frac{\partial}{\partial \xi_n}, \qquad i=\sqrt{-1},$$

which act in the Euclidean space $R^n$. We fix an arbitrary $\mathscr{P}$-matrix $p$; let its size be $t \times s$. We consider the corresponding system of differential equations with constant coefficients

$$p(iD)\,u=0, \tag{1.4}$$ [1]

consisting of $t$ equations in the unknown vector function $u$ with $s$ components. In the context of Eq. (1), we shall interpret $u$ as a column of height $s$.

Let $\Phi$ be one of the spaces introduced in §§ 2 and 3, consisting of functions or distributions, defined on some closed set or in a region of the space $R^n$. We shall assume that this space is a $\mathscr{P}$-module. Then the matrix $p(iD)$ defines a $\mathscr{P}$-mapping $p\colon \Phi^s \to \Phi^t$. The kernel of this mapping, endowed with the topology induced by $\Phi^s$, will be denoted by $\Phi_p$.

If the space $\Phi$ is not a $\mathscr{P}$-module, but there is a wider space $\Psi$, such that the operator $p(iD)$ acts from $\Phi^s$ to $\Psi^t$, then we shall say that the function $u \in \Phi^s$ is a solution of the system (1) in $\Psi$, if it belongs to the kernel of this operator, which we shall also denote by $\Phi_p$.

---

1 We denote an arbitrary operator with constant coefficients by $p(iD)$, and not by $p(D)$ purely for convenience in the application of the Fourier transform.

**1°. The representation in the spaces of smooth functions and of distributions**

**Theorem 1.** *Let* $\{N^\lambda, d^\lambda(z, D), \lambda=0, \ldots, l\}$ *be a collection of varieties and normal Noetherian operators, associated with the matrix* $p'$. *Further, let* $\{I_\alpha, -\infty<\alpha<\infty\}$ *be some decreasing family of logarithmically convex functions, and let* $\{I_\alpha\}$ *be a family of logarithmically reciprocal functions. Then for arbitrary integer* $\alpha$ *and* $q$ *every element* $u\in[\mathscr{E}^q_{I_\alpha}]^s$ *which is a solution of* (1) *in* $\mathscr{D}^*(\Omega_{I_\alpha})$ *can be written in the form*

$$\overline{(u, \phi)}=\sum_{\lambda=0}^{l} \int_{N^\lambda} d^\lambda(z, D)\, \tilde{\phi}^*(z)\, \mu^\lambda \tag{2.4}$$

*for an arbitrary* $\phi\in[S^{A-q}_{I_\alpha-A}]^s$. *Here* $\mu^\lambda$ *is a complex additive measure with support lying in the set* $N^\lambda\cap(R^n_x\times\Omega_{\mathscr{I}_{\alpha-A}})$ *and*

$$\|\mu\|_{\alpha-A}=\sum_\lambda \int (|z|+1)^{a-A}\mathscr{I}_{\alpha-A}(y)\,|\mu^\lambda|\leqq C\,\|u\|^\alpha_{I_\alpha}, \qquad \mu=(\mu^0, \ldots, \mu^l). \tag{3.4}$$

*Conversely, every functional* $u$, *given by formula* (2) *where the measures* $\mu^\lambda$ *have a finite value* $\|\mu\|^q_\alpha$, *is defined and continuous on the space* $[S^{B-q}_{I_\alpha-B}]^s$ (*that is belongs to* $[\mathscr{E}^{q-B}_{I_\alpha-B}]^s$) *and is a solution of* (1) *in* $\mathscr{D}^*(\Omega_{I_\alpha-B})$ *and*

$$\|u\|^{q-B}_{I_\alpha-B}\leqq C\|\mu\|^q_\alpha.$$

*The constants A and B depend only on the matrix p and the operators* $d^\lambda$.

*Proof.* We can suppose that $\alpha=q$ without loss of generality. We consider the sequence of spaces $S^{\mathscr{I}_\alpha}_{-\alpha}$, $-\infty<\alpha<\infty$. By Proposition 6, § 3, these spaces form a complete family $\mathscr{H}_{\mathscr{M}}$, where $\mathscr{M}$ is a complete family of majorants

$$M_\alpha(z)=(|z|+1)^\alpha\mathscr{I}_\alpha(y), \qquad \Omega_\alpha=R^n_x\times\Omega_{\mathscr{I}_\alpha}$$

of type $I$. Proposition 2, § 3, asserts that for arbitrary integer $\delta$, there is defined a continuous operator $F$: $S^{-\delta}_{I_\delta}\to S^{\mathscr{I}^*_\delta+1}_{-\delta}$, coinciding with the Fourier transform on the subspace $\mathscr{D}(\Omega_{I_\delta})$, which is dense in $S^{-\delta}_{I_\delta}$. Hence, in particular, it follows that the operator $F$ transforms the action of the operator $(i\,D)^j$ into multiplication by $z^j$. Therefore, the action of the operator $p^*(i\,D)$: $[S^{-\delta+m}_{I_\delta}]^t\to[S^{-\delta}_{I_\delta}]^s$, $m=\deg p$, after the application of $F$, becomes the action of the mapping $p^*(z)$: $[S^{\mathscr{I}^*_\delta+1}_{-\delta+m}]^t\to[S^{\mathscr{I}^*_\delta+1}_{-\delta}]^s$. We remark that the anti-linear operation $*$, introduced in § 3, represents for arbitrary $\alpha$ and $q$ an anti-isomorphism of the spaces $S^{I^*_\alpha}_q$ and $S^{I_\alpha}_q$ and it carries the operator $p^*$ into $p'$. Suppose further that $d$ is a Noetherian operator corresponding to the matrix $p'$, with components that are operators $d^\lambda$ (see 2°, § 5, Chapter IV). The continuity of the mapping (3.5), Chapter IV, implies that the operator $d$ is continuous from $[\mathscr{H}_{\delta+1}]^s$ to $\mathscr{H}_\gamma\{p', d\}$, where $\gamma=\delta+\kappa+2$. All this implies that we may consider the sequence of continuous mappings

$$\begin{aligned}[][S_{I_\delta}^{-\delta}]^s/p^*[S_{I_\delta}^{-\delta+m}]^t &\xrightarrow{F} [S_{-\delta}^{\mathscr{I}_{\delta+1}^*}]^s/p^*[S_{-\delta+m}^{\mathscr{I}_{\delta+1}^*}]^t \\ &\xrightarrow{*} [\mathscr{H}_{\delta+1}]^s/p'[\mathscr{H}_{\delta+1-m}]^t \xrightarrow{d} \mathscr{H}_\gamma\{p',d\}.\end{aligned} \tag{4.4}$$

Accordingly, the composition of these mappings is a continuous antilinear mapping, which we denote by $E$.

Let $u$ be a functional defined by formula (2) with a finite value of $\|\mu\|_\gamma$. This implies that the right side of (2) is a continuous functional over the functions $d\tilde{\phi}^*\in\mathscr{H}_\gamma\{p',d\}$. The continuity of the operator $E$ implies that $u$ is a continuous functional on the factor-space appearing in the left side of (4), that is, the functional $u$ is continuous on $[S_{I_\delta}^{-\delta}]^s$ and vanishes on the subspace $p^*[S_{I_\delta}^{-\delta+m}]^t$. Accordingly, $u\in[\mathscr{E}_{I_\delta}^{\delta}]^s$, and the functional $pu$ as an element of the space $[\mathscr{E}_{I_\delta}^{\delta-m}]^t$ is equal to zero. Thus the second assertion of the theorem is proved.

We shall prove the first assertion. To this end, we construct the operator $E^{-1}$, the inverse of $E$. By Theorem 2, § 5, and the supplement to it, there exists an integer constant $a$ depending only on $p$ and $d$, such that for arbitrary integer $\gamma$ the operator

$$d^{-1}\colon\quad \mathscr{H}_\gamma\{p',d\}\to[\mathscr{H}_{\gamma+a}]^s/[\mathscr{H}_{\gamma+a}]^s\cap\operatorname{Ker} d, \tag{5.4}$$

is defined, continuous and inverse to the operator $d$, and the subspace $[\mathscr{H}_{\gamma+a}]^s\cap\operatorname{Ker} d$ is contained in $p'[\mathscr{H}_{\gamma+2a-m}]^t$. This last assertion implies that the operator (5) can be extended to an operator acting in the factor-space $[\mathscr{H}_{\gamma+2a}]^s$ (factorized with respect to the subspace $p'[\mathscr{H}_{\gamma+2a-m}]^t$). Further, according to Proposition 2, § 3, for an arbitrary integer $\beta$, we can define the operator $F^{-1}\colon S_{-\beta}^{\mathscr{I}_\beta^*}\to S_{I_\alpha}^{-\alpha}$, $\alpha=\beta+\nu+1$ inverse to the operator $F$, introduced above. We see from relation (12.3) that the operator $F^{-1}$ transforms the operation of multiplication by the matrix $p^*(z)$ into the action of the operator $p^*(iD)$. Thus we may consider the sequence of continuous operators

$$\begin{aligned}\mathscr{H}_\gamma\{p',d\} &\xrightarrow{d^{-1}} [\mathscr{H}_{\gamma+a}]^s/[\mathscr{H}_{\gamma+a}]^s\cap\operatorname{Ker} d\to[\mathscr{H}_{\gamma+2a}]^s/p'[\mathscr{H}_{\gamma+2a-m}]^t \\ &\xrightarrow{*} [S_{-\beta}^{\mathscr{I}_\beta^*}]^s/p^*[S_{-\beta+m}^{\mathscr{I}_\beta^*}]^t \xrightarrow{F^{-1}} [S_{I_\alpha}^{-\alpha}]^s/p^*[S_{I_\alpha}^{-\alpha+m}]^t,\end{aligned} \tag{6.4}$$

where the integer $\gamma$ is arbitrary, and $\alpha=\beta+\nu+1=\gamma+2a+\nu+1$. The composition of these mappings will be denoted by $E^{-1}$.

By hypothesis, the function $u$ belongs to $[\mathscr{E}_{I_\alpha}^{\alpha}]^s$, that is, it is continuous on the space $[S_{I_\alpha}^{-\alpha}]^s$. The fact that $u$ is a solution of (1) in $\mathscr{D}(\Omega_{I_\alpha})$ means that the functional $pu$, which is defined and continuous on the space $[S_{I_\alpha}^{-\alpha+m}]^t$, is equal to zero on the subspace $[\mathscr{D}(\Omega_{I_\alpha})]^t$. Since this subspace is dense in $[S_{I_\alpha}^{-\alpha+m}]^t$, it follows that $pu=0$, that is, the functional $u$

vanishes on the subspace $p^*[S_{I_\alpha}^{-\alpha+m}]^t$. Therefore we may regard $u$ as a continuous functional on the factor-space appearing in the right side of (6). The continuity of the anti-linear mapping $E^{-1}$ implies that we may construct a continuous linear functional $v$ on $\mathscr{H}_\gamma\{p', d\}$, which acts by the formula $(v, \psi) = \overline{(u\ E^{-1}\psi)}$. We now assume that the numbers $\gamma$ appearing in (4) and (6) coincide. Then the composition $E^{-1}E$ is an identity operator, and therefore, for an arbitrary function $\phi \in [S_{I_\delta}^{-\delta}]^s$, we have

$$\overline{(u, \phi)} = \overline{(u, E^{-1}E\phi)} = (v, E\phi) = (v, d\tilde{\phi}^*). \tag{7.4}$$

We represent the functional $v$ in the form of an integral with some measure. To this end, we recall that the norm in the space $\mathscr{H}_\gamma\{p', d\}$ is expressed by the formula:

$$\|f\|_\gamma = \max_\lambda \sup_{N^\lambda \cap \Omega_\gamma} \frac{|f^\lambda(z)|}{M_\gamma(z)}.$$

Let $C^\lambda$, $\lambda = 0, \ldots, l$ be the space of continuous vector functions defined on the set $2N^\lambda \cap \Omega_\gamma$, the number of components of which is equal to the number of elements in the column $d^\lambda$ for which the norm $\|F\| = \sup\{|F(z)|, z \in N^\lambda \cap \Omega_\gamma\}$ is finite. The mapping

$$\mathscr{H}_\gamma\{p', d\} \ni f \to \left(\frac{f^0}{M_\gamma}, \ldots, \frac{f^l}{M_\gamma}\right) \in \oplus C^\lambda$$

is an isometric imbedding of the space $\mathscr{H}_\gamma\{p', d\}$ in the direct sum $\oplus C^\lambda$. According to the Hahn-Banach theorem, the functional $v$ can be extended from $\mathscr{H}_\gamma\{p', d\}$ to the whole space $\oplus C^\lambda$ with conservation of the norm. In view of the general theorem [2] the functional $v$, extended to $\oplus C^\lambda$, can be written as an integral

$$(v, (F_0, \ldots, F_l)) = \sum_\lambda \int F_\lambda \mu_\lambda,$$

where $\mu_\lambda$, $\lambda = 0, \ldots, l$ are certain complex, bounded, additive measures, concentrated in the sets $N^\lambda \cap \Omega_\gamma$, and $\sum_\lambda \int |\mu_\lambda| = \|v\|$. Writing

$$\mu^\lambda = \frac{\mu_\lambda}{M_\gamma}, \quad \lambda = 0, \ldots, l,$$

we may express the functional $v$ in the following form:

$$(v, f) = \sum_\lambda \int f^\lambda(z)\, \mu^\lambda, \quad f = \{f^\lambda\} \in \mathscr{H}_\gamma\{p', d\}, \tag{8.4}$$

where

$$\sum_\lambda \int M_\gamma |\mu^\lambda| = \sum_\lambda \int |\mu_\lambda| = \|v\|. \tag{9.4}$$

2 See, for example, Danforth and Schwartz [1].

Combining (7) and (8), we obtain the representation (2). The inequality (3) follows from (9), and from the continuity of the operator $E^{-1}$. □

**2°. Remarks and corollaries.** The representation (2) becomes simpler when the space $\mathscr{E}^{\alpha}_{I_\alpha}$ consists of sufficiently smooth functions and, to be precise, if $\alpha \geqq A+\nu$, where $\nu=[n/2]+1$. If this inequality is satisfied, then by Proposition 3, §3, the space $S^{A-\alpha}_{I_\alpha-A}$ contains all the functionals $\delta(\xi-\eta)$ with $\eta \in \Omega_{I_\alpha-A}$. Let $e$ be the unit matrix of order $s$. We substitute in (2) the matrix $\phi=\delta(\xi-\eta)\,e$, that is, we substitute one after another each of the columns of this matrix. We obtain the row

$$\overline{(u, \delta(\xi-\eta)\,e)}=\sum_{\lambda}\int_{N^\lambda} d^\lambda(z, D)\exp(z, -i\eta)\,\mu^\lambda = \sum_{\lambda}\int_{N^\lambda} d^\lambda(z, -i\eta)\exp(z, -i\eta)\,\mu^\lambda. \tag{10.4}$$

The left side of this equation represents the row formed by the components of the vector function $u$. Since the vector $u$ is interpreted as a column, this row is the transpose of this column, that is, it is equal to $u'$. Therefore the Eq. (10) yields a representation of the solution $u$, if the matrices $d^\lambda(z, -i\eta)$ in the right side are replaced by the transposed matrices $(d^\lambda(z, -i\eta))'$. Thus, Eq. (10) shows that the function $u$ in the region $\Omega_{I_\alpha-A}$ is representable as the sum of integrals with certain measures $\mu^\lambda$ over the manifold of exponential polynomials $(d^\lambda(z, -i\eta))' \cdot \exp(z, -i\eta)$, which are solutions of the system (1). When $\alpha$ is arbitrary, the representation (2) can also be given an "exponential" form. To do this, we note that for arbitrary $z \in \Omega_{I_\alpha-A}$ the functional

$$\phi \to \overline{d^\lambda(z, D)\,\tilde{\phi}^*(z)},$$

defined on $[S^{A-\alpha}_{I_\alpha-A}]^s$, corresponds to the functions $d^\lambda(z, -i\xi)\exp(z, -i\xi)$. Therefore, Eq. (2) can be rewritten as:

$$(u, \phi)=\sum_{\lambda}\int_{N^\lambda} (d^\lambda(z, -i\xi)\exp(z, -i\xi), \phi)\,\mu^\lambda,$$

where the left and righthand sides are to be understood as functionals on the space $[S^{A-\alpha}_{I_\alpha-A}]^s$.

Let us realize the representation (2) for the spaces of distributions defined on convex compacts.

**Corollary 1.** *Let $K$ and $K' \supset\supset K$ be arbitrary convex compacts in $R^n$. For an arbitrary integer $\alpha$, every function $u \in [\mathscr{E}^{\alpha}_{K'}]^s$ which is a solution of (1) in $\mathscr{D}_{K'}$ can be expressed in the form (2), for an arbitrary function $\phi \in [\mathscr{D}^*_K]^s$, and*

$$\|\mu\|^{\alpha-A}_K=\sum_{\lambda}\int_{N^\lambda} (|z|+1)^{\alpha-A}\,\mathscr{I}_K(y)\,|\mu^\lambda| \leqq C\,\|u\|^{\alpha}_{K'}. \tag{11.4}$$

*Conversely, every functional, defined by formula* (2), *with a finite* $\|\mu\|_{K'}^{\alpha}$, *belongs to the space* $[\mathscr{E}_K^{\alpha-B}]^s=[(S_K^{B-\alpha})^*]^s$ *and is a solution of* (1) *in* $\mathscr{D}_K^*$, *and* [3]

$$\|u\|_K^{\alpha-B}\leqq C\,\|\mu\|_{K'}^{\alpha}.$$

*Proof.* We construct a strictly increasing sequence of convex compacts $K_\alpha$, $-\infty<\alpha<\infty$, satisfying the conditions $K\subset\cap K_\alpha$ and $\cup K_\alpha\subset K'$. As we know (see 2°, § 3) the functions $I_{K_\alpha}$ form a decreasing family of logarithmically convex functions. For an arbitrary integer $\alpha$, the identity imbeddings $\mathscr{D}_K^\alpha\to S_{I_\alpha}^\alpha\to\mathscr{D}_{K'}^\alpha$ are defined and continuous. Moreover, the inequalities $\mathscr{I}_K(y)\leqq\mathscr{I}_\alpha(y)\leqq\mathscr{I}_{K'}(y)$ are satisfied. Therefore, Corollary 1 follows from Theorem 1 applied to the family $\{I_{K_\alpha}\}$. □

**Corollary 2.** *Let* $\Omega$ *be a region in* $R^n$, *and* $\kappa\subset\Omega$ *a convex compact. For an arbitrary integer* $q$ *every solution of* (1) *in* $\mathscr{D}^*(\Omega)$ *can be written in the form* $u=u_0+v$, *where* $u_0$ *is a solution of* (1) *in* $\mathscr{D}^*(R^n)$, *and*

$$v\in[\mathscr{E}_\kappa^q]^s,\quad \textit{and}\quad \|v\|_\kappa^q\leqq\frac{1}{q}.$$

*Proof.* Let $\kappa'$, $K$ and $K'$ be convex compacts such that $\kappa\subset\subset\kappa'\subset\subset K\subset\subset K'\subset\Omega$. Since $u$ is a distribution in $\Omega$, it belongs to the space $[\mathscr{E}_{K'}^\alpha]^s$ for some integer $\alpha$. Accordingly, we may apply Corollary 1 to the function $u$ and we derive the representations (2)–(11).

Suppose, further, that $\tau$ and $\sigma$ are arbitrary positive numbers. We denote by $C_{\tau,\sigma}$ the region in $C^n$ in which the inequality $|y|<\tau\ln(|z|+1)+\sigma$ is satisfied. We choose some continuous function $h_{\tau,\sigma}$ in $C^n$, having values that lie between zero and one, equal to one on $C_{\tau,\sigma}$ and equal to zero outside $C_{\tau,\sigma+1}$. We write

$$\overline{(u_0,\phi)}=\sum_\lambda\int d^\lambda(z,D)\,\tilde\phi^*(z)\,h_{\tau,\sigma}(z)\,\mu^\lambda,$$

$$\overline{(v,\phi)}=\sum_\lambda\int d^\lambda(z,D)\,\tilde\phi^*(z)\,[1-h_{\tau,\sigma}(z)]\,\mu^\lambda$$

Let us establish the convergence of these intervals in $[\mathscr{D}_K^*]^s$; we shall then have proved that $u=u_0+v$ in this space.

We shall suppose that $K$ contains the origin of coordinates. Then Eq. (11) implies that the integral

$$\int(|z|+1)^k\exp(R|y|)\,h_{\tau,\sigma}(z)\,|\mu^\lambda|$$

converges absolutely for arbitrary $\lambda$, $\tau$, $\sigma$ and $R$, if $k$ is so chosen that $k+\tau R\leqq\alpha-A$. Therefore, the converse assertion of Corollary 1 implies

3 It is easy to verify that this assertion remains true provided $\|\mu\|_K^\alpha<\infty$.

that $u_0$ is a distribution and satisfies (1) in an arbitrary sphere of the form $|\xi|<R$, and therefore, over the whole space $R^n$.

We now fix an arbitrary integer $q$, and we write $\rho=\rho(\kappa', CK)$. We consider the following inequality:

$$\int (|z|+1)^{q+B} \mathscr{I}_{\kappa'}(y)[1-h_{\tau,\sigma}]\,|\mu^\lambda| \leqq \int \frac{(|z|+1)^{q+B}}{\exp(\rho|y|)} \mathscr{I}_K(y)[1-h_{\tau,\sigma}]\,|\mu^\lambda|.$$

We choose the number $\tau$ so that $q+B-\rho\tau \leqq \alpha - A$. Then the right side converges because of the inequality (11). The converse assertion of Corollary 1 implies that from the finiteness of the left side it follows that the functional $v$ belongs to $[\mathscr{E}_\kappa^q]^s$ and satisfies the inequality

$$\|v\|_\kappa^q \leqq C\,\|[1-h_{\tau,\sigma}]\,\mu\|_{\kappa'}^{q+B}.$$

For $\sigma\to\infty$, the function $h_{\tau,\sigma}$ is bounded and tends to unity in every bounded region of the space $C^n$. Therefore, the right side of our inequality tends to zero. It follows that the norm of the functional $v$ can be made less than $1/q$ by a suitable choice of $\sigma$. □

**3°. The representation in the spaces of ultra-distributions**

**Theorem 2.** *Let $\{I_\alpha\}$ be a decreasing family of logarithmically convex functions, and let $\{\mathscr{I}_\alpha\}$ be a family of logarithmically reciprocal functions. Suppose further that $\{b_\alpha\}$ and $\{R_\alpha\}$ are sequences of functions satisfying the conditions of Proposition 5, § 3. Then for arbitrary integer $\alpha$ every functional $u\in[(S_{I_\alpha}^{b_\alpha})^*]^s$ which is a solution of the system (1) in $(S_{I_\alpha}^{b_\alpha-m})^*$, can be written in the form (2) for arbitrary $\phi\in[S_{I_{\alpha-A}}^{b_{\alpha-A}}]^s$, and*

$$\|\mu\|_{\alpha-A} = \sum_\lambda \int R_{\alpha-A}(z)\,\mathscr{I}_{\alpha-A}(y)\,|\mu^\lambda| \leqq C\,\|u\|_\alpha, \tag{12.4}$$

*where $\|u\|_\alpha$ is the norm of $u$ as an element of $[(S_{I_\alpha}^{b_\alpha})^*]^s$.*

*Conversely, every functional given by formula (2) in which the quantity $\|\mu\|_{\alpha+B}$ is finite, is continuous on the space $[S_{I_\alpha}^{b_\alpha}]^s$ and is a solution of (1) in $(S_{I_\alpha}^{b_\alpha-m})^*$, and*

$$\|u\|_\alpha \leqq C\,\|\mu\|_{\alpha+B}. \tag{13.4}$$

*The constants $A$ and $B$ depend only on the operators $p$ and $d^\lambda$.*

Let us first establish the validity of the expression "every functional $u\in[(S_{I_\alpha}^{b_\alpha})^*]^s$ which is a solution of (1) in $(S_{I_\alpha}^{b_\alpha-m})^*$," that is, let us show that the operator $p(iD)$ carries the space $[(S_{I_\alpha}^{b_\alpha})^*]^s$ into $[(S_{I_\alpha}^{b_\alpha-m})^*]^t$. In fact, in accordance with the Remark 6°, § 3, a differential operator of the form $D^j$ acts continuously from $S_{I_\alpha}^{b_\alpha}$ to $S_{I_\alpha}^{b_\alpha+|j|}$. Therefore, the operator $p^*(iD)$: $[S_{I_\alpha}^{b_\alpha-m}]^t \to [S_{I_\alpha}^{b_\alpha}]^s$ is defined and continuous. Then the conjugate operator acts from $[(S_{I_\alpha}^{b_\alpha})^*]^s$ to $[(S_{I_\alpha}^{b_\alpha-m})^*]^t$, which is what we were to prove.

The proof of Theorem 2 makes use of Propositions 5 and 6, § 3 and is altogether similar to the proof of Theorem 1. ▯

**4°. The Fourier transform of analytic functions.** Analytic functions are solutions of a homogeneous Cauchy-Riemann system. Since this system has constant coefficients, Theorem 1 is applicable. From it we now derive an exponential representation for analytic functions, taking account of their growth at infinity.

Let $n$ be an even number: $n=2m$. In the space $R_\xi^n$ we introduce the structure of an $m$-dimensional complex space, by setting up the complex variables $\zeta_j=\xi_j+i\,\xi_{m+j}$, $j=1,\ldots,m$. We write $\xi'=(\xi_1,\ldots,\xi_m)$, $\xi''=(\xi_{m+1},\ldots,\xi_n)$ and $\zeta=(\zeta_1,\ldots,\zeta_m)=\xi'+i\,\xi''$. In the conjugate space $C_z^n$, which appears in the representation (2), we write $z'=(z_1,\ldots,z_m)$, $z''=(z_{m+1},\ldots,z_n)$; $z'=x'+i\,y'$, $z''=x''+i\,y''$.

In the space $R_\xi^n$ we consider a homogeneous Cauchy-Riemann system

$$2i\frac{\partial u}{\partial\bar\zeta_j}=\left(i\frac{\partial u}{\partial\xi_j}\right)+i\left(i\frac{\partial u}{\partial\xi_{m+j}}\right)=0,\qquad j=1,\ldots,m. \tag{14.4}$$

**Corollary 3.** *Let $\{I_\alpha\}$ be a decreasing family of logarithmically convex functions, and let $\{\mathscr{I}_\alpha\}$ be a family of logarithmically reciprocal functions. For arbitrary $\alpha$ every function holomorphic in $\Omega_{I_\alpha}$ satisfying the inequality*

$$|u(\zeta)|\leqq C I_\alpha(\xi), \tag{15.4}$$

*can be represented in the form*

$$u(\zeta)=\int\exp(\zeta,z'')\,\tau, \tag{16.4}$$

*where $\tau$ is some measure in $C_{z''}^m$ such that*

$$\|\tau\|_\beta=\int\mathscr{I}_\beta(x'',-y'')|\tau|<\infty,\qquad \beta=\alpha-\text{const}. \tag{17.4}$$

*Conversely, every function $u$ admitting the representation* (16) *with $\|\tau\|_\alpha<\infty$ is analytic in $\Omega_{I_\alpha}$ and satisfies the inequality* (15).

*Proof.* Let $p$ be a matrix corresponding to the system (14). The matrix $p'(z)$ has the form

$$p'(z)=(z_1+i\,z_{m+1},\ldots,z_m+i\,z_n).$$

Hence, it follows that the ideal $p'\mathscr{P}^m\subset\mathscr{P}$ consists of polynomials of the form

$$f=(z_1+i\,z_{m+1})f_1+\cdots+(z_m+i\,z_n)f_m,\qquad f_1,\ldots,f_m\in\mathscr{P}.$$

It is easy to see that such polynomials are characterized by the fact that they vanish on the irreducible variety $N=\{z\colon z'=-i\,z''\}$, associated with the matrix $p'$. It follows that the ideal $p'\mathscr{P}^m$ is prime. Therefore, Proposition 3, § 4, Chapter IV implies that we may choose the variety $N$ and the operator $d(z, D)\equiv 1$ as the set of the varieties and of the normal Noetherian operators associated to the matrix $p'$.

The bound (15) and the inequality

$$I_\alpha(\xi)\leqq C\exp(-\varepsilon_{\alpha-1}|\xi|)\, I_{\alpha-1}(\xi)$$

imply that the function $u$ belongs to the space $\mathscr{E}^0_{I_{\alpha-1}}$ and is a solution of the system (1) in $\mathscr{D}^*(\Omega_{I_{\alpha-1}})$. It therefore follows from Theorem 1 that this functional admits the representation (2) for all functions $\phi\in\mathscr{D}(\Omega_{I_{\alpha-1-A}})$, and that the quantity

$$\int (|z|+1)^{-A}\,\mathscr{I}_{\alpha-1-A}(y)|\mu| \qquad (\mu=\mu^0) \tag{18.4}$$

is finite.

Since the variety $N$ is defined by the equations

$$x'=y'', \qquad y'=-x'', \tag{19.4}$$

the inequality $|z|\leqq 2|y|$ holds on it, and therefore $(|z|+1)^q=o\big(\exp(\varepsilon_{\alpha-2-A}|y|)\big)$ for arbitrary $q$. Thus, the finiteness of (18) implies the finiteness of the integral

$$\int (|z|+1)^{B+\nu}\,\mathscr{I}_{\alpha-2-A}(y)|\mu|, \tag{20.4}$$

where $B$ is the constant appearing in Theorem 1. By the converse assertion of Theorem 1, the functional $u$ is continuous in the norm $\|\cdot\|_{I\beta}^{-\nu}$, $\beta=\alpha-2-A-B$, and, accordingly, can be continued from the dense subspace $\mathscr{D}(\Omega_{I_\beta})$ to the whole space $S_{I_\beta}^{-\nu}$. Therefore Remark 2° implies that the function $u$ admits the exponential representation (10).

Taking account of the fact that $d(z, D)\equiv 1$ and remembering Eq. (19), which defines the variety $N$, this representation can be rewritten in the form

$$u(\zeta)=\int_N \exp(z, -i\,\xi)\,\mu=\int \exp(\zeta, -z'')\,\mu. \tag{21.4}$$

We set $\tau(z'')=\mu(i\,z'', -z'')$. Then (21) implies (16), and the finiteness of the quantity (20) implies (17).

We shall prove the converse assertion. Making use of the inequality connecting the function $I_\alpha$ and the logarithmic inverse $\mathscr{I}_\alpha$, we obtain the inequality

$$|\exp(\zeta, z'')|=\exp\big(\xi, (x'', -y'')\big)\leqq I_\alpha(\xi)\,\mathscr{I}_\alpha(x'', -y'').$$

Hence in view of the finiteness of $\|\tau\|_\alpha$, we have

$$|u(\zeta)|\leqq\int|\exp(\zeta, z'')|\,|\tau|\leqq I_\alpha(\xi)\int\mathscr{I}_\alpha(x'', -y'')\,|\tau|=C I_\alpha(\xi).$$

Accordingly, the integral (16) converges absolutely in the region $\Omega_{I_\alpha}$ and so represents a function $u(\zeta)$ holomorphic in this region. □

Making use of Corollary 4, we can obtain an exponential representation of an entire function of order not more than one whose indicatrix belongs to a given convex compact $K \subset C^m$. To this end, we construct a strictly decreasing sequence of convex compacts $K_\alpha$, $-\infty < \alpha < \infty$, such that $\cap K_\alpha = K$. The sequence of functions $I_\alpha = \mathscr{I}_{K_\alpha}$ is obviously a decreasing family of logarithmically convex functions, and the logarithmically reciprocal functions have the form $\mathscr{I}_\alpha = I_{K_\alpha}$. Therefore, an entire function of order not more than one whose indicatrix belongs to $K$, is characterized by the fact that it satisfies the inequality (15) for an arbitrary $\alpha$. By Corollary 4, such functions can be represented for arbitrary $\alpha$ in the form (16) with a finite value of

$$\int I_{K_\alpha}(x'', -y'')|\tau|.$$

This latter condition means that $\tau$ is a bounded measure in $C^m_{z''}$, whose support is contained in an arbitrarily small neighborhood of the compact $K^*$, symmetric to $K$ with respect to $R^m_{x''}$.

## § 5. Hypoelliptic operators

The exponential representation of the solutions of homogeneous systems obtained in the preceding section is useful for the study of local properties of these solutions. Local properties of solutions of systems are completely characterized by the variety $N$ associated with the matrix $p'$ in the sense of 1°, § 1, Chapter IV. We recall the simplest characteristics of the variety $N$: it consists of points $z \in C^n$ at which the rank of the matrix $p(z)$ is less than $s$. On the other hand, $N$ is the union of the varieties $N^\lambda$ that make up the ensemble associated with the matrix $p'$. Accordingly, in the representation (2.4) the integration is carried out precisely over the variety $N$, and this explains the influence that $N$ has on the properties of the solutions of the system (1.4).

**1°. Fundamental properties of hypoelliptic operators.** We consider the following characteristic of the variety $N$:

$$m(r) = \inf\{|y|: z = x + iy \in N, |z| = r\}.$$

Let us suppose that $N$ is not bounded. Then the function $m(r)$ is defined for sufficiently large $r$. Since the variety $N$ is algebraic, $m(r) \sim m_0 r^\gamma$ as $r \to \infty$ for some $m_0 \geqq 0$ and rational $\gamma$[4]. If the real part $N \cap R^n_x$ of the variety $N$ is unbounded, then $m(r) \equiv 0$ for sufficiently large $r$, and accordingly, $m_0 = 0$. In this case we set $\gamma = -\infty$. If $N \cap R^n_x$ is bounded, then

4 See Gorin [1].

$m_0 > 0$, and the quantity $\gamma$, defined uniquely in this case, is clearly not greater than unity. If $m(r) \to \infty$ as $r \to \infty$, then $\gamma$ is positive.

From the definition of $\gamma$ it follows that when $\gamma > -\infty$, the inequality

$$|z| \leqq B(|y|+1)^{1/\gamma} \tag{1.5}$$

holds on $N$ for some sufficiently large $B > 0$. The inequality holds also when $N$ is bounded if we set $\gamma = \infty$, $1/\gamma = 0$, and $B = \max\{|z|, z \in N\}$. The number $\gamma$ so defined will be called the index of hypoellipticity. If $\gamma > 0$, the variety $N$ is said to be hypoelliptic.

Let us turn our attention to the system (1.4). Both the operator $p$ and the system are called hypoelliptic if for an arbitrary region $\Omega \subset R^n_\xi$ all the solutions of (1.4) in the space $[\mathscr{D}^*(\Omega)]^s$ are infinitely differentiable functions. It turns out that the operator $p$ is hypoelliptic if and only if $m(r) \to \infty$ as $r \to \infty$, that is, if and only if $\gamma$ is positive. The following theorem states a more precise result.

**Theorem 1.** *Suppose that $\gamma > 0$. Then, for an arbitrary region $\Omega \subset R^n$ the natural imbedding*

$$\mathscr{E}_p(\Omega) \to \mathscr{D}^*_p(\Omega) \tag{2.5}$$

*is an isomorphism, and for an arbitrary bounded subregion $\omega \subset\subset \Omega$ the mapping of the restriction*

$$\mathscr{E}_p(\Omega) \to \mathscr{E}_p(\omega) \tag{3.5}$$

*is compact.*

We recall that in accordance with the notation of § 4, $\mathscr{E}_p(\Omega)$ and $\mathscr{D}^*_p(\Omega)$ are subspaces in $[\mathscr{E}(\Omega)]^s$ and $[\mathscr{D}^*(\Omega)]^s$, consisting of solutions of the system (1.4).

*Proof of Theorem* 1. We shall suppose, to begin with, that the region $\Omega$ is convex. We prove the first assertion. The continuity of the imbedding (2) is self-evident. Let us show that the inverse mapping is defined and is also continuous. First, we describe the topology in the spaces appearing in (2). As we established in § 2, sets of the form $\{u\colon \|u\|^q_K \leqq \varepsilon$ where $K \subset \Omega$ is a compact, $q$ is a positive integer, and $\varepsilon$ a positive number, form a fundamental system of neighborhoods of zero in the space $\mathscr{E}(\Omega)$ and, therefore, in the spaces $\mathscr{E}_p(\Omega)$. In the space $\mathscr{D}^*_p(\Omega)$, a fundamental system of neighborhoods of zero is formed by the polars of the bounded sets in $[\mathscr{D}(\Omega)]^s$, and as follows from Proposition 6, § 2, we need consider only bounded sets of the form

$$\mathscr{B} = \left\{\phi\colon \|\phi\|^b_{K'} \leqq \frac{1}{\varepsilon'}\right\},$$

where $K' \subset \Omega$ is a compact, $b = b(\eta)$ is a positive function, and $\varepsilon' > 0$ is some number. Thus, to show that (2) is an isomorphism, we need only for arbitrary $K, q$ and $\varepsilon$, find $K', b, \varepsilon'$ such that every element $u$ of the space $\mathscr{D}_p^*(\Omega)$ belonging to the polar of the set $\mathscr{B}$, is infinitely differentiable in $\Omega$ and $\|u\|_K^q < \varepsilon$.

We fix $\beta > 1$ to be not less than $1/\gamma$. We further choose some strictly increasing sequence of convex compacts $\{K_\alpha, -\infty < \alpha < \infty\}$, tending to $\Omega$, and such that $K \subset K_0$, and we choose a strictly increasing sequence of numbers $\{a_\alpha, -\infty < \alpha < \infty\}$, larger than $a = B\rho^{-\beta}$, where $B$ is the constant in (1), and $\rho = \rho(K_2, CK_3)$. We consider the increasing sequence of functions

$$b_\alpha(\eta) = a_\alpha^\eta \eta^{\eta\beta}, \qquad R_\alpha(z) = \exp\left(-\frac{\beta}{e}\left|\frac{z}{a_\alpha}\right|^{1/\beta}\right), \qquad -\infty < \alpha < \infty.$$

The calculations made in 8°, § 3, imply that these sequences satisfy the conditions of Proposition 6, § 3. Therefore, we may apply Theorem 2, § 4, to these sequences and to the decreasing family of logarithmically convex functions $I_\alpha = I_{K_\alpha}$. We set $b = b_{A+3}$ and $K' = K_{A+3}$, where $A$ is the constant in the theorem cited.

Let $u$ be an arbitrary element of the space $\mathscr{D}_p^*(\Omega)$, belonging to the polar of the set $\mathscr{B}$. It is clear that the functional $u$ is defined and continuous on the space $[\mathscr{D}_{K_{A+3}}^{b_{A+3}}]^s$, that its norm in the conjugate space $[\mathscr{D}_{K_{A+3}}^{b_{A+3}}]^s$ does not exceed $\varepsilon'$, and that it is a solution of system (1.4) in $(\mathscr{D}_{K_{A+3}}^{b_{A+3-m}})^*$. By Theorem 2, § 4, the functional $u$ has the representation (2.4), which is valid for all the functions $\phi \in [\mathscr{D}_{K_3}^{b_3}]^s$ and we have from (12.4)

$$\|\mu\|_3 = \sum_\lambda \int R_3(z)\,\mathscr{I}_3(y)\,|\mu^\lambda| \leq C\,\|u\| \leq C\varepsilon'. \tag{4.5}$$

Using this representation, we now find a bound for the functional $u$[5]:

$$|(u, \phi)| \leq \sup_N \frac{|d(z, D)\,\tilde{\phi}^*(z)|}{(|z|+1)^{q+\kappa}\,\mathscr{I}_2(y)} \sup_N \frac{(|z|+1)^{q+\kappa}\,\mathscr{I}_2(y)}{R_3(z)\,\mathscr{I}_3(y)}\,\|\mu\|_3, \quad \phi \in [\mathscr{D}_{K_3}^{b_3}]^s. \tag{5.5}$$

Since the coefficients of the differential operators $d(z, D)$ are polynomials of order not exceeding $\kappa$, the first factor on the right side can be bounded from above by the quantity

$$C \sup \frac{|\tilde{\phi}^*(z)|}{(|z|+1)^q\,\mathscr{I}_1(y)} = C\,\|\tilde{\phi}\|_{-q}^{K_1}.$$

It follows from the calculations made in proving Proposition 2, § 3, that the right side does not exceed $C'\|\phi\|^{-q}$, if supp $\phi \subset K_0$. We now estimate the second factor on the right side of (5). It follows from the

5 The symbol $d(\ldots)$ denotes the vector $(d^0, \ldots, d^l)$.

choice of $a$ that on the variety $N$

$$\frac{(|z|+1)^{q+\kappa}\mathscr{I}_2(y)}{R_3(z)\mathscr{I}_3(y)} \leqq (|z|+1)^{q+\kappa}\exp\left(\left|\frac{z}{a}\right|^{1/\beta}\right)\exp(-\rho|y|) \leqq C_q.$$

The third factor on the right side of (5) is not larger than $C\varepsilon'$, because of the inequality (4). Thus, we arrive at the inequality

$$|(u,\phi)| \leqq \varepsilon' C_q' \|\phi\|^{-q}, \qquad \phi\in[\mathscr{D}_{K_0}^{b_3}]^s. \tag{6.5}$$

It follows that the functional $u$ is continuous in the norm $\|\cdot\|^{-q}$ on the functions belonging to the space $[\mathscr{D}_{K_0}^{b_3}]^s$. Proposition 8, § 2, implies that this space is dense in $[\mathscr{D}_{K_0}^{-q}]^s$. Therefore, the functional $u$ can be extended by continuity to the whole space $[\mathscr{D}_{K_0}^{-q}]^s$ and, accordingly, coincides with some element of the space $[\mathscr{E}_{K_0}^q]^s$. Setting $\varepsilon'=\varepsilon/C_q'$, we find from the inequality (6) that $\|u\|_K^q \leqq \varepsilon$. There remains only to remark that since $q$ and the compact $K$ were chosen arbitrarily, $u$ belongs to an arbitrary one of the spaces $[\mathscr{E}_K^q]^s$, that is, it is infinitely differentiable in $\Omega$. Thus the first assertion of the theorem has been proved.

We now establish the second assertion. The condition $\omega\subset\subset\Omega$ implies that the sequence of compacts $K_\alpha$ can be so chosen that $\omega\subset K_0$. The inequality (6) being true for all functionals $u\in\mathscr{B}^0$, implies that for arbitrary $q$, $\|u\|_{K_0}^q \leqq C_q''$. The set of functions in $\mathscr{E}_p(\omega)$, satisfying these inequalities, is bounded and, therefore, relatively compact, since $\mathscr{E}(\omega)$ is an $\mathscr{F}\mathscr{S}$-space. On the other hand, the polars of the sets of the type $\mathscr{B}$ form a fundamental system of neighborhoods of zero in the space $\mathscr{D}^*(\Omega)$ and, therefore in the space $\mathscr{E}_p(\Omega)$. We have thus proved that the mapping (3) is compact.

We now prove the theorem for an arbitrary region $\Omega$. We choose some locally finite covering $U=\{U_\nu\}$ of the region $\Omega$, consisting of convex domains. By Proposition 9, § 2, the topology of the space $\mathscr{D}^*(\Omega)$ coincides with the topology induced by ${}^0\mathscr{D}^*(U)$. Therefore, the topology in $\mathscr{D}_p^*(\Omega)$ coincides with that induced by the space ${}^0\mathscr{D}_p^*(U)$, which is the kernel of the mapping $p$: $[{}^0\mathscr{D}^*(U)]^s\to[{}^0\mathscr{D}^*(U)]^t$. Similarly, the topology in $\mathscr{E}_p(\Omega)$ coincides with that induced by ${}^0\mathscr{E}_p(U)$. Then for arbitrary $\nu$ we have the isomorphism $\mathscr{E}_p(U_\nu)\cong\mathscr{D}_p^*(U_\nu)$. It follows that the spaces ${}^0\mathscr{E}_p(U)$ and ${}^0\mathscr{D}_p^*(U)$ are isomorphic and, accordingly, so are the subspaces $\mathscr{E}_p(\Omega)$ and $\mathscr{D}_p^*(\Omega)$. With this, the first assertion of the theorem is proved for an arbitrary region $\Omega$.

We now prove the second assertion. Let $N$ be large enough so that

$$\omega\subset\subset\bigcup_1^N U_\nu.$$

In every domain $U_\nu$, $\nu=1,\ldots,N$, we choose a subdomain $V_\nu \subset\subset U_\nu$ so that

$$\bigcup_1^N V_\nu=\omega.$$

Since all the domains $U_\nu$ are convex, it follows from what we have proved earlier that for arbitrary $\nu$ the mapping of the restriction $\mathscr{E}_p(U_\nu)\to\mathscr{E}_p(V_\nu)$ is compact. Accordingly, the mapping of the direct products

$${}^0\mathscr{E}_p(U)\cong\prod_\nu \mathscr{E}_p(U_\nu)\to\prod_{\nu=1}^N \mathscr{E}_p(V_\nu)\cong{}^0\mathscr{E}_p(V), \qquad V=\{V_\nu\},$$

is also compact, since the product $\prod \mathscr{E}_p(V_\nu)$ is finite. Therefore, the mapping of the subspaces $\mathscr{E}_p(\Omega)\to\mathscr{E}_p(\omega)$ of these direct products is also compact. ▯

*Remark* 1. In the proof of the bound (6), we used only the fact that $u$ is a continuous functional on the space $(\mathscr{D}_{K_{A+3}}^{b_{A+3}-m})^*$, and a solution of (1.4) in $[\mathscr{D}_{K_{A+3}}^{b_{A+3}}]^s$. On the other hand, the bound (6) holds for arbitrary $q$, and we may conclude that $u$ is infinitely differentiable in $K_0$. The spaces $\mathscr{D}_{K_\alpha}^{b_\alpha}$ form an increasing family. We denote by $\mathscr{D}^{\beta,a}(\Omega)$ the inductive limit of this family. Let $u$ belong to the space $[(\mathscr{D}^{\beta,a}(\Omega))^*]^s$ and let it be a solution of (1.4). Then for arbitrary $\alpha$, $u$ can be regarded as a continuous functional on $[\mathscr{D}_{K_{\alpha+A+3}}^{b_{\alpha+A+3}}]^s$, and a solution of (1) in $(\mathscr{D}_{K_{\alpha+A+3}}^{b_{\alpha+A+3}-m})^*$. Hence, as we have noted, it follows that $u$ is an infinitely differentiable function within $K_\alpha$. Since $\alpha$ is arbitrary, $u$ is infinitely differentiable over the whole region $\Omega$.

*Remark* 2. The condition $\gamma>0$ is necessary for the spaces $\mathscr{E}_p(\Omega)$ and $\mathscr{D}_p(\Omega)$ to coincide for at least one non-empty region $\Omega$. We shall now establish this fact, by proving the stronger assertion: if every continuous function in $R^n$ which is a solution of the system (1.4) in $\mathscr{D}^*(R^n)$ has a continuous first derivative, then $\gamma>0$.

We denote by $C^q$, where $q=0,1$, the space of functions in $R^n$ which are continuous if $q=0$ and have continuous first derivatives if $q=1$. We endow $C^0$ with the topology of compact convergence and $C^1$ with the topology of compact convergence of functions and their first derivatives. Then $C^q$, $q=0,1$ are $\mathscr{F}$-spaces. The subspace $C_p^q\subset[C^q]^s$ consisting of solutions of (1.4) in $\mathscr{D}^*(R^n)$ is closed and consequently is an $\mathscr{F}$-space also. By hypothesis the imbedding maps $C_p^1$ onto $C_p^0$. It follows from Proposition 7, §1 that this mapping is an isomorphism. Therefore for every $u\in C_p^1$ the inequality

$$\max_{|j|=1}|D^j u(0)|\leqq C\sup_{|\xi|\leqq R}|u(\xi)| \tag{7.5}$$

holds with some constants $C$ and $R$.

Let $z=x+iy$ be an arbitrary point of the set $N$. As we noted earlier, the rank of the matrix $p(z)$ is less than $s$. Therefore, for the chosen point $z$ there exists a non-zero vector $a \in C^s$, orthogonal to all the rows of $p(z)$. It follows that the vector function $a \exp(z, -i\xi)$ is a solution of the system (1.4). Substituting this function in the inequality (7) we obtain

$$|z| \leqq C \exp(R|y|),$$

where the constant $C$ is independent of $z$. It is clear from this inequality that as $|z| \to \infty$, so also $|y| \to \infty$. From this we derive the desired inequality $\gamma > 0$. ▯

**2°. Dependence of the smoothness of solutions of hypoelliptic systems of equations on their growth at infinity.** The following property of entire analytic functions is well-known: If the function $u(\zeta)$ is of bounded growth at infinity, so also are the derivatives $D^j u(\zeta)$ as $|j| \to \infty$. The solutions of homogeneous hypoelliptic systems have a similar property as is shown by the following theorem

**Theorem 2.** *Let $p$ be a hypoelliptic operator with index $\gamma$, let $I(\xi)$ be a spherically symmetric logarithmically convex function, and let $\mathscr{I}(y) = i(|y|)$ be the logarithmic reciprocal. Then every solution of the system* (1.4), *defined in the region $\Omega_I$ and satisfying the inequality*

$$|u(\xi)| \leqq C I((1+\varepsilon)\xi), \tag{8.5}$$

*for arbitrary $\varepsilon > 0$, also satisfies the inequality*

$$|D^j u(\xi)| \leqq C b^{|j|} \mathscr{B}\left(\frac{|j|}{\gamma}\right) I((1+\varepsilon)\xi)$$

*with arbitrary $\varepsilon > 0$. Here $\mathscr{B}(\eta)$ is the logarithmic reciprocal of* $i(\exp t)$.

*Proof.* We choose an arbitrary $\varepsilon > 0$ and some strictly decreasing sequence of positive numbers $\{\varepsilon_\alpha, -\infty < \alpha < \infty\}$, such that $\lim_{\alpha \to -\infty} \varepsilon_\alpha = \varepsilon$. Without loss of generality, we may suppose that $I(0) = 1$. We consider the sequence of functions

$$I_\alpha(\xi) = I((1+\varepsilon_\alpha)\xi), \qquad -\infty < \alpha < \infty.$$

We shall show that this sequence is a decreasing family of logarithmically convex functions. Since the function $\ln I(\xi)$ is spherically symmetric, convex, and equal to zero at the origin of coordinates, we have for arbitrary $\xi'$ and $\xi''$

$$\ln I(\xi') + \ln I(\xi'') \leqq \ln I(\xi' + \xi''). \tag{9.5}$$

It is clear from the construction of the functions $I_\alpha$ that for arbitrary $\alpha$ we can find a positive number $\delta$ such that $I_{\alpha+1}((1+\delta)\,\xi)=I_\alpha(\xi)$. Therefore, it follows from (9) that

$$I_{\alpha+1}(\xi)\,I_{\alpha+1}(\delta\,\xi)\leqq I_{\alpha+1}((1+\delta)\,\xi)=I_\alpha(\xi). \tag{10.5}$$

Since the function $I$ is logarithmically convex, we have from (5.3) that $I_{\alpha+1}(\delta\,\xi)\geqq c\exp(\delta'\,|\xi|)$ for some positive $\delta'$. Therefore, we conclude from (10) that the sequence of functions $I_\alpha$ satisfies inequality $I'$, § 3.

The function $I$, being logarithmically convex, is defined and continuous in some neighborhood of the origin of coordinates. It follows that for some sufficiently small $\delta''$ we have

$$\sup_{|\xi'-\xi|\leqq\delta''} I_{\alpha+1}(\xi')\leqq C\leqq C\,I_\alpha(\xi) \tag{11.5}$$

for all $\xi$ such that $|\xi|<\delta''$. For the remaining points $\xi$, we have the inequality $|(1+\delta)\,\xi|\geqq|\xi|+\delta\,\delta''$. Then from (10)

$$\sup_{|\xi'-\xi|\leqq\delta\delta''} I_{\alpha+1}(\xi')\leqq I_\alpha(\xi).$$

From this and from (11) there follows the inequality II', § 3.

Since by hypothesis the function $u$ satisfies the inequality (8) for arbitrary $\varepsilon>0$, we may set $\varepsilon=\varepsilon_{-1}$ in this inequality. Thus, the function $u$ belongs to $\mathscr{E}^0_{I_0}$ and is a solution of the system (1.4) in $\mathscr{D}^*(\Omega_{I_0})$. Applying Theorem 1, § 4, we obtain the representation (2.4) for all functions $\phi\in[S^A_{I-A}]^s$, and

$$\|\mu\|_{-A}=\sum_\lambda\int(|z|+1)^{-A}\,\mathscr{I}_{-A}(y)\,|\mu^\lambda|<\infty.$$

We choose an arbitrary $q\geqq 0$, set $v=[n/2]+1$, and making use of (2.4), we estimate the functional $u$:

$$\begin{aligned}|(u,\phi)|\leqq&\sup_N\frac{|d(z,D)\,\tilde{\phi}^*(z)|}{(|z|+1)^{q+v+\kappa}\,\mathscr{I}_{-A-2}(y)}\sup_N\frac{(|z|+1)^{v+\kappa+A}\,\mathscr{I}_{-A-2}(y)}{\mathscr{I}_{-A-1}(y)}\\ &\times\sup_N\frac{(|z|+1)^q\,\mathscr{I}_{-A-1}(y)}{\mathscr{I}_{-A}(y)}\,\|\mu\|_{-A}.\end{aligned} \tag{12.5}$$

The first factor on the right side does not exceed

$$C\sup\frac{|\tilde{\phi}^*(z)|}{(|z|+1)^{q+v}\,\mathscr{I}_{-A-3}(y)}=C\,\|\tilde{\phi}\|^{\mathscr{I}^*_{-A-3}}_{-q-v},$$

and the constant $C$ does not depend on $q$. Since the functions $\mathscr{I}_\alpha$ form an increasing family of logarithmically convex functions,

$$\exp(\delta\,|y|)\,\mathscr{I}_{-A-2}(y)\leqq\mathscr{I}_{-A-1}(y)$$

for some $\delta>0$. This inequality, taken together with (1), shows that the second factor on the right side of (12) is finite.

We now estimate the third factor. Since the function $\mathscr{I}$ is spherically symmetric, and logarithmically convex, and since $\mathscr{I}_{\alpha+1}(y)=\mathscr{I}_\alpha((1+\delta)\,y)$ for some $\delta>0$, we conclude by analogy with the functions $I_\alpha$ that

$$\mathscr{I}_{-A-1}(y)\, i(\delta'(|y|+1))\leqq C\mathscr{I}_{-A}(y),$$

where $\delta'$ is sufficiently small. It follows that the third factor on the right side of (12) does not exceed

$$C\sup_N \frac{(|z|+1)^q}{i(\delta'(|y|+1))}\leqq C\cdot(2B)^q \sup \frac{(|y|+1)^{q/\gamma}}{i(\delta'(|y|+1))},$$

where the constant $C$ does not depend on $q$. Making the substitution $t=\ln(\delta'(|y|+1))$ in the right side, we conclude that it does not surpass

$$C\, b^q \sup \frac{\exp\left(t\dfrac{q}{\gamma}\right)}{i(\exp t)}=C\, b^q \mathscr{B}\left(\frac{q}{\gamma}\right).$$

Returning to (12), we arrive at the inequality

$$|(u,\phi)|\leqq C\, b^q \mathscr{B}\left(\frac{q}{\gamma}\right) \|\tilde{\phi}\|_{-q-v}^{\mathscr{I}^*_{A-3}}, \qquad \phi\in[\mathscr{D}(\Omega_{I_0})]^s, \tag{13.5}$$

in which the constant $C$ likewise is independent of $q$. According to Proposition 2, § 3, the righthand side does not exceed $C_q\|\phi\|_{I'}^{-q-v}$, $I'=I_{-A-4}$, and therefore the functional $u$ is continuous in the norm $\|\cdot\|_{I'}^{-q-v}$ on the functions belonging to $[\mathscr{D}(\Omega_{I'})]^s$. Since $[\mathscr{D}(\Omega_{I'})]^s$ is dense in $[S_{I'}^{-q-v}]^s$, the functional $u$ can be continued to the whole space $[S_{I'}^{-q-v}]^s$, the inequality (13) being conserved. Thus, we may substitute the function $\delta^j(\xi-\eta)$ in this inequality, with an arbitrary $\eta\in\Omega_{I'}$ and $|j|\leqq q$. On the left, we have the quantity $|D^j u(\eta)|$. To estimate the right side, we remark that

$$\|\widetilde{\delta^j(\xi-\eta)}\|_{-q-v}^{\mathscr{I}^*_{A-3}}=\sup\frac{|z^j \exp(i z,\eta)|}{(|z|+1)^{q+v}\mathscr{I}_{-A-3}(-y)}$$

$$\leqq \sup\frac{\exp(-y,\eta)}{\mathscr{I}_{-A-3}(-y)}=I_{-A-3}(\eta)\leqq I((1+\varepsilon)\,\xi).$$

Hence, finally

$$|D^j u(\eta)|\leqq C\, b^q \mathscr{B}\left(\frac{q}{\gamma}\right) I((1+\varepsilon)\,\xi), \quad \xi\in\Omega_{I'}, \quad |j|\leqq q, \quad q=0,1,2,\ldots. \quad \square$$

**3°. Examples**

**Example 1.** Let $I(\xi)=\exp(A|\xi|^{1/\alpha})$ where $0<\alpha\leqq 1$, and $A>0$. According to the calculations given in 8°, § 3, the logarithmic reciprocal of this function is equal to $\exp(A'|y|^{1/(1-\alpha)})$, and therefore, $\mathscr{B}(t)=B^t t^{t(1-\alpha)}$. It follows from Theorem 2 that an arbitrary solution of the system (1.4) which does not exceed a function of the form $C\exp(A|\xi|^{1/\alpha})$ admits the bound

$$|D^j u(\xi)|\leqq C B_1^{|j|}|j|^{|j|\frac{1-\alpha}{\gamma}}\exp(A_1|\xi|^{1/\alpha}).$$

**Example 2.** Let $\Omega$ be a sphere in $R^n$ with center at the origin of coordinates. Every solution of the system (1.4) in $\Omega$ satisfies the inequality (8), with an arbitrary $\varepsilon>0$ and $I=I_{\bar{\Omega}}$. In this case $\mathscr{I}(y)=\exp(A|y|)$ and $\mathscr{B}(t)=B^t t^t$. Therefore, by Theorem 2, the function $u$ on an arbitrary compact $K\subset\Omega$ satisfies the inequality

$$|D^j u(\xi)|\leqq C B_1^{|j|}|j|^{|j|\frac{1}{\gamma}}.$$

**4°. Elliptic operators.** An operator $p$ and the system (1.4) are said to be elliptic if for an arbitrary region $\Omega$ every solution of (1.4) which is a distribution in $\Omega$ can be extended as an analytic function in some $n$-dimensional complex neighborhood of the region $\Omega$.

**Corollary 1.** *The operator p is elliptic if and only if its index of hypoellipticity is equal to one.*

*Proof.* Let $\gamma=1$. If $\Omega$ is a sphere with center at the origin of coordinates, it follows from Example 2 of Subsection 3° that every generalized solution $u$ in this sphere admits on every compact $K\subset\Omega$ the bound

$$|D^j u(\xi)|\leqq C B^{|j|}|j|^{|j|}.$$

This bound, as is easily seen, guarantees the analyticity of the function $u$. With this the sufficiency is established. We remark that the neighborhood in which $u$ is continued as an analytic function depends solely on $B$ and, therefore, depends only on $p$, $\Omega$ and $K$, but not on $u$.

Let us prove the necessity. To do this we shall prove a more general assertion: if every continuous solution of (1.4) in $R^n$ is infinitely differentiable and satisfies the inequality

$$|D^j u(0)|\leqq C|j|^{|j|\beta}$$

for some $\beta>1$, then for the operator $p$, $\gamma\geqq 1/\beta$. Arguing as we did in the proof of Remark 2, we arrive at the inequality

$$\sup_j\frac{|D^j u(0)|}{|j|^{|j|\beta}}\leqq C\sup_{|\xi|\leqq R}|u(\xi)|,$$

which is valid for an arbitrary solution of (1.4) in $R^n$. Substituting a function $a\exp(z, -i\,\xi)\in\mathscr{E}_p(R^n)$ where $z\in N$, we obtain

$$\sup_j \frac{|z^j|}{|j|^{|j|\beta}} \leqq C\exp(R|y|), \qquad z\in N.$$

From the calculations connected with (23.3) it follows that the left side is not less than

$$C\exp\left(\frac{|z|}{A}\right)^{1/\beta},$$

whence we obtain

$$\exp\left(\frac{|z|}{A}\right)^{1/\beta} \leqq C\exp(R|y|),$$

which validates the inequality (1) with $\gamma \geqq 1/\beta$. □

**5°. Partially hypoelliptic operators.** A partially hypoelliptic operator is one corresponding to systems (1.4) with solutions which are in some sense infinitely differentiable with respect to some of their variables. At the moment, we shall describe two classes of partially hypoelliptic operators, corresponding to the two following distinct definitions of the infinite differentiability of distributions with respect to a portion of their variables.

Let $\xi=(\xi_1,\ldots,\xi_n)$ be a fixed system of coordinates in $R^n_\xi$ and let $\xi'=(\xi_1,\ldots,\xi_m)$, $\xi''=(\xi_{m+1},\ldots,\xi_n)$ be a fixed division of these variables $\xi$ into two groups. We denote by $R_{\xi'}$ and by $R_{\xi''}$ the corresponding coordinate subspaces $R^n_\xi$, and by $z'=(z_1,\ldots,z_m)$ and $z''=(z_{m+1},\ldots,z_n)$ the corresponding groups of dual variables.

Let $\Omega'$ and $\Omega''$ be regions in the spaces $R_{\xi'}$ and $R_{\xi''}$ respectively. Let $f$ be a distribution in the region $\Omega=\Omega'\times\Omega''\subset R^n$. Corresponding to each function $\psi\in\mathscr{D}(\Omega'')$, we can define the distribution $(f,\psi)_{\xi''}\in\mathscr{D}^*(\Omega')$, by the formula

$$((f,\psi)_{\xi''},\phi)=(f,\phi\times\psi), \qquad \phi\in\mathscr{D}(\Omega'). \tag{14.5}$$

It is not difficult to verify that the functional $(f,\psi)_{\xi''}$ is continuous on $\mathscr{D}(\Omega')$, that is, it is a distribution in $\Omega'$. If $K'\subset\Omega'$ and $K''\subset\Omega''$ are compacts and the distribution $f$ belongs to $\mathscr{E}^{-q}_{K'\times K''}$ for some $q$, then in the scalar product $(f,\phi\times\psi)$, we can substitute arbitrary functions $\phi\in\mathscr{D}^q_{K'}$ and $\psi\in\mathscr{D}^q_{K''}$. Hence it is not difficult to conclude that for arbitrary functions $\psi\in\mathscr{D}^q_{K''}$, formula (14) defines a distribution $(f,\psi)_{\xi''}$, belonging to $\mathscr{E}^{-q}_{K'}$.

**Definition.** We shall say that the distribution $f$ is weakly infinitely differentiable with respect to $\xi'$, if for an arbitrary function $\psi\in\mathscr{D}(\Omega'')$ the functional $(f,\psi)_{\xi''}$ is an infinitely differentiable function.

We shall say that the distribution $f$ is strongly infinitely differentiable with respect to $\xi'$, if for arbitrary compacts $K' \subset \Omega'$ and $K'' \subset \Omega''$, there exists an integer $q$, such that $f \in \mathscr{E}^{-q}_{K' \times K''}$ and for arbitrary functions $\psi \in \mathscr{D}^q_{K''}$, the functional $(f, \psi)_{\xi''}$ is an infinitely differentiable function within $K'$.

Let $\Omega$ be an arbitrary region in $R^n$. We shall say that the distribution $f$, defined in $\Omega$, is weakly (strongly) infinitely differentiable in the region if the restriction of $f$ on an arbitrary subregion of the form $\Omega' \times \Omega'' \subset \Omega$, has the same property, where $\Omega' \subset R_{\xi'}$, and $\Omega'' \subset R_{\xi''}$.

It is easy to see that the properties of weak and strong infinite differentiability are local: if either property holds in the neighborhood of an arbitrary point of the region $\Omega$, it holds over the whole region $\Omega$.

It is evident that an arbitrary function that is infinitely differentiable with respect to $\xi'$ in the ordinary sense, is strongly infinitely differentiable with respect to $\xi'$, and that an arbitrary finite sum of derivatives with respect to $\xi''$ of such functions is strongly infinitely differentiable. It is possible to show that every strongly infinitely differentiable function can be written locally in the form of such a finite sum. The class of weakly infinitely differentiable functions is essentially wider. We shall not go more deeply into this problem, since our task is more specialized, namely the study of systems of the type (1.4) having solutions that are weakly or strongly infinitely differentiable with respect to $\xi'$.

The operator $p$ and the system (1.4) will be said to be weakly (strongly) hypoelliptic in the variables $\xi'$, if for an arbitrary region $\Omega$ all the solutions of (1.4) belonging to $[\mathscr{D}^*(\Omega)]^s$, are weakly (strongly) infinitely differentiable with respect to $\xi'$.

**6°. A description of the partially hypoelliptic operators.** We consider the following characteristic:

$$\check{m}(r) = \inf\{|y| : z \in N, \ |z'| = r\}.$$

If the projection of $N$ on the coordinate subspace $C_{z'}$ is not bounded, the function $\check{m}(r)$ is defined for sufficiently large $r$, and accordingly, is equivalent to $\check{m}_0 r^{\check{\gamma}}$ as $r \to \infty$ for some $\check{m}_0 \geqq 0$ and rational $\check{\gamma}$. We set $\check{\gamma} = -\infty$, if $\check{m}_0 = 0$. If the projection of $N$ on $C_{z'}$ is bounded, we write $\check{\gamma} = \infty$. It is easily seen that for arbitrary $\check{\gamma} > -\infty$ the inequality

$$|z'| \leqq B(|y| + 1)^{1/\check{\gamma}} \tag{15.5}$$

holds on the variety $N$ for some $B > 0$.

The function

$$\check{m}(r) = \inf\{|y'| + |z''| : z \in N, \ |z'| = r\}.$$

has similar properties. If the projection of $N$ on $C_{z'}$ is unbounded, this function is defined for sufficiently large $r$ and is equivalent to $\hat{m}_0 r^{\check{\gamma}}$ as $r \to \infty$. We set $\hat{\gamma} = -\infty$, if $\hat{m}_0 = 0$, and we set $\hat{\gamma} = \infty$, if the projection of $N$ on $C_{z'}$ is bounded. For arbitrary $\gamma > -\infty$, we have the inequality

$$|z'| \leqq B(|y'| + |z''| + 1)^{1/\check{\gamma}}, \qquad B > 0. \tag{16.5}$$

on the variety $N$.

The following theorem defines weakly and strongly hypoelliptic operators.

**Theorem 3.** *The operator $p$ is strongly hypoelliptic in the variables $\xi'$, if and only if $\check{\gamma} > 0$.*

*The operator $p$ is weakly hypoelliptic in the variables $\xi'$, if and only if $\check{\gamma} > 0$.*

*Proof.* We shall establish the sufficiency only. Since infinite differentiability in either the strong or the weak sense is a local property, we have only to show that it holds for solutions of the systems (1.4) in regions of the form $\Omega = \Omega' \times \Omega''$, where $\Omega'$ and $\Omega''$ are spheres in $R_{\xi'}$ and $R_{\xi''}$ with centers at the origin of coordinates and of radius $\rho$, where $\rho$ is an arbitrary positive number. For an arbitrary $\varepsilon$, $0 < \varepsilon < \rho$, we shall denote by $K'_\varepsilon$ and $K''_\varepsilon$ closed spheres concentric with $\Omega'$ and $\Omega''$ of radius $\rho - \varepsilon$. We write $K_\varepsilon = K'_\varepsilon \times K''_\varepsilon$.

Let $u$ be an arbitrary distribution in $\Omega$ which satisfies (1.4). For arbitrary $\varepsilon > 0$, the function $u$ belongs to the space $[\mathscr{E}^q_{K_\varepsilon}]^s$ for some $q$. We fix an arbitrary $\varepsilon > 0$ not exceeding $\rho/5$, and we apply to $u$ Corollary 1 of § 4. This corollary says that the functional $u$ admits the representation (2.4) for an arbitrary function $\phi \in [\mathscr{D}_{K_{2\varepsilon}}]^s$, and that the inequality

$$\begin{aligned} \|\mu\| &= \sum_\lambda \int (|z|+1)^{q-A} \mathscr{I}_{K_{2\varepsilon}}(y) \, |\mu^\lambda| \\ &\leqq \sum_\lambda \int (|z|+1)^{q-A} \exp[(\rho - 2\varepsilon)(|y'| + |y''|)] \, |\mu^\lambda| \leqq C \, \|u\|^q_{K_\varepsilon} \end{aligned}$$

holds.

We fix an arbitrary non-negative integer $r$, not less than $A - q$. We suppose that $\check{\gamma} > 0$. The inequality (15) implies that for an arbitrary integer $k$, we have $|z'|^k = o(\exp(\varepsilon|y'| + |y''|))$. From this we deduce the inequality

$$\int (|z'|+1)^k \exp[(\rho - 3\varepsilon)(|y'| + |y''|)] \left| \frac{\mu^\lambda}{(|z''|^2+1)^r} \right| \leqq C_k \, \|\mu\| . \tag{17.5}$$

Let us now consider the collection of functionals

$$(v^\lambda, \phi) = \int \tilde{\phi}^*(z) \frac{\mu^\lambda}{(|z''|^2+1)^r}, \qquad \phi \in [\mathscr{D}_{K_{4\varepsilon}}]^{e_\lambda}; \ \lambda = 0, \dots, l$$

($e_\lambda$ is the number of elements in the column $d^\lambda$). The inequality (17), with $k=0$, implies that all of the functionals $v^\lambda$ are continuous in the norm $\|\tilde{\phi}^*\|_0^{K_{3\varepsilon}}$ and therefore – by Proposition 2, § 3 – they are continuous in the norm $\|\phi\|^0$. Consequently $v^\lambda$ belongs to $[\mathscr{E}^0_{K_{4\varepsilon}}]^{e_\lambda}$, i.e. it is a square-summable function on $K_{4\varepsilon}$. Since (17) holds for arbitrary $k$, all the derivatives of the $v^\lambda$ with respect to $\xi'$ have a similar property.

Taking account of the fact that $(|z''|^2+1)^r=\sum\binom{r}{j}|z''|^{2j}$, we have

$$\begin{aligned}(u,\phi)&=\sum_{\lambda,j}\left(v^\lambda, F^{-1}\left[\binom{r}{j}|z''|^{2j}d^\lambda(z,D)\tilde{\phi}^*\right]^*\right)\\ &=\sum_{\lambda,j}\left(v^\lambda,\binom{r}{j}|iD_{\xi''}|^{2j}\bar{d}^\lambda(iD,i\xi)\phi\right),\quad |iD_{\xi''}|^2=-\sum_{m+1}^{n}\frac{\partial^2}{\partial\xi_k^2}.\end{aligned}\tag{18.5}$$

In the right side we carry all the differentiations with respect to $\xi'$ from the functions $\phi$ to $v^\lambda$ by an integration by parts. This yields

$$(u,\phi)=\sum_{|j|\leqq 2r+\kappa}\int w_j(\xi)\,D^j_{\xi''}\phi(\xi)\,d\xi,\tag{19.5}$$

where the $w_j$ are functions belonging to $[\mathscr{E}^0_{K_{4\varepsilon}}]^s$ together with all their derivatives with respect to $\xi'$. Then for arbitrary $\psi\in\mathscr{D}^{2r+\kappa}_{K''_{4\varepsilon}}$, the functional $(u,\psi)_{\xi''}$ coincides on $K'_{4\varepsilon}$ with the function

$$\sum_{|j|\leqq 2r+\kappa}\int w_j(\xi',\xi'')\,D^j_{\xi''}\psi(\xi'')\,d\xi'',\tag{20.5}$$

which is infinitely differentiable with respect to $\xi'$. Since $\varepsilon$ is arbitrary, $u$ is strongly infinitely differentiable with respect to $\xi'$. Therefore the condition $\check{\gamma}>0$ is sufficient for $p$ to be strongly hypoelliptic in $\xi'$.

Now suppose that $\hat{\gamma}>0$. The inequality (16) implies that for arbitrary non-negative $k$

$$|z'|^k=O\big(\exp(\varepsilon|y'|)(|z''|+1)^{k/\hat{\gamma}}\big),$$

and, therefore, the inequality (17) follows if $r\geqq A-q+(k/\hat{\gamma})$. Choosing $r$ to satisfy this condition, we arrive at (18), in which the $v^\lambda$ are functions whose derivatives with respect to $\xi'$, up to order $k$, belong to $[\mathscr{E}^0_{K_{4\varepsilon}}]^{e_\lambda}$. Completing the integration by parts, we carry all the differentiations with respect to $\xi'$ (up to an order not exceeding $\kappa$) to the functions $v^\lambda$. Then we obtain (19), in which all the derivatives with respect to $\xi'$ up to the order $k-\kappa$ of the functions $w_j$, belong to $[\mathscr{E}^0_{K_{4\varepsilon}}]^s$. It follows that for arbitrary $\psi\in\mathscr{D}^{2r+\kappa}_{K''_{4\varepsilon}}$ the function (20) belongs to the space $[\mathscr{E}^{k-\kappa}_{K'_{4\varepsilon}}]^s$. Therefore, if $\psi\in\mathscr{D}_{K''_{4\varepsilon}}$, the function (20) is infinitely differentiable inside $K'_{4\varepsilon}$. So, $u$ is weakly infinitely differentiable with respect to $\xi'$. This proves the sufficiency of the condition $\hat{\gamma}>0$. □

*Remark.* Let $u$ be an arbitrary distribution in the region $\Omega' \times \Omega''$ which satisfies the weakly hypoelliptic system (1.4). The weak infinite differentiability allows us to define the restriction of $u$ on an arbitrary subspace of the form $\xi' = \xi'_0$, where $\xi'_0 \in \Omega'$. This restriction, which we denote by $u_0$, is a distribution in $\Omega''$ and is defined by

$$(u_0, \psi) = (u, \psi)_{\xi''}|_{\xi' = \xi'_0}. \tag{21.5}$$

We shall show that the left side is in fact a continuous functional on the subspace $\mathscr{D}(\Omega'')$ (with values in $C^s$). We fix an arbitrary $\varepsilon > 0$ and we suppose that the function $\psi$ belongs to the space $\mathscr{D}_{K_{4\varepsilon}}$. Then, as we found in proving Theorem 3, the left side of (21) is equal to the function (20) at the point $\xi' = \xi'_0$. Suppose that $k \geqq \kappa + \nu$. Then by the construction of the functions $w_j$,

$$\sum_j \|w_j\|^{\nu}_{K_{4\varepsilon}} \leqq C \sum_\lambda \|v^\lambda\|^{\nu+\kappa}_{K_{4\varepsilon}} \leqq C' \|\mu\| \leqq C'' \|u\|^q_{K_\varepsilon}.$$

Hence

$$\begin{aligned} |(u_0, \psi)| &\leqq C \|\psi\|^{2r+\kappa} \sup_{K_{4\varepsilon}} \sum_j |w_j| \\ &\leqq C' \|\psi\|^{2r+\kappa} \sum_j \|w_j\|^{\nu}_{K_{4\varepsilon}} \leqq C'' \|\psi\|^{2r+\kappa} \|u\|^q_{K_\varepsilon}. \end{aligned}$$

We have now shown that the functional $u_0$ belongs to the space $[\mathscr{E}^{-2r-\kappa}_{K_{4\varepsilon}}]^s$, and the mapping $u \to u_0$ is continuous in the norms $\|u\|^q_{K_\varepsilon}$ and $\|u_0\|^{-2r-\kappa}_{K_{4\varepsilon}}$. Since $\varepsilon > 0$ is arbitrary, we conclude that $u_0$ is a distribution in $\Omega''$, and the mapping $[\mathscr{D}^*(\Omega)]^s \ni u \to u_0 \in [\mathscr{D}^*(\Omega'')]^s$ is continuous.

For arbitrary $j \in Z^m_+$ we consider the functional

$$(u_j, \psi) = D^j_{\xi'}(u, \psi)_{\xi''}|_{\xi' = \xi'_0}$$

which is the restriction of the corresponding derivatives of the distribution $u$ for $\xi' = \xi'_0$. This functional is a distribution in $\Omega''$ and the mapping $u \to u_j$ is continuous.

We remark that the functional $u$ is an integral of its restriction (21), that is, the equation

$$(u, \phi) = \int (u, \phi(\xi'_0, \xi''))_{\xi''}|_{\xi' = \xi'_0} \, d\xi'_0.$$

holds. To prove this, we choose $\varepsilon > 0$ small enough so that $\operatorname{supp} \phi \subset K_{4\varepsilon}$. Then the action of the functional $(u, \cdot)_{\xi''}$ is given by (20), and the action of $u$ is given by (19), and this yields the result we want. In particular, if $(u, \cdot)_{\xi''} \equiv 0$, $u = 0$.

### 7°. Examples

**Example 1.** Let us suppose that the operator $p$ contains differentiations with respect only to the variables $\xi'$, that is, $p(iD) = p(iD_{\xi'})$, and that the operator $p(iD_{\xi'})$ — considered as an operator in the space $R_{\xi'}$ — is hypoelliptic. So on the variety $N' \subset C_{z'}$ associated with the matrix $p'(z')$, the inequality $|z'| \leqq B(|y'|+1)^{1/\gamma}$ holds. If we look at the matrix $p'$ in the space $R^n$, the associated variety is $N' \times C_{z''}$. In this variety, we have the inequality $|z'| \leqq B(|y|+1)^{1/\gamma}$, and accordingly, the operator $p$ in the space $R^n_\xi$ is strongly hypoelliptic with respect to the variables $\xi'$.

**Example 2.** Let $s=t=m=1$, and let $p(z)$ be an arbitrary polynomial which admits an expansion in powers of $z_1$

$$p(z) = \sum_{j=0}^{\mu} p_j(z'')\, z_1^{\mu-j}, \qquad \mu = \deg p, \tag{22.5}$$

in which $p_0(z'')$ is a constant different from zero. We shall show that the corresponding operator $p(iD)$ is weakly hypoelliptic in $\xi_1$. In fact, in the variety $N$ we have the inequality $|z_1| \leqq B(|z''|+1)^d$, where

$$d = \max_j \frac{\deg p_j}{j}\,[6]$$

from which we derive the inequality (16) with $\hat{\gamma} = 1/d$, and this implies the weak hypoellipticity of the operator $p$ with respect to $\xi_1$.

**Example 3.** Suppose again that $s=t=1$, and let

$$p(iD) = \frac{\partial^2}{\partial \xi_1^2} - \sum_2^n \frac{\partial^2}{\partial \xi_j^2}$$

be the wave operator. The variety associated with the matrix $p'$ is defined by the equation $z_1^2 = \sum_2^n z_j^2$. We write $\xi' = (\xi_2, \ldots, \xi_n)$, $\xi'' = \xi_1$ and $z' = (z_2, \ldots, z_n)$, $z'' = z_1$. Comparing the real parts in the equation of the variety, we obtain the relation $|x'|^2 - |y'|^2 = |x''|^2 - |y''|^2$ from which we may derive the inequality

$$|x'|^2 \leqq |y'|^2 + |z''|^2. \tag{23.5}$$

This shows that the operator $p$ is weakly hypoelliptic in the variables $\xi'$. We note that this inequality holds for any change of the coordinate system in $C^n_z$, for which the new axis $\tilde{x}_1$ remains within the cone $|x''| = |x'|$, provided that we admit a constant multiplier on the left side of (23). Therefore the operator $p$ remains weakly hypoelliptic with respect to $\tilde{\xi}'$ for an arbitrary rotation in $R^n_\xi$, that keeps the new axis of $\tilde{\xi}_1$ within the cone $|\xi''| = |\xi'|$. In particular, every distribution which satisfies the wave equation has a restriction on an arbitrary time-like line.

---

6 See, for example, Gel'fand and Šilov [3].

**8°. Smoothing operators for the solutions of the system** (1.4). Let $h$: $\mathscr{P}^s \to \mathscr{P}^s$ be an arbitrary $\mathscr{P}$-matrix. We shall say that an operator $h(iD)$ smooths the solutions of the system (1.4) if, for any region $\Omega \subset R^n_\xi$, any solution $u \in [\mathscr{D}^*(\Omega)]^s$, compact $K \subset \Omega$, and any non-negative integer $\alpha$, we can find an integer $\beta$, such that the function $h^\beta(iD)\,u$ belongs to $[\mathscr{E}^\alpha_K]^s$. The following theorem characterizes all smoothing operators.

**Theorem 4.** *The operator h is a smoothing operator if and only if for arbitrary* $R>0$

$$|h(z)| \to 0 \quad \text{when} \quad z \in N,\ |y| \leqq R,\ |z| \to \infty.$$

The proof is left to the reader.

## § 6. Uniqueness of solutions of the Cauchy problem

**1°. Statement of the problem.** Let $L$ be an arbitrary linear subspace in $R^n$, with dimensionality between 0 and $n-1$. We intend to formulate an analog of the Cauchy problem for the system (1.4) with initial conditions given on $L$. Since we want to include in our considerations the solutions of this system which are distributions, we must require that these solutions have restrictions on $L$. We therefore choose a system of coordinates in $R^n$ such that the subspace $L$ coincides with the coordinate subspace corresponding to the variables $\xi'' = (\xi_{m+1}, \ldots, \xi_n)$, where $n-m = \dim L$. Setting $\xi' = (\xi_1, \ldots, \xi_m)$, we assume that the operator $p$ is weakly hypoelliptic in the $\xi'$. Then as we showed in § 5, any distribution in an arbitrary neighborhood of zero $\Omega$ which satisfies (1.4), and also any of its derivatives, has a restriction on $\Omega'' = \Omega \cap L$ which is a distribution.

**Definition.** Let $\Omega$ be some neighborhood of zero in $R^n$, and let $\Phi$ be a subspace of $\mathscr{D}^*(\Omega)$. We shall say that the Cauchy problem for the system (1.4) with initial conditions on $L$ has a unique solution in $\Phi$, if the conditions

$$u \in \Phi_p, \quad D^i_{\xi'}\, u|_L = 0\ \forall\, i \tag{1.6}$$

imply that $u \equiv 0$. We have denoted by $\Phi_p$ the subspace in $[\Phi]^s$, consisting of solutions of (1.4) in $\mathscr{D}^*(\Omega)$.

What we usually mean by the uniqueness of the Cauchy problem is a special case of our present definition. In the usual form of the Cauchy problem we are dealing with a finite number of initial conditions. For example, if the system (1.4) consists of one equation with one unknown function, and $m=1$, and the subspace $L$ is not characteristic with respect to the operator $p$, then on the subspace there are given $\mu = \deg p$ initial conditions $\dfrac{\partial^i u}{\partial \xi_1^i}$, $i = 0, \ldots, \mu-1$.

Since the subspace $L$ is non-characteristic, the operator $p_0$ in the expansion

$$p=\sum_{0}^{\mu} p_j(D_{\xi''})\frac{\partial^{\mu-j}}{\partial\xi_1^{\mu-j}}$$

is a constant different from zero. Consequently the equation $pu=0$ and the conditions $\left.\frac{\partial^i u}{\partial\xi_1^i}\right|_L \equiv 0$, $i=0,\ldots,\mu-1$ imply that $\left.\frac{\partial^i u}{\partial\xi_1^i}\right|_L =0$ for all $i$.

**2°. A theorem on the uniqueness of a solution of the Cauchy problem**

**Theorem.** *Suppose that on the variety $N$ the inequality*

$$|z'|\leqq C(|z''|^{1/\beta}+|y'|^{1/\gamma}+1) \tag{2.6}$$

*holds with $0<\gamma\leqq 1$, $0<\beta<1$. Then the Cauchy problem with initial conditions given on $L$ has only one solution in the space of distributions defined on the region $\Omega=\Omega'\times R_{\xi'}$ and belonging to the space $[\mathscr{E}_I^q]^s$, where*

$$I=I(\xi)=I'(\xi')\,I''(\xi'')$$

*and*

$$\Omega'=\begin{cases} R_{\xi'},\\ \{\xi'\colon |\xi'|<A'\};\end{cases}$$

$$I'(\xi')=\begin{cases}\exp(A'|\xi'|^{1/(1-\gamma)}), & \gamma<1,\\ I_{\Omega'}(\xi'), & \gamma=1;\end{cases} \tag{3.6}$$

$$I''(\xi'')=\exp(A''|\xi''|^{1/(1-\beta)}),$$

*$A'$, and $A''$ are arbitrary positive constants and $q$ is an arbitrary integer.*

We remark that the inequality (2) holds on the variety $N$ if and only if $p$ is weakly hypoelliptic in $\xi'$. If $\gamma=1$, the theorem implies in particular that every ordinary solution of (1.4) which vanishes on $L$, and is defined in the band $|\xi'|<A'$ and increases at infinity not faster than $C\exp(A''|\xi''|^{1/(1-\beta)})$, vanishes identically. If $\gamma<1$, then a similar assertion holds in the class of functions defined in $R^n$ and not exceeding

$$C\exp(A'|\xi'|^{1/(1-\gamma)}+A''|\xi''|^{1/(1-\beta)}).$$

The proof of the theorem is based on the first example of 3°, § 5. If we postulate that the operator $p$ is hypoelliptic, the calculations that we made in that example show that every solution of (1.4) in $R^n$, which increases at infinity no faster than $C\exp(A|\xi|^{1/(1-\gamma)})$ ($\gamma$ is the index of hypoellipticity), satisfies the inequality

$$|D^j u(\xi)|\leqq C B^{|j|}\,|j|^{|j|}$$

everywhere, that is, it is analytic in $R^n$. If all the derivatives of such a function vanish at some point, it vanishes identically. Hence, we may conclude that the Cauchy problem for hypoelliptic systems (1.4) with initial conditions on a zero dimensional subspace $L$ has a unique solution increasing no faster than $C\exp(A|\xi|^{1/(1-\gamma)})$.

We remark that the considerations we have just gone through represent a particular case of our theorem, namely, that for which $m=0$. In the general case, the operator $p$ is only weakly hypoelliptic in the variables $\xi'$. Given an arbitrary function $\psi\in\mathscr{D}(R_{\xi''})$ we can consider the quantity $(u,\psi)_{\xi''}$, which is an infinitely differentiable function of $\xi'$. As we shall prove, this function has a property similar to that which we observed for the solutions of hypoelliptic systems, namely, that under the conditions of bounded growth at infinity formulated in the theorem, the function $(u,\psi)_{\xi''}$ is analytic for some dense subspace of functions $\psi$. The initial conditions (1) imply that $(u,\psi)_{\xi''}\equiv 0$, whence $u=0$.

**3°. Proof of the theorem.** We fix an arbitrary $\varepsilon>0$ and we write $\check{I}'(\xi')=I'((1+\varepsilon)\,\xi')$, $\quad \check{I}''(\xi'')=I''((1+\varepsilon)\,\xi'')$. The computations of 8°, § 3 show that the logarithmic reciprocals of $\check{I}'$ and $\check{I}''$ are respectively equal to

$$\mathscr{I}'(y')=\exp(a'|y'|^{1/\gamma}),\qquad \mathscr{I}''(y'')=\exp(a''|y''|^{1/\beta})$$

for some positive values of $a'$ and $a''$. We choose some strongly decreasing sequence of numbers $\{\varepsilon_\alpha,\ -\infty<\alpha<\infty\}$, lying between 0 and $\varepsilon$. We write

$$I'_\alpha(\xi')=I'((1+\varepsilon_\alpha)\,\xi'),\qquad I''_\alpha(\xi'')=I''((1+\varepsilon_\alpha)\,\xi'').$$

Since the functions $I'$ and $I''$ are logarithmically convex and spherically symmetric in $R_{\xi'}$ and $R_{\xi''}$ respectively, the argument given in the proof of Theorem 2, § 5, implies that the functions $I'_\alpha$ and $I''_\alpha$ form a decreasing family of logarithmically convex functions. It follows that the functions $I_\alpha(\xi)=I'_\alpha(\xi')\,I''_\alpha(\xi'')$ form a decreasing family of logarithmically convex functions in $R^n$. We remark that for an arbitrary $\alpha$, we have $I(\xi)\leqq I_\alpha(\xi)\leqq \check{I}(\xi)=\check{I}'(\xi')\,\check{I}''(\xi'')$, and the functions $I$, $I_\alpha$, and $\check{I}$ are derivable one from the other with the aid of some dilation in $R^n$ with center at the origin of coordinates. It follows that the logarithmic reciprocals are interconnected in a similar way. In particular, for arbitrary $\alpha$, the function $\mathscr{I}_\alpha$ – that is, the logarithmic reciprocal with respect to $I_\alpha$ – has the form

$$\exp[(a'+\delta')|y'|^{1/\gamma}+(a''+\delta'')|y''|^{1/\beta}]$$

for some positive $\delta'$ and $\delta''$.

Let $u$ be an arbitrary solution of (1.4) in $\Omega$, belonging to the space $[\mathscr{E}^q_I]^s$. The inequality $I\leqq I_q$ implies that $u\in[\mathscr{E}^q_{I_q}]^s$. It therefore follows from

Theorem 1 of § 4 that the functional $u$ admits the representation (2.4) in which $\phi$ is an arbitrary function in the space $[S_{I_{q-A}}^{A-q}]^s$ and

$$\sum_{\lambda} \int (|z|+1)^{q-A} \exp[(a'+\delta')|y'|^{1/\gamma} + (a''+\delta'')|y''|^{1/\beta}]\,|\mu^{\lambda}| < \infty \tag{4.6}$$

for some positive $\delta'$ and $\delta''$.

Let $\phi$ and $\psi$ be arbitrary functions belonging respectively to the spaces $S_{\tilde{I}'}^{q'}$ and $[S_{\tilde{I}''}^{q''}]^s$ for arbitrary $q'$ and $q''$. Proposition 2 of § 3 implies that the Fourier transforms satisfy the inequalities

$$|\tilde{\phi}^*(z)| \leqq C\|\phi\|_{\tilde{I}'}^{q'} (|z'|+1)^{-q'} \exp[(a'+\varepsilon')|y'|^{1/\gamma}], \tag{5.6}$$

$$|\tilde{\psi}^*(z)| \leqq C\|\psi\|_{\tilde{I}''}^{q''} (|z''|+1)^{-q''} \exp[(a''+\varepsilon'')|y''|^{1/\beta}] \tag{6.6}$$

with arbitrary integers $q'$ and $q''$ and positive $\varepsilon'$ and $\varepsilon''$. The product $\phi\times\psi$ clearly belongs to the space $[S_{\tilde{I}}^{q}]^s$ for arbitrary $q$ and, therefore, can be substituted into the representation (2.4) yielding

$$\overline{(u, \phi\times\psi)} = \sum_{\lambda} \int d^{\lambda}(z, D)\, \tilde{\phi}^*(z')\, \tilde{\psi}^*(z'')\, \mu^{\lambda}. \tag{7.6}$$

Then the inequalities (4), (5), and (6) allow us to derive a bound for the absolute value of the right side:

$$C \sup_{N} \frac{|d^{\lambda}(z, D)\, \tilde{\phi}^*(z')\, \tilde{\psi}^*(z'')|}{(|z|+1)^{q-A} \exp[(a'+\delta')|y'|^{1/\gamma} + (a''+\delta'')|y''|^{1/\beta}]}$$

$$\leqq C' \|\phi\|_{\tilde{I}'}^{q'} \|\psi\|_{\tilde{I}''}^{q''} \sup_{N} \frac{(|z'|+1)^{-q'}(|z''|+1)^{-q''}}{(|z|+1)^{q-A-\kappa}} \times \frac{\exp[(\alpha'+\varepsilon')|y'|^{1/\gamma} + (a''+\varepsilon'')|y''|^{1/\beta}]}{\exp[(a'+\delta')|y'|^{1/\gamma} + (a''+\delta'')|y''|^{1/\beta}]}. \tag{8.6}$$

(We recall that the coefficients of the operators $d^{\lambda}(z, D)$ are polynomials of order not exceeding $\kappa$.) Since the positive numbers $\varepsilon'$ and $\varepsilon''$ in this inequality are arbitrary, we can impose the conditions $\varepsilon' \leqq \delta'/2$. Then the second factor on the right side following the supremum symbol does not exceed $\exp(-\varepsilon'|y'|^{1/\gamma})$. The quantity $q'$ is to be left arbitrary and $q''$ is to be set equal to the smallest integer $q''(q')$ satisfying the inequality

$$q'' \geqq -q + A + \kappa + \frac{1}{\beta}(-q + A + \kappa - q').$$

Then, because of (2) the upper bound on the right side of (8) is finite. Hence

$$|((u, \psi)_{\xi''}, \phi)| = |(u, \phi\times\psi)| \leqq C\|\phi\|_{\tilde{I}'}^{q'} \|\psi\|_{\tilde{I}''}^{q''}.$$

This inequality implies that both sides of Eq. (7) are defined and continuous in the space $\mathscr{D}(\Omega_{\bar{I}'})\times[\mathscr{D}(\Omega_{\bar{I}''})]^s$ and are continuous in the norm $\|\phi\|_{\bar{I}'}^{q'}+\|\psi\|_{\bar{I}''}^{q''}$. Therefore, Eq. (7) can be extended to the space $S_{\bar{I}'}^{q'}\times[S_{\bar{I}''}^{q''}]^s$. In particular, for arbitrary $j\in Z_+^m$ we can substitute in (7) the functions $\phi=\delta^j(\xi'-\eta')$, where $\eta'\in\Omega_{\bar{I}'}$, and the arbitrary function $\psi\in[S_{\bar{I}''}^{q''}]^s$, where $q''=q''(-|j|-\nu)$. By an obvious transformation we arrive at the equation

$$D_{\xi'}^{j}(u,\psi)_{\xi''}|_{\xi'=\eta'}=\sum_{\lambda}\int d^{\lambda}(z,D)\exp(z',-i\eta')(-iz')^j\tilde{\psi}^*(z'')\mu^{\lambda}.$$

We find a bound for the right side by postulating that the Fourier transform of $\psi$ satisfies the inequality

$$|\tilde{\psi}^*(z'')|\leqq c\exp(-\delta|x''|^{1/\beta}+\delta|y''|^{1/\beta}) \tag{9.6}$$

for an arbitrary $\delta>0$. We have

$$\begin{aligned}
&|D_{\xi'}^{j}(u,\psi)_{\xi''}|_{\xi'=\eta'}|\\
&\qquad\leqq C\sup_N\frac{|d^{\lambda}(z,D)\exp(z',-i\eta')(-iz')^j\tilde{\psi}^*(z'')|}{(|z|+1)^{q-A}\exp[(a'+\delta')|y'|^{1/\gamma}+(a''+\delta'')|y''|^{1/\beta}]}\\
&\qquad\leqq C(|j|+1)^{\kappa}(|\eta'|+1)^{\kappa}\\
&\qquad\quad\times\sup_N\frac{(|z|+1)^{r+|j|}\exp(y',\eta')\exp[-\delta|x''|^{1/\beta}+\delta|y''|^{1/\beta}]}{\exp[(a'+\delta')|y'|^{1/\gamma}+(a''+\delta'')|y''|^{1/\beta}]},\\
&\qquad\qquad r=\kappa+A-q.
\end{aligned} \tag{10.6}$$

The quantity on the right side under the supremum sign does not exceed

$$\begin{aligned}
(|z|+1)^{r+|j|}\exp[&|y'|\,|\eta'|-(a'+\delta')|y'|^{1/\gamma}\\
&-(a''+\delta''-\delta)|y''|^{1/\beta}-\delta|x''|^{1/\beta}].
\end{aligned} \tag{11.6}$$

We require that $\delta\leqq\delta''$ and that the point $\eta'$ belong to the region $\omega'$, which is set equal to the space $R_{\xi'}$ if $\gamma<1$, and to the sphere $|\xi'|<a'$ if $\gamma=1$. It then follows, in view of the inequality (2), that the quantity inside the square brackets in (11) does not exceed $-\delta_0|z|$ for some $\delta_0>0$. Therefore, the upper bound on the righthand side of (10) does not exceed

$$C\sup(|z|+1)^{r+|j|}\exp(-\delta_0|z|).$$

The computations given in 8°, § 3 imply that this upper bound is not larger than

$$CB^{r+|j|}(r+|j|)^{r+|j|}\leqq C_1B_1^{|j|}|j|^{|j|}.$$

From this inequality and (10), we conclude that the function $(u,\psi)_{\xi''}$ is analytic in the region $\omega'$.

By hypothesis, the function $u$ vanishes on $L$ together with all its derivatives with respect to $\xi'$. This means that

$$D^j_{\xi'}(u,\psi)_{\xi''}|_{\xi'=0}=0 \tag{12.6}$$

for an arbitrary function $\psi\in[\mathscr{D}(\Omega_{\bar{I}''})]$. As we proved earlier, the functional (12) is defined on the space $[S^{q''}_{\bar{I}''}]^s$, $q''=q''(-|j|-v)$ and, therefore, vanishes, since in view of (12) it vanishes on a dense subspace $[\mathscr{D}(\Omega_{\bar{I}''})]^s$. We note that an arbitrary function $\psi$, satisfying the inequality (9) will itself belong to the space $[S^{q''}_{\bar{I}''}]^s$. Therefore, the function $(u,\psi)_{\xi''}$ is analytic in $\omega'$, and also satisfies the conditions (12). Hence, $(u,\psi)_{\xi''}=0$ in $\omega'$ for an arbitrary $\psi$ that satisfies (9).

We shall now show that any function $\chi\in[\mathscr{D}(\Omega_{\bar{I}''})]^s$ can be approximated in the norm $\|\cdot\|^{q''}_{\bar{I}''}$, $q''=q''(-v)$ by the functions satisfying (9). Since the functional $(u,\psi)_{\xi''}|_{\xi'=\eta'}$ is continuous in this norm, it will follow that $u\equiv 0$ in $\omega'\times R_{\xi''}$. Since $\beta<1$, there exists a function $\psi_1$,[7] satisfying (9), and having an integral equal to unity over $R_{\xi''}$. We consider the sequence of functions $\psi_k(\xi)=k^{n-m}\psi_1(k\xi)$, $k=1,2,\ldots$. Each of these functions satisfies (9) for arbitrary $\delta>0$. The formula

$$F[\psi_k(\xi''-\eta'')]=\exp(z'',i\eta'')\,F[\psi_k(\xi'')]$$

implies that an arbitrary translation of the function $\psi_k$ also satisfies (9) for arbitrary $\delta>0$. Hence it follows that for an arbitrary $\chi\in[\mathscr{D}(\Omega_{\bar{I}''})]^s$ the convolution $\chi*\psi_k$ has the same property. We remark that the integral of the function $|I''\psi_k|^2$, taken over the complement of an arbitrary neighborhood of zero, tends to zero as $k\to\infty$, while the integral of $\psi_k$ over $R_{\xi''}$ is equal to unity for arbitrary $k$. Hence it follows that $\|\chi*\psi_k-\chi\|^0_{\bar{I}''}\to 0$. Therefore, for arbitrary $j$ the difference $D^j_{\xi''}(\chi*\psi_k)-D^j_{\xi''}\chi=D^j\chi*\psi_k-D^j\chi$ tends to zero in the norm $\|\cdot\|^0_{\bar{I}''}$. Consequently the difference $\chi*\psi_k-\chi$ tends to zero in the norm $\|\cdot\|^{q''}_{\bar{I}''}$, which is what we were to prove.

Therefore, we have established that the function $u$ is equal to zero in the region $\omega'\times R_{\xi''}$. If $\gamma<1$, (3) implies that $\omega'=R_{\xi'}$, whence $u\equiv 0$. We consider the case $\gamma=1$. Then by hypothesis the domain of definition of $u$ is the band $|\xi'|<A'$, while $\omega'$ is the band $|\xi'|<a'$. The construction of the function $\exp(a'|y'|)$ implies that it is the logarithmic reciprocal of a function which is equal to unity in the sphere $|\xi'|<\frac{A'}{1+\varepsilon}$, and infinite outside this sphere. Hence $a'=\frac{A'}{1+\varepsilon}$. Since $\varepsilon>0$ is arbitrary we have $u\equiv 0$ over the whole domain of definition. □

7 See Gel'fand and Šilov [2].

### 4°. Examples

**Example 1.** Let $s=t$, $m=1$. Then the reduced order of the system (1.4) is the smallest $p_0$ such that on the variety $N$, the inequality

$$|z'| \leqq C(|z''|^{p_0}+1).$$

is satisfied. If $p_0>1$, then the inequality (2) is satisfied, with $\gamma=1$, and $\beta=1/p_0$. Then the theorem we have just proved guarantees the uniqueness of the solution of the Cauchy problem in the space of distributions defined in the band $|\xi'|<A'$ and increasing at infinity not faster than $\exp(A''|\xi''|^{(p_0/p_0-1)})$. Here $A'>0$ and $A''>0$ are arbitrary. G. N. Zolotarev showed in [2] that the exponent $p_0/(p_0-1)$ is the largest that will guarantee the uniqueness of the solution of the Cauchy problem. It is reasonable to suppose that the exponents $1/\beta$ and $1/\gamma$ in the general theorem are also the best possible, but this question has not been investigated.

Chapter VII

# Inhomogeneous Systems

In this chapter we shall take up the solubility of general inhomogeneous systems of equations in an arbitrary region of Euclidian space and we shall also investigate the possibility of approximating the solutions of homogeneous systems by exponential polynomials. In § 7, we give an exact and invariant formulation of these problems in the form of relationships between the functional space and the $\mathscr{P}$-module corresponding to the given system of equations. In § 8, we shall prove an important theorem on the solubility of arbitrary inhomogeneous systems in convex regions. In §§ 9–12, we shall arrive at a number of necessary and sufficient conditions for the solubility of inhomogeneous systems and for the possibility of approximations in non-convex regions.

## § 7. Solubility of inhomogeneous systems. $M$-convexity

We again suppose that $p$ is an arbitrary $\mathscr{P}$-matrix of size $t \times s$. We consider the corresponding inhomogeneous system of equations

$$p(iD)\, u = w. \tag{1.7}$$

Here $w$ is a known vector-function with $t$ components. We are to find the solubility conditions for this system without imposing on the solution $u$ either initial or boundary conditions. Since the matrix $p$ is arbitrary, its rows may admit differential relationships. The solubility of (1) clearly requires that the vector $w$ satisfy these relationships, whatever they may be.

Theorem 1 of this section implies that the local solubility of the system (1) is guaranteed by relationships of the form $r(ID)\,p(iD)=0$, where $r'\in\mathscr{P}^t$. In order to write down all the relationships of this form, we proceed as follows: We choose a $\mathscr{P}$-matrix $p_1$ such that the sequence

$$\mathscr{P}^{t_2} \xrightarrow{p_1'} \mathscr{P}^t \xrightarrow{p'} \mathscr{P}^s, \tag{2.7}$$

is exact, where $p_1'$ is the transpose of $p_1$. The exactness of the sequence implies that $p'\,p_1'=0$, whence $p_1\,p=0$. Therefore, if the system (1) is soluble, we have

$$p_1(iD)\,w=0. \tag{3.7}$$

We shall show that every vector $r$ satisfying the equation $r\,p=0$, is a linear combination of rows of the matrix $p_1$ with coefficients from $\mathscr{P}$. In fact, $r\,p=0$ implies that $p'\,r'=0$. Therefore, the exactness of (2) implies $r'\in p_1'\,\mathscr{P}^{t_2}$, that is $r'=p_1'\,F$, where $F\in\mathscr{P}^{t_2}$. Hence $r=F\,p_1$, which is what we were to show. Therefore, an arbitrary relation of the form $r(iD)\,w=0$, where $r\,p=0$, follows from Eq. (3).

The task of finding the solubility conditions of the system (1) we now formulate as follows: to determine the functional spaces in which the condition (3) is sufficient for the solubility of the system (1) in the same space. Now we give the exact definitions.

### 1°. *M*-convexity

**Definition 1.** Let $\Phi$ be a space of distributions and also a topological $\mathscr{P}$-module. Further, let $M$ be a finite $\mathscr{P}$-module and let

$$M_*\colon \cdots \to \mathscr{P}^{t_{k+1}} \xrightarrow{p_k'} \mathscr{P}^{t_k} \to \cdots \xrightarrow{p_1'} \mathscr{P}^t \xrightarrow{p'} \mathscr{P}^s \to M \to 0 \tag{4.7}$$

be a free resolution of this module. We shall say that the space (module) $\Phi$ is $M$-convex if the sequence

$$0 \to \Phi_p \to \Phi^s \xrightarrow{p} \Phi^t \xrightarrow{p_1} \cdots \to \Phi^{t_k} \xrightarrow{p_k} \Phi^{t_{k+1}} \to \cdots, \tag{5.7}$$

is exact; we shall denote this by $\mathrm{Hom}(M,\Phi)$. We recall that the exactness of (5) means that it is algebraically exact, and all the mappings in it are homomorphisms. By Proposition 7 of § 3, Chapter I, the $M$-convexity of the space $\Phi$ does not depend on the choice of the free resolution $M_*$.

The $M$-convexity of the space $\Phi$ implies that for an arbitrary right side $w$, belonging to the space $\Phi_{p_1}$, there exists a solution $u$ of the system (1) and the mapping

$$\Phi_{p_1} \ni w \to u \in \Phi^s / \Phi_p \tag{6.7}$$

is continuous. The $M$-convexity of the space $\Phi$ amounts to this, that the same property holds also for the matrices $p_1, p_2, \ldots$.

In the general case, the sequence (5) is merely semi-exact and therefore we can consider the factor-modules

$$\operatorname{Hom}(M, \Phi) = \operatorname{Ker} p = \Phi_p,$$

$$\operatorname{Ext}^i(M, \Phi) = \Phi_{p_i} / p_{i-1} \Phi^{t_{i-1}}, \qquad i = 1, 2, \ldots \ (t_1 = t,\ t_0 = s).$$

By Proposition 7 of § 3, Chapter I, these modules depend, up to a $\mathscr{P}$-isomorphism, only on $M$ and $\Phi$. More exactly, the following assertion is true: Let

$$\cdots \to \mathscr{P}^{\tau_{k+1}} \xrightarrow{\pi_k'} \mathscr{P}^{\tau_k} \to \cdots \xrightarrow{\pi_1'} \mathscr{P}^{\tau} \xrightarrow{\pi'} \mathscr{P}^{\sigma} \to M \to 0$$

be another free resolution of the module $M$. Then for arbitrary $i = 0, 1, 2, \ldots$ there exist $\mathscr{P}$-matrices $f_i$ and $g_i$ such that the mappings

$$\Phi_{p_i} / p_{i-1} \Phi^{t_{i-1}} \underset{\check{f}_i}{\overset{\check{g}_i}{\rightleftarrows}} \Phi_{\pi_i} / \pi_{i-1} \Phi^{\tau_{i-1}} \qquad (\pi_0 = \pi,\ p_0 = p),$$

associated with them are defined and are isomorphisms.

We shall often use the abbreviation $\Phi_M$ for the modules $\operatorname{Hom}(M, \Phi)$. This means that $\Phi_M$ is a space obtained by identifying all the spaces $\Phi_p$, where $\operatorname{Coker} p' \cong M$, with the aid of the isomorphisms $\hat{f}_0$ and $\hat{g}_0$.

By an exponential polynomial in $R_\xi^n$ we shall mean any finite sum of the form

$$f(\xi) = \sum_j f_j(\xi) \exp(z_j, -i\xi), \qquad z_j \in C^n,\ f_j(\xi) \in \mathscr{P}_\xi. \tag{7.7}$$

We denote by $E$ the space of all exponential polynomials in $R_\xi^n$ with the discrete topology. An arbitrary differential operator with constant coefficients generates a linear operator in the space $E$. It follows that $E$ is a $\mathscr{P}$-module.

**Definition 2.** We suppose that the space $\Phi$ contains $E$. Then for an arbitrary finite $\mathscr{P}$-module $M$, the space $E_M$ is a subspace in $\Phi_M$. We shall say that the space $\Phi$ is strongly $M$-convex with respect to the given finite $\mathscr{P}$-module $M$, if it is $M$-convex and the subspace $E_M$ is dense in $\Phi_M$.

Now using the definitions 1 and 2, we can formulate the solubility problem for the system (1) in the following way:

*The $M$-convexity problem in the wide sense*

Suppose given a finite $\mathscr{P}$-module $M$. We are to describe the class of (strongly) $M$-convex functional spaces.

For the spaces $\mathscr{D}^*(\Omega)$ and $\mathscr{E}(\Omega)$ it can be formulated as follows:

*The restricted M-convexity problem*

For a given finite $\mathscr{P}$-module $M$, we are to define a class of regions $\Omega$, for which the space $\mathscr{D}^*(\Omega)(\mathscr{E}(\Omega))$ is (strongly) $M$-convex.

Sections §§7–12 are mainly devoted to the restricted $M$-convexity problem.

**2°. Examples**

**Example 1.** Let $s=t$, and let the matrix $p(z)$ not be identically singular. It is clear that $p_1(z)\equiv 0$. Therefore the space $\Phi$ is $M$-convex if and only if the sequence

$$\Phi^s \xrightarrow{p} \Phi^t \to 0,$$

is exact, that is, if the system (1) is soluble in $\Phi^s$ for an arbitrary righthand side $w\in\Phi^t$ and if the mapping (6) is continuous.

**Example 2.** Let $d$ be the exterior differential operator in $R^n$. It operates on the $k$-th order differential form ($k=0, 1, 2, \ldots$) according to the formula

$$d\colon\ u=\sum_{i_1<\cdots<i_k} u_{i_1,\ldots,i_k}\, d\xi^i{}_1\wedge\cdots\wedge d\xi_{i_k}\to w$$

$$=\sum_{i_1,\ldots,i_k,j}\frac{\partial u_{i_1,\ldots,i_k}}{\partial\xi_j}\, d\xi_j\wedge d\xi_i\wedge\cdots\wedge d\xi_{i_k}.$$

If the coefficients $u_{i_1,\ldots,i_k}$ and $w_{i_1,\ldots,i_{k+1}}$ of the forms $u$ and $w$ are written in a fixed order, we obtain vector functions $\check{u}$ and $\check{w}$ with $\binom{n}{k}$ and $\binom{n}{k+1}$ components respectively. The action of $d$ on the form $u$ corresponds to the action of a differential operator $d_k(D)$ with constant coefficients on the vector $\check{u}$. In particular

$$d_0(D)=\mathrm{grad}, \qquad d_{n-1}(D)=\mathrm{div}.$$

Let us now consider the sequence of $\mathscr{P}$-mappings

$$0\to\mathscr{P}\xrightarrow{d_0}\mathscr{P}^n\xrightarrow{d_1}\cdots\to\mathscr{P}^{\binom{n}{k}}\xrightarrow{d_k}\mathscr{P}^{\binom{n}{k+1}}\to\cdots\to\mathscr{P}^n\xrightarrow{d_{n-1}}\mathscr{P}.$$

Since the square of the exterior differential operator is equal to zero, we have $d_{k+1}d_k=0$ for arbitrary $k\geqq 0$, and therefore, the sequence is semi-exact. It is easy to show that this sequence is exact (see for example, Proposition 2, § 13). Furthermore, the transpose matrix $d_k'$ coincides with the matrix $d_{n-1-k}$ up to the sign and the numbering of the rows and columns (*ibid.*). It follows that the sequence

$$\begin{array}{c} 0 \to \mathscr{P} \xrightarrow{d'_{n-1}} \mathscr{P}^n \xrightarrow{d'_{n-2}} \cdots \to \mathscr{P}^{\binom{n}{n-k}} \xrightarrow{d'_{n-1-k}} \mathscr{P}^{\binom{n}{n-1-k}} \to \cdots \\ \cdots \to \mathscr{P}^n \xrightarrow{d'_0} \mathscr{P} \end{array} \tag{8.7}$$

coincides with the preceding sequence and is therefore a free resolution of the module Coker $d'_0 = \mathscr{P}/d'_0\, \mathscr{P}^n$. Therefore, by definition, the convexity of the space $\Phi$ with respect to the module Coker $d'_0$ means the exactness of the sequence

$$\begin{array}{c} 0 \to \Phi_{d_0} \to \Phi \xrightarrow{d_0} \Phi^n \xrightarrow{d_1} \cdots \to \Phi^{\binom{n}{k}} \xrightarrow{d_k} \Phi^{\binom{n}{k+1}} \to \cdots \\ \cdots \to \Phi^n \xrightarrow{d_{n-1}} \Phi \to 0, \end{array} \tag{9.7}$$

where $\Phi_{d_0}$ is a subspace in $\Phi$, consisting of the functions that satisfy the system of equations $d_0 u = 0$, that is, the subspace consisting of locally constant functions.

In other words, the convexity of the space $\Phi$ with respect to the module Coker $d'_0$ means that for arbitrary $k$, $0 \leqq k \leqq n$, the system of equations $d_k u = w$ is always soluble in $\Phi^{\binom{n}{k}}$ under the condition that $w \in \Phi^{\binom{n}{k+1}}$ and $d_{k+1} w = 0$, and that the solution $u$, as an element of the factor-space $\Phi^{\binom{n}{k}}/\Phi_{d_k}$, depends continuously on $w$.

The space $\Phi^{\binom{n}{k}}$ will be interpreted as the space of differential forms of order $k$, with coefficients from $\Phi$. Then the system of equations that we took under consideration above can be written: $du = w$. Thus if $\Phi$ is convex with respect to Coker $d'_0$, the cohomologies of the differential forms with coefficients from $\Phi$ are equal to zero, and the exterior differential operator acts in the space of such forms as a homomorphism. In the general case, the cohomology space of differential forms of order $k \geqq 0$ with coefficients in $\Phi$ is equal to the space $\operatorname{Ext}^k(\text{Coker } d'_0, \Phi)$.

If we break the sequence (9) at the term $\mathscr{P}^{\binom{n}{k}}$, we obtain a free resolution of the module Coker $d'_k$. Therefore, the convexity of the space $\Phi$ with respect to this module implies that the sequence (9) is exact, beginning with the term $\Phi^{\binom{n}{k}}$.

**Example 3.** Let $n$ be an even number: $n = 2m$. Writing $\zeta_j = \xi_j + i\,\xi_{m+j}$, $j = 1, \ldots, m$ we introduce into $R^n$ the structure of the complex space $C^m$. We expand the operator $d$ into the sum of two operators $'d$ and $''d$, which act according to the formulae

$$'d\colon\ \omega \to \sum d\zeta_j \wedge \frac{\partial \omega}{\partial \zeta_j}, \qquad ''d\colon\ \omega \to \sum d\bar{\zeta}_j \wedge \frac{\partial \omega}{\partial \bar{\zeta}_j}.$$

As in Example 2, for arbitrary $k=0, 1, \ldots$ we can represent the action of the operator $''d$ on the form $\omega$ of type $(0, k)$ as the applying of some $\mathscr{P}$-matrix $''d_k(D)$ to the vector-function having components that are the coefficients of the form $\omega$. It is not difficult to see that the operator $''d_k$ — up to the numbering of the rows and columns — can be obtained from the operator $d_k$, acting in $R^m$, by the substitution

$$\frac{\partial}{\partial \xi_j} \to \frac{\partial}{\partial \bar{\zeta}_j}, \qquad j=1, \ldots, m.$$

Starting from this observation, it is not difficult to establish the exactness of the sequence

$$0 \to \mathscr{P} \xrightarrow{''d'_{m-1}} \mathscr{P}^m \xrightarrow{''d'_{m-2}} \cdots \xrightarrow{''d'_1} \mathscr{P}^m \xrightarrow{''d'_0} \mathscr{P} \to \operatorname{Coker} ''d'_0 \to 0. \tag{10.7}$$

The sequence $\operatorname{Hom}(M_*, \Phi)$ has the form

$$0 \to \Phi_{''d_0} \to \Phi \xrightarrow{''d_0} \Phi^m \xrightarrow{''d_1} \cdots \xrightarrow{''d_{m-2}} \Phi^m \xrightarrow{''d_{m-1}} \Phi \to 0,$$

where $\Phi_{''d_0}$ is the space of holomorphic functions belonging to $\Phi$.

We conclude that the space $\operatorname{Ext}^k(\operatorname{Coker} ''d'_0, \Phi)$ is the space of cohomologies of order $k$ of the $''d$-differential form with coefficients in $\Phi$.

**Example 4.** We now consider a class of operators containing those that appeared in Examples 2 and 3. Let $p$ be a $\mathscr{P}$-column, i.e. a $\mathscr{P}$-matrix of size $t \times 1$, and $p_1, \ldots, p_t$ its components. We suppose that the variety $N(\operatorname{Coker} p')$, i.e. the variety of common roots of the polynomials $p_1, \ldots, p_t$, has a dimension which does not exceed $n-t$. (In fact this inequality implies the equality $\dim N(\operatorname{Coker} p')=n-t$.) Under these conditions, as we shall prove in Proposition 2, §13, the module $\operatorname{Coker} p'$ admits the following free resolution

$$\begin{aligned} 0 \to \mathscr{P} \xrightarrow{d_0(p)} \mathscr{P}^t \xrightarrow{d_1(p)} \cdots \to \mathscr{P}^{\binom{t}{k}} \xrightarrow{d_k(p)} \mathscr{P}^{\binom{t}{k+1}} \to \cdots \\ \cdots \to \mathscr{P}^t \xrightarrow{d_t(p)} \mathscr{P} \to \operatorname{Coker} p' \to 0, \end{aligned} \tag{11.7}$$

where $d_k(p)$, $k=0, \ldots, t-1$ is a matrix obtained from $d_k(D)$ (Example 2, corresponding to the space $R^t$) by the substitution $\dfrac{\partial}{\partial \xi_i} \to p_i$, $i=1, \ldots, t$.

The matrix $p$ satisfies the given condition, if for instance, the operators $p_1, \ldots, p_t$ act on different groups of variables. These conditions are satisfied in particular by the matrices $d_0$ and $''d_0$. The sequences (8) and (10) are therefore exact, since they are particular cases of the sequence (11) for $p=d_0$, and $p=''d_0$.

### 3°. General properties of $M$-convex spaces

**Proposition 1**

I. *The space $E_p$ is dense in $\Phi_p$ if*

$$[\Phi^*]^s \cap E_p^0 = p^* [\Phi^*]^t, \tag{12.7}$$

*where $[\Phi^*]^s \cap E_p^0$ is a subspace in $[\Phi^*]^s$, orthogonal to $E_p$.*

II. *If the space $\Phi$ is strongly $M$-convex, the sequence*

$$\cdots \to [\Phi^*]^{t_{k+1}} \xrightarrow{p_k^*} [\Phi^*]^{t_k} \to \cdots \xrightarrow{p_1^*} [\Phi^*]^t \xrightarrow{p^*} [\Phi^*]^s \cap E_p^0 \to 0 \tag{13.7}$$

*is algebraically exact.*

III. *Let $\Phi$ be an $\mathcal{F}\mathcal{S}$-space or be conjugate to such a space. It will be $M$-convex if and only if one of the three following equivalent conditions is fulfilled.*

a) *The sequence* (5) *is algebraically exact;*

b) *The sequence*

$$\cdots \to [\Phi^*]^{t_{k+1}} \xrightarrow{p_k^*} [\Phi^*]^{t_k} \to \cdots \xrightarrow{p_1^*} [\Phi^*]^t \xrightarrow{p^*} [\Phi^*]^s \tag{14.7}$$

*is algebraically exact, and the image of the operator $p^*$ is closed;*

c) *The sequence* (14) *is exact.*

*The space $\Phi$ is strongly $M$-convex if and only if the sequence* (13) *is (algebraically) exact.*

*Proof.* We shall establish the first proposition. We suppose that $E_p$ is not dense in $\Phi_p$. Then there exists a functional $\phi \in [\Phi^*]^s$ vanishing on $E_p$, but not vanishing identically on $\Phi_p$. Then $\phi \in [\Phi^*]^s \cap E_p^0$, and therefore, by (12) $\phi \in p^* [\Phi^*]^t$. It follows that $\phi$ vanishes on $\Phi_p$. This however is a contradiction and therefore $E_p$ is dense in $\Phi_p$.

We prove the second proposition. Let $\Phi$ be strongly $M$-convex. If the sequence (5) is exact, then so is the sequence

$$0 \to \Phi^s/\Phi_p \xrightarrow{p} \Phi^t \xrightarrow{p_1} \cdots.$$

By the first assertion of Proposition 8, § 1, the sequence of conjugate mappings is algebraically exact. This implies that (13) is exact, provided that we can show that the space conjugate to $\Phi^s/\Phi_p$ as a linear space is isomorphic with $[\Phi^*] \cap E_p^0$. In fact, making use of the canonical mapping $\Phi^s \to \Phi^s/\Phi_p$, we can think of an arbitrary functional on $\Phi^s/\Phi_p$ as an element of $[\Phi^*]^s$, vanishing on $\Phi_p$, and vice versa. Since $E_p$ is dense in $\Phi_p$, the functional $\phi$ vanishes on $\Phi_p$ if and only if $\phi \in E_p^0$. This proves that $(\Phi^s/\Phi_p)^* = [\Phi^*]^s \cap E_p^0$.

We now prove the third assertion. It follows from the second part of Proposition 8, § 1 that the conditions a), b), c) and the assertion: $\Phi$ is $M$-convex, are equivalent.

We shall show finally that the following three conditions are equivalent: (1) the space $\Phi$ is strongly $M$-convex, (2) the sequence (13) is algebraically exact, and (3) the sequence (13) is exact. The implication (1) ⇒ (2) follows from the assertion II, § 1. Since $[\Phi^*]^s \cap E_p^0$ is a closed subspace in $[\Phi^*]^s$, the implication (2) ⇒ (3) follows from the fact that conditions b) and c) are equivalent. Finally the implication (3) ⇒ (1) follows from the fact that c) implies that $\Phi$ is $M$-convex and that assertion I is true. □

**Proposition 2.** *Let $\Omega$ be the union of an increasing sequence of spaces $\Omega_\alpha, \alpha=1, 2, \ldots$, such that each of the spaces $\mathscr{E}(\Omega_\alpha)$ is $M$-convex. We suppose that for arbitrary $\alpha$ the space $\mathscr{E}_M(\Omega_{\alpha+1})$ is dense in $\mathscr{E}_M(\Omega_\alpha)$. Then the space $\mathscr{E}(\Omega)$ is also $M$-convex and the space $\mathscr{E}_M(\Omega)$ is dense in $\mathscr{E}_M(\Omega_\alpha)$ for arbitrary $\alpha$.*

*If for arbitrary $\alpha$ the space $\mathscr{E}(\Omega_\alpha)$ is strongly $M$-convex, then $\mathscr{E}(\Omega)$ is also strongly $M$-convex.*

*Proof.* The set of spaces $\mathscr{E}(\Omega_\alpha)$ and of restriction mappings $\mathscr{E}(\Omega_{\alpha+1}) \to \mathscr{E}(\Omega_\alpha)$ forms a decreasing family of l.t.s. which we denote by $\{\mathscr{E}(\Omega_\alpha)\}$. For an arbitrary $\mathscr{P}$-matrix $q$ we may also consider the decreasing family $\{\mathscr{E}_q(\Omega_\alpha)\}$ constructed in a like manner. We fix a free resolution (4.7) of the module $M$ and for every $i=0, 1, 2, \ldots$ we consider the sequence of decreasing families

$$0 \to \{\mathscr{E}_{p_i}(\Omega_\alpha)\} \to \{[\mathscr{E}(\Omega_\alpha)]^{t_i}\} \xrightarrow{p_i} \{\mathscr{E}_{p_{i+1}}(\Omega_\alpha)\} \to 0, \qquad (15.7)$$

where $p_0=p$. Since for arbitrary $\alpha$ the space $\mathscr{E}(\Omega_\alpha)$ is $M$-convex, we find on removing the curly brackets in (15) an exact sequence of l.t.s. Therefore, the sequence of decreasing families (15) is also exact.

We shall verify the fact that the conditions of Proposition 11, § 1 are satisfied. Condition a) is satisfied since $\mathscr{E}_{p_i}(\Omega_\alpha)$ is a closed subspace of the $\mathscr{F}$-space $[\mathscr{E}(\Omega_\alpha)^{t_i}]$. We shall verify condition b). If $i=0$, the space $\mathscr{E}_{p_i}(\Omega_{\alpha+1})$ is dense in $\mathscr{E}_{p_i}(\Omega_\alpha)$ by hypothesis.

However, when $i>0$ the fact that the $\mathscr{E}(\Omega_\alpha)$ are $M$-convex implies that $\mathscr{E}_{p_i}(\Omega_\alpha)=p_{i-1}[\mathscr{E}(\Omega_\alpha)]^{t_{i-1}}$ for all $\alpha$. But since the space $\mathscr{E}(\Omega_{\alpha+1})$ is clearly dense in $\mathscr{E}(\Omega_\alpha)$, the space $p_{i-1}[\mathscr{E}(\Omega_{\alpha+1})]^{t_{i-1}}$ is dense in $p_{i-1}[\mathscr{E}(\Omega_\alpha)]^{t_{i-1}}$. Thus, we are allowed to apply Proposition 11 of § 1 which in turn implies that the sequence of projective limits of the family in (15) is exact.

We now determine the projective limits. Since every compact $K \subset \Omega$ belongs to some region $\Omega_\alpha$, we have $\varprojlim \mathscr{E}(\Omega_\alpha) \cong \mathscr{E}(\Omega)$. It follows that $\varprojlim \mathscr{E}_q(\Omega_\alpha) \cong \mathscr{E}_q(\Omega)$ for an arbitrary $\mathscr{P}$-matrix $q$. Then the limiting sequence is given by:

$$0 \to \mathscr{E}_{p_i}(\Omega) \to [\mathscr{E}(\Omega)]^{t_i} \xrightarrow{p_i} \mathscr{E}_{p_{i+1}}(\Omega) \to 0.$$

It is exact for an arbitrary $i \geqq 0$, and this implies the $M$-convexity of the space $\mathscr{E}_M$.

The space $\mathscr{E}_M(\Omega)$, being the projective limit of the spaces $\mathscr{E}_M(\Omega_\alpha)$ is (by Proposition 2, § 1) dense in everyone of them.

Now, suppose that all of the spaces $\mathscr{E}(\Omega_\alpha)$ are strongly $M$-convex. Then $\mathscr{E}_M(\Omega_{\alpha+1})$ is dense in $\mathscr{E}_M(\Omega_\alpha)$ since $\mathscr{E}_M$ is dense in each of these spaces. Therefore, the space $\mathscr{E}(\Omega)$ is $M$-convex. An arbitrary function $u \in \mathscr{E}_M(\Omega)$ can be approximated by functions in $E_M$ in the topology of each of the spaces $\mathscr{E}_M(\Omega_\alpha)$ and, therefore, in the topology of their projective limit $\mathscr{E}_M(\Omega)$. Thus $E_M$ is dense in $\mathscr{E}_M(\Omega)$. □

### 4°. $M$-convexity of the space $E$

**Proposition 3**

*The space $E$ is strongly $M$-convex for any $\mathscr{P}$-module $M$.*

*Proof.* We have only to show that for an arbitrary $\mathscr{P}$-matrix $p$ and a matrix $p_1$ such that the sequence (2) is exact, the sequence

$$E^s \xrightarrow{p} E^t \xrightarrow{p_1} E^{t_2}. \tag{16.7}$$

is also exact. Here for convenience we set $p = p(D)$ and $p_1 = p_1(D)$ where $D = \left(\frac{\partial}{\partial \xi_1}, \ldots, \frac{\partial}{\partial \xi_n}\right)$. If an exponential polynomial of the form (7) belongs to the kernel of the mapping $p_1$ in (16), and all the points $z_j$ appearing in (7) are distinct, then every term on the right side of (7) also belongs to the kernel of $p_1$. Therefore, to establish the exactness of (16), we need only show that for arbitrary $Z \in C^n$ the sequence

$$E_Z^s \xrightarrow{p} E_Z^t \xrightarrow{p_1} E_Z^{t_2}, \tag{17.7}$$

is exact, here $E_Z$ is the space of exponential polynomials of the form $f(\xi) \times \exp(Z, -i\xi)$. We may limit ourselves to the case $Z = 0$, since the general case reduces to the special case by a suitable translation of the matrices $p(z)$ and $p_1(z)$. We may, therefore, suppose that in (17) $Z = 0$. We note that $E_0$ is the space of all polynomials in $\xi = (\xi_1, \ldots, \xi_n)$ with complex coefficients.

We consider the space $\mathscr{S}$ of all formal power series with complex coefficients in $n$ variables $z = (z_1, \ldots, z_n)$. We set up a duality relation between $E_0$ and $\mathscr{S}$ via the bilinear form

$$(f, \phi) = \sum f_i \phi_i, \quad f = \sum \frac{f_i}{i!} \xi^i \in E_0, \quad \phi = \sum \phi_i z^i \in \mathscr{S}. \tag{18.7}$$

Here we look on every element of the space $E_0$ as a linear functional on $\mathscr{S}$. It is easy to see that every linear functional on $\mathscr{S}$ can be written in the form (18), that is, the space $E_0$ is isomorphic with the space conjugate to $\mathscr{S}$.

We shall show that the operator $D^j$, which acts in $E_0$, is conjugate to the operation of multiplication by $z^j$ in $\mathscr{S}$. In fact, for arbitrary $f \in E_0$ and $\phi \in \mathscr{S}$, we have

$$(f, z^j \phi) = \sum f_i (z^j \phi)_i = \sum f_i \phi_{i-j} = \sum f_{i+j} \phi_i = \sum (D^j f)_i \phi_i = (D^j f, \phi),$$

which is what we were to prove. It follows that the operators in the sequences (17) are conjugate to the operators in the sequence

$$\mathscr{S}^{t_2} \xrightarrow{p_1} \mathscr{S}^t \xrightarrow{p'} \mathscr{S}^s \quad (p = p(z),\, p_1 = p_1(z)). \tag{19.7}$$

Since $\mathscr{S}$ is a flat $\mathscr{P}$-module, the exactness of the sequence (2) implies the algebraic exactness of (19). Therefore, Proposition 8, § 1, implies that (17) is exact, which is what we were to prove. □

We now obtain a characterization of the spaces $E_p$, which will be useful to us later.

**Proposition 4.** *Let $p$ be an arbitrary $\mathscr{P}$-matrix of size $t \times s$, and let $\mathscr{D}(z) = \sum \xi^i \mathscr{D}_j(z, \delta)$ be a $p'$-operator. Then for arbitrary $z$ and $j$ the exponential polynomial*

$$\mathscr{D}_j'(z, -i\,\xi) \exp(z, -i\,\xi) \tag{20.7}$$

*belongs to the space $E_p$, and the set of all such polynomials (with the exception of the polynomial that vanishes identically) forms a basis in this space. Here $\mathscr{D}_j'(z, i\,\xi)$ denotes the matrix obtained from the transpose of $D_j'(z, \delta)$ by replacing the functionals $\delta^{(\alpha)}$ by the monomials $(-i\,\xi)^\alpha/\alpha!$.*

*Proof.* It follows from the nature of a $p'$-operator that for an arbitrary fixed $z$ the function (20) fails to vanish identically only when

$$j \in Z_+^n \setminus \mathscr{I}(z),$$

where $I(z)$ is the basis set at the given point $z$, and the functions (20) with $j \in Z_+^n \setminus \mathscr{I}(z)$ are linearly independent. It follows that the set of all functions (20) with $z \in N(p)$, $j \in Z_+^n \setminus \mathscr{I}(z)$ are linearly independent. It therefore remains only to prove that an arbitrary element of the space $E_p$ is a linear combination of functions of the type (20) and that these functions themselves belong to $E_p$.

An arbitrary element of the space $E^s$ will be written in the form $f = \sum f_\alpha(-i\,\xi) \exp(z_\alpha, -i\,\xi)$ where all the $z_\alpha$ are distinct, and $f_\alpha(-i\,\xi)$ are the columns of height $s$, formed from polynomials in $-i\,\xi$. The function $f$ will belong to $E_p$ if and only if for an arbitrary function

$\phi \in [\mathscr{D}(R^n)]^s$ the scalar product $(p(iD)f, \phi)$ is equal to zero. This equation can be written as follows:

$$0=(p(iD)f,\phi)=\Big(\sum_\alpha f_\alpha(-i\xi)\exp(z_\alpha,-i\xi),\, p^*(iD)\phi\Big)$$
$$=\sum_\alpha\big(\exp(z_\alpha,-i\xi), f_\alpha^*(i\xi)\,p^*(iD)\phi\big)=\sum_\alpha \overline{f_\alpha^*(i\xi)\,p^*(iD)\phi}\,|_{z=\bar z_\alpha}$$
$$=\sum_\alpha f_\alpha^*(D)\,p^*(z)\,\tilde\phi\,|_{z=\bar z_\alpha}=\sum_\alpha \overline{f_\alpha'(D)\,p'(z)\,\tilde\phi^*(z)}\,|_{z=z_\alpha}$$
$$=\sum_\alpha \overline{f_\alpha\big(\delta(z-z_\alpha)\big)\,p'(z)\,\tilde\phi^*(z)}.$$

Since the values of arbitrary derivatives of the function $\tilde\phi$ at the different points $z_\alpha$ are mutually independent, the vanishing of the right side implies the vanishing of each of its terms. In other words, for arbitrary $\alpha$ the functional $f_\alpha'(\delta(z-z_\alpha))$ vanishes on all functions of the form $p'(z)\,\tilde\phi(z)$. It is clear that at an arbitrary point $z_\alpha$ the Taylor series expansion of the function $\tilde\phi$ may have an arbitrarily determined portion of finite but arbitrarily great length. It follows that the functional $f_\alpha'(\delta)$ vanishes on the whole space $p'(z_\alpha)\mathscr{S}^t$, where $\mathscr{S}$ is the space of formal power series in $n$ variables. But, as follows from the property of $p'$-operators, every functional vanishing on $p'(z_\alpha)\mathscr{S}^t$ is a linear combination of the functionals $\mathscr{D}_j(z_\alpha,\delta)$. Therefore, the exponential polynomial $f_\alpha(-i\xi)\exp(z_\alpha-i\xi)$ is a linear combination of exponential polynomials of the form

$$\mathscr{D}_j'(z_\alpha,-i\xi)\exp(z_\alpha,-i\xi),\qquad j\in Z_+^n.$$

This implies the assertion we are seeking to prove. □

## § 8. *M*-convexity in convex regions

In this section we shall show that for an arbitrary convex region $\Omega$, an arbitrary one of the spaces $\mathscr{E}(\Omega)$, $\mathscr{D}^{*F}(\Omega)$, $\mathscr{D}^*(\Omega)$ is strongly $M$-convex for every finite $\mathscr{P}$-module $M$. We shall also show that the spaces $\mathscr{E}_I$ and $S_I^*$ are strongly $M$-convex, where $I$ is an arbitrary family of logarithmically convex functions (the precise conditions on $I$ are formulated in Theorem 5). These results justify the term "$M$-convexity[1]."

In this section we need the following terminology: Let $\Psi$ be a functional space, $\Phi$ be a subspace such that the operator $p$ acts from $\Phi^s$ to $\Psi^t$. We shall say that the function $u\in\Phi^s$ satisfies the system (1.7) in $\Psi$, if the right side $w$ belongs to $\Psi^t$ and coincides with $p\,u$.

1 The reason for using this name is given in the section "Notes."

### 1°. $M$-convexity of the space $\mathscr{D}^*(\Omega)$

**Theorem 1.** *Let $\Omega \subset R^n$ be a convex region. Then for an arbitrary finite $\mathscr{P}$-module $M$, the space $\mathscr{D}^*(\Omega)$ is strongly $M$-convex.*

*Proof.* We first establish the $M$-convexity of the space $\mathscr{D}^*(\Omega)$. For this it is sufficient to show that for an arbitrary matrix $p$ and for a matrix $p_1$ such that the sequence (2.7) is exact, the mapping

$$\mathscr{D}^*_{p_1}(\Omega) \to [\mathscr{D}^*(\Omega)]^s / \mathscr{D}^*_p(\Omega), \tag{1.8}$$

which is the inverse of the operator $p$, is defined and continuous.

**Lemma 1.** *Let $w$ be an arbitrary element of the space $\mathscr{D}^*_{p_1}(\Omega)$, and $K \subset\subset K'$ convex compacts belonging to $\Omega$.*

I. *There exists a distribution $u \in [\mathscr{D}^*_K]^s$, which is a solution of the system* (1.7) *in $\mathscr{D}^*_K$, such that* $\deg_K u \leqq \deg_{K'} w + a$, *and the constant $a$ depends only on the operator $p$.*

II. *For an arbitrary solution $u \in [\mathscr{D}^*_{K'}]^s$ of the system* (1.7) *and an arbitrary $r > 0$ and $\varepsilon > 0$, we can find a distribution $u' \in [\mathscr{D}^*(\Omega)]^s$ satisfying the system* (1.7) *in $\Omega$ and such that*

$$u' - u \in [\mathscr{E}^r_K]^s \; u \; \|u' - u\|^r_K \leqq \varepsilon.$$

*Proof of the Lemma.* We choose a convex compact $K''$ such that $K \subset\subset K'' \subset\subset K'$. Since the sequence (2.7) is exact, Proposition 2, § 4, Chapter IV, implies that the submodule $p'_1 \mathscr{P}^{t_2} \subset \mathscr{P}^t$ is primary, while $C^n$ and $p'$ are respectively the associated variety and normal Noetherian operator. By hypothesis, the function $w$ satisfies the homogeneous system of equations $p_1(iD)\, w = 0$ and belongs to the space $[\mathscr{E}^\alpha_{K'}]^t$, where $\alpha = -\deg_{K'} w$. Applying Corollary 1, § 4 to this function and to the compacts $K'' \subset\subset K'$ we obtain the representation

$$\overline{(w, \phi)} = \int_{C^n} p'(z)\, \tilde{\phi}^*(z)\, \mu, \qquad \phi \in [\mathscr{D}^*_{K''}]^t, \tag{2.8}$$

in which

$$\|\mu\| = \int_{C^n} (|z|+1)^{\alpha - A}\, \mathscr{I}_{K''}(y)\, |\mu| \leqq C\, \|w\|^\alpha_{K'}. \tag{3.8}$$

Let us consider the functional

$$\overline{(u, \psi)} = \int \tilde{\psi}^*(z)\, \mu, \qquad \psi \in [\mathscr{D}^*_K]^s. \tag{4.8}$$

The inequality (3) and Proposition 2, § 3 imply that

$$|(u, \psi)| \leqq \|\tilde{\phi}^*\|^{K''}_{A-\alpha}\, \|\mu\| \leqq C\, \|\psi\|^{A-\alpha}_K\, \|w\|^\alpha_{K'}. \tag{5.8}$$

Therefore, the functional $u$ belongs to the space $[\mathscr{E}_K^{\alpha-A}]^s$, that is, $\deg_K u \leqq A-\alpha = A+\deg_{K'} w$. From (2) and (4), we obtain

$$\overline{(p(iD)\,u,\phi)} = \overline{(u, p^*(iD)\,\phi)} = \int p'(z)\,\tilde{\phi}^*(z)\,\mu = \overline{(w,\phi)}$$

for an arbitrary function $\phi \in [\mathscr{D}_K^*]^t$, that is $p(iD)\,u = w$. This proves the first assertion of the Lemma.

We now prove the second assertion. Let $u_1 \in [\mathscr{D}_{K'}^*]^s$ be an arbitrary solution of the system (1.7) in $\mathscr{D}_{K'}^*$. We choose a strongly increasing sequence of convex compacts $K_\alpha$, $-\infty < \alpha < \infty$, tending to $\Omega$, and such that $K_0 = K$ and $K_1 = K'$. We fix an arbitrary $\varepsilon > 0$ and integer $r \geqq 0$. We construct the sequence of functions $u'_\alpha \in [\mathscr{D}_{K_\alpha}^*]^s$, $\alpha = 1, 2, \ldots$, satisfying the conditions

1) $u'_1 = u_1$,

2) $p(iD)\,u'_\alpha = w$ in $\mathscr{D}_{K_\alpha}^*$,

3) $\|u'_{\alpha+1} - u'_\alpha\|_{K_{\alpha-1}}^r \leqq \dfrac{\varepsilon}{2^\alpha}$, $\quad \alpha = 1, 2, \ldots$.

This sequence will be constructed by induction on $\alpha$. The function $u'_1 = u_1$ satisfies the conditions 1) and 2). We shall suppose that the functions $u'_1, \ldots, u'_\alpha$ have already been constructed. Using the first assertion of the Lemma, we find a function $u_{\alpha+1} \in [\mathscr{D}_{K_{\alpha+1}}^*]^s$ which is a solution of the system (1.7). The difference $u_{\alpha+1} - u_\alpha$ satisfies the homogeneous system (1.4) in $\mathscr{D}_{K_\alpha}^*$. Applying Corollary 2, § 4 to it, we write it in the form of a sum $v_0 + v$, where $v_0$ is the distribution which is a solution of (1.4) in $R^n$, and $v \in [\mathscr{E}_{K_{\alpha-1}}^r]^s$, while $\|v\|_{K_{\alpha-1}}^r \leqq \varepsilon/2^\alpha$. We set $u'_{\alpha+1} = u_{\alpha+1} - v_0$. Condition 2) is satisfied in an obvious way. That the inequality 3) is satisfied follows from the fact that $u'_{\alpha+1} - u'_\alpha = v$ in $\mathscr{D}_{K_{\alpha+1}}^*$. Thus the induction has been completed and the desired sequence $\{u'_\alpha\}$ has been constructed. It follows from inequality 3) that this is a fundamental sequence in the norm $\|\cdot\|_{K_\alpha}^r$ on an arbitrary one of the compacts $K_\alpha$, and, therefore, converges in an arbitrary one of the spaces $[\mathscr{D}_{K_\alpha}^*]^s$. Since all these spaces are complete, the sequence $\{u'_\alpha\}$ has a limit $u$ in each of them. Since the functional $u$ belongs to all the spaces $[\mathscr{D}_{K_\alpha}^*]^s$, it is a distribution in the region $\Omega$, and 2) implies that $p(iD)\,u = w$. The conditions 1) and 3) imply that $\|u - u_1\|_K^r \leqq \varepsilon$. But this proves the second assertion of the Lemma. □

We now prove the theorem. It follows from Lemma 1, in particular, that for an arbitrary right side $w \in \mathscr{D}_{p_1}^*(\Omega)$, there exists a solution of the system (1.7) belonging to $[\mathscr{D}^*(\Omega)]^s$, that is, the mapping (1) is defined. It remains to show that it is continuous. This means that for an arbitrary neighborhood of zero $U$ in $[\mathscr{D}^*(\Omega)]^s$, we can find a neighborhood $U'$ of zero in $\mathscr{D}_{p_1}^*(\Omega)$ such that for an arbitrary function $w \in U'$ there exists a solution of the system (1.7) $u \in U$. By definition of the topology in the space $[\mathscr{D}^*(\Omega)]^s$, a fundamental system of neighborhoods of zero in that space is formed by the polars of the bounded sets in $[\mathscr{D}(\Omega)]^s$. By Proposition 6, § 2, it is sufficient to consider only the bounded sets of the form

$$\mathscr{B}=\left\{\phi\in[\mathscr{D}(\Omega)]^s,\ \|\phi\|_K^b\leqq\frac{1}{\varepsilon}\right\},$$

where $\varepsilon$ is a positive number, $K$ is a convex compact belonging to $\Omega$, and $b=b(\eta)$ is some positive function defined for $\eta\geqq 0$. Our aim is to construct a set $\mathscr{B}'\in[\mathscr{D}(\Omega)]^t$ of the same form, and such that for an arbitrary function $w\in\mathscr{D}^*_{p_1}(\Omega)$, belonging to the polar of $\mathscr{B}'$, there exists a solution $u\in[\mathscr{D}^*(\Omega)]^s$ of the system (1.7), belonging to the polar of $\mathscr{B}$. By constructing such a set $\mathscr{B}'$, we shall establish the continuity of the mapping (10).

Without loss of generality, we may suppose that the function $b$ satisfies the inequality $b(\eta)\geqq B^n\eta^{n\beta}$, where $B>0$, and $\beta>1$, since the larger the function $b$, the smaller the polar of the set $\mathscr{B}$. Further, we construct some increasing family of logarithmically convex functions $\{b_\alpha(\eta),\ -\infty<\alpha<\infty\}$, defined in $R^1$, and satisfying the conditions of Proposition 5, § 3, and finally, such that $b(\eta)\leqq b_\alpha(\eta)$ for all $\alpha$ (see Remark 7°, § 3).

**Lemma 2.** *Every ultra-distribution* $v\in[(\mathscr{D}_{K_c}^{b_c})^*]^s$, $c=A+B$, *which is a solution of* (1.2) *in* $(\mathscr{D}_{K_c}^{b_c-m})^*$, *can be approximated by elements of the space* $E_p$ *in the norm of the space* $[(\mathscr{D}_K^b)^*]^s$.

*Proof.* We apply to the functional $v$ the representation given in Theorem 2, § 4. The measures $\mu^\lambda$ in this representation have a finite norm $\|\mu\|_B$, and therefore, they can be approximated in this norm by measures $\mu_0^\lambda$, which are finite linear combinations of delta-functions, concentrated on the variety $N^\lambda$. Let $\mu_0^\lambda=\sum c_i^\lambda\,\delta(z-z_i^\lambda)$, where $z_i^\lambda\in N^\lambda$. If we substitute these measures in the representation (2.4), we obtain the functional

$$\begin{aligned}\overline{(v_0,\phi)}&=\sum_\lambda\int d^\lambda(z,D)\,\tilde{\phi}^*(z)\,\mu_0^\lambda=\sum_{j,\lambda}c_j^\lambda\,d^\lambda(z_j^\lambda,D)\,\tilde{\phi}^*(z_j^\lambda)\\&=\sum_{j,\lambda}\left([c_j^\lambda\,d^\lambda(z_j^\lambda,-i\,\xi)]'\exp(z_j^\lambda,-i\,\xi),\bar{\phi}\right).\end{aligned}$$

It is clear from this equation that this functional is equal to

$$\sum[c_j^\lambda\,d^\lambda(z_j^\lambda,-i\,\xi)]'\exp(z_j^\lambda,-i\,\xi)$$

and therefore belongs to $E_p$. Since the measures $\mu^\lambda$ approximate the measures $\mu^\lambda$ in the norm $\|\mu\|_B$, we have by Theorem 2, § 4, that the functional $v_0$ approximates the functional $v$ in the norm of the space $[(\mathscr{D}_K^b)^*]^s$. □

We now turn to the proof of the theorem. We apply Theorem 2, § 4, to the family of majorants $\{I_{K_\alpha}\}$, $\{b_\alpha\}$, and to the system $p_1(iD)\,w=0$.

Since $w$ is a distribution in $\Omega$, it belongs to the space $[(\mathscr{D}_{K'}^{b'})^*]^t$, where $b'=b_{A+c+v}$, $K'=K_{A+c+v}$, and $v=[n/2]+1$. Therefore, the functional $w$ can be written in the form (2.4), and

$$\|\mu\|_{c+v} \leqq C\|w\|', \tag{6.8}$$

where $\|w\|'$ is the norm of $w$ as a functional over the space $[\mathscr{D}_{K'}^{b'}]^t$. We consider the functional

$$\overline{(u', \psi)} = \int \tilde{\psi}^*(z)\,\mu, \qquad \psi \in [\mathscr{D}_{K_c}^{b_c}]^s.$$

By the inequality (6) and Proposition 5, § 3, we have

$$|(u', \psi)| \leqq \|\tilde{\psi}^*\|_{R_{c+v}}^{K_{c+v}} \|\mu\|_{c+v} \leqq C\|\psi\|_{K_c}^{b_c} \|w\|'. \tag{7.8}$$

It follows that the functional $u'$ belongs to $[(\mathscr{D}_{K_c}^{b_c})^*]^s$, and it is clear from the construction that it satisfies the system (1.7) in $(\mathscr{D}_{K_c}^{b_c-m})^*$.

It follows from the inequality (7) that $\|u'\| \leqq c_0\|w\|'$. The desired set $\mathscr{B}'$ is now defined as follows: $\mathscr{B}' = \left\{\phi\colon \|\phi\|_{K'}^{b'} \leqq \frac{1}{\varepsilon'}\right\}$, $\varepsilon' = \frac{\varepsilon}{2c_0}$. Then if $w$ belongs to the polar of $\mathscr{B}'$, these bounds imply that the functional $u'$ belongs to the polar of $2\mathscr{B}$.

Finally, suppose that $u_0 \in [\mathscr{D}^*(\Omega)]^s$ is some solution of the system (1.7) in $\mathscr{D}^*(\Omega)$. The difference $u_0 - u'$ belongs to $[(\mathscr{D}_{K_c}^{b_c})^*]^s$ and is a solution of the system (1.4) in $(\mathscr{D}_{K_c}^{b_c-m})^*$. Therefore, it follows from Lemma 2 that it can be approximated by functions from $E_p$ in the norm of the space $[(\mathscr{D}_K^b)^*]^s$. In particular, we can find a function $v \in E_p$ such that the difference $u_0 - u' - v$ belongs to the polar of the set $2\mathscr{B}$. Hence it follows that the function $u_0 - v \in [\mathscr{D}^*(\Omega)]^s$ belongs to the polar of the set $\mathscr{B}$ and is the solution of the system (1.7) in $\mathscr{D}^*(\Omega)$. With this, the continuity of the mapping (1) is proved.

Let $u$ be an arbitrary generalized solution of (1.4) in $\mathscr{D}^*(\Omega)$. It follows from Lemma 2 that the function $u$ can be approximated by functions from $E_p$ in the norm of the space $[(\mathscr{D}_K^b)]^s$. In particular, we can find a function $v \in E_p$, such that the difference $u - v$ belongs to the polar of $\mathscr{B}$. Therefore, the functions belonging to $E_p$ are dense in $\mathscr{D}_p^*(\Omega)$ in the topology of that space. The proof of Theorem 1 is complete. ☐

**Corollary 1.** *Let $\Omega$ be a convex region and $K \subset\subset K' \subset \Omega$ be convex compacts. Then for an arbitrary right side $w \in \mathscr{D}_{p_1}^*(\Omega)$, we can find a solution $u \in [\mathscr{D}^*(\Omega)]^s$ of the system* (1.7) *such that*

$$\deg_K u \leqq \deg_{K'} w + a^2,$$

*where the constant a depends only on the operator p.*

2 We postulate that $\deg_{K^i} w > -\infty$; see Remark 3°.

In fact, in accordance with the first assertion of Lemma 1, we can find a solution $u_0 \in [\mathscr{D}^*_{K''}]^s$ satisfying this inequality. By the second assertion, we can find a solution $u \in [\mathscr{D}^*(\Omega)]^s$ such that $\|u-u_0\|^r_K \leqq 1$, where $r \geqq -\deg_K u_0$, whence $\deg_K u \leqq \deg_K u_0$. ◻

**2°. $M$-convexity of the spaces $\mathscr{D}^{*F}(\Omega)$ and $\mathscr{E}(\Omega)$**

**Theorem 2.** *Let $\Omega$ be a convex region. Then for an arbitrary finite $\mathscr{P}$-module $M$ the space $\mathscr{D}^{*F}(\Omega)$ is strongly $M$-convex. Moreover, for an arbitrary $w \in \mathscr{D}^{*F}_{p_1}(\Omega)$, we can find a solution $u$ of the system* (1.7), *satisfying the inequality*

$$\deg_\Omega u \leqq \deg_\Omega w + d$$

*where $a$ is a constant depending only on $p$.*

*Proof.* We begin with the following Lemma.

**Lemma 3.** *For arbitrary $\alpha$, we can approximate any distribution $u \in [\mathscr{E}^\alpha_{K_\alpha}]^s$, which is a solution of the system* (1.4) *in $\mathscr{D}^*_{K_\alpha}$, by functions from $E_p$ in the norm $\|\cdot\|^{\alpha-c}_{K_\alpha-c}$, where $c$ is a constant depending only on $p$.*

The proof of this Lemma is based on Corollary 1, § 4, and is altogether analogous to the proof of Lemma 2. We leave it to the reader. ◻

We can suppose that $q = \deg_\Omega w > -\infty$ since the other case is contained in Theorem 3.

We now construct the sequence of distributions $u'_\alpha \in [\mathscr{E}^{-q-a}_{K_\alpha}]^s$, satisfying the conditions

$$1) \quad p(iD)\,u'_\alpha = w \ \text{ in } \mathscr{D}^*_{K_\alpha}, \qquad 2) \quad \|u'_{\alpha+1} - u'_\alpha\|^{-q-b}_{K_\alpha-c} \leqq \frac{1}{2^\alpha},$$

where the constants $a$ and $b$ depend only on $p$. By Lemma 1, there exists for arbitrary $\alpha$ a distribution $u_\alpha \in [\mathscr{E}^{-q-a}_{K_\alpha}]^s$, satisfying condition 1). We write $u'_1 = u_1$. We assume that the distributions $u'_1, \ldots, u'_\alpha$ have been constructed. We consider the difference $u_{\alpha+1} - u'_\alpha \in [\mathscr{E}^{-q-a}_{K_\alpha}]^s$. This, obviously, satisfies the system (1.7) in $\mathscr{D}^*_{K_\alpha}$. Therefore by Lemma 3 it can be approximated by functions belonging to $E_p$ in the norm $\|\cdot\|^{-q-b}_{K_\alpha-c}$, $b = a + c$. In particular, we shall find a function $v \in E_p$, such that

$$\|u_{\alpha+1} - u'_\alpha - v\|^{-q-b}_{K_\alpha-c} \leqq \frac{1}{2^\alpha}.$$

It is evident that the distribution $u'_{\alpha+1} = u_{\alpha+1} - v$ satisfies conditions 1) and 2).

Continuing the induction, we construct an infinite sequence of distributions $u'_\alpha$, satisfying these conditions. It follows from the inequality 2) that this sequence converges in each of the spaces $[\mathscr{E}^{-q-b}_{K_\alpha}]^s$, $\alpha = 1, 2, \ldots$. Therefore, the limiting distribution $u$ belongs to $\mathscr{D}^{*F}(\Omega)$ and satisfies the inequality $\deg_\Omega u \leqq q+b$. It follows from 1) that it is a solution of the system (1.7) in $\mathscr{D}^*(\Omega)$. With this the second assertion of the theorem is proved.

It then follows, in particular, that the mapping

$$\mathscr{D}^{*F}_{p_1}(\Omega) \to [\mathscr{D}^{*F}(\Omega)]^s / \mathscr{D}^{*F}_p(\Omega),$$

which is inverse to the operator $p$, is defined. The continuity follows from the continuity of the mapping (1), and with it, the $M$-convexity of the space $\mathscr{D}^{*F}(\Omega)$. The fact that $E_p$ is dense in $\mathscr{D}^{*F}_p(\Omega)$ follows from the fact that it is dense in $\mathscr{D}^*_p(\Omega)$. □

**Theorem 3.** *For an arbitrary convex region $\Omega$ and finite module $M$, the space $\mathscr{E}(\Omega)$ is strongly $M$-convex.*

*Proof.* For arbitrary $\alpha$ we denote by $(E_p)_\alpha$ the closure of $E_p$ in the space $[\mathscr{E}^\alpha_{K_\alpha}]^s$. We denote by $\{[\mathscr{E}^\alpha_{K_\alpha}]^s\}$ a decreasing family of spaces $[\mathscr{E}^\alpha_{K_\alpha}]^s$, $\alpha = 1, 2, \ldots$, and their restriction mappings. The symbols $\{(\mathscr{E}^\alpha_{K_\alpha})_{p_1}\}$ and $\{(E_p)_\alpha\}$ are similarly defined. We consider the sequence of families

$$0 \to \{(E_p)_\alpha\} \to \{[\mathscr{E}^\alpha_{K_\alpha}]^s\} \xrightarrow{p} \{(\mathscr{E}^\alpha_{K_\alpha})_{p_1}\} \to 0. \tag{8.8}$$

and we prove that it is exact. We shall need only the exactness of the last term and the algebraic exactness of the second term.

We fix an arbitrary integer $\alpha$ and we apply Lemma 1 to the compact $K = K_\alpha$ and to some region $\omega \supset K$, belonging to $K_{\alpha+a}$ (we suppose that $a > 0$, and therefore, $K \subset\subset K_{\alpha+a}$, and so such a region exists). By the Lemma, we can find for any function $w \in (\mathscr{E}^{\alpha+a}_{K_{\alpha+a}})_{p_1}$ a function $u \in [\mathscr{E}^\alpha_{K_\alpha}]^s$, satisfying the system (1.7) in the $\mathscr{D}^*_{K_\alpha}$. Since the distribution $pu - w$ belongs to $[\mathscr{E}^{\alpha-m}_{K_\alpha}]^t$ and vanishes on the subspace $[\mathscr{D}_{K_\alpha}]^t$, which is dense in $[\mathscr{D}^{m-\alpha}_{K_\alpha}]^t$, it is equal to zero, that is, $u$ is a solution of (1.7) in $\mathscr{E}^{\alpha-m}_{K_\alpha}$. Thus we have constructed an operator

$$p^{-1}\colon\ (\mathscr{E}^{\alpha+a}_{K_{\alpha+a}})_{p_1} \to [\mathscr{E}^\alpha_{K_\alpha}]^s / (\mathscr{E}^\alpha_{K_\alpha})_p$$

such that the composition $pp^{-1}\colon (\mathscr{E}^{\alpha+a}_{K_{\alpha+a}})_{p_1} \to (\mathscr{E}^{\alpha-m}_{K_\alpha})_{p_1}$ is a restriction mapping. The continuity of the operator $p^{-1}$ follows from the inequality (5). With this the exactness of (8) in the final term has been proved.

By Lemma 3 for arbitrary $\alpha$ the space $E_p$ is dense in $(\mathscr{E}^\alpha_{K_\alpha})_p$ in the norm $\|\cdot\|^{\alpha-c}_{K_{\alpha-c}}$. The inclusion relation $(\mathscr{E}^\alpha_{K_\alpha})_p \subset (E_p)_{\alpha-c}$, follows from this; it implies that (8) is algebraically exact in the second term. Thus the exactness of (8) is established. Since $(\mathscr{E}_{p_\alpha})$ is a Frechet space and obviously dense in $(E_p)_{\alpha-1}$, we can apply Proposition 11, § 1, to the sequence (8) and derive the exactness of the limit sequence

$$[\mathscr{E}(\Omega)]^s \xrightarrow{p} \mathscr{E}_{p_1}(\Omega) \to 0.$$

Since $p$ was chosen at the outset to be an arbitrary $\mathscr{P}$-matrix, we have proved that the space $\mathscr{E}(\Omega)$ is $M$-convex.

As we noted earlier, every function belonging to $(\mathscr{E}^{\alpha}_{K_\alpha})_p$ can be approximated by functions from $E_p$ in the norm $\|\cdot\|^{\alpha-c}_{K_\alpha-c}$. Therefore an arbitrary function belonging to $\mathscr{E}_p(\Omega)$ can be approximated by functions from $E_p$ in an arbitrary one of the norms $\|\cdot\|^{\alpha}_{K_\alpha}$. This proves that the space $E_p$ is dense in $\mathscr{E}_p(\Omega)$. ▯

### 3°. A sharpening of the results of 1° and 2°

**Corollary 2.** *Let $\Omega$ and $\omega \subset \Omega$ be arbitrary convex regions. Then the spaces $\mathscr{D}^*(\Omega)$ and $\mathscr{E}(\Omega)$, with the topology induced by $\mathscr{D}^*(\omega)$ and $\mathscr{E}(\omega)$, respectively, are strongly $M$-convex for an arbitrary module $M$.*

*Proof.* Let $\check{\mathscr{D}}^*(\Omega)$ be the space $\mathscr{D}^*(\Omega)$, with the topology induced by $\mathscr{D}^*(\omega)$. For an arbitrary $\mathscr{P}$-matrix $p$, the space $E_p$ is dense in $\mathscr{D}^*_p(\omega)$, by Theorem 1. But since $E_p$ is contained in $\check{\mathscr{D}}^*_p(\Omega)$, it is also dense in that space.

There remains to show that the space $\check{\mathscr{D}}^*(\Omega)$ is $M$-convex. It is sufficient to show that the mapping (1), constructed in Theorem 1, is continuous in the weakened topologies. In other words, we have to show that for an arbitrary neighborhood of zero $U$ in $[\mathscr{D}^*(\omega)]^s$, we can find a neighborhood of zero $W$ in $\mathscr{D}^*_{p_1}(\omega)$, such that for an arbitrary distribution $w \in W \cap \mathscr{D}^*_{p_1}(\Omega)$ there exists a distribution $u \in U \cap [\mathscr{D}^*(\Omega)]^s$, which is a solution of the system (1.7).

Since Theorem 1 is true in the region $\omega$, we can find, corresponding to an arbitrary neighborhood $U'$ of zero in $[\mathscr{D}^*(\omega)]$, a neighborhood of zero $W'$ in $\mathscr{D}^*_{p_1}(\omega)$, such that for every distribution $w \in W'$, there exists a distribution $u' \in U'$, satisfying the system (1.7) in $\omega$. We now suppose that the distribution $w$ belongs also to $\mathscr{D}^*_{p_1}(\Omega)$. Then the distribution $u'$ can be approximated by distributions $u \in [\mathscr{D}^*(\Omega)]^s$, satisfying the system (1.7) in $\Omega$, with respect to the topology of the space $[\mathscr{D}^*(\omega)]^s$. To construct such an approximation, we have only to find some distribution $u_0 \in [\mathscr{D}^*(\Omega)]^s$, satisfying the system (1.7) in $\Omega$, and to approximate the difference $u' - u_0$, which is a solution of the homogeneous system (1.4) in $\omega$, by elements of the space $E_p$.

We now choose one such distribution $u \in [\mathscr{D}^*(\Omega)]^s$, sufficiently close to $u'$, so that $u - u' \in U'$. Then $u|_\omega = u' + (u - u') \in 2U'$. Putting $U' = \frac{1}{2}U$, and $W = W'$, we obtain the inclusion relationship $u \in U \cap [\mathscr{D}^*(\Omega)]^s$ if $w \in W$. But this proves that the space $\check{\mathscr{D}}^*(\Omega)$ is strongly $M$-convex.

Strong $M$-convexity of the space $\mathscr{E}(\Omega)$, in the topology induced by $\mathscr{E}(\omega)$, can be demonstrated in exactly the same way. ▯

**Corollary 3.** *Let $\Omega$ be an arbitrary region, and let $p$ be a hypoelliptic operator. Then every solution $u \in [\mathscr{D}^*(\Omega)]^s$ of the system* (1.7) *is an infinitely differentiable function in $\Omega$ if the right side is.*

*Proof.* Let $\xi_0$ be an arbitrary point of the region $\Omega$, and let $\omega$ be some convex neighborhood of $\xi_0$, belonging to $\Omega$. The restriction $w$ on $\omega$ belongs to $[\mathscr{E}(\omega)]^t$. Since we assume that the system (1.7) has at least one solution in $[\mathscr{D}^*(\Omega)]^s$, the function $w$ satisfies the system $p_1(iD)\,w=0$. Therefore, it follows from Theorem 3 that there exists a function $u_0\in[\mathscr{E}(\omega)]^s$, which is a solution of the system (1.7). The difference $u-u_0$ is the solution of the homogeneous system (1.4) in the region $\omega$, and therefore, is infinitely differentiable, since $p$ is a hypoelliptic operator. It follows that the distribution $u$ is also infinitely differentiable in $\omega$. Since the point $\xi_0$ was chosen arbitrarily, the distribution $u$ is infinitely differentiable in the whole region $\Omega$. □

*Remark.* Let $\Omega$ be a convex region, and $K\subset\Omega$ a convex compact. Suppose that the function $w\in\mathscr{D}^*_{p_1}(\Omega)$ is infinitely differentiable within $K$. By Theorem 3 we can find an infinitely differentiable function defined inside $K$, which is a solution of the system (1.7). On the other hand, by Corollary 1, we can find a solution of this system belonging to $[\mathscr{D}^*(\Omega)]^s$, that has derivatives on an arbitrary given compact $\kappa\subset\subset K$ up to an arbitrary order specified in advance. However, generally speaking, we cannot find a solution defined in the whole region $\Omega$ which is infinitely differentiable in the neighborhood of a given point $\xi_0\in\Omega$. The equation

$$p\,u=\left(\frac{\partial}{\partial\xi_1}+i\frac{\partial}{\partial\xi_2}\right)u=\delta, \tag{9.8}$$

provides an example. Here $u$ is a distribution of three variables, and $\delta$ is the delta-function. Since $p_1=0$ (see, for example, 1, 2°, § 7), the condition $p_1\,\delta=0$ is trivially satisfied. We shall show that for an arbitrary point $\xi_0$ lying in the intersection of the sphere $\Omega$, with center at the origin of coordinates and radius $\varepsilon$ ($\varepsilon>0$ is arbitrary), and the plane $\xi_3=0$, there exists no solution of (9) belonging to $\mathscr{D}^*(\Omega)$ which is infinitely differentiable in the neighborhood of $\xi_0$.

Let us suppose the opposite. Let $u$ be such a solution. We note that the operator $p$ is strongly hypoelliptic in the variables $\xi'=(\xi_1,\xi_2)$. Therefore, on every line $\xi'=\text{const}$, the restriction function $u$ is defined, and this restriction depends analytically on $\xi_1+i\,\xi_2$. By a theorem of Zerner[3], this implies that the function $u$, being infinitely differentiable *a priori* in the neighborhood of the point $\xi_0$, is also infinitely differentiable in some neighborhood of the origin of coordinates, with the exception of the origin itself. The existence of such a solution of (9) implies that the operator $p$ is hypoelliptic[4], whereas in fact, it is not. This contradiction proves our assertion.

3 See Zerner [1].

4 This is easy to show if we begin with the Theorem of the Mean; see Šilov [2].

On the other hand, Corollary 3 implies that any distribution in $\Omega$ which satisfies the hypoelliptic system (1.7) is infinitely differentiable within $K$ if the right side is so. It is not known, however, whether systems other than hypoelliptic have this property.

**4°. The role of convexity of the region**

**Theorem 4**

I. *In order that the space $\mathscr{D}^*(\Omega)$ or $\mathscr{E}(\Omega)$ be M-convex for an arbitrary finite $\mathscr{P}$-module M, it is necessary and sufficient that every connected component of the region $\Omega$ be convex.*

II. *In order that the space $\mathscr{D}^*(\Omega)$ or $\mathscr{E}(\Omega)$ be strongly M-convex for an arbitrary finite $\mathscr{P}$-module M, it is necessary and sufficient that the region $\Omega$ be convex.*

*Proof.* For an arbitrary region $\omega \subset R^n$, we denote by $\Phi(\omega)$ the space $\mathscr{D}^*(\omega)$ or $\mathscr{E}(\omega)$. Let $\Omega = \cup \Omega_\alpha$ be a decomposition of the region $\Omega$ into connected components. Then the space $\Phi(\Omega)$ is topologically isomorphic to the direct product of the spaces $\Phi(\Omega_\alpha)$. If each of the regions $\Omega_\alpha$ is convex, then each of the spaces $\Phi(\Omega_\alpha)$ is $M$-convex, that is, the sequence (5.7) is exact with $\Phi = \Phi(\Omega_\alpha)$. Accordingly, the sequence (5.7) is exact for the space $\Phi = \Phi(\Omega)$, also. But this proves the $M$-convexity of the space $\Phi(\Omega)$.

We shall now prove that the condition is necessary: every component $\Omega_\alpha$ is convex. We suppose that some one of the components is not convex. In this case, we can find a sequence of equal and parallel segments $l_k$, $k=1, 2, \dots$, belonging to the domain $\Omega$, and tending to some segment $l$, which belongs to $\Omega$, with the exception of its midpoint. Choosing the system of coordinates in $R^n$ in a suitable way, we locate the midpoint of the segment $l$ at the origin of coordinates and its endpoints at the points $\xi^\pm = (\pm 1, 0, \dots, 0)$. We denote by $\xi_k$, $k=1, 2, \dots$, the midpoints of the segments $l_k$.

We now consider the differential operator $p(iD) = \dfrac{\partial}{\partial \xi_1}$. We shall show that for some function $w \in \mathscr{E}(\Omega)$, there exists no solution of (1.7) belonging to $\mathscr{D}^*(\Omega)$. Since $p_1 = 0$, this will complete the proof of the first assertion of the theorem. For every point $\xi_k$, we choose the neighborhood $U_k$ to be such that the sum $U_k + l_k$ belongs to $\Omega$, and the sequence $\{U_k\}$ tends to the origin of coordinates. For arbitrary $k$, we choose the function $\phi_k \in \mathscr{D}(U_k)$, satisfying the conditions $\phi_k \geqq 0$ and $\int \phi_k \, d\xi = 1$. We write

$$\phi'_k = \frac{1}{\|\phi\|^k} \phi_k, \qquad k = 1, 2, \dots$$ [5].

5 The norm $\|\cdot\|^k$ was defined in § 2.

Since the regions $\xi^{\pm}+U_k$ belong for large $k$ to some compact $K\subset\Omega$ the functions $\phi'_k(\xi-\xi^{\pm})$ form a bounded set in $\mathscr{D}(\Omega)$.

Let $w$ be some infinitely differentiable function in $\Omega$, and let $u$ be a distribution which satisfies (1.7) in $\Omega$. The Newton-Leibnitz formula yields

$$\left(u, \phi'_k(\xi-\xi^+)\right)-\left(u, \phi'_k(\xi-\xi^-)\right)=\int \int_{-1}^{1} w(t\,\xi^+ +\eta)\,dt\,\phi'_k(\eta)\,d\eta\,. \quad (10.8)$$

Since the functions $\phi'_k(\xi-\xi^{\pm})$ form a bounded set, the left side of this equation is bounded. On the other hand, the region $U_k$, which belongs to the domain of integration of the right side, tends to the origin of coordinates, which is a limit point of $\Omega$. Therefore, the function $w\in\mathscr{E}(\Omega)$ can be chosen so that

$$\int_{-1}^{1} w(t\,\xi^+ +\eta)\,dt>k\,\|\phi\|^k, \qquad \eta\in U_k.$$

Since $\int\phi'_k\,d\eta=1/\|\phi\|^k$, the right side of (10) is not less than $k$. But this contradicts the fact that the left side of (10) is bounded.

Let us turn to the second assertion of the theorem. The sufficiency of the condition that $\Omega$ be convex follows from Theorems 1 and 3. We shall prove the necessity. Because of the first assertion, it is sufficient to show that $\Omega$ is connected. Let us assume that it is not connected. Let $p=d_0$ (see Example 2, 2°, § 7). The solutions of the corresponding homogeneous system (1.4) in $\Omega$ are functions that are constant on each of the connected components of $\Omega$. On the other hand, the space $E_p$ consists of functions that are constant throughout the whole of $R^n$. It is clear that one cannot approximate a solution of the system (1.4) using such functions, if the solution is equal to unity on one of the connected components of $\Omega$, and to zero on the remainder. This proves the second assertion of the theorem. □

### 5°. Corollaries concerning the solutions of homogeneous systems. The space $\Phi_p$ as a function of the module Coker $p'$

**Corollary 4.** *If the region $\Omega$ is convex, the spaces $\mathscr{E}(\Omega)$, $\mathscr{D}^{*F}(\Omega)$ and $\mathscr{D}^*(\Omega)$ are injective, and the space $\mathscr{E}^*(\Omega)$ and the space $\mathscr{D}(\Omega)$, with the discrete topology, are flat $\mathscr{P}$-modules*[6].

*Proof.* It follows from Theorems 1 and 3 that if the sequence (2.7) is exact, then so is the sequence

$$\Phi^s \xrightarrow{\ p\ } \Phi^t \xrightarrow{\ p_1\ } \Phi^{t_2},$$

where $\Phi$ is an arbitrary one of the spaces $\mathscr{E}(\Omega)$, $\mathscr{D}^{*F}(\Omega)$, $\mathscr{D}^*(\Omega)$. By Proposition 9, § 3, Chapter I, this means that the module $\Phi$ is injective.

6 It can be shown that the space $\mathscr{D}(\Omega)$ is a flat $\mathscr{P}$-module also, with its own topology.

Proposition 8, § 1 implies that the sequence

$$\Phi^{t_2} \xrightarrow{p_1'} \Phi^t \xrightarrow{p'} \Phi^s$$

is algebraically exact for $\Phi = \mathscr{D}(\Omega)$ and exact for $\Phi = \mathscr{E}^*(\Omega)$. This proves the second assertion. ▯

**Corollary 5.** *Let* $M = \mathscr{P}^s / p' \mathscr{P}^t$ *and* $M_i = \mathscr{P}^{s_i} / p_i' \mathscr{P}^{t_i}$, $i = 1, \ldots, k$, *be finite* $\mathscr{P}$*-modules and let* $\phi_i$: $M \to M_i$ *be mappings associated with certain* $\mathscr{P}$*-matrices* $f_i$: $\mathscr{P}^s \to \mathscr{P}^{s_i}$, $i = 1, \ldots, k$, *such that* $\cap \operatorname{Ker} \phi_i = 0$. *Then for an arbitrary convex region* $\Omega$, *every element of* $\mathscr{D}_p^*(\Omega)$ *can be written in the form* $\sum f_i(i\,D)\, u_j$, *where* $u_j \in \mathscr{D}_{p_j}^*(\Omega)$.

*Proof.* We consider the sequence

$$0 \to M \xrightarrow{\phi} \oplus M_j,$$

where the mapping $\phi$ carries the element $x$ into $(\phi_1(x), \ldots, \phi_k(x))$. It follows from the condition $\cap \operatorname{Ker} \phi_j = 0$, that the sequence is exact. Applying to it the functor $\operatorname{Hom}(\cdot, \mathscr{D}^*(\Omega))$, we obtain the sequence

$$\bigoplus_j \mathscr{D}_{p_j}^*(\Omega) \xrightarrow{f} \mathscr{D}_p^*(\Omega) \to 0,$$

in which the mapping $f$ carries the element $(u_1, \ldots, u_k)$ into $\sum f_j'(i\,D)\, u_j$. By Corollary 4, this sequence is also exact, which proves the assertion. ▯

The considerations given in 1°, § 7, show in particular that for an arbitrary $\mathscr{P}$-module $\Phi$, the module $\Phi_p$, consisting of the solutions of the homogeneous system (1.4) that belong to $\Phi^s$, depends essentially only on the module Coker $p'$. If the $\mathscr{P}$-matrices $p$ and $\pi$ are such that Coker $p' \cong$ Coker $\pi'$, then as we observed in 1°, § 7, there exist $\mathscr{P}$-matrices $f_0$ and $g_0$ such that the associated mappings $\check{f}_0$: $\Phi_\pi \to \Phi_p$ and $\check{g}_0$: $\Phi_p \to \Phi_\pi$ establish an isomorphism between these modules. It follows that an arbitrary property of the module $\Phi_p$ which is conserved under the application of a $\mathscr{P}$-matrix, is common among all modules $\Phi_\pi$ with Coker $\pi' \cong$ Coker $p'$. We shall often use the notation $\Phi_M$ for these modules, where $M =$ Coker $p'$.

In particular, the fact that $\Phi_p$ consists of infinitely differentiable (analytic) functions is conserved under $\mathscr{P}$-isomorphisms and therefore depends only on the module Coker $p'$. In consequence, the following definition is correct.

**Definition 3.** A finite $\mathscr{P}$-module $M$ is said to be hypoelliptic (elliptic) if it is equal to Coker $p'$, where $p$ is a hypoelliptic (elliptic) operator; in other words, if for an arbitrary region $\Omega$, the space $\mathscr{D}_p^*(\Omega)$ consists only of infinitely differentiable (analytic) functions. In view of the results of § 5 $M$ is hypoelliptic if and only if the associated variety $N$ is hypoelliptic.

This definition is useful, in particular, for the solution of the following problem. Suppose given a $\mathscr{P}$-matrix $q: \mathscr{P}^s \to \mathscr{P}^\tau$. The operator $p$ is said to be hypoelliptic (elliptic) with respect to the operator $q$, if for an arbitrary region $\Omega$, every function $u \in \mathscr{D}_p^*(\Omega)$ has the property that $q(iD)u$ is infinitely differentiable (analytic).

We now reformulate the definition so that it is invariant with respect to the module $M = \operatorname{Coker} p'$. The mapping $q': \mathscr{P}^\tau \to \mathscr{P}^s$, defined by the transposed matrix $q'$, is extended to the mapping:

$$q': \quad \mathscr{P}^\tau \to M. \tag{11.8}$$

Applying the functor $\operatorname{Hom}(\cdot, \mathscr{D}^*(\Omega))$, we obtain the mapping

$$\mathscr{D}_M^*(\Omega) \xrightarrow{q} [\mathscr{D}^*(\Omega)]^\tau. \tag{12.8}$$

Our definition can now be formulated as follows: we shall say that the mapping (11) is hypoelliptic (elliptic) if the image of the mapping (12) consists only of infinitely differentiable (analytic) functions. It is easy to see that this property of the mapping (11) is equivalent to saying that the operator $p$ is hypoelliptic (elliptic) with respect to $q$.

**Corollary 6.** *The mapping* (11) *is hypoelliptic (elliptic) if and only if its image is a hypoelliptic (elliptic) module.*

*Proof.* We construct the $\mathscr{P}$-matrix $r$ so that the sequence

$$\mathscr{P}^\sigma \xrightarrow{r'} \mathscr{P}^\tau \xrightarrow{q'} M \tag{13.8}$$

is exact. In the definition of relatively hypoelliptic (elliptic) operators, we may limit ourselves to convex regions. Let $\Omega$ be an arbitrary convex region. By Corollary 4, the module $\mathscr{D}^*(\Omega)$ is injective, and therefore, applying the functor $\operatorname{Hom}(\cdot, \mathscr{D}^*(\Omega))$ to (13), we again obtain the exact sequence

$$\mathscr{D}_M^*(\Omega) \xrightarrow{q} [\mathscr{D}^*(\Omega)]^\tau \xrightarrow{r} [\mathscr{D}^*(\Omega)]^\sigma.$$

Because it is exact, the image of the mapping (12) coincides with $\mathscr{D}_r^*(\Omega)$. Therefore the mapping (11) is hypoelliptic (elliptic) if and only if the operator $r$ is hypoelliptic (elliptic). However, as we saw earlier, this property of the operator $r$ is really a property of the module $\operatorname{Coker} r'$. Since (13) is exact, we have the isomorphism $\operatorname{Coker} r' = \operatorname{Coim} q' \cong \operatorname{Im} q'$. Therefore, the hypoellipticity (ellipticity) of the operator $r$ is equivalent to the hypoellipticity of the image of the mapping (11). ▯

Let us now consider a problem connected with the local properties of the solutions of homogeneous undetermined systems. We shall say that the operator $p$ and the module $M = \operatorname{Coker} p'$ are virtually hypoelliptic, if every distribution satisfying the system (1.4) in the neighborhood of zero in $\Omega$, is the sum of a solution defined in $\Omega$ and infinitely differentiable near zero, and a solution with support contained in an arbitrarily small neighborhood of zero. If dim $M<n$, there is no solution of the system (1.4) with compact non-empty support (see § 14). Therefore, the assertion that $M$ is a virtually hypoelliptic module of dimension less than $n$ is equivalent to saying that $M$ is a hypoelliptic module.

Let us characterize the virtually hypoelliptic modules. Let $M$ be an arbitrary module. We find a reduced primary decomposition of its null submodules. Let $M_0$ be an $n$-dimensional component of this decomposition. Then the module $M_0$ does not depend on the choice of the primary representation, since it may be defined otherwise: it coincides with the set of elements $\phi \in M$, having the property that $f\phi=0$, for some polynomial $f \in \mathscr{P}$, which is not zero.

**Corollary 7.** *The module $M$ is virtually hypoelliptic if and only if the module $M_0$ is hypoelliptic; in other words, if and only if the union of the varieties $N^\lambda$, associated with $M$ and distinct from $C^n$, is hypoelliptic.*

*Proof.* We choose the $\mathscr{P}$-matrix $q$ so that the sequence

$$\mathscr{P}^r \xrightarrow{q} \mathscr{P}^s \xrightarrow{p} \mathscr{P}^t.$$

is exact. Proposition 1, § 4, Chapter IV, implies that the kernel of the mapping $q'\colon \mathscr{P}^s \to \mathscr{P}^r$ is an $n$-dimensional primary component $\mathfrak{p}_0$ of the submodule $p'\mathscr{P}^t \subset \mathscr{P}^s$. The submodule $\mathfrak{p}_0/p'\mathscr{P}^t \subset M$ coincides with $M_0$ and, therefore, we have the exact sequence

$$0 \to M_0 \to M \xrightarrow{q'} \mathscr{P}^r.$$

Let $\Omega$ be a convex neighborhood of zero. Since the module $\mathscr{D}^*(\Omega)$ is injective its exactness implies the exactness of the sequence

$$[\mathscr{D}^*(\Omega)]^r \xrightarrow{q} \mathscr{D}^*_M(\Omega) \to \mathscr{D}^*_{M_0}(\Omega) \to 0.$$

Hence every element of $\mathscr{D}^*_m(\Omega)$ which is a solution of (1.4) coincides, up to a distribution of the form $q\,v$, where $v \in [\mathscr{D}^*(\Omega)]^r$, with an element of $\mathscr{D}^*_{M_0}(\Omega)$ which is also a solution of (1.4). Every distribution of the form $q\,v$ can be represented as the sum of a distribution of the same type vanishing in the neighborhood of zero and a distribution with support in an arbitrarily small neighborhood of zero. It follows that every element of $\mathscr{D}^*_M(\Omega)$ can be made infinitely differentiable in the neighborhood of zero by the addition of an element of the same space with a compact support contained in the given neighborhood of zero, if and only if the elements of $\mathscr{D}^*_{M_0}(\Omega)$ have the same property. In other words, the modules $M$ and $M_0$ may be virtually hypoelliptic only simultaneously. But since dim $M_0<n$, the virtual hypoellipticity of the module $M$ is equivalent to the hypoellipticity of $M_0$, and this is what we were to prove. □

### 6°. Examples

**Example 1.** In Corollary 6 we set $s=t=\tau=1$. We determine the conditions under which the image of the mapping (11) is a hypoelliptic module. Since this image is isomorphic to $\operatorname{Coim} q'$, it is hypoelliptic if and only if $\operatorname{Ker} q'=p_0\mathscr{P}$, where $p_0$ is a hypoelliptic operator. On the other hand, as is easily seen, $p_0$ is the product of all the factors of the polynomial $p$ that do not divide the polynomial $q$. Thus, we may assert that the operator $p$ will be hypoelliptic (elliptic), relative to $q$, if and only if every irreducible divisor of the polynomial $p$, which is not also a divisor of $q$, is a hypoelliptic (elliptic) operator.

**Example 2.** In Corollary 5, we set $M=\mathscr{P}/p\mathscr{P}, M_i=\mathscr{P}/p_i\mathscr{P}$. We assume that the polynomial $p$ is divisible by any of the polynomials $p_i$. Then the identity mapping $f_i'\colon \mathscr{P}\to\mathscr{P}$ is extended to a mapping of the modules $\phi_i\colon M\to M_i$. The condition $\cap \operatorname{Ker}\phi_i=0$ means that $p$ is the least common multiple of the polynomials $p_i$. By Corollary 5, every solution of the equation $pu=0$ in the convex region $\Omega$ can be written in the form of a sum $\sum u_i$, where $u_i$ is a solution of the equation $p_i u_i=0$ in $\Omega$.

### 7°. $M$-convexity in the spaces $\mathscr{E}_I$ and $S_I^*$

**Theorem 5.** *Let $I=\{I_\alpha\}$ be an arbitrary decreasing family of logarithmically convex functions satisfying the conditions $\Omega_{I_\alpha}=\Omega_{\mathscr{I}_\alpha}=R^n$, for arbitrary $\alpha$, where $\mathscr{I}_\alpha$ is the logarithmic reciprocal of $I_\alpha$.*

I. *Let $p$ be an arbitrary $\mathscr{P}$-matrix. Let $\mathscr{D}=\sum \xi^j\mathscr{D}_j(z,\delta)$, be a $p'$-operator, let $\{N^\lambda\}$ be the set of varieties associated with the matrices $p'$, and let $B\subset C^n$ be a set such that for arbitrary $\lambda$ the intersection $B\cap N^\lambda$ is nonempty. Then the linear combinations of exponential polynomials of the form*

$$\mathscr{D}_j'(z,-i\xi)\exp(z,-i\xi),\qquad j\in Z_+^n,\qquad z\in B \tag{14.8}$$

*are dense in the spaces $(\mathscr{E}_I)_p$ and $(S_I^*)_p$.*

II. *For an arbitrary finite $\mathscr{P}$-module the spaces $\mathscr{E}_I$ and $S_I^*$ are $M$-convex.*

We observe that all the exponential polynomials (14) belong to $E_p$ (see Proposition 4, § 7). Therefore, it follows from the second assertion that $E_p$ is dense in $(\mathscr{E}_I)_p$ and $(S_I^*)_p$, so that the spaces $\mathscr{E}_I$ and $S_I^*$ are strongly $M$-convex for an arbitrary module $M$.

*Proof of the theorem.* We begin with a general observation. The spaces $X_{-\alpha}=\mathscr{E}_{I_\alpha}^\alpha$, $-\infty<\alpha<\infty$, and the identity imbedding $e_\alpha\colon \mathscr{E}_{I_\alpha}^\alpha\to\mathscr{E}_{I_{\alpha-1}}^{\alpha-1}$ form a family of linear topological spaces. The limits of this family are, by 5°, § 3, equal to

$$\mathscr{E}_I=\varprojlim_{\alpha\to\infty}\{\mathscr{E}_{I_\alpha}^\alpha, e_\alpha\},\qquad S_I^*=\varinjlim_{\alpha\to-\infty}\{\mathscr{E}_{I_\alpha}^\alpha, e_\alpha\}.$$

Since the mappings $e_\alpha$ are biunique, the space $\mathscr{E}_I$ can be (algebraically) identified with the intersection of all the spaces $\mathscr{E}^{\alpha}_{I_\alpha}$, and the space $S^*_I$ with their union. For every $\alpha$, we denote by $(\mathscr{E}^{\alpha}_{I_\alpha})_p$ the subspace in $[\mathscr{E}^{\alpha}_{I_\alpha}]^s$, consisting of distributions satisfying the system (1.4) in $\mathscr{E}^{\alpha-m}_{I_\alpha}$, that is, the distributions $u$ such that the functional $pu$ vanishes on $[S^{m-\alpha}_{I_\alpha}]^t$. (We recall that the spaces $\mathscr{E}^{\alpha}_{I_\alpha}$ and $S^{-\alpha}_{I_\alpha}$ are mutually conjugate.) It follows from the condition $\Omega_{I_\alpha}=R^n$ that the space $\mathscr{D}(R^n)$ is dense in $S^{m-\alpha}_{I_\alpha}$, and, therefore, the assertion that $pu=0$ on the space $[S^{m-\alpha}_{I_\alpha}]^t$ is equivalent to the assertion that $pu=0$ on the subspace $[\mathscr{D}(R^n)]^t$. Therefore, every function belonging to $[\mathscr{E}^{\alpha}_I]^s$ and satisfying the system (1.4) in $\mathscr{D}^*(R^n)$, satisfies it also in $\mathscr{E}^{\alpha-m}_{I_\alpha}$, and conversely.

Since the space $\mathscr{D}(\mathrm{R}^n)$ is dense in each of the spaces $S^{-\alpha}_{I_\alpha}$, it is also dense in the limit spaces

$$S_I=\varprojlim S^{-\alpha}_{I_\alpha},\qquad \mathscr{E}^*_I=\varinjlim S^{-\alpha}_{I_\alpha}.$$

Therefore $(\mathscr{E}_I)_p$ is the space of all functions in $[\mathscr{E}_I]^s$, satisfying the system (1.4) in $\mathscr{D}^*(R^n)$, and a similar statement is true for to the spaces $(S^*_I)_p$. It follows from all this that $(\mathscr{E}_I)_p$ is the intersection of all the spaces $(\mathscr{E}^{\alpha}_{I_\alpha})_p$, and $(S^*_I)_p$ is their union.

Let us prove the first assertion of the theorem from the beginning. We denote by $E^B_p$ the space of linear combinations of functions of the form (14). The condition $\Omega_{\mathscr{I}_\alpha}=R^n$ means that the function $I_\alpha$ grows at infinity faster than an arbitrary power and therefore the space $\mathscr{E}^{\alpha}_{I_\alpha}$ contains $E$, and the space $(\mathscr{E}^{\alpha}_{I_\alpha})_p$ contains $E_p$ and in particular, $E^B_p$.

We fix the integer $\alpha$ arbitrarily and we choose an arbitrary function $\phi\in[S^{-\alpha}_{I_\alpha}]^s$, which annihilates all the functionals in the space $E^B_p\subset[\mathscr{E}^{\alpha}_{I_\alpha}]^s$, that is

$$(\overline{\mathscr{D}'_j(z,-i\xi)\exp(z,-i\xi),\phi)}=\overline{\mathscr{D}_j(z,D)\tilde{\phi}^*(z)}=0 \tag{15.8}$$

for arbitrary $j$ and arbitrary $z\in B$. Proposition 2, § 3, implies that $\tilde{\phi}^*\in[S^{\mathscr{I}_{\alpha+1}}_{-\alpha}]^s$. By Proposition 6, § 3, the ensemble of all the spaces $S^{\mathscr{I}_\alpha}_{-\alpha}$, $-\infty<\alpha<\infty$, is a complete family $\mathscr{H}_{\mathscr{M}}$, corresponding to the family of majorants $\mathscr{M}$ of type $\mathscr{I}$. We may apply Corollary 3, § 4, Chapter IV to this family, and infer from (15) that $\mathscr{D}(z)\,\phi^*(z)\equiv 0$. Making use of Theorem 2 and the supplement to § 5, Chapter IV, we conclude that $\tilde{\phi}^*\in p'[S^{\mathscr{I}_{\alpha+a}}_{-\alpha-a}]^t$, where $a$ is a constant not depending on $\alpha$. Applying the inverse Fourier transform and using Proposition 2, § 3, we obtain the inclusion relation $\phi\in p^*[S^{-\alpha-b+m}_{I_{\alpha+b}}]^t$ for some constant $b>0$. It follows that the function $\phi$ annihilates all the functionals in the space $(\mathscr{E}^{\alpha+b}_{I_{\alpha+b}})_p$.

We have thus shown that every functional $\phi$ on the space $[\mathscr{E}^{\alpha}_{I_\alpha}]^s$ which vanishes on the subspace $E^B_p$, vanishes also on the subspace $(\mathscr{E}^{\alpha+b}_{I_{\alpha+b}})_p$. This means that $E^B_p$ is dense in $(\mathscr{E}^{\alpha+b}_{I_{\alpha+b}})_p$ in the norm $\|\cdot\|^{\alpha}_{I^{-1}_\alpha}$. Since $(\mathscr{E}_I)_p$ is the intersection of all the spaces $(\mathscr{E}^{\alpha}_{I_\alpha})_p$, an arbitrary function in

$(\mathscr{E}_I)_p$ can be approximated by functions from $E_p^B$ in any of the norms $\|\cdot\|_{I_\alpha^{-1}}^\alpha$, that is, $E_p^B$ is dense in $(\mathscr{E}_I)_p$.

As we have shown, the space $(S_I^*)_p$ is the union of the spaces $(\mathscr{E}_{I_\alpha}^\alpha)_p$. Therefore, any element from this space can be approximated by functions from $E_p^B$ in some one of the norms $\|\cdot\|_{I_\alpha^{-1}}^\alpha$. Every such norm is stronger than the topology of the space $(S_I^*)_p$, and therefore $E_p^B$ is dense in $(S_I^*)_p$. This proves the second assertion of the theorem.

We now pass to the second assertion. Again, suppose that $p$ is an arbitrary $\mathscr{P}$-matrix and $p_1$ is a matrix such that the sequence (2.7) is exact. We consider the sequence of families

$$0 \to \{(E_p)_\alpha\} \to \{[\mathscr{E}_{I_\alpha}^\alpha]^s\} \xrightarrow{p} \{(\mathscr{E}_{I_\alpha}^\alpha)_{p_1}\} \to 0, \tag{16.8}$$

where $(E_p)_\alpha$, $-\infty<\alpha<\infty$, is the closure of $E_p$ in $(\mathscr{E}_{I_\alpha}^\alpha)_p$. We shall show that this sequence is exact. We shall need only to prove the exactness in the last term and algebraic exactness in the second term.

The exactness in the last term stems from the following assertion: every function $w \in (\mathscr{E}_{I_\alpha}^\alpha)_{p_1}$ can be represented in the form $pu$, where $u \in [\mathscr{E}_{I_{\alpha-c}}^{\alpha-c}]^s$ and $\|u\|_{I_\alpha^{-1}c}^{\alpha-c} \leqq C\|w\|_{I_\alpha^{-1}}^\alpha$, while the constant $c$ does not depend on $\alpha$. The proof of this assertion is altogether similar to the proof of the first assertion of Lemma 1, 1°, and rests on Theorem 1, § 4. This proof we leave to the reader.

We showed above that the space $E_p^B \subset E_p$ is dense in $(\mathscr{E}_{I_{\alpha+b}}^{\alpha+b})_p$ in the norm $\|\cdot\|_{I_\alpha^{-1}}^\alpha$. This implies that $(\mathscr{E}_{I_{\alpha+b}}^{\alpha+b})_p \subset (E_p)_\alpha$, which in turn implies the algebraic exactness of (16) in the second term. Thus the exactness of (16) is proved. We observe that this sequence satisfies the conditions of Proposition 10, § 1 and of the second assertion of Proposition 11, § 1. Therefore, passing to the inductive and projective limits in it, we obtain two exact sequences

$$[S_I^*]^s \xrightarrow{p} \varinjlim (S_{I_\alpha}^\alpha)_{p_1} \to 0, \qquad [\mathscr{E}_I]^s \xrightarrow{p} \varprojlim (\mathscr{E}_{I_\alpha}^\alpha)_{p_1} \to 0. \tag{17.8}$$

The projective limit $\varinjlim (\mathscr{E}_{I_\alpha}^\alpha)_{p_1}$ has for elements those that make up the intersection of all the spaces $(\mathscr{E}_{I_\alpha}^\alpha)_{p_1}$ and it therefore coincides with the space $(\mathscr{E}_I)_{p_1}$, in accordance with the observation made at the outset of the proof of this theorem. Therefore, the exactness of the second sequence in (17) implies the equation $p[\mathscr{E}_I]^s = (\mathscr{E}_I)_{p_1}$. In a similar way, the exactness of the first sequence in (17) implies that $p[S_I^*]^s = (S_I^*)_{p_1}$. Since $p$ was chosen from the outset to be an arbitrary $\mathscr{P}$-matrix, these relationships imply that for the spaces $\Phi = \mathscr{E}_I, S_I^*$, the sequence (5.7) is always algebraically exact.

We remarked in 5°, § 3, that $\mathscr{E}_I$ and $S_I$ are $\mathscr{F}\mathscr{S}$-spaces. Therefore, by Proposition 1, § 7, the algebraic exactness of the sequences (5.7) with $\Phi = \mathscr{E}_I, S_I^*$ implies their exactness. □

*Remark.* Theorem 5 implies, in particular, that the subspace $E_p^B$[7] is dense in $E_p$ in the topology of any one of the spaces $(\mathscr{E}_I)_p$. This topology is stronger than the topology of any of the spaces $\mathscr{E}_p(\Omega)$ and $\mathscr{D}_p^*(\Omega)$. Therefore, for an arbitrary region $\Omega$, the assertion that $E_p$ is dense in $\mathscr{E}_p(\Omega)$ or in $\mathscr{D}_p^*(\Omega)$ is equivalent to the assertion that $E_p^B$ is dense in $\mathscr{E}_p(\Omega)$ or $\mathscr{D}_p^*(\Omega)$.

## § 9. The connection between $M$-convexity and the properties of a sheaf of solutions of a homogeneous system

**1°. Theorem on isomorphism.** Let $\Omega$ be a region in $R^n$. We denote by $\Phi(\Omega)$ either of the spaces $\mathscr{D}^*(\Omega)$ or $\mathscr{E}(\Omega)$. We denote by $\wp$ a sheaf in $R^n$, which correlates to $\Omega$ the space $\Phi(\Omega)$, that is, either of the sheaves $\mathscr{D}^*\colon \Omega \sim\to \mathscr{D}^*(\Omega)$ or $\mathscr{E}\colon \Omega\sim\to\mathscr{E}(\Omega)$. Let $M$ be a finite $\mathscr{P}$-module. By $\wp_M$ we denote the sheaf which correlates to the region $\Omega$ the space of solutions of the system (1.4) (where $\mathscr{P}^s/p'\,\mathscr{P}^t \cong M$) belonging to $[\Phi(\Omega)]^s$, and to the pair of regions $\omega\subset\Omega$ the restriction mapping $r_\Omega^\omega\colon \Phi_M(\Omega)\to \Phi_M(\omega)$. The spaces $\Phi_M(\Omega)$ and the mappings $r_\Omega^\omega$ do in fact form a sheaf, since the action of the differential operator $p$ is local.

Let $\Omega$ be a region in $R^n$, and let $U=\{U_i\}$ be some locally finite open covering of this region. In § 2, we introduced the sequence of spaces ${}^\nu\Phi(U)$, $\nu=0,1,2,\ldots$, formed by the cochain on this covering with coefficients in the space $\Phi(U_{i_0}\cap\cdots\cap U_{i_\nu})$. We recall that the topological space ${}^\nu\Phi(U)$ coincides with the direct product $\prod_{i_0<\cdots<i_\nu}\Phi(U_{i_0}\cap\cdots\cap U_{i_\nu})$. We construct the spaces ${}^\nu\Phi_M(U)$, $\nu=0,1,2,\ldots$ formed by the cochains on $U$ with coefficients in the spaces $\Phi_M(U_{i_0}\cap\cdots\cap U_{i_\nu})$, by a similar process, and we define the coboundary operator $\partial^\nu\colon {}^\nu\Phi_M(U)\to{}^{\nu+1}\Phi_M(U)$. Therefore, we may consider the complex

$$0\to\Phi_M(\Omega)\to{}^0\Phi_M(U)\xrightarrow{\partial^0}{}^1\Phi_M(U)\xrightarrow{\partial^1}\cdots. \tag{1.9}$$

The spaces $H^\nu(U,\wp_M)=\operatorname{Ker}\partial^\nu/\operatorname{Im}\partial^{\nu-1}$ $(\partial^{-1}=0)$ are said to be cohomologies of this complex.

Let $V$ be a covering of the subregion $\omega\subset\Omega$ inscribed in the covering $U$. Then for arbitrary $\nu\geqq 0$ there is a defined a restriction mapping $r_U^V\colon {}^\nu\Phi_M(U)\to{}^\nu\Phi_M(V)$, which commutes with the operator $\partial^\nu$.

**Theorem 1.** *Let $M$ be an arbitrary finite $\mathscr{P}$-module and let $U$ be a locally finite convex (open) covering of the region $\Omega$.*

---

7 This space consists of linear combinations of the functions (14).

I. *We have the isomorphism*

$$H^{\nu}(U, \mathscr{P}_M) \cong \operatorname{Ext}^{\nu}(M, \Phi(\Omega)), \qquad \nu = 0, 1, 2, \ldots .^{8} \tag{2.9}$$

*The sequence* (1) *is exact if and only if the space* $\Phi(\Omega)$ *is* $M$*-convex.*

II. *Let* $V$ *be a locally finite convex covering of the subregion* $\omega \subset \Omega$ *inscribed in* $U$. *Then the diagram*

$$\begin{array}{ccc} H^{\nu}(V, \mathscr{P}_M) & \cong & \operatorname{Ext}^{\nu}(M, \Phi(\omega)) \\ {\scriptstyle r_U^V}\uparrow & & \uparrow{\scriptstyle \rho_\Omega^\omega} \\ H^{\nu}(U, \mathscr{P}_M) & \cong & \operatorname{Ext}^{\nu}(M, \Phi(\Omega)), \end{array}$$

*in which the rows contain isomorphisms of the type* (2), *and the columns contain restriction mappings, is commutative.* (The mapping $\rho_\Omega^\omega$ corresponds to the restriction mapping $\Phi(\Omega) \to \Phi(\omega)$.)

*Proof.* Let (4.7) be a free resolution of the module $M$. We consider the following commutative diagram

$$\begin{array}{ccccccccc}
 & & \vdots & & \vdots & & & & \\
 & & \partial^1\uparrow & & \partial^1\uparrow & & \vdots & & \\
0 & \longrightarrow & {}^1\Phi_M(U) & \longrightarrow & [{}^1\Phi(U)]^s & \xrightarrow{p} & & \cdots & \\
 & & \partial^0\uparrow & & \partial^0\uparrow & & \partial^0\uparrow & & \vdots \\
0 & \longrightarrow & {}^0\Phi_M(U) & \longrightarrow & [{}^0\Phi(U)]^s & \xrightarrow{p} & [{}^0\Phi(U)]^t & \xrightarrow{p_1} & \cdots \\
 & & \uparrow & & \uparrow & & \uparrow & & \uparrow \\
0 & \longrightarrow & \Phi_M(\Omega) & \longrightarrow & [\Phi(\Omega)]^s & \xrightarrow{p} & [\Phi(\Omega)]^t & \xrightarrow{p_1} & [\Phi(\Omega)]^{t_2} \longrightarrow \cdots \\
 & & \uparrow & & \uparrow & & \uparrow & & \uparrow \\
 & & 0 & & 0 & & 0 & & 0
\end{array} \tag{3.9}$$

The exactness of all the columns of this diagram, excluding the leftmost, follows from Proposition 9, § 2. We next establish the exactness of all the rows except the bottom one. Let the cochain $\phi = \sum \phi_{i_0, \ldots, i_\nu} U_{i_0} \wedge \cdots \wedge U_{i_\nu} \in [{}^{\nu}\Phi(U)]^{t_j}$ belong to the kernel of the operator $p_j$. This means that $p_j \phi_{i_0, \ldots, i_\nu} = 0$ in $U_{i_0} \cap \cdots \cap U_{i_\nu}$, for an arbitrary array of indices $i_0, \ldots, i_\nu$. Since every region $U_{i_0} \cap \cdots \cap U_{i_\nu}$ is convex, we know by Theorems 1

8 Here and in what follows the symbol $\Phi(\Omega)$ will mean either of the two spaces $\mathscr{D}^*(\Omega)$, $\mathscr{E}(\Omega)$. However, writing $\Phi(\Omega) = \mathscr{E}(\Omega)$ in the right side of (2), we must write $\mathscr{P} = \mathscr{E}$ on the left side. If $\Phi(\Omega) = \mathscr{D}^*(\Omega)$ then $\mathscr{P} = \mathscr{D}^*$. A similar two-valued meaning will be assigned to all of our succeeding statements.

and 3 of § 8, that we can find a function $\psi_{i_0,\ldots,i_\nu} \in [\Phi(U_{i_0} \cap \cdots \cap U_{i_\nu})]^{t_{j-1}}$ such that $\phi_{i_0,\ldots,i_\nu} = p_{j-1} \psi_{i_0,\ldots,i_\nu}$ and the mapping

$$\Phi_{p_j}(U_{i_0,\ldots,i_\nu}) \ni \phi_{i_0,\ldots,i_\nu} \to \psi_{i_0,\ldots,i_\nu} \in [\Phi(U_{i_0} \cap \cdots \cap U_{i_\nu})]^{t_{j-1}} / \operatorname{Ker} p_{j-1} \quad (4.9)$$

is continuous. Since the function $\phi_{i_0,\ldots,i_\nu}$ depends by definition skew-symmetrically on its indices, we may suppose that the function $\psi_{i_0,\ldots,i_\nu}$ is also skew-symmetric in $i_0, \ldots, i_\nu$. Therefore, the functions $\psi_{i_0,\ldots,i_\nu}$ are the coefficients of some cochain $\psi \in [{}^\nu\Phi(U)]^{t_{j-1}}$. The continuity of the mapping (4) for arbitrary $i_0, \ldots, i_\nu$ implies the continuity of the mapping

$${}^\nu\Phi_{p_j}(U) \ni \phi \to \psi \in [{}^\nu\Phi(U)]^{t_{j-1}} / \operatorname{Ker} p_{j-1},$$

which is inverse to the operator $p_{j-1}$. That proves our assertion. We now break the diagram into fragments, each of which is a particular case of Diagram (α) of Theorem 1, § 2, Chapter I. Then that theorem implies the isomorphism (2).

We establish the second assertion of the theorem. Together with the diagram (3), we shall consider a similar diagram (3′) formed from the spaces ${}^\nu\Phi_M(V)$, $[{}^\nu\Phi(V)]^{t_j}$, $\nu \geqq -1$, $j \geqq 0$, and the mappings $\partial^\nu$ and $p_j$. By hypothesis, the covering $V$ is inscribed in $U$ and, therefore, the restriction mappings

$${}^\nu\Phi_M(U) \to {}^\nu\Phi_M(V), \qquad [{}^\nu\Phi(U)]^{t_j} \to [{}^\nu\Phi(V)]^{t_j}.$$

are defined. These mappings commute with the mappings $\partial^\nu$ and $p_j$ of diagrams (3) and (3′). Therefore, the second assertion of the theorem follows from Theorem 2, § 2, Chapter I. □

**Corollary 1.** *We have the algebraic isomorphism*

$$\operatorname{Ext}^\nu(M, \Phi(\Omega)) \cong H^\nu(\Omega, \mathscr{P}_M), \qquad \nu = 0, 1, 2, \ldots, \quad (5.9)$$

*where $H^\nu(\Omega, \mathscr{P}_M)$ is the cohomology space of the sheaf $\mathscr{P}_M$ in the region $\Omega$.*

*Proof.* By definition, the space $H^\nu(\Omega, \mathscr{P}_M)$ is the inductive limit of the spaces $H^\nu(U, \mathscr{P}_M)$ and the restriction mappings $r_U^V$ on the set of all open coverings of the region $\Omega$. Since $\Omega$ is a paracompact topological space, we can inscribe a locally finite convex covering in every open covering of $\Omega$. Therefore, the locally finite convex coverings form a cofinal subset of the set of all open coverings of $\Omega$. Consequently, in view of Proposition 1, § 1, $H^\nu(\Omega, \mathscr{P}_M)$ is the inductive limit of the spaces $H^\nu(U, \mathscr{P}_M)$, taken over the set of all locally finite convex coverings of $\Omega$. Assertion II of Theorem 1 implies that for any two locally finite convex coverings $U$

and $V$ of the region $\Omega$, such that $V$ is inscribed in $U$, the following diagram:

$$\begin{array}{ccc} & \rightarrow & H^\nu(V, \mathscr{P}_M) \\ \mathrm{Ext}^\nu(M, \Phi(\Omega)) & & \uparrow r_U^V \\ & \rightarrow & H^\nu(U, \mathscr{P}_M) \end{array}$$

is defined and commutative. Therefore, in the isomorphism $\mathrm{Ext}^\nu(M, \Phi(\Omega)) \cong H^\nu(U, \mathscr{P}_M)$, we may pass to the inductive limit with respect to such coverings $U$. In the limit we obtain the isomorphism (5). □

**Corollary 2.** *Let $M$ be a hypoelliptic module (see 5°, § 8). Then the mappings*

$$\mathrm{Ext}^\nu(M, \mathscr{E}(\Omega)) \to \mathrm{Ext}^\nu(M, \mathscr{D}^*(\Omega)), \qquad \nu = 0, 1, 2, \ldots, \tag{6.9}$$

*corresponding to the imbedding $\mathscr{E}(\Omega) \to \mathscr{D}^*(\Omega)$, are isomorphisms. The spaces $\mathscr{E}(\Omega)$ and $\mathscr{D}^*(\Omega)$ are simultaneously (strongly) $M$-convex or not.*

*Proof.* We shall consider two particular cases of diagram (3) corresponding to the spaces $\Phi(\Omega) = \mathscr{E}(\Omega)$ and $\Phi(\Omega) = \mathscr{D}^*(\Omega)$. The imbedding $\mathscr{E}(\Omega) \to \mathscr{D}^*(\Omega)$ generates the mapping of the first of these diagrams in the second. Applying Theorem 2, § 2, of Chapter I, we obtain the commutative diagram

$$\begin{array}{ccc} \mathrm{Ext}^\nu(M, \mathscr{D}^*(\Omega)) & \cong & H^\nu(U, \mathscr{D}_M^*) \\ \uparrow & & \uparrow \\ \mathrm{Ext}^\nu(M, \mathscr{E}(\Omega)) & \cong & H^\nu(U, \mathscr{E}_M). \end{array} \tag{7.9}$$

It follows from Theorem 1, § 5 that the mappings

$${}^\nu\mathscr{E}_M(U) \to {}^\nu\mathscr{D}_M^*(U), \qquad \nu = -1, 0, 1, \ldots, \tag{8.9}$$

are isomorphisms. This implies that the right vertical mapping in diagram (7) is an isomorphism. The commutativity then implies that the mapping (6) is also an isomorphism.

Let us consider two particular cases of the sequence (1), corresponding to the spaces $\Phi(\Omega) = \mathscr{E}(\Omega)$ and $\mathscr{D}^*(\Omega)$. In view of the fact that the imbeddings (8) are isomorphisms, these two sequences are isomorphic. It follows that the spaces $\mathscr{E}(\Omega)$ and $\mathscr{D}^*(\Omega)$ are $M$-convex or not, in unison. Since the spaces $\mathscr{E}_M(\Omega)$ and $\mathscr{D}_M^*(\Omega)$ are isomorphic, the subspace $E_M$ is dense or not in these spaces simultaneously. □

**2°. Examples**

**Example 1.** Let $M = \mathrm{Coker}\, d_0'$. As we proved in Example 2, 2°, § 7, $\mathrm{Ext}^\nu(M, \Phi(\Omega))$ is the cohomology space of differential forms with coefficients in $\Phi(\Omega)$. On the other hand, $\mathscr{P}_M$ is the sheaf $\mathscr{C}$ of locally constant functions. Therefore the isomorphism of Corollary 1 can be written as:

$$\operatorname{Ext}^\nu(M, \Phi(\Omega)) \cong H^\nu(\Omega, \mathscr{C}).$$

This isomorphism is a particular case of a theorem of de Rham, which refers to a region $\Omega$ in Euclidian space.

**Example 2.** Let $n=2m$ and $M=\operatorname{Coker} {}''d_0'$ (see Example 3, 2°, § 7). For an arbitrary region $\Omega$, the space $\Phi_M(\Omega)$ is the space $\mathscr{H}(\Omega)$ of functions that are analytic in $\Omega$. Accordingly, $\mathscr{P}_M = \mathscr{H}$, where $\mathscr{H}$ is the sheaf of analytic functions. It follows from Corollary 1 that we have the isomorphism $\operatorname{Ext}^\nu(M, \mathscr{P}(\Omega)) \cong H^\nu(\Omega, \mathscr{H})$. If $\Phi(\Omega)=\mathscr{E}(\Omega)$, it is a particular case of a theorem of Dolbeault. Thus, by Proposition 1, § 7, for $M$-convexity of the space $\mathscr{E}(\Omega)$, it is necessary and sufficient that the equations $H^\nu(\Omega, \mathscr{H})=0$, $\nu \geqq 1$, be satisfied. Since the operator ${}''d_0$ is elliptic, the module $M$ is also elliptic. Therefore by Corollary 2, these equations are necessary and sufficient for the $M$-convexity of the space $\mathscr{D}^*(\Omega)$. On the other hand, as is known[9], the equations $H^\nu(\Omega, \mathscr{H})=0$, $\nu \geqq 1$, are necessary and sufficient in order that $\Omega$ be a domain of holomorphy. Thus, $M$-convexity of the space $\mathscr{E}(\Omega)$ or $\mathscr{D}^*(\Omega)$ is equivalent to the statement that $\Omega$ is a domain of holomorphy.

As we showed in 4°, § 4, the collection of varieties associated with $M$ consists of one variety containing the origin of coordinates. Therefore, by the remark in 7°, § 8, the condition: $E_M$ is dense in $\Phi_M(\Omega)$, is equivalent to the condition: the space $E_M^B$ is dense in $\Phi_M(\Omega)$, where $B$ is the set consisting only of the origin of coordinates. As is easily seen, $E_M^B$ is the space of all polynomials in the variables $\zeta_j = \xi_j + i\,\xi_{m+j}$. Thus, the condition: $E_M$ is dense in $\Phi_M(\Omega)$, is equivalent to the statement that the polynomials in $\zeta_j$ are dense in the space $\mathscr{H}(\Omega)$ (in the topology of $\Phi_M(\Omega)$, which coincides with the natural topology of $\mathscr{H}(\Omega)$). This latter condition means that $\Omega$ is a Runge domain of the first type. Thus, we conclude that strong $M$-convexity of $\Phi(\Omega)$ is equivalent to the condition that $\Omega$ be a Runge domain.

**3°. Convexity relative to a zero-dimensional module.** Let $M$ be a zero-dimensional $\mathscr{P}$-module. Then by Proposition 4, § 4, Chapter IV, it has a finite dimension $l(M)$ as a linear space over the field $C$.

**Theorem 2.** *Let $M$ be a zero-dimensional finite $\mathscr{P}$-module. For an arbitrary region $\Omega$ and a locally finite convex covering $U$ of this region*

I. *All the spaces $H^\nu(U, \mathscr{P}_M)$, $\nu \geqq 0$, are separable.*

II. *We have the isomorphisms*

$$H^\nu(\Omega, \mathscr{P}_M) \cong [H^\nu(\Omega, \mathscr{C})]^{l(M)}, \qquad \nu \geqq 0, \tag{9.9}$$

*where $\mathscr{C} = \mathscr{E}_{d_0}$ is the sheaf of locally constant functions.*

9 See, for example, Serre [3].

*Proof*. By Proposition 4, § 4, Chapter IV, the variety $N(M)$ is a finite set, and the ensemble of its points $z^\lambda$, $\lambda=0, \ldots, l$, is the set of varieties associated with the module $M$. Suppose, further, that $M \cong \mathscr{P}^s/p' \mathscr{P}^t$, and that $\mathscr{D}=\sum \xi^j \mathscr{D}_j(z, \delta)$ is the $p'$-operator. At every point $z^\lambda$ only a finite number $l_\lambda$ of the operators $\mathscr{D}_j(z^\lambda, D)$ are distinct from zero, and the operators that are different from zero are linearly independent. Let $d^\lambda(z^\lambda, D)$ be a column of height $l_\lambda$, made up of these differential operators. The ensemble of the operators $d^\lambda(z^\lambda, D)$, $\lambda=0, \ldots, l$, is the set of normal Noetherian operators associated with the matrix $p'$.

Let $\omega$ be a convex region. By Corollary 1 of §4, we can write an arbitrary distribution $u \in \Phi_p(\omega)$ on an arbitrary compact $\kappa \subset \omega$ in the form of a sum

$$\begin{aligned} u(\xi) &= \sum_\lambda \left(d^\lambda(z^\lambda, -i\xi)\right)' \exp(z^\lambda, -i\xi)\, \mu^\lambda \\ &= \sum_{j,\lambda} \mathscr{D}'_j(z^\lambda, -i\xi) \exp(z^\lambda, -i\xi)\, \mu^\lambda_j, \end{aligned} \tag{10.9}$$

where the $\mu^\lambda$ are vectors whose components are arbitrary complex numbers $\mu^\lambda_j$. The number of these arbitrary constants is equal to the sum $\sum l_\lambda$, and by Proposition 4, §4, Chapter IV, coincides with the dimension $l(M)$.

As we observed earlier, for every $\lambda$ the differential operators $\mathscr{D}_j(z^\lambda, D)$ are linearly independent. Therefore, the ensemble of all the $l(M)$ functions $\mathscr{D}'_j(z^\lambda, -i\xi) \exp(z^\lambda, -i\xi)$ forms a linearly independent set. Then all the numbers $\mu^\lambda_j$ are uniquely defined (under the hypothesis that the compact $\kappa$ has inner points), and so do not depend on the compact $\kappa$. Therefore, the representation (10) is valid throughout the region $\omega$.

Now let $\omega$ be an arbitrary region. We choose some convex covering $V=\{V_\alpha\}$, and we write out the representation for the functions $u \in \Phi_p(\omega)$ in every region $V_\alpha$ of this covering. If two regions $V_\alpha$ and $V_\beta$ intersect, the representations (10) will coincide in their intersection, and therefore, the corresponding coefficients $\mu^\lambda_j$ will be the same in this intersection. Thus, we obtain the representation (10) for the functions belonging to $\Phi_p(\omega)$, in which the $\mu^\lambda_j$ are unique functions of the point $\xi$, locally constant in the region $\omega$, and therefore constant on every connected compact of this region.

By $\mathscr{C}(\omega)$ we denote the space of all complex valued locally constant functions in $\omega$. Correlating to the function $u$ the vectors formed from the coefficients $\mu^\lambda_j$ in the representation (10), we obtain the linear mapping

$$\Phi_p(\omega) \to [\mathscr{C}(\omega)]^{l(M)}. \tag{11.9}$$

This mapping is obviously an algebraic isomorphism. The space $\mathscr{C}(\omega)$ will be endowed with the topology of uniform convergence on every compact $\kappa \subset \omega$. Then it is easy to see that the isomorphism (11) is a topological isomorphism.

Let $U$ be the locally finite convex covering of the region $\Omega$ fixed in the hypothesis of the theorem. We consider the complex

$$0 \to \mathscr{C}(\Omega) \to {}^0\mathscr{C}(U) \xrightarrow{\partial^0} {}^1\mathscr{C}(U) \xrightarrow{\partial^1} \cdots, \tag{12.9}$$

which is a particular case of the complex (1) for $M = \operatorname{Coker} d_0'$. Here ${}^\nu\mathscr{C}(U)$, where $\nu = 0, 1, 2, \ldots$, is the space of cochains (of order $\nu$ on the covering $U$ with coefficients from the sheaf $\mathscr{C}$) which is topologically a direct product of the spaces $\mathscr{C}(U_{i_0} \cap \cdots \cap U_{i_\nu})$, isomorphic to the complex line $C$. (The isomorphism $\mathscr{C}(U_{i_0} \cap \cdots \cap U_{i_\nu}) \cong C$ is valid, since each of the regions $U_{i_0} \cap \cdots \cap U_{i_\nu}$ is connected, since it is convex.) The cohomologies of the complex (12) are by definition equal to the spaces $H^\nu(U, \mathscr{C})$, $\nu = 1, 2, \ldots$.

We fix $\nu \geqq 0$ and we apply the isomorphism (11) to all the regions of the form $\omega = U_{i_0} \cap \cdots \cap U_{i_\nu}$. Compounding the direct product of these isomorphisms, we obtain the isomorphisms ${}^\nu\Phi_p(U) \cong [{}^\nu\mathscr{C}(U)]^{l(M)}$ from which we construct the following diagram:

$$\begin{array}{ccccccc} 0 \to \Phi_p(\Omega) & \to & {}^0\Phi_p(U) & \xrightarrow{\partial^0} & {}^1\Phi_p(U) & \xrightarrow{\partial^1} & \cdots \\ \wr\| & & \wr\| & & \wr\| & & \\ 0 \to \mathscr{C}(\Omega) & \to & {}^0\mathscr{C}(U) & \xrightarrow{\partial^0} & {}^1\mathscr{C}(U) & \xrightarrow{\partial^1} & \cdots. \end{array} \tag{13.9}$$

This diagram is commutative, since every linear operation on the coefficients of the cochain $\phi \in {}^\nu\Phi_p(U)$, in particular, the application of the coboundary operator $\partial^\nu$, corresponds to the same linear operation on the corresponding coefficients $\mu_j^\lambda$. The commutativity of the diagram (13) implies that the isomorphisms between the terms of the first and second columns define isomorphisms between the corresponding cohomologies

$$H^\nu(U, \mathscr{P}_M) \cong [H^\nu(U, \mathscr{C})]^{l(M)}, \qquad \nu = 0, 1, 2, \ldots. \tag{14.9}$$

Since these isomorphisms are topological, the separability of the space $H^\nu(U, \mathscr{P}_M)$ is equivalent to the separability of the space $H^\nu(U, \mathscr{C})$. By definition $H^\nu(U, \mathscr{C}) = Z^\nu / B^\nu$, where $Z^\nu$ is the kernel of the operator $\partial^\nu$, and $B^\nu$ is the image of $\partial^{\nu-1}$ in (12). Therefore, the separability of the space $H^\nu(U, \mathscr{C})$ is in turn equivalent to the closedness of the subspace $B^\nu \subset {}^\nu\mathscr{C}(U)$.

We shall prove that this subspace is closed. Let $({}^\nu\mathscr{C}(U))'$ be the space of all linear functionals on ${}^\nu\mathscr{C}(U)$ (not necessarily continuous), and let $(B^\nu)^0$ be the subspace in $({}^\nu\mathscr{C}(U))'$, orthogonal to $B^\nu$. By a general property of linear spaces, the space $B^\nu$ itself is orthogonal to $(B^\nu)^0$, that is, it is the intersection of the kernels of all functionals belonging to $(B^\nu)^0$.

We now describe the space $({}^\nu\mathscr{C}(U))'$. We correlate to every cochain $\phi \in {}^\nu\mathscr{C}(U)$ the ensemble of its coefficients $\phi_{i_0, \ldots, i_\nu}$, and thus define an algebraic isomorphism ${}^{\nu}\mathscr{C}(U)$ and a direct product $\prod_{i_0<\cdots<i_\nu} C$ of lines. It follows that the space $({}^{\nu}\mathscr{C}(U))'$ is isomorphic to the direct sum of the same number of lines, that is, every linear functional on ${}^{\nu}\mathscr{C}(U)$ can be written in the form

$$(f,\phi)=\sum f_{i_0,\ldots,i_\nu}\phi_{i_0,\ldots,i_\nu}, \qquad f_{i_0,\ldots,i_\nu}\in C, \tag{15.9}$$

where only a finite number of the quantities $f_{i_0,\ldots,i_\nu}$ are different from zero. By definition, however, the space ${}^{\nu}\mathscr{C}(U)$ coincides with the direct product $\prod_{i_0<\cdots<i_\nu} C$ topologically as well. It follows that every functional of the form (15) is continuous on ${}^{\nu}\mathscr{C}(U)$. Thus, the subspace $B^\nu$ is the intersection of the kernels of a certain number of continuous linear functionals and is, therefore, closed. This proves the first assertion of the theorem.

We now prove the second assertion. We observe that the isomorphism (14) is compatible with the transition to an inscribed covering, that is, for every covering $V$ of the region $\Omega$ inscribed in $U$, the following diagram

$$\begin{array}{ccc} H^\nu(V,\mathscr{P}_M) & \cong & [H^\nu(V,\mathscr{C})]^{l(M)} \\ {\scriptstyle r_U^V}\uparrow & & \uparrow{\scriptstyle r_U^V} \\ H^\nu(U,\mathscr{P}_M) & \cong & [H^\nu(U,\mathscr{C})]^{l(M)} \end{array}$$

is commutative. It follows that the isomorphism (14) is conserved under a passage to the inductive limit, with respect to the set of all locally finite convex coverings of $\Omega$. Carrying out this passage, we obtain the isomorphism (9). ▯

**Corollary 3.** *Let $M$ be a zero-dimensional finite $\mathscr{P}$-module, and let $\Omega$ be a non-empty region in $R^n$. The space $\Phi(\Omega)$ is $M$-convex if and only if*

$$H^\nu(\Omega,\mathscr{C})=0, \qquad \nu\geqq 1. \tag{16.9}$$

*The space $\Phi(\Omega)$ is strongly $M$-convex if and only if*

$$H^\nu(\Omega,\mathscr{C})=0,\quad \nu\geqq 1, \qquad H^0(\Omega,\mathscr{C})\cong C.$$

*Proof.* Since every zero-dimensional module is hypoelliptic, Corollary 2 implies that we may limit ourselves to the space $\Phi(\Omega)=\mathscr{E}(\Omega)$. By Proposition 1, § 7, $\mathscr{E}(\Omega)$ is $M$-convex if and only if $\operatorname{Ext}^\nu(M,\mathscr{E}(\Omega))=0$, $\nu\geqq 1$. Corollary 1 and Theorem 2 of this section imply that this equation is equivalent to the relation (16). This proves the first assertion.

By definition, $\mathscr{E}(\Omega)$ is strongly $M$-convex if it is $M$-convex and the closure of $E_M$ in $\mathscr{E}_M(\Omega)$ coincides with $\mathscr{E}_M(\Omega)$. Proposition 4, § 7 implies that the space $E_M$ is the space of all sums of the form (10), where the $\mu_j^\lambda$ are arbitrary complex coefficients constant throughout $R^n$. We therefore have the isomorphism $E_M\cong C^{l(M)}$. Since $E_M$ is finite-dimensional and $\Omega$ is non-empty, the topology induced in $E_M$ from $\mathscr{E}_M(\Omega)$ coincides with the topology of a finite-dimensional Euclidian space. Therefore, the closure of $E_M$ in $\mathscr{E}_M(\Omega)$ is $E_M$. This means that the condition that $E_M$ is dense in $\mathscr{E}_M(\Omega)$ is equivalent to the equation $\mathscr{E}_M(\Omega)=E_M$. Since $\mathscr{E}_M(\Omega)\cong[H^0(\Omega,\mathscr{C})]^{l(M)}$, and $E_M\cong C^{l(M)}$, this equation in turn is equivalent to the isomorphism $H^0(\Omega,\mathscr{C})\cong C$. This proves the second assertion. □

The necessary conditions for $M$-convexity and strong $M$-convexity, which are given in Corollary 3, will now be generalized to an arbitrary module $M$.

### 4°. Cohomologies connected with two finite $\mathscr{P}$-modules

**Theorem 3.** *Let $M$ and $L$ be two finite $\mathscr{P}$-modules, such that*

$$\operatorname{Tor}_i(M,L)=0,\qquad i\geqq 1, \tag{17.9}$$

*and let $\Phi$ be a topological $\mathscr{P}$-module. Then the following assertions are true.*

I. *The following mappings are defined and continuous*

$$\operatorname{Ext}^k(L,\operatorname{Hom}(M,\Phi))\to\operatorname{Ext}^k(M\otimes L,\Phi),\qquad k=0,1,2,\ldots. \tag{18.9}$$

II. *If $\Phi$ is $M$-convex, these mappings are isomorphisms.*

III. *If $\Phi$ is also $M\otimes L$-convex, the space $\Phi_M$ is $L$-convex.*

*Proof.* Suppose that

$$\cdots\to\mathscr{P}^{t_2}\xrightarrow{p_1'}\mathscr{P}^{t_1}\xrightarrow{p_0'}\mathscr{P}^{t_0}\to M\to 0 \tag{19.9}$$

and

$$0\to\mathscr{P}^{s_d}\xrightarrow{q_{d-1}'}\mathscr{P}^{s_{d-1}}\to\cdots\to\mathscr{P}^{s_2}\xrightarrow{q_1'}\mathscr{P}^{s_1}\xrightarrow{q_0'}\mathscr{P}^{s_0}\to L\to 0 \tag{20.9}$$

are free resolutions of the modules $M$ and $L$. Tensoring these resolutions one by another, we obtain the commutative diagram

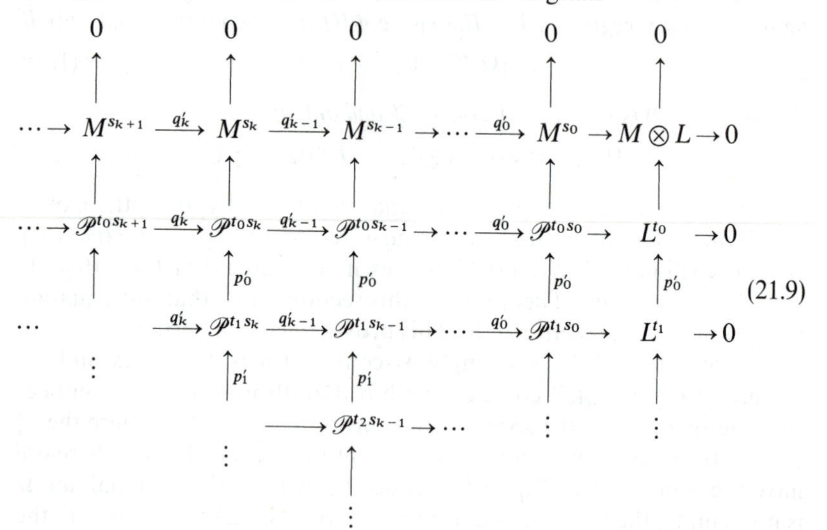

(21.9)

Here the mapping $p'_i$: $\mathscr{P}^{t_{i+1}s_j} \to \mathscr{P}^{t_i s_j}$ consists of multiplying the vector $(f_1, \dots, f_{t_{i+1}})$, $f_\alpha \in \mathscr{P}^{s_j}$ by the matrix $p'_i$, that is, it coincides with the action of the mapping $p'_i \otimes e$: $\mathscr{P}^{t_{i+1}} \otimes \mathscr{P}^{s_j} \to \mathscr{P}^{t_i} \otimes \mathscr{P}^{s_j}$ ($e$ being the identity mapping); similarly, the mapping $q'_j$ coincides with the mapping $e \otimes q'_j$: $\mathscr{P}^{t_i} \otimes \mathscr{P}^{s_{j+1}} \to \mathscr{P}^{t_i} \otimes \mathscr{P}^{s_j}$. The exactness of the top row of the diagram and the right column follows from the conditions (17). The exactness of the remaining rows and columns follows from the fact that all the modules $\mathscr{P}^k$ are flat.

We consider the free modules $\mathscr{P}_k = \bigoplus_{i+j=k} \mathscr{P}^{t_i s_j}$, $k=0, 1, 2, \dots$, and we construct the special free resolution of the module $M \otimes L$

$$\dots \to \mathscr{P}_{k+1} \xrightarrow{r'_k} \mathscr{P}_k \to \dots \to \mathscr{P}_1 \xrightarrow{r'_0} \mathscr{P}_0 \to M \otimes L \to 0, \qquad (22.9)$$

where the mappings $r'_k$ act as follows:

$$r'_k: \quad \mathscr{P}_{k+1} \ni (f_{k+1,0}, \dots, f_{0,k+1}) \to (g_{k,0}, \dots, g_{0,k}) \in \mathscr{P}_k,$$

and

$$f_{i,j} \in \mathscr{P}^{t_i s_j}, \qquad g_{ij} = p'_i f_{i+1,j} + (-1)^i q'_j f_{i,j+1}.$$

We shall show that this sequence is, in fact, a free resolution[10]. For an arbitrary vector $f \in \mathscr{P}_{k+1}$ the vector $r'_{k-1} r'_k f$ is by definition formed from the polynomials

$$\begin{aligned} h_{i,j} &= p'_i (p'_{i+1} f_{i+2,j} + (-1)^{i+1} q'_j f_{i+1,j+1}) \\ &\quad + (-1)^i q'_j (p'_i f_{i+1,j+1} + (-1)^i q'_{j+1} f_{i,j+2}) \\ &= p'_i p'_{i+1} f_{i+2,j} + q'_j q'_{j+1} f_{i,j+2} = 0. \end{aligned}$$

Thus the semi-exactness of (22) is proved. The exactness in the two last terms follows from Proposition 3, § 3, Chapter I.

It remains to show that for an arbitrary $k > 0$, the inclusion $\operatorname{Ker} r'_{k-1} \subset \operatorname{Im} r'_k$ is valid. We consider the module

$$\operatorname{Im} q'_{k-1} \cap \operatorname{Im} p'_0 / \operatorname{Im} q'_{k-1} p'_0, \qquad (23.9)$$

associated with the module $\mathscr{P}^{t_0 s_{k-1}}$, where $q'_{k-1}$ and $p'_0$ are mappings in (21). In the diagram (21), we distinguish the fragment formed of the modules $\mathscr{P}^{t_i s_j}$ where $i+j=k+1$, $k-1$. This fragment has the form of the diagram ($\gamma$) of Theorem 1, § 2, Chapter I, and all the rows and columns are exact. Therefore, the arguments of 4°, § 2, Chapter I imply that all the modules $H^i_j$ constructed for this fragment are equal to zero. In particular, the module (23) is equal to zero.

---

10 The reader who is familiar with spectral sequences can deduce this conclusion from the fact that the spectral sequence connected with the double complex (21) is degenerate (see, for example, Godement [1] Theorem 4.8.1).

Now suppose that $f=(f_{k,0},\dots,f_{0,k})$ is an arbitrary vector belonging to the kernel of the mapping $r_{k-1}$. This condition, in particular, implies that $p'_0 f_{1,k-1}=-q'_{k-1} f_{0,k}$. Both sides of this equation obviously belong to the module $\operatorname{Im} q'_{k-1} \cap \operatorname{Im} p'_0$. Therefore, the fact that the module (23) is equal to zero implies that $q'_{k-1} f_{0,k}=q'_{k-1} p'_0 g$ where $g \in \mathscr{P}^{t_1 s_k}$. Hence $q'_{k-1}(f_{0,k}-p'_0 g)=0$, and therefore the exactness of the rows of the diagram implies that $f_{0,k}-p'_0 g=q'_k h$, where $h \in \mathscr{P}^{t_0 s_{k+1}}$. We consider a new vector $f'=(f'_{k,0},\dots,f'_{0,k})=f-r'_k(0,\dots,0,g,h)$. It is clear that $f'_{0,0}=0$. Since $f'-f \in \operatorname{Im} r'_k$, the vector $f'$ also belongs to the kernel of $r'_{k+1}$. Then it follows that $p'_0 f'_{1,k-1}=p'_0 f'_{1,k-1}+q'_{k-1} f'_{0,k}=0$. Thus the exactness of the columns of the diagram (21) implies that $f'_{1,k-1}=p'_1 g'$, where $g' \in \mathscr{P}^{t_2 s_{k-1}}$. Further, we write

$$f''=(f''_{k,0},\dots,f''_{0,k})=f'-r'_k(0,\dots,0,g',0,0).$$

By construction $f''_{0,k}=f''_{1,k-1}=0$ and $f''-f' \in \operatorname{Im} r'_k$. It follows that $p'_1 f''_{2,k-2}=0$ and so on. Repeating this argument $k$ times, we obtain finally the vector $\hat{f}=(0,\dots,0)$. Thus, $f=f-\hat{f} \in \operatorname{Im} r'_k$, which is what we were to prove. This establishes the exactness of (22).

We now apply to the diagram (21) the functor $\operatorname{Hom}(\cdot,\Phi)$, and we obtain the following commutative diagram

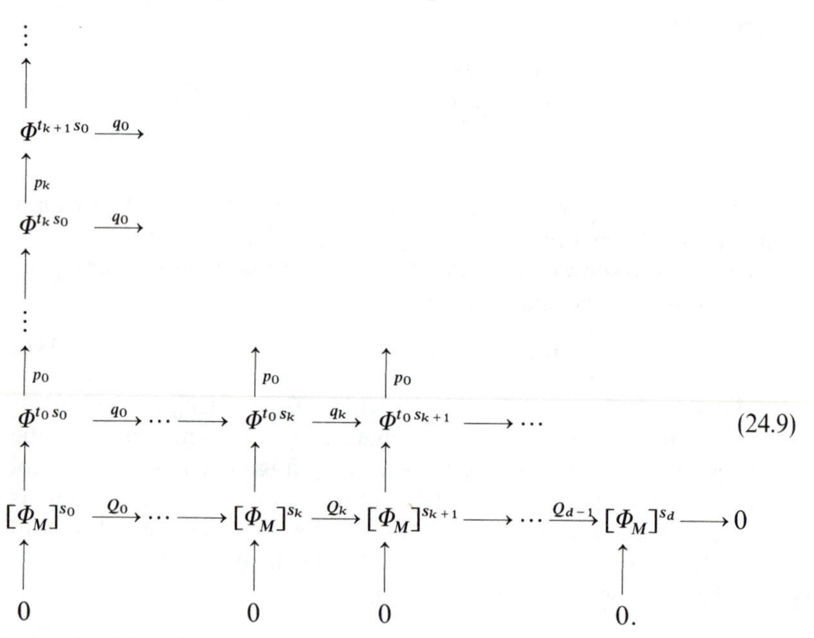

(24.9)

We have used the formula $\operatorname{Hom}(M^k,\Phi) \cong [\operatorname{Hom}(M,\Phi)]^k=[\Phi_M]^k$. Here $Q_k$ is a mapping which multiplies the vector $(u_1,\dots,u_{s_k})$, where $u_\alpha \in \Phi_M$, by the matrix $q_k$; the mappings $p_i$ and $q_j$ have a similar effect. We observe that the lowest row of this diagram represents the result of applying the functor $\operatorname{Hom}(\cdot, \Phi_M)$ to the free resolution of the module $L$. It follows that

$$\begin{aligned} &\operatorname{Hom}(L, \Phi_M) = \operatorname{Ker} Q_0, \\ &\operatorname{Ext}^k(L, \Phi_M) = \operatorname{Ker} Q_k / \operatorname{Im} Q_{k-1}, \qquad k = 1, 2, \ldots. \end{aligned} \tag{25.9}$$

If we now apply the functor $\operatorname{Hom}(\cdot, \Phi)$ to (22), we obtain the sequence

$$0 \to \Phi_{M \otimes L} \to \Phi_0 \xrightarrow{r_0} \Phi_1 \xrightarrow{r_1} \cdots \to \Phi_k \xrightarrow{r_k} \Phi_{k+1} \to \cdots,$$

in which $\Phi_k = \bigoplus_{i+j=k} \Phi^{t_i s_j}$ and the mapping $r_k$ is defined by:

$$r_k\colon\ \Phi_k \ni (\phi_{k,0}, \ldots, \phi_{0,k}) \to (\psi_{k+1,0}, \ldots, \psi_{0,k+1}) \in \Phi_{k+1},$$

where

$$\phi_{i,j} \in \Phi^{t_i s_j}, \qquad \psi_{i,j} = p_{i-1}\,\phi_{i-1,j} + (-1)^i q_{j-1}\,\phi_{i,j-1}.$$

Since the sequence (22) is a free resolution of the module $M \otimes L$, we have by definition

$$\begin{aligned} &\operatorname{Hom}(M \otimes L, \Phi) = \operatorname{Ker} r_0, \\ &\operatorname{Ext}^k(M \otimes L, \Phi) = \operatorname{Ker} r_k / \operatorname{Im} r_{k-1}, \qquad k = 1, 2, \ldots. \end{aligned} \tag{26.9}$$

Hence, in particular, it follows that the module $\operatorname{Hom}(M \otimes L, \Phi)$ coincides with the submodule in $\Phi^{t_0 s_0}$, consisting of the intersection of the kernels of the mappings $p_0$ and $q_0$. The kernel of $p_0$ is $[\Phi_M]^{s_0}$, and the operator $q_0$, acting on this submodule, coincides, by definition, with $Q_0$. Thus, $\operatorname{Hom}(M \otimes L, \Phi) \cong \operatorname{Ker} Q_0$. Thus the mapping (18), with $k=0$, is defined and will always be an isomorphism.

Let us now suppose that $k \geqq 1$. To construct the mapping (18), we construct the mapping $\operatorname{Ker} Q_k \to \operatorname{Ker} r_k$, which carries $\operatorname{Im} Q_{k-1}$ into $\operatorname{Im} r_{k-1}$. By the definition of the module $\Phi_M$, it follows that the module $[\Phi_M]^{s_k}$ coincides with the kernel of the mapping $p_0\colon \Phi^{t_0 s_k} \to \Phi^{t_1 s_k}$. Let $\phi$ be an arbitrary element of the module $\operatorname{Ker} Q_k$. If we consider it as an element of the space $\Phi^{t_0 s_k}$, it belongs to the intersection of the kernels of $p_0$ and $q_k$. Therefore, the vector $(0, \ldots, 0, \phi) \in \Phi_k$ belongs to the kernel of $r_k$. The correspondence

$$\operatorname{Ker} Q_k \ni \phi \to (0, \ldots, 0, \phi) \in \operatorname{Ker} r_k \tag{27.9}$$

is continuous. We shall suppose that the function $\phi$ belongs to $\operatorname{Im} Q_{k-1}$, that is, $\phi = q_{k-1}\psi$, where $\psi \in [\Phi_M]^{s_{k-1}}$. Then

$$(0, \ldots, 0, \phi) = r_{k-1}(0, \ldots, 0, \psi) \in \operatorname{Im} r_{k-1}.$$

Accordingly, the mapping (27) carries $\operatorname{Im} Q_{k-1}$ into $\operatorname{Im} r_{k-1}$ and therefore defines the mapping (18).

We now suppose that the space $\Phi$ is $M$-convex. This supposition implies that all the columns of the diagram (24) are exact.

We denote by $\Phi_k^i$, $i=0,\dots,k$, the subspace in $\Phi_k$ consisting of vectors of the form $\phi=(0,\dots,0,\phi_{i,k-i},\dots,\phi_{0,k})$. We remark that the operator $r_k$ acts from $\Phi_k^i$ to $\Phi_{k+1}^{i+1}$. By $\Psi_k^i$ we denote the subspace in $\Phi_k$, formed of vectors which the operator $r_k$ carries into $\Phi_{k+1}^i$. It is clear that the space $\Psi_k^i$ contains $\Phi_k^{i-1}$ and $\operatorname{Ker} r_k$.

**Lemma.** *The mapping*

$$[\Phi_M]^{s_k}/\operatorname{Im} Q_{k-1}\to\Psi_k^0/\operatorname{Im} r_{k-1}, \tag{28.9}$$

*which corresponds to the imbedding* $[\Phi_M]^{s_k}\to\Psi_k^0$, *is an isomorphism.*

*Proof.* We shall show to begin with that for arbitrary $i=0,\dots,k-1$, the mapping

$$\Phi_k^i\cap\Psi_k^0/\Phi_k^i\cap\operatorname{Im} r_{k-1}\to\Phi_k^{i+1}\cap\Psi_k^0/\Phi^{i+1}\cap\operatorname{Im} r_{k-1} \tag{29.9}$$

is an isomorphism. It is clear that it is continuous and biunique. We construct the inverse mapping. Let $\phi$ be an arbitrary element of the space $\Phi_k^{i+1}\cap\Psi_k^0$. This means that the vector $\phi$ has the form

$$(0,\dots,0,\phi_{i+1,j},\cdots,\phi_{0,k}),\qquad j=k-i-1 \text{ and } r_k\phi\in\Phi_{k+1}^0.$$

These conditions imply that $p_{i+1}\phi_{i+1,j}=0$. Since all the columns of the diagram (24) are exact, the function $\phi_{i+1,j}$ can be written in the form $p_i\psi_{i,j}$, where $\psi_{i,j}\in\Phi^{t_i s_j}$. We construct the vector $\psi=(0,\dots,0,\psi_{i,j},0,\dots,0)$ $\in\Phi_{k-1}^i$ and we write $\phi'=\phi-r_{k-1}\psi$. The construction implies that $\phi'\in\Phi_k^i$ and $r_k\phi'=r_k\phi\in\Phi_{k+1}^0$, i.e., $\phi'\in\Psi_k^0$.

The correspondence that we have set up, $\phi\to\phi'$, is the composition of the operators

$$\Phi_k^{i+1}\cap\Psi_k^0\ni\phi\to\phi_{i+1,j}\xrightarrow{p_i^{-1}}\psi_{i,j}\to\psi\to\phi'=\phi-r_k\psi\in\Phi_k^i\cap\Psi_k^0.$$

Each of these operations is a continuous mapping, but the mapping $p_i^{-1}$ is, generally, many-valued. Therefore, the composition is also a many-valued continuous mapping. The image of zero is made up of functions of the form $-r_k\psi$, belonging to $\Phi_k^i$, and it therefore belongs to the subspace $\Phi_k^i\cap\operatorname{Im} r_{k-1}$. Thus the associated mapping

$$\Phi_k^{i+1}\cap\Psi_k^0\to\Phi_k^i\cap\Psi_k^0/\Phi_k^i\cap\operatorname{Im} r_{k-1} \tag{30.9}$$

is continuous and single-valued. Since the vectors $\phi$ and $\phi'$ coincide modulo the subspace $\Phi_k^{i+1}\cap\operatorname{Im} r_{k-1}$, the mapping (30) is an inverse of the mapping (29). But this proves that the mapping (29) is an isomorphism.

We consider the mapping

$$\Phi_k^0\cap\Psi_k^0/\Phi_k^0\cap\operatorname{Im} r_{k-1}\to\Psi_k^0/\operatorname{Im} r_{k-1}, \tag{31.9}$$

which is the composition of the mappings (29) for all $i=0,\dots,k-1$. The space $\Phi_k^0\cap\Psi_k^0$ is by definition made up of vectors of the form

$\phi=(0,\ldots,0,\phi_{0,k})\in\Phi_k$ such that $r_k\phi\in\Phi^0_{k+1}$. This inclusion shows that $p_0\phi_{0,k}=0$, that is $\phi_{0,k}\in[\Phi_M]^{s_k}$. If we make the vector $\phi$ correspond to its component $\phi_{0,k}$, we obtain an isomorphic mapping $\Phi^0_k\cap\Psi^0_k\to[\Phi_M]^{s_k}$. Let us find the image of the subspace $\Phi^0_k\cap\operatorname{Im} r_{k-1}$ under this mapping. We observe, to begin with, the obvious equality $\Phi^0_k\cap\operatorname{Im} r_{k-1}=r_{k-1}\Psi^0_{k-1}$. In the isomorphism (31) we replace $k$ by $k-1$ and apply the operator $r_{k-1}$ to both sides. We obtain the isomorphism $r_{k-1}(\Phi^0_{k-1}\cap\Psi^0_{k-1})=r_{k-1}\Psi^0_{k-1}$. Using the arguments set forth above in which $k$ is replaced by $k-1$, we conclude that the operation $\phi\to\phi_{0,k-1}$ defines an isomorphism $\Phi^0_{k-1}\cap\Psi^0_{k-1}\to[\Phi_M]^{s_{k-1}}$, and that the action of the mapping $r_{k-1}$ on $\Phi^0_{k-1}\cap\Psi^0_{k-1}$ goes over into the action of the mapping $Q_{k-1}$ on $[\Phi_M]^{s_{k-1}}$. In sum then, we have obtained the isomorphism $\Phi^0_k\cap\operatorname{Im} r_{k-1}\cong\operatorname{Im} Q_{k-1}$, which is the restriction isomorphism $\Phi^0_k\cap\Psi^0_k\cong[\Phi_M]^{s_k}$ that we established earlier. It follows that the left factor-space in (31) is isomorphic to $[\Phi_M]^{s_k}/\operatorname{Im} Q_{k-1}$. □

We now take up the proof of the second assertion. We observe that the factor-space $\operatorname{Ker} r_k/\operatorname{Im} r_{k-1}$ is a subspace of the right side of the isomorphism (28), and coincides with the kernel of the mapping $r_k$, acting from the right side of (28) to $\Phi_{k+1}$. Therefore, the preimage of this subspace, under the mapping (28), is the kernel of the mapping $Q_k$, acting from $[\Phi_M]^{s_k}/\operatorname{Im} Q_{k-1}$. Thus we arrive at the isomorphism

$$\operatorname{Ker} r_k/\operatorname{Im} r_{k-1}\cong\operatorname{Ker} Q_k/\operatorname{Im} Q_{k-1}.$$

In view of (25) and (26), this implies that the mappings (18) are also isomorphisms. This proves the second assertion of the theorem.

Let us prove the third assertion. It follows from the hypothesis that for arbitrary $k\geqq 1$, the mapping

$$\Phi_k/\operatorname{Im} r_{k-1}\xrightarrow{r_k}\operatorname{Ker} r_{k+1} \tag{32.9}$$

is an isomorphism. We consider the commutative diagram:

$$\begin{array}{ccc}\Psi^0_k/\operatorname{Im} r_{k-1} & \xrightarrow{r_k} & \Phi^0_{k+1}\cap\operatorname{Ker} r_{k+1}\\ {\scriptstyle I}\uparrow & & \uparrow{\scriptstyle I'}\\ [\Phi_M]^{s_k}/\operatorname{Im} Q_{k-1} & \xrightarrow{Q_k} & \operatorname{Ker} Q_{k+1},\end{array}$$

in which $I$ is an isomorphism of the Lemma, and $I'$ is a natural imbedding (it is also, obviously, an isomorphism). We observe that the mapping $r_k$ in this diagram is a restriction isomorphism (32). Hence, it follows that $Q_k$ is also an isomorphism. With this, the $L$-convexity of the space $\Phi_M$ is proved. □

*Remark.* The fact the sequence (22) is a free resolution of the module $M \otimes L$ implies that the cohomological dimension of this module does not exceed the length of this sequence. On the other hand, the length is equal to the sum of the lengths of the resolutions (19) and (20). Since these resolutions were chosen arbitrarily, we may suppose that they have the smallest possible length, that is, that their lengths are equal to $\delta(M)$ and $\delta(L)$. Thus, we arrive at the inequality

$$\delta(M \otimes L) \leqq \delta(M) + \delta(L),$$

which holds for any two finite $\mathscr{P}$-modules satisfying the condition (17).

**Corollary 4.** *Let $\Omega$ be a convex region in $C^n$, and $\mathscr{H}(\Omega)$ the space of holomorphic functions on $\Omega$ with the topology of uniform convergence on all compact subsets of $\Omega$, having the structure of a $\mathscr{P}$-module in accordance with the formulae*

$$p\colon\ \phi(z) \to p\left(i \frac{\partial}{\partial z}\right) \phi(z), \qquad p \in \mathscr{P},\ \phi \in \mathscr{H}(\Omega).$$

*This $\mathscr{P}$-module is injective.*

*Proof.* Introducing the real structure of $R^{2n}$ in the space $C^n$, we consider the space $\mathscr{E}(\Omega)$ and the ring $\bar{\mathscr{P}}$ of polynomials with constant coefficients in the differential operators $\partial/\partial \bar{z}_j$, $j=1, \ldots, n$. The tensor product $\tilde{\mathscr{P}} = \mathscr{P} \otimes_C \bar{\mathscr{P}}$ is the ring of all differential operators with constant coefficients in $R^{2n}$. Let $L$ be a $\tilde{\mathscr{P}}$-module corresponding to the Cauchy-Riemann system, that is $L = \tilde{\mathscr{P}}/I$ where $I$ is the ideal in $\tilde{\mathscr{P}}$ generated by the elements $\partial/\partial \bar{z}_j$, $j=1, \ldots, n$. The space $\mathscr{E}_L(\Omega)$ has the same elements as $\mathscr{H}(\Omega)$, and also, as is easily seen, the same topology. Therefore we must prove that the space $\mathscr{E}_L(\Omega)$ is $M$-convex for any arbitrary finite $\mathscr{P}$-module $M$.

Let us consider the $\mathscr{P}$-module $\tilde{M} = M \otimes_C \bar{\mathscr{P}}$. It is of finite type, and Proposition 5, § 5 implies that $\operatorname{Tor}_i^{\tilde{\mathscr{P}}}(\tilde{M}, L) = 0$ for $i \geqq 1$. Since $\mathscr{E}(\Omega)$ is an injective $\tilde{\mathscr{P}}$-module Theorem 3 implies that the space $\mathscr{E}_L(\Omega)$ is $\tilde{M}$-convex. If $M_*$ is a free resolution of the $\mathscr{P}$-module $M$, then $M_* \otimes_C \bar{\mathscr{P}}$ is a free resolution of the $\tilde{\mathscr{P}}$-module $\tilde{M}$, since tensor multiplication over the field $C$ conserves exactness. Therefore the $\tilde{M}$-convexity of the space $\mathscr{E}_L(\Omega)$ implies its $M$-convexity. □

### 5°. The connection between $M$-convexity of the space $\Phi(\Omega)$ and the topology of the region $\Omega$

**Theorem 4.** *Let $M$ be an arbitrary finite $\mathscr{P}$-module of dimension $d > 0$. Then, if the space $\Phi(\Omega)$ is $M$-convex, $H^i(\Omega, \mathscr{C}) = 0$ for all $i > d$. If, moreover, the space $\Phi(\Omega)$ is strongly $M$-convex, then $H^d(\Omega, \mathscr{C}) = 0$.*

*Proof.* We construct first a finite $\mathcal{P}$-module $L$, having the following properties:

A) $\delta(L) \leqq d$,

B) $\operatorname{Tor}_i(M, L) = 0$, $i \geqq 1$,

C) the dimension of the module $M \otimes L$ is equal to zero. The fundamental ring in $\mathcal{P}$ will now be interpreted as the ring of polynomials in the complex variables $z = (z_1, \ldots, z_n)$. The existence of the module $L$ will be proved by induction on $n$. We shall suppose that a similar task for the ring $\mathcal{P}'$ of all polynomials in the variables $z_1, \ldots, z_{n-1}$ has already been carried out. We observe that in the case $n-1=0$, the task is trivial.

We begin the construction of the module $L$. If $d=0$, we may write $L = \mathcal{P}$. We shall suppose that $d > 0$. The module $M$ will be represented as a factor-module $\mathcal{P}^s/\mathfrak{p}$ and we fix some reduced primary decomposition of the submodule $\mathfrak{p}$: $\mathfrak{p} = \mathfrak{p}_0 \cap \cdots \mathfrak{p}_e$. In the space $C^n$, we choose a coordinate such that for all $\lambda$ the submodule $\mathfrak{p}_\lambda$ is normally placed and the intersection of the variety $N(\mathfrak{p}_\lambda)$ with the subspace $z_n = 0$ has dimension lower by one than the dimension of $N(\mathfrak{p}_\lambda)$. We consider the module $L_n = \mathcal{P}/z_n \mathcal{P}$. The sequence

$$0 \to \mathcal{P} \xrightarrow{z_n} \mathcal{P} \to L_n \to 0$$

is a free resolution of this module. Therefore $\delta(L_n) = 1$, and by Proposition 3, § 3, Chapter I

$$\operatorname{Tor}_1(M, L_n) \cong \mathfrak{p} \cap z_n \mathcal{P}^s / z_n \mathfrak{p}, \qquad \operatorname{Tor}_i(M, L_n) = 0, \qquad i \geqq 2.$$

We shall show that $\operatorname{Tor}_1(M, L) = 0$ also. For this, it is sufficient to show that an arbitrary element $f \in \mathfrak{p} \cap z_n \mathcal{P}^s$ belongs to $z_n \mathfrak{p}$. We have $f = z_n g \in \mathfrak{p}$, where $g \in \mathcal{P}^s$. We shall fix some $\lambda$; if the dimension of the submodule $\mathfrak{p}_\lambda$ is greater than zero, then in view of Proposition 6, § 1, Chapter IV, the inclusion relation $z_n g \in \mathfrak{p}_\lambda$ implies the inclusion $g \in \mathfrak{p}_\lambda$, since the submodule $\mathfrak{p}_\lambda$ is normally placed. If, however, $\mathfrak{p}_\lambda$ is zero-dimensional, then by hypothesis the associated variety $N(\mathfrak{p}_\lambda)$ does not intersect the subspace $Z_n = 0$. Therefore, $z_n g \in \mathfrak{p}_\lambda$ implies $g \in \mathfrak{p}_\lambda$. Thus, we have shown that $g \in \mathfrak{p}_\lambda$ for arbitrary $\lambda$, that is, $g \in \mathfrak{p}$. Hence $f = z_n g \in z_n \mathfrak{p}$, which is what we were to prove. We have now shown that

$$\operatorname{Tor}_i(M, L_n) = 0, \qquad i \geqq 1. \tag{33.9}$$

Let us consider the $\mathcal{P}'$-module $m$, which is the restriction of the module $M$ on the subspace $z_n = 0$ (see 5°, § 1, Chapter IV). By Lemma 2, § 1, Chapter IV, the variety $N(m)$ is equal to the intersection of $N(M)$ and the subspace $z_n = 0$ and, therefore, of dimension $d-1$ because of the choice of a system of coordinates in $C^n$. The induction hypothesis implies that there exists a finite $\mathcal{P}'$-module $l$, having the following property:

a) $\delta(l) \leqq d-1$, b) $\operatorname{Tor}_i(m, l)=0$ for $i \geqq 1$, and c) $\dim(m \otimes l)=0$. Suppose that

$$\begin{aligned} 0 \to \mathscr{P}'^{\tau_\delta} \to \cdots \to \mathscr{P}'^{\tau_{k+1}} \xrightarrow{q_k} \mathscr{P}'^{\tau_k} \xrightarrow{q_{k-1}} \cdots \\ \cdots \to \mathscr{P}'^{\tau_1} \xrightarrow{q_0} \mathscr{P}'^{\tau_0} \to l \to 0 \end{aligned} \tag{34.9}$$

is a free resolution of the $\mathscr{P}'$-module $l$ of length $\delta=\delta(l)$. We now replace $\mathscr{P}'$ by $\mathscr{P}$ in this sequence, and we replace $l$ by the $\mathscr{P}$-module $L'=\mathscr{P}^{\tau_0}/q_0\,\mathscr{P}^{\tau_1}$:

$$\begin{aligned} 0 \to \mathscr{P} \xrightarrow{\tau_\delta} \cdots \to \mathscr{P}^{\tau_{k+1}} \xrightarrow{q_k} \mathscr{P}^{\tau_k} \xrightarrow{q_{k-1}} \cdots \\ \cdots \to \mathscr{P}^{\tau_1} \xrightarrow{q_0} \mathscr{P}^{\tau_0} \to L' \to 0. \end{aligned} \tag{35.9}$$

We show that this sequence is exact. Let $f \in \mathscr{P}^{\tau_k}$ be an element belonging to the kernel of $q_{k-1}$. We expand it in a power series $z_n$: $f=\sum f_i z_n^i$. The equation $q_{k-1} f=0$ implies that $q_{k-1} f_i=0$ for arbitrary $i$. Since the sequence (34) is exact, we have $f_i=q_k g_i$, $g_i \in \mathscr{P}'^{\tau_{k+1}}$ whence $f=q_k g$, that is, $g=\sum z_n^i g_i \in \mathscr{P}^{\tau_{k+1}}$. This establishes the exactness of (35). Thus the sequence (35) is a free resolution of the $\mathscr{P}$-module $L'$, whence $\delta(L') \leqq d-1$.

Let us set $L=L' \otimes L_n$. Proposition 5, § 4, Chapter IV, implies that $\operatorname{Tor}_i(L', L_n)=0$, $i \geqq 1$. Therefore, by the remark in 4°, we have the inequality $\delta(L) \leqq \delta(L')+\delta(L_n) \leqq d$. This proves property A).

Let (19) be a free resolution of the module $M$. The relation (33) implies that the sequence

$$\cdots \to L_n^{t_2} \xrightarrow{p_1'} L_n^{t_1} \xrightarrow{p_0'} L_n^{t_0} \to M \otimes L_n \to 0. \tag{36.9}$$

is exact. We now remark that the operation of multiplication of the ring $\mathscr{P}$ generates a multiplication operation among the cosets in the factor-module $L_n=\mathscr{P}/z_n\mathscr{P}$. This operation converts $L_n$ into a ring isomorphic to $\mathscr{P}'$. The $\mathscr{P}$-modules $M \otimes L_n$ and $L$ are also $L_n$-modules. The ring isomorphism $L_n \cong \mathscr{P}'$ generates an isomorphism of the modules $M \otimes L_n \cong m$, $L \cong l$, and the sequence (36) can be looked on as a free resolution of the $\mathscr{P}'$-module $m$. It therefore follows from b) that the sequence

$$\cdots \to L^{t_2} \xrightarrow{p_1'} L^{t_1} \xrightarrow{p_0'} L^{t_0}.$$

is exact. If we compare this sequence with the resolution (19), we arrive at the conclusion that $\operatorname{Tor}_i(M, L)=0$, $i \geqq 1$. This proves property B).

Let us prove property C). Proposition 3, § 1, Chapter IV implies that

$$N(M \otimes L)=N(M \otimes (L' \otimes L_n))=N(M) \cap N(L') \cap N(L_n). \tag{37.9}$$

The variety $N(L_n)$ clearly coincides with the subspace $z_n=0$. By Lemma 3, § 1, Chapter IV, the intersections of this subspace with the varieties $N(M)$ and $M(L')$ are varieties $N(m)$ and $N(l)$. Therefore, the right side of (37) is equal to the variety $N(m) \cap N(l)=N(m \otimes l)$, which has dimension zero in accordance with c). This proves property C).

We now proceed to prove the theorem. We shall establish the equation

$$\mathrm{Ext}^i(L, \Phi_M)=0, \quad i>d, \qquad \Phi=\Phi(\Omega). \tag{38.9}$$

It follows from A) that a suitable free resolution of the module $L$ has a length not exceeding $d$, that is, it has the form (20). Let (24) be a diagram constructed with the aid of such a resolution. The bottom row consists of zeroes to the right of the terms $[\Phi_M]^{s_d}$. Therefore, the spaces (38), which are the cohomology spaces of this row, are equal to zero for $i>d$.

We shall establish Eq. (38) with $i=d$, on the assumption that the space $\Phi$ is strongly $M$-convex. This equation is equivalent to the following:

$$q_{d-1}[\Phi_M]^{s_{d-1}}=[\Phi_M]^{s_d}.$$

To prove that the left and right side coincide, it is sufficient to show that a) the left side is closed in the right, and b) that the left side is dense in the right. That the left side is closed is equivalent to saying that the space $\mathrm{Ext}^d(L, \Phi_M)$ is separable. It follows from Theorem 3 that this space is isomorphic to $\mathrm{Ext}^d(M \otimes L, \Phi)$. The latter, by Theorem 1, is isomorphic to the space $H^d(U, \mathscr{P}_{M\otimes L}$, where $U$ is some locally finite convex covering of the region $\Omega$. It follows from Property C) and Theorem 2 that the space $H^d(U, \mathscr{P}_{M\otimes L})$ is separable. This proves the assertion a).

We prove assertion b). Since by hypothesis the space $E_M$ is dense in $\Phi_M$, it is sufficient to establish the equation

$$q_{d-1}[E_M]^{s_{d-1}}=[E_M]^{s_d}.$$

This equation is equivalent to the relation $\mathrm{Ext}^d(L, E_M)=0$. Since by Proposition 3, § 7, the space $E$ is $M$-convex, and $M \otimes L$ is convex, Theorem 3 implies that

$$\mathrm{Ext}^d(L, E_M) \cong \mathrm{Ext}^d(M \otimes L, E)=0,$$

which is what we were to prove.

Thus, we have established the Eqs. (38). Using these, Theorem 3, Corollary 1 and Theorem 2, in turn, we obtain the chain of isomorphisms

$$\begin{aligned}\mathrm{Ext}^i(L, \Phi_M) &\cong \mathrm{Ext}^i(M \otimes L, \Phi) \cong H^i(\Omega, \mathscr{P}_{M\otimes L})\\ &\cong [H^i(\Omega, \mathscr{C})]^{l(M\otimes L)}=0, \qquad i>d \ (i \geqq d).\end{aligned} \tag{39.9}$$

Since each of the zero-dimensional modules is not equal to zero, it follows that $l(M \otimes L) \neq 0$. Therefore our theorem is proved. □

### 6°. Examples

**Example 1.** Let $p$ be a square $\mathscr{P}$-matrix of order $s$, with a non-constant determinant. The variety associated with the module $M=\mathscr{P}^s/p' \mathscr{P}^s$, is the variety of roots of the polynomials $\det p$ (see Proposition 2, § 1, Chapter IV), and, therefore, has dimension $n-1$. Then, by Theorem 4, the space $\Phi(\Omega)$ will be strongly $M$-convex only if $H^{n-1}(\Omega, \mathscr{C})=0$, that is, only if the complement of $\Omega$ in $R^n$ has no connected compact components.

**Example 2.** Let $p$ and $q$ be elements of the ring $\mathscr{P}$. We set $M=\mathscr{P}/q\mathscr{P}$ and $L=\mathscr{P}/p\mathscr{P}$. We shall find the conditions on the polynomials $p$ and $q$, for which $\operatorname{Tor}_i(M, L)=0$, $i\geqq 1$. By Proposition 3, § 3, Chapter I these relationships are equivalent to the equation $p\mathscr{P}\cap q\mathscr{P}=pq\mathscr{P}$. This equation means that every polynomial divisible by $p$ and $q$ is divisible by the product $pq$, that is, that the polynomials $p$ and $q$ are relatively prime.

We thus suppose that the polynomials $p$ and $q$ are relatively prime. Then Theorem 3 implies the following proposition. Let $\Omega$ be an arbitrary convex region. Then the equation $pu=w$, where $w\in\Phi(\Omega)$ and $qw=0$, has a solution $u\in\Phi(\Omega)$ such that $qu=0$. We shall show that the condition that $p$ and q be relatively prime is also necessary for the theorem to hold. In fact, let us suppose that $p$ and $q$ have a common divisor $r$, different from a constant, and we write $p=p_1 r$ and $q=q_1 r$. Then there exists an exponent $w\in E$, such that $qw=0$, and $q_1 w\neq 0$. In this case, the system of equations

$$pu=w, \qquad qu=0$$

are insoluble since the necessary solubility condition: $q_1 w-p_1\cdot 0=0$. is not fulfilled.

**Example 3.** We return to the notation of Example 3, 2°, § 7. We consider the $\mathscr{P}$-matrices $'d_0$ and $''d_0$, corresponding to the action of the operators $'d$ and $''d$ on the differential forms of zero order. We write

$$M=\mathscr{P}/d_0'\mathscr{P}^n, \qquad M'=\mathscr{P}/'d_0'\mathscr{P}^m, \qquad M''=\mathscr{P}/''d_0'\mathscr{P}^m.$$

The spaces $\Phi_M(\Omega)$, $\Phi_{M''}(\Omega)$ and $\Phi_{M'}(\Omega)$ are respectively, the spaces of locally constant, holomorphic, and anti-holomorphic functions in the region $\Omega$.

Proposition 3, § 3, Chapter I implies that

$$M'\otimes M''\cong\mathscr{P}/'d_0'\mathscr{P}^m+''d_0'\mathscr{P}^m.$$

The ring $\mathscr{P}$ will be interpreted as the ring of polynomials in $C^n$. Then the monomials $z_j-iz_{m+j}$, $j=1,\dots,m$, form a basis for the ideal $'d_0'\mathscr{P}^0$, and the monomials $z_j+iz_{m+j}$, $j=1,\dots,m$, form a basis for the ideal $''d_0'\mathscr{P}^m$. The sum of these ideals is an ideal in which the basis is formed by all monomials $z_j$, $j=1,\dots,n$, that is the ideal $d_0'\mathscr{P}^n$. Hence

$$M'\otimes M''\cong M, \qquad \dim M=0, \qquad l(M)=1.$$

Since the modules $M'$ and $M''$ "depend" on different groups of variables, we have by Proposition 5, § 4, Chapter IV, that $\operatorname{Tor}_j(M', M'') = 0$, $i \geqq 1$. Theorem 3, therefore, implies the existence of a continuous mapping

$$\operatorname{Ext}^i(M', \Phi_{M''}(\Omega)) \to \operatorname{Ext}^i(M, \Phi(\Omega)), \qquad i \geqq 0. \tag{40.9}$$

$\operatorname{Ext}^i(M', \Phi_{M''}(\Omega))$ is the cohomology space of the $d'$-differential-holomorphic form, and $\operatorname{Ext}^i(M, \Phi(\Omega))$ is the cohomology space of $d$-differential forms with coefficients in $\Phi(\Omega)$. It is clear from the construction of the mapping (40) that it operates in the natural way, that is, it carries the class of holomorphic $d'$-forms into the class of $d$-forms that contains it.

Let $\Omega$ be a domain of holomorphy. In this case, the space $\Phi(\Omega)$ is $M''$-convex (see Example 2, 2°), and therefore, Theorem 3 contains the following classical result: the mapping (40) is an algebraic isomorphism. Furthermore, since the module $M''$ is of dimension $m$, Theorem 4 implies that $H^i(\Omega, \mathscr{C}) = 0$ for all $i > m$. If, however, $\Omega$ is a Runge region, then since $\Phi(\Omega)$ is strongly $M''$-convex (see the same example), Theorem 4 implies that $H^m(\Omega, \mathscr{C}) = 0$ also.

## § 10. The algebraic conditions for $M$-convexity

We recall that the cohomology dimension $\delta(M)$ of a finite $\mathscr{P}$-module $M$ is the smallest length of a free resolution of it (see 7°, § 3, Chapter I). In particular, if $\delta(M) = -1$, by definition the module $M$ is equal to zero and if $\delta(M) = 0$, the module has a free resolution of the form $0 \to \mathscr{P}^s \to M \to 0$. In both cases, $t = 0$, and therefore, the corresponding system (1.7) is empty. Thus the problem of $M$-convexity for $\delta(M) < 1$ is trivial. When $\delta(M) = 1$, the problem is meaningful, and it admits an equivalent formulation in terms of a condition connecting the support of the function $\phi \in [\mathscr{E}^*(R^n)]^t$ with the support of $p'\phi$.

**1°. The condition for $M$-convexity of the space $\mathscr{E}(\Omega)$ when $\delta(M) = 1$.** The equation $\delta(M) = 1$ implies that $M$ has a free resolution of the form

$$0 \to \mathscr{P}^t \xrightarrow{p'} \mathscr{P}^s \to M \to 0,$$

where $p' \neq 0$.

**Theorem 1.** *Let $\delta(M) = 1$. Then the space $\mathscr{E}(\Omega)$ is $M$-convex if and only if the following condition is satisfied:*

*($S_\Omega$) For an arbitrary compact $K \subset \Omega$ there exists a compact $K' \subset \Omega$ such that the relation* $\operatorname{supp} p'\phi \subset K$, *where* $\phi \in [\mathscr{D}(\Omega)]^t$, *implies that* $\operatorname{supp} \phi \subset K'$.

*Proof. Sufficiency.* Suppose that ($S_\Omega$) is satisfied. We shall show that it holds also for functions in $[\mathscr{E}^*(\Omega)]^t$. We fix an arbitrary compact

$K \subset \Omega$, and we choose the compact $K_0$ to be such that $\Omega \supset K_0 \supset \supset K$. Let $\phi$ be an arbitrary function belonging to $[\mathscr{E}^*(\Omega)]^t$ such that $\operatorname{supp} p'\phi \subset K$. We choose a function $\chi \in \mathscr{D}(R^n)$ satisfying $\int \chi \, d\xi = 1$. The sequence of functions $\chi_v = v^n \chi(v\xi)$, $v = 1, 2, \ldots$ tends to the delta-function in the topology of distributions. For sufficiently large $v$ the support of the convolution $\chi_v * p'\phi = p'(\chi_v * \phi)$ belongs to $K_0$. The hypothesis $(S_\Omega)$ implies the existence of a compact $K'$, depending only on $K_0$, and such that $\operatorname{supp}(\chi_v * \phi) \subset K'$. Since $\chi_v * \phi \to \phi$ for $v \to \infty$, we conclude that $\operatorname{supp} \phi \subset K'$. But this proves that the condition $(S_\Omega)$ holds for distributions with compact supports.

Let $K_\alpha$, $\alpha = 1, 2, \ldots$ be some strictly increasing sequence of admissible compacts tending to $\Omega$, such that for arbitrary $\alpha$ the compact $(K_\alpha)'$ corresponding to $K_\alpha$ in $(S_\Omega)$ belongs to $K_{\alpha+1}$. The Frechet spaces $\mathscr{E}_{K_\alpha}$ obviously form a decreasing family. For every $\alpha$ in ${}_t[\mathscr{E}^*_{K_\alpha}]^s$ we consider the subspace $P_\alpha = [\mathscr{E}^*_{K_\alpha}]^s \cap p^*[\mathscr{E}^*(\Omega)]$ and its orthogonal complement $P_\alpha$ in the conjugate space $[\mathscr{E}_{K_\alpha}]^s$. The spaces $p_\alpha^0$ form a decreasing subfamily of the family $\{[\mathscr{E}_{K_\alpha}]^s\}$; we consider the sequence

$$0 \to \{P_\alpha^0\} \to \{[\mathscr{E}_{K_\alpha}]^s\} \xrightarrow{p} \{[\mathscr{E}_{K_\alpha}]^t\} \to 0 \tag{1.10}$$

where the mapping $p$ is generated by the differential operator $p$. We show that this sequence is exact. It is clear that $P_\alpha^0$ belongs to $(\mathscr{E}_{K_\alpha})_p$; on the other hand $(S_\Omega)$ implies that $P_\alpha \subset p^*[\mathscr{E}^*_{K_{\alpha+1}}]^t$ (here we make use of Proposition 4, §2) whence we may conclude that the image of the restriction mapping $(\mathscr{E}_{K_{\alpha+1}})_p \to (\mathscr{E}_{K_\alpha})_p$ belongs to $P_\alpha^0$. This implies the exactness of (1) in the middle term.

Let us prove the exactness of (1) in the third term. We fix $\alpha$ and the function $h \in \mathscr{D}(\operatorname{int} K_{\alpha+1})$ equal to unity on $K_\alpha$. Choosing $w \in [\mathscr{E}_{K+1}]^t$ arbitrarily, we extend the product $h w$ as zero outside of $K_{\alpha+1}$. The function so obtained belongs to $[\mathscr{E}(R^n)]^t$ and therefore by Theorem 3, §8 there exists a function $u \in [\mathscr{E}(R^n)]^s$ satisfying the system $p u = h w$. The operator $w \mapsto u|_{K_\alpha}$ (generally, many-valued) acts continuously from $[\mathscr{E}_{K_{\alpha+1}}]^t$ to $[\mathscr{E}_{K_\alpha}]^s$. The family mapping having these operators as components inverts the mapping $p$. This proves the exactness of (1).

We now show that the subspace $P_\alpha$ is closed. In view of Corollary 4, §8 the mapping

$$[\mathscr{E}^*(R^n)]^t \xrightarrow{p^*} [\mathscr{E}^*(R^n)]^s \tag{2.10}$$

is a monomorphism. Proposition 4, §2 implies that the subspace $[\mathscr{E}^*_{K_{\alpha+1}}]^t$ is closed in the left side of (2). Consequently its image is closed in the right side. The subspace $P_\alpha$ is equal to the intersection of this image with the closed subspace $[\mathscr{E}^*_{K_\alpha}]^s$ and therefore is closed in the space $[\mathscr{E}^*(R^n)]^s$. By Proposition 4, §2 we conclude that $P_\alpha$ is closed in $[\mathscr{E}^*_{K_\alpha}]^s$ also.

We shall verify that the sequence (1) satisfies the conditions of Proposition 11, §1. The condition a) follows from the fact that $P_\alpha^0$ is a closed subspace of the $\mathscr{F}$-space $[\mathscr{E}_{K_\alpha}]^s$. To prove b) it is sufficient to establish that any functional $\phi \in [\mathscr{E}_{K_\alpha}^*]^s$ which vanishes on $P_{\alpha+1}^0$ also vanishes on $P_\alpha^0$. Since the space $\mathscr{E}_{K_{\alpha+1}}^*$ is reflexive and $P_{\alpha+1}$ is closed, the functional $\phi$ belongs to $P_{\alpha+1}$ and therefore to $P_{\alpha+1} \cap [\mathscr{E}_{K_\alpha}^*]^s = P_\alpha$. Hence $\phi$ is equal to zero on $P_\alpha$.

Proposition 11, §1 implies that the sequence

$$[\mathscr{E}(\Omega)]^t \xrightarrow{p} [\mathscr{E}(\Omega)]^s \to 0$$

obtained from (1) by passage to the projective limits, is exact. This completes the proof of sufficiency.

Let us now prove the necessity of the condition. We shall suppose that the space $\mathscr{E}(\Omega)$ is $M$-convex and we fix an arbitrary compact $K \subset \Omega$. We consider the space $\Psi$, formed of the distributions $\phi \in [\mathscr{E}^*(\Omega)]^t$ for which $\operatorname{supp} p^* \phi \subset K$. We shall endow this space with a topology defined by the semi-norms $\|p^* \phi\|^k$, $k=0, 1, 2, \ldots$. We consider the bilinear form

$$(f, \phi) = \int \bar{f} \phi \, d\xi, \tag{3.10}$$

defined for functions $f \in [\mathscr{E}(\Omega)]^t$ and distributions $\phi \in \Psi$. For a fixed distribution $\phi$, this form is continuous in $f$, since $\phi$ has a compact support. On the other hand, since the space $\mathscr{E}(\Omega)$ is $M$-convex, for an arbitrary function $f$ we can find a function $u \in [\mathscr{E}(\Omega)]^s$ such that $p\,u = f$. Therefore,

$$(f, \phi) = (p\,u, \phi) = (u, p^* \phi).$$

This equation shows that the form (3) is continuous in $\phi$ in the topology of $\Psi$ for an arbitrary fixed function $f$. Thus, the form (3), which is defined on the direct product of the Frechet space $[\mathscr{E}(\Omega)]^t$ and the metrizable space $\Psi$, is separately continuous in both variables. Therefore, it is continuous[11], that is, there exist constants $k$, $C$ and a compact $K' \subset \Omega$ such that

$$\left|\int \bar{f} \phi \, d\xi\right| \leqq C \, \|p^* \phi\|^k \, \|f\|_{K'}^k.$$

This inequality implies that for an arbitrary distribution $\phi \in \Psi$ we have $\operatorname{supp} \phi \subset K'$, that is the condition $(S_\Omega)$ is fulfilled. □

**Corollary 1.** *Let $M$ be an elliptic module for which $\delta(M) = 1$. Then for an arbitrary region $\Omega \subset R^n$, the space $\Phi(\Omega)$ is $M$-convex. The space $\Phi(\Omega)$ will be strongly $M$-convex if and only if $H^{n-1}(\Omega, \mathscr{C}) = 0$, that is, if and only if the complement to $\Omega$ has no connected compact components.*

*Proof.* Corollary 2, §9, implies that both these assertions need to be proved only for the space $\Phi(\Omega) = \mathscr{E}(\Omega)$. We shall prove the first

11 See Bourbaki [1], Chapter III, §4, 1°, Proposition 2, p. 184.

assertion. For this we need to verify that the condition $S_\Omega$ is fulfilled for an arbitrary region $\Omega$. We note that the condition $\delta(M)=1$ implies that $t \leqq s$. On the other hand, since the module $M$ is elliptic, the manifold $N(M)=\{z\colon \operatorname{rank} p'(z)<s\}$ does not coincide with $C^n$. Hence $t \geqq s$, i.e., $t=s$. We have thus shown that the matrix $p'$ is square, and therefore, the variety $\{z\colon \operatorname{rank} p(z)<t\}$ coincides with $N(M)$, whence it follows that the operator $p'$ is elliptic.

Let us fix an arbitrary region $\Omega$ and a compact $K \subset \Omega$. Let $\phi \in [\mathscr{D}(\Omega)]^t$ be an arbitrary function such that $\operatorname{supp} p'\phi \subset K$. The operator $p'$ being elliptic, the function $\phi$ is analytic inside $K$. Since the support of the function $\phi$ belongs to $\Omega$, the uniqueness property of analytic functions implies that $\phi=0$ in every connected component of the region $\complement K$, intersecting $\complement \Omega$. Let $\Omega'$ be the union of these components, and $K'=\complement \Omega'$. The compact $K'$ belongs to $\Omega$, and what we have just said implies that $\operatorname{supp} \phi \subset K'$. This proves the first assertion.

Let us prove the second assertion. Since $\dim M = n-1$, the necessity of the condition $H^{n-1}(\Omega, \mathscr{C})=0$ follows from Theorem 4, § 9. We shall establish the sufficiency. By Proposition 1, § 7, it is sufficient to show that

$$p^*[\mathscr{E}^*(\Omega)]^t \supset [\mathscr{E}^*(\Omega)]^s \cap E_p^0. \tag{4.10}$$

Let $\phi$ be an arbitrary element of the right side. Since the space $\mathscr{E}(R^n)$ is strongly $M$-convex, it follows that $\phi = p^*\psi$, where $\psi \in [\mathscr{E}^*(R^n)]^t$. Let $\Gamma$ be an arbitrary connected component of the set $\complement \Omega$. Since $\operatorname{supp} p^*\phi \subset \Omega$, and the operator $p^*$ is elliptic, the function $\psi$ is analytic in the neighborhood of $\Gamma$. Since by hypothesis $\Gamma$ is connected and is not bounded, and the function $\psi$ is equal to zero for a sufficiently large $|\xi|$, we have $\psi \equiv 0$ in the neighborhood of $\Gamma$. Hence, it follows that $\operatorname{supp} \psi \subset \Omega$, that is, $\phi$ belongs to the left side of (4). But this establishes the inclusion relation (4). □

**Example.** Suppose that

$$s=t=1, \quad \text{and that} \quad p=\Delta=\sum_1^n \frac{\partial^2}{\partial \xi_j^2}$$

is a Laplace operator. The corresponding polynomial $\Delta(z)$ is equal to $-\sum z_j^2$, and therefore, the variety associated with $p$ has the form $N=\{z\colon \sum z_j^2=0\}$. It is easy to verify that on this variety, the inequality (1.5) holds with $\gamma=1$, that is, $p$ is elliptic. Accordingly, the module $M=\mathscr{P}/p\mathscr{P}$ is also elliptic, and, as is easily seen, $\delta(M)=1$. Thus, Corollary 1 is applicable to this module. The first assertion of this corollary implies that the equation $\Delta u = w$ is always soluble in $\Phi(\Omega)$ for $w \in \Phi(\Omega)$.

The second assertion of Corollary 1 implies that the space $E_p$ is dense in $\Phi_p(\Omega)$ if and only if the complement to $\Omega$ has no compact connected components. We may sharpen this result by making use of the remark of 7°, § 8. We noted there that $E_p$ is dense in $\Phi_p(\Omega)$ if and only if $E_p^B$ is dense in $\Phi_p(\Omega)$, where $B$ is an arbitrary set in $C^n$, intersecting each of the varieties $N^\lambda$ associated with $M$.

Let us determine the varieties $N^\lambda$. In the case $n=2$, the polynomial $\Delta(z)$ admits the following decomposition into irreducible factors: $\Delta(z)=-(z_1-iz_2)(z_1+iz_2)$. Therefore, the varieties $N^0=\{z\colon z_1=iz_2\}$ and $N^1=\{z\colon z_1=-iz_2\}$ form the ensemble associated with the module $M$. When $n>2$, the polynomials $\Delta(z)$ are irreducible, and accordingly, the ideal $p\mathscr{P}$ is prime, i.e., $N^0=N$, and $N^\lambda=\emptyset, \lambda>0$. Thus, for arbitrary $n>1$ each of the varieties $N^\lambda$, associated with $M$, passes through the origin of coordinates. Therefore, for $B$ we may always choose a set consisting solely of the origin of coordinates.

By definition, the corresponding space $E_p^B$ is formed of all the polynomials $u$ satisfying the condition $\Delta u=0$, that is, of all harmonic polynomials. Thus, we have arrived at the following classical result: *in order that the harmonic polynomials be dense in the space $\Phi_p(\Omega)$ of harmonic functions in a region $\Omega$, it is necessary and sufficient that the complement of the region $\Omega$ have no connected compact components.*

**2°. The condition of strong $M$-convexity connected with the hyperbolicity of the module $M$.** We recall, to begin with, the definition of an improper point of an algebraic variety. Let $N$ be an algebraic variety in $C^n$. The space $C^n$ will be imbedded in $C^{n+1}$ by means of the mapping $z\to(1,z)$. Let $H(N)$ be the set of all homogeneous polynomials in $C^{n+1}$ vanishing on $N$. An arbitrary point of the form $(0,z)$, $z\in C^n$, at which all the polynomials belonging to $H(N)$ vanish, will be called improper points of the variety $N$. The following assertion yields an analytic characterization of the improper points of the variety.

**Proposition 1**[12]**.** *Let $(0,w)$ be an improper point of the variety $N$. Then there exists an algebraic function $z(\zeta)$ with values in $N$, defined and holomorphic for $|\zeta|>a, \zeta\in C$, satisfying the following conditions:*

a) $z(\zeta)=\zeta^\mu w+\kappa(\zeta)$ *for some natural* $\mu$,

b) $|\kappa(\zeta)|\leqq c(|\zeta|^\mu+1)^h$ *for some* $h<1$.

We now formulate the concept of a hyperbolic variety.

**Definition.** Let $(\xi',\xi'')$ be some division of the variables $\xi$ into two groups, in which the group $\xi'$ is non-empty; let $(z',z'')$ be the corresponding division of the dual variables. We shall say that the algebraic variety $N\subset C_z^n$ is hyperbolic in the variables $\xi'$, if for an arbitrary improper point $(0,w',w'')$ belonging to this variety, the condition $\operatorname{Im} w''=0$ implies that $\operatorname{Im} w'=0$. Let $L$ be a subspace linear in $R^n$, distinct from $R^n$.

12 For the proof, see Matsuura [1].

We shall say that the variety $N$ is hyperbolic with respect to $L$, if it is hyperbolic with respect to the variables $\xi'$ so chosen that $L$ is the manifold of solutions of the system $\xi'=0$.

**Theorem 2.** *Let $\Omega$ be a finite or countable union of convex regions $U_\nu$, satisfying the following condition: for an arbitrary variety $N^\lambda$, associated with the given module $M$, we can find a hyper-subspace $L$ in $R^n$, with respect to which the variety $N^\lambda$ is not hyperbolic, and such that the projections of the regions $U_\nu$ on $L^\perp$ are pairwise disjoint. Then the space $\Phi(\Omega)$ is strongly $M$-convex.*

*Proof.* We shall suppose that the variety $N(M)=\cup N^\lambda$ is empty. Then the module $M$ is equal to zero, and therefore, the theorem is true in an obvious way. We now suppose that the variety $N(M)$ is non-empty. Then the hypothesis of the theorem is meaningful and it implies that the regions $U_\nu$ are pairwise disjoint. Therefore, the $M$-convexity of the space $\Phi(\Omega)$ follows from Theorem 4, § 8.

It remains to show that $E_M$ is dense in $\Phi_M(\Omega)$. Proposition 1, § 7 implies that for this, we have only to establish the inclusion relation

$$p^*[\Phi^*(\Omega)]^t \supset [\Phi^*(\Omega)]^s \cap E_p^0 \tag{5.10}$$

($p$ is a matrix from the resolution (4.7) of the module $M$). Let $\phi$ be an arbitrary element of the right side of (5). Since the regions $U_\nu$ are pairwise disjoint, the function $\phi$ is uniquely determined as a sum $\sum \phi_\nu$, where $\phi_\nu \in [\Phi^*(U_\nu)]^s$. We shall show that $\phi_\nu \in E_p^0$ for arbitrary $\nu$.

We fix an arbitrary variety $N^\lambda$, associated with $M$. It follows from the conditions of the theorem that there exists a linear manifold $L \subset R^n$ of dimension $n-1$, separating the regions $U_1$ and $U'=\sum_{\nu>1} U_\nu$, with respect to which $N^\lambda$ is not hyperbolic. In $R^n$, we choose a system of coordinates such that the subspace $\xi_1=0$ coincides with $L$, and the region $U_1(U')$ lies in the half-space $\xi_1>0\,(\xi_1<0)$. Then the support of the function $\phi_1\ \phi'=\sum_{\nu>1}\phi_\nu$ lies in the half-space $\xi_1>0$, respectively $\xi_1<0$. Therefore, Proposition 2, § 3 implies that the Fourier transforms $\chi^+=\tilde{\phi}_1$ and $\chi^-=\tilde{\phi}'$ satisfy the inequalities

$$|\chi^{(\pm)}(z)| \leqq C(|z|+1)^q \times \begin{cases} \exp((\mp)\,\varepsilon\, y_1 + A|y''|), & y_1>0\ (y_1<0);\\ \exp((\mp)\,a\, y_1 + A|y''|), & y_1<0\ (y_1>0);\end{cases} \tag{6.10}$$

where $\varepsilon=\rho(\operatorname{supp}\phi, L)$, $y_1=\operatorname{Im} z_1$, $y''=\operatorname{Im}(z_2,\dots,z_n)$, and $q$ and $A$ are certain positive numbers.

On the other hand, since the function $\phi$ belongs to $[\Phi^*(R^n)]^s \cap E_p^0$, and the space $\Phi(R^n)$ is strongly $M$-convex, it can be written in the form

$\phi = p^* \psi$, where $\psi \in [\Phi^*(R^n)]^t$. It follows that

$$\chi^+(z) + \chi^-(z) = p^*(z)\,\tilde{\phi}(z). \tag{7.10}$$

Since by hypothesis the variety $N^\lambda$ is not hyperbolic, relative to the subspace $L$, it has at least one improper point $(0, w_1, w'')$ such that $\operatorname{Im} w'' = 0$, but $\operatorname{Im} w_1 \neq 0$. Let $z(\zeta) = \zeta^\mu(w_1, w'') + \kappa(\zeta)$ be an algebraic curve belonging to $N^\lambda$, constructed with the aid of Proposition 1. Suppose, further, that $\mathscr{D} = \sum \eta^i \mathscr{D}_i(z, \delta)$ is a $p^*$-operator. We consider the sequence of functions $\chi_i^\pm(\zeta) = \mathscr{D}_i(z(\zeta), D)\,\chi^\pm(z(\zeta))$. By Theorem 1, § 4, Chapter II, for arbitrary $i$ the coefficients of the functional $\mathscr{D}_i(z, \delta)$ are rational everywhere regular functions on the set of the form $N_k \setminus N_{k+1}$, where $\{N_k\}$ is some algebraic stratification. Since the curve $z = z(\zeta)$ is defined and algebraic for $|\zeta| > a$, it belongs altogether to one of the sets $N_k \setminus N_{k+1}$ for $|\zeta| > a'$. Therefore, the coefficients of the operators $\mathscr{D}_i(z(\zeta), D)$ are holomorphic and algebraic for $|\zeta| > a'$ and, therefore, they grow at infinity no faster than some power of $|\zeta|$. Then for arbitrary $i$, we have the inequality

$$|\chi_i^\pm(\zeta)| \leqq C(|\zeta|+1)^{q_i} \sup_{|z - z(\zeta)| \leqq 1} |\chi^\pm(z)|, \qquad i \in Z_+^n. \tag{8.10}$$

Making use of (6) and b) we obtain the bound

$$|\chi_i^+(\zeta)| \leqq C \exp(A'|\zeta|^\mu), \tag{9.10}$$

which holds for arbitrary $i$. We estimate the function $\chi_i^+$ for real values of $\zeta^\mu$. Suppose for definiteness that $\operatorname{Im} w_1 > 0$. We shall assume to begin with that $\zeta^\mu > 0$. From (6), (8), and b), we have

$$\begin{aligned} |\chi_i^+(\zeta)| &\leqq C(|\zeta|+1)^{q_i'} \exp(-\varepsilon \operatorname{Im} w_1\, \zeta^\mu + A|\kappa(\zeta)|) \\ &\leqq C' \exp\left(-\frac{\varepsilon}{2} \operatorname{Im} w_1\, \zeta^\mu\right). \end{aligned}$$

Now suppose $\zeta^\mu < 0$. From (7) we derive the equation $\chi_i^+ \equiv -\chi_i^-$, and, therefore, again from (6), (8), b), we have

$$|\chi_i^+(\zeta)| = |\chi_i^-(\zeta)| \leqq C \exp\left(\frac{\varepsilon}{2} \operatorname{Im} w_1\, \zeta^\mu\right).$$

Thus, for all real values of $\zeta^\mu$, we have

$$|\chi_i^+(\zeta)| \leqq C \exp\left(-\frac{\varepsilon}{2} \operatorname{Im} w_1\, |\zeta|^\mu\right).$$

This inequality, together with (9), shows, via the Phràgmen-Lindelöf Theorem, that $\chi_i^+ \equiv 0$ for arbitrary $i$, that is $\mathscr{D}(z)\,\chi^+(z) \equiv 0$ on the curve

$z=z(\zeta)$, belonging to the variety $N^\lambda$. Since $\lambda$ was chosen arbitrarily from the beginning, we have shown that on every one of the varieties $N^\lambda$, associated with the module $M$, we can find at least one point at which $\mathscr{D}(z)\,\chi^+(z)=0$. By Corollary 4, § 4, Chapter IV, this implies that $\mathscr{D}(z)\,\chi^+(z)\equiv 0$ on $N(M)$, that is, the function $\chi^+$ is orthogonal to all the functions of the form $\mathscr{D}_i(z,\delta)$. But this means that its inverse Fourier transform $\phi_1$ is orthogonal to all the exponential polynomials belonging to the space $E_p$ (see Proposition 4, § 7), that is,

$$\phi_1\in[\Phi^*(U_1)]^s\cap E_p^0.$$

We conclude that the function $\phi'=\phi-\phi_1$ is also orthogonal to the space $E_p$. Representing it in the form of a sum of the functions $\phi_2$ and $\phi''=\sum_{\nu>2}\phi_\nu$, we find in a similar fashion that the function $\phi_2$ is orthogonal to $E_p$, and so on. Thus, we have proved that for arbitrary $\nu$,

$$\phi_\nu\in[\Phi^*(U_\nu)]^s\cap E_p^0.$$

Since all the regions $U_\nu$ are convex, it follows from § 8 that $\phi_\nu=p^*\,\psi_\nu$, where $\psi_\nu\in[\Phi^*(U_\nu)]^t$. Setting $\psi=\sum\psi_\nu$, we obtain $\phi=p^*\,\psi\in p^*[\Phi^*(\Omega)]^t$. This proves the inclusion relation (5). □

*Remark.* We can show that when $\Omega$ is the union of two non-intersecting convex regions $U_1$ and $U_2$, the condition for strong $M$-convexity obtained in Theorem 2, is also necessary. In fact, the following assertion is true: In order that the space $\Phi(\Omega)$ be strongly $M$-convex, it is necessary and sufficient that for arbitrary $\lambda$ the variety $N^\lambda$ be non-hyperbolic with respect to some linear manifold $L^\lambda$ of dimension $n-1$, separating $U_1$ and $U_2$. When there are more than two regions $U_\nu$, the condition of Theorem 2 is no longer necessary. Suppose for example $n=2$, and $\Omega$ is the exterior of three rays, $\phi=0, \frac{2\pi}{3}, \frac{4\pi}{3}$ in the plane. Then by Corollary 1, the space $\Phi(\Omega)$ is strongly $M$-convex for an arbitrary elliptic module $M$. The hypothesis of Theorem 2 is not fulfilled, since no line divides the connected components of the region $\Omega$.

**3°. Homological conditions for *M*-convexity.** The conditions in question amount to this: a given finite module $L$ can be included in an exact sequence of the form

$$0\to L\to\mathscr{P}^{t_{k-1}}\xrightarrow{p'_{k-2}}\mathscr{P}^{t_{k-2}}\to\cdots\xrightarrow{p'_1}\mathscr{P}^{t_1}\xrightarrow{p'_0}\mathscr{P}^{t_0}. \quad (10.10)$$

In § 13, we shall find the conditions on $L$, under which such a sequence can be constructed. We show here that the longer the sequence, that is, the larger the number $k$, the wider the class of regions $\Omega$ for which the sequence $\Phi(\Omega)$ is (strongly) $L$-convex.

**Theorem 3.** *We assume that the module $L$ can be included in an exact sequence of the form* (10), *of length $k$, and that the region $\Omega$ has a locally finite convex covering $U$ consisting of regions, having no more than $k+1$-fold mutual intersections. Then*

I. *The space $\Phi(\Omega)$ is $L$-convex.*

II. *Let $\Omega'$ be a region having a convex covering $U'$ inscribed in $U$. Then the image of the restriction mapping $\Phi_L(\Omega) \to \Phi_L(\Omega')$ is dense in $\Phi_L(\Omega')$.*

*Proof.* Suppose that

$$\cdots \xrightarrow{p'_{k+1}} \mathscr{P}^{t_{k+1}} \xrightarrow{p'_k} \mathscr{P}^{t_k} \to L \to 0$$

is a free resolution of the module $L$. We join it to the sequence (10). To this end, we consider the mapping $p'_{k-1}$, the composition of the mappings $\mathscr{P}^{t_k} \to L \to \mathscr{P}^{t_{k-1}}$. It is easy to verify that this mapping makes the sequence

$$\cdots \to \mathscr{P}^{t_{k+1}} \xrightarrow{p'_k} \mathscr{P}^{t_k} \xrightarrow{p'_{k-1}} \mathscr{P}^{t_{k-1}} \to \cdots \xrightarrow{p'_1} \mathscr{P}^{t_1} \xrightarrow{p'_0} \mathscr{P}^{t_0} \to M \to 0 \quad (11.10)$$

exact. Here we have written $M = \operatorname{Coker} p'_0$; thus this sequence is a free resolution of $M$.

Let us consider the following commutative diagram

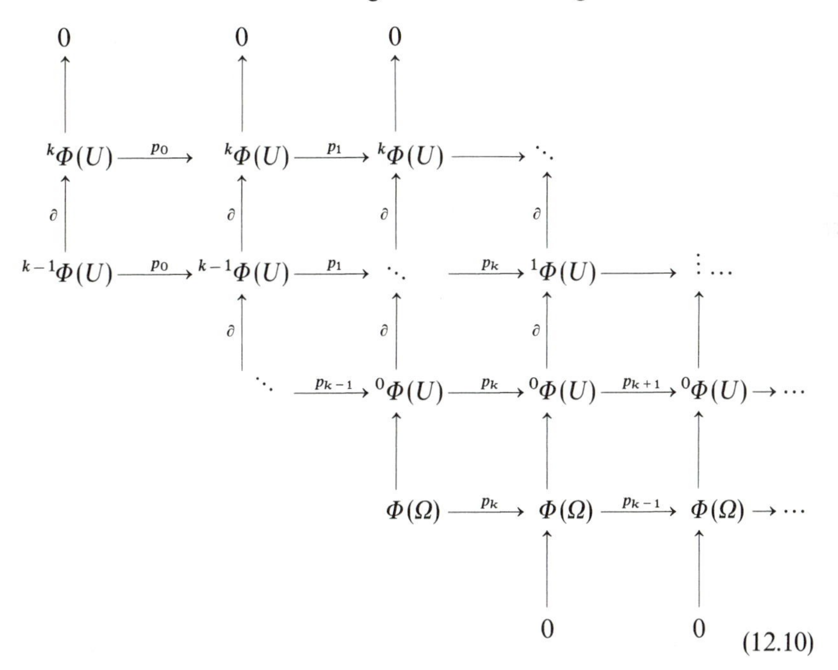

(12.10)

where for simplicity we have omitted the brackets of the form $[\dots]^{t_i}$. It is a particular case of the diagram (3.9) (from which the left column has been removed), since the hypothesis of our theorem implies that ${}^{k+1}\Phi(U)=0$. In the course of the proof of Theorem 1, § 9, we showed that all the columns and all the rows of this diagram except the lowest were exact. Therefore, on the basis of Theorem 1, § 2, Chapter I, we conclude that the bottom row of (12) is also exact. This proves that $\Phi(\Omega)$ is convex.

We now prove the second assertion of the theorem. Since the region $\Omega'$ belongs to $\Omega$, and the covering $U'$ is inscribed in $U$, the restriction mappings $\Phi(\Omega)\to\Phi(\Omega')$, ${}^{v}\Phi(U)\to{}^{v}\Phi(U')$ are defined. By these mappings the topological spaces $\Phi(\Omega')$ and ${}^{v}\Phi(U')$ induce the topologies on the spaces $\Phi(\Omega)$ and ${}^{v}\Phi(U)$. The latter with the induced topology will be denoted by $\check{\Phi}(\Omega)$ and ${}^{v}\check{\Phi}(U)$. Let us consider the diagram (12), obtained from diagram (12) when we replace the $\Phi(\Omega)$ and ${}^{v}\Phi(U)$ by $\check{\Phi}(\Omega)$ and ${}^{v}\check{\Phi}(U)$. We shall prove that the space $\check{\Phi}(\Omega)$ is $M$-convex. To this end, we prove that all the rows and columns of the diagram (12) are exact.

Let $U=\{U_i\}$, $U'=\{U_i'\}$. For arbitrary $i_0,\dots,i_v$, we denote by $\check{\Phi}(U_{i_0}\cap\cdots\cap U_{i_v})$ the space $\Phi(U_{i_0}\cap\cdots\cap U_{i_v})$, in which we have introduced the topology induced by $\Phi(U_{i_0}'\cap\cdots\cap U_{i_v}')$. The space ${}^{v}\check{\Phi}(U)$ is clearly the direct product of the spaces $\check{\Phi}(U_{i_0}\cap\cdots\cap U_{i_v})$. Since by hypothesis all the regions $U_i$ and $U_i'$ are convex, it follows from Corollary 2, § 8, that each of the spaces $\check{\Phi}(U_{i_0}\cap\cdots\cap U_{i_v})$ is $M$-convex. It then follows that the spaces ${}^{v}\check{\Phi}(U)$, $v\geqq 0$ are also $M$-convex. In view of the exactness of (11), we have now proved the exactness of all the rows of the diagram (12), except the lowest.

The set of spaces $\check{\Phi}(\Omega)$, $\check{\Phi}(U_{i_0}\cap\cdots\cap U_{i_v})$ is a presheaf on the covering $U$. This presheaf, clearly, satisfies the conditions of Proposition 9, § 2. But this proposition implies that in the diagram (12), all the columns are exact. Thus, in diagram (12), Theorem 1, § 2, Chapter I, is applicable, and implies that the lowest row is exact, that is, the space $\check{\Phi}(\Omega)$ is $M$-convex.

Since $L=\operatorname{Coker} p_k'$, we have $\Phi_L(\Omega')=\Phi_{p_k}(\Omega')$. Suppose that $u$ is an arbitrary element of $\Phi_{p_k}(\Omega')$. Since the space $\Phi(\Omega)$ is dense in $\Phi(\Omega')$, we can find an element $v\in[\Phi(\Omega)]^{t_k}$, belonging to an arbitrarily small neighborhood of $u$ in the topology of $[\Phi(\Omega')]^{t_k}$. Since the operator $p_k$ is continuous, and $p_k u=0$, the element $p_k v$ is arbitrarily near to zero in the topology of the space $[\Phi(\Omega)]^{t_{k+1}}$. The $M$-convexity of $\check{\Phi}(\Omega)$ implies that the mapping $p_k$: $[\check{\Phi}(\Omega)]^{t_k}\to[\Phi(\Omega)]^{t_{k+1}}$ is a homomorphism. Therefore, the element $v$ is arbitrarily close in the topology of $[\check{\Phi}(\Omega)]^{t_k}$ to the subspace $\Phi_{p_k}(\Omega)$. Therefore, the element $u$ is also arbitrarily close to the subspace $\Phi_{p_k}(\Omega)$, which is what we were to prove. □

We observe that Theorems 1 and 3 of § 8 can be looked on as particular cases of the theorem we have just proved, corresponding to the case $k=0$.

**Theorem 4.** *We suppose that the conditions of Theorem 3 are satisfied, and we impose the following conditions also: for an arbitrary variety $N^\lambda$, associated with the module $M$ that appears in* (11), *there exists a hypersubspace $L^\lambda$ in $R^n$, with respect to which $N^\lambda$ is not hyperbolic, and such that for arbitrary $i_1, \ldots, i_k$ the projections of the regions*

$$V_{i_0} = U_{i_0} \cap U_{i_1} \cap \cdots \cap U_{i_k}, \qquad i_0 \neq i_1, \ldots, i_k$$

*on $(L^\lambda)^\perp$ are pairwise disjoint. Then the space $\Phi(\Omega)$ is strongly $L$-convex.*

*Proof.* The $L$-convexity of $\Phi(\Omega)$ follows from Theorem 3. Let us consider the commutative diagram

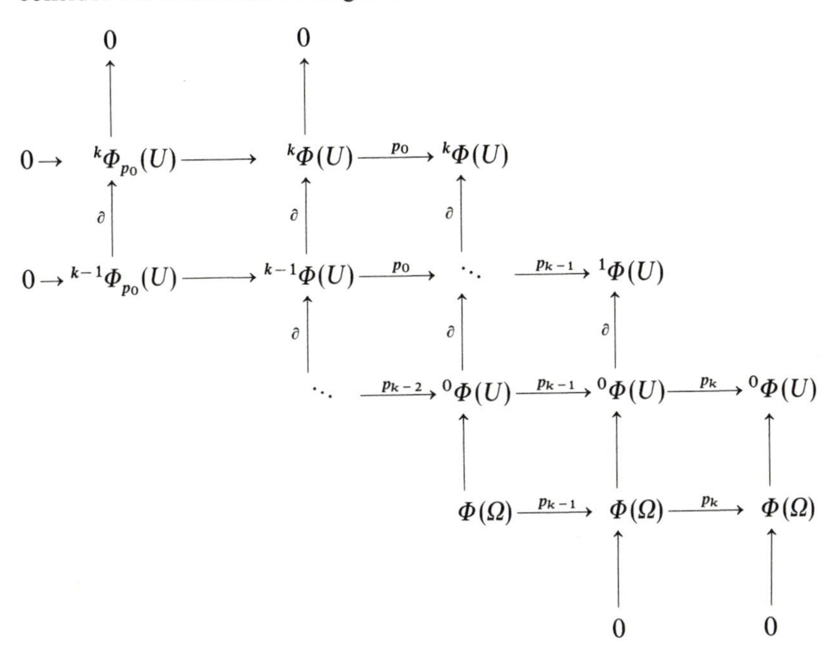

which is obtained by extending diagram (12) one step to the left. By what we have proved already, all the columns of this diagram are exact, except the leftmost, and all the rows except the lowest. Therefore, we may derive from Theorem 1, § 2, Chapter I, the isomorphism

$$^k\Phi_p(U)/\partial^{k-1}\Phi_p(U) \cong \Phi_{p_k}(\Omega)/p_{k-1}[\Phi(\Omega)]^{t_{k-1}}, \qquad p = p_0. \qquad (13.10)$$

We shall show that the factor-space on the left is absolutely inseparable, that is, the only open set in it is the space itself. Let $u = \sum u_{i_0, \ldots, i_k} \cdot U_{i_0} \wedge \cdots \wedge U_{i_k}$ be an arbitrary element of the space $^k\Phi_p(U)$. We fix arbitrarily the indices $i_1, \ldots, i_k$, and we consider the components $v_{i_0} = u_{i_0, \ldots, i_k}$ of the function $u$. Each of them is defined and satisfies the homogeneous system (1.4) in the region $V_{i_0}$. Since these regions are pairwise disjoint, the function $v_{i_0}$ defines a single-valued function $v$ in the region $V=\cup V_{i_0}$, belonging to the space $\Phi_p(V)$. The hypothesis of our theorem implies that the covering of the region $V$, formed by the regions $V_{i_0}$, satisfies the conditions of Theorem 2 with respect to the module $M$. Therefore, by Theorem 2, the space $\Phi_p(V)$ is strongly $M$-convex. Therefore the function $v$ can be approximated by exponential polynomials $v' \in E_p$. Since the indices $i_1, \ldots, i_k$ were arbitrarily chosen, we may assert that the cochain $u$ can be approximated by cochains of the form $u'=\sum v'_{i_1,\ldots,i_k} U_{i_0} \wedge \cdots \wedge U_{i_k}$, where all the $v'_{i_1,\ldots,i_k} \in E_p$. The cochains $u'$ of this form belong to $\partial^{k-1}\Phi_p(U)$. We have thus shown that the subspace $\partial^{k-1}\Phi_p(U)$ is dense in ${}^k\Phi_p(U)$, and therefore, the left factor-space in (13) is absolutely inseparable.

Since the isomorphism (13) is topological, the factor-space on the left is also absolutely inseparable, and therefore, the subspace $p_{k-1}[\Phi(\Omega)]^{t_{k-1}}$ is dense in $\Phi_{p_k}(\Omega)$. Since the exponential polynomials are dense in $\Phi(\Omega)$, the exponential polynomials of the form $p_{k-1} f, f \in E^{t_{k-1}}$ are dense in $p_{k-1}[\Phi(\Omega)]^{t_{k-1}}$ and, therefore, in $\Phi_{p_k}(\Omega)$. The exponential polynomials of this form, obviously, belong to $E_{p_k}$. Therefore, the space $E_{p_k}$ is dense in $\Phi_{p_k}(\Omega)$. □

**Corollary 1.** *Let the region $\Omega$ admit a locally finite convex covering, consisting of regions that have no more than $k+1$-fold intersections. Then for an arbitrary finite $\mathscr{P}$-module $M$*

$$\operatorname{Ext}^i(M, \Phi(\Omega))=0, \quad i>k. \tag{14.10}$$

*If, moreover, the module $M$ satisfies the conditions of Theorem 4, the space $\operatorname{Ext}^k(M, \Phi(\Omega))$ is absolutely inseparable.*

*Proof.* Let (11) be a free resolution of the module $M$. Then the module $L=\operatorname{Coker} p'_k$ satisfies the conditions of Theorem 3. Therefore, the lowest row of the diagram (12) is exact, and this implies equation (14). If the module $M$ satisfies the conditions of Theorem 4, then as we showed in the proof of that theorem, both the spaces in (13) are absolutely inseparable. There remains only to observe that the right space is $\operatorname{Ext}^k(M, \Phi(\Omega))$. □

## § 11. Geometrical conditions of $M$-convexity

In this section, we prove a theorem which contains the geometric conditions on the functional space $\Phi$, under which the equation $\operatorname{Ext}^i(M, \Phi)=0$, $i>k$ holds for an arbitrary hypoelliptic $\mathscr{P}$-module $M$. This result may also be treated as a condition for $L$-convexity of the space $\Phi$, where $L$ is the module connected with $M$ by the sequences (10.10)–(11.10). It is clear from these sequences that for arbitrary $i>k$, the isomorphism (2) holds.

**1°. $k$-convexity of a region.** Let $k$ be some integer lying between 1 and $n$. A function $h(\xi)$ defined in some region $\Omega$, and belonging to the class $C^2$ in this region, will be called $k$-convex if it is real and if at every point $\xi \in \Omega$, its Hessian $\operatorname{Hess} h = \left\{ \frac{\partial^2 h}{\partial \xi_i \, \partial \xi_j} \right\}$ has no less than $k$ positive eigenvalues. We shall say that the region $\Omega$ is $k$-convex if we can define in it some continuous function $h$ which is $k$-convex outside some compact $K \subset \Omega$, and such that for arbitrary real $c$ the set $K_c = \{\xi \in \Omega, h(\xi) \leqq c\}$ is compact. We shall say that the region $\Omega$ is completely $k$-convex, if it is $k$-convex and $K = \varnothing$.

It is easy to prove that every convex region is $n$-convex.

**2°. $M$-cohomologies in $k$-convex regions**

**Theorem 1.** *Let $M$ be a hypoelliptic module, and let $k$, $0 \leqq k \leqq n$ be some integer.*

I. *If the region $\Omega$ is $(n-k)$-convex, the spaces* $\operatorname{Ext}^i(M, \Phi(\Omega))$, $i > k$, *are finite dimensional.*

II. *If the region $\Omega$ is completely $(n-k)$-convex, then* $\operatorname{Ext}^i(M, \Phi(\Omega)) = 0$ *for $i > k$.*

Let us take note of some necessary facts. If the module $M$ is hypoelliptic, Corollary 2, §9 implies the isomorphism

$$\operatorname{Ext}^i(M, \mathscr{D}^*(\Omega)) \cong \operatorname{Ext}^i(M, \mathscr{E}(\Omega)), \qquad i \geqq 0.$$

Therefore, it is sufficient to prove the theorem only for the space $\Phi(\Omega) = \mathscr{E}(\Omega)$. Let

$$\cdots \xrightarrow{p_{k+1}} \mathscr{P}^{t_{k+1}} \xrightarrow{p'_k} \mathscr{P}^{t_k} \xrightarrow{p'_{k-1}} \cdots \xrightarrow{p'_1} \mathscr{P}^{t_1} \xrightarrow{p'_0} \mathscr{P}^{t_0} \to M \to 0 \qquad (1.11)$$

be a free resolution of the module $M$. We consider the module $L = \operatorname{Coker} p'_k$. It is clear that the mappings $\ldots, p'_{k+1}, p'_k$ in the sequence (1) form a free resolution of the module $L$. Hence, it is obvious that for an arbitrary $\mathscr{P}$-module $\Phi$

$$\operatorname{Ext}^i(L, \Phi) \cong \operatorname{Ext}^{k+i}(M, \Phi), \qquad i \geqq 1. \qquad (2.11)$$

The remaining mappings in the sequence (1) form an exact sequence (10.10).

We take note of the following transparent geometrical lemma.

**Lemma 1.** *In $R^n$, there exists a countable system of closed parallelepipeds $\pi = \{\pi_\alpha\}$, having the following properties:*

I. *For arbitrary $\varepsilon > 0$, those $\pi_\alpha$ having a diameter less than $\varepsilon$ form a covering of $R^n$.*

II. *An arbitrary set of* $k$ *parallelepipeds* $\pi_\alpha$, *of which no one contains any of the others, intersects in a set of codimension not less than* $k-1$. (The empty set will be given an arbitrary negative dimension.)

**Lemma 2.** *We set* $\xi'=(\xi_1,\dots,\xi_m)$ *and* $\xi''=(\xi_{m+1},\dots,\xi_n)$, *where* $m=n-k$. *Let* $\omega$ *be a region in* $R^n$, *such that its section by an arbitrary subspace of the form* $\xi''=\text{const}$ *is a convex region. Then the space* $\mathscr{E}(\omega)$ *is strongly L-convex.*

For the proof, we construct an increasing sequence of regions tending to $\omega$, in each of which Theorem 3, § 10 is applicable. Let $K$ be an arbitrary compact belonging to $\omega$. We choose some compact $K'$ such that $K\subset\subset K'\subset\omega$, and such that the intersection of $K'$ with every subspace of the form $\xi''=\text{const}$ is convex. We write $\rho=\min\{\rho(K,\complement K'),\rho(K',\complement\omega)\}$. Making use of Lemma 1, we cover the coordinate subspace $R^k_{\xi''}$ with a system of parallelepipeds $\{\pi_\alpha\}$, the diameter of which does not exceed $\rho/4$, having not more than $k+1$-fold mutual intersections. For every $\alpha$ we choose a convex neighborhood $u_\alpha$ of the parallelepiped $\pi_\alpha$, the diameter of which does not exceed $\rho/2$, and such that the regions $u_\alpha$ intersect in no more than $k+1$-fold fashion.

Further, suppose that $\kappa$ is a projection of $K'$ on $R^k_{\xi''}$. If the given region $u_\alpha$ intersects $\kappa$, we choose some point $\xi''_\alpha\in u_\alpha\cap\kappa$ and we consider the region $v_\alpha$, which is the intersection of $\operatorname{int}K'$ with the subspace $\xi''=\xi''_\alpha$. If, however, $u_\alpha\cap\kappa=\emptyset$, then we write $v_\alpha=\emptyset$. The regions $w_\alpha=v_\alpha\times u_\alpha$ are convex, they belong to $\omega$, they have no more than $k+1$-fold mutual intersections, and they form a finite covering of the region $\omega_K=\cup\, w_\alpha$, containing $K$. The regions $R^m_{\xi'}\times u_\alpha$ are also convex, they intersect in no more than $k+1$-fold fashion, and they form a locally finite covering of the space $R^n$. Therefore, on the basis of Theorem 3, § 10, we conclude that the space $\mathscr{E}(\omega_K)$ is $L$-convex, and the subspace $\mathscr{E}_L(R^n)$ is dense in $\mathscr{E}_L(\omega_K)$. Since the subspace $E_L$ is dense in $\mathscr{E}_L(R^n)$, it is dense also in $\mathscr{E}_L(\omega_K)$, and therefore, the space $\mathscr{E}(\omega_K)$ is strongly $L$-convex.

It follows from the construction of the regions $\omega_K$ that $K\subset\omega_K\subset\subset\omega$. Since the compact $K\subset\omega$ was chosen arbitrarily, we may apply this construction to the sequence of compacts $K_\nu$, $\nu=1,2,\dots$, tending to $\omega$, and so constructed that for an arbitrary $\nu$ we have $\omega_{K_\nu}\subset K_{\nu+1}$. The sequence of regions $\omega_{K_\nu}$ is an increasing sequence and it tends to $\omega$. Since for arbitrary $\nu$, the space $\mathscr{E}(\omega_{K_\nu})$ is strongly $L$-convex, Proposition 2, § 7, implies that the space $\mathscr{E}(\omega)$ is also strongly $L$-convex, which is what we were to prove. ▯

**3°. Proof of Theorem 1.** We shall suppose that the region $\Omega$ is $n-k$-convex, and $h(\xi)$ is a corresponding function continuous in $\Omega$, and $n-k$-convex in $\Omega\setminus K$, where $K\subset\Omega$ is some compact. By hypothesis, for an arbitrary real $c$, the set $K_c$, consisting of the points of $\Omega$ at which

$h(\xi)\leqq c$, is compact. We denote by $\Omega_c$ the subregion of $\Omega$, in which $h(\xi)<c$. As $c\to\infty$, the regions $\Omega_c$ form an increasing sequence tending to $\Omega$.

For convenience in the argument, we add to the function $h$ a constant such that we have the inequality $\sup_K h<0$.

**Lemma 3.** *For arbitrary* $c\geqq 0$, *we can find a positive number* $\varepsilon=\varepsilon(c)$ *and a finite sequence of regions* $\Omega^\alpha$ *such that*

$$\Omega_c=\Omega^0\subset\Omega^1\subset\cdots\subset\Omega^\alpha\subset\cdots\subset\Omega^A=\Omega_{c+\varepsilon}. \tag{3.11}$$

*and having the following properties: For arbitrary* $\alpha$, $0<\alpha\leqq A$, *the restriction mapping*

$$r^i_\alpha:\ \operatorname{Ext}^i(L,\mathscr{E}(\Omega^\alpha))\to\operatorname{Ext}^i(L,\mathscr{E}(\Omega^{\alpha-1}))$$

A) *for* $i>1$ *is an algebraic isomorphism;*

B) *for* $i=1$ *is an algebraic epimorphism, and its kernel is an absolutely inseparable space;*

C) *for* $i=0$ *has as cokernel an absolutely inseparable space if the space* $\operatorname{Ext}^1(L,\mathscr{E}(\Omega^\alpha))$ *is separable.*

*Proof.* The condition imposed on the function $h$ means that in the neighborhood of an arbitrary point $\xi\in\Omega\setminus K$, we can choose a system of coordinates $\eta=(\eta',\eta'')$, where $\eta'\in R^m$ $(m=n-k)$, and $\eta''\in R^k$, such that the matrix

$$\operatorname{Hess}_{\eta'} h(\xi)=\left\{\frac{\partial^2 h(\xi)}{\partial\eta_i\,\partial\eta_j},\ 1\leqq i,j\leqq m\right\}$$

is positive definite at this point. Since the second derivatives of the function $h$ are continuous in $\Omega\setminus K$, this property of the matrix $\operatorname{Hess}_{\eta'} h$ is preserved in some neighborhood $U_\xi$ of the point $\xi$. Since $K\subset\Omega_0$, in the covering of the region $\Omega\setminus K$ formed by the regions $U_\xi$, $\xi\in\Omega\setminus K$, we can inscribe a sequence of spheres $U^\alpha$, $\alpha=1,2,\ldots$, locally finite in $\Omega$, and such that its union contains $\Omega\setminus\Omega_0$.

Suppose, further, that $e_\alpha$, $\alpha=1,2,\ldots$, is a sequence of non-negative functions in $R^n$ belonging to the class $C^2$ and satisfying the conditions

(i) for arbitrary $\alpha$ $\operatorname{supp} e_\alpha\subset U^\alpha$,

(ii) $\sum e_\alpha>0$ in $\Omega\setminus\Omega_c$,

(iii) the functions $e_\alpha$ are small enough so that for arbitrary $\alpha$ and $\beta$, the matrix

$$\operatorname{Hess}_{\eta'} h_\alpha(\xi),\quad\text{where}\quad h_\alpha=h-\sum_1^\alpha e_i,$$

is positive definite for all $\xi\in U^\beta$, where $\eta=(\eta',\eta'')$ is a local system of coordinates corresponding to the sphere $U^\beta$.

Further suppose that $A$ is the largest of the numbers $\alpha$ such that the intersection $K_c \cap U^\alpha$ is not empty. For every $\alpha$, $0 \leqq \alpha \leqq A$, we consider the region

$$\Omega_c^\alpha = \{\xi \in \Omega, h_\alpha(\xi) < c\}.$$

From (ii) it follows that the function $h_A$ is strictly less than $h$ in the neighborhood of the set $K_c \setminus \Omega_c$. This implies that $K_c \subset \Omega_c^A$. Consequently we have the relation $\Omega_{c+\varepsilon} \subset \Omega_c^A$ for sufficiently small positive $\varepsilon = \varepsilon(c)$. We shall show that the regions $\Omega^\alpha = \Omega_c^\alpha \cap \Omega_{c+\varepsilon}$, $\alpha = 0, \ldots, A$, satisfy the conditions of the Lemma. In fact, the relation (3) clearly holds.

We take note of two properties of the regions that we have constructed:

a) For arbitrary $\alpha$, $\Omega^\alpha \setminus \Omega^{\alpha-1} \subset U^\alpha$,

b) For arbitrary $\alpha$ and $\beta$, the space $\mathscr{E}(\Omega^\alpha \cap U^\beta)$ is strongly $L$-convex.

Property a) follows from (i), since $h_\alpha - h_{\alpha-1} = -e_\alpha$. Let us verify property b). We choose $\alpha$ and $\beta$ arbitrarily. By construction, the matrices $\mathrm{Hess}_{\eta'} h$ and $\mathrm{Hess}_{\eta'} h_\alpha$ are positive definite within the sphere $U^\beta$. Therefore, the intersections of the regions $\Omega_{c+\varepsilon} \cap U^\beta$ and $\Omega_c^\alpha \cap U^\beta$ with an arbitrary manifold of the form $\eta'' = \mathrm{const}$ are convex. Therefore, the intersection of the regions $\Omega^\alpha \cap U^\beta = \Omega_c^\alpha \cap \Omega_{c+\varepsilon} \cap U^\beta$ with such a manifold is also convex. Therefore property b) follows from Lemma 2.

From a) we derive the equation $\Omega^{\alpha-1} \cup (\Omega^\alpha \cap U^\alpha) = \Omega^\alpha$, where $\alpha$ is an arbitrary number lying between 1 and $A$. It follows that the regions $\Omega^{\alpha-1}$ and $\Omega^\alpha \cap U^\alpha$ form a covering of the region $\Omega^\alpha$. The sequence (19.2), when applied to this covering and the sheaf $\mathscr{E}$ is written as follows:

$$0 \to \mathscr{E}(\Omega^\alpha) \xrightarrow{\rho} \mathscr{E}(\Omega^{\alpha-1}) \oplus \mathscr{E}(\Omega^\alpha \cap U^\alpha) \xrightarrow{\sigma} \mathscr{E}(\Omega^{\alpha-1} \cap U^\alpha) \to 0.$$

The mapping $\rho$ carries the function $\phi$ into the pair $(\phi', \phi'')$, where $\phi'$ and $\phi''$ are restrictions of $\phi$ on $\Omega^{\alpha-1}$ and $\Omega^\alpha \cap U^\alpha$, and the mapping $\sigma$ carries the pair $(\phi, \psi)$ into the difference $\phi' - \psi'$, where $\phi'$ and $\psi'$ are restrictions of $\phi$ and $\psi$ on $\Omega^{\alpha-1} \cap U^\alpha$. Proposition 9, §2 implies that this sequence is exact as a sequence of mappings of topological $\mathscr{P}$-modules. Therefore, we may write the corresponding algebraically exact sequence for the functor $\mathrm{Hom}(L, \cdot)$:

$$\begin{aligned} \cdots \xrightarrow{\delta^{i-1}} \mathrm{Ext}^i(L, \mathscr{E}(\Omega^\alpha)) \xrightarrow{\rho^*} \mathrm{Ext}^i(L, \mathscr{E}(\Omega^{\alpha-1})) \\ \oplus \mathrm{Ext}^i(L, \mathscr{E}(\Omega^\alpha \cap U^\alpha)) \xrightarrow{\sigma^*} \mathrm{Ext}^i(L, \mathscr{E}(\Omega^{\alpha-1} \cap U^\alpha)) \xrightarrow{\delta^i} \cdots, \end{aligned} \tag{4.11}$$

in which all the mappings are continuous (see Proposition 8, § 3, Chapter I). It follows from b) that $\mathrm{Ext}^i(L, \mathscr{E}(\Omega^\beta \cap U^\alpha)) = 0$ for all $i > 0, \alpha, \beta$. Therefore, the sequence (4) may be divided into the fragments

$$0 \longrightarrow \mathrm{Ext}^i(L, \mathscr{E}(\Omega^\alpha)) \xrightarrow{r_\alpha^i} \mathrm{Ext}^i(L, \mathscr{E}(\Omega^{\alpha-1})) \to 0, \quad i = 2, 3, \ldots, \tag{5.11}$$

and

$$\begin{aligned}0 &\longrightarrow \mathscr{E}_L(\Omega^\alpha) \xrightarrow{\rho^*} \mathscr{E}_L(\Omega^{\alpha-1}) \oplus \mathscr{E}_L(\Omega^\alpha \cap U^\alpha) \\ &\xrightarrow{\sigma^*} \mathscr{E}_L(\Omega^{\alpha-1} \cap U^\alpha) \xrightarrow{\delta^0} \operatorname{Ext}^1(L, \mathscr{E}(\Omega^\alpha)) \xrightarrow{r^1_\alpha} \operatorname{Ext}^1(L, \mathscr{E}(\Omega^{\alpha-1})) \to 0.\end{aligned} \tag{6.11}$$

The algebraic exactness of (5) implies that for $i > 1$, the mapping $r^i_\alpha$ is an algebraic isomorphism, and the algebraic exactness of (6) implies that $r^1_\alpha$ is an algebraic isomorphism. By b) the space $E_L$ is dense in $\mathscr{E}_L(\Omega^{\alpha-1} \cap U^\alpha)$. On the other hand, $E_L$ belongs to the image of the mapping $\sigma^*$ in (6), therefore the image of $\sigma^*$ is dense in $\mathscr{E}_L(\Omega^{\alpha-1} \cap U^\alpha)$, that is, the cokernel of $\sigma^*$ is an absolutely inseparable space. The mapping $\delta^0$, being continuous, acts continuously from the space Coker $\sigma^*$ to the space Ker $r^1_\alpha$. Therefore, the space Ker $r^1_\alpha$ is also absolutely inseparable[13]. Thus, the properties A) and B) of the mappings $r^i_\alpha$ have been established.

We now prove property C). We shall suppose that the space $\operatorname{Ext}^1(L, \mathscr{E}(\Omega^\alpha))$ is separable. A separable space cannot have any inseparable subspace. Therefore, B) implies that $\operatorname{Ker} r^1_\alpha = 0$. We consider the commutative diagram

$$\begin{array}{ccccccc}
0 & & 0 & & & & \\
\uparrow & & \uparrow & & & & \\
\mathscr{E}_L(\Omega^\alpha) & \xrightarrow{r^0_\alpha} & \mathscr{E}_L(\Omega^{\alpha-1}) & & \longrightarrow & 0 & \\
\uparrow & & \uparrow \pi & & & \uparrow & \\
\mathscr{E}_L(\Omega^\alpha) & \xrightarrow{\rho^*} & \mathscr{E}_L(\Omega^{\alpha-1}) \oplus \mathscr{E}_L(\Omega^\alpha \cap U^\alpha) & & \xrightarrow{\sigma^*} & \mathscr{E}_L(\Omega^{\alpha-1} \cap U^\alpha) & \longrightarrow 0 \\
 & & \uparrow i & & & \uparrow -R & \\
 & & \mathscr{E}_L(\Omega^\alpha \cap U^\alpha) & & \longrightarrow & \mathscr{E}_L(\Omega^\alpha \cap U^\alpha) & \longrightarrow 0
\end{array}$$

in which $R$ is a restriction mapping, and $i$ and $\pi$ are mappings acting according to the formulae

$$i: \ \phi \to (0, \phi), \qquad \pi: \ (\phi, \psi) \to \phi.$$

The left column and the lowest row of this diagram, which consist of identity operators, are exact. The exactness of the second column is obvious. Since $\operatorname{Ker} r^1_\alpha = 0$, the algebraic exactness of (6) implies the algebraic exactness of the second row. Since this row is formed of Frechet spaces, it is also topologically exact by Proposition 8, § 1. Thus, we may apply to our diagram Theorem 1, § 2, Chapter I, which implies the topological isomorphism

$$\operatorname{Coker} r^0_\alpha \cong \operatorname{Coker} R. \tag{7.11}$$

13 The image of an absolutely inseparable space under a continuous mapping is again absolutely inseparable. This can be verified immediately.

Since the space $\mathscr{E}_L(\Omega^\alpha \cap U^\alpha)$ contains $E_L$, which by b) is dense in $\mathscr{E}_L(\Omega^{\alpha-1} \cap U^\alpha)$, the image of the mapping $R$ is dense in $\mathscr{E}_L(\Omega^{\alpha-1} \cap U^\alpha)$. Therefore, the space Coker $R$ is absolutely inseparable. Taking account of the isomorphism (7), we conclude that the space Coker $r_\alpha^0$ is also absolutely inseparable, which is what we were to prove. □

We consider the numerical sequence

$$c_1=0, \quad c_2=c_1+\varepsilon(c_1), \ldots, c_{\lambda+1}=c_\lambda+\varepsilon(c_\lambda), \ldots,$$

where $\varepsilon(c)$ is the function constructed in Lemma 3. We shall show that this function can be chosen so that $c_\lambda \to \infty$. In fact, the number $\varepsilon(c)$ is defined by the condition that the region $\Omega_{c+\varepsilon}$ belongs to $\Omega_c^A$. The construction of $\Omega_c^A$ makes it obvious that this region contains a $\delta$-neighborhood of the compact $K_c$, where the quantity $\delta=\delta(c)>0$ can be assumed to be a non-increasing function of $c$. The number $\varepsilon(c)$ can be made to satisfy the condition: $\varepsilon(c)$ is the largest of the numbers $\varepsilon$ for which the region $\Omega_{c+\varepsilon}$ belongs to the region $\omega_c$, which is a $\delta(c)$-neighborhood of the compact $K_c$. We shall show that in this case $c_\lambda \to \infty$. Let us assume the opposite.

Then, $c_\lambda \to c_* < \infty$, and therefore, for arbitrary $\lambda$, we can find a point $\xi_\lambda \subset \Omega_{c_*}$, not lying in $\omega_{c_\lambda}$. Since $\bar{\Omega}_{c_*}$ is compact, the sequence $\{\xi_\lambda\}$ has some limit point $\xi_* \in \Omega_{c_*}$. On the other hand, the union of the compacts $K_{c_\lambda}$ is equal to $\Omega_{c_*}$. Therefore, the point $\xi_*$ belongs to a $\delta(c_*)$-neighborhood of some compact $K_{c_\lambda}$ and therefore, belongs to the region $\omega_{c_\lambda}$ for all sufficiently large $\lambda$. We have thus obtained a contradiction with the relationship $\xi_\lambda \in \omega_{c_\lambda}$. It follows that $c_\lambda \to \infty$, which is what we were to prove.

**Lemma 4.** *Let $\omega$ and $\upsilon \supset\supset \omega$ be arbitrary bounded regions in $R^n$, such that the restriction mapping*

$$\operatorname{Ext}^i(M, \mathscr{E}(\upsilon)) \to \operatorname{Ext}^i(M, \mathscr{E}(\omega)) \tag{8.11}$$

*is an algebraic epimorphism. Then the space* $\operatorname{Ext}^i(M, \mathscr{E}(\omega))$ *is separable and finite dimensional.*

*Proof.* Let $V=\{V_\nu\}$ be some locally finite convex covering of the region $\upsilon$, and $U=\{U_\nu\}$ be a locally finite convex covering of the region $\omega$, such that for some $\delta>0$ the covering formed of $\delta$-neighborhoods of the regions $U_\nu$ is inscribed in $V$. We denote by $Z_M^i(V)$ the kernel of the coboundary operator $\partial^i$: ${}^i\mathscr{E}_M V) \to {}^{i+1}\mathscr{E}_M(V)$. We give a similar meaning to the expression $Z_M^i(U)$, and we consider the mapping

$$\mathscr{R}\colon Z_M^i(V) \oplus {}^{i-1}\mathscr{E}_M(U) \to Z_M^i(U), \qquad i>k, \tag{9.11}$$

which is equal to the sum $R+\Delta$, where $R$ and $\Delta$ are operators defined by the formulae

$$R: \begin{cases} Z^i_M(V) \xrightarrow{\rho} Z^i_M(U) \\ {}^{i-1}\mathscr{E}_M(U) \longrightarrow ; \end{cases} \qquad \Delta: \begin{cases} Z^i_M(V) \longrightarrow 0 \\ {}^{i-1}\mathscr{E}_M(U) \xrightarrow{\partial^{i-1}} Z^i_M(U), \end{cases}$$

and $\rho$ is a restriction mapping. Since by hypothesis, the mapping (8) is an algebraic epimorphism, Theorem 1, § 9, implies that the restriction mapping $H^i(V, \mathscr{E}_M) \to H^i(U, \mathscr{E}_M)$ is also an algebraic epimorphism. It follows that the mapping $\mathscr{R}$ has the same property.

Let us now consider the restriction operator $\rho: {}^i\mathscr{E}_M(V) \to {}^i\mathscr{E}_M(U)$. We may look on it as a mapping acting from the direct product

$$\Pi\, \mathscr{E}_M(V_{\nu_0} \cap \cdots \cap V_{\nu_i})$$

to the direct product $\Pi\, \mathscr{E}_M(U_{\nu_0} \cap \cdots \cap U_{\nu_i})$. The mapping of direct products differs from zero only on those factors $\mathscr{E}_M(V_{\nu_0} \cap \cdots \cap V_{\nu_i})$ for which the intersection $V_{\nu_0} \cap \cdots \cap V_{\nu_i} \cap \omega$ is not empty (the number of such factors is finite, since $\omega \subset \subset v$) and it acts on these factors as a restriction operator on a subregion of the form $U_{\mu_0} \cap \cdots \cap U_{\mu_i}$, which together with its $\delta$-neighborhood is bounded and belongs to $V_{\nu_0} \cap \cdots \cap V_{\nu_i}$. By Theorem 1, § 5, every such restriction operator is compact. This implies the compactness of the operator $\rho$ and, therefore, of the operator $R$.

Thus, $\Delta = \mathscr{R} - R$, where $\mathscr{R}$ is an algebraic epimorphism, and $R$ is a compact mapping. Since all the spaces appearing in (9) are Frechet spaces, we may apply to $\Delta$ the Schwartz Theorem[14], which implies that the space

$$\operatorname{Coker} \Delta = H^i(U, \mathscr{E}_M) \cong \operatorname{Ext}^i(M, \mathscr{E}(\omega))$$

is separable and finite dimensional. □

**Corollary 1.** *Let $c$ be an arbitrary, non-negative number, and let $\varepsilon = \varepsilon(c)$ be the constant defined in Lemma 3. Then for arbitrary $i > k$, the restriction mapping*

$$R^i\colon\ \operatorname{Ext}^i(M, \mathscr{E}(\Omega_{c+\varepsilon})) \to \operatorname{Ext}^i(M, \mathscr{E}(\Omega_c))$$

*is a topological isomorphism, and the restriction mapping*

$$R_{i-1}\colon\ \mathscr{E}_{p_{i-1}}(\Omega_{c+\varepsilon}) \to \mathscr{E}_{p_{i-1}}(\Omega_c)$$

*has an image dense in $\mathscr{E}_{p_{i-1}}(\Omega_c)$.*

*Proof.* We fix an arbitrary $i > 0$ and we consider the restriction mapping

$$\operatorname{Ext}^i(L, \mathscr{E}(\Omega_{c+\varepsilon})) \to \operatorname{Ext}^i(L, \mathscr{E}(\Omega_c)). \tag{10.11}$$

14 See Schwartz, [2], Part II, § 7.

It is a composition of the operators $r_\alpha^i$ with $\alpha = A, A-1, \ldots, 1$, which are algebraically epimorphic in view of Lemma 3. Therefore, (10) is also an algebraic epimorphism. We further choose $\beta$ arbitrarily, $0 \leqq \beta \leqq A$, we set $\varepsilon' = \varepsilon(c + \varepsilon(c))$ and we consider the sequence of operators

$$\operatorname{Ext}^i(L, \mathscr{E}(\Omega_{c+\varepsilon+\varepsilon'})) \to \operatorname{Ext}^i(L, \mathscr{E}(\Omega_{c+\varepsilon})) \to \operatorname{Ext}^i(L, \mathscr{E}(\Omega^\beta)).$$

The first of these coincides with (10) when $c$ is replaced with $c+\varepsilon$ and, is therefore, algebraically epimorphic. The second is the composition of operators $r_\alpha^i$ with $\alpha = A, A-1, \ldots, \beta+1$ and, is, therefore, also algebraically epimorphic. Therefore, the composition of these has the same property. Since $\Omega^\beta \subset\subset \Omega_{c+\varepsilon+\varepsilon'}$, we conclude on the basis of Lemma 4 that the space $\operatorname{Ext}^i(L, \mathscr{E}(\Omega^\beta))$ is separable.

By definition, $\operatorname{Ext}^i(L, \mathscr{E}(\Omega^\beta))$ is a factor-space of the $\mathscr{F}$-space $\mathscr{E}_{p_{k+i}}(\Omega^\beta)$ with respect to its subspace $p_{k+i-1}[\mathscr{E}(\Omega^\beta)]^{t_{k+i-1}}$. Since the factor-space is separable, the subspace is closed. Therefore, the subspace and the factor-space are Frechet spaces. By Lemma 3, the mapping $r_\alpha^1$ and the mappings $r_\alpha^i$ with $i > 1$ are algebraic isomorphisms and, therefore, by Proposition 7, § 1, are topological isomorphisms. Thus, the mapping (10), which is their composition, is also an isomorphism. The mapping $R^{i+k}$ is isomorphic to (10) in view of (2), and, is, therefore a topological isomorphism.

We shall now show that for arbitrary $\alpha$ and $i \geqq k$, the image of the mapping

$$\rho_\alpha^i \colon \mathscr{E}_{p_i}(\Omega^\alpha) \to \mathscr{E}_{p_i}(\Omega^{\alpha-1})$$

is dense in $\mathscr{E}_{p_i}(\Omega^{\alpha-1})$. If $i > k$, then the fact that $r_\alpha^i$ is an epimorphism implies that the image of $\rho_\alpha^i$, together with the subspace $p_{i-1}[\mathscr{E}(\Omega^{\alpha-1})]^{t_{i-1}}$ is equal to the space $\mathscr{E}_{p_i}(\Omega^{\alpha-1})$. But the subspace $\mathscr{E}(\Omega^\alpha)$ is, clearly, dense in $\mathscr{E}(\Omega^{\alpha-1})$, and, therefore, $p_{i-1}[\mathscr{E}(\Omega^\alpha)]^{t_{i-1}} \subset \operatorname{Im} \rho_\alpha^i$ is dense in $p_{i-1}[\mathscr{E}(\Omega^{\alpha-1})]^{t_{i-1}}$. It follows that $\operatorname{Im} \rho_\alpha^i$ is dense in $\mathscr{E}_{p_i}(\Omega^{\alpha-1})$.

If $i = k$, this assertion follows from Property C) of Lemma 3.

Since the operator $R_i$ is the composition of the operators $\rho_\alpha^i$ with $\alpha = A, A-1, \ldots, 1$, its image is dense in $\mathscr{E}_{p_i}(\Omega_c)$. □

**4°. Completion of the proof of Theorem 1.** We now fix an arbitrary $i > k$ and $\lambda$ and we consider the sequence of linear topological spaces

$$0 \to \mathscr{E}_{p_{i-1}}(\Omega_{c\lambda}) \to [\mathscr{E}(\Omega_{c\lambda})]^{t_{i-1}} \xrightarrow{p_{i-1}} p_{i-1}[\mathscr{E}(\Omega_{c\lambda})]^{t_{i-1}} \to 0. \quad (11.11)$$

It is, clearly, algebraically exact. As we noted above, $p_{i-1}[\mathscr{E}(\Omega_{c\lambda})]^{t_{i-1}}$ is a Frechet space. Therefore, (11) is an algebraically exact sequence of Frechet spaces. Thus, it is exact.

When $\lambda$ runs over the sequence of natural numbers, the terms of (11) run over a decreasing family of linear topological spaces, with respect to restriction mappings. In view of the corollary, the space $\mathscr{E}_{p_{i-1}}(\Omega_{c_{\lambda+1}})$ is dense in $\mathscr{E}_{p_{i-1}}(\Omega_{c_\lambda})$. Therefore, by Proposition 11, § 1, the sequence (11) permits us to pass to the projective limit while preserving its exactness. Since $c_\lambda \to \infty$ for $\lambda \to \infty$, we have $\Omega_{c_\lambda} \to \Omega$, and therefore, the projective limit of the sequence of spaces $\mathscr{E}(\Omega_{c_\lambda})$ coincides with $\mathscr{E}(\Omega)$. Then the limiting sequence may be expressed as follows:

$$0 \to \mathscr{E}_{p_{i-1}}(\Omega) \to [\mathscr{E}(\Omega)]^{t_{i-1}} \xrightarrow{p_{i-1}} \varprojlim_{\lambda} p_{i-1}[\mathscr{E}(\Omega_{c_\lambda})]^{t_{i-1}} \to 0.$$

Its exactness implies that

$$\varprojlim_{\lambda} p_{i-1}[\mathscr{E}(\Omega_{c_\lambda})]^{t_{i-1}} \cong p_{i-1}[\mathscr{E}(\Omega)]^{t_{i-1}}. \tag{12.11}$$

is an algebraic isomorphism.

In a similar way, we pass to the limit in the sequence

$$0 \to p_{i-1}[\mathscr{E}(\Omega_{c_\lambda})]^{t_{i-1}} \to \mathscr{E}_{p_i}(\Omega_{c_\lambda}) \to \operatorname{Ext}^i(M, \mathscr{E}(\Omega_{c_\lambda})) \to 0$$

(the space $p_{i-1}[\mathscr{E}(\Omega_{c_{\lambda+1}})]^{t_{i-1}}$, as we have seen, is dense in $p_{i-1}[\mathscr{E}(\Omega_{c_\lambda})]^{t_{i-1}}$). Taking account of (12), we obtain in the limit the exact sequence

$$0 \to p_{i-1}[\mathscr{E}(\Omega)]^{t_{i-1}} \to \mathscr{E}_{p_i}(\Omega) \to \varprojlim_{\lambda} \operatorname{Ext}^i(M, \mathscr{E}(\Omega_{c_\lambda})) \to 0.$$

The exactness of this sequence implies that

$$\varprojlim_{\lambda} \operatorname{Ext}^i(M, \mathscr{E}(\Omega_{c_\lambda})) \cong \operatorname{Ext}^i(M, \mathscr{E}(\Omega)). \tag{13.11}$$

is a topological isomorphism.

In accordance with the corollary, all the restriction mappings

$$R^i\colon \operatorname{Ext}^i(M, \mathscr{E}(\Omega_{c_{\lambda+1}})) \to \operatorname{Ext}^i(M, \mathscr{E}(\Omega_{c_\lambda})),$$

which are used in the left side of (13), are isomorphisms. Therefore, this projective limit is isomorphic to an arbitrary one of the spaces $\operatorname{Ext}^i(M, \mathscr{E}(\Omega_{c_\lambda}))$. Hence, finally

$$\operatorname{Ext}^i(M, \mathscr{E}(\Omega_0)) \cong \operatorname{Ext}^i(M, \mathscr{E}(\Omega)), \qquad i > k. \tag{14.11}$$

In view of Lemma 4, the space $\operatorname{Ext}^i(M, \mathscr{E}(\Omega_0))$ is finite dimensional, and, therefore the space $\operatorname{Ext}^i(M, \mathscr{E}(\Omega))$ is also finite dimensional, which is what we were to prove.

Let us now suppose that the region $\Omega$ is completely $n-k$-convex. Then $K = \varnothing$, and, therefore, we may suppose that $h > 0$ in $\Omega$. Hence $\Omega_0 = \varnothing$ and, accordingly, the left side of (14) is equal to zero. Therefore, the right side is also equal to zero. $\square$

*Remark.* We may deduce from the proof of Theorem 2 certain facts about the spaces $\operatorname{Ext}^i(M, \mathscr{E}(\Omega))$ in those cases also when $M$ is a non-hypoelliptic module. In fact, we note that the arguments of the theorem for $i>k+1$ do not depend on Lemma 4 and, therefore, are true for an arbitrary module $M$. Thus, the algebraic isomorphism (14) with $i>k+1$ holds under the assumption only that the region $\Omega$ is $n-k$-convex. To establish this isomorphism for $i=k+1$, it is sufficient to know that the spaces $\operatorname{Ext}^{k+1}(M, \mathscr{E}(\omega))$ with $\omega \subset \Omega$ are all separable. Thus, we may formulate

**Corollary 2.** *Let $M$ be an arbitrary finite $\mathscr{P}$-module, and let $\Omega$ be an $n-k$-convex region. Then for arbitrary $i>k+1$, the algebraic isomorphism (14) is valid. If all the spaces $\operatorname{Ext}^{k+1}(M, \mathscr{E}(\omega))$ with $\omega \subset \Omega$ are separable, then this isomorphism holds also for $i=k+1$.*

*If the region $\Omega$ is completely $n-k$-convex, then $\operatorname{Ext}^i(M, \mathscr{E}(\Omega))=0$ for all $i>k+1$, and, in the second case, $i>k$.*

## § 12. Operators of the form $p(D_{\bar{\zeta}})$ in domains of holomorphy

Let $n$ be an even number: $n=2m$. In the space $R^n_\xi$ we define the structure of a complex space $C^m_\xi$, writing $\zeta_j = \xi_j + i\,\xi_{m+j}$, $j=1, \ldots, m$. We consider the differential operators

$$D_\zeta = \left(\frac{\partial}{\partial \zeta_1}, \ldots, \frac{\partial}{\partial \zeta_m}\right), \quad D_{\bar{\zeta}} = \left(\frac{\partial}{\partial \bar{\zeta}_1}, \ldots, \frac{\partial}{\partial \bar{\zeta}_m}\right).$$

The polynomials in the operators $\dfrac{\partial}{\partial \zeta_j}$, $j=1, \ldots, m$, or in the operators $\dfrac{\partial}{\partial \bar{\zeta}_j}$, $j=1, \ldots, m$, with complex coefficients, form subrings $\mathscr{P}'$ and $\mathscr{P}''$ of the ring $\mathscr{P}$. A finite $\mathscr{P}$-module $M$ will be said to be holomorphic if $M \cong \operatorname{Coker} p'$ where $p$ is some $\mathscr{P}''$-matrix, that is, a matrix formed from polynomials in $D_{\bar{\zeta}}$. In particular, the module corresponding to the Cauchy-Riemann system of equations in $C^m$ is holomorphic. In this section we shall show that for holomorphic $\mathscr{P}$-modules, the domains of holomorphy have the same roles as convex regions have for arbitrary $\mathscr{P}$-modules.

### 1°. Formulation of the theorem

**Theorem 1.** *Let $\Omega \subset C^m_\zeta$ be an arbitrary domain of holomorphy. Then for an arbitrary holomorphic $\mathscr{P}$-module $M$, the spaces $\mathscr{E}(\Omega)$ and $\mathscr{D}^*(\Omega)$ are $M$-convex. If $\Omega$ is a Runge region, then these spaces are strongly $M$-convex.*

The basic property of a domain of holomorphy that is needed in the proof is the following: every domain of holomorphy in $C^m$ is a Stein manifold, and therefore, admits a holomorphic, regular, proper imbedding into the space $C^{2m+1}$ [15]. That the mapping is proper means that the pre-image of every compact in $C^{2m+1}$ is a compact in $\Omega$. The regularity of the mapping amounts to the fact that the Jacobian at every point $\zeta \in \Omega$ is equal to $m$.

For convenience, we set $\nu = 2m+1$. The variables in the space $C^\nu$ will be denoted by $\lambda = (\lambda_1, \ldots, \lambda_\nu)$. The mapping $\Omega \to C^\nu_\lambda$ described above, will be denoted by $\lambda$. The image of the point $\zeta$ under this mapping will be denoted by $\lambda(\zeta) = (\lambda_1(\zeta), \ldots, \lambda_\nu(\zeta))$. That the mapping $\lambda$ is holomorphic amounts to saying that all the functions $\lambda_1(\zeta), \ldots, \lambda_\nu(\zeta)$ are holomorphic in $\Omega$.

Let us consider the space $C^{m+\nu}$, which is a direct product of the spaces $C^m_\zeta$ and $C^\nu_\lambda$. We denote by $\pi$ the projection operator from $C^{m+\nu}$ on $C^m_\zeta$, and by $\Lambda$ the mapping from $\Omega$ to $C^{m+\nu}$ which carries the point $\zeta$ into the point $(\zeta, \lambda(\zeta))$. The operators $\Lambda$ and $\pi$ set up a biholomorphic homeomorphism between the region $\Omega$ and the manifold $\Lambda(\Omega)$, defined by the system of equations $\lambda = \lambda(\zeta)$, $\zeta \in \Omega$. It is easy to see that the manifold $\Lambda(\Omega)$ is a closed set in $C^{m+\nu}$.

For every region $U$, belonging to the space $C^m$ or to the space $C^{m+\nu}$, we denote by $\Phi(U)$ any one of the three spaces $\mathscr{H}(U)$, $\mathscr{E}(U)$, $\mathscr{D}^*(U)$, where $\mathscr{H}(U)$ is the space of holomorphic functions in $U$, with its topology induced by $\mathscr{E}(U)$. Let $U$ be an arbitrary region in $C^{m+\nu}$, and let $V$ be the projection on $C^m_\zeta$ of its intersection with $\Lambda(\Omega)$. We now consider the operator

$$\Lambda^*: \Phi(U) \ni \phi(\zeta, \lambda) \to \phi(\zeta, \lambda(\zeta)) \in \Phi(V),$$

which carries the function $\phi$ into its restriction on $\Lambda(\Omega)$, looked on as a function of $\zeta$. Clearly, this operator is defined and continuous if $\Phi(U) = \mathscr{H}(U)$, and $\Phi(V) = \mathscr{H}(V)$ or $\Phi(U) = \mathscr{E}(U)$, and $\Phi(V) = \mathscr{E}(V)$.

We denote by $\mathscr{Q}$ the ring of polynomials with complex coefficients in the operators $D_\lambda$, $D_{\bar{\lambda}}$. The ring $\mathscr{R}$ of all differential operators in $C^{m+\nu}$ with complex coefficients is equal to the tensor product $\mathscr{P} \underset{C}{\otimes} \mathscr{Q}$ over the field $C$. We denote by $\mathscr{M}$ the $\mathscr{R}$-module $M \underset{C}{\otimes} \mathscr{Q} \cong \mathscr{R}^s / p' \mathscr{R}^t$. We consider the Cauchy-Riemann system in the variables $\lambda$:

$$\frac{\partial u}{\partial \bar{\lambda}_1} = \cdots = \frac{\partial u}{\partial \bar{\lambda}_\nu} = 0. \tag{1.12}$$

We denote by $L$ the $\mathscr{Q}$-module corresponding to this system, and by $\mathscr{L}$ the $\mathscr{R}$-module $\mathscr{P} \underset{C}{\otimes} L$ corresponding to the same system, considered as being in the space $C^{m+\nu}$.

---

15 See Narasimhan [1]. In fact, for the proof of Theorem 1, it is sufficient to use a most elementary property of domains of holomorphy: such a domain is the limit of an increasing sequence of Weil regions. However, the use of the theorems already formulated in the text allows us to simplify the proof.

If $U$ is a region in $C^{m+\nu}$, then $\Phi_{\mathscr{L}}(U)$ is the space of functions holomorphic in $\lambda$, belonging to $\Phi(U)$, and $\Phi_{\mathscr{M}\otimes\mathscr{L}}(U)$ is a subspace of it, consisting of the solutions of the system $p(D_\zeta)u=0$ (here, and later, the symbol $\otimes$ denotes the tensor product over the ring $\mathscr{R}$). We can now formulate the second fundamental theorem.

**Theorem 2.** *If $M$ is a holomorphic $\mathscr{P}$-module, the sequence*

$$\Phi_{\mathscr{M}\otimes\mathscr{L}}(C^{m+\nu}) \xrightarrow{\Lambda^*} \Phi_M(\Omega) \to 0 \tag{2.12}$$

*is defined and exact. Here $\Phi(\Omega)$ is one of the spaces $\mathscr{E}(\Omega)$ or $\mathscr{D}^*(\Omega)$, and $\Phi(C^{m+\nu})$ is correspondingly, either $\mathscr{E}(C^{m+\nu})$ or $\mathscr{D}^*(C^{m+\nu})$.*

### 2°. Four lemmas

**Lemma 1.** *Let $U$ be an arbitrary region in $\Omega\times C^\nu$. If the function $f\in\Phi_{\mathscr{L}}(U)$ vanishes for $\lambda_1=0$, there exists a uniquely defined function $g\in\Phi_{\mathscr{L}}(U)$, such that $f=\lambda_1 g$. The correspondence $f\to g$ is continuous.*

*Proof.* For the spaces $\Phi(U)=\mathscr{H}(U)$, $\mathscr{E}(U)$, the assertion is self-evident. We shall prove it for the space $\mathscr{D}^*(U)$. We assume, to begin with, that the region $U$ is a direct product of the regions $V\subset\Omega$ and $W\subset C^\nu$. Every $f\in\mathscr{D}^*(U)$ and $\phi\in\mathscr{D}(V)$ can be put into correspondence with the distribution $(f,\phi)_\xi\in\mathscr{D}^*(W)$ by the equation

$$((f,\phi)_\xi,\psi)=(f,\phi\times\psi), \qquad \psi\in\mathscr{D}(W).$$

This correspondence defines a continuous mapping

$$\mathscr{D}^*(U)\to\mathrm{Hom}_C(\mathscr{D}(V),\mathscr{D}^*(W)), \tag{3.12}$$

where $\mathrm{Hom}_C(\mathscr{D}(V),\mathscr{D}^*(W))$ is the space of all continuous linear mappings from $\mathscr{D}(V)$ to $\mathscr{D}^*(W)$, with the topology of bounded convergence. According to the theorem on the kernel[16], the mapping (3) is a topological isomorphism.

The subspace $\mathscr{D}^*_{\mathscr{L}}(U)\subset\mathscr{D}^*(U)$ is characterized by the fact that the image of every function of this subspace under the mapping (3) is an operator carrying $\mathscr{D}(V)$ into the subspace $\mathscr{H}(W)\subset\mathscr{D}^*(W)$, consisting of analytic functions. In other words, the restriction mapping (3) on $\mathscr{D}^*_{\mathscr{L}}(U)$ acts according to the formula

$$\mathscr{D}^*_{\mathscr{L}}(U)\to\mathrm{Hom}_C(\mathscr{D}(V),\mathscr{H}(W)). \tag{4.12}$$

It follows from Theorem 1, § 5, that the topology of the space $\mathscr{H}(W)$ coincides with the topology induced by $\mathscr{D}^*(W)$. Therefore, the topology of the space $\mathrm{Hom}_C(\mathscr{D}(V),\mathscr{H}(W))$ is that induced by $\mathrm{Hom}_C(\mathscr{D}(V),\mathscr{D}^*(W))$. Thus, the fact that the mapping (3) is an isomorphism implies that the mapping (4) is also an isomorphism.

16 See, for example, Schwartz [3].

Let us suppose that the function $f \in \mathcal{D}^*_{\mathcal{L}}(U)$ vanishes on the subspace $\lambda_1 = 0$. Then for an arbitrary function $\phi \in \mathcal{D}(V)$, the analytic function $(f, \phi)_\xi \in \mathcal{H}(W)$ also vanishes for $\lambda_1 = 0$. Since the lemma is true for the subspace $\mathcal{H}(W)$, the quotient $g_\phi = \lambda_1^{-1}(f, \phi)_\xi$ also belongs to $\mathcal{H}(W)$. The mapping $(f, \phi)_\xi \to g_\phi$ is continuous. We have thus constructed a mapping $\mathcal{D}(V) \ni \phi \to g_\phi \in \mathcal{H}(W)$, which, as an element of the space $\mathrm{Hom}_C(\mathcal{D}(V), \mathcal{H}(W))$, is continuous in $f$. Since (4) is an isomorphism, there corresponds to this mapping some function $g \in \mathcal{D}^*_{\mathcal{L}}(U)$. It follows from its construction that it is unique, continuous in $f$ and that $\lambda_1 g = f$. This proves the lemma for regions of the form $U = V \times W$.

Now suppose that $U$ is an arbitrary region in $\Omega \times C^\nu$. We construct a locally finite covering $\mathcal{U}$ of this region, consisting of regions of the form $U_k = V_k \times W_k$. We consider the function $f \in \mathcal{D}^*_{\mathcal{L}}(U)$, which vanishes for $\lambda_1 = 0$, as an element of the space of cochains ${}^0\mathcal{D}^*_{\mathcal{L}}(\mathcal{U})$. Using the particular case of the lemma proved above, we can represent the function $f$ in the form $\lambda_1 g$, where the cochain $g$ belongs to ${}^0\mathcal{D}^*_{\mathcal{L}}(\mathcal{U})$, is uniquely defined, and is continuous in $f$. We conclude from the equation $\partial^0 f = \partial^0 \lambda_1 g = \lambda_1 \partial^0 g = 0$ that $\partial^0 g = 0$. Since the sequence (25.2) is exact with $\Phi = \mathcal{D}^*$, it follows that the cochain $g$ belongs to $\mathcal{D}^*_{\mathcal{L}}(U)$ and that as an element of this space it depends continuously on $f$. □

Again, let $U$ be an arbitrary region in $C^{m+\nu}$. For every $k = 0, 1, \dots, \nu$, we denote by $\Phi^k_{\mathcal{L}}(U)$ the space of differential forms typified by

$$\sum \phi_{i_1, \dots, i_k} \, d\lambda_{i_1} \wedge \cdots \wedge d\lambda_{i_k}, \qquad \phi_{i_1, \dots, i_k} \in \Phi_{\mathcal{L}}(U).$$

We consider also the operator $l\colon \Phi^k_{\mathcal{L}}(U) \to \Phi^{k+1}_{\mathcal{L}}(U)$, which consists of the multiplication by the differential form

$$l = \sum_1^\nu [\lambda_j - \lambda_j(\zeta)] \, d\lambda_j.$$

It is clear that $l\,l = 0$.

**Lemma 2.** *Let $U$ be a region in $C^{m+\nu}$, satisfying the following condition: $U$ is the direct product of the convex region $V = \Lambda^{-1}(U) \subset C^m_\zeta$ and $W \subset C^\nu_\lambda$. Then the sequence*

$$0 \to \Phi^0_{\mathcal{L}}(U) \xrightarrow{l} \Phi^1_{\mathcal{L}}(U) \xrightarrow{l} \cdots \to \Phi^{\nu-1}_{\mathcal{L}}(U) \xrightarrow{l} \Phi^\nu_{\mathcal{L}}(U) \xrightarrow{\Lambda^\circledast} \Phi(V) \to 0 \qquad (5.12)$$

*is exact, where $\Lambda^\circledast$ is an operator acting according to the formula*

$$\Lambda^\circledast\colon \ \phi \, d\lambda_1 \wedge \cdots \wedge d\lambda_\nu \to \Lambda^* \phi.$$

(In the course of the proof, we shall show that this operator is, in fact, defined for all three types of spaces denoted by $\Phi(U)$.)

*Proof.* In the region $\Omega \times C^\nu$ we make the following holomorphic substitution of variables:

$$\zeta \to \zeta, \qquad \lambda \to \lambda' = \lambda - \lambda(\zeta). \qquad (6.12)$$

We rewrite the sequence (5) in the new system of coordinates, and by $U'$ we denote the image of the region $U$. It is easy to see that the change of variables (6) defines an isomorphic mapping of the space $\Phi(U')$ on the space $\Phi(U)$. Since the system of equations (1) does not change its form, the subspace $\Phi_{\mathscr{L}}(U')$ goes over under this isomorphism into $\Phi_{\mathscr{L}}(U)$. Therefore, for arbitrary $k$, we have the isomorphism $\Phi_{\mathscr{L}}^k(U') \cong \Phi_{\mathscr{L}}^k(U)$. The operator $l$ is transformed into the operator $l'$, whose action consists in multiplying by the form $l' = \sum \lambda'_j \, d\lambda'_j$. Finally, the operator $\Lambda^{\circledast}$ is transformed into an operator which acts according to the formula

$$\Lambda': \; \phi(\zeta, \lambda') \, d\lambda'_1 \wedge \cdots \wedge d\lambda'_\nu \to \phi(\zeta, 0).$$

Thus, the sequence (5) is isomorphically transformed into the sequence

$$\begin{aligned} 0 \to \Phi_{\mathscr{L}}^0(U') \xrightarrow{l'} \Phi_{\mathscr{L}}^1(U') \xrightarrow{l'} \cdots \\ \cdots \to \Phi_{\mathscr{L}}^{\nu-1}(U') \xrightarrow{l'} \Phi_{\mathscr{L}}^\nu(U') \xrightarrow{\Lambda'} \Phi(V) \to 0. \end{aligned} \tag{7.12}$$

When $\Phi(U') = \mathscr{H}(U')$ or $\mathscr{E}(U')$, the operator $\Lambda'$ is clearly defined and continuous. We shall prove that it is so for the space $\Phi(U') = \mathscr{D}^*(U')$. In the system (1), the variables $\lambda$ are to be replaced by $\lambda'$ and we note that this system is strongly hypoelliptic in $\lambda'$, with the exponent $\check{\gamma} = 1$ (see § 5). Accordingly, by Theorem 3 of § 5, every distribution $u$ defined in $U'$ which satisfies this system, has a restriction on the subspace $\lambda' = 0$ which is a distribution in the region $V$. In accordance with the remark following Theorem 3, § 5, this restriction, as an element of the space $\mathscr{D}^*(V)$, depends continuously on $u$. We have now established that the operator $\Lambda': \mathscr{D}_{\mathscr{L}}^*(U) \to \mathscr{D}^*(V)$ is defined and continuous. It follows that the operator $\Lambda^{\circledast}$ in (5) is also defined and continuous.

To establish the exactness of the sequence (5), it is sufficient to establish the exactness of (7). We denote by $Z^k(U')$, $k = 0, \ldots, \nu - 1$, the kernel of the operator $l': \Phi_{\mathscr{L}}^k(U') \to \Phi_{\mathscr{L}}^{k+1}(U')$ and by $Z^\nu(U')$ the kernel of $\Lambda'$. The space $Z^0(U')$ consists of all the functions $f$, belonging to $\Phi_{\mathscr{L}}(U')$, such that $\lambda'_j f = 0$, $j = 1, \ldots, \nu$. Therefore, $Z^0(U') = 0$, when $\Phi(U') = \mathscr{H}(U')$ or $\mathscr{E}(U')$. When $\Phi(U') = \mathscr{D}^*(U')$, this follows from the isomorphism (4). We now construct the continuous (single-valued) operators

$$l'^{-1}: \; Z^k(U') \to \Phi_{\mathscr{L}}^{k-1}(U'), \qquad k = 1, \ldots, \nu \tag{8.12}$$

and the operator

$$\Lambda'^{-1}: \; \Phi(V) \to \Phi_{\mathscr{L}}^\nu(U'),$$

which are inverse to the operators $l'$ and $\Lambda'$ respectively. The fact that they exist implies the exactness of the sequence (7) and therefore, of the sequence (5).

We begin by constructing the operator $\Lambda'^{-1}$. An arbitrary function $\phi \in \Phi(V)$ can be continued in $U'$, as a function which is constant in $\lambda$.

The extension $\tilde{\phi}$, so obtained, clearly belongs to $\Phi_{\mathscr{L}}(U')$ and depends continuously on $\phi$. The mapping $\phi \to \tilde{\phi}\, d\lambda'_1 \wedge \cdots \wedge d\lambda'_v$ is the desired inverse of $\Lambda'$.

We now construct the operators $l'^{-1}$. We observe that this task depends only on the region $U'$, and does not depend on $\Omega$ or $\Lambda$. Therefore, we may use induction on $v$. We shall assume that operators similar to $l'^{-1}$ have been constructed for an arbitrary region $\hat{U}$ in $C^{m+v-1}$. We note that when $v=1$, this postulate is clearly verified.

For an arbitrary form $\phi \in \Phi^k_{\mathscr{L}}(U')$, where $0 \leqq k < v$, we denote by $\hat{\phi}$ its restriction on the subspace $\lambda'_1 = 0$. The correspondence $\phi \to \hat{\phi}$ defines a continuous mapping $\Phi^k_{\mathscr{L}}(U') \to \Phi^k_{\hat{\mathscr{L}}}(\hat{U})$, where $\hat{U} \subset C^{m+v-1}$ is the intersection of $U'$ with the subspace $\lambda'_1 = 0$, and $\hat{\mathscr{L}}$ is a module corresponding to the Cauchy-Riemann system in the variables $\lambda'_2, \ldots, \lambda'_v$. We write

$$\hat{l} = \sum_2^v \lambda'_j\, d\lambda'_j.$$

Suppose $k < v$. We consider an arbitrary form $\phi \in \Phi^k_{\mathscr{L}}(U')$, belonging to the kernel of the operator $l'$. The equation $l' \wedge \phi = 0$, obviously, implies that $\hat{l} \wedge \hat{\phi} = 0$. By the induction hypothesis, the form $\hat{\phi}$ can be written as $\hat{\phi} = \hat{l} \wedge \hat{\psi}$, where $\hat{\psi} = \hat{l}^{-1}\, \hat{\phi} \in \Phi^{k-1}_{\hat{\mathscr{L}}}(\hat{U})$, and the operator $\hat{l}^{-1}$ is continuous. We continue the form $\hat{\psi}$ to the whole region $U$, by extending all its coefficients as functions constant with respect to $\lambda'_1$, and we consider the difference $\phi' = \phi - l' \wedge \psi$, where $\psi$ is the just determined extension of $\hat{\psi}$. The restriction of the form $\phi'$ on the subspace $\lambda'_1 = 0$ is equal to zero, and therefore, using Lemma 1, we may write it in the form

$$\phi' = d\lambda'_1 \wedge \psi' + \lambda'_1\, \chi + \lambda'_1\, d\lambda'_1 \wedge \theta, \tag{9.12}$$

where $\psi'$ and $\theta$ belong to $\Phi^{k-1}_{\mathscr{L}}(U')$, and $\chi \in \Phi^k_{\mathscr{L}}(U')$. None of the three forms $\psi'$, $\theta$, and $\chi$ contain $d\lambda'_1$, and the form $\psi'$ does not depend on $\lambda'_1$. Starting from the fact that division by $\lambda'_1$ in $\Phi_{\mathscr{L}}(U')$ is a single-valued operation, by virtue of Lemma 1, it is easy to convince oneself that the representation (9) under the conditions on $\psi'$, $\theta$, and $\chi$ is unique, and the forms $\psi'$, $\theta$, and $\chi$ depend continuously on $\phi$.

The equation $l' \wedge \phi' = 0$ implies that

$$d\lambda'_1 \wedge [\lambda'^2_1\, \chi - \hat{l} \wedge \psi' - \lambda'_1\, \hat{l} \wedge 0] + \lambda'_1\, \hat{l} \wedge \chi = 0.$$

Since the forms $\lambda'_1\, \hat{l} \wedge \chi$ and $\lambda'^2_1\, \chi - \hat{l} \wedge \psi' - \lambda'_1\, \hat{l} \wedge \theta$ do not contain $d\lambda'_1$, we have

$$\hat{l} \wedge \psi' = \lambda'^2_1\, \chi - \lambda'_1\, \hat{l} \wedge \theta. \tag{10.12}$$

Since the form $\psi'$ does not depend on $\lambda'_1$, we obtain $\hat{l} \wedge \psi' = 0$. Now using the induction hypothesis, we find that $\psi' = \hat{l} \wedge \eta$, where $\eta = \hat{l}^{-1}\, \psi'$ is a form belonging to $\Phi^{k-2}_{\mathscr{L}}(U')$ and independent of $\lambda'_1$. On the other hand, from Eq. (10), again, we find that $\lambda'_1\, \chi = \hat{l} \wedge \theta$, since $\lambda'_1$ is not a divisor of zero in $\Phi_{\mathscr{L}}(U')$. Substituting this expression in (9), we obtain

$$\phi' = d\lambda'_1 \wedge \hat{l} \wedge \eta + \hat{l} \wedge \theta + \lambda'_1\, d\lambda'_1 \wedge \theta = l' \wedge [\theta - d\lambda_1 \wedge \eta].$$

The mapping

$$Z^k(U') \ni \phi \to \theta - d\lambda'_1 \wedge \eta + \psi \in \Phi^{k-1}_{\mathscr{L}}(U'),$$

as is obvious from its construction, is continuous and is the inverse of $l'$.

To complete the induction, we must construct the operator (8) in the case $k = \nu$. Let the form $\phi^* = \phi\, d\lambda'_1 \wedge \cdots \wedge d\lambda'_\nu$ belong to the kernel of the operator $\Lambda'$, that is, $\phi(\zeta, 0) \equiv 0$. We consider the form

$$\hat{\phi} = \phi(\zeta; 0, \lambda'_2, \ldots, \lambda'_\nu)\, d\lambda'_2 \wedge \cdots \wedge d\lambda'_\nu.$$

Since the function $\phi(\zeta; 0, \lambda'_2, \ldots, \lambda'_\nu)$ vanishes on the subspace $\lambda'_2 = \cdots = \lambda'_\nu = 0$, the induction hypothesis implies that $\hat{\phi} = \hat{l} \wedge \psi$, where $\psi \in \Phi^{\nu-2}_{\mathscr{L}}(U')$, and the mapping $\phi^* \to \psi$ is a continuous operator. We consider the difference

$$\phi' = \phi^* - d\lambda'_1 \wedge \hat{l} \wedge \psi = [\phi(\zeta, \lambda'_1, \ldots, \lambda'_\nu) - \phi(\zeta; 0, \lambda'_2, \ldots, \lambda'_\nu)]\, d\lambda'_1 \wedge \cdots \wedge d\lambda'_\nu.$$

Since the function $\phi(\zeta, \lambda'_1, \ldots, \lambda'_\nu) - \phi(\zeta; 0, \lambda'_2, \ldots, \lambda'_\nu)$ vanishes on $\lambda'_1 = 0$, it follows from Lemma 1 that the form $\chi = \dfrac{\phi'}{\lambda'_1\, d\lambda'_1}$ belongs to $\Phi^{\nu-1}_{\mathscr{L}}(U')$ and depends continuously on $\phi'$. Setting $\psi' = \chi - d\lambda'_1 \wedge \psi$, we obtain

$$\phi^* = \phi' + d\lambda'_1 \wedge \hat{l} \wedge \psi = \lambda'_1\, d\lambda'_1 \wedge \chi - \hat{l} \wedge d\lambda'_1 \wedge \psi = l' \wedge \psi'.$$

The operator $\phi^* \to \psi'$ is clearly continuous, and, as we have shown, it is inverse to $l'$. This completes the induction, and concludes the proof of the lemma. $\square$

The result of Lemma 2 can be put into somewhat different form. Every differential form in the space $\Phi^k_{\mathscr{L}}(U)$ will be written as a column whose elements are the coefficients of the form. This sets up an isomorphism $\Phi^k_{\mathscr{L}}(U) \cong [\Phi_{\mathscr{L}}(U)]^{\binom{\nu}{k}}$. After this transformation the operator $l\colon \Phi^k_{\mathscr{L}}(U) \to \Phi^{k+1}_{\mathscr{L}}(U)$ is transformed into the operation of multiplication by some matrix $l_k$ whose elements are functions having the form $\pm[\lambda - \lambda_j(\zeta)]$ or are zeroes. Then the sequence (5) can be rewritten as

$$\begin{aligned} 0 \to \Phi_{\mathscr{L}}(U) &\xrightarrow{l_0} [\Phi_{\mathscr{L}}(U)]^\nu \xrightarrow{l_1} \cdots \\ &\cdots \to [\Phi_{\mathscr{L}}(U)]^\nu \xrightarrow{l_{\nu-1}} \Phi_{\mathscr{L}}(U) \xrightarrow{\Lambda^*} \Phi(V) \to 0. \end{aligned} \tag{11.12}$$

In particular, if we make the substitution $\Phi(U) = \mathscr{H}(U)$, we obtain the exact sequence

$$\begin{aligned} 0 \to \mathscr{H}(U) &\xrightarrow{l_0} [\mathscr{H}(U)]^\nu \xrightarrow{l_1} \cdots \\ &\cdots \to [\mathscr{H}(U)]^\nu \xrightarrow{l_{\nu-1}} \mathscr{H}(U) \xrightarrow{\Lambda^*} \mathscr{H}(V) \to 0. \end{aligned} \tag{12.12}$$

The defect of the sequences (11) and (12) is that they were constructed differently for different regions $U$. In the following lemma we obtain similar exact sequences that are constructed uniformly for all the $U$.

**Lemma 3.** *Let $\square$ be an arbitrary open cube in $C^{m+\nu}$, defined as the direct product of the open cubes $\Delta\subset C^m$ and $\nabla\subset C^\nu$. There exists a finite sequence $F_1, \ldots, F_k, \ldots$ of matrices holomorphic in $C^{m+\nu}$ such that for any convex region $U\subset\square$ there is defined an exact sequence*

$$0\to\Phi_{\mathscr{L}}(U)\to\Phi_{\mathscr{L}}(U)\xrightarrow{\Lambda^*}\Phi(V)\to 0, \tag{13.12}$$

*provided $U$ either satisfies the conditions of Lemma 2 or does not intersect $\Lambda(\Omega)$. In the latter case we set $\Phi(V)=0$.*

*Proof.* Let $\mathscr{H}$ be a sheaf of germs of holomorphic functions in $C^{m+\nu}$, and let $\mathscr{H}^\Lambda$ be a subsheaf consisting of the germs of functions vanishing on the manifold $\Lambda(\Omega)$. The sheaf $\mathscr{H}^\Lambda$ is a coherent analytic sheaf in $C^{m+\nu}$. Therefore, we may write down a resolution for the restriction $\mathscr{H}^\Lambda_\square$ of this sheaf on $\square$, in the following form:

$$0\to\cdots\to[\mathscr{H}_\square]^{\alpha_k}\xrightarrow{F_k}[\mathscr{H}_\square]^{\alpha_{k-1}}\to\cdots\to[\mathscr{H}_\square]^{\alpha_1}\xrightarrow{F_1}\mathscr{H}^\Lambda_\square\to 0, \tag{14.12}$$

where $\mathscr{H}_\square$ is the restriction of the sheaf $\mathscr{H}$ on $\square$, and $F_1, \ldots, F_k, \ldots$ are matrices consisting of functions analytic in $C^{m+\nu}$ [17].

Since the region $U$ is convex, it is a domain of holomorphy. Therefore, by the theory of coherent sheaves, we can make a transition in the sequence (16) from the sheaves to the spaces of their sections on $U$, without losing exactness [18]. Therefore, the sequence of linear spaces

$$\begin{aligned}0\to\cdots\to[\mathscr{H}(U)]^{\alpha_k}\xrightarrow{F_k}[\mathscr{H}(U)]^{\alpha_{k-1}}\to\cdots\\ \cdots\to[\mathscr{H}(U)]^{\alpha_1}\xrightarrow{F_1}\mathscr{H}(U)\xrightarrow{\Lambda^*}\mathscr{H}(V)\to 0\end{aligned} \tag{15.12}$$

is algebraically exact.

We remark that the space $\mathscr{H}(U)$ is a commutative ring with a unit relative to the operation of multiplication by functions. This ring will be denoted by $A$. In the space $[\mathscr{H}(U)]^k=A^k\, k=1, 2, \ldots$, there is defined a natural structure of an $A$-module. We shall define the structure of an $A$-module in the space $\mathscr{H}(V)$, by prescribing the action of an element $\phi\in A$ via the formula

$$\phi f=\phi(\zeta, \lambda(\zeta))\, f(\zeta), \qquad f\in\mathscr{H}(V).$$

Thus, the terms of the sequences (12), (15) are $A$-modules, and the mappings are $A$-mappings.

17 See, for example, Canning, Rossi [1].
18 See, for example, Canning, Rossi [1].

In a similar fashion, we introduce the structures of $A$-modules in the spaces $[\Phi_{\mathscr{L}}(U)]^k$ and $\Phi(V)$. It is clear that for arbitrary $k$, we have a topological $A$-isomorphism

$$[\mathscr{H}(U)]^k \underset{A}{\otimes} \Phi_{\mathscr{L}}(U) \cong [\Phi_{\mathscr{L}}(U)]^k.$$

The exactness of the sequence (11) in the last term implies that

$$\mathscr{H}(V) \underset{A}{\otimes} \Phi_{\mathscr{L}}(U) \cong \Phi(V).$$

Therefore tensoring the sequences (12) and (15) by the $A$-module $\Phi_{\mathscr{L}}(U)$, we obtain the sequences (11) and (13). Since (12) and (15) are free resolutions of the finite $A$-module $\mathscr{H}(V)$, it follows from Proposition 2, § 3, Chapter I that the exactness of (11) implies the exactness of (13) for any region $U$ satisfying the conditions of Lemma 2. If $U$ does not intersect $\Lambda(\Omega)$ the sequence (15) is a free resolution of the trivial $A$-module $\mathscr{H}(V)=0$. In that case the same proposition implies again the exactness of (13). ▯

**Lemma 4.** *The sequence*

$$0 \to \cdots \to [\Phi_{\mathscr{L}}(\square)]^{\alpha_k} \xrightarrow{F_k} [\Phi_{\mathscr{L}}(\square)]^{\alpha_{k-1}} \to \cdots$$
$$\cdots \to [\Phi_{\mathscr{L}}(\square)]^{\alpha_1} \xrightarrow{F_1} \Phi_{\mathscr{L}}(\square) \xrightarrow{\Lambda^*} \Phi(\omega) \to 0, \qquad \omega = \Lambda^{-1}(\square)$$

*is exact.*

*Proof.* We cover the cube $\square$ with a locally finite covering $U$, consisting of regions of the form $U_\alpha = V_\alpha \times W_\alpha$, where $V_\alpha \subset C^m_\zeta$, and $W_\alpha \subset C^\nu_\lambda$, each of which satisfies the conditions of Lemma 3. We denote by $V$ a covering of the region $\omega$, consisting of regions $\Lambda^{-1}(U_\alpha)$. We consider the following commutative diagram

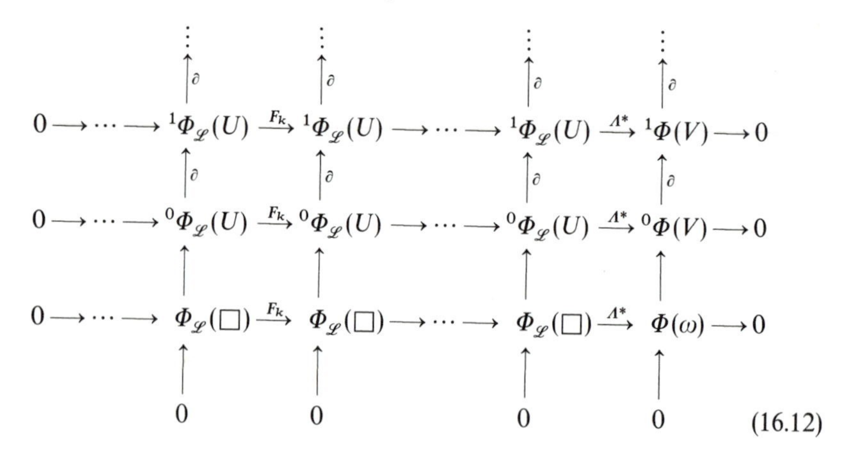

$$\begin{array}{ccccccccccc}
 & & \vdots & & \vdots & & & & \vdots & & \vdots \\
 & & \uparrow\partial & & \uparrow\partial & & & & \uparrow\partial & & \uparrow\partial \\
0 \to \cdots \to & & {}^1\Phi_{\mathscr{L}}(U) & \xrightarrow{F_k} & {}^1\Phi_{\mathscr{L}}(U) & \to \cdots \to & & & {}^1\Phi_{\mathscr{L}}(U) & \xrightarrow{\Lambda^*} & {}^1\Phi(V) \to 0 \\
 & & \uparrow\partial & & \uparrow\partial & & & & \uparrow\partial & & \uparrow\partial \\
0 \to \cdots \to & & {}^0\Phi_{\mathscr{L}}(U) & \xrightarrow{F_k} & {}^0\Phi_{\mathscr{L}}(U) & \to \cdots \to & & & {}^0\Phi_{\mathscr{L}}(U) & \xrightarrow{\Lambda^*} & {}^0\Phi(V) \to 0 \\
 & & \uparrow & & \uparrow & & & & \uparrow & & \uparrow \\
0 \to \cdots \to & & \Phi_{\mathscr{L}}(\square) & \xrightarrow{F_k} & \Phi_{\mathscr{L}}(\square) & \to \cdots \to & & & \Phi_{\mathscr{L}}(\square) & \xrightarrow{\Lambda^*} & \Phi(\omega) \to 0 \\
 & & \uparrow & & \uparrow & & & & \uparrow & & \uparrow \\
 & & 0 & & 0 & & & & 0 & & 0
\end{array} \qquad (16.12)$$

in which we have omitted the square brackets $[\ldots]^{\alpha_k}$. By hypothesis, $\square$ is a convex region, and $U$ is a convex covering of this region. Since the space $\Phi(\square)$ is $\mathscr{L}$-convex in view of the results contained in §8, all the columns of this diagram, with the exception of the last, are exact in accordance with Theorem 1, §9. The exactness of the last column follows from Proposition 9, §2.

By the construction, each of the regions $U_\alpha$ satisfies the conditions of Lemma 3, and therefore, an arbitrary finite intersection $U_{\alpha_0}\cap\cdots\cap U_{\alpha_k}$ of these regions also satisfies the conditions of this lemma. Therefore, Lemma 3 implies that every sequence of the form

$$0\to\cdots\to[\Phi_{\mathscr{L}}(U_{\alpha_0}\cap\cdots\cap U_{\alpha_k})]^{\alpha_1}\xrightarrow{F_1}$$
$$\to\Phi_{\mathscr{L}}(U_{\alpha_0}\cap\cdots\cap U_{\alpha_k})\xrightarrow{\Lambda^*}\Phi(V_{\alpha_0,\ldots,\alpha_k})\to 0,$$
$$V_{\alpha_0,\ldots,\alpha_k}=\Lambda^{-1}(U_{\alpha_0}\cap\cdots\cap U_{\alpha_k})$$

is exact. From this, we conclude that in the diagram (16), all the rows are exact, with the exception of the lowest. Applying Theorem 1, §2, Chapter I, we establish the exactness of the lowest row. $\square$

**3°. Completion of the proof of Theorems 1 and 2.** We shall show that the module $\mathscr{M}$ has a free resolution of the form (4.7), in which $p, p_1, p_2, \ldots$ are $\mathscr{P}''$-matrices. By hypothesis $M\cong\operatorname{Coker}p'$, where $p$ is some $\mathscr{P}''$-matrix. We write down a free resolution for the $\mathscr{P}''$-module

$$M''=\mathscr{P}''^{s}/p'\mathscr{P}''^{t}:\ 0\to\mathscr{P}''^{t_\delta}\xrightarrow{p'_{\delta-1}}\mathscr{P}''^{t_{\delta-1}}\to\cdots\xrightarrow{p'_2}\mathscr{P}''^{t_2}\xrightarrow{p'_1}$$
$$\xrightarrow{p'_1}\mathscr{P}''^{t}\xrightarrow{p'}\mathscr{P}''^{s}\to M''\to 0. \tag{17.12}$$

In this sequence, $p, p_1, p_2, \ldots, p_{\delta-1}$ are by construction $\mathscr{P}''$-matrices. The sequence (17) is now multiplied in tensor fashion by the ring $\mathscr{P}'$ over the field $C$, and we then perform a tensor multiplication by $\mathscr{Q}$ over the same field. Since

$$\mathscr{P}''\underset{C}{\otimes}\mathscr{P}'\underset{C}{\otimes}\mathscr{Q}=\mathscr{R},\qquad M''\underset{C}{\otimes}\mathscr{P}'\underset{C}{\otimes}\mathscr{Q}=\mathscr{M}$$

we obtain the exact sequence:

$$0\to\mathscr{R}^{t_\delta}\xrightarrow{p'_{\delta-1}}\mathscr{R}^{t_{\delta-1}}\to\cdots\xrightarrow{p'_2}\mathscr{R}^{t_2}\xrightarrow{p'_1}\mathscr{R}^{t}\xrightarrow{p'}\mathscr{R}^{s}\to\mathscr{M}\to 0\ [19],$$

which is the desired resolution of the module $\mathscr{M}$. Using this resolution, we construct a diagram (see p. 366) in which we have omitted the brackets $[\ldots]^{\alpha_k t_i}$.

Since all the matrices $F_k$ are holomorphic, and the operators $p, p_1, \ldots$ contain differentiations with respect to the variable $\zeta$ only, we have

19 The sequence is exact since the functor of a tensor product over a field is exact.

for arbitrary $k$ and $j$ the relation $p_j F_k = F_k p_j$. It follows that multiplication by the matrix $F_k$ carries a vector consisting of solutions of the system $p(D_\zeta)u=0$, into a vector consisting of solutions of the same system. Therefore, the mappings $F_k$ in the bottom row of the diagram (18)

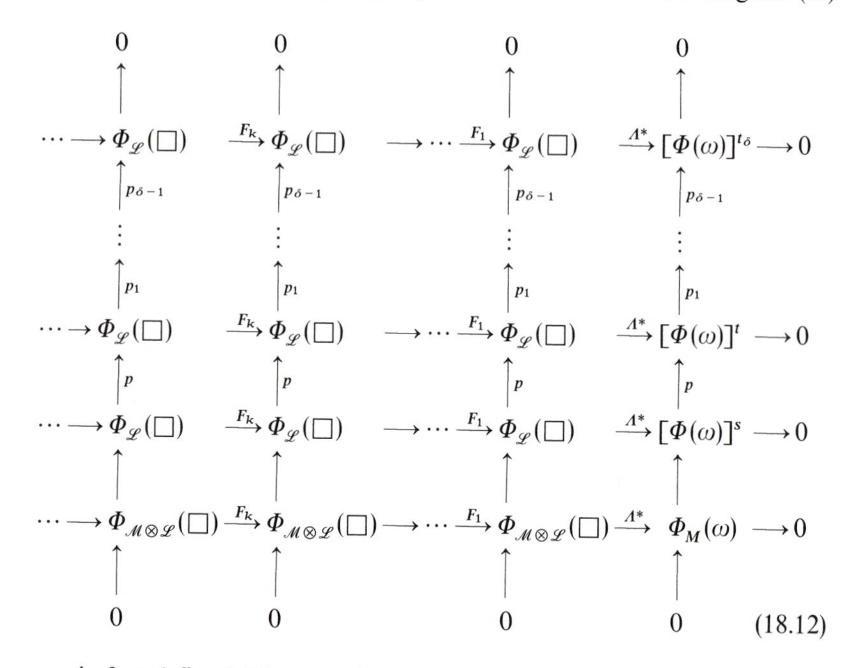

(18.12)

are in fact defined. The equations $p_j F_k = F_k p_j$ imply that the diagram (18) is commutative in all its rows and all its columns except the last.

We shall show that it is commutative in the last column also. For arbitrary $j=1, \ldots, m$ and arbitrary functions $f$, holomorphic in $\square$ with respect to the variables $\lambda$, we have

$$\frac{\partial}{\partial \bar{\zeta}_j} \Lambda^*(f) = \frac{\partial}{\partial \bar{\zeta}_j} f(\zeta, \lambda(\zeta))$$

$$= \left\{ \frac{\partial f}{\partial \bar{\zeta}_j} + \sum_i \frac{\partial f}{\partial \lambda_i} \frac{\partial \lambda_i(\zeta)}{\partial \bar{\zeta}_j} + \sum_i \frac{\partial f}{\partial \bar{\lambda}_j} \frac{\partial \bar{\lambda}_i}{\partial \bar{\zeta}_j} \right\} \Bigg|_{\lambda = \lambda(\zeta)} = \Lambda^* \left( \frac{\partial f}{\partial \bar{\zeta}_j} \right).$$

Repeated application of this equation yields an arbitrary equation $p_j \Lambda^*(f) = \Lambda^*(p_j f)$, $j = 0, 1, 2, \ldots$, and these equations imply that (18) is commutative in the last column. This proves that diagram (18) is commutative as a whole.

Lemma 4 implies the exactness of all the rows of this diagram, except the lowest. We shall prove the exactness of the columns. Since the operators $p(D_\zeta)$ and $\frac{\partial}{\partial\bar{\lambda}_1}, \ldots, \frac{\partial}{\partial\bar{\lambda}_\nu}$, act on different groups of variables, Proposition 5, § 4, Chapter IV implies that $\mathrm{Tor}_i(\mathcal{M}, \mathcal{L})=0$ for all $i \geqq 1$. Since the region $\square$ is convex, Theorem 3, § 9, implies that the space $\Phi_{\mathcal{L}}(\square)$ is $\mathcal{M}$-convex. It follows that all the columns of (18) are exact except the rightmost. The rightmost column is exact in the term $\Phi_M(\omega)$ and is algebraically exact in the term $[\Phi(\omega)]^s$. Applying Theorem 1, § 2, Chapter I, we arrive at the conclusion that the bottom row and the righthand column are exact.

We now prove the first assertion of Theorem 1. Let $\square$ run over all increasing sequences of open cubes tending to $C^{m+\nu}$. Since the mapping $\Lambda$ is proper, the region $\omega=\Lambda^{-1}(\square)$ runs over some increasing sequence of regions with compact closures tending to $\Omega$. The exactness of the right column of diagram (18) implies that for an arbitrary region $\omega$ in this sequence, the space $\Phi(\omega)$ is $M$-convex. In particular, the space $\mathscr{E}(\omega)$ is $M$-convex. The exactness of the bottom row of (18) in the last term implies that every function $u \in \mathscr{E}_M(\omega)$ can be written in the form $\Lambda^*(u^*)$, where $u^* \in \mathscr{E}_{\mathcal{M}\otimes\mathcal{L}}(\square)$. Since the region $\square$ is convex, the function $u^*$ can be approximated by functions $e^* \in E_{\mathcal{M}\otimes\mathcal{L}}$ in the topology of the space $\mathscr{E}_{\mathcal{M}\otimes\mathcal{L}}(\square)$. Since the operator $\Lambda^*$ is continuous, we may use functions of the form $\Lambda^*(e^*)$ to approximate the function $u$ in the topology of $\mathscr{E}_M(\omega)$. Since all functions of the form $\Lambda^*(e^*)$ belong to $\mathscr{E}_M(\Omega)$, we have proved that the space $\mathscr{E}_M(\Omega)$ is dense in $\mathscr{E}_M(\omega)$, and this fact, together with the $M$-convexity of all the spaces $\mathscr{E}(\omega)$ implies that the space $\mathscr{E}(\Omega)$ is also $M$-convex (see Proposition 2, § 7).

To establish the $M$-convexity of $\mathscr{D}^*(\Omega)$, we need the following assertion.

**Corollary 1.** *Let $\kappa$ be some compact belonging to $\omega$. Then for an arbitrary integer $q$, every distribution $u \in \mathscr{D}^*_M(\omega)$ can be written in the form of a sum $u_0+v$, where $u_0 \in \mathscr{D}^*_M(\Omega)$, and $v \in [\mathscr{E}^q_\kappa]^s$, while $\|v\|^q_\kappa \leqq \frac{1}{q}$.*

*Proof.* Let $K \subset \square$ be a convex compact containing the set $\Lambda(\kappa)$. The exactness of the bottom row of the diagram (21) in the last term implies that the function $u$ can be represented in the form $\Lambda^*(u^*)$, where $u^* \in \mathscr{D}^*_{\mathcal{M}\otimes\mathcal{L}}(\square)$. Applying Corollary 2, § 4, to the distribution $u^*$, we write it as a sum $u^*_0+v^*$, where $u^*_0 \in \mathscr{D}^*_{\mathcal{M}\otimes\mathcal{L}}(C^{m+\nu})$ and $v^* \in [\mathscr{E}^q_K]^s$, and $\|v^*\|^q_K \leqq \frac{1}{q}$. We have $u=u_0+v$, where $u_0=\Lambda^*(u^*_0) \in \mathscr{D}^*_{\mathcal{M}}(\Omega)$ and $v=\Lambda^*(v^*) \in [\mathscr{E}^q_\kappa]^s$. Since the mapping $\Lambda^*\colon [\mathscr{E}^q_K]^s \to [\mathscr{E}^q_\kappa]^s$ is continuous, and the norm of the function $v^*$ is arbitrarily small, the norm of $v$ can also be made arbitrarily small. $\square$

We shall prove the $M$-convexity of the space $\mathscr{D}^*(\Omega)$. Since every holomorphic $\mathscr{P}_\zeta$-module has a free resolution of the form (4.7), in which all the $p, p_1, \ldots$ are $\mathscr{P}''$-matrices, it is sufficient to prove that there exists a continuous operator

$$\mathscr{D}^*_{p_1}(\Omega) \to [\mathscr{D}^*(\Omega)]^s / \mathscr{D}^*_p(\Omega), \tag{19.12}$$

which is inverse to $p$. Here, $p$ is an arbitrary $\mathscr{P}''$-matrix and $p_1$ is a $\mathscr{P}''$-matrix such that the sequence (2.7) is exact.

Let $B$ be an arbitrary bounded set in $[\mathscr{D}(\Omega)]^s$. We choose a region $\omega \subset \Omega$, for which the right column of (18) is exact, and which is sufficiently large so that $B$ belongs to the space $[\mathscr{D}(\omega)]^s$ and is bounded in it. The exactness of the right column of (18) implies that we can find a bounded set $B'$ in $[\mathscr{D}(\omega)]^t$ such that for an arbitrary function $w \in \mathscr{D}^*_{p_1}(\omega)$, belonging to the polar of $B'$, we can find a function $u' \in [\mathscr{D}^*(\omega)]^s$, belonging to the polar of $B$, and such that $p\,u' = w$.

We shall now suppose that $w \in \mathscr{D}^*_{p_1}(\Omega)$. Then, relying on the fact that the space $\mathscr{D}^*(\omega')$ is $M$-convex for some sequence of regions $\omega'$, tending to $\Omega$, and relying on Corollary 1, we may apply the arguments of Lemma 1, § 8, to prove that the function $u'$ can be approximated by functions $u \in [\mathscr{D}^*(\Omega)]^s$, satisfying the same system $p\,u = w$. In particular, we may choose a function $u \in [\mathscr{D}^*(\Omega)]^s$, satisfying this system, and not exceeding the value 2, on the set $B$. The correspondence $w \to u$ defines the desired continuous operator (19). Thus the first assertion of Theorem 1 is proved.

We prove the second assertion. Let $z \in C^n$ and $l \in C^{2\nu}$ be variables dual to $\zeta$ and $\lambda$. Suppose, further, that $\{N^\mu_M \subset C^n_z, d^\mu\}$ is a set of algebraic varieties and normal Noetherian operators associated with the module $M$. As we know (see 4°, § 4), the set of algebraic varieties and normal Noetherian operators associated with the module $L$ consists of the manifold $N_L$, defined by the system of equations $l' = -i\,l''$, where $l' = (l_1, \ldots, l_\nu)$, $l'' = (l_{\nu+1}, \ldots, l_{2\nu})$, and $l = (l', l'')$, and the operator $d \equiv 1$. By Proposition 5, § 4, Chapter IV, the variety $N^\mu_M \times N_L \subset C^n \times C^{2\nu}$ and the operators $d^\mu$ form a set, associated with the $\mathscr{R}$-module $\mathscr{M} \otimes \mathscr{L}$. This implies that the linear combinations of exponential polynomials of the form

$$d^\mu(z, -i\,\xi)\exp(z, -i\,\xi)\exp(l, -i\,\lambda), \qquad z \in N^\mu_M,\ l \in N_L,$$

are dense in $\Phi_{\mathscr{M} \otimes \mathscr{L}}(\square)$ (Corollary 1, § 4). Therefore, the exactness of the lowest row of diagram (21) in the last term implies that the linear combinations of functions of the form

$$d^\mu(z, -i\,\xi)\exp(z, -i\,\xi)\exp\big(l, -i\,\lambda(\zeta)\big) \tag{20.12}$$

are dense in $\Phi_M(\omega)$.

We shall suppose that $\Omega$ is a Runge region. Then every function of the form $\exp(l, -i\lambda(\xi))$ can be approximated by polynomials in $\zeta$ in the topology of $\mathscr{H}(\Omega)$. Therefore, functions of the form (20) can be approximated by exponential polynomials of the form

$$d^{\mu}(z, -i\xi)\exp(z, -i\xi)\,h(\zeta)\in E_M, \tag{21.12}$$

where $h$ is a polynomial. It follows that the linear combinations of functions of the form (21) are dense in $\Phi_M(\omega)$. Since $\omega$ can be chosen arbitrarily close to $\Omega$, we have proved that the linear combinations of functions of the form (21) are dense in $\Phi_M(\Omega)$. This completes the proof of Theorem 1. □

We shall now prove Theorem 2. We choose some increasing sequence of open cubes of the form $\square_k = \Delta_k \times V_k$, $k=1, 2, \ldots$, tending to the form $C^{m+\nu}$. All the preceding arguments are applicable to each of the cubes $\square = \square_k$.

Let $v$ be an arbitrary element of $\Phi_M(\Omega)$. What we have already said implies that for arbitrary $k$, it can be written in the form $v = \Lambda^*(u_k)$, where $u_k \in \Phi_{\mathscr{M}\otimes\mathscr{L}}(\square_k)$ is a many-valued, continuous function of $v$. The difference $u_2 - u_1$ belongs to $\Phi_{\mathscr{M}\otimes\mathscr{L}}(\square_1)$ and is annihilated by the operator $\Lambda^*$. Accordingly, the exactness of the bottom row of (21) with $\square = \square_1$ implies that it can be represented in the form

$$u_2 - u_1 = F_1 w_1, \qquad w_1 \in [\Phi_{\mathscr{M}\otimes\mathscr{L}}(\square_1)]^{\alpha_1}.$$

We fix an arbitrary integer $q$ and positive $E$. Since the domain $\square_1$ is convex, it follows from Corollary 2, § 4, that $w_1$ can be written as a sum $w_1' + w_1''$, where $w_1' \in [\Phi_{\mathscr{M}\otimes\mathscr{L}}(C^{m+\nu})]^{\alpha_1}$, and $w_1'' \in [\mathscr{E}^q_{\square_0}]^{\alpha_1 s}$, and $\|w_1''\|^q_{\square_0} < \varepsilon/2$. We set $u_2' = u_2 - F_1 w_1'$. Applying a similar argument to the difference $u_3 - u_2'$, we write it in the form of a sum $F_1 w_2' + F_1 w_2''$, where $w_2' \in [\Phi_{\mathscr{M}\otimes\mathscr{L}}(C^{m+\nu})]^{\alpha_1}$, where $w_2'' \in [\mathscr{E}^q_{\square_1}]^{\alpha_1 s}$ and $\|w_2''\|^q_{\square_1} < \varepsilon/4$. We further write $u_3' = u_3 - F_1 w_2'$ and so on.

Continuing this argument, we construct a sequence of elements $u_k' \in \Phi_{\mathscr{M}\otimes\mathscr{L}}(\square_k)$, $k=1, 2, \ldots$; $u_1' = u_1$ such that for arbitrary $k$

$$u_{k+1}' = u_k' + F_1 w_k'' \quad \text{and} \quad \|w_k''\|^7_{\square_{k-1}} < \frac{\varepsilon}{2^k}.$$

Thus, this sequence has a limit in each of the spaces $\Phi_{\mathscr{M}\otimes\mathscr{L}}(\square_k)$ and, therefore, has a limit $u$ in the space $\Phi_{\mathscr{M}\otimes\mathscr{L}}(C^{m+\nu})$. Since for arbitrary $k$ we have $\Lambda^*(u_k') = v$ in the regions $\Lambda^{-1}(\square_k)$, we also have $\Lambda^*(u) = v$.

In the cube $\square_0$

$$u = u_1 + F_1[w_1'' + w_2'' + \cdots].$$

The distribution in square brackets belongs $[\mathscr{E}^q_{\square_0}]^{\alpha_1 s}$ and does not exceed $\varepsilon$ in the norm $\|\cdot\|^q_{\square_0}$. Since $\varepsilon>0$ and $q$ are arbitrary, it follows that $u$ can be made arbitrarily close to $u_1$ in the topology of the space $\Phi_{\mathscr{M}\otimes\mathscr{L}}(\square_0)$. Since $u_1$ is a continuous function of $v$ and the cube $\square_0$ can be chosen to be arbitrarily large, we conclude that the many-valued mapping

$$\Phi_M(\Omega)\ni v\to u\in\Phi_{\mathscr{M}\otimes\mathscr{L}}(C^{m+\nu})$$

is continuous. Since $\Lambda^*(u)=v$, these mappings are inverses with respect to the operator $\Lambda^*$ in (2). □

Combining this theorem (or the exactness of the lowest row of (18)) with Corollary 1, § 4, we obtain

**Corollary 2.** *Let $\Omega$ be a domain of holomorphy and $K\subset C^{m+\nu}$ a compact. Every distribution $u\in\mathscr{D}^*_M(\Omega)$ can be written in the form*

$$(u,\phi)=\sum_\mu \int_{N^\mu_M\times C^\nu_{l''}} (d^\mu(z,-i\,\xi)\exp[(z,-i\,\xi)+(l'',\lambda(\zeta))],\phi)\,\rho_\mu,$$

*where $\rho_\mu$ are measures concentrated on the varieties $N^\mu_M\times C^\nu_{l''}$, and*

$$\sum_\mu \int (|z|+|l''|+1)^q\,\mathscr{I}_K(y,\operatorname{Re} l'',-\operatorname{Im} l'')\,|\rho_\mu|<\infty$$

*for some integer $q$. If the function $u$ is infinitely differentiable, then $q$ can be made arbitrarily small.*

**Corollary 3.** *If $\Omega$ is a domain of holomorphy, the spaces $\mathscr{E}(\Omega)$ and $\mathscr{D}^*(\Omega)$ are injective, and $\mathscr{E}^*(\Omega)$ and $\mathscr{D}(\Omega)$, endowed with the discrete topology, are flat $\mathscr{P}''$-modules.*

Corollary 3 is proved in a fashion similar to the proof of Corollary 4, § 8.

Chapter VIII

# Overdetermined Systems

The concepts of determined and overdetermined systems will be introduced in § 14. In §§ 14 and 15 we present a series of results connected with these concepts: theorems on the possibility of extensions of solutions of homogeneous systems defined in the neighborhood of the boundary of regions within these regions, a theorem on the continuous dependence of the solutions on their values on portions of the boundary, and theorems on the extension of smoothness, and some theorems on uniqueness.

In §13 we collect the homological results needed for this chapter and the preceding chapter. In particular, Theorem 1, §13, establishes a connection between the dimension of the module $M$ and the vanishing of the module $\operatorname{Ext}^i(M, \mathscr{P})$. This result unifies two different approaches to the study of overdetermined systems: the direct approach which uses the properties of the characteristic variety $N$ (Theorems 1 and 2, §15) and the cohomological approach which uses the modules $\operatorname{Ext}^i(M, \mathscr{P})$ (other theorems in §15).

## §13. Concerning the modules $\operatorname{Ext}^i(M, \mathscr{P})$

The modules $\operatorname{Ext}^i(M, \mathscr{P})$ will play a significant role in the next two sections. In view of this, we now establish some general properties and some ways for computing these modules.

**1°. The modules $\operatorname{Ext}^i(M, \mathscr{P})$ and the dimension of the module $M$.** We shall say that a finite $\mathscr{P}$-module $M$ has at $z \in C^n$, the dimension $d = \dim_z M$, if the highest dimension of an irreducible component of the variety $N(M)$ containing the point $z$ is equal to $d$.

**Theorem 1.** *In order that the dimension of the module $M$ at the point $z$ be less than $n-k$, where $0 \leqq k \leqq n$, it is necessary and sufficient that the point not belong to the varieties*

$$N(\operatorname{Ext}^i(M, \mathscr{P})), \qquad i = 0, \ldots, k.$$

*Proof.* We shall consider the case $k=0$ separately because of its simplicity. As we proved in 4°, §3, Chapter I, the module $\operatorname{Ext}^0(M, \mathscr{P})$ is isomorphic to the kernel $\mathscr{P}_p$ of the mapping $p$: $\mathscr{P}^s \to \mathscr{P}^t$. If $\mathscr{P}_p \neq 0$, then $N(\mathscr{P}_p) = C^n$. Thus, the assertion under proof reduces to the following: to show that the condition $N(M) \neq C^n$ is equivalent to $\mathscr{P}_p = 0$.

On the other hand, as we proved in §1, Chapter IV, the variety $N(M)$ is defined by the condition rank $p(z) < s$. We shall suppose that the module $\mathscr{P}_p$ is not empty. Then there exists a vector $F \in \mathscr{P}^s$, which is not equal to zero, and which is such that $pF = 0$. And therefore, rank $p(z) < s$ identically, that is, $N(M) = C^n$. Conversely, if $N(M) = C^n$, then rank $p(z) < s$ throughout $C^n$. We can then find a vector $F \in \mathscr{P}^s$, distinct from zero, and such that $pF = 0$. This vector can be constructed, for example, as follows: we consider a normal Noetherian operator $d^0$, associated with the $n$-dimensional primary component of the submodule $p' \mathscr{P}^t \subset \mathscr{P}^s$. Since $d^0$ is a normal operator, it contains no differentiations, that is, it is a $\mathscr{P}$-matrix. Let $d_j^0 \neq 0$ be a row of this $\mathscr{P}$-matrix, and let $(d_j^0)'$ be the transposed matrix, that is, a column belonging to $\mathscr{P}^s$. Since $d^0$ is Noetherian, we have $d_j^0 p' = 0$, whence $p(d_j^0)' = 0$, that is, the column $(d_j^0)'$ is the one we were looking for.

Now suppose that $k>0$. We shall establish the sufficiency. Let the dimension of the module $M$ at the point $z$ be $n-h$. We are to show that $h>k$. We begin by assuming the opposite: suppose that $h \leqq k$. We choose a variety $N_0$, from the set associated with $M$ of dimension $n-h$ and containing the point $z$. In $C^n$ we choose a system of coordinates such that the variety $N_0$ is normally placed. Then, by definition, $N_0$ belongs to the variety of the roots of the system of equations

$$q_1(z_1, w) = \cdots = q_h(z_h, w) = 0, \qquad w = (z_{h+1}, \ldots, z_n), \tag{1.13}$$

where $q_i(z_i, w)$ is a polynomial having unity as the coefficient of the highest power of $z_i$. We shall suppose that $\partial q_i/\partial z_i$ does not vanish identically on $N_0$. We denote by $N_*$ the intersection of $N_0$ with the variety of the roots of the polynomials $\partial q_i/\partial z_i$, $i=1, \ldots, h$.

We assume that the point $z$ does not belong to $N_*$. Let $\mathcal{O} = \mathcal{O}_z$ be a ring of functions holomorphic in the neighborhood of $z$. We prove that for arbitrary $j=2, \ldots, h$, we have

$$q_j \mathcal{O} \cap (q_1 \mathcal{O} + \cdots + q_{j-1} \mathcal{O}) = q_j (q_1 \mathcal{O} + \cdots + q_{j-1} \mathcal{O}). \tag{2.13}$$

Since by hypothesis, the derivatives $\partial q_i/\partial z_i$ are different from zero at $z$, the change of coordinates in the neighborhood of $z$ defined by the formula

$$z_j \to z_j' = q_j(z_j, w), \qquad j=1, \ldots, h; \; w \to w,$$

is regular. It follows that the ideal $q_1 \mathcal{O} + \cdots + q_{j-1} \mathcal{O}$ in the ring $\mathcal{O}$ is the set of functions vanishing for $z_1' = \cdots = z_{j-1}' = 0$. Therefore, if a function of the form $z_j' \phi$, where $\phi \in \mathcal{O}$, belongs to this ideal, $\phi$ belongs to it also. This implies equation (2).

We consider the $\mathcal{O}$-modules $\mathcal{L}_j = \mathcal{O}/q_i \mathcal{O}, j=1, \ldots, h$. The cohomological dimension of each of these is equal to one. Therefore, we derive from (2), by virtue of Proposition 3, §3, Chapter I, the equation

$$\mathrm{Tor}_1^{\mathcal{O}} \left( \mathcal{L}_j, \bigotimes_1^{j-1} \mathcal{L}_i \right) = 0.$$

Using this equation and the remark of §9, we conclude by induction on $j$ that the cohomological dimension of the module

$$\bigotimes_1^{j} \mathcal{L}_1$$

does not exceed $j$. In particular, the cohomological dimension of the module

$$\mathcal{L} = \bigotimes_1^{h} \mathcal{L}_i = \mathcal{O}/(q_1 \mathcal{O} + \cdots + q_h \mathcal{O})$$

does not exceed $h$. Suppose that

$$0 \to \mathcal{O}^{\tau_h} \to \mathcal{O}^{\tau_{h-1}} \to \cdots \to \mathcal{O}^{\tau_1} \to \mathcal{O}^{\tau_0} \to \mathscr{L} \to 0 \tag{3.13}$$

is a free resolution of this $\mathcal{O}$-module of length $h$.

We consider the commutative diagram

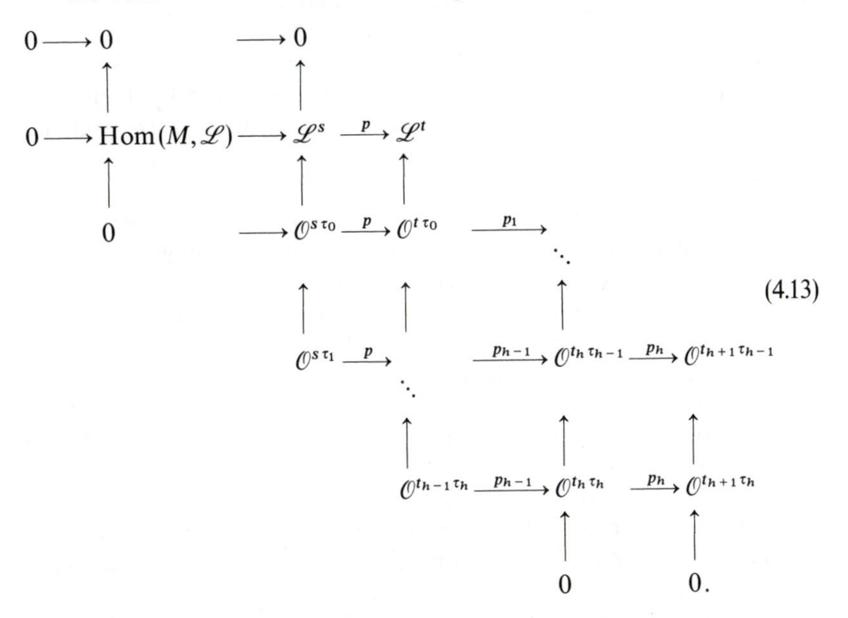

(4.13)

The exactness of (3) implies the exactness of all the columns in the diagram beginning with the second. As for the rows, since $\mathcal{O}$ is a flat $\mathscr{P}$-module, we may write the chain of isomorphisms

$$\operatorname{Ext}^i(M,\mathscr{P}) \otimes \mathcal{O} = \mathscr{P}_{p_i}/p_{i-1}\mathscr{P}^{t_{i-1}} \otimes \mathcal{O} \cong \mathcal{O}_{p_i}/p_{i-1}\mathcal{O}^{t_{i-1}} = \operatorname{Ext}^i(M,\mathcal{O}).$$

Comparing the left and right sides, we obtain

$$\operatorname{Ext}^i(M,\mathscr{P}) \otimes \mathcal{O} \cong \operatorname{Ext}^i(M,\mathcal{O}). \tag{5.13}$$

Since by hypothesis $z$ does not belong to $N(\operatorname{Ext}^i(M, \mathscr{P}))$ for $i=0, \dots, k$, the left side is equal to zero for $i=0, \dots, k$. Hence, we conclude that for these values of $i$, the right side is equal to zero, whence we derive the exactness of the sequence

$$0 \to \mathcal{O}^s \xrightarrow{p} \mathcal{O}^t \xrightarrow{p_1} \cdots \to \mathcal{O}^{t_{k-1}} \xrightarrow{p_{k-1}} \mathcal{O}^{t_k} \xrightarrow{p_k} \mathcal{O}^{t_{k+1}}. \tag{6.13}$$

Since, by hypothesis, $h \leqq k$, we conclude that all the rows of (4) except the uppermost are exact. The uppermost row is exact in view of the relation (19.3), Chapter I. Applying Theorem 1, § 2, Chapter I, we arrive at the conclusion that the leftmost column is exact, that is

$$\operatorname{Hom}(M, \mathscr{L})=0. \tag{7.13}$$

We shall now show that this equation, in fact, does not hold, which leads to a contradiction with the postulate that $h \leqq k$. Let $d_0$ be a normal Noetherian operator belonging to the set associated with the submodule $p' \mathscr{P}^t \subset \mathscr{P}^s$, corresponding to the previously fixed variety $N_0$. We shall assume that it is constructed by the method given in § 3, Chapter IV. Then its rows are obtained from the coefficients of the corresponding $p'$-operator by multiplication with some polynomial. As we know, the null coefficient of a $p'$-operator is a functional of zero order. Therefore, the corresponding row $\partial$ of the matrix $d_0$ is a differential operator of order zero, and therefore, the transposed column $\partial'$ belongs to $\mathscr{P}^s$.

In accordance with the property of Noetherian operators, we have $d_0 p'=0$ on $N_0$, whence $p \partial'|_{N_0}=0$. By hypothesis, the variety of roots of the system (1) is regular in the neighborhood of $z$. Since it has dimension $n-h$ and contains the $n-h$-dimensional irreducible variety $N_0$, it coincides with $N_0$ in the neighborhood of $z$. Thus, the vector $p \partial'$ is equal to zero on the variety (1) in the neighborhood of $z$, whence $p \partial' \in q_1 \mathscr{O}+\cdots+q_h \mathscr{O}$. Accordingly, the vector $\partial'$ belongs to the kernel of the mapping $p: \mathscr{L}^s \to \mathscr{L}^t$, but is not equal to zero as an element of $\mathscr{L}^s$, since the zero coefficient of a $p'$-operator does not vanish in $N(M)$. Thus, $\partial'$ is a non-zero element of the module $\operatorname{Hom}(M, \mathscr{L})$, which contradicts (7), and therefore, $h>k$.

To complete the proof of sufficiency, there remains to consider the case when $z \in N_*$. Since the variety $N_*$ belongs to the irreducible variety $N_0$, but does not coincide with it, it is nowhere dense in $N_0$. Therefore, an arbitrary neighborhood of the point $z$ intersects $N_0 \setminus N_*$. On the other hand, by hypothesis, the point $z$ does not belong to the varieties $N(\operatorname{Ext}^i(M, \mathscr{P}))$, $i=0, \ldots, k$, and, accordingly, a sufficiently small neighborhood $U$ does not intersect them. Therefore, we can apply to an arbitrary point $\zeta \in U \cap (N_0 \setminus N_*)$ the arguments set forth above, which lead to the contradiction with the assumption $h \leqq k$. This completes the sufficiency proof.

We shall now establish the necessity, that is, we shall show that $\dim_z M<n-k$ implies that $z$ does not belong to the varieties

$$N(\operatorname{Ext}^i(M, \mathscr{P})), \qquad i=0, \ldots, k.$$

The latter condition, by Proposition 2, § 1, Chapter IV and by (5) is equivalent to the relationships

$$\operatorname{Ext}^i(M, \mathscr{O})=0, \qquad i=0, \ldots, k. \tag{8.13}$$

Let us consider the $\mathcal{O}$-module $\mathscr{M} = M \otimes \mathcal{O}$ and establish the isomorphism

$$\mathrm{Ext}^i_{\mathscr{P}}(M, \mathcal{O}) \cong \mathrm{Ext}^i_{\mathcal{O}}(\mathscr{M}, \mathcal{O}). \tag{9.13}$$

Since the $\mathscr{P}$-module $\mathcal{O}$ is flat, if we perform a tensor multiplication of the resolution (4.7) by this module, we obtain an exact sequence which is a free resolution of the $\mathcal{O}$-module $\mathscr{M}$. It follows that the factor-module $\mathcal{O}_{p_i}/p_{i-1}\,\mathcal{O}^{t_{i-1}}$, corresponding to the $i$-th term of the sequence (6) coincides with the left and right sides of (9). This establishes (9) as an isomorphism.

We now prove (8). Suppose to begin with that $k=n$. We have $\dim_z M = -1$, that is, $z$ does not belong to $N(M)$. Then by Proposition 2, §1, Chapter IV, $\mathscr{M}=0$, whence $\mathrm{Ext}^i_{\mathcal{O}}(\mathscr{M}, \mathcal{O})=0$ for all $i \geqq 0$. Then, using the isomorphism (9), we obtain (8).

For convenience we shall suppose that $z$ is the origin of coordinates. We now set $k=n-1$. Then $\dim_0 M \leqq 0$, and therefore, the dimension of $\mathscr{M}$ as a linear space over the field $C$ is finite. We prove (8) by induction on this dimension, which we shall denote by $l_0(M)$. We remark that when $l_0(M)=0$, we have $\dim_0 M = -1$, and therefore, as we have just proved, relation (8) is valid. We shall establish it for a special module $M$. This module will have the form $\mathscr{P}/\mathscr{I}$, where $\mathscr{I}$ is the ideal in $\mathscr{P}$ having $z_1, \ldots, z_n$ as basis. This basis satisfies the conditions of Proposition 2 of this section and therefore, $\mathrm{Ext}^i(M, \mathscr{P})=0$, $i=0, \ldots, n-1$. These imply relation (8) for the module $\mathscr{P}/\mathscr{I}$.

Now let $M$ be an arbitrary $\mathscr{P}$-module with $\dim_0 M=0$ and $l_0(M) \geqq 1$. Since the origin of coordinates belongs to $N(M)$, we can find an element $F \in \mathrm{M}$, different from zero, and such that $f\,F=0$ for an arbitrary polynomial $f \in \mathscr{I}$. Let $M_0$ be a submodule of $M$, consisting of elements of the form $f\,F$, $f \in \mathscr{P}$. The mapping $f \to f\,F$ defines an isomorphism $M_0 \cong \mathscr{P}/\mathscr{I}$, and therefore, (8) is proved for the module $M_0$.

Let us consider the exact sequence

$$0 \to M_0 \otimes \mathcal{O} \to M \otimes \mathcal{O} \to M/M_0 \otimes \mathcal{O} \to 0.$$

Its terms are finite-dimensional spaces over the field $C$, and therefore the dimension of the middle term is equal to the sum of the dimensions of the surrounding terms, that is, $l_0(M) = l_0(M_0) + l_0'(M/M_0)$. Since $M_0 \cong \mathscr{P}/\mathscr{I}$, we have $l_0(M_0) = l_0(\mathscr{P}/\mathscr{I}) = 1$, whence $l_0(M/M_0) = l_0(M) - 1$. Therefore (8) is proved for the modules $M/M_0$ by the induction hypothesis. Thus, in the exact sequence

$$\cdots \to \mathrm{Ext}^i(M/M_0, \mathcal{O}) \to \mathrm{Ext}^i(M, \mathcal{O}) \to \mathrm{Ext}^i(M_0, \mathcal{O}) \to \cdots$$

we have $\mathrm{Ext}^i(M_0, \mathcal{O}) = \mathrm{Ext}^i(M/M_0, \mathcal{O}) = 0$ for all $i \leqq n-1$. From this we obtain $\mathrm{Ext}^i(M, \mathcal{O})=0$ for all $i \leqq n-1$. This proves (8) for the case $k=n-1$.

In the general case, the proof of (8) will be carried out by induction on the number $\dim_0 M$, on the supposition that $\dim_0 M>0$ (if $\dim_0 M\leqq 0$, relation (8) has already been proved). Suppose that

$$\mathfrak{p}=\mathfrak{p}_0\cap\cdots\cap\mathfrak{p}_l \tag{10.13}$$

is a reduced primary decomposition of the submodule $\mathfrak{p}=p'\mathscr{P}^t\subset\mathscr{P}^s$. We denote by $\mathfrak{p}^0$ the intersection of all the zero dimensional modules $\mathfrak{p}_\lambda$, and by $\mathfrak{p}'$ the intersection of the remaining modules $\mathfrak{p}_\lambda$. We set $M'=\mathscr{P}^s/\mathfrak{p}'$ and we consider the exact sequence

$$0\to\mathfrak{p}'/\mathfrak{p}\xrightarrow{i}M\to M'\to 0,$$

in which $i$ is an imbedding. This gives rise to the exact sequence for the functor Ext

$$\begin{aligned}\cdots\to\operatorname{Ext}^{i-1}(\mathfrak{p}'/\mathfrak{p},\mathcal{O})\to\operatorname{Ext}^i(M',\mathcal{O})\to&\\ \to\operatorname{Ext}^i(M,\mathcal{O})\to\operatorname{Ext}^i(\mathfrak{p}'/\mathfrak{p},\mathcal{O})\to\cdots&\end{aligned} \tag{11.13}$$

The self-evident inclusion $\mathfrak{r}(\mathfrak{p}'/\mathfrak{p})\supset\mathfrak{r}(\mathfrak{p}^0)$ implies that the module $\mathfrak{p}'/\mathfrak{p}$ is zero-dimensional[1]. Relation (8) for this module and the fact that (11) is exact implies the isomorphism $\operatorname{Ext}^i(M',\mathcal{O})\cong\operatorname{Ext}^i(M,\mathcal{O})$, $i\leqq n-1$. Thus to verify relation (8) for the module $M$, it is sufficient to verify it for the module $M'$. Our construction implies that in the reduced primary decomposition of the submodule $\mathfrak{p}'\subset\mathscr{P}^s$ there are no zero-dimensional components and, therefore, without loss of generality, we may suppose that in the initial decomposition (10), there are no zero-dimensional components.

In $C^n$ we choose a system of coordinates, such that all the varieties $N(\mathfrak{p}_\lambda)$ will be normally placed. Since the dimension of each of them is greater than zero, their intersection with an arbitrary subspace of the form $z_n=\zeta$ is a variety of dimension one less. Therefore, it follows from the proof of Theorem 4, § 9, that

$$\operatorname{Tor}_i(M,L_\zeta)\equiv 0,\quad i\geqq 1,\qquad\text{where}\qquad L_\zeta=\mathscr{P}/(z_n-\zeta)\mathscr{P}. \tag{12.13}$$

Let $\mathscr{P}'$ be a ring of polynomials and $\mathcal{O}'$ be a ring of functions in the variables $z_1,\dots,z_{n-1}$ holomorphic at zero, with the structure of the $\mathscr{P}'$-modules. We consider the $\mathscr{P}'$-module $M_\zeta=M\otimes L_\zeta$. Lemma 2, § 1, Chapter IV, implies that $N(M_\zeta)$ is equal to the intersection of $N(M)$ with the manifold $z_n=\zeta$. It follows that $\dim_0 M_0=\dim_0 M-1$. Therefore, by the induction hypothesis

$$\operatorname{Ext}^i(M_0,\mathcal{O}')=0,\qquad i\leqq k. \tag{13.13}$$

1 Here we shall suppose that $\mathfrak{p}'\neq\mathfrak{p}$.

Here and later for brevity we shall write $\mathrm{Ext}^i(M_\zeta, \mathcal{O}')$ instead of $\mathrm{Ext}^i_{\mathscr{P}'}(M_\zeta, \mathcal{O}')$.

From the exactness of (4.7) and from (12) we infer the exactness of the sequence

$$\cdots \xrightarrow{p'_{1,\zeta}} \mathscr{P}'^{t} \xrightarrow{p'_\zeta} \mathscr{P}'^{s} \to M_\zeta \to 0,$$

where $p_\zeta, p_{1,\zeta}, \ldots$ are restrictions of the matrices $p, p_1, \ldots$ on the subspace $z_n = \zeta$. Thus, we have obtained a free resolution for the $\mathscr{P}'$-module $M_\zeta$. Therefore, the relation (13) implies the exactness of the sequence

$$\mathcal{O}'^{t_{i-1}} \xrightarrow{p_{i-1,0}} \mathcal{O}'^{t_i} \xrightarrow{p_{i,0}} \mathcal{O}'^{t_{i+1}}, \qquad i=0, \ldots, k \ (p_0 = p,\ p_{-1} = 0). \quad (14.13)$$

We have now to infer the exactness of the sequence

$$\mathcal{O}^{t_{i-1}} \xrightarrow{p_{i-1}} \mathcal{O}^{t_i} \xrightarrow{p_i} \mathcal{O}^{t_{i+1}}, \qquad i=0, \ldots, k. \quad (15.13)$$

Let $F$ be an arbitrary element of the kernel of the mapping $p_i$ in (15). The exactness of (14) implies that $F|_{z_n=0} = p_{i-1}\, g$, where $g \in \mathcal{O}'^{t_{i-1}}$. We look on the function $g$ as a function of all of the arguments $z$, and as a constant with respect to $z_n$. The difference $F - p_{i-1}\, g$ vanishes on $z_n = 0$, and therefore can be written in the form $z_n F'$, where $F' \in \mathscr{C}^{t_i}$, that is, $F \in p_{i-1}\, \mathcal{O}^{t_{i-1}} + z_n\, \mathcal{O}^{t_i}$. The equation $p_i\, z_n\, F' = 0$ implies that $p_i\, F' = 0$, and therefore the function $F'$ can be written in the form $p_{i-1}\, g' + z_n\, F''$, where $g \in \mathcal{O}^{t_{i-1}}$, $F'' \in \mathcal{O}^{t_i}$. Therefore, $F \in p_{i-1}\, \mathcal{O}^{t_{i-1}} + z_n^2\, \mathcal{O}^{t_i}$ and so on. We have thus proved that $F \in p_{i-1}\, \mathcal{O}^{t_{i-1}} + z_n^k\, \mathcal{O}^{t_i}$ for arbitrary $k = 1, 2, \ldots$. It follows that the $p_{i-1}$-operator $\mathscr{D}$ vanishes on $F$, and therefore $F \in p_{i-1}\, \mathcal{O}^{t_{i-1}}$. □

**2°. Corollaries**

**Corollary 1.** *In order that* $\dim M < n-k$, *where* $0 \leqq k \leqq n$, *it is necessary and sufficient that*

$$\mathrm{Ext}^i(M, \mathscr{P}) = 0, \qquad i = 0, \ldots, k.$$

*Proof.* The finite $\mathscr{P}$-module $L$ is equal to zero if and only if $N(L) = \varnothing$. On the other hand, for the dimension of an algebraic variety $N$ to be less than $n-k$, it is necessary and sufficient that at every point $z$, $\dim_z N < n-k$. Combining these remarks with Theorem 1, we obtain the desired results. □

**Corollary 2.** *For an arbitrary finite* $\mathscr{P}$-*module* $M$

$$N(M) = \bigcup_{i=0}^{n} N\big(\mathrm{Ext}^i(M, \mathscr{P})\big).$$

For the proof it is sufficient to set $k = n$ in the Theorem. □

**3°. The dimension of the modules $\mathrm{Ext}^i(M, \mathscr{P})$**

**Theorem 2.** *For an arbitrary finite* $\mathscr{P}$-*module* $M$

$$\dim \mathrm{Ext}^i(M, \mathscr{P}) \leqq n-i, \qquad i = 0, 1, 2, \ldots.$$

*Proof.* We return to the notation used in the proof of Theorem 1. We shall represent the module $M$ in the form $\mathscr{P}^s/\mathfrak{p}$, and we write the reduced primary decomposition (10) of the submodules $\mathfrak{p} \subset \mathscr{P}^s$ and we consider the module $\mathfrak{p}'$, which is equal to the intersection of all the non-zero-dimensional components of the decomposition (10). Setting $M' = \mathscr{P}^s/\mathfrak{p}'$, we replace $\mathscr{O}$ by $\mathscr{P}$ in the sequence (11) and we obtain:

$$\begin{aligned} \cdots \to \operatorname{Ext}^{i-1}(\mathfrak{p}'/\mathfrak{p}, \mathscr{P}) \to \operatorname{Ext}^i(M', \mathscr{P}) \to \operatorname{Ext}^i(M, \mathscr{P}) \\ \to \operatorname{Ext}^i(\mathfrak{p}'/\mathfrak{p}, \mathscr{P}) \to \cdots. \end{aligned} \tag{16.13}$$

Since the module $\mathfrak{p}'/\mathfrak{p}$ is zero-dimensional, Corollary 1 implies that $\operatorname{Ext}^i(\mathfrak{p}'/\mathfrak{p}, \mathscr{P}) = 0$ for all $i \leqq n-1$. Then (16) implies the isomorphisms

$$\operatorname{Ext}^i(M', \mathscr{P}) \cong \operatorname{Ext}^i(M, \mathscr{P}), \qquad i \leqq n-1,$$

and the exact sequence

$$0 \to \operatorname{Ext}^n(M', \mathscr{P}) \to \operatorname{Ext}^n(M, \mathscr{P}) \to \operatorname{Ext}^n(\mathfrak{p}'/\mathfrak{p}, \mathscr{P}) \to 0.$$

Since this sequence is exact, we have

$$N(\operatorname{Ext}^n(M, \mathscr{P})) = N(\operatorname{Ext}^n(M', \mathscr{P})) \cup N(\operatorname{Ext}^n(\mathfrak{p}'/\mathfrak{p}, \mathscr{P})).$$

By Corollaries 1 and 2, the varieties $N(\operatorname{Ext}^n(\mathfrak{p}'/\mathfrak{p}, \mathscr{P}))$ and $N(\mathfrak{p}'/\mathfrak{p})$ coincide. Since the module $\mathfrak{p}'/\mathfrak{p}$ is zero-dimensional, it follows that $N(\operatorname{Ext}^n(\mathfrak{p}'\mathfrak{p}, \mathscr{P}))$ is a zero-dimensional variety.

All this implies that if the theorem is proved for the module $M'$, it is proved for the module $M$. We may therefore assume that in the decomposition (10) there are no zero-dimensional components.

The proof of the theorem will now be carried out by induction on $n$. Let $\mathscr{O}'$ be the ring of functions in the variables $z_1, \ldots, z_{n-1}$ that are holomorphic in the neighborhood of a point $z' \in C^{n-1}$, and let $\mathscr{O}$ be the ring of functions in the variables $z_1, \ldots, z_n$, holomorphic in the neighborhood of the point $(z', \zeta) \in C^n$. These rings will be looked on as modules over the rings $\mathscr{P}'$ and $\mathscr{P}$, respectively. Suppose that $z'$ does not belong to the variety associated with $\operatorname{Ext}^i(M_\zeta, \mathscr{P}')$. Then $\operatorname{Ext}^i(M_\zeta, \mathscr{O}') = 0$. Since in (10) there are no zero-dimensional components, we infer the exactness of the sequence (15) (see the proof of Theorem 1), which is equivalent to the equation $\operatorname{Ext}^i(M, \mathscr{O}) = 0$. This equation means that the point $(z', \zeta)$ does not belong to the variety associated with $\operatorname{Ext}^i(M, \mathscr{P})$. Thus, we have established the implication

$$z' \,\bar{\in}\, N(\operatorname{Ext}^i(M_\zeta, \mathscr{P}')) \Rightarrow (z', \zeta) \,\bar{\in}\, N(\operatorname{Ext}^i(M, \mathscr{P})),$$

whence

$$N(\operatorname{Ext}^i(M_\zeta, \mathscr{P}')) \supset N(\operatorname{Ext}^i(M, \mathscr{P})) \cap \{z_n = \zeta\}. \tag{17.13}$$

As follows from the induction hypothesis, for arbitrary $\zeta \in C^1$ the dimension of the variety $N(\operatorname{Ext}^i(M_\zeta, \mathscr{P}'))$ does not exceed $n-1-i$. The inclusion (17) implies that the dimension of the variety $N(\operatorname{Ext}^i(M, \mathscr{P}))$ does not exceed $n-i$. $\square$

The result which we shall now establish is a sharpening of Corollary 2.

**Corollary 3.** *For arbitrary $k$, $0 \leqq k \leqq n$, the variety $\bigcup_{i \leqq k} N(\operatorname{Ext}^i(M, \mathscr{P}))$ coincides with the union of all irreducible components of the variety $N(M)$, whose dimension is not less than $n-k$.*

*Proof.* Let $N^\lambda$, $\lambda = 0, \ldots, l$, be the set of varieties associated with $M$. The assertion to be proved is equivalent to the equation

$$\bigcup_{\dim N^\lambda \geqq n-k} N^\lambda = \bigcup_{i \leqq k} N(\operatorname{Ext}^i(M, \mathscr{P})).$$

We shall prove this equation. The inclusion $\subset$ follows from Corollary 2, since the union of all the varieties $N^\lambda$ is $N(M)$, and the varieties $N(\operatorname{Ext}^i(M, \mathscr{P}))$ with $i > k$ has a dimension less than $n-k$.

We shall prove the inclusion $\supset$. Suppose that $M \cong \mathscr{P}^s/\mathfrak{p}$, and (10) is a reduced primary decomposition of $\mathfrak{p}$. By $\check{\mathfrak{p}}$ we denote the intersection of all the primary components of this decomposition having dimension not less than $n-k$, and by $\hat{\mathfrak{p}}$, the intersection of the remaining components. Thus we have the exact sequence

$$0 \to \check{\mathfrak{p}}/\mathfrak{p} \to M \to \check{M} \to 0, \qquad \check{M} = \mathscr{P}^s/\check{\mathfrak{p}},$$

from which we derive the exact sequence

$$\begin{aligned} \cdots \to \operatorname{Ext}^{i-1}(\check{\mathfrak{p}}/\mathfrak{p}, \mathscr{P}) \to \operatorname{Ext}^i(\check{M}, \mathscr{P}) \to \operatorname{Ext}^i(M, \mathscr{P}) \\ \to \operatorname{Ext}^i(\check{\mathfrak{p}}/\mathfrak{p}, \mathscr{P}) \to \cdots. \end{aligned} \tag{18.13}$$

Since $\mathfrak{p} = \check{\mathfrak{p}} \cap \hat{\mathfrak{p}}$, we have $\mathfrak{r}(\check{\mathfrak{p}}/\mathfrak{p}) \supset \mathfrak{r}(\hat{\mathfrak{p}})$, and, therefore, the dimension of the module $\check{\mathfrak{p}}/\mathfrak{p}$ is less than $n-k$. Thus Corollary 1 implies $\operatorname{Ext}^i(\check{\mathfrak{p}}/\mathfrak{p}, \mathscr{P}) = 0$, $i = 0, \ldots, k$. Using this equation in (18), we obtain the isomorphism

$$\operatorname{Ext}^i(\check{M}, \mathscr{P}) \cong \operatorname{Ext}^i(M, \mathscr{P}), \qquad i = 0, \ldots, k.$$

Combining this fact with Corollary 2, we obtain the desired inclusion

$$\bigcup_{\dim N^\lambda \geqq n-k} N^\lambda = N(\check{M}) \supset \bigcup_{i \leqq k} N(\operatorname{Ext}^i(\check{M}, \mathscr{P})) = \bigcup_{i \leqq k} N(\operatorname{Ext}^i(M, \mathscr{P})). \quad \square$$

**4°. The modules $E_i(M)$.** In this subsection we shall establish an effective criterion for deciding whether a given module $M$ can be included in an exact sequence of the form (10.10). This criterion is formulated in terms of the modules $E_i(M)$, which we are about to define. For a given finite $\mathscr{P}$-module $M$, we consider the canonical mapping

$$j_M\colon M \to \operatorname{Hom}(\operatorname{Hom}(M, \mathscr{P}), \mathscr{P})$$

into its "second associated" module. This mapping is constructed as follows: to an element $f \in M$ there corresponds the mapping $\operatorname{Hom}(M, \mathscr{P}) \to \mathscr{P}$, which carries the homomorphism $\phi \in \operatorname{Hom}(M, \mathscr{P})$ into its value $\phi(f)$ at this element. We write

$$E_0(M) = \operatorname{Ker} j_M, \qquad E_1(M) = \operatorname{Coker} j_M$$

and, further,

$$E_i(M) = \operatorname{Ext}^{i-1}(\operatorname{Hom}(M, \mathscr{P}), \mathscr{P}), \qquad i = 2, 3, \ldots.$$

We establish rules for the calculation of these modules. The module $E_0(M)$ admits the following simple description: it is equal to the $n$-dimensional primary component of the null submodule of the module $M$ [2]. The proof of this assertion is left to the reader.

**Proposition 1.** *Suppose that $M \cong \operatorname{Coker} p'$. Then*

$$E_i(M) \cong \operatorname{Ext}^{i+1}(\operatorname{Coker} p, \mathscr{P}), \qquad i = 0, 1, 2, \ldots. \tag{19.13}$$

*Proof.* Throughout this subsection we will use the following abbreviations:

$$H(M) = \operatorname{Hom}(M, \mathscr{P}), \qquad E^i(M) = \operatorname{Ext}^i(M, \mathscr{P}), \qquad i \geqq 0.$$

Suppose that

$$\cdots \xrightarrow{q_3} \mathscr{P}^{s_3} \xrightarrow{q_2} \mathscr{P}^{s_2} \xrightarrow{q_1} \mathscr{P}^{s} \xrightarrow{p} \mathscr{P}^{t} \to \operatorname{Coker} p \to 0$$

is a free resolution of the module $\operatorname{Coker} p$. Since $H(M) \cong \operatorname{Ker}\{p\colon \mathscr{P}^s \to \mathscr{P}^t\}$, the mappings $\ldots q_3, q_2, q_1$ of this sequence form a free resolution of $H(M)$. If we apply to these resolutions the functor $\operatorname{Hom}(\cdot, \mathscr{P})$, we obtain two sequences

$$\begin{gathered} 0 \to H(\operatorname{Coker} p) \to \mathscr{P}^t \xrightarrow{p'} \mathscr{P}^s \xrightarrow{q'_1} \mathscr{P}^{s_2} \xrightarrow{q'_2} \mathscr{P}^{s_3} \xrightarrow{q'_3} \cdots, \\ 0 \to H(H(M)) \xrightarrow{q_1^*} \mathscr{P}^{s_2} \xrightarrow{q'_2} \mathscr{P}^{s_3} \xrightarrow{q'_3} \cdots, \end{gathered} \tag{20.13}$$

which coincide beginning with the term $\mathscr{P}^{s_2}$ ($q_1^*$ is the mapping associated with $q'_1$). From these we derive the isomorphisms $E^{i+2}(\operatorname{Coker} p) \cong E^i(H(M))$, $i \geqq 1$. And this proves (19) for all $i \geqq 2$.

---

2 We have met this module in § 8 as a description of the virtually hypoelliptic operators.

We now consider the diagram

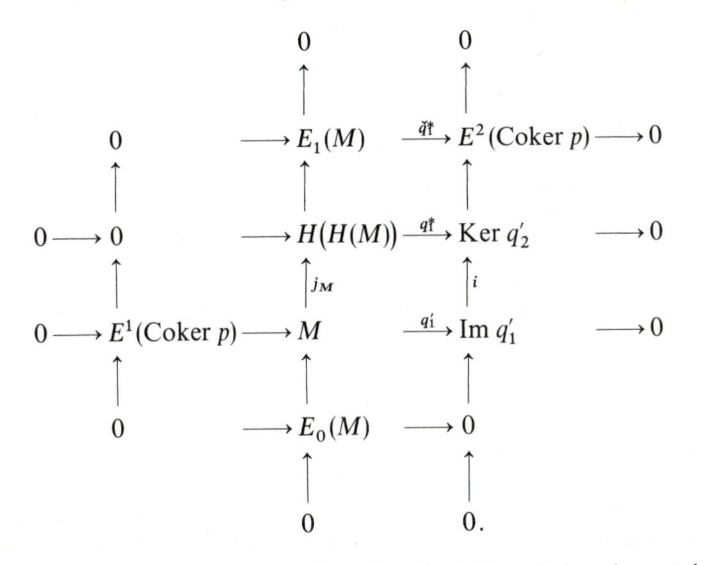

Its central square is commutative, that is, $q_1^* j_M = i q_1'$, where $i$ is an imbedding. Therefore, there is defined a mapping $\check{q}_1^*$ associated to $q_1^*$. Thus, the diagram is commutative and the first and third rows, as well as the second and third columns are exact. Applying Theorem 1, §2, Chapter I twice – first to the lower left corner and then to the upper right corner, we establish the isomorphism $E_0(M) \cong E^1(\mathrm{Coker}\, p)$ and we prove that $\check{q}_1^*$ is also an isomorphism. □

The criterion that we spoke about earlier is formulated as follows.

**Theorem 3.** *In order that the module M be included in an exact sequence of the form*

$$0 \to M \xrightarrow{q_1} \mathscr{P}^{s_2} \xrightarrow{q_2} \mathscr{P}^{s_3} \xrightarrow{q_3} \cdots \xrightarrow{q_{k-1}} \mathscr{P}^{s_k} \xrightarrow{q_k} \mathscr{P}^{s_{k+1}}, \qquad (21.13)$$

*it is necessary and sufficient that*

$$E_i(M) = 0, \qquad i = 0, 1, \ldots, k-1. \qquad (22.13)$$

By Proposition 1, the necessary and sufficient condition can be formulated as: $\mathrm{Ext}^i(\mathrm{Coker}\, p, \mathscr{P}) = 0$, $i = 1, \ldots, k$.

*Proof. Sufficiency.* Suppose that the equations $E_i(M) = 0$, $i = 0, \ldots, k-1$ are satisfied. We shall suppose to begin with that $k = 1$. Then by the definition of $E_0(M)$ the mapping $j_M\colon M \to H(H(M))$ is a monomorphism. The mapping $q_1^*$ in (20) is also a monomorphism (the exactness of (20) in the first two terms follows from the property of the functor

Hom). Therefore, the composition of these mappings $q_1^* j_M \colon M \to \mathscr{P}^{s_2'}$ is also a monomorphism.

Now suppose that $k>1$. Then the sequence (20) is exact in all terms beginning with $\mathscr{P}^{s_k}$ and $H(H(M)) \cong M$, which implies the exactness of (21).

*Necessity*. The case $k=1, 2$. We shall suppose to begin with that in the sequences (21) $k=1$. We write $\Pi = \mathscr{P}^{s_2}$, $L = \operatorname{Coker} q_1$ and we consider the exact sequence

$$0 \to M \xrightarrow{q_1} \Pi \xrightarrow{\alpha} L \to 0,$$

in which $\alpha$ is a canonical mapping. Applying the functor $\operatorname{Hom}(\cdot, \mathscr{P})$, we obtain the exact sequence

$$\begin{aligned} 0 \to H(L) \xrightarrow{\alpha^*} \Pi \xrightarrow{q_1^*} H(M) \to E^1(L) \to 0 \to \cdots \\ \to E^i(L) \to 0 \to E^i(M) \to E^{i+1}(L) \to 0 \to \cdots \end{aligned} \tag{23.13}$$

(where for an arbitrary mapping $\beta$, we write $\beta^* = \operatorname{Hom}(\beta, \mathscr{P})$). Applying the functor $\operatorname{Hom}(\cdot, \mathscr{P})$ again to the three terms of this sequence, we obtain the commutative diagram

$$\begin{array}{ccccccc} 0 \longrightarrow & H(H(M)) & \xrightarrow{q_1^{**}} & \Pi & \xrightarrow{\alpha^{**}} & H(H(L)) & \\ & \uparrow{\scriptstyle j_M} & & \| & & \uparrow{\scriptstyle j_L} & \\ 0 \longrightarrow & M & \xrightarrow{q_1} & \Pi & \xrightarrow{\alpha} & L & \longrightarrow 0. \end{array} \tag{24.13}$$

Its commutativity implies that $q_1 = q_1^{**} j_M$. Since $q_1$ is a monomorphism, $j_M$ is also a monomorphism, that is, $E_0(M)=0$, which is what we were to prove.

We shall now assume that $k=2$. Then by what we have already proved, $j_L$ is a monomorphism. Hence, since the first row of (24) is semi-exact, we have

$$\operatorname{Im} q_1^{**} \subset \operatorname{Ker} \alpha^{**} = \operatorname{Ker} j_L \alpha = \operatorname{Ker} \alpha = \operatorname{Im} q_1,$$

whence $\operatorname{Im} q_1^{**} \subset \operatorname{Im} q_1$. Since $q_1$ establishes the isomorphism $M \cong Im\, q_1$, and $q_1^{**}$ establishes the isomorphism $H(H(M)) \cong \operatorname{Im} q_1^{**}$ (we shall prove this latter assertion later), we conclude that $j_M$ is an isomorphism, that is, $E_1(M)=0$.

We now turn to the sequence (23). We write $Q = \operatorname{Im} q_1^*$. We have two exact sequences

$$0 \to H(L) \xrightarrow{\alpha^*} \Pi \to Q \to 0,$$

$$0 \to Q \to H(M) \to E^1(L) \to 0,$$

and these imply the exactness of two other sequences

$$0 \to H(Q) \to \Pi \xrightarrow{\alpha^{**}} H(H(L)) \to E^1(Q) \to 0 \to \cdots \\ \cdots \to E^i(Q) \to 0 \to E^i(H(L)) \to E^{i+1}(Q) \to 0 \to \cdots, \tag{25.13}$$

$$0 \to H(E^1(L)) \to H(H(M)) \to H(Q) \to E^1(E^1(L)) \to \cdots \\ \cdots \to E^i(E^1(L)) \to E^i(H(M)) \to E^i(Q) \to E^{i+1}(E^1(L)) \to \cdots. \tag{26.13}$$

By Theorem 2, $\dim E^1(L) < n$, and therefore, by Theorem 1, $H(E^1(L)) = 0$. Thus the mapping $H(H(M)) \to H(Q)$ in (26) is a monomorphism. The mapping $H(Q) \to \Pi$ in (25) is also a monomorphism. Thus the composition of these mappings, which is equal to $q_1^{**}$, is also a monomorphism, whence $H(H(M)) \cong \operatorname{Im} q_1^{**}$, which is what we were to prove.

*The general case.* The relation (22) will now be proved, together with the inequality

$$\dim E^i(M) \leqq n - i - k, \qquad i \geqq 1, \tag{27.13}$$

by induction on $k$. For $k=0$, the inequalities follow from Theorem 2. If $k \geqq 1$, we may suppose that for $k-1$ the equations (27) have been proved. Since the module $L$ can be included in an exact sequence of the form (21) of length $k-1$, it follows from our assumption that for arbitrary $i \geqq 1$, we have $\dim E^i(L) \leqq n-i-k+1$. The exactness of (23) implies the isomorphisms $E^i(M) \cong E^{i+1}(L)$, $i \geqq 1$, which together with the inequality for $\dim E^i(L)$ imply (27). This establishes the inequality (27).

We shall now prove (22) by induction on $k$ for $k > 2$. By substituting $L$ in (27) in place of $M$, we obtain in particular $\dim E^1(L) \leqq n-k$, whence by Theorem 1, $E^i(E^1(L)) = 0$, $i = 1, \ldots, k-1$. Taking account of these relations in (26), we obtain the isomorphisms

$$E_i(M) = E^{i-1}(H(M)) \cong E^{i-1}(Q), \qquad i = 2, \ldots, k-1. \tag{28.13}$$

Since by the induction hypothesis $E_i(L) = 0$, $i = 0, \ldots, k-2$, we have $H(H(L)) \cong L$ and, therefore, the mapping $\alpha^{**}$ in (25) coincides with $\alpha$ and is thus an epimorphism. Hence, $E^1(Q) = 0$. On the other hand, if $k > 3$, the exactness of (25) implies that $E^{i+1}(Q) \cong E^i(H(L)) = 0$ for $i = 1, \ldots, k-3$. This proves that the right sides of (28) are equal to zero. This brings us to (22). ☐

**Corollary 4.** *In order that the module $M$ be imbeddable in an exact sequence of the form* (21) *of length $k$, it is sufficient that it be representable in the form $\mathscr{P}^s / p' \mathscr{P}^t$, where the matrix $p$ satisfies the condition*

$$\dim\{z : \operatorname{rang} p(z) < t\} < n - k.$$

*Proof.* The condition just formulated, by Theorem 1, implies the equations $\operatorname{Ext}^i(\operatorname{Coker} p, \mathscr{P}) = 0$, $i = 0, \ldots, k$. Taking account of Proposition 1, we obtain $E_i(M)=0$, $i=0,\dots,k-1$, whence by Theorem 3, we derive the desired results. □

### 5°. Calculation of the modules $\mathrm{Ext}^i(M,\mathscr{P})$ in certain particular cases

**Proposition 2.** *Let $M=\mathrm{Coker}\, p'$, where $p$ is the column $(p_1,\dots,p_t)$, and $\dim M\leqq n-t$. Then*

I. *The module $M$ admits the following free resolution (the complex of Koszul):*

$$0\to\mathscr{P}\xrightarrow{\delta_0(p)}\mathscr{P}^t\xrightarrow{\delta_1(p)}\cdots\to\mathscr{P}^{\binom{t}{k}}\xrightarrow{\delta_k(p)}\mathscr{P}^{\binom{t}{k+1}}\to\cdots$$
$$\cdots\to\mathscr{P}^t\xrightarrow{\delta_{t-1}(p)}\mathscr{P}\to M\to 0, \tag{29.13}$$

*in which the mapping $\delta_k(p)$, $k=0,\dots,t-1$ acts according to the formula*

$$\{f_{i_1,\dots,i_k}\}\mapsto\{g_{i_0,\dots,i_k}=\sum(-1)^j p_{i_j} f_{i_0,\dots,\hat{i}_j,\dots i_k}\}$$

*where the components of a vector of the module $\mathscr{P}^{\binom{t}{k}}$ are indexed by all the sets of integers $i_1,\dots,i_k$ such that $1\leqq i_1<\dots<i_k\leqq t$. In particular, $\delta_0(p)=p$, $\delta_{t-1}(p)=p'$.*

II. *We have the isomorphisms*

$$\mathrm{Ext}^i(M,\mathscr{P})=0,\quad i\neq t,\qquad \mathrm{Ext}^t(M,\mathscr{P})\cong M. \tag{30.13}$$

*Proof.* It is easy to verify that the sequence (29.13) is semi-exact. We denote this complex by $K$ and we write $Z^k=\mathrm{Ker}\,\delta_k(p)$, $B^k=\mathrm{Im}\,\delta_{k-1}(p)$, $0\leqq k<t$, where $B^0=0$ and $B^t=M$. We consider also the complex $K_z=K\otimes\mathcal{O}_z$ for an arbitrary point $z\in C^n$ whose mappings $\Delta_k$, $k=0,\dots,t-1$ acts by the same formulae as the mappings $\delta_k(p)$. We set $Z^k_z=\mathrm{Ker}\,\Delta_k$, $B^k_z=\mathrm{Im}\,\Delta_{k-1}$, $0\leqq k<t$ where $B^0_z=0$. According to the property of tensor multiplication we have $B^k_z\cong B^k\otimes\mathcal{O}_z$, $0\leqq k\leqq t$. Since $\mathcal{O}_z$ is a flat $\mathscr{P}$-module we have also $Z^k_z=Z^k\otimes\mathcal{O}_z$ and

$$Z^k/B^k\otimes\mathcal{O}_z\cong Z^k_z/B^k_z,\qquad 0\leqq k<t. \tag{31.13}$$

Now we verify that the right side is equal to zero when $z\bar{\in}N(M)$. In that case at least one of the components of the column $p$, for example $p_1$, does not vanish at $z$. Let $\phi=\{\phi_{i_1,\dots,i_k}\}$ be an element of $\mathcal{O}_z^{\binom{t}{k}}$ such that $\Delta_k\phi=0$. We set

$$\psi=\{\psi_{i_2,\dots,i_k}\}\in\mathcal{O}_z^{\binom{t}{k-1}},$$

where $\psi_{i_2,\dots,i_k}=p_1^{-1}\phi_{1,i_2,\dots,i_k}$ if $i_2>1$ and $\psi_{i_2,\dots,i_k}=0$ if $i_2=1$. It is easy to verify that $\Delta_{k-1}\psi=\phi$, which proves our assertion. By (31) and Proposition 2, §1, Chapter IV it follows that

$$N(Z^k/B^k)\subset N(M),\qquad 0\leqq k<t. \tag{32.13}$$

Let us prove part I of the proposition. If $t>n$, we have $N(M)=\varnothing$, hence (32) implies that $Z^k=B^k$, $0 \leqq k<t$. Now we suppose that $t=n$. From the exact sequence

$$0 \to Z^k \to \mathscr{P}^{\binom{t}{k}} \to B^{k+1} \to 0 \qquad 0 \leqq k<n$$

we infer the isomorphisms

$$E^{j+1}(Z^k) \cong E^j(B^{k+1}), \qquad j \geqq 1, \tag{33.13}$$

(we use the abbreviation of 4°). In view of (32) $\dim Z^k/B^k \leqq 0$, hence Corollary 1 implies that $E^j(Z^k/B^k)=0$, $j<n$. Therefore we conclude from the exact sequence $0 \to B^k \to Z^k \to Z^k/B^k \to 0$ that

$$\begin{aligned} &E^j(B^k) \cong E^j(Z^k), \qquad j<n, \\ &E^n(Z^k)/E^n(B^k) \cong E^n(Z^k/B^k). \end{aligned} \tag{34.13}$$

Combining this with (33) we find

$$E^n(Z^k) \cong E^{n-1}(B^{k+1}) \cong E^{n-1}(Z^{k+1}) \cong \cdots \cong E^k(M).$$

Corollary 1 implies that the right side is equal to zero if $0 \leqq k<n$. Hence it follows from the second isomorphism (34) that $E^n(Z^k/B^k)=0$. Consequently by Corollary 1 we can conclude that $Z^k=B^k$, $0 \leqq k<n$.

In the case $t<n$ we use an induction on $n-t$. We choose a system of coordinates in $C^n$ such that $N(M)$ is normally placed and we consider the $\mathscr{P}$-module $L_\zeta=\mathscr{P}/(z_n-\zeta)\mathscr{P}$ where $\zeta \in C$. The arguments at the end of 1° show that the complex $K \otimes L_\zeta$ consists of the $\mathscr{P}'$-module $M_\zeta \cong \operatorname{Coker} p'_\zeta$, of the free $\mathscr{P}'$-modules and of the mappings $\delta_k(p_\zeta)$, $k=0, \ldots, t-1$ where $p_\zeta$ is the restriction of the column $p$ on the variety $z_n=\zeta$. Since $\dim M_\zeta \leqq n-1-t$, the induction hypothesis implies that for any $\zeta$ the complex $K \otimes L_\zeta$ is an exact sequence. Repeating the arguments at the end of 1° we infer that the sequence $K$ itself is exact.

Now we prove part II of the proposition. We consider a pairing

$$\mathscr{P}^{\binom{t}{k}} \times \mathscr{P}^{\binom{t}{t-k}} \to \mathscr{P}$$

which acts by the formula

$$(\{f_{i_1, \ldots, i_k}\}, \{g_{j_1, \ldots, j_{t-k}}\}) \mapsto \sum_{i_1, \ldots, i_k, j_1, \ldots, j_{t-k}} \varepsilon f_{i_1, \ldots, i_k} g_{j_1, \ldots, j_{t-k}}$$

where the sum is taken over all the transpositions of the set $(1, 2, \ldots, t)$ and $\varepsilon$ is the sign of the transposition. It is easy to verify that the pairing defines the isomorphisms

$$\operatorname{Hom}(\mathscr{P}^{\binom{t}{k}}, \mathscr{P}) \cong \mathscr{P}^{\binom{t}{t-k}}, \qquad \operatorname{Hom}(\delta_k(p), \mathscr{P}) \cong \pm \delta_{t-k}(p).$$

It follows that, applying the functor $\operatorname{Hom}(\cdot, \mathscr{P})$ to the sequence (29), we obtain the sequence

$$0 \to \operatorname{Hom}(M, \mathscr{P}) \to \mathscr{P} \xrightarrow{\pm\delta_0(p)} \mathscr{P}^t \xrightarrow{\pm\delta_1(p)} \cdots \\ \cdots \to \mathscr{P}^t \xrightarrow{\pm\delta_{t-1}(p)} \mathscr{P} \to 0, \tag{35.13}$$

which, up to its sign, coincides with (30), except for the first and last terms. Therefore, the sequence (35) is exact in all terms except the last, and the kernel of $\delta_0(p)$ is equal to zero. Hence, we derive the isomorphism (31). □

**Proposition 3.** *Let* $M = \operatorname{Coker} p'$, *where* $p$ *is the column* $(p_1, \ldots, p_t)$. *Then*

$$\operatorname{Ext}^1(M, \mathscr{P}) \cong \mathscr{P}/\Delta\mathscr{P},$$

*where* $\Delta$ *is the greatest common divisor of the polynomials* $p_1, \ldots, p_t$.

*Proof.* We recall that

$$\operatorname{Ext}^1(M, \mathscr{P}) \cong \mathscr{P}_{p_1}/p\,\mathscr{P},$$

where the $\mathscr{P}$-matrix $p_1: \mathscr{P}^t \to \mathscr{P}^{t_2}$ is such that $\mathscr{P}_{p'} = p_1' \mathscr{P}^{t_2}$. Let $\check{p}$ be a row consisting of the polynomials $\check{p}_\tau = p_\tau/\Delta$, $\tau = 1, \ldots, t$. Since $\mathscr{P}_{\check{p}'} = \mathscr{P}_{p'}$ the sequence

$$\mathscr{P}^{t_2} \xrightarrow{p_1'} \mathscr{P}^t \xrightarrow{\check{p}'} \mathscr{P} \to \operatorname{Coker} \check{p}' \to 0$$

is exact and therefore can be continued to a free resolution of the module $\operatorname{Coker} \check{p}'$. Since the greatest common divisor of the polynomials $\check{p}_1, \ldots, \check{p}_t$ is a constant, the variety of their common roots (associated with $\operatorname{Coker} \check{p}'$) has dimension less than $n-1$. Therefore, by Theorem 1, we have $\operatorname{Ext}^1(\operatorname{Coker} p', \mathscr{P}) = 0$, which proves the exactness of the sequence

$$\mathscr{P} \xrightarrow{\check{p}} \mathscr{P}^t \xrightarrow{p_1} \mathscr{P}^{t_2}.$$

This in turn implies that $\mathscr{P}_{p_1} = \check{p}\,\mathscr{P}$, and therefore, $\operatorname{Ext}^1(M, \mathscr{P}) \cong \check{p}\,\mathscr{P}/p\,\mathscr{P}$. The mapping $\mathscr{P} \ni f \to p\,f \in \check{p}\,\mathscr{P}$ defines an isomorphism

$$\mathscr{P}/\Delta\mathscr{P} \cong \check{p}\,\mathscr{P}/p\,\mathscr{P}.$$

Since the right side coincides with $\operatorname{Ext}^1(M, \mathscr{P})$ it is the isomorphism we were seeking. □

## § 14. The extension of solutions of homogeneous systems

We now describe all systems of the form (1.4), having the following property: Every solution defined in the neighborhood of the boundary of a convex bounded region can be continued within the region as a solution of the same system. It turns out that such systems are characterized by the condition $\operatorname{Ext}^1(M,\mathscr{P})=0$, where $M$ is the $\mathscr{P}$-module corresponding to the system in question. We shall, moreover, obtain a number of more exact theorems on the possibility of extensions of solutions of the system (1.4).

**1°. Determined and overdetermined systems**

**Definition.** The operator $p$ and the module $M=\mathscr{P}^s/p'\mathscr{P}^t$ will be said to be determined if $\operatorname{Hom}(M,\mathscr{P})=0$, and underdetermined otherwise. We shall say that the operator $p$ and the module $M$ are overdetermined if

$$\operatorname{Hom}(M,\mathscr{P})=\operatorname{Ext}^1(M,\mathscr{P})=0.$$

In view of Corollary 1, § 13, the determined modules $M$ are characterized by the inequality $\dim M<n$, and the overdetermined modules by the inequality $\dim M<n-1$.

As an example, we may consider the operators $d_i$, $i=0,\dots,n-1$, from Example 2, § 7. It is easy to see that all the operators $d_i$ with $i>0$ are underdetermined, and the operator $d_0$ is determined, and when $n>1$ is overdetermined.

We shall now prove that the determined operators are characterized by the fact that the corresponding system (1.4) has no solution with a compact support. We shall further prove that for the overdetermined operators every system (1.4), defined in the neighborhood of the boundary of a convex region, has a unique extension throughout the whole region.

**Proposition 1.** *In order that the system* (1.4) *have no non-zero solution which is a distribution with compact support, it is necessary and sufficient that the operator $p$ be determined.*

*Proof.* By (19.3), Chapter I, the module $\operatorname{Hom}(M,\mathscr{P})$ is isomorphic to the kernel $\mathscr{P}_p$ of the mapping $p\colon \mathscr{P}^s\to\mathscr{P}^t$. We shall assume that the operator $p$ is underdetermined, that is, this kernel is different from zero. Let $F$ be a non-zero element of the module $\mathscr{P}_p$, and let $\delta$ be the delta-function in $R^n$. The distribution $F(iD)\,\delta$ is different from zero, has a compact support, and is a solution of the system (1.4). This proves the necessity.

Let us prove the sufficiency. Since $\operatorname{Hom}(M,\mathscr{P})\cong\mathscr{P}_p$, and the $\mathscr{P}$-module $\mathscr{E}^*(R^n)$ is flat (see Corollary 4, § 8), the sequence

$$0\to\operatorname{Hom}(M,\mathscr{P})\otimes\mathscr{E}^*(R^n)\to[\mathscr{E}^*(R^n)]^s\xrightarrow{\;p\;}[\mathscr{E}^*(R^n)]^t$$

is exact. In this sequence, the kernel of the operator $p$ is the space of all solutions of the system (1.4) having compact supports. If $\operatorname{Hom}(M,\mathscr{P})=0$, this space consists of the single function which is identically zero. □

**2°. Characterization of the irremovable singularities of solutions.** Let $\Omega$ be a region in $R^n$, and $K\subset\Omega$ be a compact. We shall say that the distribution $u$ is a solution of the system (1.4) in $\Omega$, having a singularity on $K$, if $u \in \mathscr{D}_M^*(\Omega \setminus K)$. In the space $\mathscr{D}_M^*(\Omega \setminus K)$ we consider the subspace $\widehat{\mathscr{D}_M^*(\Omega)}$, consisting of limits of sequences of the form $\{u_\alpha \in \mathscr{D}_M^*(\Omega)\}$, which are stable outside an arbitrary neighborhood of $K$. We shall say that the singularity of the function $u \in \mathscr{D}_M^*(\Omega \setminus K)$ is removable, if $u$ belongs to $\widehat{\mathscr{D}_M^*(\Omega)}$. The factor-space $\mathscr{D}_M^*(\Omega \setminus K)/\mathscr{D}_M^*(\Omega)$ characterizes the set of irremovable singularities of the solution of (1.4) in $\Omega$ that belong to $K$.

Let $u$ be an infinitely differentiable solution of (1.4) in $\Omega \setminus K$, i.e., $u \in \mathscr{E}_M(\Omega \setminus K)$. We shall assume that the singularity of $u$ on the compact $K$ is removable, i.e., $u \in \widehat{\mathscr{D}_M^*(\Omega)}$. Then by Example 1, § 15, there exists a sequence of functions $u_\alpha \in \mathscr{E}_M(\Omega)$ tending to $u$, and stable outside any neighborhood of $K$, that is, the function $u$ belongs to the subspace $\mathscr{E}_M(\Omega)$, constructed in the same way as the subspace $\widehat{\mathscr{D}_M^*(\Omega)}$. Thus, the factor-space

$$\mathscr{E}_M(\Omega \setminus K)/\widehat{\mathscr{E}_M(\Omega)}$$

characterizes the irremovable singularities, belonging to $K$, of infinitely differentiable solutions of (1.4).

Both the factor-spaces that we have under consideration will be denoted in the same way:

$$\Phi_M(\Omega \setminus K)/\widehat{\Phi_M(\Omega)}. \tag{1.14}$$

Our immediate aim is to describe them. We first obtain some conditions for an element of $\widehat{\Phi_M(\Omega)}$ to belong to $\Phi_M(\Omega)$.

**Proposition 2.** *If $M$ is a determined module, we have $\widehat{\Phi_M(\Omega)} = \Phi_M(\Omega)$. If $K$ is a finite set, the inclusion relation*

$$\widehat{\mathscr{D}_M^*(\Omega)} \cap [\mathscr{D}^*(\Omega)]^s = \mathscr{D}_M^*(\Omega),$$

*holds, where the left and the right sides are looked on as subspaces in $\mathscr{D}_M^*(\Omega \setminus K)$.*

*Proof.* Let $\{u_\alpha \in \Phi_M(\Omega)\}$ be a sequence stable outside any neighborhood of $K$. Without loss of generality, we shall suppose that $u_1 \equiv u_2 \equiv \cdots$ outside some compact $\mathscr{K} \subset \Omega$. Then for arbitrary $\alpha$ the function $u_\alpha - u_{\alpha+1}$ has a compact support and satisfies the system (1.4). If $M$ is a definite module, by Proposition 1, we have $u_\alpha \equiv u_{\alpha+1}$ and, therefore, $u = \lim u_\alpha \equiv u_1 \in \Phi_M(\Omega)$.

Let us prove the second assertion. Clearly, we may limit ourselves to the case when $K=0$, where 0 is the origin of coordinates. Let $u$ belong to the intersection $\widehat{\mathscr{D}_M^*(\Omega)} \cap [\mathscr{D}^*(\Omega)]^s$. Then $pu$ is a distribution whose support is the origin of coordinates. Accordingly, $pu = F(iD)\,\delta$, where $F \in \mathscr{P}^t$. On the other hand, $pu = p(u-u_1) \in p[\mathscr{E}^*(R^n)]^s$. Therefore, the polynomial $F$ belongs to the space $p[\mathscr{E}^*(R^n)]^s$ and because of Proposition 4, § 1, Chapter II, it has the form $pG$, where $G \in \mathscr{P}^s$. It follows that the function $u - G(iD)\delta$ satisfies the system (1.4) in $\Omega$ and, therefore, belongs to the space $\mathscr{D}_M^*(\Omega)$. Hence $u \in \mathscr{D}_M^*(\Omega)$, if $\mathscr{D}_M^*(\Omega)$ is considered as a subspace in $\mathscr{D}_M^*(\Omega \setminus O)$. □

Let us consider some spaces of entire functions. Let $\mathscr{K}$ be a convex compact in $R^n$. We denote by $\mathscr{E}^{\mathscr{K}+}$ the space of entire functions in $C^n$, characterized by the fact that each of them for arbitrary $\varepsilon > 0$ satisfies the inequality

$$|\phi(z)| \leqq C(|z|+1)^b \exp(\varepsilon |y|)\, \mathscr{I}_{\mathscr{K}}(-y) \tag{2.14}$$

for some $b$ and $C$, depending on $\varepsilon$. The subspace in $\mathscr{E}^{\mathscr{K}+}$, consisting of functions $\phi$, which satisfy this inequality for arbitrary $\varepsilon > 0$ and $b > -\infty$, will be denoted by $S^{\mathscr{K}+}$. Let $q$ be a $\mathscr{P}$-matrix. We denote by $\mathscr{E}^{\mathscr{K}+}\{q\}$ and $S^{\mathscr{K}+}\{q\}$ the spaces of $q$-functions holomorphic in $C^n$, which satisfy the inequality (2) for arbitrary $\varepsilon > 0$ and $b$, depending on $\varepsilon$, or respectively, for arbitrary $\varepsilon > 0$ and $b > -\infty$. Let $L$ be a finite $\mathscr{P}$-module. Using the construction of 4°, § 5, Chapter IV, we can set up a natural isomorphism between the spaces $\mathscr{E}^{\mathscr{K}+}\{q\}$, for which $\operatorname{Coker} q' \cong L$. By $\mathscr{E}^{\mathscr{K}+}\{L\}$ we denote the space obtained by identifying all the $\mathscr{E}^{\mathscr{K}+}\{q\}$, for which $\operatorname{Coker} q' \cong L$. The symbol $S^{\mathscr{K}+}\{L\}$ has a similar meaning.

**Theorem 1.** *Let $M$ be a finite $\mathscr{P}$-module. Let $K$ be a convex compact, and let $\Omega$ be a neighborhood of it. Then there exists an operator*

$$B\colon\ \mathscr{D}_M^*(\Omega \setminus K)/\widehat{\mathscr{D}^*(\Omega)} \to \mathscr{E}^{K+}\{\operatorname{Ext}^1(M, \mathscr{P})\}, \tag{3.14}$$

*which establishes an isomorphism between these spaces. The restriction*

$$B\colon\ \mathscr{E}_M(\Omega \setminus K)/\widehat{\mathscr{E}_M(\Omega)} \to S^{K+}\{\operatorname{Ext}^1(M, \mathscr{P})\} \tag{4.14}$$

*of this operator is defined. It is also an isomorphism.* (Here and later, all the spaces will be endowed with the discrete topology.)

*Proof.* Let (4.7) be a free resolution of the module $M$. By definition

$$\operatorname{Ext}^1(M, \mathscr{P}) = \mathscr{P}_{p_1}/p\mathscr{P}^s, \qquad \mathscr{P}_{p_1} = \operatorname{Ker}\{p_1\colon \mathscr{P}^t \to \mathscr{P}^{t_2}\}. \tag{5.14}$$

Therefore, the sequence

$$0 \to \operatorname{Ext}^1(M, \mathscr{P}) \to \mathscr{P}^t/p\mathscr{P}^s \xrightarrow{p_1} \mathscr{P}^{t_2} \tag{6.14}$$

is defined and exact.

Let $K$ be an arbitrary convex compact in $R^n$ with a non-empty interior int $\mathscr{K}$. We choose a strongly increasing sequence $\{\mathscr{K}_\alpha\}$ of convex compacts tending to $\mathscr{K}$. We consider the family of majorants

$$\mathscr{M} = \{M_\alpha(z) = (|z|+1)^\alpha \mathscr{I}_{\mathscr{K}_\alpha}(-y),\ \alpha = 1, 2, \ldots\}.$$

Using this family, we can construct for an arbitrary finite $\mathscr{P}$-module $L$ the corresponding family of spaces $\mathscr{E}^{\mathscr{K}}\{L\}=\mathscr{H}_{\mathscr{M}}\{L\}$. Proposition 6, §3, implies that $\mathscr{M}$ is a family of majorants of type $\mathscr{I}$. Therefore, by the basic theorem of Chapter IV, we have that the functor $L\sim\to\mathscr{E}^{\mathscr{K}}\{L\}$ is exact. Then, applying this functor to the sequence (6), we obtain an exact sequence of families. Before writing out this sequence, we note that in view of the same theorem, we have the isomorphism $\mathscr{E}^{\mathscr{K}}\{\mathscr{P}^t/p\mathscr{P}^s\}\cong[\mathscr{E}^{\mathscr{K}}]^t/p[\mathscr{E}^{\mathscr{K}}]^s$ where $\mathscr{E}^{\mathscr{K}}=\mathscr{E}^{\mathscr{K}}\{\mathscr{P}\}$. Therefore, the exact sequence that we now have in mind can be written as:

$$0\to\mathscr{E}^{\mathscr{K}}\{\mathrm{Ext}^1(M,\mathscr{P})\}\to[\mathscr{E}^{\mathscr{K}}]^t/p[\mathscr{E}^{\mathscr{K}}]^s\xrightarrow{p_1}[\mathscr{E}^{\mathscr{K}}]^{t_2}. \tag{7.14}$$

For an arbitrary module $L$, we denote by $\mathscr{E}^{\mathscr{K}-}\{L\}$ the inductive limit of the family $\mathscr{E}^{\mathscr{K}}\{L\}$. Let us pass to this limit in the sequence (7). By Proposition 10, § 1, the sequence of spaces

$$0\to\mathscr{E}^{\mathscr{K}-}\{\mathrm{Ext}^1(M,\mathscr{P})\}\to[\mathscr{E}^{\mathscr{K}-}\}^t/p[\mathscr{E}^{\mathscr{K}-}]^s\xrightarrow{p_1}[\mathscr{E}^{\mathscr{K}-}]^{t_2} \tag{7'.14}$$

that we so obtained is exact. Since it is exact, the space $\mathscr{E}^{\mathscr{K}-}\{\mathrm{Ext}^1(M,\mathscr{P})\}$, in which we are interested, is isomorphic to the kernel of the mapping $p_1$. Let us write out this kernel. Let $\partial=\{\partial_\lambda,\lambda=0,\ldots,l\}$ be a normal Noetherian operator associated with the matrix $p$. By Proposition 1, § 4, Chapter IV, we may write $\partial_0=p_1$, where $\partial_0$ is a component of the operator $\partial$, corresponding to the $n$-dimensional primary component of the submodule $p\mathscr{P}^s\subset\mathscr{P}^t$.

Using Theorem 2, § 5, Chapter IV, and again using Proposition 10, § 1, we arrive at the conclusion that the mapping

$$[\mathscr{E}^{\mathscr{K}-}]^t/p[\mathscr{E}^{\mathscr{K}-}]^s\xrightarrow{\partial}\mathscr{E}^{\mathscr{K}-}\{p\}$$

is an isomorphism. Therefore, the kernel of $p_1$ in (7′) may be identified with the subspace $\mathscr{E}^{\mathscr{K}-}\{p\}$, consisting of $p$-functions for which the component corresponding to the operator $\partial_0$ is equal to zero. We denote this subspace by $\mathscr{E}^{\mathscr{K}-}_{p_1}\{p\}$. We have thus established the fact that the operator

$$\mathscr{E}^{\mathscr{K}-}\{\mathrm{Ext}^1(M,\mathscr{P})\}\xrightarrow{\partial}\mathscr{E}^{\mathscr{K}-}_{p_1}\{p\} \tag{8.14}$$

is an isomorphism.

Let us now construct the operator $B$. Let $\mathscr{K}$ be a convex compact such that $\mathscr{K}\supset\supset K$. We choose a function $\alpha\in\mathscr{D}(\Omega)$, equal to unity in some neighborhood $V$ of the compact $K$, and having a support belonging to $\operatorname{int}\mathscr{K}$. Let $u$ be an arbitrary element of the space $\mathscr{D}^*_M(\Omega\setminus\mathrm{K})$. The function $p(iD)\alpha u$ is, clearly, equal to zero in the region $V\setminus K$, and it may, therefore, be extended from $\Omega\setminus K$ to $\Omega$, if we set it equal to zero on $K$. We denote by $E$ the operator. Since $\operatorname{supp} E p\alpha u\subset\operatorname{supp}\alpha$,

$$E p\alpha u\in[\mathscr{E}^*(\operatorname{int}\mathscr{K})]^t.$$

Let $\widetilde{Ep\alpha u}$ be the Fourier transform of $Ep\alpha u$. Applying the operator $-\partial$, we obtain the desired mapping

$$B\colon u \to -\partial \widetilde{Ep\alpha u}.$$

This mapping carries the space $\mathscr{D}_M^*(\Omega\setminus K)$ into the space of $p$-functions holomorphic in $C^n$. We shall show that it sets up the isomorphism (3). First, we shall show that it does not depend on the choice of the compact $\mathscr{K}$, nor on the function $\alpha$. Let $\alpha$ and $\alpha'$ be two arbitrary functions on $\mathscr{D}(\Omega)$, which are equal to unity in some neighborhood $V\supset K$. Since the function $(\alpha-\alpha')u$ vanishes on $V\setminus K$, the operator $E$ is applicable to it. Accordingly,

$$Ep\alpha u - Ep\alpha' u = Ep(\alpha-\alpha')u = pE(\alpha-\alpha')u.$$

Since $pE(\alpha-\alpha')u\in p[\mathscr{E}^*(\Omega)]^s$, we have $\partial\widetilde{Ep\alpha u}-\partial\widetilde{Ep\alpha' u}=0$, which is what we were to prove.

Since the support of the function $Ep\alpha u$ belongs to int $\mathscr{K}$, the properties of the Fourier transform (see Propsoition 2, § 3) imply that the function $\widetilde{Ep\alpha u}$ belongs to the space $[\mathscr{E}^{\mathscr{K}-}]^t$ and therefore, $-\partial\widetilde{Ep\alpha u}\in \mathscr{E}^{\mathscr{K}-}\{p\}$. We note that

$$p_1(z)\,\widetilde{Ep\alpha u} = p_1(\widetilde{iD)\,Ep\alpha u} = \widetilde{Ep_1 p\alpha u} = 0,$$

since $p_1 p=0$. It follows that the function $Bu=-\partial\widetilde{Ep\alpha u}$ belongs to the subspace $\mathscr{E}_{p_1}^{\mathscr{K}-}\{p\}$. Taking account of the isomorphism (8), we may look on $Bu$ as an element of the space $\mathscr{E}^{\mathscr{K}-}\{\mathrm{Ext}^1(M,\mathscr{P})\}$. Since this function does not depend on the compact $\mathscr{K}$, it belongs to the intersection of all of the spaces $\mathscr{E}^{\mathscr{K}-}\{\mathrm{Ext}^1(M,\mathscr{P})\}$ with $\mathscr{K}\supset\supset K$. This intersection, clearly, coincides with $\mathscr{E}^{K+}\{\mathrm{Ext}^1(M,\mathscr{P})\}$. Thus, we have constructed the linear operator

$$B\colon \mathscr{D}_M^*(\Omega\setminus K)\to \mathscr{E}^{K+}\{\mathrm{Ext}^1(M,\mathscr{P})\}.$$

We now determine the kernel of this operator. Let $Bu=0$, that is, $\partial\widetilde{Ep\alpha u}=0$. We choose a strongly decreasing sequence $\{K_\nu\}$ of convex compacts tending to $K$ [3]. For every $\nu=1,2,\ldots$ we find a function $\alpha_\nu\in\mathscr{D}(\Omega)$, equal to unity in the neighborhood of $K$, and such that $\operatorname{supp}\alpha_\nu\subset\subset K_\nu$. Then $\operatorname{supp} Ep\alpha_\nu u\subset\subset K_\nu$ and, therefore, $Ep\alpha_\nu u\in[\mathscr{E}^{K_\nu-}]^t$. The equation $\partial\widetilde{Ep\alpha_\nu u}=\partial\widetilde{Ep\alpha u}=0$ implies, by Theorem 2, § 5, Chapter IV, that $\widetilde{Ep\alpha_\nu u}\in p[\mathscr{E}^{K_\nu-}]^s$, whence

$$\widetilde{Ep\alpha_\nu u}=p\,v_\nu,\qquad v_\nu\in[\mathscr{E}^*(\operatorname{int}K_\nu)]^s.$$

It follows that $p(\alpha_\nu u-v_\nu)=0$ in the region $\Omega\setminus K$. We now consider the distribution $u_\nu=E(1-\alpha_\nu)u+v_\nu$. It coincides with $u$ outside of $K_\nu$ and

3 That is, $K_1\supset\supset K_2\supset\supset\cdots$ and $\cap K_\nu=K$.

is a solution of the system (1.4) in $\Omega$, since $u_\nu = u - (\alpha_\nu u - v_\nu)$, and in the neighborhood of $K$, we have $u_\nu = v_\nu$. Thus, the sequence of distributions $u_\nu \in \mathscr{D}_M^*(\Omega)$ stabilizes to the function $u$ outside any neighborhood of $K$. It follows that $u \in \widehat{\mathscr{D}_M^*(\Omega)}$. But this shows that $\operatorname{Ker} B \subset \widehat{\mathscr{D}_M^*(\Omega)}$.

Let us now establish the converse inclusion relationship. Suppose that $u \in \mathscr{D}_M^*(\Omega)$. Then

$$E\, p\, \alpha\, u = p\, \alpha\, u \in p[\mathscr{E}^*(\Omega)]^s,$$

whence $\partial \widetilde{E\, p\, \alpha\, u} = 0$. Let $u'$ be an arbitrary distribution coinciding with $u$ outside some neighborhood $V \subset\subset \Omega$ of the compact $K$. Choosing the function $\alpha \in \mathscr{D}(\Omega)$ to be equal to unity on $V$, we obtain

$$\partial \widetilde{E\, p\, \alpha\, u'} = \partial \widetilde{E\, p\, \alpha\, u} = 0.$$

It follows that an arbitrary distribution in the space $\widehat{\mathscr{D}_M^*(\Omega)}$ belongs to the kernel of the operator $B$. We have thus established the equation $\operatorname{Ker} B = \widehat{\mathscr{D}_M^*(\Omega)}$. It follows that the mapping (3) is biunique.

Let us now construct the inverse mapping. Let $F$ be an arbitrary element of the space $\mathscr{E}^{K+}\{\operatorname{Ext}^1(M, \mathscr{P})\}$. This space is the intersection of the spaces $\mathscr{E}^{K_\nu -}\{\operatorname{Ext}^1(M, \mathscr{P})\}$, $\nu = 1, 2, \ldots$. Taking account of the isomorphism (8), in which $\mathscr{K} = K_\nu$, $\nu = 1, 2, \ldots$, we may think of $F$ as an element of an arbitrary one of the spaces $\mathscr{E}_{p_1}^{K_\nu -}\{p\}$, $\nu = 1, 2, \ldots$. Then Theorem 2, § 5, Chapter IV, implies that for arbitrary $\nu$ we may find a function $\psi_\nu \in [\mathscr{E}^{K_\nu -}]^t$ such that $F = \partial \psi_\nu$. Since by hypothesis the component of the function $F$, corresponding to the operator $\partial_0$, is equal to zero, we have $p_1(z)\, \psi_\nu = 0$. Let $\phi_\nu$ be the inverse Fourier transform of the function $\psi_\nu$. By Theorem 2, § 3, we have $\phi_\nu \in [\mathscr{E}^*(R^n)]^t$, and $\operatorname{supp} \phi_\nu \subset K_\nu$ and $p_1(i D)\, \phi_\nu = 0$. Taking account of this latter relationship, and of Theorem 1, § 8, we may find a distribution $u_\nu \in [\mathscr{D}^*(R^n)]^s$ such that $p(i D)\, u_\nu = \phi_\nu$.

Let us now construct the sequence of distributions $u'_\nu \in [\mathscr{D}^*(R^n)]$, $\nu = 1, 2, \ldots$, such that for an arbitrary $\nu$, we have $u'_{\nu+1} = u'_\nu$ outside $K_\nu$ and $p\, u'_\nu = \phi_\nu$, and $u'_1 = u_1$. Let us agree that the first $\nu$ components of this sequence have been already constructed. Then we construct the function $u'_{\nu+1}$. Since for arbitrary $\nu$ we have $\partial \psi_\nu = F$ and $\psi_\nu \in [\mathscr{E}^{K_\nu -}]^t$, it follows that $\partial(\psi_{\nu+1} - \psi_\nu) = 0$. Applying Theorem 2, § 5, Chapter IV, to the family $\mathscr{H}_{\mathscr{M}} = \mathscr{E}^{K_\nu}$, we find that $\psi_{\nu+1} - \psi_\nu \in p[\mathscr{E}^{K_\nu -}]^s$, whence $\phi_{\nu+1} - \phi_\nu = p\, \rho_\nu$, where $\rho_\nu$ is some distribution with its support in $K_\nu$. Then $p(u_{\nu+1} - u'_\nu) = \phi_{\nu+1} - \phi_\nu = p\, \rho_\nu$, and we see that the distribution $v_\nu = u_{\nu+1} - u_\nu - \rho_\nu$ is a solution of the system (1.4) throughout $R^n$. Therefore the distribution $u'_{\nu+1} = u_{\nu+1} - v_\nu = u'_\nu + \rho_\nu$ is the one that we are seeking. And with this, the desired sequence $\{u'_\nu\}$ has been constructed.

Since this sequence is stable outside any neighborhood of $K$, there exists in the space $\mathscr{D}_M^*(\Omega\setminus K)$ a limiting distribution $u=\lim u'_\nu$. We shall show that the mapping $F\to u$ is the inverse of the operator $B$. Without loss of generality, we may suppose that $K_1\subset\Omega$. We choose a function $\alpha$ to be equal to unity in some neighborhood $V_1$ of the compact $K_1$. And we consider the function $p\alpha u_1$. In the region $V_1$ it is equal to $\phi_1$, and in $\Omega\setminus V_1$ it coincides with $p\alpha u$, since $u=u_1$ outside $K_1$. Therefore $p\alpha u_1= E p\alpha u+\phi_1$. Hence

$$Bu=-\partial\widetilde{E p\alpha u}=-\partial p\alpha u_1+\partial\tilde{\phi}_1=\partial\tilde{\phi}_1=\partial\psi_1=F,$$

which is what we were to prove. We have now shown that the mapping (3) is an isomorphism.

We now take up the mapping (4). Let $\mathscr{K}$ be an arbitrary convex compact with a non-empty interior. By analogy with the space $\mathscr{E}^{\mathscr{K}-}$, we consider the space $S^{\mathscr{K}-}$, which is the union of the spaces $S^{\mathscr{K}_\nu+}$, $\nu=1,2,\ldots$ ($\{\mathscr{K}_\nu\}$ is a strongly increasing sequence of convex compacts tending to $\mathscr{K}$). In the construction of the isomorphism (3) we made use of the fact that $\mathscr{E}^{\mathscr{K}-}=\varinjlim \mathscr{H}_{\mathscr{M}}$, where $\mathscr{M}$ is a family of majorants of type $\mathscr{I}$. The space $S^{\mathscr{K}-}$ cannot be represented in this form, but it is equal to $\cup \varinjlim \mathscr{H}_{\mathscr{M}}$, where the union runs over all sets of families $\mathscr{M}$, having the form

$$\mathscr{M}=\{M_\alpha(z)=R_\alpha(z)\,\mathscr{I}_{\mathscr{K}_\alpha}(-y),\ \alpha=1,2,\ldots\},$$

and $\{R_\alpha\}$ is an arbitrary sequence of functions, satisfying the conditions of Proposition 5, § 3. The remaining arguments that we used above apply with the obvious changes to the mapping (4). ▯

**3°. Remark and corollaries**

*Remark.* The operator $B$, constructed in Theorem 1, is independent of the compacts $K$ and of the region $\Omega$ in the following sense. Let $\kappa$ be a convex compact, containing $K$, and let $\omega$ be a neighborhood of it, belonging to $\Omega$. Then the diagram

$$\begin{array}{ccc} \mathscr{D}_M^*(\omega\setminus\kappa)/\widehat{\mathscr{D}_M^*(\omega)} & \xrightarrow{B} & \mathscr{E}^{\kappa+}\{\mathrm{Ext}^1(M,\mathscr{P})\} \\ \uparrow r_\omega^\Omega & & \uparrow i \\ \mathscr{D}_{\mathscr{M}}^*(\Omega\setminus K)/\widehat{\mathscr{D}_{\mathscr{M}}^*(\Omega)} & \xrightarrow{B} & \mathscr{E}^{K+}\{\mathrm{Ext}^1(M,\mathscr{P})\} \end{array} \tag{9.14}$$

is commutative; here $i$ is the identity imbedding, and $r_\omega^\Omega$ is a restriction mapping.

**Corollary 1.** *Let $\kappa=K$. Then the mapping $r_\omega^\Omega$ in* (9) *is an isomorphism.*

*Proof.* If $\kappa=K$, the imbedding $i$ is an isomorphism. And therefore the commutativity of (9) implies that $r_\omega^\Omega$ is an isomorphism. ▯

**Corollary 2.** *Let $K$ be a non-empty convex compact, and $\Omega$ be a neighborhood of it. In order that the equation $\mathscr{D}_M^*(\Omega\setminus K)=\widehat{\mathscr{D}_M^*(\Omega)}$ should hold, that is, in order that an arbitrary singularity of the solution of* (1.4) *on $K$ be removable, it is necessary and sufficient that* $\operatorname{Ext}^1(M,\mathscr{P})=0$.

*Proof.* The sufficiency of the condition $\operatorname{Ext}^1(M,\mathscr{P})=0$ follows from Theorem 1.

Let us prove the necessity of the condition. Without loss of generality we may suppose that the compact $K$ contains the origin of coordinates. Then the ring $\mathscr{P}$ is a subspace in $\mathscr{E}^{K+}$, so that the $p$-functions of the form $\partial F$, where $F\in\mathscr{P}^t$, belong to $\mathscr{E}^{K+}\{p\}$. Therefore, the Noetherian operator $\partial$ defines a mapping

$$\mathscr{P}_{p_1}/p\,\mathscr{P}^s \xrightarrow{\partial} \mathscr{E}_{p_1}^{K+}\{p\}. \tag{10.14}$$

This mapping is a monomorphism. If $\partial F=0$, where $F\in\mathscr{P}^t$, Proposition 4, § 1, Chapter II, implies that $F=p\,G$, $G\in\mathscr{P}^s$. Applying the isomorphisms (5) and (8), we transform the monomorphism (10) to the form

$$\operatorname{Ext}^1(M,\mathscr{P})\to\mathscr{E}^{K+}\{\operatorname{Ext}^1(M,\mathscr{P})\}.$$

Therefore, $\operatorname{Ext}^1(M,\mathscr{P})\neq 0$ implies that $\mathscr{E}^{K+}\{\operatorname{Ext}^1(M,\mathscr{P})\}\neq 0$. This proves the necessity. □

**Corollary 3.** *Let us again suppose that $K$ is a convex compact and $\Omega$ is a neighborhood of it. In order that every distribution satisfying the system* (1.4) *in $\Omega\setminus K$ have a unique extension in $\Omega$ which also satisfies* (1.4), *it is necessary and sufficient that the operator $p$ be overdetermined.*

*Proof. Necessity.* Corollary 2 implies the necessity of the condition $\operatorname{Ext}^1(M,\mathscr{P})=0$. The uniqueness of the extension requires that the system (1.4) have no solution with a support belonging to $K$. By Proposition 1, this means that $\operatorname{Hom}(M,\mathscr{P})=0$. Therefore, $p$ is an overdetermined operator.

*Sufficiency.* Let $p$ be overdetermined. By Corollary 2, for every solution $u\in\mathscr{D}_M^*(\Omega\setminus K)$, and for every compact $\mathscr{K}\supset\supset K$, there exists a solution $u_{\mathscr{K}}\in\mathscr{D}_M^*(\Omega)$, coinciding with $u$ in $\Omega\setminus\mathscr{K}$. For any two compacts $\mathscr{K}$, $\mathscr{L}\supset K$, belonging to $\Omega$, the difference $u_{\mathscr{K}}-u_{\mathscr{L}}$ is a solution of the system (1.4) with a compact support. By Proposition 1, we have $u_{\mathscr{K}}\equiv u_{\mathscr{L}}$, and therefore, $u_{\mathscr{K}}$ coincides with $u$ in $\Omega\setminus K$ and is uniquely defined by $u$. □

**4°. One-point irremovable singularities.** We now limit ourselves to a narrower problem: We shall determine the irremovable singularities, at the origin of coordinates, of those solutions of the system (1.4) for which we know that they can be extended in the neighborhood of the origin of coordinates as distributions.

Let $O$ be the origin of coordinates in $R^n$, and let $\Omega$ be a neighborhood of it. The solutions of (1.4) in $\Omega \setminus O$, which are extendable in the neighborhood of $O$ as distributions, form the space $\mathscr{D}_M^*(\Omega \setminus O) \cap [\mathscr{D}^*(\Omega)]^s$. The irremovable singularities of such solutions form a space, which is the image of the natural mapping

$$\mathscr{D}_M^*(\Omega \setminus O) \cap [\mathscr{D}^*(\Omega)]^s \to \mathscr{D}_M^*(\Omega \setminus O)/\widehat{\mathscr{D}_M^*(\Omega)}.$$

The kernel of this mapping is equal to the intersection $\widehat{\mathscr{D}_M^*(\Omega)} \cap [\mathscr{D}_M^*(\Omega)]^s$ and, therefore, coincides with $\mathscr{D}_M^*(\Omega)$, by Proposition 2. This means that the space of irremovable singularities of the kind that we are now studying can be identified with the factor-space

$$\mathscr{D}_M^*(\Omega \setminus O) \cap [\mathscr{D}^*(\Omega)]^s/\mathscr{D}_M^*(\Omega). \tag{11.14}$$

**Theorem 2.** *The operator $B$ establishes an isomorphism of the space* (11) *and the space* $\mathrm{Ext}^1(M, \mathscr{P})$ *(looked on as a subspace in* $\mathscr{E}^{O+}\{\mathrm{Ext}^1(M, \mathscr{P})\}$*).*

*Proof.* Let $u$ be an arbitrary distribution belonging to

$$\mathscr{D}_M^*(\Omega \setminus O) \cap [\mathscr{D}^*(\Omega)]^s.$$

Since it is extendable in $\Omega$ as a distribution, we have $p\alpha u = Ep\alpha u + pu$. Therefore $Bu = -\partial \widetilde{Ep\alpha u} = \partial \widetilde{pu}$. Since the distribution $pu$ is concentrated at the origin of coordinates, it is equal to $F(iD)\delta$, where $F \in \mathscr{P}^t$, and $p_1 F = 0$. Hence $\widetilde{pu} = F(z)$ and $p_1(z)F(z) = 0$. Then the function $Bu$ is equal to $\partial F$ and, accordingly, belongs to the image of the mapping (10), i.e., $Bu \in \mathrm{Ext}^1(M, \mathscr{P})$.

Conversely, let $\partial F$ be an arbitrary $p$-function, belonging to the image of the mapping (10). Then $F \in \mathscr{P}^t$, and $p_1 F = 0$. Then Theorem 1, § 8, implies that the system of equations $pu = F(iD)\delta$ has a solution in $R^n$. This solution clearly belongs to the space $\mathscr{D}_M^*(\Omega \setminus O) \cap [\mathscr{D}^*(\Omega)]^s$. □

**Corollary 4.** *In order that the inclusion*

$$\mathscr{E}_M(\Omega \setminus O) \cap [\mathscr{D}^*(\Omega)]^s \subset \mathscr{D}_M^*(\Omega), \tag{12.14}$$

*be valid, it is necessary and sufficient that none of the varieties associated with the module* $\mathrm{Ext}^1(M, \mathscr{P})$ *be hypoelliptic* (see § 5).

We emphasize that both sides of (12) are to be looked on as subspaces of $\mathscr{D}_M^*(\Omega \setminus O)$, that is, every distribution concentrated at the origin of coordinates is equal to zero as an element of the spaces in (12). Thus, the inclusion relation (12) shows that every infinitely differentiable solution of (1.4), defined in $\Omega \setminus O$, and extendable in $\Omega$ as a distribution, has an extension in $\Omega$ as a solution of (1.4).

*Proof. Sufficiency.* Suppose that $p\mathscr{P}^s = \mathfrak{p}_0 \cap \cdots \cap \mathfrak{p}_l$ is a reduced primary decomposition of the submodule $p\mathscr{P}^s \subset \mathscr{P}^t$ and that $N_\lambda$, $\partial_\lambda$, $\lambda =$

$0, \ldots, l$, are the varieties and the normal Noetherian operators associated with the modules $\mathfrak{p}_\lambda$, and suppose that $\mathfrak{p}_0 = \mathscr{P}_{p_1}$, and that $\partial_0 = p_1$. Then $\mathfrak{p}_1 \cap \cdots \cap \mathfrak{p}_l$ is a reduced primary decomposition of the submodule $p\mathscr{P}^s \subset \mathscr{P}_{p_1}$ and $M_1 \cap \cdots \cap M_l$, where $M_\lambda = \mathfrak{p}_\lambda / p\mathscr{P}^s$ is a reduced primary decomposition of zero in the module $\mathscr{P}_{p_1}/p\mathscr{P}^s$ (see Proposition 1, § 1, Chapter IV). Since the variety $N_\lambda$, $\lambda > 0$ is associated with the submodule $\mathfrak{p}_\lambda \subset \mathscr{P}_{p_1}$, it is associated also with the submodule $M_\lambda \subset \mathscr{P}_{p_1}/p\mathscr{P}^s$. Accordingly, the varieties $N_\lambda$, $\lambda = 1, \ldots, l$, form the collection associated with the module $\mathscr{P}_{p_1}/p\mathscr{P}^s = \mathrm{Ext}^1(M, \mathscr{P})$.

Let us suppose that none of the varieties $N_\lambda (\lambda > 0)$ are hypoelliptic. By Theorem 1 the operator $B$ carries the space $\mathscr{E}_M(\Omega \setminus O) \cap [\mathscr{D}^*(\Omega)]^s$ into $S^{O+}\{\mathrm{Ext}^1(M, \mathscr{P})\}$, and by Theorem 2 carries it into $\mathrm{Ext}^1(M, \mathscr{P})$. Therefore, for an arbitrary element $u$ belonging to this space, $Bu = \partial F$, where $F \in \mathscr{P}_{p_1}$, and the $p$-function $\partial F$ decreases faster than an arbitrary power of $|z|$ in an arbitrary band of the form $|y| \leq C$. This means that every component $\partial_\lambda F$ of $p$-functions decreases faster than an arbitrary power of $|z|$ in the intersection of $N_\lambda$ and this band. Since by hypothesis the varieties $N_\lambda$ are not hypoelliptic, the intersection of each with some such band is unbounded, and therefore, the condition that was formulated above concerning the decrease of the functions is meaningful.

Let us choose an arbitrary $\lambda > 0$ and consider the function

$$m(r) = \inf\{|\partial_\lambda(z+\zeta, D)\, F(z+\zeta)|,\ z+\zeta \in N_\lambda,\ |z| = r,\ |\zeta| \leq 1\}.$$

We shall suppose that $\partial_\lambda F \not\equiv 0$. Since the function $\partial_\lambda F$ is a polynomial, we have $m(r) \sim c\, r^\alpha$ as $r \to \infty$ for some $c \geq 0$ and $\alpha$ [4]. On the other hand, what we said above implies that the function $m(r)$ decreases at infinity faster than an arbitrary power of $r$. It follows that $m(r) \equiv 0$. Therefore $\partial_\lambda F \equiv 0$ on $N_\lambda$ in the neighborhood of some point $z \in N_\lambda$ and, so $\partial_\lambda F \equiv 0$ on $N_\lambda$, since this variety is irreducible. Thus $\partial F = 0$, whence by Theorem 2, we have $u \in \mathscr{D}_M^*(\Omega)$. This proves the sufficiency of our condition.

We shall now prove the necessity. We suppose that one of the varieties $N_\lambda$, for example, $N_l$, is hypoelliptic. We construct a function belonging to the left side of (12) and not belonging to the right side.

By a theorem due to Lech [5] we can find a polynomial $h \in \mathscr{P}$, which vanishes on $N_l$, and such that the variety of its roots is also hypoelliptic. By Theorem 1, § 8, the equation $h(iD)\, v = \delta$ has a solution in $R^n$. Corollary 3 of the same section implies that the distribution $v$ is infinitely differentiable away from the origin of coordinates. Suppose, further, that $F$ is an element of $\mathscr{P}^t$, belonging to all the $\mathfrak{p}_\lambda$ for $0 \leq \lambda < l$ but not belonging to $\mathfrak{p}_l$. (Such an element exists since by hypothesis none of the modules $\mathfrak{p}_\lambda$

4 See Gorin [1].
5 See Lech [1].

contains the intersection of the others.) Since the polynomial $h$ vanishes on $N_l$, it belongs to the radical of $\mathfrak{p}_\lambda$, and therefore, we have for some natural $\rho$, $h^\rho F \in p\mathscr{P}^s$. Without loss of generality, we may suppose that $\rho=1$ (if $\rho>1$, then we can go back to the beginning and replace $h$ by $h^\rho$). Hence $hF=pG$, where $G\in\mathscr{P}^s$. Setting $u=G(iD)\,v$, we obtain

$$p\,u=p\,G\,v=h\,F\,v=F\,h\,v=F(i\,D)\,\delta.$$

Thus the distribution $u$ belongs to $[\mathscr{D}^*(R^n)]^s$, is infinitely differentiable and is the solution of the system (1.4) away from origin of coordinates; therefore, it belongs to the left side of (12). We shall show that it does not belong to the right side. Let us suppose that it does. Then it coincides in $\Omega\setminus O$ with some distribution $u'\in\mathscr{D}^*_M(\Omega)$. The difference $u-u'$ is concentrated at the origin of coordinates and, therefore, has the form $G'(i\,D)\,\delta$, $G'\in\mathscr{P}^s$. Hence

$$F\,\delta=p\,u=p(u-u')=p\,G'\,\delta,$$

that is, $F=p\,G'\in p\,\mathscr{P}^s$, which contradicts the way in which $F$ was chosen. □

**5°. Examples**

**Example 1.** Let $M=\mathscr{P}/p\mathscr{P}$, where $p$ is a non-zero element of the ring $\mathscr{P}$. By Proposition 2, § 13, $\operatorname{Ext}^1(M,\mathscr{P})\cong M$. Therefore, the irreducible components of the variety $N$ of the roots of the polynomial $p$ form a collection of varieties associated with $\operatorname{Ext}^1(M,\mathscr{P})$. Therefore, Corollary 4 in this case reads as follows: Any distribution in a neighborhood $\Omega$ of the origin of coordinates, which is an infinitely differentiable solution of the system (1.4) in $\Omega\setminus O$, will be extendable in $\Omega$ as a solution of the same equation, if and only if none of the irreducible components of the variety $N$ is hypoelliptic.

**Example 2.** The operator

$$p=\frac{\partial}{\partial\xi_1}\Delta-1,\qquad \Delta=\sum_1^n\frac{\partial^2}{\partial\xi_j^2}.$$

is not hypoelliptic, and the corresponding variety is irreducible, since the polynomial $p$ is irreducible. Therefore by Corollary 4 every infinitely differentiable solution of the system (1.4) in $\Omega\setminus O$, which is a distribution in $\Omega$, is continuable in $\Omega$ as a solution of this equation. We shall now show that the requirement that the solution be infinitely differentiable cannot be replaced by a requirement that it be differentiable up to a finite order, no matter how large that order. To this end, we construct solutions of (1.4) in $R^n\setminus O$ of an arbitrarily high degree of smoothness, which are distributions in $R^n$, but have irremovable singularities at the origin of coordinates.

Let $u \in \mathscr{D}^*(R^n)$ be a solution of the equation $pu = \delta$. Let $K$ be an arbitrary compact belonging to $R^n \setminus O$. Since

$$\Delta \frac{\partial u}{\partial \xi_1} = u$$

in the neighborhood of $K$, we know by a property of the Laplace operator that

$$\deg_K \frac{\partial u}{\partial \xi_1} \leqq \deg_L u - 2$$

for every compact $L \supset \supset K$. Hence

$$\deg_K \frac{\partial^2 u}{\partial \xi_1^2} \leqq \deg_L u - 4$$

and so on. Thus, for arbitrary integer $q$ we can find an integer $k$ such that the function $\partial^k u / \partial \xi_1^k$ has continuous derivatives on $K$ up to order $q$. (This result can be derived from Theorem 4, § 5 as well.) It is clear that this is a solution of (1.4) in $R^n \setminus O$, but cannot be continued in $R^n$ as a solution of (1.4).

**Example 3.** We shall now show that Corollary 4 ceases to be valid if we consider the irremovable singularities, at the origin of coordinates, of all infinitely differentiable solutions of (1.4) in $R^n \setminus O$.

Suppose that $n = 2$, and

$$p = \frac{\partial^2}{\partial \xi_1^2} - i \frac{\partial^3}{\partial \xi_2^3}.$$

The polynomial $p$ is clearly irreducible, and the variety $N = \{z \colon z_1^2 = z_2^3\}$ associated with it is not hypoelliptic, since it contains the unbounded real part $\{x \colon x_1^2 = x_2^3\}$. We construct a solution of the corresponding equation (1.4), which is infinitely differentiable away from zero, and has an irremovable singularity at the origin. In view of Theorem 1, it is sufficient to construct a non-zero element of the space $S^{O+}\{\mathrm{Ext}^1(M, \mathscr{P})\}$.

Let us characterize this space. It is clear that $p_1 = 0$. Therefore $S^{O+}\{\mathrm{Ext}^1(M, \mathscr{P})\} = S^{O+}\{p\}$. Since the polynomial $p$ is irreducible, $\partial \equiv 1$ is a normal Noetherian operator, associated with $p$ (see Proposition 3, § 4, Chapter IV). Therefore, in this case, the concept of a $p$-function holomorphic in $C^n$ has a simple meaning: It is a function given on $N$, which is extendable as a holomorphic function in the neighborhood of an arbitrary point $z \in N$. The fact that a holomorphic $p$-function $f$ belongs to the space $S^{O+}\{p\}$ means that for arbitrary integer $k$ it satisfies the inequality

$$|f(z)| \leqq C(|z|+1)^{-k} \exp\left(\frac{1}{k}|y|\right) \qquad (z \in N). \tag{13.14}$$

Let $\psi(\lambda)$ be an even, entire function in $C^1$, which is not identically zero and satisfies the inequality

$$|\psi(\lambda)| \leqq \rho(|\lambda|^3+1)\exp(|\operatorname{Im}\lambda|), \tag{14.14}$$

where $\rho(t)$ is some positive function decreasing monotonely as $t\to\infty$ faster than an arbitrary power of $t$. (We may, for instance, choose $\psi$ to be the Fourier transform of an arbitrary even function $\phi\in\mathscr{D}(R^1)$ with a support in $(-1,1)$.) Since the function $\psi(\lambda)$ is even, it is equal to $\psi_0(\lambda^2)$, where $\psi_0$ is some entire function. We consider a holomorphic $p$-function $f(z_1,z_2)=\psi_0(z_2)$, $(z_1,z_2)\in N$. It is, clearly, not identically zero, and we shall show that it satisfies the inequality (13).

We first establish the inequality

$$|\lambda^3|+|\lambda^2| \leqq 2(|\lambda|^3+1), \qquad |\operatorname{Im}\lambda| \leqq C(|\operatorname{Im}\lambda^3|^{\frac{1}{3}}+|\operatorname{Im}\lambda^2|^{\frac{1}{2}}), \tag{15.14}$$

where $C$ is some sufficiently large number. The first inequality follows from the fact that $|\lambda|^2 \leqq |\lambda|^3+1$. Since $\operatorname{Im}\lambda^k=|\lambda|^k\sin k\phi$, where $\phi=\arg\lambda$, the second inequality follows from the fact that for arbitrary $\phi$ and sufficiently large $C$, we have $C\,|\sin\phi| \leqq C(|\sin 3\phi|^{\frac{1}{3}}+|\sin 2\phi|^{\frac{1}{2}})$. Let $\lambda$ be any of the roots $\sqrt{z_2}$. Then on the variety $N\, z_1=\pm\lambda^3$, and $z_2=\lambda^2$, and therefore, (14) and (15) imply the inequality

$$\begin{aligned}|f(z_1,z_2)|=|\psi(\lambda)| &\leqq \rho(\tfrac{1}{2}[|\lambda|^3+|\lambda|^2])\exp(C[|\operatorname{Im}\lambda^3|^{\frac{1}{3}}+|\operatorname{Im}\lambda^2|^{\frac{1}{2}}])\\ &\leqq C\rho(\tfrac{1}{2}|z|)\exp(C\,|\operatorname{Im} z|^{\frac{1}{2}}),\end{aligned}$$

from which (13) follows for arbitrary $k=1,2,\ldots$. Hence $f\in S^{O+}\{p\}$. This proves the non-triviality of $S^{O+}\{\operatorname{Ext}^1(M,\mathscr{P})\}$.

**Example 4.** We shall find an explicit expression for the operator $B$ when $n=2$, and $K\supset O$ and

$$p=\frac{1}{2}\left[\frac{\partial}{\partial\xi_1}+i\frac{\partial}{\partial\xi_2}\right]=\frac{\partial}{\partial\bar\zeta}.$$

The polynomial $p$ is irreducible, the variety $N=N(p)$ is given by the equations $z_1=-i\,z_2$, and we may assume $\partial\equiv 1$.

The operator $B$ may be applied to functions of the form $\zeta^{-j-1}$, $j=0,1,2,\ldots$. Since these are infinitely differentiable outside $K$, we can choose $\alpha$ to be the characteristic function of some compact $\mathscr{K}\supset\supset K$. We shall suppose that the boundary of $\mathscr{K}$ is smooth. Since the operator $p$ is of first order, we have $E\,p\,\alpha\,\zeta^{-j-1}=\zeta^{-j-1}\,p\,\alpha$. Hence

$$\begin{aligned}B\zeta^{-j-1}&=-\partial\widetilde{p\,\alpha\,\zeta^{-j-1}}=-\int\exp[(z_1,i\,\xi_1)+(z_2,i\,\xi_2)]\,\zeta^{-j-1}\frac{\partial\alpha}{\partial\bar\zeta}\,d\xi_1\,d\xi_2\Big|_N\\ &=-\int\exp(\zeta,z_2)\,\zeta^{-j-1}\frac{\partial\alpha}{\partial\bar\zeta}\,d\xi_1\,d\xi_2.\end{aligned}$$

Since the distribution $\partial\alpha/\partial\bar{\zeta}$ is concentrated on $\partial\mathcal{K}$ we may introduce into the integrand on the right side a factor $h\in\mathcal{D}(R^2)$, which is equal to unity in the neighborhood of $\partial\mathcal{K}$ and which vanishes in the neighborhood of the origin of coordinates. Applying Stokes's formula, we obtain

$$\begin{aligned} Bu &= -\frac{1}{2i}\int h(\xi)\exp(\zeta, z_2)\,\zeta^{-j-1}\frac{\partial\alpha}{\partial\bar{\zeta}}\,d\bar{\zeta}\,d\zeta \\ &= \frac{1}{2i}\int_{\mathcal{K}}\frac{\partial}{\partial\bar{\zeta}}\left[h\exp(\zeta, z_2)\,\zeta^{-j-1}\right]d\bar{\zeta}\,d\zeta \\ &= \frac{1}{2i}\int_{\partial\mathcal{K}}\exp(\zeta, z_2)\,\zeta^{-j-1}\,d\zeta = \frac{\pi}{j!}\,z_2^j. \end{aligned}$$

Thus the operator $(1/\pi)B$ carries the function $\zeta^{-j-1}$ into $(1/j!)\,z_2^j$.

Since this operator is linear and continuous, it carries an arbitrary series of the form $\sum_{j\geqq 0} a_j\,\zeta^{-j-1}$, which converges outside of $K$, into the series

$$\sum\frac{a_j}{j!}\,z_2^j.$$

We note that the series $\sum a_j\,\zeta^{-j-1}$ is the Borel transform of the series

$$\sum\frac{a_j}{j!}\,z_2^j.$$

Therefore, the operator $(1/\pi)B$ represents the transform inverse to the Borel transform. Thus, by Theorem 1, the Borel transform sets up an isomorphism of the space $\mathscr{E}^{K+}\{p\}$ and the space of series of the form $\sum a_j\,\zeta^{-j-1}$, convergent in $R^2\backslash K$. We note that functions of this type are holomorphic in $R^2\backslash K$, and tend to zero at infinity. Conversely, every function which is holomorphic in $R^2\backslash K$ and which tends to zero at infinity, may be expanded in a series of this form.

Since the operator $p$ is hypoelliptic, we have $\mathcal{D}_M^*(\Omega\backslash K)=\mathscr{E}_M(\Omega\backslash K)$ and therefore, the spaces $\mathscr{E}^{K+}\{p\}$ and $S^{K+}\{p\}$ coincide. This implies that the functions belonging to these spaces are characterized by the fact that they satisfy the inequality

$$|f(z)|\leqq C_\varepsilon\exp(\varepsilon\,|y|)\,\mathscr{I}_K(-y_1, -y_2)=C_\varepsilon\exp(\varepsilon\,|z_2|)\,\mathscr{I}_K(x_2, -y_2), \qquad z\in N,$$

for arbitrary $\varepsilon>0$. Hence, it is obvious that the space $\mathscr{E}^{K+}\{p\}$ coincides with the space of functions that are entire in $C_{z_2}^1$ of order no higher than one, and which have indicator diagrams belonging to a compact $K^*$, symmetric to $K$ with respect to the axis $\xi_1$. Thus, we have arrived at the well-known theorem of Polya: The Borel transform sets up an isomorphism of the space of entire functions of order no higher than one, having indicator diagrams belonging to $K^*$, and the space of functions that are holomorphic in $R^2 \setminus K$, and tend to zero at infinity.

### 6°. A generalization of a duality theorem of Grothendieck

**Corollary 5.** *Let the module* $\mathrm{Ext}^1(M, \mathscr{P})$ *be elliptic. Then the bilinear form* $\langle v, u\rangle = (\bar{v}, E\, p\, \alpha\, u)$, *defined on* $\mathscr{E}_{p'}(K) \times \mathscr{D}_p^*(\Omega \setminus K)$, *transforms* $\mathscr{E}_{\mathrm{Ext}^1(M,\mathscr{P})}(K)$ *into the space of continuous functionals on* $\mathscr{D}_M^*(\Omega \setminus K)/\widehat{\mathscr{D}_M^*(\Omega)}$.

*The same bilinear form transforms* $\mathscr{D}^*_{\mathrm{Ext}^1(M,\mathscr{P})}(K)$ *into the space of continuous functionals on* $\mathscr{E}_M(\Omega \setminus K)/\widehat{\mathscr{E}_M(\Omega)}$.

*Proof.* We shall establish only the first assertion, since the second is proved in the same way. By definition every element $v \in \mathscr{E}_{p'}(K)$ belongs to the space $\mathscr{E}_{p'}(U)$, where $U$ is some convex neighborhood of the compact $K$. Therefore the form $\langle v, u\rangle$ is defined, if the function $\alpha$ is chosen in such a way that the support of its gradient belongs to $U \setminus K$. For any two functions $\alpha$ and $\alpha'$ satisfying this condition, we have

$$(\bar{v}, E\, p\, \alpha\, u) - (\bar{v}, E\, p\, \alpha' u) = (\bar{v}, E\, p(\alpha - \alpha')\, u)$$
$$= (\bar{v}, p\, E(\alpha - \alpha')\, u) = (\overline{p' v}, E(\alpha - \alpha')\, u) = 0,$$

since $p' v = 0$. It follows that the form $\langle v, u\rangle$ does not depend on the choice of the function $\alpha$ and, therefore, it is linear in both arguments. We note, also, that this form is continuous in the second argument.

Let $\{N_\lambda, \partial_\lambda\}$ be the set of varieties and normal Noetherian operators associated with the matrix $p$, with $N_0 = C^n$, and $\partial_0 = p_1$ (see the proof of Theorem 1). By Theorem 1, § 4, for any integer $q$ every function $v \in \mathscr{E}_{p'}(K)$ can be written in the form

$$(\bar{v}, \phi) = \sum_\lambda \int_{N_\lambda} \partial_\lambda(z, D)\, \tilde{\phi}(z)\, \mu_\lambda, \tag{I}$$

where the $\mu_\lambda$ are measures with finite integrals

$$\int_{N_\lambda} (|z| + 1)^q \mathscr{I}_{\mathscr{K}}(-y)\, |\mu_\lambda|, \tag{II}$$

and $\mathscr{K}$ a compact neighborhood of $K$. Applying Formula (I), we rewrite our bilinear form as

$$\langle v, u\rangle = \sum_\lambda \int_{N_\lambda} \partial_\lambda\, \widetilde{E\, p\, \alpha\, u}\, \mu_\lambda = -\sum_\lambda \int_{N_\lambda} (B\, u)_\lambda\, \mu_\lambda, \tag{III}$$

where $(B\, u)_\lambda$, $\lambda = 1, \ldots, l$, are the components of the $p$-function $B\, u$. Changing the function $v$, we may obtain as the $\mu_\lambda$ arbitrary measures with finite integrals (II). Therefore, if some distribution $u \in \mathscr{D}_p^*(\Omega \setminus K)$ causes the form $\langle v, u\rangle$ to vanish for all $v$, we have $B\, u = 0$ and conversely. By Theorem 1, $B\, u = 0$ if and only if $u \in \widehat{\mathscr{D}_p^*(\Omega)}$. Thus, we may suppose that the second argument in our form is an element of the factor-space

$$\mathscr{D}_p^*(\Omega\setminus K)/\widehat{\mathscr{D}_p^*(\Omega)}; \tag{IV}$$

and that our form is non-singular with respect to this argument.

Let us now consider the second argument. We note that the form vanishes if $v\in p_1'[\mathscr{E}(K)]^{t_2}$. In fact, suppose that $v=p_1' w$; then

$$\langle v,u\rangle=(\overline{p_1' w}, E\,p\,\alpha\,u)=(\overline{w}, p_1\,E\,p\,\alpha\,u)=(\overline{w}, E\,p_1\,p\,\alpha\,u)=0,$$

since $p_1\,p=0$. We shall show that the converse is true: if $\langle v,u\rangle=0$ for all $u$, then $v\in p_1'[\mathscr{E}(K)]^{t_2}$.

We specify $\mathscr{P}$-matrices $q$ and $e$ such that the sequence

$$0\to\mathscr{P}^\tau/e'\,\mathscr{P}^\sigma\xrightarrow{q}\mathscr{P}_{p^1}/p\,\mathscr{P}^s\to 0. \tag{V}$$

is exact. Setting $L=\mathscr{P}^t/\mathscr{P}_{p_1}$, we obtain two more exact sequences

$$0\to L\xrightarrow{p_1}\mathscr{P}^{t_2};\qquad 0\to\mathscr{P}_{p_1}/p\,\mathscr{P}^s\to\mathscr{P}/p\,\mathscr{P}^s\to L\to 0.$$

Since the $\mathscr{P}$-module $\mathscr{E}(K)$ is injective and the relation $\mathscr{P}_{p_1}/p\,\mathscr{P}^s\cong \operatorname{Ext}^1(M,\mathscr{P})$ is an isomorphism, we obtain the exact sequences

$$\begin{gathered} 0\to\mathscr{E}_{\operatorname{Ext}^1(M,\mathscr{P})}(K)\xrightarrow{q'}\mathscr{E}_e(K)\to 0,\\ [\mathscr{E}(K)]^{t_2}\xrightarrow{p_1'}\mathscr{E}_L(K)\to 0;\\ 0\to\mathscr{E}_L(K)\to\mathscr{E}_p{}'(K)\to\mathscr{E}_{\operatorname{Ext}^1(M,\mathscr{P})}(K)\to 0.\end{gathered} \tag{VI}$$

Combining these, we find another exact sequence

$$p_1'[\mathscr{E}(K)]^{t_2}\to\mathscr{E}_{p'}(K)\xrightarrow{q'}\mathscr{E}_e(K)\to 0. \tag{VII}$$

The function $q'v$ belongs to $\mathscr{E}_e(K)$ and is, therefore, analytic, since the module $\operatorname{Coker} e'\cong\operatorname{Ext}^1(M,\mathscr{P})$ is elliptic. On the other hand, for an arbitrary distribution $\phi\in[\mathscr{E}_K^*]^\tau$ the distribution $q\,\phi$ is annihilated by the operator $p_1$ since (V) is exact. Therefore, the function $q\,\tilde{\phi}\in[\mathscr{E}^{K+}]^t$ is annihilated by the matrix $p_1(z)$, and we have $\partial q\,\tilde{\phi}\in\mathscr{E}_{p_1}^{K+}\{p\}$. By Theorem 1, we have $\partial\widetilde{q\phi}=-B\,u$ for some distribution $u\in\mathscr{D}_p^*(\Omega\setminus K)$, whence

$$(\overline{q' v},\phi)=(\overline{v},q\,\phi)=\sum_\lambda\int\partial_\lambda q\,\tilde{\phi}\,\mu_\lambda=-\sum_\lambda\int(B\,u)_\lambda\,\mu_\lambda=\langle v,u\rangle=0.$$

But this equation implies that the analytic function $q'v$ vanishes on $K$ together with all its derivatives. Therefore, the function $q'v$ vanishes in the neighborhood of $K$, i.e., it belongs to the kernel of the mapping $q'$ in (VII). Since this sequence is exact, we have $v\in p_1'[\mathscr{E}(K)]^{t_2}$, which is what we were to prove.

From all this it follows that the form $\langle v, u\rangle$ is nonsingular, if we consider its first argument as an element of the factor-space

$$\mathscr{E}_{p'}(K)/p_1'[\mathscr{E}(K)]^{t_2}. \tag{VIII}$$

Since our form is continuous in the second argument, we have imbedded the factor-space (VIII) in the space of continuous linear functionals on (IV). We shall now verify the fact that this imbedding is an epimorphism.

We note that the variety $\bigcup_{\lambda\geq 1} N_\lambda$ is elliptic since it is associated with the module $\mathrm{Ext}^1(M, \mathscr{P})$. Hence it is easy to conclude that the space $\mathscr{E}_{p_1}^{K+}\{p\}$ coincides with the space of $p$-functions holomorphic in $C^n$ and having zero $\partial_0$-components, and satisfying the condition that all the norms

$$\|f\|_{\mathscr{K}} = \max_{\lambda\geq 1} \sup_{N_\lambda} \frac{|f_\lambda(z)|}{\exp\left(\mathscr{I}_{\mathscr{K}}(-y)\right)}, \tag{IX}$$

are finite, where $\mathscr{K}\supset\supset K$. In $\mathscr{E}_{p_1}^{K+}\{p\}$ we introduce the topology defined by the ensemble of these norms. Returning to the proof of Theorem 1, it is easy to detect the fact that the operator $B$ establishes a topological isomorphism between the factor-space (IV) and the space $\mathscr{E}_{p_1}^{K+}\{p\}$, with this topology. Therefore, every continuous functional $V$ on (IV) can be considered to be a continuous functional on $\mathscr{E}_{p_1}^{K+}\{p\}$. The family of norms (IX) yields a topology of countable type and therefore $V$ is continuous in one of these norms. Further, applying an argument similar to the proof of Theorem 1 of § 4, we write the functional $V$ in the form of an integral

$$(V, f) = \sum_\lambda \int_{N_\lambda} f_\lambda\, \mu_\lambda,$$

where the measures $\mu_\lambda$ have finite integrals of the form (II). Substituting these measures in the right side of (I), we obtain the corresponding function $v\in\mathscr{E}_{p'}(K)$, which in view of (III) coincides with the functional $V$.

We have thus proved that (VIII) is the space of all continuous linear functionals on (IV). It remains to remark that because (VI) and (VII) are exact, the factor-space (VIII) coincides with $\mathscr{E}_{\mathrm{Ext}^1(M,\mathscr{P})}(K)$. □

**7°. Special results on the extension of solutions**

**Theorem 3.** *Let $L$ be a subspace of $R^n$ of dimension $n-m$, $0<m<n$. We suppose that the module $M$ satisfies the following conditions:*

a) $\mathrm{Ext}^i(M, \mathscr{P})=0$, $i=1, \ldots, m$;

b) *None of the varieties associated with the module* $\mathrm{Ext}^{m+1}(M, \mathscr{P})$ *is hyperbolic with respect to $L$. Then for an arbitrary compact $\kappa\subset L$ and for an arbitrary region $\omega\subset\subset\kappa$ every solution of* (1.4), *defined in the neighbor-*

*hood of the compact $\kappa\setminus\omega$, can be extended in the neighborhood of the compact $\kappa$ as a solution of the same system.*

*Proof.* Let $u$ be a solution of the system (1.4), defined in the neighborhood of $\kappa\setminus\omega$. We choose the region $\omega'\subset\subset\omega$ and the convex neighborhood $v$ of the origin of coordinates in the subspace $L^{\perp}$ in such a way that the distribution $u$ is defined and is a solution of (1.4) in the neighborhood of $\bar{v}\times(\kappa\setminus\omega')$. We further choose a function $\alpha\in\mathscr{D}(R^n)$, to be equal to one in the neighborhood of $\bar{v}\times(\kappa\setminus\omega)$, and zero in the neighborhood of $\bar{v}\times\bar{\omega}'$. The distribution $\alpha u$ is equal to zero near $v\times\omega'$ and, can, therefore, be extended to have the value zero in $v\times\omega'$. This extension will be denoted by $\check{u}$. We have $\operatorname{supp} p\check{u}\subset v\times(\omega\setminus\omega')$.

Suppose that $\Delta\subset v$ is an $n$-dimensional tetrahedron containing the origin of coordinates in $L^{\perp}$. Its boundary $\partial\Delta$ is the union of faces $\Delta_j$, $j=1,\ldots,m+1$, of dimension $m-1$ and having no more than $m$-fold mutual intersections. Let $V_j$, $j=1,\ldots,m+1$ be a convex neighborhood of the faces, also having no more than $m$-fold mutual intersections, and contained in $v$. We construct the tetrahedron $\Delta'$, a dilation of $\Delta$, such that $\Delta\subset\subset\Delta'\subset v$, and near enough to $\Delta$ so that $\Delta'\setminus\Delta\subset V=\cup V_j$. We choose the function $\beta\in\mathscr{D}(R^n)$ with support in $\Delta'\times L$, equal to one in the neighborhood $\Delta\times L$. It is clear that

$$\operatorname{supp} p_1\,\beta\, p\,\check{u}\subset\Omega=V\times\omega. \tag{16.14}$$

Let (4.7) be a free resolution of the module $M$. We consider the $\mathscr{P}$-module $M_i=\operatorname{Coker} p_i$, $i=1, m$. We shall show that Theorem 4, § 10 is applicable to the module $M_m$ and the region $\Omega$. Applying the functor $\operatorname{Hom}(\cdot,\mathscr{P})$ to the resolution (4.7), we obtain the sequence

$$0\to M_1\xrightarrow{p_2}\mathscr{P}^{t_3}\xrightarrow{p_3}\mathscr{P}^{t_4}\to\cdots\xrightarrow{p_{m-1}}\mathscr{P}^{t_m}\xrightarrow{p_m}\mathscr{P}^{t_{m+1}}\xrightarrow{p_{m+1}}\mathscr{P}^{t_{m+2}}.$$

The condition a) with $i=2,\ldots,m$ implies that this sequence is exact in the terms $M_1$, $\mathscr{P}^{t_3},\ldots,\mathscr{P}^{t_m}$. Thus the module $M_1$ is included in an exact sequence of the form (10.10) with $k=m-1$. Further, applying the arguments used in the proof of Corollary 4 to the matrix $p=p_m$, we find that the set of varieties associated with the matrix $p_m$, is equal to the set associated with the module

$$\operatorname{Ext}^{m+1}(M,\mathscr{P})=\mathscr{P}_{p_{m+1}}/p_m\,\mathscr{P}^{t_m},$$

plus the variety associated with the submodule $\mathscr{P}_{p_{m+1}}\subset\mathscr{P}^{t_{m+1}}$, which is equal to $C^n$ (or to the empty set, if $p_{m+1}=0$).

We now note that the regions $U_j=V_j\times\omega$, $j=1,\ldots,m+1$, are convex, and have no more than $m$-fold mutual intersections, and form a finite covering of the region $\Omega$. Let $N_\lambda$ be a variety belonging to the collection associated with the module $M_m$, that is, associated with the matrices $p_m$. What we have already proved implies that $N_\lambda$ belongs to the collection associated with the module $\mathrm{Ext}^{m+1}(M, \mathscr{P})$, else $N_\lambda = C^n$. In the first case, condition b) implies the existence of a hypersubspace $L_\lambda \supset L$, with respect to which $N_\lambda$ is not hyperbolic. It is clear that the projection of an arbitrary region of the form $U_{j_1} \cap \cdots \cap U_{j_m}$ on $L_\lambda^\perp$ is equal to the projection of the region $V_{j_1} \cap \cdots \cap V_{j_m}$. Since the regions $V_j$ can be chosen to be arbitrarily close to the faces of the tetrahedron $\Delta$, their intersections of the form $V_{j_1} \cap \cdots \cap V_{j_m}$ can be made arbitrarily close to the vertices of this tetrahedron. We may suppose that the projections of the vertices of the tetrahedron $\Delta$ on $L_\lambda^\perp$ are distinct: if this is not so, we have only to carry out an arbitrarily small rotation of $\Delta$. Then by a suitable choice of the neighborhoods of $V_j \supset \Delta_j$ we find that the projections of the intersections $V_{j_1} \cap \cdots \cap V_{j_m}$ on $L_\lambda^\perp$ are pairwise disjoint, that is, for a given variety $N_\lambda$ the condition prescribed in Theorem 4, § 10 is satisfied.

The variety $N_\lambda = C^n$ is not hyperbolic with respect to an arbitrary hyper-subspace in $R^n$ and, therefore, the condition of Theorem 4, § 10 is satisfied as well. Thus, all the conditions of the theorem are satisfied and, therefore, the space $\mathscr{E}(\Omega)$ is strictly $M_1$-convex. Hence, by Proposition 1, § 7 we conclude that

$$[\mathscr{E}^*(\Omega)]^{t_2} \cap p_1[\mathscr{E}^*(R^n)]^t = p_1[\mathscr{E}^*(\Omega)]^t. \tag{17.14}$$

The inclusion relation (16) implies that the function $p_1 \beta p \check{u}$ belongs to the lefthand side. It therefore belongs to the right side also, that is, $p_1 \beta p \check{u} = p_1 v$, where $v \in [\mathscr{E}^*(\Omega)]^t$. Hence $p_1(\beta p \check{u} - v) = 0$. The condition a) with $i=1$ implies that the sequence

$$\mathscr{P}^s \xrightarrow{p} \mathscr{P}^t \xrightarrow{p_1} \mathscr{P}^{t_2}.$$

is exact. Since the region $U = v \times \omega$ is convex, the $\mathscr{P}$-module $\mathscr{E}^*(U)$ is flat. Therefore, the exactness of this sequence implies the equation $\mathscr{E}^*_{p_1}(U) = p[\mathscr{E}^*(U)]^s$. Since $\mathrm{supp}(\beta p \check{u} - v) \subset U$, we find that $\beta p \check{u} - v = p w$, where $w \in [\mathscr{E}^*(U)]^s$. Thus, the function $\check{u} - w$ is the desired extension of the function $u$. □

**Theorem 4.** *Let $\Omega$ be a convex region, let $K \subset \Omega$ be a compact, and let $\Sigma_i$, $i=1, \ldots, m$, where $0<m<n$, be closed half-spaces in $R^n$. We shall suppose, further, that the conditions $\mathrm{Ext}^i(M, \mathscr{P}) = 0$, $i=1, \ldots, m$, are satisfied. Then for an arbitrary solution $u$ of the system* (1.4), *defined in $\Omega \setminus (K \cup \Sigma)$, where $\Sigma = \cup \Sigma_i$, we can find a solution of* (1.4) *defined in $\Omega \setminus \Sigma$, coinciding with $u$ in $\Omega \setminus (K' \cup \Sigma)$, where $K' \subset \Omega$ is also compact.*

*If the module $M$ is elliptic, the convexity condition for the region $\Omega$ can be replaced by a weaker condition: the set $\complement(\Omega \cup \Sigma)$ has no bounded connected components.*

*Proof.* We denote by $\mathscr{E}^*(\Omega, \Sigma)$ the subspace of $\mathscr{D}^*(\Omega\setminus\Sigma)$, consisting of distributions vanishing on a set of the form $\Omega\setminus(K'\cup\Sigma)$, where $K'\subset\Omega$ is a compact.

**Lemma.** *If the conditions of the theorem are fulfilled, and $\Omega$ is a convex region, the sequence*

$$[\mathscr{E}^*(\Omega, \Sigma)]^s \xrightarrow{p} [\mathscr{E}^*(\Omega, \Sigma)]^t \xrightarrow{p_1} [\mathscr{E}^*(\Omega, \Sigma)]^{t_2} \tag{18.14}$$

*is exact*[6].

*Proof of the lemma.* Without loss of generality, we may suppose that the half-spaces $\Sigma_i$, $i=1, \ldots, m$, are in a general position. In $R^n$ we choose a system of coordinates such that the set $L=\cap\Sigma_i$ coincides with the subspace $\xi'=(\xi_1, \ldots, \xi_m)=0$. We denote by $B$ the intersection of $R^n\setminus\Sigma$ with the subspace $\xi''=(\xi_{m+1}, \ldots, \xi_n)=0$. Making use of Lemma 1, § 11, we cover the region $B$ with a locally finite covering, formed from closed parallelepipeds $\pi_j$, $j=1, 2, \ldots$, with sides parallel to the coordinate axis and intersecting in lots of no more than $m+1$ and further intersecting only in boundary points. Corresponding to each parallelepiped $\pi_j$, we choose a convex neighborhood $V_j$, close enough to $\pi_j$, so that

$$\pi_{j_0}\cap\cdots\cap\pi'_{j_\nu}=\varnothing$$

always implies that $V_{j_0}\cap\cdots\cap V_{j_\nu}=\varnothing$.

We consider the covering $U$ of the region $\Omega\setminus\Sigma$ consisting of the regions $U_j=(V_j\times L)\cap\Omega$. The convexity of $\Omega$ implies the convexity of this covering[7]. On the covering $U$ we consider the precosheaf $\mathscr{E}^*$, consisting of the spaces $\mathscr{E}^*(\omega)$, where $\omega \neq \Omega\setminus\Sigma$, and the space $\mathscr{D}^*(\Omega\setminus\Sigma)$. We denote by $S$ the ensemble of all sets in $R^n$ with compact closures contained in the region $\Omega$. It is clear that $S$ is a family of supports in the sense of 11°, § 2. We easily see that the precosheaf $\mathscr{E}^*$ and the family of supports $S$ satisfy the conditions of Proposition 10, § 2, which implies that the sequence

$$0\to {}^m\mathscr{E}^*_S(U) \xrightarrow{\partial_m} {}^{m-1}\mathscr{E}^*_S(U)\to\cdots\to {}^0\mathscr{E}^*_S(U)\xrightarrow{\partial_0}\mathscr{E}^*(\Omega, \Sigma)\to 0, \tag{19.14}$$

is exact. Here ${}^\nu\mathscr{E}^*_S(U)$, $\nu=0, 1, \ldots, m$, is the space of cochains of order $\nu$ on the covering $U$ with coefficients in the space $\mathscr{E}^*(U_{j_0}\cap\cdots\cap U_{j_\nu})$, whose supports belong to the family $S$.

---

6 In this lemma we ignore the topology of the linear spaces; the exactness of a sequence therefore means algebraic exactness.

7 This is the only occasion for the use of the convexity of the region $\Omega$. Thus our theorem is proved for arbitrary regions $\Omega$ such that all the $U_j$ are convex.

Let us consider the commutative diagram

$$\begin{array}{cccccccccc}
0 & & 0 & & & & & & & \\
\uparrow & & \uparrow & & & & & & & \\
\mathscr{E}^*(\Omega,\Sigma) & \xrightarrow{p} & \mathscr{E}^*(\Omega,\Sigma) & \xrightarrow{p_1} & \mathscr{E}^*(\Omega,\Sigma) & & & & & \\
\uparrow{\scriptstyle \partial_0} & & \uparrow{\scriptstyle \partial_0} & & \uparrow{\scriptstyle \partial_0} & & & & & \\
{}^0\mathscr{E}_S^*(U) & \xrightarrow{p} & {}^0\mathscr{E}_S^*(U) & \xrightarrow{p_1} & & & & & & \\
& & \uparrow{\scriptstyle \partial_1} & & \ddots & & & & & \\
& & & & & \xrightarrow{p_m} & p_m\,{}^{m-2}\mathscr{E}_S^*(U) & & & \\
& & & & \uparrow{\scriptstyle \partial_{m-1}} & & \uparrow{\scriptstyle \partial_{m-1}} & & & \\
& & \ddots & \xrightarrow{p_{m-1}} & {}^{m-1}\mathscr{E}_S^*(U) & \xrightarrow{m\,p_m} & p_m\,{}^{m-1}\mathscr{E}_S^*(U) & \longrightarrow & 0 & \\
& & & & \uparrow{\scriptstyle \partial_m} & & \uparrow{\scriptstyle \partial_m} & & & \\
& & & & {}^m\mathscr{E}_S^*(U) & \xrightarrow{p_m} & p_m\,{}^m\mathscr{E}_S^*(U) & \longrightarrow & 0 &
\end{array} \tag{20.14}$$

in which we have omitted brackets of the form $[\dots]^{t_k}$. The columns of this diagram, with the exception of the last, are exact, because (19) is exact.

The conditions of the theorem imply that the sequence

$$\mathscr{P}^s \xrightarrow{p} \mathscr{P}^t \xrightarrow{p_1} \mathscr{P}^{t_2} \xrightarrow{p_2} \cdots \xrightarrow{p_{m-1}} \mathscr{P}^{t_m} \xrightarrow{p_m} \mathscr{P}^{t_{m+1}}. \tag{21.14}$$

is exact. An arbitrary region of the form $\omega = U_{j_0} \cap \cdots \cap U_{j_\nu}$ is convex and therefore, the $\mathscr{P}$-module $\mathscr{E}^*(\omega)$ is flat, and this implies that the sequence

$$[\mathscr{E}^*(\omega)]^s \xrightarrow{p} [\mathscr{E}^*(\omega)]^t \xrightarrow{p_1} \cdots \to [\mathscr{E}^*(\omega)]^{t_m} \xrightarrow{p_m} p_m[\mathscr{E}^*(\omega)]^{t_m} \to 0. \tag{22.14}$$

is exact. The exactness of this sequence implies the exactness of all the rows of (20), except the topmost.

We shall prove that the righthand column is exact. Let $\phi$ be an arbitrary element of the space $p_m[{}^{m-1}\mathscr{E}_S^*(U)]^{t_m}$, such that $\partial_{m-1}\phi = 0$. In view of the exactness of (19), we have $\phi = \partial_m \psi$, where

$$\psi = \sum \psi_{j_0,\dots,j_m} U_{j_0} \wedge \cdots \wedge U_{j_m} \in [{}^m\mathscr{E}_S^*(U)]^{t_{m+1}}.$$

We shall show that all the distributions $\psi_{j_0,\dots,j_m}$ belong to $p_m[\mathscr{E}^*(R^n)]^{t_m}$. Let us fix an arbitrary set of indices $j_0, \dots, j_m$ and choose a polygon $l$ consisting of segments of the form $\pi_{i_1} \cap \cdots \cap \pi_{i_m}$, which join the point

$\pi_{j_0} \cap \cdots \cap \pi_{j_m}$, and some other point of the form $\pi_{k_0} \cap \cdots \cap \pi_{k_m}$, lying on the boundary of the projection supp $\psi$ on $B$. Let us write out the sequence of coefficients of the cochain $\psi$, corresponding to the vertices of the polygon $l$:

$$\psi_{k_0, \ldots, k_m}, \ldots, \psi_{i_0, \ldots, i_m}, \ldots, \psi_{j_0, \ldots, j_m}. \tag{23.14}$$

The relation $\partial_m \psi = \phi$ implies that the coefficient of the cochain $\phi$ corresponding to an arbitrary segment of the polygon $l$ is equal to the algebraic sum of the two neighboring distributions in (23) corresponding to the vertices of this segment. Since each coefficient of the cochain $\phi$ belongs to the space $p_m[\mathscr{E}^*(R^n)]^{t_m}$, and the first term in (23) is equal to zero because of the choice of $k_0, \ldots, k_m$, all the terms in the sequence (23) also belong to this space, whence $\psi_{j_0, \ldots, j_m} \in p_m[\mathscr{E}^*(R^n)]^{t_m}$.

We now choose some convex region $\omega_{j_0, \ldots, j_m} \subset U_{j_0} \cap \cdots \cap U_{j_m}$, containing the support of $\psi_{j_0, \ldots, j_m}$. The results of § 8 imply the equation

$$[\mathscr{E}^*(\omega_{j_0, \ldots, j_m})]^{t_{m+1}} \cap p_m[\mathscr{E}^*(R^n)]^{t_m} = p_m[\mathscr{E}^*(\omega_{j_0, \ldots, j_m})]^{t_m}.$$

By what we have already proved, the distribution $\psi_{j_0, \ldots, j_m}$ belongs to the left side, and therefore, belongs to the right side also, that is

$$\psi_{j_0, \ldots, j_m} = p_m \chi_{j_0, \ldots, j_m}, \quad \text{where} \quad \chi_{j_0, \ldots, j_m} \in [\mathscr{E}^*(\omega_{j_0, \ldots, j_m})]^{t_m}.$$

Thus, $\psi = p_m \chi$, where $\chi$ is the cochain whose coefficients are the distributions $\chi_{j_0, \ldots, j_m}$. Since by hypothesis supp $\psi \in S$, we know that for a suitable choice of the regions $\omega_{j_0, \ldots, j_m}$ we can cause supp $\chi$ to belong to $S$ also. Then $\psi \in p_m[{}^m\mathscr{E}_S^*(U)]^{t_m}$, which implies that the right column of diagram (20) is exact.

Applying Theorem 1, § 2, Chapter I to diagram (20), we find that the topmost row is exact. □

We now prove the theorem. Let $\check{u}$ be a distribution in $\Omega \setminus \Sigma$, coinciding with $u$ in $\Omega \setminus (K' \cup \Sigma)$, where $K' \subset \Omega$ is a compact. The distribution $p\check{u}$, obviously, belongs to the kernel of the operator $p_1$ in (18). Since the sequence (18) is exact, it follows that $p\check{u} = pv$ for some distribution $v \in [\mathscr{E}^*(\Omega, \Sigma)]^s$. The distribution $\check{u} - v$ is the desired continuation of $u$.

Let us now suppose that $M$ is an elliptic module, that is, $p$ is an elliptic operator and the region $\Omega$ is such that the set $R^n \setminus (\Omega \cup \Sigma)$ has no bounded connected components. In the sequence (18), we set $\Omega = R^n$. In view of our lemma, the sequence so obtained is exact. The distribution $p\check{u}$ belongs to the kernel of the operator $p_1$ of this sequence. Therefore $p\check{u} = pv$, where $v \in [\mathscr{E}^*(R^n, \Sigma)]^s$. The fact that

$$\operatorname{supp} p v = \operatorname{supp} p \check{u} \subset K' \setminus \Sigma,$$

implies that supp $v \subset K'' \setminus \Sigma$, where $K''$ is a compact belonging to $\Omega$, since the function $v$ is analytic in $R^n \setminus (K''' \cup \Sigma)$, where $K'''$ is a compact in $R^n$. It follows that the distribution $\check{u}-v$ is the desired continuation of $u$. ▯

**Theorem 5.** *The assertion of Theorem 4 is valid under the following hypotheses: $n=2m$, $\Omega$ is a holomorphy domain in $C^m=R^n$, and $M$ is a holomorphic $\mathscr{P}$-module* (see § 12), *and* $\operatorname{Ext}^i(M,\mathscr{P})=0$, $i=1,\ldots,m$.

*Proof.* Let (4.7) be a free resolution of the module $M$, consisting of $\mathscr{P}''$-matrices. We establish the exactness of the sequence (22). To do this, we consider the regions $U_j=(V_j\times L)\cap\Omega$. Since each of the regions $V_j\times L$ is convex, it is a domain of holomorphy. Therefore, the intersection $(V_j\times L)\cap\Omega$ is also a domain of holomorphy. Thus for arbitrary $j_0,\ldots,j_v$ the region $\omega=U_{j_0}\cap\cdots\cap U_{j_v}$ is also a domain of holomorphy, and therefore, by Corollary 3, § 12, the $\mathscr{P}''$-module $\mathscr{E}^*(\omega)$ is flat. This implies the exactness of the sequence (22). The remaining arguments in the proof of Theorem 4 apply without change. ▯

**8°. The uniqueness of the extension of the solutions ot a homogeneous system.** Making use of the method applied in Theorems 3 and 4 we now obtain two theorems which yield sufficient conditions for the uniqueness of solutions of the system (1.4).

**Theorem 6.** *Let $L$ be a subspace in $R^n$, of dimension $n-m$, $0\leqq m<n$. Let the following conditions be satisfied:*

a) $\operatorname{Ext}^i(M,\mathscr{P})=0$, $i=0,\ldots,m-1$, *that is,* $\dim M\leqq n-m$ (see § 13);

b) *an arbitrary variety from the set associated with the module* $\operatorname{Ext}^m(M,\mathscr{P})$, *is not hyperbolic with respect to $L$. Then every distribution defined and satisfying* (1.4) *in the neighborhood of $L$ and equal to zero in the neighborhood of $L\setminus\omega$, where $\omega\subset L$ is a bounded region, is equal to zero in the neighborhood of $L$.*

*Proof.* Let $\kappa\subset L$ be some convex compact such that $\omega\subset\subset\kappa$ and let $\beta$ and $\Omega$ have the same meaning as in the proof of Theorem 3. We have $\operatorname{supp} p\,\beta\,u\subset\Omega$. We observe that the matrix $p$ satisfies the same conditions as the matrix $p_1$ in Theorem 3. Taking account of the fact that for the matrix $p_1$ we have established the Eq. (17), we are entitled to write down the equation

$$[\mathscr{E}^*(\Omega)]^t\cap p[\mathscr{E}^*(R^n)]^s=p[\mathscr{E}^*(\Omega)]^s.$$

Since the distribution $p\,\beta\,u$ belongs to the left side of this equation, it belongs to the right side as well, whence $p\,\beta\,u=p\,v$, where $v\in[\mathscr{E}^*(\Omega)]^s$. Accordingly, $p(\beta\,u-v)=0$, and therefore $\beta\,u=v$, since $\operatorname{Hom}(M,\mathscr{P})=0$. Since $\bar{\Omega}\cap L=\varnothing$, the distribution $v$ is equal to zero in the neighborhood of $L$. We have thus proved that the distribution $u$ is equal to zero in the neighborhood of $L$. ▯

**Theorem 7.** *Let $K$ be a compact, and let $\Sigma_i$, $i=1, \ldots, m$, be arbitrary closed half-spaces in $R^n$. We shall suppose that the conditions* $\operatorname{Ext}^i(M, \mathscr{P})=0$, $i=0, \ldots, m-1$ *are satisfied, that is,* $\dim M \leqq n-m$. *Then every distribution which satisfies* (1.4) *in $R^n \setminus \Sigma$, where $\Sigma = \cup \Sigma_i$, and equal to zero in $R^n \setminus (K \cup \Sigma)$, is equal to zero in $R^n \setminus \Sigma$ also.*

*Proof.* The conditions of the theorem imply the exactness of the sequence

$$0 \to \mathscr{P}^s \xrightarrow{p} \mathscr{P}^t \xrightarrow{p_1} \mathscr{P}^{t_2} \to \cdots \xrightarrow{p_{m-2}} \mathscr{P}^{t_{m-1}} \xrightarrow{p_{m-1}} \mathscr{P}^{t_m}.$$

Starting from this sequence in the same way as from (21), and repeating the arguments given in the lemma, we can establish the exactness of the sequence

$$0 \to [\mathscr{E}^*(R^n, \Sigma)]^s \xrightarrow{p} [\mathscr{E}^*(R^n, \Sigma)]^t. \tag{24.14}$$

Let $u$ be a solution of the system (1.4), satisfying the conditions of the theorem. Then $u \in [\mathscr{E}^*(R^n, \Sigma)]^s$ and $p u = 0$ in $R^n \setminus \Sigma$. Therefore, $u$ belongs to the kernel of the operator $p$ in (24). But the exactness of (24) implies that $u \equiv 0$ in $R^n \setminus \Sigma$. □

**9°. Examples**

**Example 5.** Let $p$ be a non-zero element of the ring $\mathscr{P}$. By Proposition 2, § 13, we have $\operatorname{Hom}(M, \mathscr{P})=0$ and $\operatorname{Ext}^1(M, \mathscr{P}) \cong M$. Let $L$ be a subspace in $R^n$ of dimension $n-1$. Then Theorem 6 reads as follows: if no irreducible component of the variety of the roots of the polynomial $p$ is hyperbolic with respect to $L$, then every solution of (1.4) defined in the neighborhood of $L$ and equal to zero in the neighborhood of $L \setminus \omega$, is equal to zero in the neighborhood of the whole subspace $L$.

**Example 6.** Let $n=2m$, and $p = {}''d_0$ be an operator corresponding to a Cauchy-Riemann system in $R^n = C^m$. The module $M = \operatorname{Coker} p'$ is elliptic, and by Proposition 2, § 13, $\operatorname{Ext}^i(M, \mathscr{P})=0$, $i=0, \ldots, m-1$. Therefore, Theorem 4 implies the following proposition: Let $\Sigma$ be the union of $m-1$ closed halfspaces in $C^m$. Let $\Omega$ be a bounded region such that the set $\mathsf{C}(\Omega \cup \Sigma)$ has no bounded connected components, that is, let the set $\partial\Omega \setminus \Sigma$ be connected. Then an arbitrary holomorphic function defined in the neighborhood of $\partial\Omega \setminus \Sigma$ can be continued to a holomorphic function in $\Omega \setminus \Sigma$.

## § 15. The influence of boundary values on the behavior of the solutions within a region

We now establish another property of determined and overdetermined systems of type (1.4): Every solution of such a system, defined in a region $\Omega$, depends continuously on its values near some portion of the boundary of $\Omega$; the less the dependence, the greater the "degree of overdetermination" of the system. The role of the "degree of overdetermination" will again be played by the module $\operatorname{Ext}^i(M, \mathscr{P})$.

**1°. The continuous dependence of the solution of a determined system on its values in the neighborhood of the boundary.** As we know (see § 14), every solution of a determined system (1.4), given in a region $\Omega$, and equal to zero in the neighborhood of the boundary, is identically equal to zero. In fact, a stronger assertion is valid: the solution of such a system depends continuously on its values in $\Omega \setminus K$, where $K \subset \Omega$ is an arbitrary compact.

**Theorem 1.** *Let $M$ be a determined module* (that is, $\operatorname{Hom}(M, \mathscr{P}) = 0$), *let $K$ be an arbitrary compact in $R^n$, let $\Omega$ be a neighborhood of it, and let $\Delta$ be a compact such that $\partial K \subset\subset \Delta$. Then for an arbitrary integer $q$, every distribution in $\Omega$ which satisfies* (1.4) *and belongs to $[\mathscr{E}_\Delta^{q+A}]^s$, belongs also to the space $[\mathscr{E}_K^q]^s$ and satisfies the inequality*

$$\|u\|_K^q \leqq C \|u\|_\Delta^{q+A}, \tag{1.15}$$

*where $A$ is a constant depending only on $p$.*

*Proof.* Let $\alpha$ be an infinitely differentiable function in $\Omega$ with a support belonging to $\Delta \cup K$, and equal to unity in the neighborhood of $K$. The support of the distribution $p \alpha u$ belongs to $\Delta$, and therefore, we have the inequality

$$\|p \alpha u\|_{\Delta \cup K}^q \leqq C \|u\|_\Delta^{q+m}, \qquad m = \deg p, \qquad -\infty < q < \infty. \tag{2.15}$$

Let $K_q$, $-\infty < q < \infty$, be a strictly decreasing sequence of convex compacts containing $\Delta \cup K$. We consider the decreasing sequence of spaces $\mathscr{D}_{K_q}^q$, $-\infty < q < \infty$. By Proposition 2, § 3 the Fourier transform is a continuous mapping from $\mathscr{D}_{K_q}^q$ to $S_q^{K_q^*+1}$, and the inverse Fourier transform maps $S^{K_q^*}$ continuously into $\mathscr{D}_{K_{q+2}}$. On the other hand, the sequence $\mathscr{H}_q = S_-^{K^*_{-q}}$ is a family $\mathscr{H}_{\mathscr{M}}$, corresponding to the family of majorants $\mathscr{M}$ of type $\mathscr{I}$. Therefore, by Theorem 2, § 5, Chapter IV, and the Supplement to that theorem, we know that corresponding to an arbitrary integer $q$ there is defined a continuous mapping

$$p[\mathscr{H}_{q-m}]^s \to [\mathscr{H}_{q+a}]^s / [\mathscr{H}_{q+a}]^s \cap \operatorname{Ker} p,$$

inverse to the operator $p$, where $a$ is a constant depending only on $p$. Combining this fact with the properties of the Fourier transform noted above, and arguing by analogy with the proof of Theorem 1, §4, we arrive at the following result: For arbitrary $q$ there is defined a continuous operator

$$p^{-1}: \quad p[\mathscr{D}_{K_{q+m}}^{q+m}]^s \to [\mathscr{D}_{K_{q-b}}^{q-b}]^s / [\mathscr{D}_{K_{q-b}}^{q-b}]^s \cap \operatorname{Ker} p, \tag{3.15}$$

inverse to $p$. The left side has the topology induced by $[\mathscr{D}^q_{K_q}]^t$, and the constant $b$ again depends only on $p$.

We now remark that the distribution $p\,\alpha\,u$ belongs to the left side of (3), and therefore the operator $p^{-1}$ is applicable to it. Since this operator is continuous, we know that for every $\varepsilon>0$ we can find an element $\phi\in p^{-1}(p\,\alpha\,u)$ such that

$$\|\phi\|^{q-b}_{K_{q-b}} \leqq C\,\|p\,\alpha\,u\|^q_{K_q}+\varepsilon, \qquad C=\|p^{-1}\|. \tag{4.15}$$

Since the support of the distribution $p\,\alpha\,u$ belongs to $\Delta$, the first term on the right side is equal to $C\|p\,\alpha\,u\|^q_\Delta$. From the equation $p\,\phi=p\,\alpha\,u$ it follows that $\phi=\alpha\,u$, since $p$ is a determined operator. Therefore, on the right side of (4) we may set $\varepsilon=0$. Combining (4) and (2), we obtain

$$\|\alpha\,u\|^{q-b}_{K_{q-b}} \leqq C\,\|p\,\alpha\,u\|^q_\Delta \leqq C'\,\|u\|^{q+m}_\Delta.$$

The left side is not less than $\|u\|^{q-b}_K$. Therefore, replacing $q-b$ by $q$, we arrive at (1). □

**2°. A theorem on continuous dependence for overdetermined systems**

**Theorem 2.** *Let the module $M$ satisfy the conditions* $\dim M \leqq n-k$ *for some* $0<k\leqq n$ (which is equivalent to the condition $\operatorname{Ext}^i(M,\mathscr{P})=0$, $i=0,\ldots,k-1$, by § 13). *Suppose further that $\Pi$ is a bounded polyhedron in $R^n$, and that $\Gamma_{n-k}$ is a compact containing the neighborhood of an $n-k$-dimensional skeleton of the polyhedron $\Pi$, and let $\Omega$ be a neighborhood of $\Pi\cup\Gamma_{n-k}$. Then for an arbitrary integer $q$, every distribution in $\Omega$ which satisfies* (1.4), *and belongs to $[\mathscr{E}^{q+A}_{\Gamma_{n-k}}]^s$, belongs also to the space $[\mathscr{E}^q_\Pi]^s$ and satisfies the inequality*

$$\|u\|^q_\Pi \leqq C\,\|u\|^{q+A}_{\Gamma_{n-k}}, \tag{5.15}$$

*where $A$ is a constant depending only on $p$.*

Let us first note an elementary fact.

**Lemma.** *For an arbitrary polynomial $q\in\mathfrak{r}(M)$ we infer from* (1.4) *the inequality $Q(i\,D)\,u=0$, where $Q$ is a diagonal matrix of size $s\times s$, which contains in its diagonal a sufficiently high power of the operator $q$.*

*Proof of the lemma.* By definition of the ideal $\mathfrak{r}(M)$ every polynomial $q\in\mathfrak{r}(M)$, when raised to a sufficiently high power, annihilates an arbitrary element of the module $M$. It follows that for sufficiently large $\rho$, all the columns $q^\rho\,e_\sigma$, $\sigma=1,\ldots,s$ ($e_\sigma$ are the columns of the unit matrix) belong to the submodule $p'\,\mathscr{P}^t$, that is, $q^\rho\,e_\sigma=p'\,F_\sigma$, $F_\sigma\in\mathscr{P}^t$. Hence $e'_\sigma\,q^\rho=F'_\sigma\,p$. Therefore $p(i\,D)\,(u_1,\ldots,u_s)=0$ implies $q^\rho(i\,D)\,u_\sigma=0$ for all $\sigma$. □

*Proof of the theorem.* The case $k=1$ is contained in Theorem 1. Let us suppose that $k>1$. Let $\kappa$ be an arbitrary $n-k+1$-dimensional face of the polyhedron $\Pi$, and $L$ be a linear manifold of the same dimension containing this space. In $R^n$ we choose a system of coordinates such that the axes of $\xi_k, \ldots, \xi_n$ lie in $L$. Without loss of generality, we may suppose that the variety $N(M)$ is normally placed in the corresponding system of coordinates in $C^n$. In fact, as we noted in § 1, Chapter IV, those systems of coordinates in $C^n$, in which a given algebraic variety is not normally placed, form a nowhere dense subset in the set of all systems of coordinates. Therefore, we may always find a polyhedron $\Pi' \supset \Pi$, as close as we please to $\Pi$, and having faces that satisfy the conditions we have just prescribed.

So, $N(M)$ is normally placed in the system of coordinates fixed above. Then there exists a polynomial $q(z_k, \ldots, z_n) \not\equiv 0$, and vanishing on $N(M)$. This means that it belongs to the ideal $\mathfrak{r}(M)$, and therefore, by our lemma, every solution of the system (1.4) satisfies also the system $Q(i D_{\xi''})\, u = 0$, where $Q$ is a determined operator containing differentiations with respect to the variables $\xi'' = (\xi_k, \ldots, \xi_n)$ only. Let us write $\xi' = (\xi_1, \ldots, \xi_{k-1})$. Let us further choose a closed sphere $S$ with a center at the origin of coordinates, and an admissible compact $\delta \subset L$, containing a neighborhood of the boundaries of the face $\kappa$, which is sufficiently small so that $S + \kappa \subset \Omega$ and $S + \delta \subset \Gamma_{n-k}$.

For every function $\phi \in \mathscr{D}_S$ we consider the convolution $u * \phi = (u(\eta), \phi(\xi - \eta))$. In view of the choice of the sphere $S$, the convolution $u * \phi$ is defined in an $n$-dimensional neighborhood of the compact $\kappa \cup \delta$, it is infinitely differentiable, and it satisfies in this neighborhood the system (1.4); it therefore satisfies the system $Q(i D_{\xi''})\, (u * \phi) = 0$. It follows that the function $v(\xi'') = (u * \phi)|_{\xi' = 0}$, is defined in some $n-k+1$-dimensional neighborhood $\omega \subset L$ of the compact $\kappa$, and in the subspace $L$ satisfies the determined system of equations $Q(i D_{\xi''})\, v = 0$. Applying Theorem 1 to the function $v$, and also making use of Proposition 3, § 3, we obtain the chain of inequalities

$$\begin{aligned}
\sup_{\xi'' \in \kappa} |(u(\eta), \phi(\xi'' - \eta))| &= \sup_{\kappa} |v(\xi'')| \leqq C\, \|v\|_\kappa^{\nu} \\
&\leqq C'\, \|v\|_\delta^{\nu + a} \leqq C'' \max_{|i| \leqq \nu + a} \sup_{\delta} |D_{\xi''}^i\, v| \leqq C'' \max_{|i| \leqq \nu + a} \sup_{\delta} |u * D^i \phi| \\
&\leqq C''\, \|u\|_{\Gamma_{n-k}}^{q + \nu + a} \max_{|i| \leqq \nu + a} \sup_{\xi'' \in \delta} \|D_{\xi''}^i\, \phi(\xi'' - \eta)\|^{-(q + \nu + a)} \\
&\leqq C'''\, \|u\|_{\Gamma_{n-k}}^{q + \nu + a}\, \|\phi\|^{-q},
\end{aligned}$$

where $\nu = \left[\dfrac{n}{2}\right] + 1$. It follows that for an arbitrary function $\psi \in \mathscr{D}_{\kappa + S}$, which is a translation of a function $\phi \in \mathscr{D}_S$, we have the inequality

$$|(u, \psi)| \leqq C\, \|u\|_{\Gamma_{n-k}}^{q + \nu + a}\, \|\psi\|^{-q}. \tag{6.15}$$

Since $\kappa$ is compact, a finite number of translations of the sphere $S$ will cover the compact $K \subset \Omega$, containing a neighborhood of $\kappa$. Using a partition of the identity subordinate to this covering, we can represent an arbitrary function $\psi \in \mathscr{D}_{\kappa+S}$ as a sum $\sum \psi_i$ with a fixed number of terms, which are functions belonging to $\mathscr{D}_S$ and continuously depend on $\psi$. If we apply the inequality (6) to each of these, we obtain

$$\|u\|_{\kappa+S}^{q} \leqq C \|u\|_{\Gamma_{n-k}}^{q+\nu+a}.$$

By summing over all the $n-k+1$-dimensional faces of $\kappa$, we obtain

$$\|u\|_{\Gamma_{n-k+1}}^{q} \leqq C \|u\|_{\Gamma_{n-k}}^{q+\nu-a},$$

where $\Gamma_{n-k+1}$ is a compact containing a neighborhood of the $n-k+1$-dimensional skeleton of the polyhedron $\Pi$. In a similar fashion, we can establish an inequality of the form $\|u\|_{\Gamma_{i+1}}^{q} \leqq C \|u\|_{\Gamma_i}^{q+b}$ for arbitrary $i > n-k$. Combining all these inequalities, we arrive at (5). $\square$

**Corollary 1.** *Let the conditions of Theorem* 2 *be satisfied. Then every distribution defined and satisfying* (1.4) *in the neighborhood of* $\Pi$, *and analytic in the neighborhood of the* $n-k$*-dimensional skeleton of* $\Pi$, *is analytic in the neighborhood of* $\Pi$.

*Proof.* We choose an open neighborhood $\Omega$ of the polyhedron $\Pi$ and a closed neighborhood $\Gamma_{n-k}$ of its $n-k$-dimensional skeleton so that the solution $u$ is defined in $\Omega$ and analytic in the neighborhood $\Gamma_{n-k}$. An arbitrary derivative $D^i u$ of this solution belongs to the space $[\mathscr{E}_{\Gamma_{n-k}}^{\nu+A}]^s$, where $\nu = \left[\frac{n}{2}\right] + 1$, and it satisfies the inequality

$$\|D^i u\|_{\Gamma_{n-k}}^{\nu+A} \leqq C B^{|i|} i!.$$

By Theorem 2 all the distributions $D^i u$ belong to $[\mathscr{E}_{\Pi}^{\nu}]^s$ and satisfy a similar set of inequalities

$$\|D^i u\|_{\Pi}^{\nu} \leqq C' B^{|i|} i!.$$

In view of Proposition 3, § 3, the fact that the left sides are finite implies that all the distributions $D^i u$ are continuous in int $\Pi$, and their absolute value does not exceed $C'' B^{|i|} i!$. It follows that the function $u$ is analytic in the neighborhood of $\Pi$. $\square$

**3°. Extension of smoothness.** From now on the differentiability of a distribution will mean infinite differentiability.

Theorem 2 implies, in particular, that if the conditions $\operatorname{Ext}^i(M, \mathscr{P}) = 0$, $i = 0, \ldots, k$, are satisfied, then every solution defined in the neighborhood of the polyhedron $\Pi$, and differentiable in the neighborhood of the $n-k$-dimensional skeleton $\gamma_{n-k}$, is differentiable in the neighborhood of the whole polyhedron. We shall now show that this property (even in a stronger form) is conserved when we weaken the assumption relative to the module $M$.

**Theorem 3.** *Let the module $M$ be such that all the modules* $\operatorname{Ext}^i(M, \mathscr{P})$, *$i=0, \ldots, k$, are hypoelliptic. Suppose further that $\Omega$ is a convex region, $K \subset \Omega$ is a compact, and $\Sigma_i$, $i=0, \ldots, k$, – are closed half-spaces in $R^n$ and that $\Sigma = \cup \Sigma_i$. Then every distribution which satisfies* (1.4) *in $\Omega \setminus \Sigma$, and is differentiable in $\Omega \setminus (K \cup \Sigma)$, is differentiable in $\Omega \setminus \Sigma$.*

*Proof.* The condition imposed on the module $M$ will now be formulated in another way. By Corollary 3, § 13, the union $H$ of all irreducible components of the variety $N(M)$ having dimension not less than $n-k$, coincides with $\bigcup_{i \leqq k} N(\operatorname{Ext}^i(M, \mathscr{P}))$. Each of the varieties $N(\operatorname{Ext}^i(M, \mathscr{P}))$, $i \leqq k$, is by hypothesis hypoelliptic, and therefore the union of these components $H$ is also hypoelliptic. Therefore the hypothesis of the theorem can be formulated as follows: The variety $N(M)$ is the union of a hypoelliptic variety $H$ and a variety $v$, of dimension less than $n-k$.

We denote by $\mathfrak{p}_H$ the intersection of all the primary components $\mathfrak{p}_\lambda$ of the submodule $p' \mathscr{P}^t \subset \mathscr{P}^s$ such that $\dim N(\mathfrak{p}_\lambda) \geqq n-k$, and we denote by $\mathfrak{p}_v$ the intersection of the remaining prime components of this submodule. We write $M_H = \mathscr{P}^s / \mathfrak{p}_H$ and $M_v = \mathscr{P}^s / \mathfrak{p}_v$. We have $N(M_H) = H$, and $N(M_v) = v$, and therefore the module $M_H$ is hypoelliptic and the dimension of the module $M_v$ is less than $n-k$. The identity mapping $\mathscr{P}^s \to \mathscr{P}^s$ generates the mappings $M \to M_H$ and $M \to M_v$. Since $p' \mathscr{P}^t = \mathfrak{p}_H \cap \mathfrak{p}_v$, the intersection of the kernels of these mappings is equal to zero. Therefore Corollary 5, § 8 implies that every distribution $u$ which satisfies (1.4) in the convex region $\Omega \setminus \Sigma$, can be represented as a sum $u_H + u_v$, where $u_H \in \mathscr{D}^*_{M_H}(\Omega \setminus \Sigma)$ and $u_v \in \mathscr{D}^*_{M_v}(\Omega \setminus \Sigma)$. Since the module $M_H$ is hypoelliptic, the distribution $u_H$ is differentiable in $\Omega \setminus \Sigma$. Accordingly, the distribution $u_v = u - u_H$ is differentiable in $\Omega \setminus (K \cup \Sigma)$.

Thus, our task is reduced to showing that an arbitrary distribution $u_v \in \mathscr{D}^*_{M_v}(\Omega \setminus \Sigma)$, which is differentiable in $\Omega \setminus (K \cup \Sigma)$, has the same property throughout the whole region. In other words, the proof of the theorem for the module $M$ can be reduced to the proof of the theorem for the module $M_v$, which has dimension less than $n-k$. In order to conserve our former notation, we shall prove the theorem for the initial module $M$, supposing now that its dimension is less than $n-k$.

We first prove a lemma. Let $\Upsilon$ and $\upsilon \subset \Upsilon$ be arbitrary convex regions in $R^n$. We denote by $\hat{\mathscr{E}}^*(\Upsilon)$ the subspace in $\mathscr{E}^*(\Upsilon)$, consisting of distributions $\phi$ such that $\operatorname{sing\,supp} \phi \subset \upsilon$.

**Lemma.** *For an arbitrary $\mathscr{P}$-matrix $p$*

$$[\hat{\mathscr{E}}^*(\Upsilon)]^t \cap p[\mathscr{E}^*(\Upsilon)]^s = p[\hat{\mathscr{E}}^*(\Upsilon)]^s.$$

*Proof of the lemma.* Let $\phi$ be an arbitrary element of the left side. We choose two strictly increasing sequences of convex compacts $\kappa_\alpha$, $K_\alpha$, $\alpha=1, 2, \ldots$, tending respectively to $v$ and to $\Upsilon$ such that $\operatorname{supp}\phi\subset K_1$, and sing $\operatorname{supp}\subset\kappa_1$. We further choose a function $e\in\mathscr{D}_\kappa$, which is equal to unity in the neighborhood of $\kappa_1$. Let $d=\{d^\lambda\}$ be a Noetherian operator associated with the matrix $p$. The inclusion relation $\phi\in p[\mathscr{E}^*(\Upsilon)]^s$ implies the equation

$$F \stackrel{\text{def.}}{=\!=\!=} \widetilde{d\,e\,\phi} = d\widetilde{(e-1)\,\phi}.$$

Since $\operatorname{supp} e\,\phi\subset\kappa_2$, we have $e\,\phi\in[\mathscr{D}^q_{\kappa_2}]^t$ for some integer $q$. Hence, $\widetilde{e\phi}\in[S_q^{\kappa_3^*}]^t$ and, therefore, $\widetilde{d\,e\,\phi}\in S^{\kappa_3^*}_{q-\kappa}\{p\}$, where $\kappa$ is the highest power of $z$ appearing in the operator $d^\lambda$. Therefore

$$|F(z)|\leqq C(|z|+1)^{-q+\kappa}\exp(\mathscr{I}_{\kappa_3}(-y)), \qquad z\in N(p). \tag{7.15}$$

On the other hand, the function $(e-1)\,\phi$ is differentiable, so that its Fourier transform satisfies the inequality

$$|\widetilde{(e-1)\,\phi}|\leqq r(z)\exp(\mathscr{I}_{K_1}(-y)), \tag{8.15}$$

where $r(z)\to 0$ for $|z|\to\infty$ faster than an arbitrary power of $|z|$. Therefore the function $F$ satisfies the inequality

$$|F(z)|\leqq r'(z)\exp(\mathscr{I}_{K_2}(-y)), \tag{9.15}$$

where $r'(z)\to 0$ for $|z|\to\infty$ faster than an arbitrary power of $|z|$.

We now show that the inequalities (7) and (9) imply an inequality of the form

$$|F(z)|\leqq R(z)\exp(\mathscr{I}_{\kappa_4}(-y)), \tag{10.15}$$

where $R(z)\to 0$ as $|z|\to\infty$ also faster than an arbitrary power of $|z|$. In fact, suppose that $|y|\leqq\frac{1}{2l}\ln\frac{1}{r'(z)}$, where $l=\sup\{|\xi|,\ \xi\in\kappa_4\}+\sup\{|\xi|,\ \xi\in K_2\}$. Then $\exp(\mathscr{I}_{\kappa_4}(-y)-\mathscr{I}_{K_2}(-y))\geqq\sqrt{r'(z)}$ and, accordingly, by (9)

$$|F(z)|\leqq\sqrt{r'(z)}\exp(\mathscr{I}_{\kappa_4}(-y)).$$

We now suppose that $|y|\geqq\frac{1}{2l}\ln\frac{1}{r'(z)}$. Then $\exp(-\varepsilon|y|)\leqq[r'(z)]^{\varepsilon/2l}$, where $\varepsilon=\frac{1}{2}\rho(\kappa_3, C\kappa_4)$ and, accordingly, (7) implies that

$$\begin{aligned}|F(z)|&\leqq C(|z|+1)^{-q+\kappa}\exp(-\varepsilon|y|)\exp(\mathscr{I}_{\kappa_4}(-y))\\&\leqq C(|z|+1)^{-q+\kappa}\,r'(z)^{\varepsilon/2l}\exp(\mathscr{I}_{\kappa_4}(-y)).\end{aligned}$$

With this we have proved the inequality (10).

This inequality in turn proves that $F \in S_R^{\kappa_4^*}\{p\}$. We construct an increasing sequence of functions $R_\alpha(z)$, $\alpha=1, 2, \ldots$, which satisfy the conditions of Proposition 5, § 3, and such that $r(z)+R(z) \leqq R_1(z)$. Since the family $\mathscr{H}_{\mathscr{M}}=\{S_{R_\alpha}^{\kappa_\alpha^*}\}$ corresponds to a family of majorants $\mathscr{M}$ of type $\mathscr{I}$, we may apply Theorem 2, § 5, Chapter IV. This theorem shows that $F=d\psi$, where $\psi \in [S_{R_\alpha}^{\kappa_\alpha^*}]^t$ for some $\alpha$. Hence $d[\widetilde{(1-e)\,\phi}+\psi]=0$. The inequality (8) allows us to conclude that the function $\widetilde{(1-e)}\,\phi+\psi$ belongs to $[S_{R_\alpha}^{K_\alpha^*}]^t$. Therefore the theorem just cited implies that

$$\widetilde{(1-e)}\,\phi+\psi=p\,\chi, \qquad \chi \in [S_{R_\beta}^{K_\beta^*}]$$

for some $\beta$. In a similar fashion, we show that $d[\widetilde{e\phi}-\psi]=0$, where $\widetilde{e\phi}-\psi \in [S_{-q}^{\kappa_\alpha^*}]^t$ for some $q$. Applying our theorem again to the family of spaces $\{S_{-\alpha}^{\kappa_\alpha^*}, \alpha=1, 2, \ldots\}$, we find that

$$\widetilde{\alpha\phi}-\psi=p\,\omega, \qquad \omega \in [S_{-\beta}^{\kappa_\beta^*}]^s.$$

Carrying out the inverse Fourier transform, we obtain

$$\phi=(e\,\phi-\tilde{\psi})+[(1-e)\,\phi+\tilde{\psi}]=p(\tilde{\chi}+\tilde{\omega}),$$

$$\tilde{\chi} \in [\mathscr{D}(\Upsilon)]^s, \qquad \tilde{\omega} \in [\mathscr{E}^*(\upsilon)]^s.$$

Hence $\phi \in p[\hat{\mathscr{E}}^*(\Upsilon)]^s$. □

**Corollary 2.** *The space $\hat{\mathscr{E}}^*(T)$, when endowed with the discrete topology, is a flat $\mathscr{P}$-module.*

The proof is left to the reader.

**4°. Completion of the proof of Theorem 3.** By hypothesis the dimension of the module $M$ is less than $n-k$. Therefore by Corollary 1, § 13 we have the equation

$$\operatorname{Ext}^i(M, \mathscr{P})=0, \qquad i=0, \ldots, k. \tag{11.15}$$

**Inductive proposition.** *Let us suppose that the module $M$ satisfies the conditions (11) for $i=1, \ldots, k$. Let $T$ be a convex region, and let $\Sigma_i'$, $i=0, \ldots, k$, be open halfspaces in $R^n$. Then for an arbitrary distribution $\phi \in p[\mathscr{E}^*(R^n)]^s$ with support in $T$, differentiable in the neighborhood of $R^n \setminus \Sigma'$ $(\Sigma'=\bigcup \Sigma_i')$, we can find a distribution $\psi$ with support in $T$, which is differentiable in the neighborhood of $R^n \setminus \Sigma'$, and such that $\phi=p\,\psi$.*

This assertion will be proved by induction on $k$. For $k=0$ it follows from our lemma, if we put $\upsilon=\Upsilon \cap \Sigma_0$. We shall show it for an arbitrary $k \geqq 1$, on the assumption that it has already been proved for $k-1$.

We choose a free resolution (4.7) of the module $M$. The module $M_1=\operatorname{Coker} p_1'$ satisfies the relations

$$\operatorname{Ext}^i(M_1, \mathscr{P}) \cong \operatorname{Ext}^{i+1}(M, \mathscr{P})=0, \qquad i=1, \ldots, k-1,$$

and therefore the operator $p_1$ satisfies the inductive proposition when $k$ is replaced by $k-1$.

Let $\phi$ be an arbitrary distribution satisfying the conditions of the inductive proposition. We choose a closed half-space $\Sigma_0''\subset\Sigma_0'$ such that the distribution $\phi$ is differentiable in the neighborhood of

$$R^n\backslash(\Sigma_0''\cup\hat{\Sigma}), \qquad \text{where } \hat{\Sigma}=\bigcup_1^k \Sigma_i'.$$

We choose a function $\alpha\in\mathscr{E}(R^n)$, equal to unity in the neighborhood of $R^n\backslash\Sigma_0'$ and zero in the neighborhood of $\Sigma_0''$. The product $\alpha\phi$ is differentiable in the neighborhood of $R^n\backslash\hat{\Sigma}$. It follows that the distribution $p_1\alpha\phi$ is also differentiable in the neighborhood $R^n\backslash\hat{\Sigma}$, and that its support belongs to $(\Sigma_0'\backslash\Sigma_0'')\cap\Upsilon$. Consequently the distribution $p_1\alpha\phi$ satisfies the induction hypothesis, with $k-1$ in place of $k$, $p_1$ in place of $p$, and the region $(\Sigma_0'\backslash\Sigma_0'')\cap\Upsilon$ in place of $\Upsilon$. Therefore, we can find a distribution $\psi'$ with support in $(\Sigma_0'\backslash\Sigma_0'')\cap\Upsilon$, which is differentiable in the neighborhood of $R^n\backslash\hat{\Sigma}$, such that $p_1\alpha\phi=p_1\psi'$.

Since the module $\mathscr{E}^*(R^n)$ is flat, the equation $\mathrm{Ext}^1(M,\mathscr{P})=0$ implies that the sequence

$$[\mathscr{E}^*(R^n)]^s \xrightarrow{p} [\mathscr{E}^*(R^n)]^t \xrightarrow{p_1} [\mathscr{E}^*(R^n)]^{t_2}. \tag{12.15}$$

is algebraically exact. Since $p_1(\alpha\phi-\psi')=0$, the distribution $\alpha\phi-\psi'$ belongs to the kernel of the operator $p_1$ in this sequence and therefore can be written in the form

$$\alpha\phi-\psi'=p\chi, \qquad \chi\in[\mathscr{E}^*(R^n)]^s. \tag{13.15}$$

Since the distribution $\alpha\phi-\psi'$ is differentiable in the neighborhood of $R^n\backslash\hat{\Sigma}$, and its support belongs to $\Upsilon\backslash\Sigma_0''$, we can apply the induction hypothesis to it, with $k-1$ in place of $k$. It follows that the distribution $\chi$ in (13) can be chosen to be differentiable in the neighborhood of $R^n\backslash\hat{\Sigma}$ and to have a support belonging to $\Upsilon\backslash\Sigma_0''$.

The equation $p_1\phi=0$ implies that $p_1[(1-\alpha)\phi+\psi']=0$. Since the support of the distribution $(1-\alpha)\phi+\psi'$ belongs to the convex region $\Upsilon\cap\Sigma_0'$, we have

$$(1-\alpha)\phi+\psi'=p\omega, \qquad \omega\in[\mathscr{E}^*(\Upsilon\cap\Sigma_0')]^s. \tag{14.15}$$

The existence of a distribution $\omega$ with these properties follows from the exactness of the sequence (12), in which $R^n$ is replaced by $\Upsilon\cap\Sigma_0'$. Combining (13) and (14), we obtain

$$\phi=p(\chi+\omega).$$

The distribution $\psi=\chi+\omega$ is differentiable in the neighborhood of $R^n\backslash\Sigma'$, and its support belongs to $\Upsilon$; it is thus the distribution we are seeking. This completes the proof of the inductive proposition. □

Let us now turn to the proof of the theorem itself. Let $u$ be a distribution satisfying (1.4) in $\Omega\setminus\Sigma$ and differentiable in $\Omega\setminus(K\cup\Sigma)$. For every $i=0,\ldots,k$ we choose an arbitrary open half-space $\Sigma_i'\supset\Sigma_i$ and we write $\Sigma'=\cup\Sigma_i'$. We further choose a function $\beta\in\mathscr{D}(\Omega)$, equal to unity in the neighborhood of $K$. The distribution $p\beta u$ belongs to $p[\mathscr{E}^*(\Omega)]^s$ and is differentiable in the neighborhood of $R^n\setminus\Sigma'$. Applying the inductive proposition to it with $\Upsilon=\Omega$, we arrive at a distribution $\psi\in[\mathscr{E}^*(\Omega)]^s$, which is differentiable in the neighborhood of $R^n\setminus\Sigma'$, and such that $p\beta u=p\psi$. The distribution $\check{u}=(1-\beta)u+\psi$ is differentiable, it satisfies the system (1.4) in the neighborhood of $\Omega\setminus\Sigma'$, and it coincides with $u$ in $\Omega\setminus(K'\cup\Sigma')$, where $K'\subset\Omega$ is compact. The difference $u-\check{u}$ is a solution of the same system and is equal to zero in $\Omega\setminus(K'\cup\Sigma')$, and, therefore, we can extend it to be equal to zero $R^n\setminus(K'\cup\Sigma')$, again obtaining a solution of the system (1.4). Now using all the relations (5), we apply Theorem 7, § 14, with $m=k+1$. This implies that $u\equiv\check{u}$, that is, the distribution $u$ is itself differentiable in the neighborhood of $\Omega\setminus\Sigma'$. Since the half-spaces $\Sigma_i'\supset\Sigma_i$ were chosen arbitrarily, we infer that the distribution $u$ is differentiable in the whole region $\Omega\setminus\Sigma$. ▯

**5°. Corollaries.** We now obtain some corollaries of Theorem 3 which are analogous to Corollary 1 and Theorem 2.

**Corollary 3.** *Suppose that the module $M$ satisfies the conditions of Theorem* 3, *that $L$ is an $(n-k)$-dimensional subspace in $R^n$, that $\kappa$ is a convex compact, and that $\sigma$ is a closed half-space belonging to $L$. Then every distribution which satisfies* (1.4) *in the neighborhood of $\kappa\setminus\sigma$, and is differentiable in the neighborhood of $\partial\kappa\setminus\sigma$, is differentiable in the neighborhood of $\kappa\setminus\omega$.*

*Proof.* Let $U$ be a solution of (1.4), defined in the region $\Omega\supset\kappa\setminus\sigma$, and differentiable in the regions $\omega\supset\partial\kappa\setminus\sigma$. We now choose the closed half-spaces $\Sigma_i$, $i=0,\ldots,k$ in $R^n$ to be such that the set $R^n\setminus(\Sigma\cup\omega)$ has a connected, bounded component $G$, such that $\kappa\setminus\sigma\subset\omega\cup G\subset\Omega$. The function $u$ satisfies (1.4) in the region $(\omega\cup G)\setminus\Sigma$ and is differentiable in $\omega\setminus\Sigma$, and the region $\omega\cup G$, containing $\kappa\setminus\sigma$, can be assumed to be convex. It follows from Theorem 3 that the distribution $u$ is differentiable in the region $\omega\cup G)\setminus\Sigma\supset\kappa\setminus\sigma$. ▯

**Corollary 4.** *Let the module $M$ satisfy the conditions of Theorem* 3, *let $\Pi$ be a bounded polyhedron, and let $\Sigma_0$ be an open half-space in $R^n$. Then every distribution which satisfies* (1.4) *in a neighborhood of $\Pi\setminus\Sigma_0$, and is differentiable in the neighborhood of $\gamma_{n-k}\setminus\Sigma_0$* (here $\gamma_{n-k}$ is the skeleton of dimension $n-k$), *is differentiable in the neighborhood $\Pi\setminus\Sigma_0$.*

*Proof.* Let $\kappa$ be a face of the polyhedron $\Pi$, of dimension $n-j$, where $1\leqq j\leqq k$, and let it be supposed known that the solution $u$ is differentiable in the neighborhood of $\partial\kappa\backslash\Sigma_0$. Then Corollary 3 implies that $u$ is differentiable also in the neighborhood of $\kappa\backslash\Sigma_0$. Applying this argument to the faces of the polyhedron $\Pi$, of dimension $n-k$, then to the faces of dimension $n-k+1$ and so forth, and then, finally, to those of dimension $n-1$, we prove that the distribution is differentiable in the neighborhood of $\partial\Pi\backslash\Sigma_0$. We choose the neighborhood $\Omega\supset\Pi$ and the compact $K\subset\Omega$ so that the distribution $u$ is defined in the neighborhood of $\Omega\backslash\Sigma_0$ and differentiable in the neighborhood of $\Omega\backslash(K\cup\Sigma_0)$. Suppose that $u'$ is a distribution in $\Omega$, coinciding with $u$ in the neighborhood of $\Omega\backslash\Sigma_0$. We choose a function $\beta\in\mathscr{D}(\Omega)$, to be equal to unity in the neighborhood of $K$. The distribution $p\,\beta u$ is differentiable in the neighborhood of $R^n\backslash\Sigma_0$ and has a compact support. Therefore, our lemma implies that there exists a distribution $\psi$ with a compact support, which is differentiable in the neighborhood of $R^n\backslash\Sigma_0$, and such that $p\,\beta\,u=p\,\psi$. Hence $p(\beta\,u-\psi)=0$ and therefore $\beta\,u\equiv\psi$, since $p$ is a determined operator. Therefore the function $u$, coinciding with $u'$ in the neighborhood of $R^n\backslash\Sigma_0$, is differentiable in the neighborhood of this set. □

A result similar to that of Theorem 3 is valid for underdetermined operators, if the solution of the corresponding system (1.4) is looked on as being determined up to a solution with a "compact carrier."

**Corollary 5.** *Let the module $M$ satisfy the conditions* $\operatorname{Ext}^i(M,\mathscr{P})=0$, $i=1,\ldots,k$. *Then in the notation of Theorem 3 the following assertion is valid: Every distribution which satisfies* (1.4) *in the neighborhood of $\Omega\backslash\Sigma$ (all the $\Sigma_i$ are open), and is differentiable in the neighborhood of $\Omega\backslash(K\cup\Sigma)$, is the sum of a solution that is differentiable in the neighborhood of $\Omega\backslash\Sigma$ and a distribution in $\Omega\backslash\Sigma$ that is equal to zero outside an arbitrary predetermined convex neighborhood of $K$.*

*Proof.* Let $\omega$ be an arbitrary convex neighborhood of the compact $K$. We return to the last part of the proof of Theorem 3. The function $\beta$ will be subjected to the condition $\operatorname{supp}\beta\subset\omega$. Then $\operatorname{supp} p\,\beta\,u\subset\omega$ and therefore, when we apply the inductive proposition with $\Upsilon=\omega$, we obtain a distribution $\psi$ with a support in $\omega$. Therefore the distribution $\check{u}=(1-\beta)\,u+\psi$, is differentiable in the neighborhood of $\Omega\backslash\Sigma$, and coincides with $u$ in $\Omega\backslash(\Sigma\cup\omega$. The decomposition $u=\check{u}+(u-\check{u})$ is the one we are looking for. □

The result of Corollary 5, which yields a condition on the extension of smoothness for underdetermined operators, is weaker than Theorem 3, since the condition $\operatorname{Ext}^i(M,\mathscr{P})=0$ is significantly stronger than the condition: the module $\operatorname{Ext}^i(M,\mathscr{P})$ is hypoelliptic. We do not know, however, whether in Theorem 5 the condition $\operatorname{Ext}^i(M,\mathscr{P})=0$ can be replaced by the latter and more natural condition. We know only that the condition $\mathrm{Ext}^1(M, \mathscr{P})=0$ in Corollary 5 can be weakened, in that we can replace it by: $\mathrm{Ext}^1(M, \mathscr{P}) \cong \mathrm{Coker}\, h$, where $h$ is a hypoelliptic operator.

**6°. Examples**

**Example 1.** Let $p$ be an arbitrary operator. Corollary 5 implies the following assertion: Every solution of (1.4), defined in the neighborhood of $\Omega \setminus \Sigma_0$ ($\Sigma_0$ is an open half space), and differentiable in the neighborhood of $\Omega \setminus (K \cup \Sigma_0)$ – where $K$ is a convex compact belonging to $\Omega$ – is the sum of a solution which is differentiable in the neighborhood of $\Omega \setminus \Sigma_0$, and a solution with a support in an arbitrary compact $\mathscr{K} \supset \supset K$. If the regions $\Omega$ and $\Sigma_0$ do not intersect, we arrive at the conclusion which we used in 2°, §14: If the singularity on $K$ of a solution that is differentiable in $\Omega \setminus K$ is removable, it is removable in the class of solutions differentiable in $\Omega$.

**Example 2.** Let $p$ be a determined operator. Then Corollary 4 implies that every solution of (1.4), given in $\Omega \setminus \Sigma_0$ ($\Omega$ is an arbitrary region, and $\Sigma_0$ is a closed half-space), and differentiable in $\Omega \setminus (K \cup \Sigma_0)$, is differentiable in $\Omega \setminus \Sigma_0$. This result cannot be improved if we have to deal with the entire class of determined operators. In fact, as Zerner [2] and Hörmander [1] have shown, there always exists a solution of (1.4) in $R^n$ such that the set sing supp $u$ coincides with the given bicharacteristics of the operator $p(s=t=1)$. (It is clear that an arbitrary line can be a bicharacteristic of some such operator.)

**Example 3.** Let the matrix $p'$ have the form $(p_1, \ldots, p_t)$, in which not all of the $p_\tau$ are zero. Then Proposition 3, §13, implies that $\mathrm{Hom}(M, \mathscr{P})=0$, and $\mathrm{Ext}^1(M, \mathscr{P}) \cong \mathscr{P}/\varDelta\mathscr{P}$, where $\varDelta$ is the greatest common divisor of the polynomials $p_1, \ldots, p_t$. If the polynomial $\varDelta$ is hypoelliptic, Theorem 3 holds for the operator $p$ with $k=1$, and if $\varDelta=\mathrm{const}$, then Theorem 2 holds with $k=1$. More generally, if the variety $N(p)$ consisting of the common roots of the polynomials $p_1, \ldots, p_t$ is the union of a hypoelliptic variety and a variety of dimension less than $n-k$, the operator $p$ satisfies Theorem 3 in its full generality. If the variety $N(p)$ itself has dimension less than $n-k$, then we are in the conditions of Theorem 2.

# Notes

*Chapter I*. This seems to be the first occasion on which classes of equivalent families of modules have been studied as objects in a category. Theorem 1, § 2 is not altogether new: Its second assertion, in the case when the category is exact, was formulated by V.G. Boltyanskii and M.M. Postnikov in an editorial note to a book by H. Cartan and S. Eilenberg [1].

*Chapter II*. The fundamental results of the chapter – Theorems 1 and 2, § 4 – are new, except for the particular case $s=t=1$, which is Weierstrass's "Preparatory Lemma."

*Chapter III*. The method used in this chapter to prove that $\mathcal{M}$-cohomology is trivial stems from a paper by the author [4]. Complex measures in $C^n$ were investigated, generated by the null functionals on the space $Z$ of entire functions with bounds on their growth at infinity of the form $O(M_\alpha)$, where $M_\alpha(z)$ are functions of the form $(|z|+1)^{-q}\exp(b\,|\mathrm{Im}\, z|)$. One of the results obtained in the paper can be formulated as follows: Every measure $\mu$ with finite integrals $\int M_\alpha|\mu|$, which generates a null functional on $Z$, can be written in the form $\sum \frac{\partial}{\partial \bar{z}_i}\mu_i$, where the $\mu_i$ are measures with finite integrals $\int M_\alpha|\mu_i|$. This result is closely connected with $\mathcal{M}$-cohomology, since the proposition dual to the theorem on the construction of a measure-generating null functional asserts the triviality of the first group of cohomologies of analytic cocycles with bounded growth of the form $O(M_\alpha)$.

The triviality of the $\mathcal{M}$-cohomology for majorants $M_\alpha$ of that type is contained also in a paper by B. Malgrange [1] in which he investigated an equivalent problem: The triviality of the cohomology of differential forms in $C^n$ which satisfy the same condition of bounded growth. Later L. Hörmander [1] obtained a very general result: The triviality of the $\mathcal{M}$-cohomologies with bounded growth of the form $O(\exp\phi(z))$, where $\phi(z)$ is a plurisubharmonic function. Hörmander's result essentially supplements Theorem 1, § 5, but it does not contain this theorem, since it is not applicable, for example, to a family of majorants of the form $R(z)=O((|z|+1)^{-q})$ where $M_\alpha(z)=R(z)\exp(a\,|\mathrm{Im}\, z|)$, for arbitrary $q>0$ (since in this case the function $\ln M_\alpha(z)$ is not plurisubharmonic).

*Chapter IV.* A theorem similar to Theorem 2, § 5, was formulated first by L. Ehrenpreis [1] as the "Fundamental Principle." The content of Ehrenpreis's theorem consists in determining the image and the cokernel of the mapping $p: \Phi^s \to \Phi$, where $p$ is a polynomial matrix of dimension $1 \times s$, and $\Phi$ is the space of entire functions in $C^n$ that satisfy a condition of bounded growth at infinity of some not altogether clearly defined nature. He introduces a system of differential operators $\partial_k$ with constant coefficients, which are similar to the Noetherian operators constructed in § 4; each of the operators $\partial_k$ acts from $\Phi$ to the space $\Phi(V_k)$ of analytic functions on some submanifold $V_k \subset C^n$ with the same bounds on their growth as $\Phi$ has. It is asserted that the kernel of the operator $\oplus \partial_k$ is $p\Phi^s$, and its image is a closed subspace in $\oplus \Phi(V_k)$.

It was later found (Palamodov [5]), that Ehrenpreis's theorem can hold only under supplementary conditions, for example, if $s=1$. The fact is that the first assertion of the theorem: the existence of the differential operators $\partial_k$ with constant coefficients such that $\cap \operatorname{Ker} \partial_k = p\Phi^s$, is not in general substantiable (see Example 4°, § 4). This error, which is of algebraic nature, does not invalidate some of the consequences derived by Ehrenpreis in [1], which relate to the properties of differential operators with constant coefficients. In particular, some of his assertions dealing with hypoelliptic and hyperbolic operators are true inasmuch as they essentially use a more approximate theorem analogous to the "Fundamental Principle."

In the period 1960–1962, the author found a theorem which differs inessentially from Theorem 2, § 5 (see [4–7]).

*Chapter V.* § 1. The concept of a regularly increasing family was introduced by B.M. Makarov [1], under the name of a regular inductive limit. Propositions 3 and 4 are due to him. The Schwartz spaces were introduced by A. Grothendieck [1] on the basis of an analysis of the properties of the space $\mathscr{E}(R^n)$, which was introduced and carried through by L. Schwartz [1]. Proposition 6 is due to Grothendieck [1].

§ 2. Subsections 1°–5° contain an exposition of the elementary aspects of the theory of distributions on the basis of the theory of spaces of type $H$, appearing here under a notation that seems to be more natural in the context. The use of the concept of cosheaf in the theory of distributions appears to be new.

§ 3. The logarithmically convex functions and logarithmic duality (according to Jung) in that context, were first used by B.L. Gurevič [1], who perfected a construction introduced by I.M. Gel'fand and D.E. Šilov [1]. Proposition 2 and 5 are variants of known theorems of Paley-Wiener-Schwartz (see Schwartz [1]) and of Gel'fand-Šilov [2]. Proposition 3 is a variant of a known imbedding theorem of S.L.Sobolev [1].

*Chapter VI*. The theorem on the representation of solutions of a homogeneous system of differential equations with constant coefficients was first formulated by L. Ehrenpreis in 1960 [1]. This theorem was obtained as a consequence of his "Fundamental Principle," and referred to an arbitrary system of equations, with an arbitrary number of equations in one unknown function. Independently of Ehrenpreis, the author obtained a similar equation in the scalar case (one equation and one unknown function), for solutions belonging to a special class of functions. This theorem was published in [3]. In a later paper, the author [4] extended the admissible class of functions to the space of all distributions and established a particular case of Ehrenpreis's theorem dealing with one equation. In the general case, Ehrenpreis's theorem turns out to be untrue for the same reason as his "Fundamental Principle" turns out to be untrue: the absence of Noetherian operators with constant coefficients; see § 4, Chapter IV and the remarks on this chapter. The correct form of the exponential representation was found by the author in [5], [6], [7] for a system of general form. A less precise theorem of the same kind was obtained independently by Malgrange [1].

Theorems 1 and 2, § 4 give the exponential representation in the most important cases. We note that the author's notes (5), (14) contain more precise results, particularly the exponential representation for solutions-distributions in the whole convex domain.

The particular case of Corollary 3, in which $m=1$, $I_\alpha=\mathscr{I}_{K_\alpha}$ where $K_\alpha$ are convex compacts, is due to A. Martineau [1]. Another method for representing the solution was introduced earlier by M.S. Agranovič [5], [4].

There is a series of papers on a closely related question — the exponential representation of slowly growing solutions. The first result belongs to L. Schwartz [1], who showed that in the scalar case ($s=t=1$), when the only root of the characteristic equation is at the origin, every solution of the homogeneous equation belonging to $S'$ is a polynomial. This result was generalized by G.E. Šilov [1], who determined the solutions of the same class of equations which grow at infinity no faster than $C\exp(c|\xi|^\gamma)$, for some $C, c>0$ and $\gamma<1$. The exponential representation of slowly growing solutions of arbitrary systems of equations in one unknown function was obtained by the author [8].

§ 5. The class of elliptic operators was distinguished by I.G. Petrovskiĭ [1], who proved that all the possible solutions of a linear differential equation are analytic if and only if the equation is elliptic. The regularity of weak solutions was first studied by Weyl [1]; he established the smoothness of all weak solutions of the Laplace equation (Weyl's Lemma).

L. Schwartz [1] called attention to a general problem: to determine all equations having only infinitely differentiable solutions. In the class of scalar equations with constant coefficients, this problem was solved by L. Hörmander in [2] (see also the review by Šilov [2]). Extensive classes of hypoelliptic scalar equations with variable coefficients were described in papers by Mizohata [1], Malgrange [2], Hörmander [3], [4], F. Browder [1], F. Trèves [1], [2]. L.R. Volevič [1], [2] studied certain classes of hypoelliptic systems with variable coefficients.

Hypoelliptic systems of general form, with constant coefficients, were described by Hörmander [5], who relied on an algebraic theorem of Lech [1]. Hypoelliptic systems of equations with constant coefficients were also studied by Matzuura [1], [2].

The most general, and in principle the simplest method for studying local properties of solutions of systems with constant coefficients is the use of exponential representations. This method was first applied by Ehrenpreis in [1] (see the remarks on Chapter IV), and has also been applied by the author in [9], [10], [7]. The exposition in § 5 is based on this method. Theorem 1, together with Remark 2, contains the result of Hörmander that we have mentioned, on hypoelliptic systems with constant coefficients.

Hörmander obtained in [2], [5] a bound on the growth of the derivatives $D^j u$ of solutions of hypoelliptic equations of the form $O\left(\Gamma\left(\frac{|j|}{\gamma}\right)\right)$ for $|j|\to\infty$. V.V. Grušin [1] observed that the exponent $1/\gamma$ in this bound can be decreased if one imposes a bound on the growth of solutions for $|\xi|\to\infty$ of the form $O(\exp(c|\xi|^\beta))$. In [2] he published a proof of this fact, without the use of the theorem on exponential representation. Theorem 2, § 4 generalizes and sharpens these results.

The concepts of weak and strong partial regularity of distributions were introduced by Schwartz [2]. The class of weakly hypoelliptic scalar equations with respect to part of the variables (under another name) was introduced and described by Görding and Malgrange in [1], and by Ehrenpreis in [2]. Matzuura [3] transferred this result to general systems with constant coefficients. Strongly hypoelliptic operators with respect to part of the variables were studied by Ehrenpreis in [2]. E.A. Gorin [1] obtained a more detailed result in the same direction. Other classes of operators, hypoelliptic in part of the variables were studied by Görding and Malgrange [1], by Ehrenpreis [5], [2], Palamodov [9], Gorin [1], [2], Frieberg [1], [2], and V.I. Burenkov [1].

The existence of smoothing operators was observed by Gorin and V.V. Grušin [1] and was studied by them in [2] also. Theorem 4 generalizes a result obtained in these papers.

Ehrenpreis [5], [2], Hörmander [8] have investigated the hypoellipticity of convolution operators. Bengel [1] established the analyticity of all solutions of elliptic equations in the class of hyperfunctions. Another proof of this fact was given by Komatsu [1].

§ 6. The problem of the uniqueness classes of solutions of the Cauchy problem has a history. The first papers in this direction are due to Holmgren [1], A.N. Tihonov [1], and S. Tacklind [1] who determined the uniqueness classes of the Cauchy problem for the heat equation and similar equations. Tacklind [1] found the conditions on the function $\phi(x)$ that were necessary and sufficient for the functions $f(x)$ subject to the condition $f(x)=O(\exp\phi(x))$ to form a uniqueness class for the heat equation.

The first general results here were found by Petrovskiĭ [2], who considered square systems of equations with constant coefficients or coefficients depending only on $t$, which are soluble with respect to the highest derivative in $t$. He showed that the functions that are bounded with respect to space-variables form a uniqueness class for an arbitrary system of this type. V.E. Lyance [1] and L. Schwartz [3] strengthened this theorem, replacing the class of functions bounded with respect to $x$ by a class of slowly increasing functions. I.M. Gel'fand and G.E. Šilov [1] found uniqueness classes for the system considered by Petrovskiĭ, containing exponentially increasing functions. In a more precise form, these results were published in [3]. The uniqueness theorem of Gel'fand-Šilov, which relates to systems with constant coefficients, is a particular case of the theorem of §6; it is formulated in Example 1. The theorem of § 6 is due to the author.

The uniqueness classes for the Cauchy problem for parabolic equations with variable coefficients were determined by O.A. Ladiženskaya [1], and for general parabolic systems by S.D. Eĭdel'man [1], [2]. G.N. Zolotraev [1] generalized the result of Tacklind mentioned above, to systems that are parabolic in the sense of Petrovskiĭ. The study of uniqueness classes for the solution of the Cauchy problem was also taken up by Ya.I. Žitomirskii [1], [2]. In [3] he obtained a very precise result on the uniqueness process for the solution of the Cauchy problem in the case where there is one space-variable.

In the review by Malgrange [10] a series of papers are cited on the uniqueness of solutions of the Cauchy problem for non-characteristic subspaces.

*Chapter VII.* §7. The term "$p$-convexity" was first used by Hörmander in [6]. However, this term was used not for a functional space but for an open set in $R^n$ – namely, the open set $\Omega$ was called by Hörmander $p$-convex ($p$ being a scalar operator with constant coefficients), if it satisfied the condition $(S_\Omega)$ of Theorem 1, § 10. This condition $(S_\Omega)$ first appeared, however, in Malgrange's thesis [3], in which he showed that it is equivalent to each of two theorems on the solubility of the nonhomogeneous equations $p\mathscr{E}(\Omega)=\mathscr{E}(\Omega)$ and $\mu\mathscr{D}'^F(\Omega)=\mathscr{D}'^F(\Omega)$. Hörmander introduced in [6] the term "strongly $p$-convex open set" which refers to some stronger condition of the type $(S_\Omega)$ and is equivalent to the relation $p\mathscr{D}'(\Omega)=\mathscr{D}'(\Omega)$. The term "$M$-convex domain" as applied to an arbitrary finite $\mathscr{P}$-module $M$ was first used by Malgrange [4]. His meaning is close to our own: an open set $\Omega$ is $M$-convex in the sense of Malgrange if and only if $\mathscr{E}(\Omega)$ is an $M$-convex space in our sense. The term "$M$-convex space" introduced by the author is connected with the fact that the solubility of a nonhomogeneous system depends not only on the open set in which it is defined, but on the class of functions in which the solution lies (see Hörmander [6], [10]).

The narrow problem of $M$-convexity was first formulated by Malgrange in [4]. There is an interesting variant of the narrow problem of $M$-convexity, formulated in the same paper by Malgrange.

§ 8. The general problem of the solubility of nonhomogeneous equations arose out of the classical problem of constructing fundamental solutions in the large. In the works of Fredholm [1], Zeylon [1], Herglotz [1], and Petrovskiĭ [3] fundamental solutions were constructed for special classes of operators with constant coefficients. One of the first achievements of the general theory of equations with constant coefficients due to Ehrenpreis [3] and Malgrange [3] was the proof that there exists a fundamental solution for an arbitrary scalar operator. This result was obtained as a consequence of more general theorems on solubility, of the form $p\Phi=\Phi$. Malgrange established in [3] a theorem for the spaces $\Phi=\mathscr{E}(\Omega)$ and $\mathscr{D}'^F(\Omega)$, where $\Omega$ is an arbitrary convex domain in $R^n$. Ehrenpreis obtained in [3] the same results for the case $\Omega=R^n$. In [4] Ehrenpreis established a similar theorem for the space $\mathscr{D}'(R^n)$. Malgrange [5] generalized this result to the space $\mathscr{D}'(\Omega)$. In his thesis [3], Malgrange showed that for all the spaces $\Phi$ that we have listed and for an arbitrary scalar operator $p$, the exponential polynomials are dense in $\Phi_p$. Ehrenpreis [4] also established a theorem on solubility for certain spaces $\Phi$ of entire functions. The solubility of the scalar equation in $\mathscr{D}'(\Omega)$ was studied in the already cited work of Hörmander [6], and also by Neymark [1].

Malgrange [3] and Hörmander [7] investigated the problem of the best possible local properties of fundamental solutions. Hörmander and Trèves proposed an explicit construction for the fundamental solution based on a multidimensional contour of integration. Trèves [3], [4], M. Zerner [3], and T. Shirota [1] considered the problem of constructing a fundamental solution for an equation depending on a parameter. Agranovič [1]–[4], who perfected Shirota's method, gave an explicit construction for the solutions in some of the theorems on solubility that we have described above, and he established certain new solubility theorems. Palamodov [1], [2], found that in some functional spaces theorems on solubility and uniqueness of the solution hold simultaneously, provided $p(z) \neq 0$ for $\operatorname{Im} z = 0$.

A theorem on the solubility of general systems of equations, when a certain compatibility condition holds, was first developed by Ehrenpreis [1] as a consequence of his "Fundamental Principle." For the spaces $\mathscr{E}(\Omega)$, $\mathscr{D}'(\Omega)$, where $\Omega$ is an arbitrary convex open set, and for similar spaces, this theorem was obtained by the author in [5], and independently by Malgrange [4], [1], for the spaces $\mathscr{E}(\Omega)$ and $\mathscr{D}'(\Gamma)$ where $\Gamma$ is an arbitrary convex compact.

Malgrange [4] was the first to note that if $p\mathscr{E}(\Omega) = \mathscr{E}(\Omega)$ for an arbitrary scalar operator $p \neq 0$ is to hold, the connected open set $\Omega$ must be convex.

The homological form of the theorem on the solubility of general systems was suggested by Malgrange [4]. Corollary 5, and the additive representation of solutions (Example 2), yields solutions to Hadamard's problem on the factorization of the operator $p$. Hadamard's problem in non-convex open sets was studied by Matzuura [4]. Corollary 6 generalizes some results of Grušin [3] on $Q$-hypoelliptic operators formulated in Example 1.

Theorem 5 belongs to the author. The second assertion of this theorem, in view of the remark made at the end, contains an earlier result of Malgrange [6], [3].

For scalar operators, the theorem on solubility in the space $S'$ was obtained by Hörmander [9] and by S. Lojasiewicz [1], [2]. For systems of general form, a similar result was obtained by Malgrange [7]; see also Palamodov [8].

Theorems on the solubility and on exponential approximations for convolution equations were obtained by Malgrange [3], Ehrenpreis [2], [7], Hörmander [6], and Martinau [2].

The solubility of equations for operators with variable coefficients was studied by Hörmander [12], [13], [10]. In [13] he distinguished a wide class of operators with variable coefficients, for which the nonhomogeneous equation is almost always insoluble. The same question for equations of first order was studied by L. Nirenberg and by Trèves [1], and for systems of equations Matsumura [1].

§ 9. Corollary 1 (its algebraic part) was first formulated by Malgrange [4]. Theorems 2 and 4 were published by the author in [11]. The algebraic part of Theorem 3 is the well-known associativity formula for homological algebra. Theorem 4 generalizes a theorem of Serre [2] on the topology of Runge domains (see Example 3). The method of proof uses an argument due to Serre.

§ 10. Theorem 1 is due to Malgrange [3]. It should be noted that for scalar operators with a real principal part, Malgrange [8] obtained sufficient and nearly necessary geometric conditions on $\Omega$ for the fulfillment of the condition $(S_{\Omega})$ (see also Hörmander [10]). Hörmander [6], [10] obtained similar conditions for the existence of solutions in the case $\mathscr{D}'(\Omega)$. Corollary 1 belongs, in essence, to Malgrange. Theorems 2, 3, and 4 are due to the author.

§ 11. Theorem 1 is an analog of the theorem on finiteness of Andreotti-Grauert. The proof of Theorem 1 in general outline follows the arguments of Andreotti-Grauert [1], who used some ideas of Ehrenpreis.

A theorem of the Andreotti-Grauert type was obtained by Malgrange [11]. Hörmander [1] obtained a number of very precise existence and approximation theorems for the operator $''d$. In [11] Hörmander found geometric conditions on the open set which guarantee the solubility of first order systems of equations with variable coefficients and one unknown variable under very general assumptions.

§ 12. Theorem 1 strengthens the results of Malgrange [12]. Malgrange established this theorem only for the space $\mathscr{E}(\Omega)$. Theorem 2 is due to the author.

*Chapter VIII*. § 13. This section does not contain new results but it is difficult to give all the exact references. Theorems 1 and 2 are contained in the "grade-theory" of D. Rees [1], who inferred these from the theorem of Lasker-Maccauley. Contrariwise, we obtain the latter in the form of Proposition 2 as a corollary. Theorem 3 is due to J. Jans [1]. The paper of H. Bass [1] contains a proof of this theorem which is near ours.

§ 14. The first of the theorems on the possibility of extending the solution from the neighborhood of the boundary into the open set was formulated by Ehrenpreis in [6]. A stronger result was obtained by Malgrange [4], [14] and by the author with V.D. Grušin [12] as a consequence of theorems on the solubility of general systems. Theorems 1 and 2 for scalar operators were established by Grušin [4], who used the work of the author [4]. Corollary 5 generalizes a duality theorem of Grothendieck [2], which refers only to scalar operators.

Problems relating to irremovable singularities of smooth solutions were studied by L.A. Čudov [1], Grušin [3], [4], and by E.A. Gorin and Grušin [2]. Corollary 4 generalizes the results of these authors. Example 3 belongs to Grušin [4]. Grušin found in [4] a broad, sufficient condition for the absence of irremovable singularities of infinitely differentiable solutions.

The particular case of Theorem 6 formulated in Example 5 of 9° belongs to F. John [1] (the case of the non-hyperbolic operator) and to B. Brodda [1] (the case of the characteristic subspace); A.A. Gorin and

E. A. Gorin [1] studied an adjacent question: the existence of solutions whose Cauchy conditions have compact supports. The result formulated in Example 6 is a theorem of Osgood-Brown (see, for example, Fuks [1]).

Theorems 3, 4, 5, 6, and 7 belong to the author. Results close to these are contained in papers by the author [12], [10]. In connection with the problem of extension, Malgrange [13] studied the cohomology of the sheaf of solutions $\mathscr{E}_M$ with compact supports.

§ 15. The fact that every solution of a scalar equation which is differentiable in a neighborhood of the boundary turns out to be differentiable in the interior, was noted for the first time by Agranovič [1]. A stronger result contained in Example 2 of 7° was obtained by Malgrange [5], John [2], and see also Grušin [5]. Theorem 2 – 3, Corollaries 3 and 4 are due to the author. Similar results were published by the author in [10]. The problems on the extension of analyticity were studied by Boman [1].

The work of Zerner [2] and Hörmander [10] should be noted, in which solutions of a scalar equation are constructed, where the support of the singularity coincides with a given bicharacteristic. Shirota [2] finds necessary and sufficient conditions that differentiability should be extendable across a hyper-plane. A precise theorem on the extendability of differentiability of solutions was found by Grušin [6]. Problems on the extendability of smoothness of solutions of equations with variable coefficients have been studied by Shirota [3] and by Hörmander [10].

# Bibliography

Agranovič, M. S.

[1] Neskol'ko teorem ob uravneniyah v častnyh proizvodnyh s postoyannymi koefficientami. Dokl. Akad. Nauk SSSR **128**, No. 3, 439 – 442 (1959). (Theorems on partial differential equations with constant coefficients.)

[2] Ob analitičeskih rešeniyah uravneniĭ v častnyh proizvodnyh s postoyannymi koefficientami. Dokl. Akad. Nauk SSSR **124**, No. 6, 1183 – 1186 (1959). (Analytic solutions of partial differential equations with constant coefficients.)

[3] Suščestvovanie rešeniĭ uravneniĭ v častnyh proizvodnyh s postoyannymi koefficientami v nekotoryh klassah funkciĭ. Vestnik MGU, No. 3, 3 – 13 (1959). (The existence of solutions of partial differential equations with constant coefficients in certain classes of functions.)

[4] Ob uravneniyah v častnyh proizvodnyh s postoyannymi koefficientami. Uspehi Mat. Nauk **16**, No. 2 (98), 27 – 93 (1958). (On partial differential equations with constant coefficients.)

[5] Obščie rešeniya differencial'no-raznostnyh uravneniĭ s postoyannymi koefficientami. Dokl. Akad. Nauk SSSR **123**, No. 1, 9 – 12 (1958). (The general solution of differential-difference equations with constant coefficients.)

Andreotti, A., Grauert, H.

[1] Théorèmes de finitude pour la comologie des espaces complexes. Bull. Soc. Math. France **90**, 193 – 259 (1962).

Bass, H.

[1] On the ubiquity of Gorenstein rings. Math. Z. **82**, No. 1, 8 – 28 (1963).

Bengel, G.

[1] a) Sur une extension de la théorie des hyperfonctions; b) Régularité des solutions hyperfonctions d'une équation elliptique. C. R. Acad. Sci. **262** (1966).

Boman, J.

[1] On the propagation of analyticity of solutions of differential equations with constant coefficients. Ark. Mat. **5**, No. 7, 271 – 279 (1964).

Bourbaki, N.

[1] Topologičeskie vektornye prostranstva, translated from the French. Moscow: IL 1959. (Espaces vectoriels topologiques. Paris 1953 – 1955.)

Brodda, B.

[1] On uniqueness theorems for differential equations with constant coefficients. Math. Scand. **9**, 55 – 68 (1961).

Browder, F.

[1] Regularity theorems for solutions of partial differential equations with variable coefficients. Proc. Nat. Acad. Sci. U.S.A. **43**, No. 2, 234 – 236 (1957).

Burenkov, V. I.

[1] O beskonečnoi differenciruemosti i analytičnosti ubyvayuščih na beskonečnosti rešeniĭ uravneniĭ s postoyannymi koefficientami. Dokl. Akad. Nauk SSSR **174**, No. 5, 1007 – 1010 (1967). (Infinite differentiability and analyticity of solutions of equations with constant coefficients, decreasing at infinity.)

Cartan, H., Eilenberg, S.

[1] Gomologičeskaya algebra, translated form the French. Moscow: IL 1960. (Homological Algebra. Princeton 1956.)

Čudov, L. A.

[1] Ob osobennostyah rešeniĭ lineĭnyh differencial'nyh uravneniĭ v častnyh proizvodnyh s postoyannymi koefficientami. Dokl. Akad.Nauk SSSR **125**, No. 3, 504 – 507 (1959). (Singularities of the solutions of linear partial differential equations with constant coefficients.)

Dieudonné, J., Schwartz, L.

[1] La dualité dans les espaces (F) et (DF). Ann. Inst. Fourier **1**, 61 – 101 (1949).

Dunford, N., Schwartz, J.

[1] Lineĭnye operatory, translated from the English. Moscow: IL 1962. (Linear Operators, Part I. Interscience 1958.)

Ehrenpreis, L.

[1] A fundamental principle for systems of linear differential equations with constant coefficients and some of its applications. Proc. Int. symp. on linear spaces, Jerusalem, 1960.

[2] Solutions of some problems of division IV. Amer. J. Math. **82**, No. 3, 522 – 588 (1960).

[3] Solutions of some problems of division I. Amer. J. Math. **76**, No. 4, 883 – 903 (1954).

[4] Solutions of some problems of division III. Amer. J. Math. **78**, No. 4, 685 – 715 (1956).

[5] General theory of elliptic equations. Proc. Nat. Acad. Sci. **42**, No. 1, 39 – 41 (1956).

[6] A new proof and an extension of Hartog's theorem. Bull. Amer. Soc. **67**, No. 5, 507 – 509 (1961).

[7] Solutions of some problems of division II. Amer. J. Math. **77**, No. 2, 286 – 292 (1955).

Eĭdel'man, S. D.

[1] Ocenki rešeniĭ paraboličeskih sistem i nekotorye ih priloženiya. Mat. Sb. **33**, No. 3, 359 – 382 (1953). (Bounds for the solutions of parabolic systems and some applications of them.)

[2] Paraboličeskie sistemy, "Nauka", 1964. (Parabolic systems.)

Fredholm, J.

[1] Sur l'integrale fondamentale d'une équation differentielle elliptique à coefficients constants. Rend. Circ. Mat. Palermo **25**, 346 – 351 (1908).

Friberg, J.

[1] Partially hypoelliptic differential equations of finite type. Math. Scand. **9**, 22 – 42 (1961).

[2] Estimates for partially hypoelliptic differential operators. Lund 1963.

Fuks, B. A.

[1] Vvedenie v teoriyu analitičeskih funkciĭ mnogih kompleksnyh peremennyh. Moscow: Fizmatgiz 1962. (Introduction to the theory of analytic functions of several complex variables.)

Gårding, L., Malgrange, B.

[1] Opérateurs différentiels partiellement hypoelliptiques et partiellement elliptiques. Math. Scand. **9**, 5 – 21 (1961).

Gel'fand, I. M., Šilov, G. E.

[1] Preobrazovaniya Fur'e bystro rastuščih funkciĭ i voprosy edinstvennosti rešeniya zadači Koši. Uspehi Mat. Nauk **8**, No. 6, 3–54 (1953). (Fourier transforms of functions of rapid growth and the uniqueness of solutions of the Cauchy problem.)

[2] Prostranstva osnovnyh i obobščennyh funkciĭ ("Obobscennye funkcii", vyp. 2). Moscow: Fizmatgiz 1958. (Spaces of fundamental and generalized functions ["Generalized Functions", 2nd issue].)

[3] Nekotorye voprosy teorii differencialnyh uravneniĭ ("Obobščennye funkcii", vyp. 3). Moscow: Fizmatgiz 1958. (Certain questions in the theory of differential equations ["Generalized Functions", 3rd issue].)

Godement, R.

[1] Algebraičeskaya topologiya i teoriya pučkov, translated from the French. Moscow: IL 1961. (Topologie algebrique et théorie des faisceaux. Paris: Hermann 1958.)

Gorin, A. A., Gorin, E. A.

[1] O razrešimosti zadači Koši s finitnymi načal'nymi dannymi. Diff. Uravnenija **1**, No. 12, 1640–1646 (1965). (Solubility of the Cauchy problem with finite initial conditions.)

Gorin, E. A.

[1] Častično gipoelliptičeskie differencial'nye uravneniya v častnyh proizvodnyh s postoyannymi koefficientami. Sibirsk. Mat. Ž. **3**, No. 4, 500–526 (1962). (Partially hypo-elliptic partial differential equations with constant coefficients.)

[2] O kvadratičnoĭ summiruemosti rešeniĭ differencialnyh uravneniĭ v častnyh proizvodnyh s postoyannymi koefficientami. Sibirsk. Mat. Ž. **2**, No. 2, 221–232 (1961). (Quadratic summability of solutions of partial differential equations with constant coefficients.)

[3] Ob asimptotičeskih svoĭstvah mnogočlenov i algebraičeskih funkciĭ ot neskol'kih peremennyh. Uspehi Mat. Nauk **16**, No. 1 (97), 91–119 (1961). (Asymptotic properties of polynomials and algebraic functions of several complex variables.)

Gorin, E. A., Grušin, V. V.

[1] Differencial'nye uravneniya, rešeniya kotoryh sglaživayutsya pri differencirovanii. Vestnik MGU No. 2, 25–32 (1963). (Differential equations whose solutions become smooth upon differentiation.)

[2] O nekotoryh lokal'nyh teoremah dlya uravneniĭ v častnyh proizvodnyh s postoyannymi koefficientami. TMMO **14**, 200–210 (1965). (Local theorems for partial differential equations with constant coefficients.)

Grothendieck, A.

[1] Sur les espaces (F) et (DF). Summa Brasil. Math. **3**, 57–123 (1954).

[2] Sur les espaces de solutions d'une classe générale d'équations aux dérivées partielles. J. Analyse Math. **2**, 243–280 (1952/53).

Grušin, V. V.

[1] Ob odnom svoĭstve rešeniĭ gipoelliptičeskogo uravneniya. Dokl. Akad. Nauk SSSR **137**, No. 4, 768–771 (1961). (A property of solutions of hypo-elliptic equations.)

[2] Svyaz' meždu lokal'nymi i global'nymi svoĭstvami rešeniĭ gipoelliptičeskih uravneniĭ s postoyannymi koefficientami. Mat. Sb. **66** (108) No. 4, 525–550 (1965). (A relation between the local and global properties of solutions of hypo-elliptic equations with constant coefficients.)

[3] O Q-gipoelliptičeskih uravneniyah. Mat. Sb. **57** (99) No. 2, 233–240 (1962). (On $Q$-hypoelliptic equations.)

[4] O rešeniyah s isolirovannymi osobennostyami dlya uravneniĭ v častnyh proizvodnyh s postoyannymi koefficientami. TMMO **15**, 262–278 (1966). (Solutions with isolated singularities for partial differential equations with constant coefficients.)

[5] O rešeniyah differencial'nyh uravneniĭ v častnyh proizvodnyh s postoyannymi koefficientami. Dokl. Akad. Nauk SSSR **139**, No. 1, 17–19 (1961). (On solutions of partial differential equations with constant coefficients.)

[6] Rasprostranenie gladkosti reseniĭ differencial'nyh uravneniĭ glavnogo tipa. Dokl. Akad. Nauk SSSR **148**, No. 6, 1241–1244 (1963). (Extension of the smoothness of solutions of differential equations of principal type.)

Gunning, R., Rossi, H.

[1] Analytic functions of several complex variables. Prentice-hall, Inc. 1965.

Gurevič, B. L.

[1] Novye tipy prostranstv osnovnyh i obobščennyh funkciĭ i problema Koši dlya operatornyh uravneniĭ. Dissertaciya, Har'kov 1956. (New types of spaces of fundamental and generalized functions, and the Cauchy problem for operator equations. Thesis.)

Herglotz, G.

[1] Über die Integration linearer, partieller Differentialgleichungen mit konstanten Koeffizienten. Ber. Verh. Sächs. Acad. Wiss. Leipzig, Math.-Phys. Kl. **78**, 93–126, 287–318 (1926); **80**, 60–144 (1928); – Abh. Math. Sem. Univ. Hamburg **6**, 189–197 (1928).

Hervé, M.

[1] Funkcii mnogih kompleksnyh peremennyh, perev. s franc. Moskva: "Mir" 1965. (Several complex variables. Oxford University Press 1963.)

Holmgren, E.

[1] Sur les solutions quasianalytiques de l'équation de la chaleur. Ark. Mat. **18** (1924).

Hörmander, L.

[1] $L^2$-estimates and existence theorems for the $\bar{\partial}$ operator. Acta Math. **113**, No. 1–2 89–152 (1965).

[2] On the theory of general partial differential operators. Acta Math. **94**, 161–248 (1955).

[3] On interior regularity of the solutions of partial differential equations. Comm. Pure Appl. Math. **11**, No. 2, 197–218 (1958).

[4] Hypoelliptic differential operators. Ann. Inst. Fourier **11**, 477–492 (1961).

[5] Differentiability properties of solutions of systems of differential equations. Ark. Mat. **3**, No. 6, 527–535 (1958).

[6] On the range of convolution operators. Ann. Math. **76**, No. 1, 148–170 (1962).

[7] Local and global properties of fundamental solutions. Math. Scand. **5**, 27–39 (1957).

[8] Hypoelliptic convolution equations. Acta Math. **9**, 551–585 (1960).

[9] On the division of distributions by polynomials. Ark. Mat. **3**, 555–568 (1958).

[10] Linear partial differential operators. Berlin-Göttingen-Heidelberg: Springer 1963.

[11] The Frobenius-Nirenberg theorem. Ark. Mat. **5**, No. 5, 425–432 (1964).

[12] Differential operators of principal type. Math. Ann. **140**, 124–146 (1960).

[13] Differential equations without solutions. Math. Ann. **140**, 160–173 (1960).

Jans, J.-P.

[1] On finitely generated modules over Noetherian rings. Trans. Amer. Math. Soc. **106**, No. 2, 330–340 (1963).

John, F.

[1] Non admissible data for differential equations with constant coefficients, Comm. Pure Appl. Math. **10**, 391–398 (1957).

[2] Continuous dependence on data for solutions of partial differential equations with a prescribed bound. Comm. Pure Appl. Math. **13**, 551–585 (1960).

Kantorovič, L. V., Akilov, G. P.

[1] Funkcional'nyi analiz v normirovannyh prostranstvah. Moscow: Fizmatgiz 1959. (Functional analysis in normed spaces.)

Komatsu, H.

[1] Resolutions by hyperfunctions of sheaves of solutions of differential equations with constant coefficients. Stanford Univ., 1960.

Ladyženskaya, O. A.

[1] O edinstvennosti reševiya zadači Koši dlya lineĭnogo paraboličeskogo uravneniya. Mat. Sb. **27**, No. 2, 175–184 (1950). (Uniqueness of the solutions of the Cauchy problem for linear parabolic equations.)

Lech, C.

[1] A metric property of the zeros of a complex polynomial ideal. Ark. Mat. **3**, No. 6, 543–554 (1958).

Lojasiewicz, S.

[1] Division d'une distribution par une fonction analytique de variables réelles. C. R. Acad. Sci. Paris **246**, 683 (1958).

[2] Sur le problème de la division. Studia Math. **18**, No. 1, 87–136 (1959).

Lyance, V. E.

[1] O zadače Koši v oblasti funkciĭ deĭstvitel'nogo peremennogo. Ukrain. Mat. Ž. **1**, No. 4, 42–63 (1949). (On the Cauchy problem in the area of functions of a real variable.)

Makarov, B. M.

[1] Ob induktivnom predele posledovatel'nosti normirovannyh prostranstv. Dokl. Akad.Nauk SSSR **119**, No. 6, 1092–1094 (1958). (On the inductive limit of a sequence of normed spaces.)

Malgrange, B.

[1] Sur les systèmes différentiels à coefficients constants. Coll. Int. C. N. R. S., Paris 1962.

[2] Sur une classe d'operateurs différentiels hypoelliptiques. Bull. Soc. Math. France **85**, 283–306 (1957).

[3] Existence et approximation des solutions des équations aux dérivées partielles et des équations de convolution. Ann. Inst. Fourier **6**, 271–355 (1956).

[4] Systèmes différentiels à coefficients constants. Séminaire Bourbaki, Paris, 1962/63, No. 246.

[5] Sur la propagation de la régularité des solutions des équations à coefficients constants. Bull. Math. Soc. Sci. Math. Phys. R. P. Roumaine **3** (53), No. 4, 433–440 (1959).

[6] Séminaire Schwartz, Paris, 1954/55, exp. 3.

[7] Division des distributions I–IV. Séminaire Schwartz, Paris, 1959/60, exp. 21–25; Séminaire Bourbaki, Paris, 1959/60, No. 203.

[8] Sur les ouverts convexes par rapport à un operateur différentiel. C. R. Acad. Sci. Paris **254**, 614–615 (1962).

[9] Sur les équations de convolution. Rend. Sem. Univ. Torino **19**, 19–27 (1959/60).

[10] Sur l'unicité du problème de Cauchy. Rend. Sem. Mat. Fis. Milano **30**, 3–10 (1960).

[11] Some remarks on the notion of convexity for differential operators. Diff. Analysis, Bombay Coll., 1964, 163–174.

[12] Quelques problèmes de convexité pour les operateurs différentiels à coefficients constants I. Séminaire Collège de France, 1962/63, exp. 7.

[13] Quelques problèmes de convexité pour les operateurs différentiels à coefficients constants II. Séminaire Collège de France, 1962/63, exp. 7.

[14] Sur les systèmes differentiels à coefficients constants. Séminaire Collège de France, 1961/62, exp. 8.

Martineau, A.

[1] Sur les fonctionelles analytiques et la transformation de Fourier-Borel. J. Analyse Math. **9**, 1–163 (1963).

[2] Équations différentielles d'ordre infini. Séminaire de France, 1965/66.

Matzumura, M.

[1] Existence locale de solutions pour quelques systèmes d'équations aux dérivées partielles. Japan J. Math. **32**, 13–49 (1962).

Matzuura, S.

[1] On general systems of partial differential operators with constant coefficients. J. Math. Soc. Japan **13**, No. 1, 94–103 (1961).

[2] A remark on ellipticity of general systems of differential operators with constant coefficients. J. Math. Kyoto Univ. **1**, No. 1, 71–74 (1961).

[3] Partially hypoelliptic and partially elliptic systems of differential operators with constant coefficients. J. Math. Kyoto Univ. **1**, No. 2, 147–160 (1962).

[4] Factorisation of differential operators and decomposition of solutions of homogeneous equations. Osaka Math. J. **15**, No. 2, 213–231 (1963).

Mizohata

[1] Hypoellipticité des équations paraboliques. Bull. Soc. Math. France **85**, 15 (1957).

Narasimhan, R.

[1] Imbedding of holomorphically complete complex spaces. Amer. J. Math. **82**, No. 4, 917–934 (1960).

Neymark, H.

[1] On the existence of solutions of differential equations with constant coefficients. Ark. Math. **5**, 433–443 (1965).

Nirenberg, L., Trèves, F.

[1] Solvability of a first order linear partial differential equation. Comm. Pure Appl. Math. **14**, 331–351 (1963).

Palamodov, V. P.

[1] Ob usloviyah na beskonečnosti, obespečivayuščih korrektnuyu razrešimost' nekotorogo klassa uravneniya vida $p\left(i\frac{\partial}{\partial x}\right)u=f$. Dokl. Akad. Nauk SSSR **129**, No. 4, 740–743 (1959). [On the conditions at infinity that ensure the correct solubility of a certain class of equations of the form $p\left(i\frac{\partial}{\partial x}\right)u=f$.]

[2] Usloviya korrektnoĭ razrešimosti v celom nekotorogo klassa uravneniĭ s postoyannymi koefficientami. Sibirsk. Mat. Ž. **4**, No. 5, 1137–1149 (1963). (The conditions of correct solubility in the large of a certain class of equations with constant coefficients.)

[3] Ob obščem vide rešeniya odnorodnogo differencial'nogo uravneniya s postoyannymi koefficientami. Dokl. Akad. Nauk SSSR **137**, No. 4, 774–777 (1961). (On the general form of the solution of a homogeneous differential equation with constant coefficients.)

[4] Obščiĭ vid rešeniĭ lineĭnyh differencial'nyh uravneniĭ s postoyannymi koefficientami. Dokl. Akad. Nauk SSSR **143**, No. 6, 1278–1281 (1962). (The general form of the solutions of linear differential equations with constant coefficients.)

[5] O sistemah differencial'nyh uravneniĭ s postoyannymi koefficientami. Dokl. Akad. Nauk SSSR **148**, No. 3, 523–526 (1963). (On systems of differential equations with constant coefficients.)

[6] Obščie teoremy o sistemah lineĭnyh differencial'nyh uravneniĭ s postoyannymi koefficientami. Sov.-amer. simp., Novosibirsk, 1963. (General theorems on systems of linear differential equations with constant coefficients.)

[7] Struktura moduleĭ nad kol'com mnogočlenov v prostranstvah analitičeskih funkciĭ. Dissertaciya, Moscow 1965. (The structure of modules over the ring of polynomials in spaces of analytic functions.)

[8] Stroenie polinomial'nyh idealov i ih faktorprostranstv v prostranstvah beskonečno differenciruemyh funkciĭ. Dokl. Akad. Nauk SSSR **141**, No. 6, 1302–1305 (1961). (The structure of polynomial ideals and their factor spaces in spaces of infinitely differentiable functions.)

[9] K teorii gipoelliptičeskih i častično gipoelliptičeskih operatorov. Dokl. Akad. Nauk SSSR **140**, No. 5, 1015–1018 (1961). (On the theory of hypo-elliptic and partially hypo-elliptic operators.)

[10] O nedoopredelennyh i pereopredelennyh sistemah differencial'nyh uravneniĭ s postoyannymi koefficientami. Dokl. Akad. Nauk SSSR **156**, No. 6, 1288–1291 (1964). (On under-determined and over-determined systems of differential equations with constant coefficients.)

[11] O probleme M-vypuklosti. Dokl. Akad. Nauk SSSR **161**, No. 5, 1015–1018 (1965). (On the problem of M-convexity.)

[12] Polinomial'nye idealy i uravneniya v častnyh proizvodnyh. Uspehi Mat. Nauk **18**, 2 (110), 164–168 (1963). (Polynomial ideals and partial differential equations.)

[13] Funktor proektivnogo predela v kategorii topologičeskih prostranstv. Mat. Sb. **75**, No. 4, 567–603 (1968). (The functor of the projective limit in the category of topological spaces.)

[14] Zamečanie ob eksponencial'nom predstavlenii rešeniĭ differencial'nyh uravneniĭ s postoyannymi koefficientami. Mat. Sb. **76**, No. 3, 417–434 (1968). (A remark on the exponential representation of the solutions of differential equations with constant coefficients.)

Petrovskii, I. G.

[1] Sur l'analycité des solutions des systèmes d'équations différentielles. Mat. Sb. **5** (47), No. 1, 3–68 (1939).

[2] O probleme Koši dlya sistem lineĭnyh uravneniĭ s častnymi proizvodnymi v oblasti neanalitičeskih funkciĭ. Byull. MGU, sekc. A, **1**, No. 7 (1938). (On the Cauchy problem for systems of linear partial differential equations with constant coefficients.)

[3] On the diffusion of waves and the lacunas for systems of hyperbolic equations. Mat. Sb. **17** (59), 289–370 (1945).

Rees, D.

[1] The grade of an ideal or module. Proc. Cambridge Philos. Soc. **53**, No. 1, 28–42 (1957)

Schwartz, L.
[1] Théorie des distributions I, II. Paris: Hermann 1950/51.
[2] Distributions semi-regulières et changements de variables. J. Math. Pures Appl. (9) **36**, 109 – 127 (1957).
[3] Les équations d'evolution liées au produit de composition. Ann. Inst. Fourier **2**, 19 – 49 (1950).
[4] Kompleksnye analitičeskie mnogoobraziya, Elliptičeskie uravneniya s častnymi proizvodnymi. Moscow: "Mir" 1964. (Complex analytic manifolds. Elliptic partial differential equations.)
[5] Théorie des noyaux. Proc. Int. Congr. Math. **1**, 220 – 230 (1952).

Serre, J.-P.
[1] Géométrie analytique et géométrie algébrique. Ann. Inst. Fourier **6**, 1 – 42 (1955/56).
[2] Une propriété topologique des domains de Runge. Proc. Amer. Math. Soc. **6**, No. 1, 133 – 134 (1955).
[3] Quelques problèmes globaux relatifs aux variétés de Stein. Coll. Bruxelles, 1953.
[4] Lokal'naya algebra i teoriya kratnosteĭ. Matematika **7**, 5, 3 – 93 (1963). (Local algebra and the theory of multiplicity.)

Shirota, T.
[1] On solutions of a partially differential equations with a parameter. Proc. Japan. Acad. **32**, No. 6, 401 – 405 (1956).
[2] On the propagation of regularity of solutions of partial differential equations with constant coefficients. Proc. Japan Acad. **38**, No. 8, 587 – 590 (1962).
[3] On the propagation of regularity of solutions of partial differential equations. Proc. Japan Acad. **39**, No. 2, 120 – 124 (1963).

Šilov, G. E.
[1] Analog odnoĭ teoremy Loran Švarca. Izv. Vysš. Učebn. Zaved. No. 4 (23), 137 – 147 (1961). (Analogue of a theorem of Laurent Schwartz.)
[2] Lokal'nye svoĭstva reseniĭ differencial'nyh uravneniĭ v častnyh proizvodnyh s postoyannymi koefficientami. Uspehi Mat. Nauk **14**, No. 5, 3 – 46 (1959). (Local properties of the solutions of partial differential equations with constant coefficients.)

Sobolev, S. I.
[1] Nekotorye primeneniya funkcional'nogo analiza v matematičeskoi fizike. Izd-vo LGU, 1950. (Some applications of functional analysis in mathematical physics.)

Tacklind, S.
[1] Sur les classes quasianalytiques des solutions des équations aux dérivées partielles du type parabolique. Nord Acta Regial Soc. Sc. Upsal. **4**, No. 10 (1937).

Trèves, F.
[1] Opérateurs différentiels hypoelliptiques. Ann. Inst. Fourier **9**, 1 – 73 (1959).
[2] An invariant criterion of hypoellipticity. Amer. J. Math. **83**, No. 4, 645 – 668 (1961).
[3] Solutions élementaires d'équations aux dérivées partielles dépendent d'une paramètre. C. R. Acad. Sci. Paris **242**, 1250 – 1252 (1956).
[4] Lekcii po lineĭnym uravneniyam v častnyh proizvodnyh s postoyannymi koefficientami. Moscow: "Mir" 1965. (Lectures on partial differential equations with constant coefficients.)

Volevič, L. P., Paneyah, V. P.
[1] Nekotorye prostranstva obobscennyh funkciĭ i teoremy vlozeniya. Uspehi Mat. Nauk **20**, No. 1, 3 – 74 (1965). (Certain spaces of generalized functions and theorems on imbedding.)

Volevič, L. R.

[1] Lokal'nye svoĭstva reseniĭ kvaziellipticeskih sistem. Mat. Sb. **59** (101), 3 – 52 (1962). (Local properties of solutions of quasi-elliptic systems.)

[2] O gipoellipticeskih sistemah s peremennymi koefficientami. Dokl. Akad. Nauk SSSR **156**, No. 6, 1262 – 1265 (1964). (Hypoelliptic systems with variable coefficients.)

Waerden, B. L. van der

[1] Sovremennaya algebra, Part II, translated from the German. Moscow-Leningrad: Gostehizdat 1947. (Algebra, II, 4. Aufl. Berlin-Göttingen-Heidelberg: Springer 1959.)

Weyl, H.

[1] The method of orthogonal projection in potential theory. Duke Math. J. **7**, 411 – 444 (1940).

Zariski, O., Samuel, P.

[1] Kommutativnaya algebra, Vols. I, II, translated from the English. Moscow: IL 1963. (Commutative algebra, vols. I, II. van Nostrand 1958, 1960.)

Zerner, M.

[1] Théorie de Hartogs et singularités des distributions. Bull. Soc. Math. France **90**, 165 – 184 (1962).

[2] Solutions de l'équation des ondes présentant des singularities sur une droite. C. R. Acad. Sci. Paris **250**, 2980 – 2982 (1960).

[3] Solution élementaire locale d'équations aux dérivées partielles dépendent d'un paramètre, C. R. Acad. Sci. Paris **248**, 3679 (1959).

Zeylon, N.

[1] Das Fundamentalintegral der allgemeinen partiellen linearen Differentialgleichungen mit Konstanten Koeffizienten. Ark. Mat. Astr. Fys. **6**, No. 38, 1 – 32 (1911).

Žitomirskiĭ, Ya. M.

[1] Zadača Koši dlya paraboličeskih sistem lineĭnyh uravneniĭ v častnyh proizvodnyh s rastuščimi koefficientami. Izv. Vysš. Učebn. Zaved. Matematika **1** (18), 55 – 74 (1959). (The Cauchy problem for parabolic systems of partial differential equations with increasing coefficients.)

[2] Točnye klassy edinstvennosti rešeniya zadači Koši dlya uravneniya vtorogo poryadka. Dokl. Akad. Nauk SSSR **171**, No. 1, 29 – 32 (1966). (Exact uniqueness classes of solutions of the Cauchy problem for second-order equations.)

[3] Klassy edinstvennosti rešeniya zadači Koši. Uspehi Mat. Nauk **21**, No. 5, 269 – 270 (1966). (Uniqueness classes of solutions of the Cauchy problem.)

Zolotarev, G. N.

[1] Neobhodimye i dostatočnye usloviya edinstvennosti rešeniya zadači Koši dlya paraboličeskih sistem. Izv. Vysš. Učebn. Zaved. Matematika **1** (1958). (Necessary and sufficient conditions for the uniqueness of solutions of the Cauchy problem for parabolic systems.)

[2] O točnyh ocenkah dlya klassov edinstvennosti rešeniya zadači Koši dlya sistem lineĭnyh differencial'nyh uravneniĭ v častnyh proizvodnyh. Dissertaciya, Moscow 1958. (On an exact bound for the uniqueness classes of solutions of the Cauchy problem for systems of partial differential equations. Thesis.)

# Subject Index

# Index of Basic Notation

## In Chapter I



## In Chapter II



## In Chapter III



## In Chapter IV



## In Chapter V



## In Chapters VI – VIII




Universitätsdruckerei H. Stürtz AG, Würzburg

## Die Grundlehren der mathematischen Wissenschaften in Einzeldarstellungen mit besonderer Berücksichtigung der Anwendungsgebiete

---

76. Tricomi: Vorlesungen über Orthogonalreihen. DM 68,– ; US $ 18.70
77. Behnke/Sommer: Theorie der analytischen Funktionen einer komplexen Veränderlichen. DM 79,– ; US $ 21.80
78. Lorenzen: Einführung in die operative Logik und Mathematik. DM 54,– ; US $ 14.90
80. Pickert: Projektive Ebenen. DM 48,60; US $ 13.40
81. Schneider: Einführung in die transzendenten Zahlen. DM 24,80; US $ 6.90
82. Specht: Gruppentheorie. DM 69,60; US $ 19.20
84. Conforto: Abelsche Funktionen und algebraische Geometrie. DM 41,80; US $ 11.50
86. Richter: Wahrscheinlichkeitstheorie. DM 68,– ; US $ 18,70
88. Müller: Grundprobleme der mathematischen Theorie elektromagnetischer Schwingungen. DM 52,80; US $ 14.60
89. Pfluger: Theorie der Riemannschen Flächen. DM 39,20; US $ 10.80
90. Oberhettinger: Tabellen zur Fourier Transformation. DM 39,50; US $ 10.90
91. Prachar: Primzahlverteilung. DM 58,– ; US $ 16.00
93. Hadwiger: Vorlesungen über Inhalt, Oberfläche und Isoperimetrie. DM 49,80; US $ 13.70
94. Funk: Variationsrechnung und ihre Anwendung in Physik und Technik. DM 120,– ; US $ 33.00
95. Maeda: Kontinuierliche Geometrien. DM 39,– ; US $ 10.80
97. Greub: Linear Algebra. DM 39,20; US $ 9,80
98. Saxer: Versicherungsmathematik. 2. Teil. DM 48,60; US $ 13.40
99. Cassels: An Introduction to the Geometry of Numbers. DM 69,– ; US $ 19.00
100. Koppenfels/Stallmann: Praxis der konformen Abbildung. DM 69,– ; US $ 19.00
101. Rund: The Differential Geometry of Finsler Spaces. DM 59,60; US $ 16.40
103. Schütte: Beweistheorie. DM 48,– ; US $ 13.20
104. Chung: Markov Chains with Stationary Transition Probabilities. DM 56,– ; US $ 14.00
105. Rinow: Die innere Geometrie der metrischen Räume. DM 83,– ; US $ 22.90
106. Scholz/Hasenjaeger: Grundzüge der mathematischen Logik. DM 98,– ; US $ 27.00
107. Köthe: Topologische lineare Räume I. DM 78,– ; US $ 21.50
108. Dynkin: Die Grundlagen der Theorie der Markoffschen Prozesse. DM 33,80; US $ 9.30
110. Dinghas: Vorlesungen über Funktionentheorie. DM 69,– ; US $ 19.00
111. Lions: Equations différentielles opérationnelles et problèmes aux limits. DM 64,– ; US $ 17.60
112. Morgenstern/Szabó: Vorlesungen über theoretische Mechanik. DM 69,– ; US $ 19.00
113. Meschkowski: Hilbertsche Räume mit Kernfunktion. DM 58,– ; US $ 16.00
114. MacLane: Homology. DM 62,– ; US $ 15.50
115. Hewitt/Ross: Abstract Harmonic Analysis. Vol. 1: Structure of Topological Groups, Integration Theory, Group Representations. DM 76,– ; US $ 20.90
116. Hörmander: Linear Partial Differential Operators. DM 42,– ; US $ 10.50
117. O'Meara: Introduction to Quadratic Forms. DM 48,– ; US $ 13.20
118. Schäfke: Einführung in die Theorie der speziellen Funktionen der mathematischen Physik. DM 49,40; US $ 13.60
119. Harris: The Theory of Branching Processes. DM 36,– ; US $ 9.90
120. Collatz: Funktionalanalysis und numerische Mathematik. DM 58,– ; US $ 16.00
121.
122. Dynkin: Markov Processes. DM 96,– ; US $ 26.40
123. Yosida: Functional Analysis. DM 66,– ; US $ 16.50
124. Morgenstern: Einführung in die Wahrscheinlichkeitsrechnung und mathematische Statistik. DM 38,– ; US $ 10.50

125. Itô/McKean: Diffusion Processes and Their Sample Paths. DM 58,– ; US $ 16.00
126. Lehto/Virtanen: Quasikonforme Abbildungen. DM 38,– ; US $ 10.50
127. Hermes: Enumerability, Decidability, Computability. DM 39,– ; US $ 10.80
128. Braun/Koecher: Jordan-Algebren. DM 48,– ; US $ 13.20
129. Nikodým: The Mathematical Apparatus for Quantum-Theories. DM 144,– ; US $ 36.00
130. Morrey: Multiple Integrals in the Calculus of Variations. DM 78,– ; US $ 19.50
131. Hirzebruch: Topological Methods in Algebraic Geometry. DM 38,– ; US $ 9.50
132. Kato: Perturbation Theory for Linear Operators. DM 79,20; US $ 19.80
133. Haupt/Künneth: Geometrische Ordnungen. DM 68,– ; US $ 18.70
134. Huppert: Endliche Gruppen I. DM 156,– ; US $ 42.90
135. Handbook for Automatic Computation. Vol. 1/Part a: Rutishauser: Description of ALGOL 60. DM 58,– ; US $ 14.50
136. Greub: Multilinear Algebra. DM 32,– ; US $ 8.00
137. Handbook for Automatic Computation. Vol. 1/Part b: Grau/Hill/Langmaack: Translation of ALGOL 60. DM 64,– ; US $ 16.00
138. Hahn: Stability of Motion. DM 72,– ; US $ 19.80
139. Mathematische Hilfsmittel des Ingenieurs. Herausgeber: Sauer/Szabó. 1. Teil. DM 88,– ; US $ 24.20
140. Mathematische Hilfsmittel des Ingenieurs. Herausgeber: Sauer/Szabó. 2. Teil. DM 136,– ; US $ 37.40
141. Mathematische Hilfsmittel des Ingenieurs. Herausgeber: Sauer/Szabó. 3. Teil. DM 98,– ; US $ 27.00
142. Mathematische Hilfsmittel des Ingenieurs. Herausgeber: Sauer/Szabó. 4. Teil. DM 124,– ; US $ 34.10
143. Schur/Grunsky: Vorlesungen über Invariantentheorie. DM 32,– ; US $ 8.80
144. Weil: Basic Number Theory. DM 48,– ; US $ 12.00
145. Butzer/Berens: Semi-Groups of Operators and Approximation. DM 56,– ; US $ 14.00
146. Treves: Locally Convex Spaces and Linear Partial Differential Equations. DM 36,– ; US $ 9.90
147. Lamotke: Semisimpliziale algebraische Topologie. DM 48,– ; US $ 13.20
148. Chandrasekharan: Introduction to Analytic Number Theory. DM 28,– ; US $ 7.00
149. Sario/Oikawa: Capacity Functions. DM 96,– ; US $ 24.00
150. Iosifescu/Theodorescu: Random Processes and Learning. DM 68,– ; US $ 18.70
151. Mandl: Analytical Treatment of One-dimensional Markov Processes. DM 36,– ; US $ 9.80
152. Hewitt/Ross: Abstract Harmonic Analysis. Vol. II. Structure and Analysis for Compact Groups. Analysis on Locally Compact Abelian Groups. DM 140,– ; US $ 38.50
153. Federer: Geometric Measure Theory. DM 118,– ; US $ 29.50
154. Singer: Bases in Banach Spaces I. DM 112,– ; US $ 30.80
155. Müller: Foundations of the Mathematical Theory of Electromagnetic Waves. DM 58,– ; US $ 16.00
156. van der Waerden: Mathematical Statistics. DM 68,– ; US $ 18.70
157. Prohorov/Rozanov: Probability Theory. DM 68,– ; US $ 18.70
159. Köthe: Topological Vector Spaces I. DM 78,– ; US $ 21.50
160. Agrest/Maksimov: Theory of Incomplete Cylindrical Functions and their Applications. In preparation
161. Bhatia/Szegö: Stability Theory of Dynamical Systems. In preparation
162. Nevanlinna: Analytic Functions. DM 76,– ; US $ 20.90
163. Stoer/Witzgall: Convexity and Optimization in Finite Dimensions I. DM 54,– ; US $ 14.90
164. Sario/Nakai: Classification Theory of Riemann Surfaces. DM 98,– ; US $ 27.00

165. Mitrinović: Analytic Inequalities. DM 88,– ; US $ 24.20
166. Grothendieck/Dieudonné: Eléments de Géometrie Algébrique I. In preparation
167. Chandrasekharan: Arithmetical Functions. DM 58,– ; US $ 16.00
168. Palamodov: Linear Differential Operators with Constant Coefficients. DM 98,– ; US $ 27.00
169. Rademacher: Topics in Analytic Number Theory. In preparation
170. Lions: Optimal Control Systems Governed by Partial Differential Equations. DM 78,– ; US $ 21.50
171. Singer: Best Approximation in Normed Linear Spaces by Elements of Linear Subspaces. DM 60,– ; US $ 16.50
172. Bühlmann: Mathematical Methods in Risk Theory. In preparation